21. $\int \sqrt{a^2+x^2}\,dx = \frac{x}{2}\sqrt{a^2+x^2} + \frac{a^2}{2}\sinh^{-1}\frac{x}{a} + C$

22. $\int x^2\sqrt{a^2+x^2}\,dx = \frac{x(a^2+2x^2)\sqrt{a^2+x^2}}{8} - \frac{a^4}{8}\sinh^{-1}\frac{x}{a} + C$

23. $\int \frac{\sqrt{a^2+x^2}}{x}\,dx = \sqrt{a^2+x^2} - a\sinh^{-1}\left|\frac{a}{x}\right| + C$

24. $\int \frac{\sqrt{a^2+x^2}}{x^2}\,dx = \sinh^{-1}\frac{x}{a} - \frac{\sqrt{a^2+x^2}}{x} + C$

25. $\int \frac{x^2}{\sqrt{a^2+x^2}}\,dx = -\frac{a^2}{2}\sinh^{-1}\frac{x}{a} + \frac{x\sqrt{a^2+x^2}}{2} + C$

26. $\int \frac{dx}{x\sqrt{a^2+x^2}} = -\frac{1}{a}\ln\left|\frac{a+\sqrt{a^2+x^2}}{x}\right| + C$

27. $\int \frac{dx}{x^2\sqrt{a^2+x^2}} = -\frac{\sqrt{a^2+x^2}}{a^2x} + C$

28. $\int \frac{dx}{\sqrt{a^2-x^2}} = \sin^{-1}\frac{x}{a} + C$

29. $\int \sqrt{a^2-x^2}\,dx = \frac{x}{2}\sqrt{a^2-x^2} + \frac{a^2}{2}\sin^{-1}\frac{x}{a} + C$

30. $\int x^2\sqrt{a^2-x^2}\,dx = \frac{a^4}{8}\sin^{-1}\frac{x}{a} - \frac{1}{8}x\sqrt{a^2-x^2}\,(a^2-2x^2) + C$

31. $\int \frac{\sqrt{a^2-x^2}}{x}\,dx = \sqrt{a^2-x^2} - a\ln\left|\frac{a+\sqrt{a^2-x^2}}{x}\right| + C$

32. $\int \frac{\sqrt{a^2-x^2}}{x^2}\,dx = -\sin^{-1}\frac{x}{a} - \frac{\sqrt{a^2-x^2}}{x} + C$

33. $\int \frac{x^2}{\sqrt{a^2-x^2}}\,dx = \frac{a^2}{2}\sin^{-1}\frac{x}{a} - \frac{1}{2}x\sqrt{a^2-x^2} + C$

34. $\int \frac{dx}{x\sqrt{a^2-x^2}} = -\frac{1}{a}\ln\left|\frac{a+\sqrt{a^2-x^2}}{x}\right| + C$

35. $\int \frac{dx}{x^2\sqrt{a^2-x^2}} = -\frac{\sqrt{a^2-x^2}}{a^2x} + C$

36. $\int \frac{dx}{\sqrt{x^2-a^2}} = \cosh^{-1}\frac{x}{a} + C = \ln\left|x+\sqrt{x^2-a^2}\right| + C$

37. $\int \sqrt{x^2-a^2}\,dx = \frac{x}{2}\sqrt{x^2-a^2} - \frac{a^2}{2}\cosh^{-1}\frac{x}{a} + C$

38. $\int (\sqrt{x^2-a^2})^n\,dx = \frac{x(\sqrt{x^2-a^2})^n}{n+1} - \frac{na^2}{n+1}\int (\sqrt{x^2-a^2})^{n-2}\,dx, \quad n \neq -1$

39. $\int \frac{dx}{(\sqrt{x^2-a^2})^n} = \frac{x(\sqrt{x^2-a^2})^{2-n}}{(2-n)a^2} - \frac{n-3}{(n-2)a^2}\int \frac{dx}{(\sqrt{x^2-a^2})^{n-2}}, \quad n \neq 2$

40. $\int x(\sqrt{x^2-a^2})^n\,dx = \frac{(\sqrt{x^2-a^2})^{n+2}}{n+2} + C, \quad n \neq -2$

41. $\int x^2\sqrt{x^2-a^2}\,dx = \frac{x}{8}(2x^2-a^2)\sqrt{x^2-a^2} - \frac{a^4}{8}\cosh^{-1}\frac{x}{a} + C$

$\int \frac{\sqrt{x^2-a^2}}{x}\,dx = \sqrt{x^2-a^2} - a\sec^{-1}\left|\frac{x}{a}\right| + C$

$\int \frac{\sqrt{x^2-a^2}}{x^2}\,dx = \cosh^{-1}\frac{x}{a} - \frac{\sqrt{x^2-a^2}}{x} + C$

Continued overleaf.

44. $\displaystyle\int \frac{x^2}{\sqrt{x^2-a^2}}\,dx = \frac{a^2}{2}\cosh^{-1}\frac{x}{a} + \frac{x}{2}\sqrt{x^2-a^2} + C$

45. $\displaystyle\int \frac{dx}{x\sqrt{x^2-a^2}} = \frac{1}{a}\sec^{-1}\left|\frac{x}{a}\right| + C = \frac{1}{a}\cos^{-1}\left|\frac{a}{x}\right| + C$

46. $\displaystyle\int \frac{dx}{x^2\sqrt{x^2-a^2}} = \frac{\sqrt{x^2-a^2}}{a^2x} + C$

47. $\displaystyle\int \frac{dx}{\sqrt{2ax-x^2}} = \sin^{-1}\left(\frac{x-a}{a}\right) + C$

48. $\displaystyle\int \sqrt{2ax-x^2}\,dx = \frac{x-a}{2}\sqrt{2ax-x^2} + \frac{a^2}{2}\sin^{-1}\left(\frac{x-a}{a}\right) + C$

49. $\displaystyle\int (\sqrt{2ax-x^2})^n\,dx = \frac{(x-a)(\sqrt{2ax-x^2})^n}{n+1} + \frac{na^2}{n+1}\int (\sqrt{2ax-x^2})^{n-2}\,dx,$

50. $\displaystyle\int \frac{dx}{(\sqrt{2ax-x^2})^n} = \frac{(x-a)(\sqrt{2ax-x^2})^{2-n}}{(n-2)a^2} + \frac{(n-3)}{(n-2)a^2}\int \frac{dx}{(\sqrt{2ax-x^2})^{n-2}}$

51. $\displaystyle\int x\sqrt{2ax-x^2}\,dx = \frac{(x+a)(2x-3a)\sqrt{2ax-x^2}}{6} + \frac{a^3}{2}\sin^{-1}\frac{x-a}{a} + C$

52. $\displaystyle\int \frac{\sqrt{2ax-x^2}}{x}\,dx = \sqrt{2ax-x^2} + a\sin^{-1}\frac{x-a}{a} + C$

53. $\displaystyle\int \frac{\sqrt{2ax-x^2}}{x^2}\,dx = -2\sqrt{\frac{2a-x}{x}} - \sin^{-1}\left(\frac{x-a}{a}\right) + C$

54. $\displaystyle\int \frac{x\,dx}{\sqrt{2ax-x^2}} = a\sin^{-1}\frac{x-a}{a} - \sqrt{2ax-x^2} + C$

55. $\displaystyle\int \frac{dx}{x\sqrt{2ax-x^2}} = -\frac{1}{a}\sqrt{\frac{2a-x}{x}} + C$

56. $\displaystyle\int \sin ax\,dx = -\frac{1}{a}\cos ax + C$

57. $\displaystyle\int \cos ax\,dx = \frac{1}{a}\sin ax + C$

58. $\displaystyle\int \sin^2 ax\,dx = \frac{x}{2} - \frac{\sin 2ax}{4a} + C$

59. $\displaystyle\int \cos^2 ax\,dx = \frac{x}{2} + \frac{\sin 2ax}{4a} + C$

60. $\displaystyle\int \sin^n ax\,dx = \frac{-\sin^{n-1} ax\cos ax}{na} + \frac{n-1}{n}\int \sin^{n-2} ax\,dx$

61. $\displaystyle\int \cos^n ax\,dx = \frac{\cos^{n-1} ax\sin ax}{na} + \frac{n-1}{n}\int \cos^{n-2} ax\,dx$

62. (a) $\displaystyle\int \sin ax\cos bx\,dx = -\frac{\cos(a+b)x}{2(a+b)} - \frac{\cos(a-b)x}{2(a-b)} + C, \qquad a^2 \neq b^2$

(b) $\displaystyle\int \sin ax\sin bx\,dx = \frac{\sin(a-b)x}{2(a-b)} - \frac{\sin(a+b)x}{2(a+b)}, \qquad a^2 \neq b^2$

(c) $\displaystyle\int \cos ax\cos bx\,dx = \frac{\sin(a-b)x}{2(a-b)} + \frac{\sin(a+b)x}{2(a+b)}, \qquad a^2 \neq b^2$

63. $\displaystyle\int \sin ax\cos ax\,dx = -\frac{\cos 2ax}{4a} + C$

64. $\displaystyle\int \sin^n ax\cos ax\,dx = \frac{\sin^{n+1} ax}{(n+1)a} + C, \qquad n \neq -1$

This table is continued on the endpap

CALCULUS AND ANALYTIC GEOMETRY

SIXTH EDITION

GEORGE B. THOMAS, JR.
MASSACHUSETTS INSTITUTE OF TECHNOLOGY

ROSS L. FINNEY
MASSACHUSETTS INSTITUTE OF TECHNOLOGY

CALCULUS AND ANALYTIC GEOMETRY

PART II: VECTORS, FUNCTIONS OF SEVERAL VARIABLES, INFINITE SERIES, AND DIFFERENTIAL EQUATIONS

ADDISON-WESLEY PUBLISHING COMPANY
Reading, Massachusetts • Menlo Park, California • London
Amsterdam • Don Mills, Ontario • Sydney

Sponsoring Editor: *Wayne Yuhasz*
Developmental Editor: *Jeffrey Pepper*
Production Manager: *Martha K. Morong*
Production Editor: *Marion E. Howe*
Text Designer: *Margaret Ong Tsao*
Illustrators: *Carmella M. Ciampa*
Dick Morton
Cover Designer: *Richard Hannus, Hannus Design Associates*
Cover Photograph: *Charles Rotmil, Four By Five*
Art Editor: *Dick Morton*
Manufacturing Supervisor: *Ann DeLacey*

Library of Congress Cataloging in Publication Data

Thomas, George Brinton, 1914–
Calculus and analytic geometry.

Includes index.
Contents: pt. 1, Functions of one variable, analytic geometry—pt. 2. Vectors, functions of several variables, infinite series, and differential equations.
1. Calculus. 2. Geometry, Analytic. I. Finney, Ross L. II. Title.
QA303.T42 1984b 515′.15 83-15483
ISBN 0-201-16291-1 (v. 1)
ISBN 0-201-16292-X (v. 2)

Reprinted with corrections, May 1984

ISBN 0-201-16292-X
CDEFGHIJ-DO-8987654

PREFACE

Calculus, believed by many to be the greatest achievement in all mathematics, was created to meet the pressing mathematical needs of seventeenth century science. Heading the list were the needs to interrelate the accelerations, velocities, and distances traveled by moving bodies, to relate the slopes of curves to rates of change, to find the maximum and minimum values a function might take on (the greatest and least distances of a planet from its sun, for example), and to find the lengths of curves, the areas bounded by curves, the volumes enclosed by surfaces, and the centers of gravity of attracting bodies. Calculus is still the best mathematics for solving problems like these (and many others) and it is now so widely used that there is hardly a professional field that does not benefit from it in some way.

In preparing the *Sixth Edition* from the *Fifth Edition,* we have tried to make calculus easier for people to learn. The number of exercises has nearly doubled. Many of the new exercises are of a basic computational nature, designed to reinforce fundamental skills. There are more examples, three hundred new figures (all captioned), frequent formula summaries, and more (and shorter) applications for people going into engineering and science. Linear and quadratic approximations of functions are introduced early in the book, along with standard error estimates. There continue to be hand-held calculator exercises.

The widespread availability of microcomputers has made it possible for people learning calculus or using it in their professional work to graph functions in the plane or in space, to solve many equations quickly, and otherwise to implement the formulas derived by calculus. This edition therefore contains references at the end of some of the problem sets to a series of twenty-seven Apple II® microcomputer programs appropriate for students enrolled in calculus courses as well as for engineers, physicists, and others who use calculus as a tool in their work. The programs and accompanying manual are available from Addison-Wesley Publishing Company under the title *The Calculus Toolkit.* There is more information about them following this preface. It is to be emphasized, however, that one's progress through the text is in no way dependent upon having access to these or any other computer programs.

The level of rigor of the text is about the same as in earlier editions. For example, we do not prove that a function that is continuous on a closed and bounded interval has a maximum and a minimum on the interval, but we state that theorem and use it in proving the Mean Value Theorem.

Chapter 1, on the rate of change of a function, has new sections on continuity and on infinity as a limit.

Chapter 2, on derivatives, now has sections on linear (tangent line) approximations, and on inverse functions and the Picard method for finding roots.

Chapter 3, on applications of derivatives, begins with new sections on curve sketching, concavity, and asymptotes. Maxima and minima now precede related rates. The chapter concludes with a section that extends the Mean Value Theorem and develops error estimates for the standard linear and quadratic approximations of functions.

Chapter 4, on indefinite and definite integrals, begins as it has in the past with differential equations of the form $y' = f(x)$, solved by separation of variables. The chapter contains a new development of the two fundamental theorems of integral calculus, however, and devotes a separate section to the technique of integration by substitution (for both definite and indefinite integrals). It is in this section that differentials are first introduced.

Chapter 5, on applications of definite integrals, has more art, problems, and worked examples than before, and frequent formula summaries.

Chapter 6, which introduces the logarithmic, exponential, and inverse trigonometric functions, also discusses relative rates of growth of functions. It concludes with a section on applications of exponential and logarithmic functions to cooling, exponential growth, radioactive decay, and electric circuits, and a section on compound interest and Benjamin Franklin's will.

In Chapter 7, on techniques of integration, the section on improper integrals has been expanded to include comparison tests for convergence. There is also a new section on using integral tables, and the treatment of integration by parts has been moved to the beginning of the chapter.

Chapters 8 (plane analytic geometry) and 9 (hyperbolic functions) have been shortened somewhat and contain additional art and problems.

Chapter 10, on polar coordinates, is shorter than before and contains a new technique for graphing polar equations of the form $r = f(\theta)$.

The presentation of infinite sequences and series has been moved forward in the book and divided into Chapters 11 and 12. Chapter 11 is devoted to sequences and infinite series of constants, Chapter 12 to Taylor's theorem (as an extended mean value theorem) and power series. Series of complex numbers are mentioned briefly. (An introduction to complex number arithmetic and Argand diagrams appears in Appendix 8.)

Chapter 13, on vectors, begins with motion in the plane and moves from there to the study of vector algebra and geometry in space.

Chapter 14, on vector functions and their derivatives, has a new treatment of tangent vectors, velocity, and acceleration, and concludes with a section on Kepler's laws of planetary motion.

Chapter 15, on partial derivatives, has new treatments of limits of functions of two variables, continuity, surfaces, partial derivatives, chain rules, directional derivatives, linear approximation and increment estimation, maxima and minima (both constrained and free), Lagrange multipliers, exact differentials, and least squares. Computer graphics have made it possible to visualize and discuss a number of surfaces that could not have been shown in earlier editions of this book. The chapter also looks briefly at solutions of some of the important partial differential equations of physics (in connection with higher order derivatives) and has a short section on how to apply chain rules when a function's variables are not independent.

Chapter 16, on multiple integrals, contains a new introduction to the subject, along with more examples, problems, and frequent formula summaries. It concludes with a presentation of surface area based on the notion of gradient.

Chapter 17, on vector analysis, begins with vector fields, surface integrals, line integrals, and work, and concludes with Green's theorem, the divergence theorem, and Stokes's theorem. In addition to many new examples and problems, the chapter now contains a brief derivation of the continuity equation of hydrodynamics.

In Chapter 18, on ordinary differential equations, the treatment of linear second order equations with constant coefficients has been expanded to include the method of undetermined coefficients in addition to the method of variation of parameters. The chapter concludes with short sections on power series solutions, direction fields and Picard's theorem, and Euler and Runge-Kutta methods.

The appendixes include expanded sections on determinants and Cramer's rule, and on matrices and linear equations, as well as new sections on mathematical induction and number systems.

The text is available in one complete volume, which can be covered in two or three semesters, depending on the pace, or as two separate parts. Part I treats functions of one variable, analytic geometry in two dimensions, and infinite series (Chapters 1 through 12). It also contains the appendixes on determinants, matrices, and mathematical induction. Part II begins with Chapter 11, on sequences and series, and contains all subsequent chapters, including the appendixes. Both parts contain answers to odd-numbered problems.

We would particularly like to thank Jack M. Pollin and Frank R. Giordano of the Mathematics Department of the U.S. Military Academy at West Point, who generously gave us access to and guided us through their department's carefully kept examination and problem files. Some of the finest problems in this new edition have come from West Point. Likewise, and no less, we would like to thank Carroll O. Wilde of the Mathematics Department of the U.S. Naval Postgraduate School for granting us access to his department's outstanding examination files, and for his many supportive comments and suggestions during the preparation of the present edition.

We would also like to acknowledge the many helpful contributions and suggestions of our colleagues at MIT, particularly Arthur Mattuck and Frank Morgan, and the many students there and elsewhere who have taken the time to share their ideas with us.

Many helpful comments and suggestions were made by people who reviewed the *Fifth Edition* as the *Sixth Edition* was being planned:

Mark Bridger, Northeastern University
Jeff Davis, University of New Mexico
Mark Bridger, Northeastern University
Jeff Davis, University of New Mexico
Simon Hellerstein, University of Wisconsin, Madison
Stanley Lukawecki, Clemson University
Ronald Morash, University of Michigan, Dearborn
Harold Oxsen, Diablo Valley College
Walter Read, California State University, Fresno
Michael Shaughnessy, Oregon State University

We would also like to thank and acknowledge the contributions of the people who participated in the preliminary planning sessions that took place in Menlo Park, California, and Cincinnati, Ohio:

Charles Austin, California State University, Long Beach
Douglas Crawford, College of San Mateo
Daniel S. Drucker, Wayne State University
Bruce H. Edwards, University of Florida
Alice J. Kelly, University of Santa Clara
Stanley Lukawecki, Clemson University

Ronald Morash, University of Michigan, Dearborn
Michael Shaughnessy, Oregon State University
Ronald J. Stern, University of Utah
Carroll O. Wilde, Naval Postgraduate School

A great many valuable contributions to the *Sixth Edition* were made by the people who reviewed the manuscript as it developed through its various stages:

Charles Austin, California State University, Long Beach
Mark Bridger, Northeastern University
Stuart Goldenberg, California Polytechnic State University
Simon Hellerstein, University of Wisconsin, Madison
Ronald Morash, University of Michigan, Dearborn
Hiram Paley, University of Illinois at Urbana-Champaign
David F. Ullrich, North Carolina State University

Charles Austin and Mark Bridger made many additional contributions to the chapters on partial derivatives and multiple integration.

Special contributions to the book were also made through correspondence and in conversation by

Harold Diamond, University of Illinois at Urbana-Champaign
Frank D. Faulkner, Naval Postgraduate School
Nathaniel Grossman, University of California, Los Angeles
Richard W. Hamming, Naval Postgraduate School
John P. Hoyt, Franklin and Marshall College
Alice J. Kelley, University of Santa Clara
Ernest J. Manfred, U.S. Coast Guard Academy
Charles G. Moore, Northern Arizona University
Joseph J. Rotman, University of Illinois at Urbana-Champaign
Arthur C. Segal, The University of Alabama in Birmingham
G. Wayne Sullivan, Volunteer State Community College.
Joseph D. Zund, New Mexico State University

The answer manuscript for the odd-numbered problems in this edition was developed through the collaborative efforts of

Kenneth R. Ballou, University of California, Berkeley
Lynda Jo Carlson, California State University, Fullerton
J. Howard Jones, Lansing Community College
Chris J. Petti, Massachusetts Institute of Technology
Craig W. Reynolds, Massachusetts Insitute of Technology
Richard J. Palmaccio, Pine Crest School
Richard Semmler, Northern Virginia Community College
Charles Swanson, University of Minnesota

The computer-generated art that appears in Chapters 13, 15, and 18 is the patient and enthusiastic work of John Aspinall of the Plasma Fusion Center at MIT.

To each and every person who has at any time contributed helpful suggestions, comments, or criticisms, whether or not we have been able to incorporate these into the book, we say "Thank you very much."

It is a pleasure to acknowledge the superb assistance in illustration, editing, design, and composition that the staff of Addison-Wesley Publishing Company has given to the preparation of this edition.

Any errors that may appear are the responsibility of the authors. We will appreciate having these brought to our attention.

October, 1983

G. B. T., Jr.
R. L. F.

CONTENTS

PROLOGUE
WHAT IS CALCULUS? xiii

11
SEQUENCES AND INFINITE SERIES 594

11.1 Introduction 594
11.2 Sequences of numbers 595
11.3 Limits that arise frequently 604
11.4 Infinite series 608
11.5 Tests for convergence of series with nonnegative terms 619
11.6 Absolute convergence 641
11.7 Alternating series. Conditional convergence 648
Review Questions and Exercises 655
Miscellaneous Problems 655

12
POWER SERIES 657

12.1 Power series for functions 657
12.2 Taylor's theorem with remainder: Sines, cosines, and e^x 663
12.3 Further computations, logarithms, arctangents, and π 672
12.4 Indeterminate forms 679
12.5 Convergence of power series. Integration, differentiation, multiplication, and division 682
Review Questions and Exercises 694
Miscellaneous Problems 694

13
VECTORS 697

13.1 Vector components and the unit vectors **i** and **j** 697
13.2 Modeling projectile motion 703
13.3 Parametric equations in analytic geometry 709
13.4 Space coordinates 715
13.5 Vectors and distance in space 719
13.6 The scalar product of two vectors 722
13.7 The vector product of two vectors in space 729
13.8 Equations of lines, line segments, and planes 732
13.9 Products of three vectors or more 740
13.10 Cylinders 745
13.11 Quadric surfaces 748
Review Questions and Exercises 754
Miscellaneous Problems 754

14
VECTOR FUNCTIONS AND THEIR DERIVATIVES 758

14.1 Derivatives of vector functions 758
14.2 Tangent vectors, velocity, and acceleration 763
14.3 Arc length for space curves. The unit tangent vector **T** 767
14.4 Curvature and normal vectors 772
14.5 Derivatives of vector products. The tangential and normal components of **v** and **a** 779

14.6 Planetary motion and Kepler's laws 785
Review Questions and Exercises 790
Miscellaneous Problems 790

15 PARTIAL DERIVATIVES 794

15.1 Functions of two or more variables 794
15.2 Limits and continuity 799
15.3 Partial derivatives 806
15.4 Chain rules 811
15.5 Nonindependent variables 818
15.6 Gradients, directional derivatives, and tangent planes 822
15.7 Higher order derivatives. Partial differential equations from physics 837
15.8 Linear approximation and increment estimation 844
15.9 Maxima, minima, and saddle points 854
15.10 Lagrange multipliers 865
15.11 Exact differentials 876
15.12 Least squares 884
Review Questions and Exercises 888
Miscellaneous Problems 889

16 MULTIPLE INTEGRALS 894

16.1 Introduction 894
16.2 Double integrals 894
16.3 Area 904
16.4 Physical applications 906
16.5 Changing to polar coordinates 910
16.6 Triple integrals in rectangular coordinates 916
16.7 Physical applications in three dimensions 921
16.8 Integrals in cylindrical and spherical coordinates 924
16.9 Surface area 930
Review Questions and Exercises 934
Miscellaneous Problems 934

17 VECTOR ANALYSIS 938

17.1 Vector fields 938
17.2 Surface integrals 942
17.3 Line integrals and work 948
17.4 Two-dimensional fields. Flux across a plane curve 960
17.5 Green's theorem 967
17.6 The divergence theorem 975
17.7 Stokes's theorem 984
Review Questions and Exercises 992
Miscellaneous Problems 993

18 DIFFERENTIAL EQUATIONS 996

18.1 Introduction 996
18.2 Solutions 998
18.3 First order: Variables separable 999
18.4 First order: Homogeneous 1000
18.5 First order: Linear 1002
18.6 First order: Exact 1005
18.7 Special types of second order equations 1006
18.8 Linear equations with constant coefficients 1008
18.9 Linear second order homogeneous equations with constant coefficients 1009
18.10 Linear second order nonhomogeneous equations with constant coefficients 1013
18.11 Higher order linear equations with constant coefficients 1022
18.12 Vibrations 1023
18.13 Approximation methods: Power series 1028
18.14 Direction fields and Picard's theorem 1029
18.15 Numerical methods 1034
Review Questions and Exercises 1039
Miscellaneous Problems 1040

APPENDIXES A–1

1. Determinants and Cramer's Rule A–1
2. Matrices and linear equations A–9
3. Proofs of the limit theorems of Article 1.9 A–18
4. The increment and mixed-derivative theorems A–22
5. Mathematical induction A–27
6. The law of cosines and the addition formulas from trigonometry A–29
7. Formulas from elementary mathematics A–31
8. Invented number systems. Complex numbers A–36
9. Tables A–46
10. The distributive law for vector cross products A–49

AVAILABLE SUPPLEMENTS The following supplementary materials are available for use by students:

Self-Study Manual	*Maurice D. Weir* (Naval Postgraduate School)
Student Supplement	*Kenneth R. Ballou* (University of California, Berkeley)
	Charles Swanson (University of Minnesota)

LIST OF PROGRAMS

The Calculus Toolkit

Ross L. Finney Dale T. Hoffman
Judah L. Schwartz Carroll O. Wilde

- Super * Grapher
- Name That Function
- Secant Lines
- Limit Problems
- Limit Definition
- Continuity at a Point
- Derivative Grapher
- Function Evaluator/Comparer
- Parametric Equations
- Root Finder
- Picard's Fixed Point Method
- Integration
- Integral Evaluator
- Antiderivatives and Direction Fields
- Partial Fraction Integration Problems
- Conic Sections
- Sequences and Series
- Taylor Series
- Complex Number Calculator
- 3D Grapher
- Double Integral Evaluator
- Scalar Fields
- Vector Fields
- First Order Initial Value Problem
- Second Order Initial Value Problem
- Damped Oscillator
- Forced, Damped Oscillator

PROLOGUE

What Is Calculus?

Calculus is the mathematics of motion and change. Where there is motion or growth, where variable forces are at work producing acceleration, calculus is the right mathematical tool. This was true in the beginnings of the subject, and it is true today.

Calculus is used to predict the orbits of earth satellites; to design inertial navigation systems, cyclotrons, and radar systems; to explore problems of space travel; to test theories about ocean currents and the dynamics of the atmosphere. It is applied increasingly to solve problems in biology, business, economics, linguistics, medicine, political science, and psychology. Calculus is also a gateway to nearly all fields of higher mathematics.

Differential calculus deals with the problem of calculating rates of change. When we have a formula for the distance that a moving body covers as a function of time, differential calculus gives us the formulas for calculating the body's velocity and acceleration at any instant. Differential calculus also lets us answer questions about the greatest and smallest values a function can take on. What angle of elevation gives a cannon its greatest range? What is the strongest rectangular beam we can cut from a cylindrical log? When is a chemical reaction proceeding at its fastest rate?

Integral calculus considers the problem of determining a function from information about its rate of change. Given a formula for the velocity of a body as a function of time, we can use integral calculus to produce a formula that tells how far the body has traveled from its starting point at any instant. Given the present size of a population and the rate at which it grows, we can produce a formula that predicts the population's size at any future time or that estimates how big the population was at some time in the past. Given the rate at which carbon-14 disintegrates, we can produce a formula that calculates the age of a sample of charcoal from its present ratio of carbon-14 to carbon-12. The age of Crater Lake in Oregon, 6660 years, was estimated by applying such a formula to the charcoal from a tree that was killed in the volcanic eruption that formed the lake.

Modern science and engineering use both differential and integral calculus to express physical laws in precise mathematical terms and to study the consequences of these laws.

Before Sir Isaac Newton (1642–1727) was born, Johannes Kepler (1571–1630) spent twenty years studying observational data and using empirical methods to discover the three laws that now bear his name:

1. Each planet moves about the sun in an orbit that is an ellipse with one focus at the sun (Fig. P.1).
2. The line joining the planet and the sun sweeps over equal areas in equal intervals of time.
3. The squares of the periods of revolution of the planets about the sun are proportional to the cubes of their average distances from the sun.

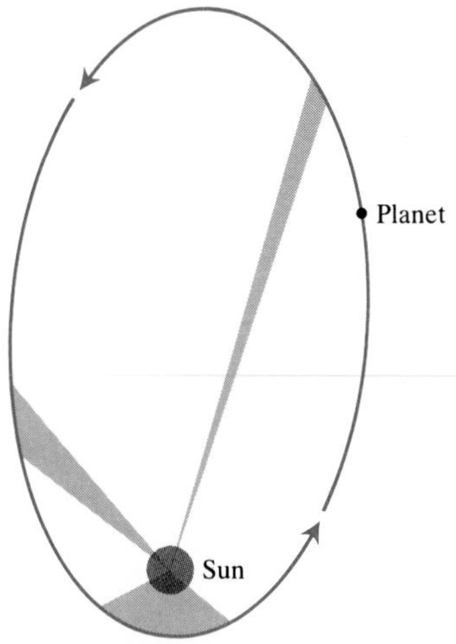

P.1 A planet moving about its sun. The shaded regions shown here have equal areas. According to Kepler's second law, the planet takes the same amount of time to traverse the curved outer boundary of each region. The planet therefore moves faster near the sun than it does farther away.

Figure P.1 gives us some idea of what Kepler's first two laws say, but what does his third law say? The earth's period of revolution about the sun is 365 days (close enough for now). Its average distance from the sun is 93 million miles. The ratio of the square of 365 to the cube of 93,000,000 is $(365)^2/(93{,}000{,}000)^3 = 1.66 \times 10^{-19}$. Kepler's third law says that if the period of revolution T of any other planet is measured in earth days, and if the planet's average distance D from the sun is measured in miles, then the numerical value of $T^2/D^3 = (\text{Period})^2/(\text{Distance})^3$ will be 1.66×10^{-19}, just as it was for the earth. The ratio is the same for every planet in the solar system. In particular, the farther a planet is from the sun, the longer it takes to go once about its orbit.

When they were first written down, there can have been no immediately apparent relationship between the measurements Kepler made of T and D for the planets he observed. He must have studied long and hard before discovering the constancy of the ratio T^2/D^3. Only a few years later, Newton was able to use calculus to derive all three of Kepler's laws from the physical assumptions we know as Newton's laws of motion and gravitation. Kepler formulated his laws for our solar system. Newton's work shows that they hold for any orbital system that obeys the laws of motion and gravitation.

The calculus we use today is the contribution of many people. Its roots can be traced to classical Greek geometry, but its invention is chiefly the work of the astronomers, mathematicians, and physicists of the seventeenth century. Among these were René Descartes (1596–1650), Bonaventura Cavalieri (1598–1647), Pierre de Fermat (1601–1665), John Wallis (1616–1703), Blaise Pascal (1623–1662), Christian Huygens (1629–1695), Isaac Barrow (1630–1677), and James Gregory (1638–1675). The work culminated in the great creations of Newton and Gottfried Wilhelm Leibniz (1646–1716).

The efforts of Newton and Leibniz changed the calculus from an appendage of Greek geometry to an independent science that had the power to handle a vast array of new scientific problems. Newton and Leibniz formulated the fundamental theorems of calculus and recognized in the particular works of their colleagues the general methods that applied to all sorts of functions. Many of their operational formulas, methods of calculation, and mathematical symbols are still in use.

The development of the calculus did not stop with the accomplishments of the seventeenth century. Great additions were made by the eighteenth-century mathematicians James Bernoulli (1654–1705) and his brother John Bernoulli (1667–1748); by Leonhard Euler (1707–1783), the key mathematical figure of the century; by Joseph-Louis Lagrange (1736–1813); and many others. The ultimate justification of the procedures of calculus was made by the mathematicians of the nineteenth century, among them Bernhard Bolzano (1781–1848), Augustin-Louis Cauchy (1789–1857), and Karl Weierstrass (1815–1897). The nineteenth century also brought spectacular developments in mathematics beyond the calculus. You can read about them in Morris Kline's magnificent book *Mathematical Thought from Ancient to Modern Times* (New York: Oxford University Press, 1972).

"The calculus was the first achievement of modern mathematics," wrote John von Neumann (1903–1957), one of the great mathematicians of the present century, "and it is difficult to overestimate its importance. I think it defines more unequivocally than anything else the inception of modern mathematics; and the system of mathematical analysis, which is its logical development, still constitutes the greatest technical advance in exact thinking."†

† *World of Mathematics*, Vol. 4 (New York: Simon and Schuster, 1960), "The Mathematician," by John von Neumann, pp. 2053–2063.

CALCULUS AND ANALYTIC GEOMETRY

11 Sequences and Infinite Series

11.1 Introduction

In this chapter and the next we deal with two related topics: *sequences* and *series*. We begin with an example that shows how a sequence of numbers can be generated by an iterative process like Picard's method for solving an equation.

EXAMPLE 1 Use Picard's method to find approximations to a solution of the equation $\sin x - x^2 = 0$.

Solution (See Art. 2.10 for a review of Picard's method.) Figure 11.1 shows graphs of portions of the curves

$$y = \sin x \qquad \text{and} \qquad y = x^2.$$

The curves intersect at the origin and between $x = 0$ and $x = 1$. To estimate the latter intersection, we rewrite the equation in the form

$$x = \sqrt{\sin x}.$$

Let $x_0 = 1$, and define $x_1, x_2, x_3, \ldots$ by the iterative formula

$$x_{n+1} = \sqrt{\sin x_n} \qquad \text{for } n = 0, 1, 2, \ldots. \tag{1}$$

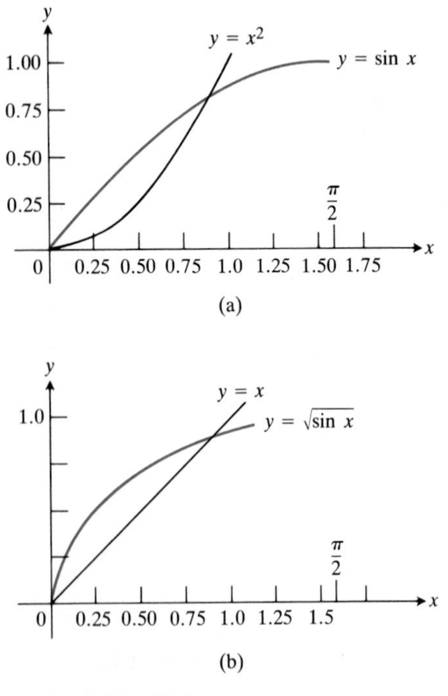

11.1 (a) The graphs of $y = x^2$ and $y = \sin x$ intersect near $x = 1$. (b) We solve $x = \sqrt{\sin x}$ by iteration with $x_0 = 1$ and $x_{n+1} = \sqrt{\sin x_n}$.

A computer or programmable calculator can easily be programmed to grind out

$$x_1 = 0.917\ldots$$
$$x_2 = 0.891\ldots$$
$$x_3 = 0.881\ldots$$

and so on. After each x_n there is a next term, x_{n+1}. In principle, the process could be continued indefinitely. In practice, we would probably stop with x_{19}, say, in this particular problem when we see that to nine decimal places, $x_{19} = x_{20} = x_{21}$. The numerical value that we got is

$$x_{19} = 0.876726216. \tag{2}$$

If we could carry out the calculations with no limit on the number of decimal places, we could get an unending sequence of numbers $x_1, x_2, x_3,$

$\ldots, x_n, \ldots$ that approach a limit that is the desired solution of $\sin x - x^2 = 0$. □

There are many examples of infinite sequences in mathematics—for example, the sequence of prime numbers:

$$2, 3, 5, 7, 11, 13, 17, 19, 23, 29, 31, \ldots.$$

In this case, we do not have a formula for the nth prime. Still simpler is the sequence of even numbers:

$$2, 4, 6, 8, \ldots, 2n, \ldots$$

where we do have a formula $x_n = 2n$ for the nth even number.

Sometimes the terms of a sequence are, or can be, formed by simple addition. For example, the decimal approximations to 1/3 are

$$0.3, \quad 0.33 = 0.3 + 0.03, \quad 0.333 = 0.3 + 0.03 + 0.003,$$

and so on. Here we could also define the terms of the sequence by saying that $x_1 = 0.3$ and $x_{n+1} = x_n + 3/10^{n+1}$.

This chapter deals primarily with sequences of constants. The next chapter will deal with sequences (and series, which are special kinds of sequences) of *functions*.

PROBLEMS

In Problems 1–5, write out in explicit form each of the terms $x_1, x_2, x_3, x_4, x_5, x_6$.

1. $x_0 = 1$, and $x_{n+1} = x_n + (\frac{1}{2})^{n+1}$

2. $x_0 = 1$, and $x_{n+1} = \dfrac{x_n}{n+1}$

3. $x_0 = 2$, and $x_{n+1} = \dfrac{x_n}{2}$

4. $x_0 = -2$, and $x_{n+1} = \dfrac{n}{n+1}x_n$

5. $x_0 = x_1 = 1$, and $x_{n+2} = x_n + x_{n+1}$. (This is called the Fibonacci sequence. The ratio $r_n = x_n/x_{n+1}$ gives a related sequence $r_1, r_2, \ldots$.)

6. If we try to solve the equation $\sin x = x^2$ by taking $x_0 = 1$ and $x_{n+1} = \sin^{-1}(x_n^2)$, what is x_1? x_2? Do you think this process will converge to an x that satisfies the given equation $\sin x = x^2$?

7. (*Calculator*) Put $x_0 = 1$ and let $x_{n+1} = x_n + \sin x_n - x_n^2$. When you get $x_{n+1} = x_n$, you have solved $\sin x_n - x_n^2 = 0$. Compare your answer with Eq. (2).

8. (*Calculator*) Use Picard's method to generate sufficiently many terms of the sequence $x_1, x_2, \ldots$, defined by $x_0 = 1$, $x_{n+1} = (\sin x_n)^{1/3}$, to estimate a root of the equation $\sin x = x^3$.

11.2 Sequences of Numbers

We continue our study of sequences of numbers with a definition.

DEFINITION 1

A *sequence* of numbers is a function whose domain is the set of positive integers.

Sequences are defined by rules the way other functions are, typical rules being

$$a(n) = n - 1, \qquad a(n) = 1 - \frac{1}{n}, \qquad a(n) = \frac{\ln n}{n^2}. \tag{1}$$

To signal the fact that the domains are restricted to the set of positive integers, it is conventional to use a letter like n from the middle of the alphabet for the independent variable instead of the x, y, z, and t used so widely in other contexts. The formulas in the defining rules, however, like the ones above, are often valid for domains much larger than the set of positive integers. This can prove to be an advantage, as we shall see at the end of this article.

The numbers in the range of a sequence are called the *terms* of the sequence, the number $a(n)$ being called the *nth term*, or the term with *index n*. For example, if $a(n) = (n + 1)/n$, then the terms are

First term	Second term	Third term		nth term		
$a(1) = 2,$	$a(2) = \frac{3}{2},$	$a(3) = \frac{4}{3},$	$\ldots,$	$a(n) = \frac{n+1}{n},$	$\ldots.$	(2)

When we use the simpler notation a_n for $a(n)$, the sequence in (2) becomes

$$a_1 = 2, \quad a_2 = \frac{3}{2}, \quad a_3 = \frac{4}{3}, \quad \ldots, \quad a_n = \frac{n+1}{n}, \quad \ldots. \tag{3}$$

To describe sequences, we often write the first few terms as well as a formula for the nth term.

EXAMPLE 1

We write	For the sequence whose defining rule is
$0, \ 1, \ 2, \ \ldots, \ n-1, \ \ldots$	$a_n = n - 1$
$1, \ \frac{1}{2}, \ \frac{1}{3}, \ \ldots, \ \frac{1}{n}, \ \ldots$	$a_n = \frac{1}{n}$
$1, \ -\frac{1}{2}, \ \frac{1}{3}, \ -\frac{1}{4}, \ \ldots, \ (-1)^{n+1}\frac{1}{n}, \ \ldots$	$a_n = (-1)^{n+1}\frac{1}{n}$
$0, \ \frac{1}{2}, \ \frac{2}{3}, \ \frac{3}{4}, \ \ldots, \ 1 - \frac{1}{n}, \ \ldots$	$a_n = 1 - \frac{1}{n}$
$0, \ -\frac{1}{2}, \ \frac{2}{3}, \ -\frac{3}{4}, \ \ldots, \ (-1)^{n+1}\left(1 - \frac{1}{n}\right), \ \ldots$	$a_n = (-1)^{n+1}\left(1 - \frac{1}{n}\right)$
$3, \ 3, \ 3, \ \ldots, \ 3, \ \ldots$	$a_n = 3$ ☐

We can, of course, use set notation to describe sequences, writing

$\{(n, a_n)\}$ The sequence whose ordered pairs are (n, a_n)

or even

$\{a_n\}$ The sequence a_n

when such an abbreviation will not cause trouble. Both choices of notation lead to graphs, for we can either plot the ordered pairs (n, a_n) in the coordinate plane, or plot the numbers a_n on a single axis (shown as a horizontal axis in Fig. 11.2). Plotting just the number a_n has the advantage

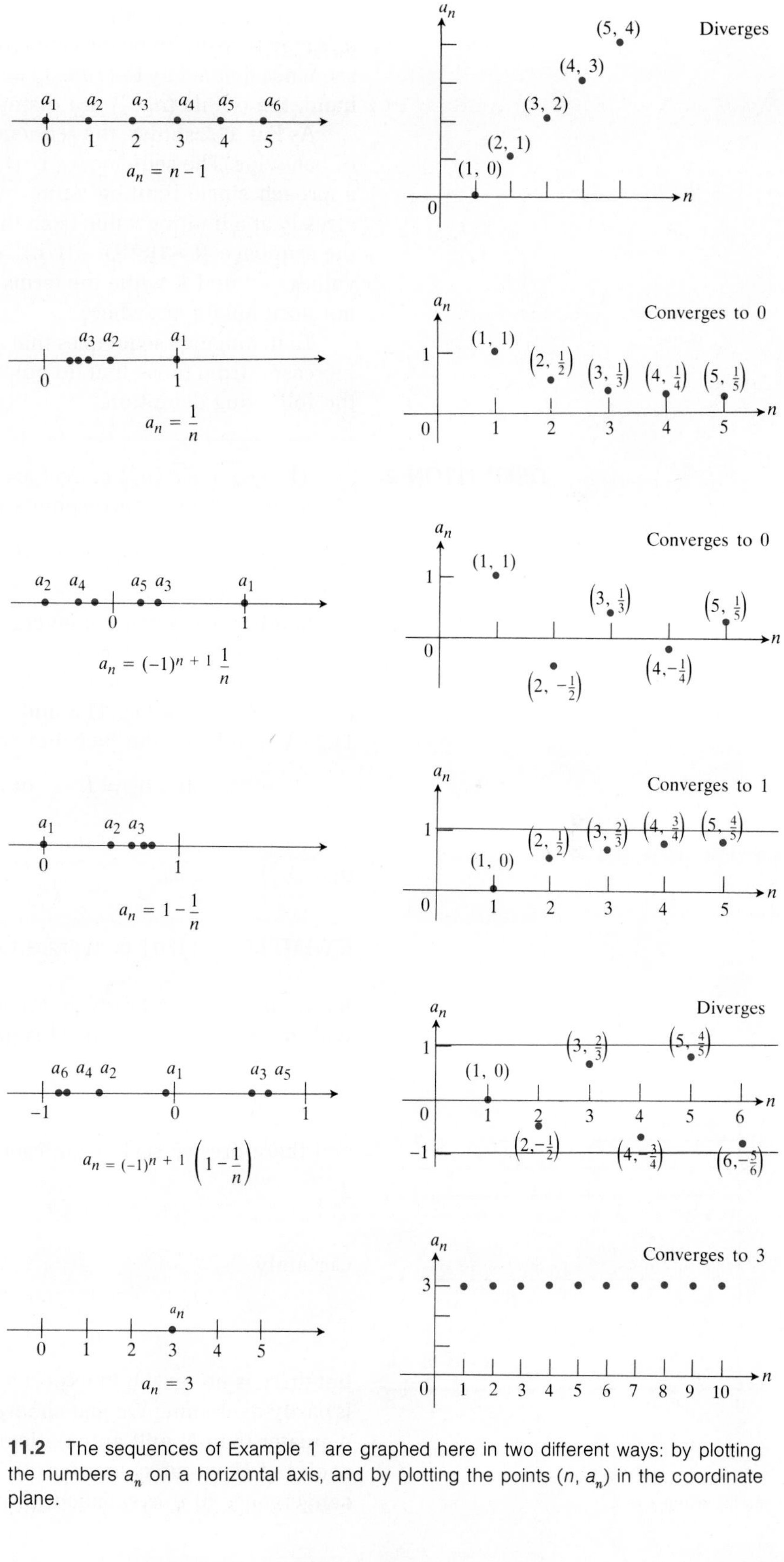

11.2 The sequences of Example 1 are graphed here in two different ways: by plotting the numbers a_n on a horizontal axis, and by plotting the points (n, a_n) in the coordinate plane.

of simplicity. A potential disadvantage, however, is the fact that several a_n's may turn out to be the same for different values of n, as they are in the sequence defined by the rule $a_n = 3$, in which every term is 3. On the other hand, the points (n, a_n) are distinct for different values of n.

As Fig. 11.2 shows, the sequences of Example 1 exhibit different kinds of behavior. The sequences $\{1/n\}$, $\{(-1)^{n+1}(1/n)\}$, and $\{1 - 1/n\}$ seem to approach single limiting values as n increases, and the sequence $\{3\}$ is already at a limiting value from the very first. On the other hand, terms of the sequence $\{(-1)^{n+1}(1 - 1/n)\}$ seem to accumulate near two different values, -1 and 1, while the terms of $\{n - 1\}$ get larger and larger and do not accumulate anywhere.

To distinguish sequences that approach a unique limiting value L, as n increases, from those that do not, we say that they *converge*, according to the following definition.

DEFINITION 2

The sequence $\{a_n\}$ converges to the number L if to every positive number ε there corresponds an index N such that

$$|a_n - L| < \varepsilon \qquad \text{for all} \quad n > N. \tag{4}$$

In other words, $\{a_n\}$ converges to L if, for every positive ε, there is an index N such that all terms after the Nth lie within a radius ε of L. Or, all but finitely many (namely the first N) terms of the sequence lie within a radius ε of L. (See Fig. 11.3 and look once more at the sequences in Fig. 11.2.) We indicate the fact that $\{a_n\}$ converges to L by writing

$$\lim_{n\to\infty} a_n = L \qquad \text{or} \qquad a_n \to L \quad \text{as } n \to \infty,$$

and we call L the *limit* of the sequence $\{a_n\}$. If no such limit exists, we say that $\{a_n\}$ *diverges*.

EXAMPLE 2 $\{1/n\}$ converges to 0.

Solution Let $\varepsilon > 0$ be given. We begin by writing down the inequality (4), with $a_n = 1/n$ and $L = 0$. This gives

$$|a_n - L| = \left|\frac{1}{n} - 0\right| = \frac{1}{n} < \varepsilon, \tag{5}$$

and therefore we seek an integer N such that

$$\frac{1}{n} < \varepsilon \qquad \text{for all} \quad n > N. \tag{6}$$

Certainly

$$\frac{1}{n} < \varepsilon \qquad \text{for all} \quad n > \frac{1}{\varepsilon}, \tag{7}$$

but there is no reason to expect $1/\varepsilon$ to be an integer. This minor difficulty is easily overcome: We just choose any integer $N > 1/\varepsilon$. Then every index n greater than N will automatically be greater than $1/\varepsilon$. In short, for this choice of N we can guarantee (6). The criterion set forth in Definition 2 for convergence to 0 is satisfied. □

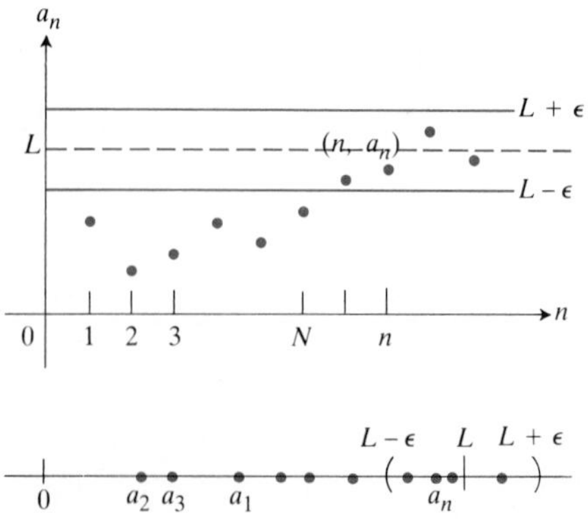

11.3 $a_n \to L$ if L is a horizontal asymptote of $\{(n, a_n)\}$. In this figure, all the a_n's after a_N lie within ϵ of L.

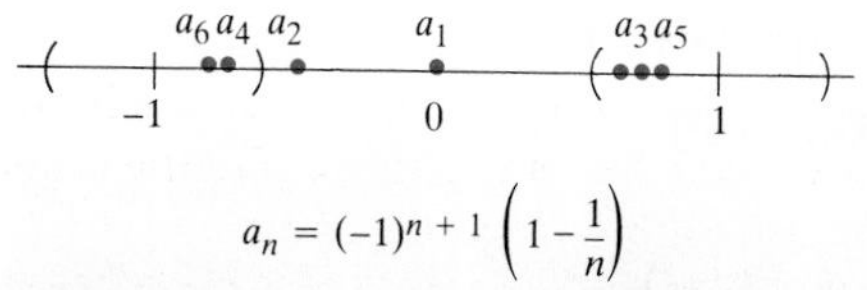

Neither the ϵ-interval about 1 nor the ϵ-interval about -1 contains a complete tail of the sequence.

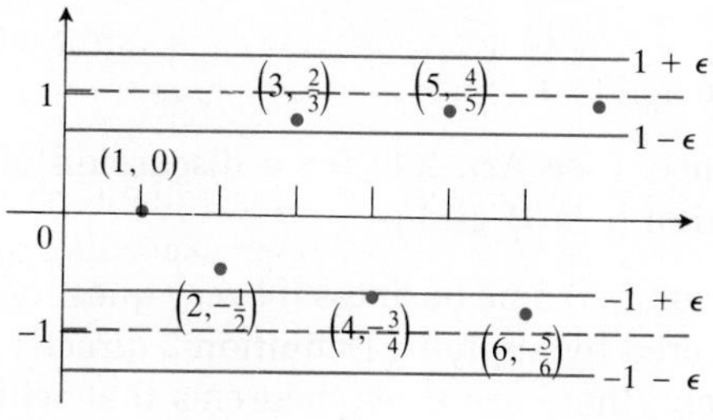

Neither of the ϵ-bands shown here contains all the points (n, a_n) from some index onward.

11.4 The sequence $\{(-1)^{n+1}[1 - (1/n)]\}$ diverges.

EXAMPLE 3 If k is any number, then the constant sequence $\{k\}$, defined by $a_n = k$ for all n, converges to k.

Solution Let $\varepsilon > 0$ be given. When we take both $a_n = k$ and $L = k$ on the left of the inequality in (4), we find

$$|a_n - L| = |k - k| = 0, \tag{8}$$

which is less than any positive ε for every $n \geq 1$. Hence $N = 1$ will work. □

EXAMPLE 4 The sequence $\{(-1)^{n+1}(1 - 1/n)\}$ diverges. To see why, pick a positive ε smaller than 1 so that the bands shown in Fig. 11.4 about the lines $y = 1$ and $y = -1$ do not overlap. Any $\varepsilon < 1$ will do. Convergence to 1 would require every point of the graph from some index on to lie inside the upper band, but this will never happen. As soon as a point (n, a_n) lies within the upper band, every alternate point starting with $(n + 1, a_{n+1})$ will lie within the lower band. Likewise, the sequence cannot converge to -1. On the other hand, because the terms of the sequence get increasingly close to 1 and -1 alternately, they never accumulate near any other value. □

A *tail* of a sequence $\{a_n\}$ is the collection of all the terms whose indices are greater than some index N; in other words, one of the sets $\{a_n | n > N\}$. Another way to say $a_n \to L$ is to say that every ε-interval about L contains a tail. The convergence (or divergence) of a sequence, as well as the limit of a convergent sequence, depends only on the tail behavior of the sequence.

As Example 4 suggests, a sequence cannot have more than one limit: there cannot be two different numbers L and L' such that every ε-interval about each one contains a complete tail. (See Problem 62.)

The behavior of the sequence $\{(-1)^{n+1}(1 - 1/n)\}$ is qualitatively different from that of $\{n - 1\}$, which diverges because it outgrows every real number L. We describe the behavior of $\{n - 1\}$ by writing

$$\lim_{n\to\infty} (n - 1) = \infty.$$

In speaking of infinity as a limit of a sequence $\{a_n\}$, we do not mean that the difference between a_n and infinity becomes small as n increases. We mean that a_n becomes numerically large as n increases.

When a sequence is defined iteratively, we may not have a formula that expresses the nth term as a function of n. For example, the sequence of numbers $\{x_n\}$ of Example 1 of Art. 11.1 appears to converge to a solution of the equation $\sin x - x^2 = 0$. But we have no explicit formula for x_n in terms of n and we don't yet have enough expertise to be sure that the sequence converges. We shall return to this point in Art. 11.6.

REMARK. Picard's method and Newton's method are familiar examples of schemes that define sequences iteratively. To return to the example of

$$\sin x - x^2 = 0,$$

we could base an iteration on the alternate form

$$x = \sin^{-1}(x^2)$$

leading to

$$x_{n+1} = \sin^{-1}(x_n^2),$$

which won't work if we start with $x_0 = 1$. (It leads to $x_1 = 1.57\ldots$, and there is no $\sin^{-1}(x_1^2)$ for such x_1.) But, when we use the inverse function $\sqrt{\sin x}$, the formula

$$x_{n+1} = \sqrt{\sin x_n}, \qquad x_0 = 1,$$

does produce a convergent sequence. (See Art. 2.10 for a discussion of criteria for convergence to a solution of $x = g(x)$.)

The study of limits would be a cumbersome business if every question about convergence had to be answered by applying Definition 2 directly, as we have done so far. Fortunately, there are three theorems that will make this process largely unnecessary from now on. The first two are practically the same as Theorems 1 and 4 in Article 1.9.

THEOREM 1

If $A = \lim_{n\to\infty} a_n$ and $B = \lim_{n\to\infty} b_n$ both exist and are finite, then

i) $\lim \{a_n + b_n\} = A + B$,

ii) $\lim \{ka_n\} = kA$ (k any number),

iii) $\lim \{a_n \cdot b_n\} = A \cdot B$,

iv) $\lim \left\{\dfrac{a_n}{b_n}\right\} = \dfrac{A}{B}$, provided $B \neq 0$ and b_n is never 0,

it being understood that all of the limits are to be taken as $n \to \infty$.

By combining Theorem 1 with Examples 2 and 3, we can proceed immediately to

$$\lim_{n\to\infty} -\frac{1}{n} = -1 \cdot \lim_{n\to\infty} \frac{1}{n} = -1 \cdot 0 = 0,$$

$$\lim_{n\to\infty} \left(1 - \frac{1}{n}\right) = \lim_{n\to\infty} 1 - \lim_{n\to\infty} \frac{1}{n} = 1 - 0 = 1,$$

$$\lim_{n\to\infty} \frac{5}{n^2} = 5 \cdot \lim_{n\to\infty} \frac{1}{n} \cdot \lim_{n\to\infty} \frac{1}{n} = 5 \cdot 0 \cdot 0 = 0,$$

$$\lim_{n\to\infty} \frac{4 - 7n^6}{n^6 + 3} = \lim_{n\to\infty} \frac{(4/n^6) - 7}{1 + (3/n^6)} = \frac{0 - 7}{1 + 0} = -7.$$

A corollary of Theorem 1 that will be useful later on is that every nonzero multiple of a divergent sequence is divergent.

COROLLARY

If the sequence $\{a_n\}$ diverges, and if c is any number different from 0, then the sequence $\{ca_n\}$ diverges.

Proof of the corollary. Suppose, on the contrary, that $\{ca_n\}$ converges. Then, by taking $k = 1/c$ in part (ii) of Theorem 1, we see that the sequence

$$\left\{\frac{1}{c} \cdot ca_n\right\} = \{a_n\}$$

converges. Thus $\{ca_n\}$ cannot converge unless $\{a_n\}$ converges. If $\{a_n\}$ does not converge, then $\{ca_n\}$ does not converge. ■

The next theorem is the sequence version of the Sandwich Theorem of Article 1.9.

THEOREM 2

> If $a_n \leq b_n \leq c_n$ for all n beyond some index N, and if $\lim a_n = \lim c_n = L$, then $\lim b_n = L$ also.

An immediate consequence of Theorem 2 is that, if $|b_n| \leq c_n$ and $c_n \to 0$, then $b_n \to 0$ because $-c_n \leq b_n \leq c_n$. We use this fact in the next example.

EXAMPLE 5

$$\frac{\cos n}{n} \to 0 \qquad \text{because} \quad 0 \leq \left|\frac{\cos n}{n}\right| = \frac{|\cos n|}{n} \leq \frac{1}{n}. \ \square$$

EXAMPLE 6

$$\frac{1}{2^n} \to 0 \qquad \text{because} \quad 0 \leq \frac{1}{2^n} \leq \frac{1}{n}. \ \square$$

EXAMPLE 7

$$(-1)^n\frac{1}{n} \to 0 \qquad \text{because} \quad 0 \leq \left|(-1)^n\frac{1}{n}\right| \leq \frac{1}{n}. \ \square$$

The application of Theorems 1 and 2 is broadened by a theorem that says that the result of applying a continuous function to a convergent sequence is again a convergent sequence. We state the theorem without proof. (See Problem 65 for an outline of the proof.)

THEOREM 3

> If $a_n \to L$ and if f is a function that is continuous at L and defined at all the a_n's, then $f(a_n) \to f(L)$.

EXAMPLE 8 Find $\lim_{n\to\infty} \sqrt{(n+1)/n}$.

Solution Let $f(x) = \sqrt{x}$ and $a_n = (n+1)/n$. Then

$$a_n \to 1 \qquad \text{and} \qquad f(a_n) = \sqrt{a_n} \to f(1) = \sqrt{1} = 1$$

because $f(x)$ is continuous at $x = 1$. □

EXAMPLE 9 Find $\lim_{n\to\infty} 2^{1/n}$.

Solution Let $f(x) = 2^x$ and $a_n = 1/n$. Then

$$a_n \to 0 \qquad \text{and} \qquad f(a_n) = 2^{1/n} \to 2^0 = 1$$

because 2^x is continuous at $x = 0$. □

THEOREM 4

> Suppose that $f(x)$ is a function defined for all $x \geq n_0$ and $\{a_n\}$ is a sequence such that $a_n = f(n)$ when $n \geq n_0$. If
>
> $$\lim_{x\to\infty} f(x) = L, \qquad \text{then} \qquad \lim_{n\to\infty} a_n = L.$$

Proof. Let $\varepsilon > 0$. Suppose that L is a finite limit such that

$$\lim_{x\to\infty} f(x) = L.$$

Then there is a number M such that

$$x > M \Rightarrow |f(x) - L| < \varepsilon.$$

Let N be an integer such that

$$N \geq n_0 \qquad \text{and} \qquad N > M.$$

Then

$$n > N \Rightarrow a_n = f(n) \qquad \text{and} \qquad |a_n - L| = |f(n) - L| < \varepsilon. \ \blacksquare$$

The proof would require modification in case $f(x) \to +\infty$ or $f(x) \to -\infty$, but in either of these cases the sequence $\{a_n\}$ would diverge.

L'Hôpital's rule can be used to determine the limits of some sequences. The next example shows how.

EXAMPLE 10 Find $\lim_{n\to\infty} (\ln n)/n$.

Solution The function $(\ln x)/x$ is defined for all $x \geq 1$ and agrees with the given sequence on the positive integers. Therefore $\lim_{n\to\infty} (\ln n)/n$ will equal $\lim_{x\to\infty} (\ln x)/x$ if the latter exists. A single application of l'Hôpital's rule shows that

$$\lim_{x\to\infty} \frac{\ln x}{x} = \lim_{x\to\infty} \frac{1/x}{1} = \frac{0}{1} = 0.$$

We conclude that $\lim_{n\to\infty} (\ln n)/n = 0$. □

When we use l'Hôpital's rule to find the limit of a sequence, we often treat n as a continuous real variable, and differentiate directly with respect to n. This saves us from having to rewrite the formula for a_n as we did in Example 10.

EXAMPLE 11 Find $\lim_{n\to\infty} (2^n/5n)$.

Solution By l'Hôpital's rule,

$$\lim_{n\to\infty} \frac{2^n}{5n} = \lim_{n\to\infty} \frac{2^n \cdot \ln 2}{5} = \infty. \ \square$$

When the terms of the sequence are ratios of polynomials in n, we have a choice of methods for finding the limit: either use l'Hôpital's rule or divide the numerator and denominator by an appropriate power of n. (For example, we could divide both numerator and denominator by the highest power of n in one or the other.) The following shows what happens when the degree of the numerator is equal to, less than, or greater than, the degree of the denominator.

EXAMPLE 12 Find the following limits as $n \to \infty$.

$$\lim \frac{n^2 - 2n + 1}{2n^2 + 5} = \lim \frac{1 - (2/n) + (1/n^2)}{2 + (5/n^2)} = \frac{1}{2},$$

$$\lim \frac{n^3 + 5n}{n^4 - 6} = \lim \frac{(1/n) + (5/n^3)}{1 - (6/n^4)} = 0,$$

$$\lim \frac{n^2 - 5}{n + 1} = \lim \frac{2n}{1} = \infty \quad \text{(l'Hôpital's rule).} \ \square$$

PROBLEMS

In Problems 1–10, write a_1, a_2, a_3, and a_4 for each sequence $\{a_n\}$. Determine which of the sequences converge and which diverge. Find the limit of each sequence that converges.

1. $a_n = \dfrac{1 - n}{n^2}$
2. $a_n = \dfrac{n}{2^n}$
3. $a_n = \left(\dfrac{1}{3}\right)^n$
4. $a_n = \dfrac{1}{n!}$
5. $a_n = \dfrac{(-1)^{n+1}}{2n - 1}$
6. $a_n = 2 + (-1)^n$
7. $a_n = \cos \dfrac{n\pi}{2}$
8. $a_n = 8^{1/n}$
9. $a_n = \dfrac{(-1)^{n-1}}{\sqrt{n}}$
10. $a_n = \sin^2 \dfrac{1}{n} + \cos^2 \dfrac{1}{n}$

In Problems 11–54, determine which of the sequences $\{a_n\}$ converge and which diverge. Find the limit of each sequence that converges.

11. $a_n = \dfrac{1}{10n}$
12. $a_n = \dfrac{n}{10}$
13. $a_n = 1 + \dfrac{(-1)^n}{n}$
14. $a_n = \dfrac{1 + (-1)^n}{n}$
15. $a_n = (-1)^n\left(1 - \dfrac{1}{n}\right)$
16. $a_n = 1 + (-1)^n$
17. $a_n = \dfrac{2n + 1}{1 - 3n}$
18. $a_n = \dfrac{n^2 - n}{2n^2 + n}$
19. $a_n = \sqrt{\dfrac{2n}{n + 1}}$
20. $a_n = \dfrac{\sin n}{n}$
21. $a_n = \sin \pi n$
22. $a_n = \sin\left(\dfrac{\pi}{2} + \dfrac{1}{n}\right)$
23. $a_n = n\pi \cos n\pi$
24. $a_n = \dfrac{\sin^2 n}{2^n}$
25. $a_n = \dfrac{n^2}{(n + 1)^2}$
26. $a_n = \dfrac{\sqrt{n} - 1}{\sqrt{n}}$
27. $a_n = \dfrac{1 - 5n^4}{n^4 + 8n^3}$
28. $a_n = \sqrt[n]{3^{2n+1}}$
29. $a_n = \tanh n$
30. $a_n = \dfrac{\ln n}{\sqrt{n}}$
31. $a_n = \dfrac{2(n + 1) + 1}{2n + 1}$
32. $a_n = \dfrac{(n + 1)!}{n!}$
33. $a_n = 5$
34. $a_n = 5^n$
35. $a_n = (0.5)^n$
36. $a_n = \dfrac{10^{n+1}}{10^n}$
37. $a_n = \dfrac{n^n}{(n + 1)^{n+1}}$
38. $a_n = (0.03)^{1/n}$
39. $a_n = \sqrt{2 - \dfrac{1}{n}}$
40. $a_n = 2 + (0.1)^n$
41. $a_n = \dfrac{3^n}{n^3}$
42. $a_n = \dfrac{\ln (n + 1)}{n + 1}$
43. $a_n = \ln n - \ln (n + 1)$
44. $a_n = \dfrac{1 - 2^n}{2^n}$
45. $a_n = \dfrac{n^2 - 2n + 1}{n - 1}$
46. $a_n = \dfrac{n + (-1)^n}{n}$
47. $a_n = \left(-\dfrac{1}{2}\right)^n$
48. $a_n = \dfrac{\ln n}{\ln 2n}$
49. $a_n = \tan^{-1} n$
50. $a_n = \sinh (\ln n)$
51. $a_n = n \sin \dfrac{1}{n}$
52. $a_n = \dfrac{2n + \sin n}{n + \cos 5n}$
53. $a_n = \dfrac{n^2}{2n - 1} \sin \dfrac{1}{n}$
54. $a_n = n\left(1 - \cos \dfrac{1}{n}\right)$
55. Show that $\lim_{n\to\infty} (n!/n^n) = 0$. (*Hint:* Expand the numerator and denominator and compare the quotient with $1/n$.)
56. (*Calculator*) The formula $x_{n+1} = (x_n + a/x_n)/2$ is the one produced by Newton's method to generate a sequence of approximations to the positive solution of $x^2 - a = 0$, $a > 0$. Starting with $x_1 = 1$ and $a = 3$, use the formula to calculate successive terms of the

sequence until you have approximated $\sqrt{3}$ as accurately as your calculator permits.

57. (*Calculator*) If your calculator has a square-root key, enter $x = 10$ and take successive square roots to approximate the terms of the sequence $10^{1/2}$, $10^{1/4}$, $10^{1/8}$, . . . , continuing as far as your calculator permits. Repeat, with $x = 0.1$. Try other positive numbers above and below 1. When you have enough evidence, guess the answers to these questions: Does $\lim_{n\to\infty} x^{1/n}$ exist when $x > 0$? Does it matter what x is?

58. (*Calculator*) If you start with a reasonable value of x_1, then the rule $x_{n+1} = x_n + \cos x_n$ will generate a sequence that converges to $\pi/2$. Figure 11.5 shows why. The convergence is rapid. With $x_1 = 1$, calculate x_2, x_3, and x_4. Find out what happens when you start with $x_1 = 5$. Remember to use radians.

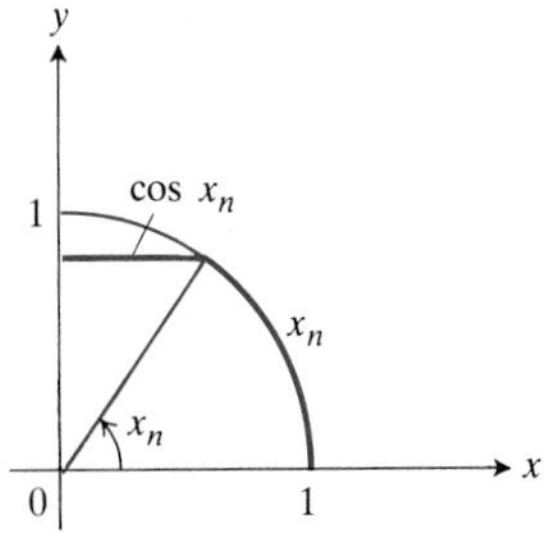

11.5 The length $\pi/2$ of the circular arc is approximated by $x_n + \cos x_n$.

59. Suppose that $f(x)$ is defined for all $0 \le x \le 1$, that f is differentiable at $x = 0$, and that $f(0) = 0$. Define a sequence $\{a_n\}$ by the rule $a_n = nf(1/n)$. Show that $\lim a_n = f'(0)$.

Use the result of Problem 59 to find the limits of the sequences in Problems 60 and 61.

60. $a_n = n \tan^{-1} \dfrac{1}{n}$ **61.** $a_n = n(e^{1/n} - 1)$

62. Prove that a sequence $\{a_n\}$ cannot have two different limits L and L'. (*Hint:* Take $\varepsilon = \frac{1}{2}|L - L'|$ in Eq. 4.)

63. Under the hypotheses of Theorem 4, prove that if $\lim_{x\to\infty} f(x) = +\infty$, then the sequence $\{a_n\}$ diverges.

64. Prove Theorem 2.

65. Prove Theorem 3. (Outline of proof. Assume the hypotheses of the theorem and let ε be any positive number. For this ε there is a $\delta > 0$ such that

$$|f(x) - f(L)| < \varepsilon \quad \text{when} \quad |x - L| < \delta.$$

For such a $\delta > 0$, there is an index N such that

$$|a_n - L| < \delta \quad \text{when } n > N.$$

What is the conclusion?)

> ***Toolkit* programs**
> Sequences and Series

11.3 Limits That Arise Frequently

Some limits arise so frequently that they are worth special attention. In this article we investigate these limits and look at examples in which they occur.

1. $\lim_{n\to\infty} \dfrac{\ln n}{n} = 0$	2. $\lim_{n\to\infty} \sqrt[n]{n} = 1$
3. $\lim_{n\to\infty} x^{1/n} = 1 \quad (x > 0)$	4. $\lim_{n\to\infty} x^n = 0 \quad (\lvert x\rvert < 1)$
5. $\lim_{n\to\infty} \left(1 + \dfrac{x}{n}\right)^n = e^x \quad$ (any x)	6. $\lim_{n\to\infty} \dfrac{x^n}{n!} = 0 \quad$ (any x)

1. $\mathbf{\lim_{n\to\infty} \dfrac{\ln n}{n} = 0}$

This limit was calculated in Example 10 of Article 11.2.

2. $\lim_{n\to\infty} \sqrt[n]{n} = 1$

Let $a_n = n^{1/n}$. Then

$$\ln a_n = \ln n^{1/n} = \frac{1}{n}\ln n \to 0, \tag{1}$$

so that, by applying Theorem 3 of Article 11.2 to $f(x) = e^x$, we have

$$n^{1/n} = a_n = e^{\ln a_n} \to e^0 = 1. \tag{2}$$

3. $\lim_{n\to\infty} x^{1/n} = 1$, **if** $x > 0$

Let $a_n = x^{1/n}$. Then

$$\ln a_n = \ln x^{1/n} = \frac{1}{n}\ln x \to 0, \tag{3}$$

because x remains fixed while n gets large. Thus, again by Theorem 3, with $f(x) = e^x$,

$$x^{1/n}\ a_n = e^{\ln a_n} \to e^0 = 1. \tag{4}$$

4. $\lim_{n\to\infty} x^n = 0$, **if** $|x| < 1$

Our scheme here is to show that the criteria of Definition 2 of Article 11.2 are satisfied with $L = 0$. That is, we show that to each $\varepsilon > 0$ there corresponds an index N so large that

$$|x^n| < \varepsilon \qquad \text{for} \quad n > N. \tag{5}$$

Since $\varepsilon^{1/n} \to 1$, while $|x| < 1$, there is an index N for which

$$|x| < \varepsilon^{1/N}. \tag{6}$$

In other words,

$$|x^N| = |x|^N < \varepsilon. \tag{7}$$

This is the index we seek, because

$$|x^n| < |x^N| \qquad \text{for} \quad n > N. \tag{8}$$

Combining (7) and (8) produces

$$|x^n| < |x^N| < \varepsilon \qquad \text{for} \quad n > N, \tag{9}$$

which is just what we needed to show.

5. $\lim_{n\to\infty}\left(1 + \frac{x}{n}\right)^n = e^x$ **(any** x**)**

Let

$$a_n = \left(1 + \frac{x}{n}\right)^n.$$

Then

$$\ln a_n = \ln\left(1 + \frac{x}{n}\right)^n = n \ln\left(1 + \frac{x}{n}\right) \to x,$$

as we can see by the following application of l'Hôpital's rule, in which we differentiate with respect to n:

$$\lim_{n\to\infty} n \ln\left(1 + \frac{x}{n}\right) = \lim_{n\to\infty} \frac{\ln(1 + x/n)}{1/n}$$

$$= \lim_{n\to\infty} \frac{\left(\dfrac{1}{1 + x/n}\right)\cdot\left(-\dfrac{x}{n^2}\right)}{-1/n^2} = \lim_{n\to\infty} \frac{x}{1 + x/n} = x.$$

Thus, by Theorem 3 of Art. 11.2, with $f(x) = e^x$,

$$\left(1 + \frac{x}{n}\right)^n = a_n = e^{\ln a_n} \to e^x.$$

6. $\displaystyle\lim_{n\to\infty} \frac{x^n}{n!} = 0$ **(any x)**

Since

$$-\frac{|x|^n}{n!} \le \frac{x^n}{n!} \le \frac{|x|^n}{n!},$$

all we really need to show is that $|x|^n/n! \to 0$. The first step is to choose an integer $M > |x|$, so that

$$\frac{|x|}{M} < 1 \qquad \text{and} \qquad \left(\frac{|x|}{M}\right)^n \to 0.$$

We then restrict our attention to values of $n > M$. For these values of n, we can write

$$\frac{|x|^n}{n!} = \frac{|x|^n}{1\cdot 2 \cdots M\cdot \underbrace{(M+1)(M+2)\cdots n}_{(n-M)\text{ factors}}}$$

$$\le \frac{|x|^n}{M!\,M^{n-M}} = \frac{|x|^n M^M}{M!\,M^n} = \frac{M^M}{M!}\left(\frac{|x|}{M}\right)^n.$$

Thus,

$$0 \le \frac{|x|^n}{n!} \le \frac{M^M}{M!}\left(\frac{|x|}{M}\right)^n.$$

Now, the constant $M^M/M!$ does not change with n. Thus the Sandwich Theorem tells us that

$$\frac{|x|^n}{n!} \to 0 \qquad \text{because} \qquad \left(\frac{|x|}{M}\right)^n \to 0.$$

REMARK. It is important to note that x is fixed and only n varies in the limits in Formulas 3 through 6, above.

A large number of limits can be found directly from the six limits we have just calculated.

EXAMPLES

1. If $|x| < 1$, then $x^{n+4} = x^4 \cdot x^n \to x^4 \cdot 0 = 0$. □
2. $\sqrt[n]{2n} = \sqrt[n]{2}\sqrt[n]{n} \to 1 \cdot 1 = 1$. □
3. $\left(1 + \dfrac{1}{n}\right)^{2n} = \left[\left(1 + \dfrac{1}{n}\right)^n\right]^2 \to e^2$. □
4. $\dfrac{100^n}{n!} \to 0$. □

5. $\dfrac{x^{n+1}}{(n+1)!} = \dfrac{x}{(n+1)} \cdot \dfrac{x^n}{n!} \to 0 \cdot 0 = 0.$ □

Still other limits can be calculated by using logarithms or l'Hôpital's rule, as in the calculations of limits (2), (3), and (5) at the beginning of this article.

EXAMPLE 6 Find $\lim_{n\to\infty} (\ln(3n+5))/n$.

Solution By l'Hôpital's rule,

$$\lim_{n\to\infty} \frac{\ln(3n+5)}{n} = \lim_{n\to\infty} \frac{3/(3n+5)}{1} = 0. \quad \square$$

EXAMPLE 7 Find $\lim_{n\to\infty} \sqrt[n]{3n+5}$.

Solution Let

$$a_n = \sqrt[n]{3n+5} = (3n+5)^{1/n}.$$

Then,

$$\ln a_n = \ln(3n+5)^{1/n} = \frac{\ln(3n+5)}{n} \to 0,$$

as in Example 6. Therefore,

$$a_n = e^{\ln a_n} \to e^0 = 1,$$

by Theorem 3 of Article 11.2. □

We know that $\ln n$ increases more slowly than n does as $n \to \infty$, because $(\ln n)/n \to 0$. But we can say much more: $\ln n$ increases more slowly than $\sqrt{n}$, $\sqrt[3]{n}$, or even $n^{0.00001}$. In fact, if c is any positive constant $(\ln n)/n^c \to 0$ as $n \to \infty$. The next example establishes this fact.

EXAMPLE 8 Show that

$$\lim_{n\to\infty} \frac{\ln n}{n^c} = 0$$

if c is any positive constant.

Solution Apply l'Hôpital's rule and get

$$\lim \frac{\ln n}{n^c} = \lim \frac{1/n}{cn^{c-1}} = \lim \frac{1}{cn^c} = 0. \quad \square$$

PROBLEMS

Determine which of the following sequences $\{a_n\}$ converge and which diverge. Find the limit of each sequence that converges.

1. $a_n = \dfrac{1+\ln n}{n}$

2. $a_n = \dfrac{\ln n}{3n}$

3. $a_n = \dfrac{(-4)^n}{n!}$

4. $a_n = \sqrt[n]{10n}$

5. $a_n = (0.5)^n$

6. $a_n = \dfrac{1}{(0.9)^n}$

7. $a_n = \left(1 + \dfrac{7}{n}\right)^n$

8. $a_n = \left(\dfrac{n+5}{n}\right)^n$

9. $a_n = \dfrac{\ln(n+1)}{n}$

10. $a_n = \sqrt[n]{n+1}$

11. $a_n = \dfrac{n!}{10^{6n}}$

12. $a_n = \dfrac{1}{\sqrt{2^n}}$

13. $a_n = \sqrt[2n]{n}$

14. $a_n = (n+4)^{1/(n+4)}$

15. $a_n = \dfrac{1}{3^{2n-1}}$

16. $a_n = \ln\left(1 + \dfrac{1}{n}\right)^n$

17. $a_n = \left(\dfrac{n}{n+1}\right)^n$

18. $a_n = \left(1 + \dfrac{1}{n}\right)^{-n}$

19. $a_n = \dfrac{\ln(2n+1)}{n}$

20. $a_n = \sqrt[n]{2n+1}$

21. $a_n = \sqrt[n]{\dfrac{x^n}{2n+1}}, \quad x > 0$

22. $a_n = \sqrt[n]{n^2}$

23. $a_n = \sqrt[n]{n^2 + n}$

24. $a_n = \dfrac{3^n \cdot 6^n}{2^{-n} \cdot n!}$

25. $a_n = \left(\dfrac{3}{n}\right)^{1/n}$

26. $a_n = \sqrt[n]{4^n n}$

27. $a_n = \left(1 - \dfrac{1}{n}\right)^n$

28. $a_n = \left(1 - \dfrac{1}{n^2}\right)^n$

29. $a_n = \dfrac{\ln(n^2)}{n}$

30. $a_n = \dfrac{(\ln n)^{200}}{n}$

31. $a_n = \dfrac{\ln n}{n^{1/n}}$

32. $a_n = \dfrac{1}{n}\displaystyle\int_1^n \dfrac{1}{x}\,dx$

33. $a_n = \displaystyle\int_1^n \dfrac{1}{x^p}\,dx, \quad p > 1$

(*Calculator*) In Problems 34–36, use a calculator to find a value of N such that the given inequality is satisfied for $n \geq N$.

34. $|\sqrt[n]{0.5} - 1| < 10^{-3}$

35. $|\sqrt[n]{n} - 1| < 10^{-3}$

36. $\dfrac{2^n}{n!} < 10^{-9}$

(*Hint:* If you do not have a factorial key, then write

$$\frac{2^n}{n!} = \left(\frac{2}{1}\right)\left(\frac{2}{2}\right)\cdots\left(\frac{2}{n}\right).$$

That is, calculate successive terms by multiplying by 2 and dividing by the next value of n.)

37. In Example 8, we assumed that obviously if $c > 0$, then $1/n^c \to 0$ as $n \to \infty$. Write out a formal proof of this fact. (*Hint:* If $\varepsilon = 0.001$ and $c = 0.04$, how large should N be in order to be sure that $|1/n^c - 0| < \varepsilon$ when $n > N$?)

Toolkit programs

Sequences and Series

11.4 Infinite Series

Infinite series are sequences of a special kind: those in which the nth term is the sum of the first n terms of a related sequence.

EXAMPLE 1 Suppose that we start with the sequence

$$1, \frac{1}{2}, \frac{1}{4}, \frac{1}{8}, \frac{1}{16}, \frac{1}{32}, \frac{1}{64}, \ldots$$

and then form a new sequence by successively adding 1, 2, 3, ... terms of the original sequence. We denote the terms of the first sequence by a_n and the terms of the second sequence by s_n. For this example, we have

$$s_1 = a_1 = 1,$$

$$s_2 = a_1 + a_2 = 1 + \frac{1}{2} = \frac{3}{2},$$

$$s_3 = a_1 + a_2 + a_3 = 1 + \frac{1}{2} + \frac{1}{4} = \frac{7}{4},$$

as the first three terms of the sequence $\{s_n\}$. □

When a sequence $\{s_n\}$ is formed in this way from a given sequence $\{a_n\}$ by the rule

$$s_n = a_1 + a_2 + \cdots + a_n = \sum_{k=1}^{n} a_k,$$

the result is called an *infinite series*. The series $\{s_n\}$ is usually denoted by

$$\sum_{k=1}^{\infty} a_k \quad \text{or} \quad \sum_{n=1}^{\infty} a_n \quad \text{or simply} \sum a_n,$$

to show how the numbers s_n are obtained from the terms a_n.

We formalize these remarks in the following definition.

DEFINITION

1. If $\{a_n\}$ is a given sequence and s_n is defined by

$$s_n = a_1 + a_2 + \cdots + a_n = \sum_{k=1}^{n} a_k,$$

then the sequence $\{s_n\}$ is called an *infinite series*.

2. Instead of $\{s_n\}$, we usually write $\Sigma_{n=1}^{\infty} a_n$ or simply $\Sigma\, a_n$ (to show how the sums s_n are related to the terms $a_1, a_2, \ldots$).
3. The number $s_n = \Sigma_{k=1}^{n} a_k$ is called the nth *partial sum* of the series $\Sigma\, a_k$.
4. The number a_n is called the nth *term* of the series (and it is also the nth term of the original sequence $\{a_n\}$).
5. The series $\Sigma\, a_n$ is said to *converge* to a number L if and only if

$$L = \lim_{n\to\infty} s_n = \lim_{n\to\infty} \sum_{k=1}^{n} a_k,$$

in which case we call L the *sum* of the series and write

$$\sum_{n=1}^{\infty} a_n = L \quad \text{or} \quad a_1 + a_2 + \cdots + a_n + \cdots = L.$$

If no such limit exists (i.e., if $\{s_n\}$ diverges), the series is said to *diverge*.

We shall illustrate one method of finding the sum of an infinite series with the repeating decimal

$$0.3333\ldots = \frac{3}{10} + \frac{3}{100} + \frac{3}{1000} + \frac{3}{10{,}000} + \cdots$$

$$s_1 = \frac{3}{10},$$

$$s_2 = \frac{3}{10} + \frac{3}{10^2},$$

$$s_n = \frac{3}{10} + \frac{3}{10^2} + \cdots + \frac{3}{10^n}.$$

We can obtain a simple expression for s_n in closed form as follows: We multiply both sides of the equation for s_n by $\frac{1}{10}$ and obtain

$$\frac{1}{10}s_n = \frac{3}{10^2} + \frac{3}{10^3} + \cdots + \frac{3}{10^n} + \frac{3}{10^{n+1}}.$$

When we subtract this from s_n, we have

$$s_n - \frac{1}{10}s_n = \frac{3}{10} - \frac{3}{10^{n+1}} = \frac{3}{10}\left(1 - \frac{1}{10^n}\right).$$

Therefore,

$$\frac{9}{10}s_n = \frac{3}{10}\left(1 - \frac{1}{10^n}\right) \quad \text{and} \quad s_n = \frac{3}{9}\left(1 - \frac{1}{10^n}\right).$$

As $n \to \infty$, $(\frac{1}{10})^n \to 0$ and

$$\lim_{n\to\infty} s_n = \frac{3}{9} = \frac{1}{3}.$$

We therefore say that the sum of the infinite series

$$\frac{3}{10^1} + \frac{3}{10^2} + \frac{3}{10^3} + \cdots + \frac{3}{10^n} + \cdots$$

is $\frac{1}{3}$, and we write

$$\sum_{n=1}^{\infty} \frac{3}{10^n} = \frac{1}{3}.$$

The decimal 0.333 . . . is a special kind of *geometric series*.

DEFINITION

A series of the form

$$a + ar + ar^2 + ar^3 + \cdots + ar^{n-1} + \cdots \tag{1}$$

is called a *geometric series*. The ratio of any term to the one before it is r.

The ratio r can be positive, as in

$$1 + \frac{1}{2} + \frac{1}{4} + \cdots + \frac{1}{2^{n-1}} + \cdots, \tag{2}$$

or negative, as in

$$1 - \frac{1}{3} + \frac{1}{9} - \cdots + (-1)^n\frac{1}{3^{n-1}} + \cdots. \tag{3}$$

The sum of the first n terms of (1) is

$$s_n = a + ar + ar^2 + \cdots + ar^{n-1}. \tag{4}$$

Multiplying both sides of (4) by r gives

$$rs_n = ar + ar^2 + \cdots + ar^{n-1} + ar^n. \tag{5}$$

When we subtract (5) from (4), nearly all the terms cancel on the right side, leaving

$$s_n - rs_n = a - ar^n,$$

or

$$(1 - r)s_n = a(1 - r^n). \tag{6}$$

If $r \neq 1$, we may divide (6) by $(1 - r)$ to obtain

$$s_n = \frac{a(1 - r^n)}{1 - r}, \qquad r \neq 1. \tag{7a}$$

On the other hand, if $r = 1$ in (4), we get

$$s_n = na, \qquad r = 1. \tag{7b}$$

We are interested in the limit as $n \to \infty$ in Eqs. (7a) and (7b). Clearly, (7b) has no finite limit if $a \neq 0$. If $a = 0$, the series (1) is just

$$0 + 0 + 0 + \cdots,$$

which converges to the sum zero.

If $r \neq 1$, we use (7a). In the right side of (7a), n appears only in the expression r^n. This approaches zero as $n \to \infty$ if $|r| < 1$. Therefore,

$$\lim_{n\to\infty} s_n = \lim_{n\to\infty} \frac{a(1 - r^n)}{1 - r} = \frac{a}{1 - r} \qquad \text{if} \quad |r| < 1. \tag{8a}$$

If we recall that $r^0 = 1$ when $r \neq 0$ we can summarize by writing

$$a + ar + ar^2 + \cdots + ar^{n-1} + \cdots = a \sum_{n=1}^{\infty} r^{n-1} = \frac{a}{1 - r}, \tag{8b}$$

or

$$a + ar + ar^2 + \cdots + ar^{n-1} + \cdots = a \sum_{n=0}^{\infty} r^{n} = \frac{a}{1 - r} \tag{8c}$$

if $0 < |r| < 1$. If $r = 0$, the series still converges, to $a/(1 - r) = a$. If $|r| > 1$, then $|r^n| \to \infty$, and (1) diverges.

The remaining case is where $r = -1$. Then $s_1 = a$, $s_2 = a - a = 0$, $s_3 = a$, $s_4 = 0$, and so on. If $a \neq 0$, this sequence of partial sums has no limit as $n \to \infty$, and the series (1) diverges.

We have thus proved the following theorem.

THEOREM 1

Geometric Series Theorem

If $|r| < 1$, the geometric series

$$a + ar + ar^2 + \cdots + ar^{n-1} + \cdots$$

converges to $a/(1 - r)$. If $|r| \geq 1$, the series diverges unless $a = 0$. If $a = 0$, the series converges to 0.

EXAMPLE 1 Geometric series with $a = \frac{1}{9}$ and $r = \frac{1}{3}$.

$$\frac{1}{9} + \frac{1}{27} + \frac{1}{81} + \cdots = \frac{1}{9}\left(1 + \frac{1}{3} + \frac{1}{3^2} + \cdots\right) = \frac{1/9}{1 - (1/3)} = \frac{1}{6}. \ \square$$

EXAMPLE 2 Geometric series with $a = 4$ and $r = -\frac{1}{2}$.

$$4 - 2 + 1 - \frac{1}{2} + \frac{1}{4} - \cdots = 4\left(1 - \frac{1}{2} + \frac{1}{4} - \frac{1}{8} + \frac{1}{16} - \cdots\right)$$

$$= \frac{4}{1 + (1/2)} = \frac{8}{3}. \ \square$$

EXAMPLE 3 A ball is dropped from a meters above a flat surface. Each time the ball hits after falling a distance h, it rebounds a distance rh, where r is a positive number less than one. Find the total distance the ball travels up and down. See Fig. 11.6.

Solution The distance is given by the series

$$s = a + 2ar + 2ar^2 + 2ar^3 + \cdots.$$

The terms following the first term form a geometric series of sum $2ar/(1 - r)$. Hence the distance is

$$s = a + \frac{2ar}{1 - r} = a\frac{1 + r}{1 - r}.$$

For instance, if a is 6 meters and $r = \frac{2}{3}$, the distance is

$$s = 6\frac{1 + \frac{2}{3}}{1 - \frac{2}{3}} = 30 \text{ m. } \square$$

EXAMPLE 4 If we take $a = 1$ and $r = x$ in the geometric series theorem, we obtain

$$1 + x + x^2 + \cdots + x^{n-1} + \cdots = \frac{1}{1 - x}, \qquad |x| < 1. \ \square \tag{9}$$

REMARK 1. We were fortunate in the case of the geometric series to have found the closed-form expressions

$$s_n = \begin{cases} a\dfrac{1 - r^n}{1 - r} & \text{when } r \neq 1, \\ na & \text{when } r = 1, \end{cases}$$

from which we could get the precise results given by Theorem 1. There are not many other types of series where such closed-form expressions are available. (The next example is one of those rare cases.) Most of the remainder of this chapter will be devoted to tests that we can apply to the individual terms a_n of the series $\Sigma\, a_n$ to determine whether the series converges or diverges, without having to calculate the partial sums s_n. It will turn out that we can do so for a great many series. If a series does converge, we still have the question of determining its sum. Chapter 12 will help to some extent in doing that, but for a great many series we will still be left with no alternative but to compute numerical values of the partial sums and use those to estimate the true sum.

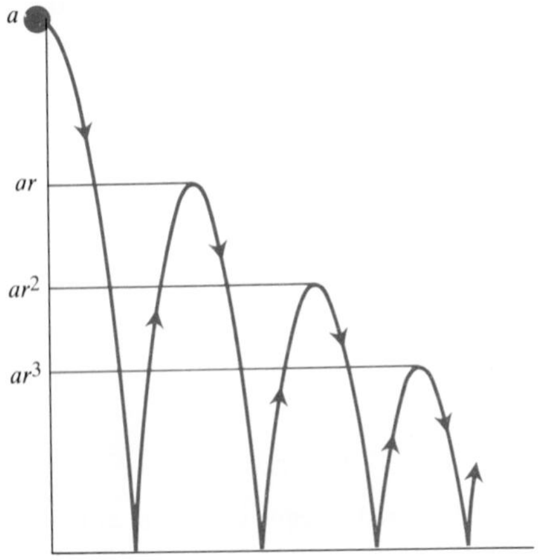

11.6 The height of each rebound is reduced by the factor r.

As noted above, the next example is another series whose sum can be found exactly by first finding a closed expression, or formula, for the kth partial sum s_k.

EXAMPLE 5 Determine whether $\sum_{n=1}^{\infty} [1/n(n+1)]$ converges. If it does, find the sum.

Solution We begin by looking for a pattern in the sequence of partial sums that might lead us to a closed expression for s_k. The key to success here, as in the integration

$$\int \frac{dx}{x(x+1)} = \int \frac{dx}{x} - \int \frac{dx}{x+1},$$

is the use of partial fractions:

$$\frac{1}{k(k+1)} = \frac{1}{k} - \frac{1}{k+1}. \tag{10}$$

This permits us to write the partial sum

$$\sum_{n=1}^{k} \frac{1}{n(n+1)} = \frac{1}{1 \cdot 2} + \frac{1}{2 \cdot 3} + \cdots + \frac{1}{k \cdot (k+1)}$$

as

$$s_k = \left(\frac{1}{1} - \frac{1}{2}\right) + \left(\frac{1}{2} - \frac{1}{3}\right) + \cdots + \left(\frac{1}{k} - \frac{1}{k+1}\right). \tag{11}$$

By removing parentheses on the right, and combining terms, we find that

$$s_k = 1 - \frac{1}{k+1} = \frac{k}{k+1}. \tag{12}$$

From this expression for s_k, we see immediately that $s_k \to 1$. Therefore the series does converge, and

$$\sum_{n=1}^{\infty} \frac{1}{n(n+1)} = 1. \ \square \tag{13}$$

There are, of course, other series that diverge besides geometric series with $|r| \geq 1$.

EXAMPLE 6

$$\sum_{n=1}^{\infty} n^2 = 1 + 4 + 9 + \cdots + n^2 + \cdots$$

diverges because the partial sums grow beyond every number L. The number $s_n = 1 + 4 + 9 + \cdots + n^2$ is greater than or equal to n^2 at each stage. $\square$

EXAMPLE 7

$$\sum_{n=1}^{\infty} \frac{n+1}{n} = \frac{2}{1} + \frac{3}{2} + \frac{4}{3} + \cdots + \frac{n+1}{n} + \cdots$$

diverges because the sequence of partial sums eventually outgrows every preassigned number: each term is greater than 1, so the sum of n terms is greater than n. □

A series can diverge without having its partial sums become large. For instance, the partial sums may oscillate between two extremes, as they do in the next example.

EXAMPLE 8 $\Sigma_{n=1}^{\infty} (-1)^{n+1}$ diverges because its partial sums alternate between 1 and 0:

$$
\begin{aligned}
s_1 &= (-1)^2 = 1, \\
s_2 &= (-1)^2 + (-1)^3 = 1 - 1 = 0, \\
s_3 &= (-1)^2 + (-1)^3 + (-1)^4 = 1 - 1 + 1 = 1,
\end{aligned}
$$

and so on. □

The next theorem provides a quick way to detect the kind of divergence that occurred in Examples 6, 7, and 8.

THEOREM 2

The nth-term Test for Divergence

If $\lim_{n\to\infty} a_n \neq 0$, or if $\lim_{n\to\infty} a_n$ fails to exist, then $\Sigma_{n=1}^{\infty} a_n$ diverges.

When we apply Theorem 2 to the series in Examples 6, 7, and 8, we find that

$\sum_{n=1}^{\infty} n^2$ diverges because $n^2 \to \infty$,

$\sum_{n=1}^{\infty} \frac{n+1}{n}$ diverges because $\frac{n+1}{n} \to 1 \neq 0$,

$\sum_{n=1}^{\infty} (-1)^{n+1}$ diverges because $\lim_{n\to\infty} (-1)^{n+1}$ does not exist.

Proof of the theorem. We prove Theorem 2 by showing that if $\Sigma\, a_n$ converges, then $\lim_{n\to\infty} a_n = 0$. Let

$$s_n = a_1 + a_2 + \cdots + a_n,$$

and suppose that $\Sigma\, a_n$ converges to S; that is

$$s_n \to S.$$

Then, corresponding to any preassigned number $\epsilon > 0$, there is an index N such that all the terms of the sequence $\{s_n\}$ after the Nth one lie between $S - (\epsilon/2)$ and $S + (\epsilon/2)$. Hence, no two of them may differ by as much as ϵ. That is, if m and n are both greater than N, then

$$|s_n - s_m| < \epsilon.$$

In particular, this inequality holds if $m = n - 1$ and $n > N + 1$, so that

$$|a_n| = |s_n - s_{n-1}| < \epsilon \qquad \text{when} \quad n > N + 1.$$

Since ϵ was any positive number whatsoever, this means that

$$\lim_{n\to\infty} a_n = 0. \blacksquare$$

In the next example, we use both Theorems 1 and 2.

EXAMPLE 9 Determine whether each series converges or diverges. If it converges, find its sum.

a) $\sum_{n=1}^{\infty} 2\left(\cos\frac{\pi}{3}\right)^n$ b) $\sum_{n=0}^{\infty} \left(\tan\frac{\pi}{4}\right)^n$

c) $\sum_{n=1}^{\infty} \frac{n}{2n+5}$ d) $\sum_{n=1}^{\infty} \frac{5(-1)^n}{4^n}$

Solution

a) $\cos \pi/3 = \frac{1}{2}$. This is a geometric series with first term $a_1 = 2(\cos \pi/3) = 1$ and ratio $r = \cos \pi/3 = \frac{1}{2}$; so the series *converges* and its *sum* is $a_1/(1 - r) = 1/(1 - \frac{1}{2}) = 2$.

b) $\tan \pi/4 = 1$. The nth term does not have 0 as its limit, so the series *diverges*.

c) $a_n = \frac{n}{2n+5}$ and $\lim_{n\to\infty} \frac{n}{2n+5} = \frac{1}{2} \neq 0$. Series (c) *diverges*.

d) This is a geometric series with first term $a_1 = -5/4$ and ratio $r = -1/4$. Series (d) *converges* and its *sum* is

$$\frac{a_1}{1-r} = \frac{-5/4}{1+\frac{1}{4}} = -1. \square$$

Because of how it is proved, Theorem 2 is often stated in the following shorter way.

THEOREM 3

If $\Sigma_{n=1}^{\infty} a_n$ converges, then $a_n \to 0$.

A Word of Caution Theorem 3 does *not* say that if $a_n \to 0$ then $\Sigma\, a_n$ converges. The series $\Sigma\, a_n$ may diverge even though $a_n \to 0$. Thus, $\lim a_n = 0$ is a necessary but not a sufficient condition for the series $\Sigma\, a_n$ to converge.

EXAMPLE 10 The series

$$1 + \underbrace{\frac{1}{2} + \frac{1}{2}}_{2\text{ terms}} + \underbrace{\frac{1}{4} + \frac{1}{4} + \frac{1}{4} + \frac{1}{4}}_{4\text{ terms}} + \cdots + \underbrace{\frac{1}{2^n} + \frac{1}{2^n} + \cdots + \frac{1}{2^n}}_{2^n\text{ terms}} + \cdots$$

diverges even though its terms form a sequence that converges to 0. $\square$

Whenever we have two convergent series we can add them, subtract them, and multiply them by constants, to make other convergent series. The next theorem gives the details.

THEOREM 4

> If $A = \Sigma_{n=1}^{\infty} a_n$ and $B = \Sigma_{n=1}^{\infty} b_n$ both exist and are finite, then
>
> i) $\sum_{n=1}^{\infty} (a_n + b_n) = A + B$,
>
> ii) $\sum_{n=1}^{\infty} ka_n = k \sum_{n=1}^{\infty} a_n = kA \qquad (k \text{ any number}).$

Proof of the theorem. Let

$$A_n = a_1 + a_2 + \cdots + a_n, \qquad B_n = b_1 + b_2 + \cdots + b_n.$$

Then the partial sums of $\Sigma_{n=1}^{\infty} (a_n + b_n)$ are

$$\begin{aligned} S_n &= (a_1 + b_1) + (a_2 + b_2) + \cdots + (a_n + b_n) \\ &= (a_1 + \cdots + a_n) + (b_1 + \cdots + b_n) \\ &= A_n + B_n. \end{aligned}$$

Since $A_n \to A$ and $B_n \to B$, we have $S_n \to A + B$. The partial sums of $\Sigma_{n=1}^{\infty} (ka_n)$ are

$$S_n = ka_1 + ka_2 + \cdots + ka_n = k(a_1 + a_2 + \cdots + a_n) = kA_n,$$

which converge to kA. ■

REMARK 2. If you think there should be two more parts of Theorem 4 to match those of Theorem 1, Article 11.2, see Problems 44–46. Also, see Problem 41, Article 11.5, and Problem 39, Article 11.7.

Part (ii) of Theorem 4 says that every multiple of a convergent series converges. A companion to this is the next corollary, which says that every *nonzero* multiple of a divergent series diverges.

COROLLARY

> If $\Sigma_{n=1}^{\infty} a_n$ diverges, and if c is any number different from 0, then the series of multiples $\Sigma_{n=1}^{\infty} ca_n$ diverges.

Proof of the corollary. Suppose, to the contrary, that $\Sigma_{n=1}^{\infty} ca_n$ actually converges. Then, when we take $k = 1/c$ in part (ii) of Theorem 4 we find that

$$\frac{1}{c} \cdot \sum_{n=1}^{\infty} ca_n = \sum_{n=1}^{\infty} \frac{1}{c} \cdot ca_n = \sum_{n=1}^{\infty} a_n$$

converges. That is, $\Sigma_{n=1}^{\infty} ca_n$ cannot converge unless $\Sigma_{n=1}^{\infty} a_n$ also converges. Thus, if $\Sigma_{n=1}^{\infty} a_n$ diverges, then $\Sigma_{n=1}^{\infty} ca_n$ must diverge. ■

REMARK 3. An immediate consequence of Theorem 4 is that if $A = \Sigma_{n=1}^{\infty} a_n$ and $B = \Sigma_{n=1}^{\infty} b_n$, then

$$\sum (a_n - b_n) = \sum a_n + \sum (-1)b_n = \sum a_n - \sum b_n = A - B. \quad (14)$$

The series $\Sigma_{n=1}^{\infty} (a_n - b_n)$ is called the *difference* of $\Sigma_{n=1}^{\infty} a_n$ and $\Sigma_{n=1}^{\infty} b_n$, while $\Sigma_{n=1}^{\infty} (a_n + b_n)$ is called their *sum*.

EXAMPLE 11

a) $\displaystyle\sum_{n=1}^{\infty} \frac{4}{2^{n-1}} = 4 \sum_{n=1}^{\infty} \frac{1}{2^{n-1}} = 4 \frac{1}{1-\frac{1}{2}} = 8,$

b) $$\sum_{n=0}^{\infty} \frac{3^n - 2^n}{6^n} = \sum_{n=0}^{\infty} \left(\frac{1}{2^n} - \frac{1}{3^n}\right) = \sum_{n=0}^{\infty} \frac{1}{2^n} - \sum_{n=0}^{\infty} \frac{1}{3^n}$$

$$= \frac{1}{1-\frac{1}{2}} - \frac{1}{1-\frac{1}{3}} = 2 - \frac{3}{2} = \frac{1}{2}. \quad \Box$$

REMARK 4. A finite number of terms can always be deleted from or added to a series without altering its convergence or divergence. If $\Sigma_{n=1}^{\infty} a_n$ converges and k is an index greater than 1, then $\Sigma_{n=k}^{\infty} a_n$ converges, and

$$\sum_{n=1}^{\infty} a_n = a_1 + a_2 + \cdots + a_{k-1} + \sum_{n=k}^{\infty} a_n. \tag{15}$$

Conversely, if $\Sigma_{n=k}^{\infty} a_n$ converges for any $k > 1$, then $\Sigma_{n=1}^{\infty} a_n$ converges and the sums continue to be related as in Eq. (15). Thus, for example,

$$\sum_{n=1}^{\infty} \frac{1}{5^n} = \frac{1}{5} + \frac{1}{25} + \frac{1}{125} + \sum_{n=4}^{\infty} \frac{1}{5^n} \tag{16}$$

and

$$\sum_{n=4}^{\infty} \frac{1}{5^n} = \sum_{n=1}^{\infty} \frac{1}{5^n} - \frac{1}{5} - \frac{1}{25} - \frac{1}{125}. \tag{17}$$

Note that while the addition or removal of a finite number of terms from a series has no effect on the convergence or divergence of the series, these operations can change the *sum* of a convergent series.

REMARK 5. The indexing of the terms of a series can be changed without altering convergence of the series. For example, the geometric series that starts with

$$1 + \frac{1}{2} + \frac{1}{4} + \cdots$$

can be described as

$$\sum_{n=0}^{\infty} \frac{1}{2^n} \quad \text{or} \quad \sum_{n=-4}^{\infty} \frac{1}{2^{n+4}} \quad \text{or} \quad \sum_{n=5}^{\infty} \frac{1}{2^{n-5}}. \tag{18}$$

The partial sums remain the same no matter what indexing is chosen, so that we are free to start indexing with whatever integer we want. Preference is usually given to an indexing that leads to a simple expression. In Example 11(b) we chose to start with $n = 0$ instead of $n = 1$, because this allowed us to describe the series we had in mind as:

$$\sum_{n=0}^{\infty} \frac{3^n - 2^n}{6^n} \quad \text{instead of} \quad \sum_{n=1}^{\infty} \frac{3^{n-1} - 2^{n-1}}{6^{n-1}}. \tag{19}$$

PROBLEMS

In Problems 1–8, find a closed expression for the sum s_n of the first n terms of each series. Then compute the sum of the series if the series converges.

1. $2 + \frac{2}{3} + \frac{2}{9} + \frac{2}{27} + \cdots + \frac{2}{3^{n-1}} + \cdots$

2. $\frac{9}{100} + \frac{9}{100^2} + \frac{9}{100^3} + \cdots + \frac{9}{100^n} + \cdots$

3. $1 + e^{-1} + e^{-2} + \cdots + e^{-(n-1)} + \cdots$

4. $1 - \frac{1}{2} + \frac{1}{4} - \frac{1}{8} + \cdots + (-1)^{n-1}\frac{1}{2^{n-1}} + \cdots$

5. $1 - 2 + 4 - 8 + \cdots + (-1)^{n-1}2^{n-1} + \cdots$

6. $\frac{1}{2 \cdot 3} + \frac{1}{3 \cdot 4} + \frac{1}{4 \cdot 5} + \cdots + \frac{1}{(n+1)(n+2)} + \cdots$

7. $\ln\frac{1}{2} + \ln\frac{2}{3} + \ln\frac{3}{4} + \cdots + \ln\frac{n}{n+1} + \cdots$

8. $1 + 2 + 3 + \cdots + n + \cdots$

9. The series in Problem 6 can be described as

$$\sum_{n=1}^{\infty} \frac{1}{(n+1)(n+2)}.$$

It can also be described as a summation beginning with $n = -1$:

$$\sum_{n=-1}^{\infty} \frac{1}{(n+3)(n+4)}.$$

Describe the series as a summation beginning with
a) $n = -2$, b) $n = 0$, c) $n = 5$.

10. a) A ball is dropped from a height of 4 m. Each time it strikes the pavement after falling from a height of h meters, it rebounds to a height of $0.75h$ meters. Find the total distance traveled up and down by the ball.
b) Calculate the total time the ball is traveling. (*Hint:* $y = \frac{1}{2}gt^2 = \frac{1}{2}(9.8)t^2$, t in seconds, y in meters.)

In Problems 11–18, find the sum of the series.

11. $\sum_{n=0}^{\infty} \frac{1}{4^n}$

12. $\sum_{n=2}^{\infty} \frac{1}{4^n}$

13. $\sum_{n=1}^{\infty} \frac{7}{4^n}$

14. $\sum_{n=0}^{\infty} (-1)^n \frac{5}{4^n}$

15. $\sum_{n=0}^{\infty} \left(\frac{5}{2^n} + \frac{1}{3^n}\right)$

16. $\sum_{n=0}^{\infty} \left(\frac{5}{2^n} - \frac{1}{3^n}\right)$

17. $\sum_{n=0}^{\infty} \left(\frac{2^n}{5^n}\right)$

18. $\sum_{n=0}^{\infty} \left(\frac{2^{n+1}}{5^n}\right)$

Use partial fractions to find the sum of the series in Problems 19–22.

19. $\sum_{n=1}^{\infty} \frac{4}{(4n-3)(4n+1)}$

20. $\sum_{n=1}^{\infty} \frac{1}{(4n-3)(4n+1)}$

21. $\sum_{n=3}^{\infty} \frac{4}{(4n-3)(4n+1)}$

22. $\sum_{n=1}^{\infty} \frac{2n+1}{n^2(n+1)^2}$

23. a) Express the repeating decimal

$$0.234\,234\,234\ldots$$

as an infinite series, and give the sum as a ratio p/q of two integers.
b) Is it true that *every* repeating decimal is a rational number p/q? Give a reason for your answer.

24. Express the decimal number

$$1.24\,123\,123\,123\ldots,$$

which begins to repeat after the first three figures, as a rational number p/q.

In Problems 25–38, determine whether each series converges or diverges. If it converges, find the sum.

25. $\sum_{n=0}^{\infty} \left(\frac{1}{\sqrt{2}}\right)^n$

26. $\sum_{n=1}^{\infty} \ln\frac{1}{n}$

27. $\sum_{n=1}^{\infty} (-1)^{n+1}\frac{3}{2^n}$

28. $\sum_{n=1}^{\infty} (\sqrt{2})^n$

29. $\sum_{n=0}^{\infty} \cos n\pi$

30. $\sum_{n=0}^{\infty} \frac{\cos n\pi}{5^n}$

31. $\sum_{n=0}^{\infty} e^{-2n}$

32. $\sum_{n=1}^{\infty} \frac{n^2+1}{n}$

33. $\sum_{n=1}^{\infty} (-1)^{n+1}n$

34. $\sum_{n=1}^{\infty} \frac{2}{10^n}$

35. $\sum_{n=0}^{\infty} \frac{2^n - 1}{3^n}$

36. $\sum_{n=1}^{\infty} \left(1 - \frac{1}{n}\right)^n$

37. $\sum_{n=0}^{\infty} \frac{n!}{1000^n}$

38. $\sum_{n=0}^{\infty} \frac{1}{x^n}$, $|x| > 1$

In Problems 39 and 40, the equalities are instances of Theorem 1. Give the value of a and of r in each case.

39. $\frac{1}{1+x} = \sum_{n=0}^{\infty} (-1)^n x^n$, $|x| < 1$

40. $\frac{1}{1+x^2} = \sum_{n=0}^{\infty} (-1)^n x^{2n}$, $|x| < 1$

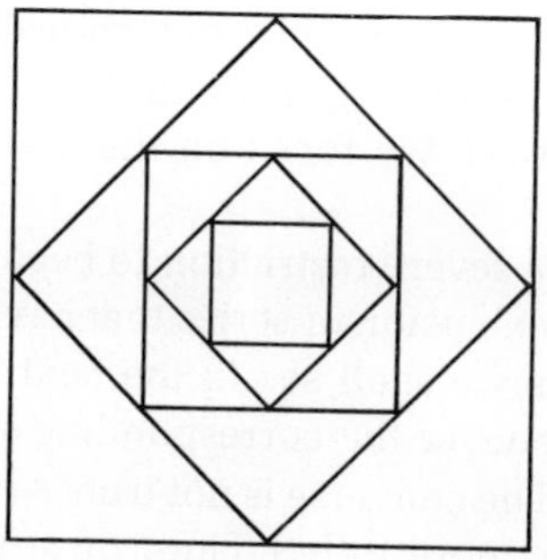

Figure 11.7

41. Figure 11.7 shows the first five of an infinite series of squares. The outermost square has an area of 4, and each of the other squares is obtained by joining the midpoints of the sides of the square before it. Find the sum of the areas of all the squares.

42. Find a closed-form expression for the nth partial sum of the series $\Sigma_{n=1}^{\infty}(-1)^{n+1}$.

43. Show by example that the term-by-term sum of two divergent series may converge.

44. Find convergent geometric series $A = \Sigma_{n=1}^{\infty} a_n$ and $B = \Sigma_{n=1}^{\infty} b_n$ that illustrate the fact that $\Sigma_{n=1}^{\infty} a_n \cdot b_n$ may converge without being equal to $A \cdot B$.

45. Show by example that $\Sigma_{n=1}^{\infty}(a_n/b_n)$ may diverge even though $\Sigma_{n=1}^{\infty} a_n$ and $\Sigma_{n=1}^{\infty} b_n$ converge and no $b_n = 0$.

46. Show by example that $\Sigma_{n=1}^{\infty}(a_n/b_n)$ may converge to something other than A/B even when $A = \Sigma_{n=1}^{\infty} a_n$, $B = \Sigma_{n=1}^{\infty} b_n \neq 0$, and no $b_n = 0$.

47. Show that if $\Sigma_{n=1}^{\infty} a_n$ converges, and $a_n \neq 0$ for all n, then $\Sigma_{n=1}^{\infty}(1/a_n)$ diverges.

48. a) Verify by long division that

$$\frac{1}{1+t} = 1 - t + t^2 - t^3 + \cdots + (-1)^n t^n + \frac{(-1)^{n+1}t^{n+1}}{1+t}.$$

b) By integrating both sides of the equation in part (a) with respect to t, from 0 to x, show that

$$\ln(1+x) = x - \frac{x^2}{2} + \frac{x^3}{3} - \frac{x^4}{4} + \cdots + (-1)^n \frac{x^{n+1}}{n+1} + R,$$

where

$$R = (-1)^{n+1}\int_0^x \frac{t^{n+1}}{1+t}\,dt.$$

c) If $x > 0$, show that

$$|R| \leq \int_0^x t^{n+1}\,dt = \frac{x^{n+2}}{n+2}.$$

(*Hint:* As t varies from 0 to x, $1 + t \geq 1$.)

d) If $x = \frac{1}{2}$, how large should n be in part (c) above if we want to be able to guarantee that $|R| < 0.001$? Write a polynomial that approximates $\ln(1+x)$ to this degree of accuracy for $0 \leq x \leq \frac{1}{2}$.

e) If $x = 1$, how large should n be in part (c) above if we want to be able to guarantee that $|R| < 0.001$?

***Toolkit* programs**

Sequences and Series

11.5

Tests for Convergence of Series with Nonnegative Terms

In this article we shall study series that do not have negative terms. The reason for this restriction is that the partial sums of these series always form increasing sequences, and increasing sequences *that are bounded from above* always converge, as we shall see. Thus, to show that a series of nonnegative terms converges, we need only show that there is some number beyond which the partial sums never go.

It may at first seem to be a drawback that this approach establishes the fact of convergence without actually producing the sum of the series in question. Surely it would be better to compute sums of series directly from nice formulas for their partial sums. But in most cases such formulas are not available, and in their absence we have to turn instead to a two-step procedure of first establishing convergence and then approximating the sum. In this article and the next, we focus on the first of these two steps.

Surprisingly enough it is not a severe restriction to begin our study of convergence with the temporary exclusion of series that have one or more negative terms, for it is a fact, as we shall see in the next article, that a series $\Sigma_{n=1}^{\infty} a_n$ will converge whenever the corresponding series of absolute values $\Sigma_{n=1}^{\infty} |a_n|$ converges. (The converse is not true. Articles 11.6 and 11.7 will deal with these matters more fully.) Thus, once we know that

$$\sum_{n=1}^{\infty} \frac{1}{n^2} = 1 + \frac{1}{4} + \frac{1}{9} + \frac{1}{16} + \frac{1}{25} + \cdots \tag{1}$$

converges, we will know that *all* of the series like

$$1 - \frac{1}{4} + \frac{1}{9} - \frac{1}{16} + \frac{1}{25} + \cdots \tag{2}$$

and

$$-1 - \frac{1}{4} + \frac{1}{9} + \frac{1}{16} - \frac{1}{25} - \cdots, \tag{3}$$

that can be obtained from (1) by changing the sign of one or more terms, also converge! We might not know at first what they converge to, but at least we know they converge, and that is a first and necessary step toward estimating their sums.

Increasing Sequences

Suppose now that $\Sigma_{n=1}^{\infty} a_n$ is an infinite series that has no negative terms. That is, $a_n \geq 0$ for every n. Then, when we calculate the partial sums s_1, s_2, s_3, and so on, we see that each one is greater than or equal to its predecessor because $s_{n+1} = s_n + a_n$. That is,

$$s_1 \leq s_2 \leq s_3 \leq \cdots \leq s_n \leq s_{n+1} \leq \cdots. \tag{4}$$

A sequence $\{s_n\}$ like the one in (4), with the property that $s_n \leq s_{n+1}$ for every n, is called an *increasing* sequence.

There are two types of increasing sequences—those that increase beyond any finite bound and those that don't. The former diverge to infinity. We turn our attention to the other kind: those that do not grow beyond all bounds. Such a sequence is said to be *bounded from above,* and any number M such that

$$s_n \leq M \qquad \text{for all } n$$

is called *an upper bound* of the sequence.

EXAMPLE 1 If $s_n = n/(n + 1)$, then 1 is an upper bound and so is any larger number, like $\sqrt{2}$, 5, 17. No number smaller than 1 is an upper bound, so that for this sequence 1 is the *least upper bound.* □

When an increasing sequence is bounded from above, we may ask, "Must it have a *least* upper bound?". The answer is yes, but we shall not prove this fact. We shall prove that *if L is the least upper bound, then the sequence converges to L.* The following argument shows why L is the limit.

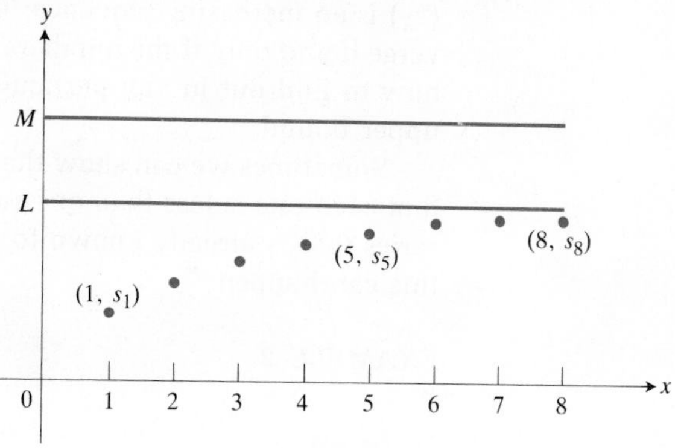

11.8 When the terms of an increasing sequence have an upper bound M, they have a limit $L \leq M$.

When we plot the points $(1, s_1), (2, s_2), \ldots, (n, s_n)$ in the xy-plane, if M is an upper bound of the sequence, all these plotted points will lie on or below the line $y = M$ (Fig. 11.8). It seems intuitively clear that there ought to be a lowest such line $y = L$. That would mean that none of the points (n, s_n) is above $y = L$, but that some do lie above any lower line $y = L - \epsilon$, if ϵ is a positive number. Then the sequence converges to L as limit, because the number L has the properties (a) $s_n \leq L$ for *all* values of n, (b) given any $\epsilon > 0$, there exists at least one integer N such that

$$s_N > L - \epsilon.$$

Then the fact that $\{s_n\}$ is an increasing sequence tells us further that

$$s_n \geq s_N > L - \epsilon \qquad \text{for all} \qquad n \geq N.$$

This means that *all* the numbers s_n, beyond the Nth one in the sequence, lie within ϵ distance of L. This is precisely the condition for L to be the limit of the sequence s_n,

$$L = \lim_{n\to\infty} s_n.$$

The facts for increasing sequences are summarized in the following theorem.

THEOREM 1

The Increasing Sequence Theorem

Let $\{s_n\}$ be an increasing sequence of real numbers. Then one or other of the following alternatives must hold:

A. The terms of the sequence are all less than or equal to some finite constant M. In this case, the sequence has a finite limit L that is also less than or equal to M. (In fact, L is the least upper bound of the sequence $\{s_n\}$.)

B. The sequence diverges to plus infinity: that is, the numbers in the sequence ultimately exceed every preassigned number.

Alternative B of the theorem is what happens when there are points (n, s_n) above any given line $y = M$, no matter how large M may be.

Let us now apply Theorem 1 to the convergence of infinite series of nonnegative numbers. If Σa_n is such a series, its sequence of partial sums $\{s_n\}$ is an increasing sequence. Therefore, $\{s_n\}$, and hence Σa_n, will converge if and only if the numbers s_n have an upper bound. The question is how to find out in any particular instance whether the s_n's do have an upper bound.

Sometimes we can show that the s_n's are bounded above by showing that each one is less than or equal to the corresponding partial sum of a series that is already known to converge. The next example shows how this can happen.

EXAMPLE 2

$$\sum_{n=0}^{\infty} \frac{1}{n!} = 1 + \frac{1}{1!} + \frac{1}{2!} + \frac{1}{3!} + \cdots \tag{5}$$

converges because its terms are all positive and less than or equal to the corresponding terms of

$$1 + \sum_{n=0}^{\infty} \frac{1}{2^n} = 1 + 1 + \frac{1}{2} + \frac{1}{2^2} + \cdots. \tag{6}$$

To see how this relationship between these two series leads to an upper bound for the partial sums of $\Sigma_{n=0}^{\infty} (1/n!)$, let

$$s_n = 1 + \frac{1}{1!} + \frac{1}{2!} + \cdots + \frac{1}{n!},$$

and observe that, for each n,

$$s_n \leq 1 + 1 + \frac{1}{2} + \frac{1}{2^2} + \cdots + \frac{1}{2^n} < 1 + \sum_{n=0}^{\infty} \frac{1}{2^n} = 1 + \frac{1}{1 - \frac{1}{2}} = 3.$$

Thus the partial sums of $\Sigma_{n=0}^{\infty} (1/n!)$ are all less than 3. Therefore, $\Sigma_{n=0}^{\infty} (1/n!)$ converges.

Just because 3 is an upper bound for the partial sums of $\Sigma_{n=0}^{\infty} (1/n!)$ we cannot conclude that the series converges to 3. The series actually converges to $e = 2.71828\ldots$. □

We established the convergence of the series in Example 2 by comparing it with a series that was already known to converge. This kind of comparison is typical of a procedure called the *comparison test* for convergence of series of nonnegative terms.

Comparison Test for Series of Nonnegative Terms

Let $\Sigma_{n=1}^{\infty} a_n$ be a series that has no negative terms.

A. Test for *convergence* of Σa_n. The series Σa_n converges if there is a convergent series of nonnegative terms Σc_n with $a_n \leq c_n$ for all $n > n_0$.

B. Test for *divergence* of $\Sigma\, a_n$. The series $\Sigma\, a_n$ diverges if there is a divergent series of nonnegative terms $\Sigma\, d_n$ with $a_n \geq d_n$ for all $n \geq n_0$.

In part (A), the partial sums of the series $\Sigma\, a_n$ are bounded from above by

$$M = a_1 + a_2 + \cdots + a_{n_0} + \sum_{n=n_0+1}^{\infty} c_n.$$

Therefore, they form an increasing sequence that has a limit L that is less than or equal to M.

In part (B), the partial sums for $\Sigma\, a_n$ are not bounded from above: if they were, the partial sums for $\Sigma\, d_n$ would be bounded by

$$M' = d_1 + d_2 + \cdots + d_{n_0} + \sum_{n=n_0+1}^{\infty} a_n.$$

That would imply convergence of $\Sigma\, d_n$. Therefore, divergence of $\Sigma\, d_n$ implies divergence of $\Sigma\, a_n$.

To apply the comparison test to a series, we do not have to include the early terms of the series. We can start the test with any index N, provided we include all the terms of the series being tested from there on.

EXAMPLE 3 The convergence of the series

$$5 + \frac{2}{3} + 1 + \frac{1}{7} + \frac{1}{2} + \frac{1}{3!} + \frac{1}{4!} + \cdots + \frac{1}{k!} + \cdots$$

can be established by ignoring the first four terms and comparing the remainder of the series from the fifth term on (the fifth term is $\frac{1}{2}$) with the convergent geometric series

$$\sum_{n=1}^{\infty} \frac{1}{2^n} = \frac{1}{2} + \frac{1}{4} + \frac{1}{8} + \cdots. \; \square$$

To apply the comparison test we need to have on hand a list of series that are known to converge and a list of series that are known to diverge. Our next example adds a divergent series to the list.

EXAMPLE 4 The *harmonic series*

$$\sum_{n=1}^{\infty} \frac{1}{n} = 1 + \frac{1}{2} + \frac{1}{3} + \frac{1}{4} + \cdots$$

diverges.

To see why, we represent the terms of the series as the areas of rectangles each of base unity and having altitudes $1, \frac{1}{2}, \frac{1}{3}, \ldots$, as in Fig. 11.9. The sum of the first n terms of the series,

$$s_n = 1 + \frac{1}{2} + \frac{1}{3} + \cdots + \frac{1}{n},$$

could be thought to represent the sum of the areas of n rectangles each of which is greater than the area underneath the corresponding portion of

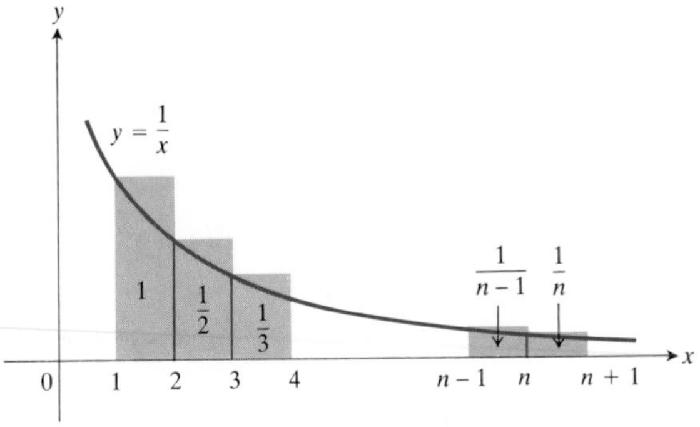

11.9 $1 + \frac{1}{2} + \frac{1}{3} + \cdots + \frac{1}{n} > \int_1^{n+1} \frac{1}{x}\,dx = \ln(n+1)$

the curve $y = 1/x$. Thus s_n is greater than the area under this curve between $x = 1$ and $x = n + 1$:

$$s_n > \int_1^{n+1} \frac{dx}{x} = \ln(n+1).$$

Therefore $s_n \to +\infty$ because $\ln(n+1) \to +\infty$. The series

$$1 + \frac{1}{2} + \frac{1}{3} + \cdots + \frac{1}{n} + \cdots$$

diverges to plus infinity. □

The harmonic series $\sum_{n=1}^{\infty}(1/n)$ is another series whose divergence cannot be detected by the nth-term test for divergence. The series diverges in spite of the fact that $1/n \to 0$.

We know that every nonzero multiple of a divergent series diverges (Corollary of Theorem 4 in the preceding article). Therefore, the divergence of the harmonic series implies the divergence of series like

$$\sum_{n=1}^{\infty} \frac{1}{2n} = \frac{1}{2} + \frac{1}{4} + \frac{1}{6} + \frac{1}{8} + \cdots$$

and

$$\sum_{n=1}^{\infty} \frac{1}{100n} = \frac{1}{100} + \frac{1}{200} + \frac{1}{300} + \frac{1}{400} + \cdots.$$

The Integral Test

In Example 4 we deduced the divergence of the harmonic series by comparing its sequence of partial sums with a divergent sequence of integrals. This comparison is a special case of a general comparison process called the *integral test,* a test that gives criteria for convergence as well as for divergence of series whose terms are positive.

Integral Test

Let the function $y = f(x)$, obtained by introducing the continuous variable x in place of the discrete variable n in the nth term of the positive series

$$\sum_{n=1}^{\infty} a_n,$$

be a decreasing function of x for $x \geq 1$. Then the series and the integral

$$\int_1^\infty f(x)\, dx$$

both converge or both diverge.

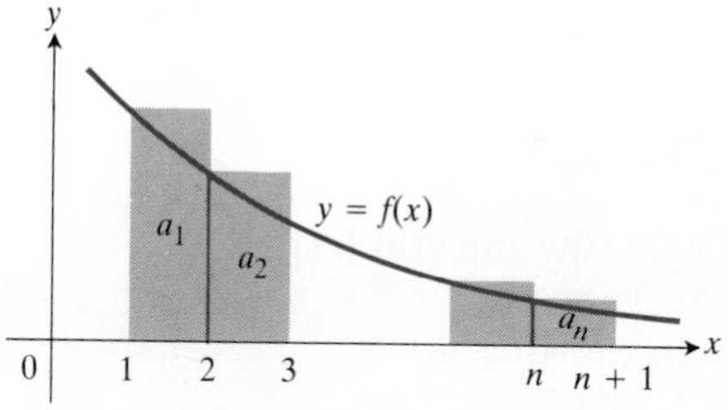

(a)

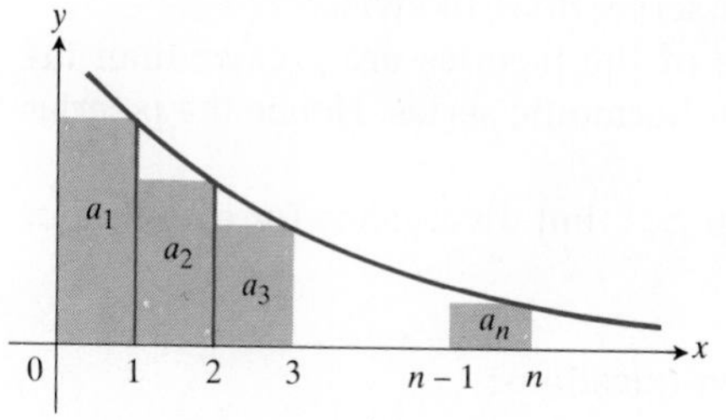

(b)

11.10 $\int_1^{n+1} f(x)\, dx \leq a_1 + a_2 + \cdots + a_n \leq a_1 + \int_1^n f(x)\, dx.$

Proof. We start with the assumption that f is a decreasing function with $f(n) = a_n$ for every n. This leads us to observe that the rectangles in Fig. 11.10(a), which have areas $a_1, a_2, \ldots, a_n$, collectively enclose more area than that under the curve $y = f(x)$ from $x = 1$ to $x = n + 1$. That is,

$$\int_1^{n+1} f(x)\, dx \leq a_1 + a_2 + \cdots + a_n.$$

In Fig. 11.10(b) the rectangles have been faced to the left instead of to the right. If we momentarily disregard the first rectangle, of area a_1, we see that

$$a_2 + a_3 + \cdots + a_n \leq \int_1^n f(x)\, dx.$$

If we include a_1, we have

$$a_1 + a_2 + \cdots + a_n \leq a_1 + \int_1^n f(x)\, dx.$$

Combining these results gives

$$\int_1^{n+1} f(x)\, dx \leq a_1 + a_2 + \cdots + a_n \leq a_1 + \int_1^n f(x)\, dx. \tag{7}$$

If the integral $\int_1^\infty f(x)\, dx$ is finite the right-hand inequality shows that $\sum_{n=1}^\infty a_n$ is also finite. But if $\int_1^\infty f(x)\, dx$ is infinite, then the left-hand inequality shows that the series is also infinite.

Hence the series and the integral are both finite or both infinite. ■

EXAMPLE 5 *The p-series.* If p is a real constant, the series

$$\sum_{n=1}^\infty \frac{1}{n^p} = \frac{1}{1^p} + \frac{1}{2^p} + \frac{1}{3^p} + \cdots + \frac{1}{n^p} + \cdots \tag{8}$$

converges if $p > 1$ and diverges if $p \leq 1$. To prove this, let

$$f(x) = \frac{1}{x^p}.$$

Then, if $p > 1$, we have

$$\int_1^\infty x^{-p}\, dx = \lim_{b\to\infty} \frac{x^{-p+1}}{-p+1}\bigg|_1^b = \frac{1}{p-1},$$

which is finite. Hence the p-series converges if p is greater than one.

If $p = 1$, we have

$$1 + \frac{1}{2} + \frac{1}{3} + \cdots + \frac{1}{n} + \cdots,$$

which we already know diverges. Or, by the integral test,

$$\int_1^\infty x^{-1}\,dx = \lim_{b\to\infty} \ln x \Big|_1^b = +\infty,$$

and, since the integral diverges, the series does likewise.

Finally, if $p < 1$, then the terms of the p-series are greater than the corresponding terms of the divergent harmonic series. Hence the p-series diverges, by the comparison test.

Thus, we have convergence for $p > 1$, but divergence for every other value of p. □

Given a series $\Sigma\, a_n$ we have two questions:

1. Does the series converge?
2. If it converges, what is its sum?

Most of this chapter is devoted to the first question. But, as a practical matter, the second question is just as important for a scientist or engineer. We digress briefly to address that question now.

Estimation of Remainders by Integrals

The difference $R_n = L - s_n$ between the sum of a convergent series and its nth partial sum is called a *remainder* or a *truncation error*. Since R_n itself is given as an infinite series, which, in principle, is as difficult to evaluate as the original series, you might think that there would be no advantage in singling out R_n for attention. But sometimes even a crude estimate for R_n can lead to an estimate of L that is closer to L than s_n is.

Suppose, for example, that we are interested in learning the numerical value of the series

$$\sum_{k=1}^{\infty} \frac{1}{k^2} = \frac{1}{1^2} + \frac{1}{2^2} + \frac{1}{3^2} + \cdots.$$

This is a p-series with $p = 2$, and hence is known to converge. This means that the sequence of partial sums

$$s_n = \frac{1}{1^2} + \frac{1}{2^2} + \cdots + \frac{1}{n^2}$$

has a limit L. If we want to know L to a couple of decimal places, we might try to find an integer n such that the corresponding *finite* sum s_n differs from L by less, say, than 0.005. Then we would use this s_n in place of L, to two decimals. If we write

$$L = \sum_{k=1}^{\infty} \frac{1}{k^2} = \frac{1}{1^2} + \frac{1}{2^2} + \cdots + \frac{1}{n^2} + \frac{1}{(n+1)^2} + \cdots,$$

we see that

$$R_n = L - s_n = \frac{1}{(n+1)^2} + \cdots.$$

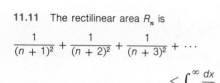

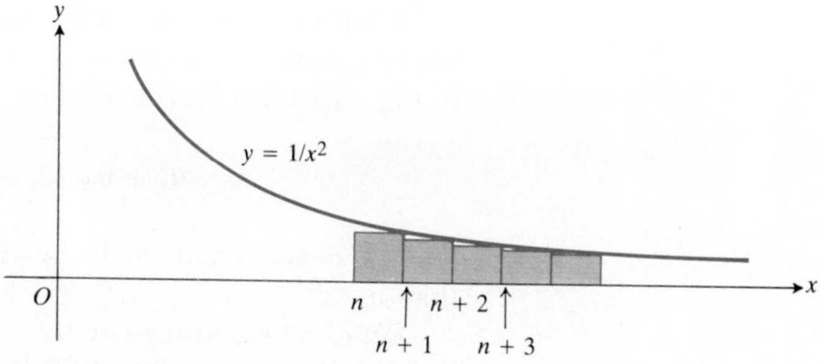

11.11 The rectilinear area R_n is

$$\frac{1}{(n+1)^2} + \frac{1}{(n+2)^2} + \frac{1}{(n+3)^2} + \cdots < \int_n^\infty \frac{dx}{x^2}.$$

We estimate the error R_n by comparing it with the area under the curve

$$y = \frac{1}{x^2}$$

from $x = n$ to ∞.

From Fig. 11.11 we see that

$$R_n < \int_n^\infty \frac{1}{x^2}\,dx = \frac{1}{n},$$

which tells us that, by taking 200 terms of the series, we can be sure that the difference between the sum L of the entire series and the sum s_{200} of these 200 terms will be less than $1/200 = 0.005$.

A somewhat closer estimate of R_n results from using the trapezoidal rule to approximate the area under the curve in Fig. 11.11. Let us write u_k for $1/k^2$ and consider the trapezoidal approximation

$$\begin{aligned} T_n &= \sum_{k=n}^{\infty} \frac{1}{2}(u_k + u_{k+1}) = \frac{1}{2}(u_n + u_{n+1}) + \frac{1}{2}(u_{n+1} + u_{n+2}) + \cdots \\ &= \frac{1}{2}u_n + u_{n+1} + u_{n+2} + \cdots = \frac{1}{2}u_n + R_n. \end{aligned}$$

Now since the curve $y = 1/x^2$ is concave upward.

$$T_n > \int_n^\infty \frac{1}{x^2}\,dx = \frac{1}{n},$$

and we have

$$R_n = T_n - \frac{1}{2}u_n > \frac{1}{n} - \frac{1}{2n^2}.$$

We now know that

$$\frac{1}{n} > R_n > \frac{1}{n} - \frac{1}{2n^2},$$

and $L = s_n + R_n$ may be estimated as follows:

$$s_n + \frac{1}{n} > L > s_n + \frac{1}{n} - \frac{1}{2n^2}. \tag{9}$$

Thus, by using $s_n + 1/n$ in place of s_n to estimate L, we shall be making an error which is numerically less than $1/(2n^2)$. By taking $n \geq 10$, this error is then made less than 0.005. The difference in time required to compute the sum of 10 terms versus 200 terms is sufficiently great to make this sharper analysis of practical importance.

What we have done in the case of this specific example may be done in any case where the graph of the function $y = f(x)$ is concave upward as in Fig. 11.11. We find that when $\int_n^\infty f(x)\,dx$ exists,

$$u_1 + u_2 + \cdots + u_n + \int_n^\infty f(x)\,dx \tag{10}$$

tends to overestimate the value of the series, but by an amount that is less than $u_n/2$.

For excellent articles on the subject of estimating remainders, see "Estimating Remainders," by R. P. Boas, *Mathematics Magazine*, Volume 51, Number 2, March 1978, pp. 83–89, and "Partial Sums of Infinite Series and How They Grow," R. P. Boas, *American Mathematical Monthly*, Volume 84, Number 4, April 1977, pp. 237–258.

The Limit Comparison Test

We resume discussion of criteria for determining the convergence or divergence of a given series $\Sigma\, a_n$. Observe that we can combine the comparison tests with the information we have about the geometric series and the various p-series. There is a more powerful form of the comparison test that is known as the *limit comparison test*, which we take up next. It is particularly handy in dealing with series in which a_n is a rational function of n. The next example will illustrate what we mean.

EXAMPLE 6 Do you think $\Sigma\, a_n$ converges? or diverges? Why?

a) $a_n = \dfrac{2n}{n^2 - n + 1}$ b) $a_n = \dfrac{2n^3 + 100n^2 + 1000}{\frac{1}{8}n^6 - n + 2}$

Discussion In determining convergence or divergence, only the tails count. And, when n is very large, the highest powers of n in both numerator and denominator are what count the most. So, in (a), we reason this way:

$$a_n = \frac{2n}{n^2 - n + 1} \quad \text{behaves about like} \quad \frac{2n}{n^2} = \frac{2}{n},$$

and we guess that $\Sigma\, a_n$ diverges by comparison with $\Sigma\, 1/n$. In (b), we reason that a_n will behave about like $2n^3/(1/8)n^6 = 16/n^3$ and we guess that the series will converge, by comparison with $\Sigma\, 1/n^3$, a p-series with $p = 3$.

To be more precise, in part (a) we can take

$$a_n = \frac{2n}{n^2 - n + 1} \quad \text{and} \quad d_n = \frac{1}{n}$$

and look at the ratio

$$\frac{a_n}{d_n} = \frac{2n^2}{n^2 - n + 1} = \frac{2}{1 - \left(\frac{1}{n}\right) + \left(\frac{1}{n^2}\right)}.$$

Clearly, as $n \to \infty$ the limit is 2:

$$\lim \frac{a_n}{d_n} = 2.$$

This means that, in particular, if we take $\varepsilon = 1$ in the definition of limit, we know there is an index N such that a_n/d_n is within 1 unit of this limit for all $n \geq N$:

$$2 - 1 \leq a_n/d_n \leq 2 + 1 \qquad \text{for} \quad n \geq N.$$

Thus $a_n \geq d_n$ for $n \geq N$. Therefore, by the comparison test, $\Sigma\, a_n$ diverges because $\Sigma\, d_n$ diverges.

In part (b), if we let $c_n = 1/n^3$, we can easily show that

$$\lim \frac{a_n}{c_n} = 16.$$

Taking $\varepsilon = 1$ in the definition of limit, we can conclude that there is an index N' such that a_n/c_n is between 15 and 17 when $n \geq N'$. Since $\Sigma\, c_n$ converges, so also does $\Sigma\, 17c_n$ and thus $\Sigma\, a_n$. □

Our rather rough guesswork paved the way for successful choices of comparison series. We make all of this more precise in the following *limit comparison test.*

Limit Comparison Test

A. Test for convergence. If $a_n \geq 0$ for $n \geq n_0$ and there is a convergent series $\Sigma\, c_n$ such that $c_n > 0$ and

$$\lim \frac{a_n}{c_n} < \infty, \tag{11}$$

then $\Sigma\, a_n$ converges.

B. Test for divergence. If $a_n \geq 0$ for $n \geq n_0$ and there is a divergent series $\Sigma\, d_n$ such that $d_n > 0$ and

$$\lim \frac{a_n}{d_n} > 0, \tag{12}$$

then $\Sigma\, a_n$ diverges.

We shall not formally prove these results. Part (A) follows easily from the fact that if (11) holds, then there is an index $N \geq n_0$ and a constant M such that $a_n < Mc_n$ when $n > N$, and $\Sigma\, Mc_n$ converges. Similarly, if (12) holds, there is an index $N' \geq n_0$ and a constant $k > 0$ such that $a_n > kd_n$ for $n \geq N'$, and $\Sigma\, kd_n$ diverges.

EXAMPLE 7 Do the following series converge, or diverge?

a) $\dfrac{3}{4} + \dfrac{5}{9} + \dfrac{7}{16} + \dfrac{9}{25} + \cdots = \displaystyle\sum_{n=1}^{\infty} \frac{2n + 1}{(n + 1)^2}$

b) $\dfrac{1}{2} + \dfrac{2}{3} + \dfrac{3}{4} + \dfrac{4}{5} + \dfrac{5}{6} + \cdots = \displaystyle\sum_{n=1}^{\infty} \frac{n}{n + 1}$

c) $\dfrac{101}{3} + \dfrac{102}{10} + \dfrac{103}{29} + \cdots = \displaystyle\sum_{n=1}^{\infty} \frac{100 + n}{n^3 + 2}$

d) $\frac{1}{1} + \frac{1}{3} + \frac{1}{7} + \cdots = \sum_{n=1}^{\infty} \frac{1}{2^n - 1}$

Solution a) Let $a_n = (2n + 1)/(n^2 + 2n + 1)$ and $d_n = 1/n$. Then

$$\sum d_n \text{ diverges} \qquad \text{and} \qquad \lim \frac{a_n}{d_n} = \lim \frac{2n^2 + n}{n^2 + 2n + 1} = 2,$$

so

$$\sum a_n \text{ diverges.}$$

b) Let $b_n = n/(n + 1)$. Then $\lim b_n = 1 \neq 0$, so $\Sigma\, b_n$ diverges, by the nth term test.

c) Let $a_n = (100 + n)/(n^3 + 2)$. When n is large, this ought to compare with $n/n^3 = 1/n^2$, so we let $c_n = 1/n^2$. Then we apply the limit comparison theorem:

$$\sum c_n \text{ converges} \qquad \text{and} \qquad \lim \frac{a_n}{c_n} = \lim \frac{n^3 + 100n^2}{n^3 + 2} = 1,$$

so

$$\sum a_n \text{ converges.}$$

d) Let $a_n = 1/(2^n - 1)$ and $c_n = 1/2^n$. (We reason that $2^n - 1$ behaves about like 2^n when n is large.) Then

$$\frac{a_n}{c_n} = \frac{2^n}{2^n - 1} = \frac{1}{1 - (1/2)^n} \to 1 \qquad \text{as} \quad n \to \infty.$$

Because $\Sigma\, c_n$ converges, we conclude that $\Sigma\, a_n$ does also. □

When we use the p-series for comparison, it is essential that p be a constant, as shown by this example.

EXAMPLE 8 Does the series

$$1^{-2} + 2^{-3/2} + 3^{-4/3} + 4^{-5/4} + \cdots + n^{-(n+1)/n} + \cdots$$

converge, or diverge?

Solution Let

$$a_n = n^{-(n+1)/n} = \frac{1}{n^{1+(1/n)}}.$$

This looks a bit like $1/n^p$ with $p = 1 + (1/n)$, which is greater than 1. But $1 + (1/n)$ isn't constant, so we need to be careful about drawing a conclusion. In fact

$$n^{1+(1/n)} = n(n)^{1/n} \qquad \text{and} \qquad (n)^{1/n} \to 1 \qquad \text{as} \quad n \to \infty.$$

Let's use the limit comparison test with $d_n = 1/n$. Then

$$\frac{a_n}{d_n} = \frac{n}{n(n)^{1/n}}$$

$$= \frac{1}{(n)^{1/n}} \to 1 \qquad \text{as} \quad n \to \infty.$$

We know that $\Sigma\, d_n$ diverges, and when we apply the limit comparison test we conclude that $\Sigma\, a_n$ also diverges. □

EXAMPLE 9 Does the series $\Sigma_{n=1}^{\infty} \ln n/n^{3/2}$ converge?

Solution Let $a_n = \ln n/n^{3/2}$ and $c_n = n^c/n^{3/2}$, where c is a positive constant that is less than $\frac{1}{2}$. For example, we might choose $c = \frac{1}{4}$. Then $c_n = 1/n^{3/2-c} = 1/n^p$ with $p = \frac{3}{2} - c > 1$. (When $c = \frac{1}{4}$, $p = \frac{5}{4}$.) Hence $\Sigma\, c_n$ converges. Because $\ln n$ goes to infinity more slowly than n^c, for any positive constant c (Article 11.3, Example 8), we have reason to believe that $\Sigma\, a_n$ also converges. To verify this hunch, we apply the limit comparison test:

$$\lim \frac{a_n}{c_n} = \lim \frac{\ln n}{n^c} = \lim \frac{1/n}{cn^{c-1}} \qquad \text{(by l'Hôpital's rule)}$$

$$= \lim \frac{1}{cn^c} = 0.$$

The given series converges. □

The Ratio Test

It is not always possible to tell whether a particular series converges by using the comparison test or the limit comparison test. We might not be able to find a series to compare it with. Nor is it always possible to use the integral test to answer whatever questions of convergence then remain unanswered. The terms of the series might not decrease as n increases, or we might not find a formula for the nth term that we can integrate. What we really need is an intrinsic test that can be applied directly to the terms of the given series $\Sigma\, a_n$. The next two tests are intrinsic and they are easy to apply. The first of these, the *ratio test*, measures the rate of growth (or decline) by examining the ratio a_{n+1}/a_n. For a geometric series, this rate of growth is a constant, and the series converges if and only if it is less than 1 in absolute value. But even if the ratio is not constant, we may be able to find a geometric series for comparison, as illustrated in the next example.

EXAMPLE 10 Let $a_1 = 1$ and define a_{n+1} to be

$$a_{n+1} = \frac{n}{2n+1} a_n.$$

Does the series $\Sigma\, a_n$ converge, or diverge?

Solution We begin by writing out a few terms of the series:

$$a_1 = 1, \qquad a_2 = \frac{1}{3}, \qquad a_3 = \frac{2}{5}a_2 = \frac{1 \cdot 2}{3 \cdot 5}, \qquad a_4 = \frac{3}{7}a_3 = \frac{1 \cdot 2 \cdot 3}{3 \cdot 5 \cdot 7}.$$

We observe that each term is somewhat less than $\frac{1}{2}$ the term before it, because $n/(2n+1)$ is less than $\frac{1}{2}$. Therefore the terms of the given series are less than or equal to the terms of the geometric series

$$1 + \left(\frac{1}{2}\right) + \left(\frac{1}{4}\right) + \cdots + \left(\frac{1}{2}\right)^{n-1} + \cdots$$

that converges to 2. So our series also converges, and we now know that its sum is less than 2. □

In *proving* the ratio test, we shall make a comparison with appropriate geometric series as in the example above, but when we *apply* it we don't actually make a direct comparison.

The Ratio Test

Let $\Sigma\, a_n$ be a series with positive terms, and suppose that

$$\lim_{n\to\infty} \frac{a_{n+1}}{a_n} = \rho \qquad \text{(Greek letter rho).}$$

Then,

a) the series *converges* if $\rho < 1$,

b) the series *diverges* if $\rho > 1$,

c) the series *may converge or it may diverge* if $\rho = 1$. (The test provides no information.)

Proof. a) Assume first that $\rho < 1$ and let r be a number between ρ and 1. Then the number

$$\varepsilon = r - \rho$$

is positive. Since

$$\frac{a_{n+1}}{a_n} \to \rho,$$

a_{n+1}/a_n must lie within ε of ρ when n is large enough, say, for all $n \geq N$. In particular,

$$\frac{a_{n+1}}{a_n} < \rho + \varepsilon = r, \qquad \text{when } n > N.$$

That is,

$$\begin{aligned} a_{N+1} &< ra_N, \\ a_{N+2} &< ra_{N+1} < r^2 a_N, \\ a_{N+3} &< ra_{N+2} < r^3 a_N, \\ &\vdots \\ a_{N+m} &< ra_{N+m-1} < r^m a_N. \end{aligned}$$

These inequalities show that the terms of our series, after the Nth term, approach zero more rapidly than the terms in a geometric series with ratio $r < 1$. More precisely, consider the series $\Sigma\, c_n$, where $c_n = a_n$ for $n = 1, 2, \ldots, N$ and $c_{N+1} = ra_N$, $c_{N+2} = r^2 a_N, \ldots, c_{N+m} = r^m a_N, \ldots$. Now $a_n \leq c_n$ for all n, and

$$\begin{aligned} \sum_{n=1}^{\infty} c_n &= a_1 + a_2 + \cdots + a_{N-1} + a_N + ra_N + r^2 a_N + \cdots \\ &= a_1 + a_2 + \cdots + a_{N-1} + a_N(1 + r + r^2 + \cdots). \end{aligned}$$

Because $|r| < 1$, the geometric series $1 + r + r^2 + \cdots$ converges, and hence so does $\Sigma\, c_n$. By the comparison test, $\Sigma\, a_n$ also converges.

b) Next, suppose $\rho > 1$. Then, from some index M on, we have

$$\frac{a_{n+1}}{a_n} > 1 \qquad \text{or} \qquad a_M < a_{M+1} < a_{M+2} < \cdots.$$

Hence, the terms of the series do not approach 0 as n becomes infinite, and the series diverges, by the nth-term test.

c) Finally, the two series

$$\sum_{n=1}^{\infty} \frac{1}{n} \quad \text{and} \quad \sum_{n=1}^{\infty} \frac{1}{n^2}$$

show that, when $\rho = 1$, some other test for convergence must be used.

$$\text{For } \sum_{n=1}^{\infty} \frac{1}{n}: \qquad \frac{a_{n+1}}{a_n} = \frac{1/(n+1)}{1/n} = \frac{n}{n+1} \to 1.$$

$$\text{For } \sum_{n=1}^{\infty} \frac{1}{n^2}: \qquad \frac{a_{n+1}}{a_n} = \frac{1/(n+1)^2}{1/n^2} = \left(\frac{n}{n+1}\right)^2 \to 1^2 = 1.$$

In both cases $\rho = 1$, yet the first series diverges while the second converges. ■

The ratio test is often effective when the terms of the series contain factorials of expressions involving n, or expressions raised to the nth power, or combinations, as in the next example. We recall that

$$n! = 1 \cdot 2 \cdot 3 \cdot \cdots \cdot n$$

implies that

$$(n+1)! = (n+1)n!$$

and that $(2(n+1))! = (2n+2)! = (2n+2)(2n+1)(2n)!$. These facts are used in parts (a) and (b) of the example. Part (b) also uses the fact that $4^{n+1}/4^n = 4$. Making use of appropriate cancellation properties of factorials and powers leads to simplified expressions for the ratio a_{n+1}/a_n.

EXAMPLE 11 Test the following series for convergence or divergence, using the ratio test.

a) $\sum_{n=1}^{\infty} \frac{n!n!}{(2n)!}$ b) $\sum_{n=1}^{\infty} \frac{4^n n!n!}{(2n)!}$ c) $\sum_{n=0}^{\infty} \frac{2^n + 5}{3^n}$

Solution a) If $a_n = n!n!/(2n)!$, then $a_{n+1} = (n+1)!(n+1)!/(2n+2)!$, and

$$\frac{a_{n+1}}{a_n} = \frac{(n+1)!(n+1)!(2n)!}{n!n!(2n+2)(2n+1)(2n)!}$$

$$= \frac{(n+1)(n+1)}{(2n+2)(2n+1)} = \frac{n+1}{4n+2} \to \frac{1}{4}.$$

The series converges because $\rho = \frac{1}{4}$ is less than 1.

b) If $a_n = 4^n n!n!/(2n)!$, then

$$\frac{a_{n+1}}{a_n} = \frac{4^{n+1}(n+1)!(n+1)!}{(2n+2)(2n+1)(2n)!} \times \frac{(2n)!}{4^n n!n!}$$

$$= \frac{4(n+1)(n+1)}{(2n+2)(2n+1)} = \frac{2(n+1)}{2n+1} \to 1.$$

Because the limit is $\rho = 1$, we cannot decide on the basis of the ratio test alone whether the series converges or diverges. However, when we note that $a_{n+1}/a_n = (2n+2)/(2n+1)$, we conclude that a_{n+1} is always greater than a_n because $(2n+2)/(2n+1)$ is always greater than 1. Therefore, all terms are greater than or equal to $a_1 = 2$ and the nth term does not go to zero as n tends to infinity. Hence, by the nth term test, the series diverges.

c) For the series $\sum_{n=0}^{\infty} (2^n + 5)/3^n$,

$$\frac{a_{n+1}}{a_n} = \frac{(2^{n+1}+5)/3^{n+1}}{(2^n+5)/3^n} = \frac{1}{3}\cdot\frac{2^{n+1}+5}{2^n+5}$$
$$= \frac{1}{3}\cdot\left(\frac{2+5\cdot 2^{-n}}{1+5\cdot 2^{-n}}\right) \to \frac{1}{3}\cdot\frac{2}{1} = \frac{2}{3}.$$

The series converges because $\rho = \frac{2}{3}$ is less than 1.

This does *not* mean that $\frac{2}{3}$ is the sum of the series. In fact,

$$\sum_{n=0}^{\infty} \frac{2^n+5}{3^n} = \sum_{n=0}^{\infty}\left(\frac{2}{3}\right)^n + \sum_{n=0}^{\infty}\frac{5}{3^n} = \frac{1}{1-\frac{2}{3}} + \frac{5}{1-\frac{1}{3}} = 10\tfrac{1}{2}. \quad \square$$

REMARK 1. While the ratio test is useful in the types of series just discussed, it is not very useful for series like p-series.

In the next example, the series is expressed in terms of powers of x. By applying the ratio test, we can learn for what values of x the series converges. For those values, the series defines a function of x. (In Chapter 12, we discuss such power series in more detail.)

EXAMPLE 12 For what values of x does the series

$$x + \frac{x^3}{3} + \frac{x^5}{5} + \frac{x^7}{7} + \cdots + \frac{x^{2n-1}}{2n-1} + \cdots$$

converge?

Solution The nth term of the series is

$$a_n = \frac{x^{2n-1}}{2n-1}.$$

We consider first the case where x is positive. Then the series is a positive series and

$$\frac{a_{n+1}}{a_n} = \frac{(2n-1)x^2}{(2n+1)} \to x^2.$$

The ratio test therefore tells us that the series converges if x is positive and less than one and diverges if x is greater than one.

Since only odd powers of x occur in the series, we see that the series simply changes sign when x is replaced by $-x$. Therefore the series also converges for $-1 < x \leq 0$ and diverges for $x < -1$. The series converges to zero when $x = 0$.

We know, thus far, that the series

converges for $|x| < 1$,
diverges for $|x| > 1$,

but we don't know what happens when $|x| = 1$. To test at $x = 1$, we apply the integral test to the series

$$1 + \frac{1}{3} + \frac{1}{5} + \frac{1}{7} + \cdots + \frac{1}{2n-1} + \cdots,$$

which we get by taking $x = 1$ in the original series. The companion integral is

$$\int_1^\infty \frac{dx}{2x-1} = \frac{1}{2}\ln(2x-1)\Big|_1^\infty = \infty.$$

Hence the series diverges to $+\infty$ when $x = 1$. It diverges to $-\infty$ when $x = -1$. Therefore the only values of x for which the given series converges are $-1 < x < 1$. □

REMARK 2. When $\rho < 1$, the ratio test is also useful in estimating the truncation error that results from using

$$s_N = a_1 + a_2 + \cdots + a_N$$

as an approximation to the sum of a convergent series of positive terms

$$S = a_1 + a_2 + \cdots + a_N + (a_{N+1} + \cdots);$$

for, if we know that

$$r_1 \le \frac{a_{n+1}}{a_n} \le r_2 \qquad \text{for} \qquad n \ge N, \tag{13}$$

where r_1 and r_2 are constants that are both less than one, then the inequalities

$$r_1 a_n \le a_{n+1} \le r_2 a_n \qquad (n = N, N+1, N+2, \ldots)$$

enable us to deduce that

$$a_N(r_1 + r_1^2 + r_1^3 + \cdots) \le \sum_{n=N+1}^{\infty} a_n \le a_N(r_2 + r_2^2 + r_2^3 + \cdots). \tag{14}$$

The two geometric series have sums

$$r_1 + r_1^2 + r_1^3 + \cdots = \frac{r_1}{1-r_1}, \qquad r_2 + r_2^2 + r_2^3 + \cdots = \frac{r_2}{1-r_2}.$$

Hence, the error

$$R_N = \sum_{n=N+1}^{\infty} a_n$$

lies between

$$a_N \frac{r_1}{1-r_1} \qquad \text{and} \qquad a_N \frac{r_2}{1-r_2}.$$

That is,

$$a_N \frac{r_1}{1-r_1} \le S - s_N \le a_N \frac{r_2}{1-r_2} \tag{15}$$

if

$$0 \le r_1 \le \frac{a_{n+1}}{a_n} \le r_2 < 1 \qquad \text{for} \quad n \ge N.$$

EXAMPLE 13 Estimate the remainder for the series

$$1 + \frac{1}{3} + \frac{1 \cdot 2}{3 \cdot 5} + \frac{1 \cdot 2 \cdot 3}{3 \cdot 5 \cdot 7} + \cdots = a_1 + a_2 + a_3 + \cdots$$

where

$$a_1 = 1 \qquad \text{and} \qquad a_{n+1} = \frac{n}{2n + 1} a_n \qquad \text{for} \quad n \geq 1.$$

Solution We see that the ratio

$$\frac{a_{n+1}}{a_n} = \frac{n}{2n + 1}$$

is less than $\frac{1}{2}$ for all n. So we can take $r_2 = \frac{1}{2}$ in (13) and (15). We also observe that the sequence $\{n/(2n + 1)\}$ is an increasing sequence, so we can take $r_2 = N/(2N + 1)$. When we substitute these values of r_1 and r_2 in (15) we get

$$a_N \frac{N}{N + 1} \leq S - s_N \leq a_N.$$

For $N = 10$, for example, the difference between the sum of all terms of the series and the sum of the first 10 terms is between $(10/11)a_{10}$ and a_{10}. For $N = 20$, the difference is between $(20/21)a_{20}$ and a_{20}. We have calculated s_{20} and a_{20} to be

$$s_{20} = 1.570795962, \qquad a_{20} = 3.80 \times 10^{-7}.$$

Therefore, R_{20} lies between 3.80×10^{-7} and $(20/21)a_{20}$, which is 3.62×10^{-7}. We estimate the sum S for the entire series by $S = s_{20} + R_{20}$ and get

$$1.570796324 \leq S \leq 1.570796342. \ \square$$

CAUTION. These calculations involve round-off errors that can easily affect the last decimal place or more. It seems safe to say that the first six decimals are correct. Scientists and engineers who need to be certain of accuracy beyond that obtained here should consult textbooks on numerical analysis.

The *n*th Root Test

We return to the question, "Does $\Sigma\, a_n$ converge?" When there is a simple formula for a_n, we can try one of the tests we already have. But, suppose that the series is the following,

$$\frac{1}{2} + \frac{2}{4} + \frac{3}{8} + \frac{1}{16} + \frac{5}{32} + \frac{1}{64} + \cdots + \frac{f(n)}{2^n} + \cdots,$$

where

$$f(n) = \begin{cases} n & \text{if } n \text{ is a prime number,} \\ 1 & \text{otherwise.} \end{cases}$$

Clearly, this is not a geometric series. The nth term approaches zero as $n \to \infty$, so we don't know that the series diverges. The integral test doesn't look promising. The ratio test produces

$$\frac{a_{n+1}}{a_n} = \frac{1}{2}\frac{f(n+1)}{f(n)} = \begin{cases} \dfrac{1}{2^n} & \text{if } n \text{ is a prime} \geq 3, \\[2mm] \dfrac{n+1}{2} & \text{if } n+1 \text{ is a prime} \geq 5. \end{cases}$$

The ratio will get very large or very close to zero and it has no limit because there are infinitely many primes. The test that will answer the question for this series is the nth root test. To apply it, we look at

$$\sqrt[n]{a_n} = \frac{\sqrt[n]{f(n)}}{2} = \begin{cases} \dfrac{\sqrt[n]{n}}{2} & \text{if } n \text{ is a prime,} \\[2mm] \dfrac{1}{2} & \text{otherwise.} \end{cases}$$

In this example, therefore, it will always be true that

$$\frac{1}{2} \leq \sqrt[n]{a_n} \leq \frac{\sqrt[n]{n}}{2}$$

and $\lim \sqrt[n]{a_n} = \frac{1}{2}$ (by the Sandwich limit theorem). Because this limit is less than 1, the nth *root test* tells us that the given series converges, as we shall now see.

The nth Root Test

Let $\Sigma\, a_n$ be a series with $a_n \geq 0$ for $n \geq n_0$, and suppose that

$$\sqrt[n]{a_n} \to \rho.$$

Then,

a) the series *converges* if $\rho < 1$

b) the series *diverges* if $\rho > 1$

c) the test is *not conclusive* if $\rho = 1$.

Proof. a) Suppose that $\rho < 1$, and choose an $\varepsilon > 0$ so small that $\rho + \varepsilon < 1$ also. Since $\sqrt[n]{a_n} \to \rho$, the terms $\sqrt[n]{a_n}$ eventually get closer than ε to ρ. In other words, there exists an index $N \geq n_0$ such that

$$\sqrt[n]{a_n} < \rho + \varepsilon \qquad \text{when} \quad n \geq N.$$

Then it is also true that

$$a_n < (\rho + \varepsilon)^n \qquad \text{for} \quad n \geq N.$$

Now,

$$\sum_{n=N}^{\infty} (\rho + \varepsilon)^n,$$

a geometric series with ratio $(\rho + \varepsilon) < 1$, converges. By comparison,

$$\sum_{n=N}^{\infty} a_n$$

converges, from which it follows that

$$\sum_{n=1}^{\infty} a_n = a_1 + \cdots + a_{N-1} + \sum_{n=N}^{\infty} a_n$$

converges.

b) Suppose that $\rho > 1$. Then, for all indices beyond some index M, we have

$$\sqrt[n]{a_n} > 1,$$

so that

$$a_n > 1 \quad \text{for} \quad n > M,$$

and the terms of the series do not converge to 0. The series therefore diverges by the nth-term test.

c) The series $\Sigma_{n=1}^{\infty} (1/n)$ and $\Sigma_{n=1}^{\infty} (1/n^2)$ show that the test is not conclusive when $\rho = 1$. The first series diverges and the second converges, but in both cases $\sqrt[n]{a_n} \to 1$. ■

EXAMPLE 14 For the series $\Sigma_{n=1}^{\infty} (1/n^n)$,

$$\sqrt[n]{\frac{1}{n^n}} = \frac{1}{n} \to 0.$$

The series converges. □

EXAMPLE 15 For the series $\Sigma_{n=1}^{\infty} (2^n/n^2)$,

$$\sqrt[n]{\frac{2^n}{n^2}} = \frac{2}{\sqrt[n]{n^2}} = \frac{2}{(\sqrt[n]{n})^2} \to \frac{2}{1^2} = 2.$$

The series diverges. □

EXAMPLE 16 For the series $\Sigma_{n=1}^{\infty} (1 - 1/n)^n = 0 + \frac{1}{4} + \frac{8}{27} + \cdots$,

$$\sqrt[n]{\left(1 - \frac{1}{n}\right)^n} = \left(1 - \frac{1}{n}\right) \to 1.$$

Because $\rho = 1$, the root test is not conclusive. However, if we apply the nth-term test for divergence, we find that

$$\left(1 - \frac{1}{n}\right)^n = \left(1 + \frac{-1}{n}\right)^n \to e^{-1} = \frac{1}{e}.$$

The series diverges. □

List of Tests

We now have seven tests for divergence and convergence of infinite series:

1. The nth-term test for divergence (applies to all series).
2. The "bounded from above" test (applies to partial sums of nonnegative series).
3. A. The comparison test for convergence (nonnegative series for $n \geq n_0$).

B. The comparison test for divergence (nonnegative series for $n \geq n_0$).

4. A. The limit comparison test for convergence (as for 3A).
   B. The limit comparison test for divergence (as for 3B).
5. The integral test (positive decreasing series).
6. The ratio test (positive series).
7. The nth root test (nonnegative series for $n \geq n_0$).

NOTE. These tests can also be applied to settle questions about the convergence or divergence of series of nonpositive or negative terms. Just factor -1 from the series in question, and test the resulting series of nonnegative or positive terms.

EXAMPLE 17

$$\sum_{n=1}^{\infty} -\frac{1}{n} = -1 \cdot \sum_{n=1}^{\infty} \frac{1}{n} \quad \text{diverges.} \ \square$$

EXAMPLE 18

$$\sum_{n=0}^{\infty} -\frac{1}{2^n} = -1 \cdot \sum_{n=0}^{\infty} \frac{1}{2^n} = -1 \cdot 2 = -2. \ \square$$

PROBLEMS

In Problems 1–30, determine whether the given series converges or diverges. In each case, give a reason for your answer.

1. $\sum_{n=1}^{\infty} \frac{1}{10^n}$

2. $\sum_{n=1}^{\infty} \frac{n}{n+2}$

3. $\sum_{n=1}^{\infty} \frac{\sin^2 n}{2^n}$

4. $\sum_{n=1}^{\infty} \frac{5}{n}$

5. $\sum_{n=1}^{\infty} \frac{n^3}{2^n}$

6. $\sum_{n=1}^{\infty} -\frac{1}{8^n}$

7. $\sum_{n=1}^{\infty} \frac{\ln n}{n}$

8. $\sum_{n=1}^{\infty} \frac{1}{n\sqrt{n}}$

9. $\sum_{n=1}^{\infty} \frac{2^n}{3^n}$

10. $\sum_{n=0}^{\infty} \frac{-2}{n+1}$

11. $\sum_{n=1}^{\infty} \frac{1}{1+\ln n}$

12. $\sum_{n=1}^{\infty} \frac{\ln n}{\sqrt{n+1}\, n}$

13. $\sum_{n=1}^{\infty} \frac{2^n}{n+1}$

14. $\sum_{n=1}^{\infty} \left(\frac{n}{3n+1}\right)^n$

15. $\sum_{n=1}^{\infty} -\frac{n^2}{2^n}$

16. $\sum_{n=1}^{\infty} \frac{1}{\sqrt{n}} \frac{(\ln n)^{10}}{n^{2/3}}$

17. $\sum_{n=1}^{\infty} \frac{1}{\sqrt{n^3+2}}$

18. $\sum_{n=1}^{\infty} \frac{1}{\sqrt[n]{2}}$

19. $\sum_{n=1}^{\infty} \frac{(n+1)(n+2)}{n!}$

20. $\sum_{n=1}^{\infty} \frac{\sqrt{n}}{n^2+1}$

21. $\sum_{n=1}^{\infty} \frac{n}{n^2+1}$

22. $\sum_{n=1}^{\infty} n^2 e^{-n}$

23. $\sum_{n=1}^{\infty} \left(1+\frac{1}{n}\right)^n$

24. $\sum_{n=1}^{\infty} \frac{1}{3^{n-1}+1}$

25. $\sum_{n=1}^{\infty} \frac{(n+3)!}{3!n!3^n}$

26. $\sum_{n=2}^{\infty} \frac{1}{n \ln n}$

27. $\sum_{n=1}^{\infty} \frac{1}{(2n+1)!}$

28. $\sum_{n=1}^{\infty} \frac{1}{(\ln 2)^n}$

29. $\sum_{n=1}^{\infty} \frac{n!}{n^n}$

30. $\sum_{n=1}^{\infty} \frac{1-n}{n \cdot 2^n}$

31. Show that the series

$$\sum_{n=1}^{\infty} \frac{1}{2n-1} = 1 + \frac{1}{3} + \frac{1}{5} + \cdots$$

diverges. (*Hint:* Compare the series with a multiple of the harmonic series.)

In Problems 32–34, find all values of x for which the given series converge. Begin with the ratio test or the root test, and then apply other tests as needed.

32. $\sum_{n=1}^{\infty}\left(\frac{x^2+1}{3}\right)^n$ **33.** $\sum_{n=1}^{\infty}\frac{x^{2n+1}}{n^2}$

34. $\sum_{n=1}^{\infty}\left(\frac{1}{|x|}\right)^n$

35. (*Calculator*) Use Inequality (9) of this article to estimate $\Sigma_{n=1}^{\infty}(1/n^2)$ with an error less than 0.005. Compare your result with the value given in Problem 36.

36. (*Calculator*) Euler discovered that

$$\sum_{n=1}^{\infty}\frac{1}{n^2}=\sum_{n=1}^{\infty}\frac{3[(n-1)!]^2}{(2n)!}=\frac{\pi^2}{6}.$$

Compute s_6 for each series. To 10 decimal places, $\pi^2/6 = 1.6449340668$.

37. (*Calculator*) Use the expression in (10) of this article to find the value of $\Sigma_{n=1}^{\infty}(1/n^4) = (\pi^4/90)$ with an error less than 10^{-6}.

38. (*Calculator*) To estimate partial sums of the divergent harmonic series, Inequality (7) with $f(x) = 1/x$ tells us that

$$\ln n < 1 + \frac{1}{2} + \cdots + \frac{1}{n} < 1 + \ln n.$$

Suppose that the summation started with $s_1 = 1$ thirteen billion years ago (one estimate of the age of the universe) and that a new term has been added every *second* since then. How large would you expect s_n to be today?

39. There are no values of x for which $\Sigma_{n=1}^{\infty}(1/nx)$ converges. Why?

40. Show that if $\Sigma_{n=1}^{\infty} a_n$ is a convergent series of nonnegative numbers then the series $\Sigma_{n=1}^{\infty}(a_n/n)$ converges.

41. Show that if $\Sigma\, a_n$ and $\Sigma\, b_n$ are convergent series with $a_n \geq 0$ and $b_n \geq 0$, then $\Sigma\, a_n b_n$ converges. (*Hint:* From some index on, $a_n b_n < a_n + b_n$.)

42. A sequence of numbers

$$s_1 \geq s_2 \geq \cdots \geq s_n \geq s_{n+1} \geq \cdots,$$

in which $s_n \geq s_{n+1}$ for every n, is called a *decreasing sequence*. A sequence $\{s_n\}$ is *bounded from below* if there is a finite constant M with $M \leq s_n$ for every n. Such a number M is called a *lower bound* for the sequence. Deduce from Theorem 1 that a decreasing sequence that is bounded from below converges, and that a decreasing sequence that is not bounded from below diverges.

43. The *Cauchy condensation test* says: Let $\{a_n\}$ be a decreasing sequence ($a_n \geq a_{n+1}$, all n) of positive terms that converges to 0. Then,

$$\sum a_n \text{ converges if and only if } \sum 2^n a_{2^n} \text{ converges.}$$

For example, $\Sigma\,(1/n)$ diverges because $\Sigma\, 2^n \cdot (1/2^n) = \Sigma\, 1$. Show why the test works.

44. Use the Cauchy condensation test of Problem 43 to show that

a) $\sum_{n=2}^{\infty}\frac{1}{n \ln n}$ diverges.

b) $\sum_{n=1}^{\infty}\frac{1}{n^p}$ converges if $p > 1$ and diverges if $p \leq 1$.

45. Pictures like the one in Fig. 11.9 suggest that, as n increases, there is very little change in the difference between the sum

$$1 + \frac{1}{2} + \cdots + \frac{1}{n}$$

and the integral

$$\ln n = \int_1^n \frac{1}{x}\,dx.$$

To explore this idea, carry out the following steps.

a) By taking $f(x) = 1/x$ in inequality (7), show that

$$\ln(n+1) \leq 1 + \frac{1}{2} + \cdots + \frac{1}{n} \leq 1 + \ln n$$

or

$$0 < \ln(n+1) - \ln n \leq 1 + \frac{1}{2} + \cdots + \frac{1}{n} - \ln n \leq 1.$$

Thus, the sequence

$$a_n = 1 + \frac{1}{2} + \cdots + \frac{1}{n} - \ln n$$

is bounded from below and from above.

b) Show that

$$\frac{1}{n+1} < \int_n^{n+1}\frac{1}{x}\,dx = \ln(n+1) - \ln n,$$

so that the sequence $\{a_n\}$ in part (a) is decreasing.

Since a decreasing sequence that is bounded from below converges (Problem 42) the numbers a_n defined in (a) converge:

$$1 + \frac{1}{2} + \cdots + \frac{1}{n} - \ln n \to \gamma.$$

The number γ, whose value is 0.5772 . . . , is called *Euler's constant*. In contrast to other special numbers like π and e, no other expression with a simple law of formulation has ever been found for γ.

46. *Logarithmic p-series.* Let p be a positive constant. Show that

$$\int_2^{\infty} \frac{dx}{x\,(\ln x)^p}$$

converges if and only if $p > 1$. (The integral does not start at 1, but at 2, because $\ln 1 = 0$.) What can you deduce about convergence or divergence of the following series?

a) $\displaystyle\sum_{n=2}^{\infty} \frac{1}{n \ln n}$ b) $\displaystyle\sum_{n=2}^{\infty} \frac{1}{n(\ln n)^{1.01}}$

c) $\displaystyle\sum_{n=5}^{\infty} \frac{n^{1/2}}{(\ln n)^3}$ d) $\displaystyle\sum_{n=3}^{\infty} \frac{1}{n \ln (n^3)}$

e) $\displaystyle\sum_{n=2}^{\infty} \frac{1}{n(\ln n)^{(n+1)/n}}$

***Tookit* programs**
Sequences and Series

11.6 Absolute Convergence

We now extend to series that have both positive and negative terms the techniques that we have developed for answering questions about the convergence of series of nonnegative numbers. The extension is made possible by a theorem that says that, if a series converges after all its negative terms have been made positive, then the unaltered series converges also.

THEOREM 1

If $\Sigma_{n=1}^{\infty} |a_n|$ converges, then $\Sigma_{n=1}^{\infty} a_n$ converges.

Proof of the theorem. For each n,

$$-|a_n| \le a_n \le |a_n|,$$

so that

$$0 \le a_n + |a_n| \le 2|a_n|.$$

If $\Sigma_{n=1}^{\infty} |a_n|$ converges, then $\Sigma_{n=1}^{\infty} 2|a_n|$ converges and, by the comparison test, the nonnegative series

$$\sum_{n=1}^{\infty} (a_n + |a_n|)$$

converges. The equality $a_n = (a_n + |a_n|) - |a_n|$ now lets us express $\Sigma_{n=1}^{\infty} a_n$ as the difference of two convergent series:

$$\sum_{n=1}^{\infty} a_n = \sum_{n=1}^{\infty} (a_n + |a_n| - |a_n|) = \sum_{n=1}^{\infty} (a_n + |a_n|) - \sum_{n=1}^{\infty} |a_n|.$$

Therefore, $\Sigma_{n=1}^{\infty} a_n$ converges. ■

An obvious corollary of Theorem 1 (which is also called the *contrapositive* form of the theorem) is

COROLLARY

If $\Sigma\, a_n$ diverges, then $\Sigma\, |a_n|$ diverges.

(See Problem 19.)

DEFINITION A series $\Sigma_{n=1}^{\infty} a_n$ is said to *converge absolutely* if $\Sigma_{n=1}^{\infty} |a_n|$ converges.

Our theorem can now be rephrased to say that *every absolutely convergent series converges.* We shall see in the next article, however, that the converse of this statement is false. Many convergent series do not converge absolutely. That is, there are many series whose convergence depends on the presence of infinitely many positive and negative terms arranged in a particular order.

Here are some examples of how the theorem can and cannot be used to determine convergence.

EXAMPLE 1 For $\sum_{n=1}^{\infty} (-1)^{n+1}\frac{1}{n^2} = 1 - \frac{1}{4} + \frac{1}{9} - \frac{1}{16} + \cdots,$

the corresponding series of absolute values is

$$\sum_{n=1}^{\infty} \frac{1}{n^2} = 1 + \frac{1}{4} + \frac{1}{9} + \frac{1}{16} + \cdots,$$

which converges because it is a p-series with $p = 2 > 1$ (Article 11.5). Therefore

$$\sum_{n=1}^{\infty} (-1)^{n+1}\frac{1}{n^2}$$

converges absolutely. Therefore

$$\sum_{n=1}^{\infty} (-1)^{n+1}\frac{1}{n^2}$$

converges. □

EXAMPLE 2 For $\sum_{n=1}^{\infty} \frac{\sin n}{n^2} = \frac{\sin 1}{1} + \frac{\sin 2}{4} + \frac{\sin 3}{9} + \cdots,$

the corresponding series of absolute values is

$$\sum_{n=1}^{\infty} \left|\frac{\sin n}{n^2}\right| = \frac{|\sin 1|}{1} + \frac{|\sin 2|}{4} + \cdots,$$

which converges by comparison with $\Sigma_{n=1}^{\infty} (1/n^2)$, because $|\sin n| \leq 1$ for every n. The original series converges absolutely; therefore it converges. □

EXAMPLE 3 For $\sum_{n=1}^{\infty} (-1)^{n+1}\frac{1}{n} = 1 - \frac{1}{2} + \frac{1}{3} - \frac{1}{4} + \cdots,$

the corresponding series of absolute values is

$$\sum_{n=1}^{\infty} \frac{1}{n} = 1 + \frac{1}{2} + \frac{1}{3} + \frac{1}{4} + \cdots,$$

which diverges. *We can draw no conclusion from this about the convergence or divergence of the original series.* Some other test must be found. In fact, the original series converges, as we shall see in the next article. □

EXAMPLE 4 The series

$$\sum_{n=1}^{\infty} (-1)^n \frac{n}{5n+1} = -\frac{1}{6} + \frac{2}{11} - \frac{3}{16} + \frac{4}{21} - \cdots$$

does not converge, by the nth-term test. Therefore, by the Corollary of Theorem 1, the series does not converge absolutely. □

REMARK. We know that $\Sigma\, a_n$ converges if $\Sigma\, |a_n|$ converges, but the two series will generally not converge to the same sum. For example,

$$\sum_{n=0}^{\infty} \left|\frac{(-1)^n}{2^n}\right| = \sum_{n=0}^{\infty} \frac{1}{2^n} = \frac{1}{1-\frac{1}{2}} = 2,$$

while

$$\sum_{n=0}^{\infty} \frac{(-1)^n}{2^n} = \frac{1}{1+\frac{1}{2}} = \frac{2}{3}.$$

In fact, when a series $\Sigma\, a_n$ converges absolutely, we can expect $\Sigma\, a_n$ to equal $\Sigma\, |a_n|$ only if none of the numbers a_n is negative.

One other important fact about absolutely convergent series is

THEOREM 2

The Rearrangement Theorem for Absolutely Convergent Series

If $\Sigma\, a_n$ converges absolutely, and $b_1, b_2, \ldots, b_n, \ldots$ is any arrangement of the sequence $\{a_n\}$, then

$$\sum b_n \text{ converges}$$

and

$$\sum_{n=1}^{\infty} b_n = \sum_{n=1}^{\infty} a_n.$$

(For an outline of the proof, see Problem 22.)

EXAMPLE 5 As we saw in Example 1, the series

$$1 - \left(\frac{1}{4}\right) + \left(\frac{1}{9}\right) - \left(\frac{1}{16}\right) + \cdots + (-1)^{n-1}\left(\frac{1}{n^2}\right) + \cdots \tag{1}$$

converges absolutely. A possible rearrangement of the terms of the series might start with a positive term, then 2 negative terms, then 3 positive terms, then 4 negative terms, and so on: after k terms of one sign, take $k+1$ terms of the opposite sign. The first ten terms of this series look like this:

$$1 - \frac{1}{4} - \frac{1}{16} + \frac{1}{9} + \frac{1}{25} + \frac{1}{49} - \frac{1}{36} - \frac{1}{64} - \frac{1}{100} - \frac{1}{144} + \cdots. \tag{2}$$

The rearrangement theorem says that both series converge to the same value. In this problem, if we had the second series to begin with, we would probably be glad to exchange it for the first, if we knew that we could. We can do even better: the sum of either series is also equal to

$$\sum_{n=1}^{\infty} \frac{1}{(2n-1)^2} - \sum_{n=1}^{\infty} \frac{1}{(2n)^2}. \tag{3}$$

(See Problem 23.) □

Multiplication of Series

Don't panic: it is possible to multiply two series, by applying the distributive law. For convenience of notation, we shall write the first series as

$$a_0 + a_1 + a_2 + \cdots + a_n + \cdots \tag{1}$$

and the second as

$$b_0 + b_1 + b_2 + \cdots + b_n + \cdots \tag{2}$$

and the product as

$$a_0b_0 + (a_0b_1 + a_1b_0) + (a_0b_2 + a_1b_1 + a_2b_0) + \cdots + (a_0b_n + a_1b_{n-1} + \cdots + a_kb_{n-k} + \cdots + a_nb_0) + \cdots. \tag{3}$$

Thus every term in the first series appears as a factor with every term in the second series, in the form a_kb_{n-k}.

EXAMPLE 6 Let

$$a_0 = 1, \quad a_1 = \frac{1}{2}, \quad \ldots, \quad a_n = \frac{1}{2^n} \tag{4}$$

and

$$b_0 = 1, \quad b_1 = \frac{1}{3}, \quad \ldots, \quad b_n = \frac{1}{3^n}. \tag{5}$$

The products in (3) are

$$a_0b_0 = 1, \quad a_0b_1 = \frac{1}{3}, \quad a_1b_0 = \frac{1}{2}, \quad a_0b_2 = \frac{1}{9}, \quad a_1b_1 = \frac{1}{6}, \quad a_2b_0 = \frac{1}{4}$$

and, in general, all terms of the form

$$a_kb_{n-k} = \frac{1}{2^k3^{n-k}}, \qquad n = 0, 1, 2, \ldots \text{ and } k = 0, 1, 2, \ldots, n. \; \square$$

The following theorem tells us that the sum of all these products is the sum of the first series, which is 2, times the sum of the second series, which is $\frac{3}{2}$ (both are geometric series: $r_1 = \frac{1}{2}$ and $r_2 = \frac{1}{3}$). By the way in which the terms of the product series are formed, we see that every number that is not divisible by a prime greater than 3 appears as a denominator, precisely once. By rearranging the terms (using Theorem 2) we can indicate the sum as

$$\frac{1}{1} + \frac{1}{2} + \frac{1}{3} + \frac{1}{4} + \frac{1}{6} + \frac{1}{8} + \frac{1}{9} + \frac{1}{12} + \frac{1}{16} + \frac{1}{18} + \frac{1}{24} + \frac{1}{32} + \cdots. \tag{6}$$

The sum of the series in (6) is 3.

We now state Theorem 3 (without proof).

THEOREM 3

Let $\Sigma\, a_n$ and $\Sigma\, b_n$ be absolutely convergent series. Define $\Sigma\, c_n$ by the equations

$$c_0 = a_0 b_0, \qquad c_1 = a_0 b_1 + a_1 b_0, \ldots, c_n = \sum_{k=0}^{n} a_k b_{n-k}. \tag{7}$$

Then $\Sigma\, c_n$ converges absolutely and

$$\sum_{n=0}^{\infty} c_n = \left(\sum_{n=0}^{\infty} a_n\right) \times \left(\sum_{n=0}^{\infty} b_n\right). \tag{8}$$

EXAMPLE 7 The sum of the reciprocals of those integers that can be expressed as products of the form $3^a 7^b$, where a and b independently take all nonnegative integer values, is

$$\frac{1}{1-\frac{1}{3}} \times \frac{1}{1-\frac{1}{7}} =$$

$$\left(1 + \frac{1}{3} + \frac{1}{9} + \cdots + \frac{1}{3^n} + \cdots\right) \times \left(1 + \frac{1}{7} + \frac{1}{49} + \cdots + \frac{1}{7^n} + \cdots\right),$$

or

$$\frac{3}{2} \times \frac{7}{6} = \frac{7}{4}. \quad \square$$

We shall have more to do with multiplication of series in Chapter 12.

Picard's Method

We conclude this section with a proof that the Picard iteration procedure based upon the iterative scheme $x_{n+1} = g(x_n)$ produces a sequence $\{x_n\}$ that converges to a solution of the equation $x = g(x)$ under suitable conditions. We shall convert the sequence $\{x_n\}$ into an infinite series, prove the absolute convergence of that series, conclude the argument by claiming the convergence of the series, and thus the convergence of $\{x_n\}$. We do this as follows. Let

$$a_0 = x_0, \qquad a_1 = x_1 - x_0, \qquad a_2 = x_2 - x_1, \ldots, a_{n+1} = x_{n+1} - x_n. \tag{9}$$

Then we also have

$$\begin{aligned} x_0 &= a_0, \\ x_1 &= a_0 + a_1, \\ x_2 &= a_0 + a_1 + a_2, \\ &\vdots \\ x_{n+1} &= a_0 + a_1 + a_2 + \cdots + a_{n+1}. \end{aligned} \tag{10}$$

Equations (10) show that if $\Sigma\, a_n$ converges, then the sequence $\{x_{n+1}\}$ also converges. Then, of course, the sequence $\{x_n\}$ converges to the same limit and this common limit satisfies the equation $x = g(x)$. Under what conditions? Those stated in the next theorem.

THEOREM 4

Picard Convergence Theorem

Let $g(x)$ be continuous on a closed interval $[a, b]$ and differentiable in the open interval (a, b). Let $x_0, x_1, x_2, \ldots$ be in (a, b) and satisfy

$$x_{n+1} = g(x_n). \tag{11}$$

If there is a constant M that is less than 1, such that

$$|g'(x)| \le M \qquad \text{for all } x \text{ in } (a, b), \tag{12}$$

then $\{x_n\}$ converges to a solution of the equation

$$x = g(x). \tag{13}$$

Proof. Using the notation of Eqs. (9) and (10) above and the Mean Value Theorem, we see that

$$a_2 = x_2 - x_1 = g(x_1) - g(x_0) = (x_1 - x_0)g'(c_1)$$

for some c_1 between x_1 and x_0. Therefore, c_1 is in (a, b) and if condition (12) is satisfied, we can conclude that

$$|a_2| \le M|x_1 - x_0| = M|a_1|.$$

In the same way,

$$a_3 = g(x_2) - g(x_1) = (x_2 - x_1)g'(c_2) = a_2 g'(c_2)$$

for some c_2 in (a, b). Thus, if (12) is satisfied,

$$|a_3| \le |a_2|M \le |a_1|M^2.$$

Continue in this manner. The conclusion will be that if the hypotheses of Theorem 4 are satisfied, then

$$\begin{aligned} |a_2| &\le M|a_1|, \\ |a_3| &\le M^2|a_1| \\ &\vdots \\ |a_n| &\le M^{n-1}|a_1|. \end{aligned}$$

The series

$$a_0 + a_1 + a_2 + \cdots + a_n + \cdots$$

converges absolutely because the series

$$|a_0| + |a_1|(1 + M + M^2 + \cdots + M^n + \cdots)$$

converges when M is a positive constant less than 1.

We conclude that the sequence $\{x_{n+1}\}$ converges to the following limit:

$$x = \lim_{N\to\infty} x_{N+1} = \lim_{N\to\infty} \sum_{k=0}^{N} a_k.$$

Because $g(x)$ is continuous at $x = \lim x_n$, it is also true that $\lim g(x_n) = g(x)$ and

$$x = \lim x_{n+1} = \lim g(x_n) = g(x). \blacksquare$$

EXAMPLE 8 In Article 11.1, we investigated the sequence generated by $x_0 = 1$, $x_{n+1} = \sqrt{\sin x_n}$. In this example,

$$g(x) = \sqrt{\sin x}$$

and we can take $a = \pi/4$, $b = \pi/2$ because all the terms of the sequence lie in this interval. The graph of g is concave downward over (a, b) because $g''(x)$ is negative there. We can therefore take M to be any positive constant less than 1 and greater than $g'(\pi/4) \approx 0.42045$. (We can do still better when we see that all terms of the sequence $\{x_n\}$ are in the interval $(a, b]$ with $a = 0.87672621$ and $b = 1$. Then we can take $M = 0.365$, which is slightly greater than $g'(a)$.)

In general, the error estimate is given by

$$|x - x_N| \le \left(\sum_{n=N+1}^{\infty} M^{n-1}\right) \times |x_1 - x_0| = \frac{M^N}{1 - M} \times |x_1 - x_0|. \quad (14)$$

With $x_0 = 1$, $x_1 = \sqrt{\sin 1} = 0.917$, $N = 20$, and $M = 0.365$, we find that x_{20} differs from the limit by less than 3×10^{-10} for the example in Article 11.1. (The numbers that we calculated involved round-off errors, so we cannot claim 9-place accuracy for our answer.) □

PROBLEMS

In Problems 1–18, determine whether the series converge absolutely. In each case give a reason for the convergence or divergence of the corresponding series of absolute values.

1. $\sum_{n=1}^{\infty} \frac{1}{n^2}$
2. $\sum_{n=1}^{\infty} \frac{1}{(-n)^3}$
3. $\sum_{n=1}^{\infty} \frac{1-n}{n^2}$
4. $\sum_{n=1}^{\infty} \left(-\frac{1}{5}\right)^n$
5. $\sum_{n=1}^{\infty} \frac{-1}{n^2 + 2n + 1}$
6. $\sum_{n=1}^{\infty} \frac{(-1)^n}{2n}$
7. $\sum_{n=1}^{\infty} \frac{\cos n\pi}{n\sqrt{n}}$
8. $\sum_{n=1}^{\infty} \frac{-10}{n}$
9. $\sum_{n=0}^{\infty} \frac{(-1)^n}{(2n)!}$
10. $\sum_{n=0}^{\infty} \frac{(-1)^n}{(2n+1)!}$
11. $\sum_{n=2}^{\infty} (-1)^n \frac{n}{n+1}$
12. $\sum_{n=1}^{\infty} \frac{-n}{2^n}$
13. $\sum_{n=1}^{\infty} (5)^{-n}$
14. $\sum_{n=1}^{\infty} \left(\frac{1}{2^n} - 1\right)$
15. $\sum_{n=1}^{\infty} \frac{(-100)^n}{n!}$
16. $\sum_{n=2}^{\infty} (-1)^n \frac{\ln n}{\ln n^2}$
17. $\sum_{n=1}^{\infty} \frac{2-n}{n^3}$
18. $\sum_{n=1}^{\infty} \left(\frac{1}{2^n} - \frac{1}{3^n}\right)$

19. Show that if $\sum_{n=1}^{\infty} a_n$ diverges, then $\sum_{n=1}^{\infty} |a_n|$ diverges.

20. Show that if $\sum_{n=1}^{\infty} a_n$ converges absolutely, then

$$\left|\sum_{n=1}^{\infty} a_n\right| \le \sum_{n=1}^{\infty} |a_n|.$$

21. Show that if $\sum_{n=1}^{\infty} a_n$ and $\sum_{n=1}^{\infty} b_n$ both converge absolutely, then so does
   a) $\sum (a_n + b_n)$ b) $\sum (a_n - b_n)$
   c) $\sum ka_n$ (k any number)

22. Prove Theorem 2. (Outline of proof. Assume the hypotheses. Let $\varepsilon > 0$. Show there is an index N_1 such that

$$\sum_{n=N_1}^{\infty} |a_n| < \frac{\varepsilon}{2}$$

and an index $N_2 \ge N_1$ such that $|s_{N_2} - L| < \varepsilon/2$, where $s_n = \sum_{k=1}^{n} a_k$. Because all of the terms $a_1, a_2, \ldots, a_{N_2}$ appear somewhere in the sequence $\{b_n\}$, there is an index $N_3 \ge N_2$ such that if $n \ge N_3$, then $\sum_{k=1}^{n} b_k - s_{N_2}$ is, at most, a sum of terms of the form a_m with $m \ge N_1$. Therefore, if $n \ge N_3$,

$$\left|\sum_{k=1}^{n} b_k - L\right| \le \left|\sum_{k=1}^{n} b_k - s_{N_2}\right| + |s_{N_2} - L|$$

$$\le \sum_{k=N_1}^{\infty} |a_k| + |s_{N_2} - L| < \varepsilon. \blacksquare$$

23. Establish the following: If $\Sigma\,|a_n|$ converges and $b_n = a_n$ when $a_n \geq 0$, while $b_n = 0$ when $a_n < 0$, then $\Sigma\, b_n$ converges. Likewise $\Sigma\, c_n$ converges, where $c_n = 0$ when $a_n \geq 0$ and $c_n = -a_n$ when $a_n < 0$. In other words, when the original series converges absolutely, its positive terms by themselves form a convergent series, and so do the negative terms. And

$$\sum_{n=1}^{\infty} a_n = \sum_{n=1}^{\infty} b_n - \sum_{n=1}^{\infty} c_n$$

because $b_n = \frac{1}{2}(a_n + |a_n|)$ and $c_n = \frac{1}{2}(-a_n + |a_n|)$.

25. In the Picard convergence theorem, if $M = 1/2$, show that the error (as estimated by inequality 14) is cut by a factor of 1/2 at each successive iteration.

24. In Example 6, where $a_n = 1/2^n$ and $b_n = 1/3^n$, let $c_n = a_0 b_n + a_1 b_{n-1} + \cdots + a_n b_0$, as in Eq. (3). Since the number $a_k b_{n-k}$ is less than or equal to $(1/2)^n$, show that $c_n \leq n + 1/2^n$ and then prove that $\Sigma\, c_n$ converges. Does it converge absolutely?

26. Let $x_0 = 1$ and define the sequence $\{x_n\}$ by the formula $x_{n+1} = \cos x_n$. Show that one could take $M = \sin 1 \approx 0.84$ in the Picard convergence theorem. If we make a fresh start with $x_0 = 0.8$, we get $x_1 = 0.6967$ and all the rest of the sequence $\{x_n\}$ will be between these numbers. What value of M can you now use?

***Toolkit* programs**

Sequences and Series

11.7 Alternating Series. Conditional Convergence

When some of the terms of a series $\Sigma\, a_n$ are positive and others are negative, the series converges if $\Sigma\,|a_n|$ converges. Thus we may apply any of our tests for convergence of nonnegative series, provided we apply them to the series of absolute values. But we do not know, when the series of absolute values *diverges*, whether the *original* series diverges or converges. If it converges, in this case, we say that it *converges conditionally.*

We shall discuss one simple case of series with mixed signs, namely, series that take the form

$$a_1 - a_2 + a_3 - a_4 + \cdots (-1)^{n+1} a_n + \cdots, \tag{1}$$

with all a's > 0. Such series are called *alternating series* because successive terms have alternate signs. Examples of alternating series are

$$1 - \frac{1}{2} + \frac{1}{3} - \frac{1}{4} + \frac{1}{5} - \frac{1}{6} + \cdots, \tag{2}$$

$$\frac{1}{\ln 2} - \frac{1}{\ln 3} + \frac{1}{\ln 4} - \frac{1}{\ln 5} + \cdots, \tag{3}$$

$$1 - \sqrt{2} + \sqrt{3} - \sqrt{4} + \cdots. \tag{4}$$

The series

$$1 - \frac{1}{2} - \frac{1}{4} + \frac{1}{6} - \frac{1}{8} - \frac{1}{10} + \frac{1}{12} \cdots \tag{5}$$

is *not* an alternating series. The signs of its terms do not alternate.

DEFINITION A sequence $\{a_n\}$ is called a *decreasing sequence* if $a_n \geq a_{n+1}$ for every n.

Examples of decreasing sequences are

$$1, \frac{1}{2}, \frac{1}{3}, \frac{1}{4}, \ldots, \quad \text{and} \quad 1, 1, 1, 1, \ldots.$$

The sequence

$$\frac{1}{3}, \frac{1}{2}, \frac{1}{6}, \frac{1}{4}, \ldots, \frac{1}{3n}, \frac{1}{2n}, \ldots$$

is *not* a decreasing sequence even though it converges to 0 from above.

One reason for selecting alternating series for study is that every alternating series whose numbers a_n form a decreasing sequence with limit 0 converges. Another reason is that whenever an alternating series converges, it is easy to estimate its sum. This fortunate combination of assured convergence and easy estimation gives us an opportunity to see how a wide variety of series behave. We look first at the convergence.

THEOREM 1

Leibniz's Theorem

$\sum_{n=1}^{\infty} (-1)^{n+1} a_n$ converges if all three of the following conditions are satisfied:

1. The a_n's are all positive,
2. $a_n \geq a_{n+1}$ for every n,
3. $a_n \to 0$.

Proof. If n is an even integer, say $n = 2m$, then the sum of the first n terms is

$$\begin{aligned} s_{2m} &= (a_1 - a_2) + (a_3 - a_4) + \cdots + (a_{2m-1} - a_{2m}) \\ &= a_1 - (a_2 - a_3) - (a_4 - a_5) - \cdots - (a_{2m-2} - a_{2m-1}) - a_{2m}. \end{aligned}$$

The first equality exhibits s_{2m} as the sum of m nonnegative terms, since each expression in parentheses is positive or zero. Hence $s_{2m+2} \geq s_{2m}$, and the sequence $\{s_{2m}\}$ is increasing. The second equality shows that $s_{2m} \leq a_1$. Since $\{s_{2m}\}$ is increasing and bounded from above it has a limit, say

$$\lim_{n\to\infty} s_{2m} = L. \tag{6}$$

If n is an odd integer, say $n = 2m + 1$, then the sum of the first n terms is

$$s_{2m+1} = s_{2m} + a_{2m+1}.$$

Since $a_n \to 0$,

$$\lim_{m\to\infty} a_{2m+1} = 0.$$

Hence, as $m \to \infty$,

$$s_{2m+1} = s_{2m} + a_{2m+1} \to L + 0 = L. \tag{7}$$

Finally, we may combine (6) and (7) and say simply

$$\lim_{n\to\infty} s_n = L. \blacksquare$$

Here are some examples of what Theorem 1 can do.

EXAMPLE 1 The *alternating harmonic* series

$$\sum_{n=1}^{\infty} (-1)^{n+1}\frac{1}{n} = 1 - \frac{1}{2} + \frac{1}{3} - \frac{1}{4} + \cdots$$

satisfies the three requirements of the theorem; therefore it converges. It converges conditionally because the corresponding series of absolute values is the harmonic series, which diverges. □

EXAMPLE 2

$$\sum_{n=1}^{\infty} (-1)^{n+1}\sqrt{n} = 1 - \sqrt{2} + \sqrt{3} - \sqrt{4} + \cdots$$

diverges by the nth-term test. □

EXAMPLE 3 Theorem 1 gives no information about

$$\frac{2}{1} - \frac{1}{1} + \frac{2}{2} - \frac{1}{2} + \frac{2}{3} - \frac{1}{3} + \cdots + \frac{2}{n} - \frac{1}{n} + \cdots.$$

The sequence $\frac{2}{1}, \frac{1}{1}, \frac{2}{2}, \frac{1}{2}, \frac{2}{3}, \frac{1}{3}, \ldots$ is not a decreasing sequence. Some other test must be found. When we group the terms of the series in consecutive pairs

$$\left(\frac{2}{1} - \frac{1}{1}\right) + \left(\frac{2}{2} - \frac{1}{2}\right) + \left(\frac{2}{3} - \frac{1}{3}\right) + \cdots + \left(\frac{2}{n} - \frac{1}{n}\right) + \cdots,$$

we see that the $2n$th partial sum of the given series is the same number as the nth partial sum of the harmonic series. Thus the sequence of partial sums, and hence the series, diverges. □

EXAMPLE 4 Does the series $\sum_{n=2}^{\infty} (-1)^n(\ln n/(n + 1))$ converge? While it is clear that this is an alternating series and $a_n = \ln n/(n + 1)$ approaches 0 as $n \to \infty$, it isn't clear at first glance that the sequence $\{a_n\}$ is a decreasing sequence. We might consider the corresponding function

$$f(x) = \frac{\ln x}{x + 1}$$

whose derivative is

$$f'(x) = \frac{(x + 1)/x - \ln x}{(x + 1)^2}.$$

The derivative is negative, and the function f decreasing, if $\ln x$ is greater than $(x + 1)/x = 1 + 1/x$. For $x = 1, 2, \ldots, 1 + (1/x) \leq 2$, so the sequence decreases when $\ln n > 2$. Since $e^2 \approx 7.4$, we conclude that the original sequence decreases for $n \geq 8$. That is, we can apply Leibniz's theorem to conclude that $\sum_{n=8}^{\infty} (-1)^n(\ln n/(n + 1))$ converges. Therefore the original series also converges. □

We use the following graphical interpretation of the partial sums to gain added insight into the way in which an alternating series converges

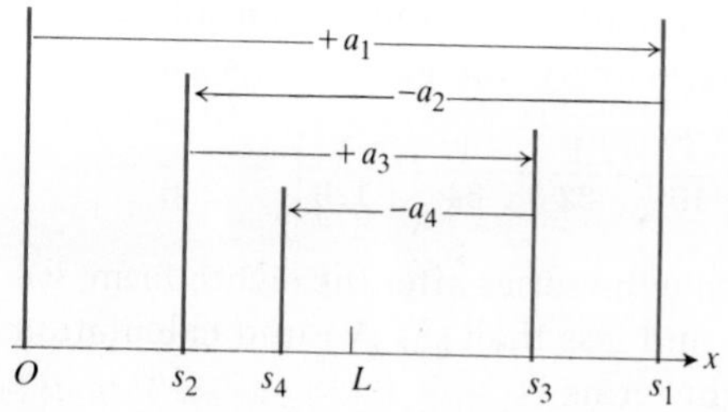

11.12 The partial sums of an alternating series that satisfies the hypotheses of Leibniz's theorem straddle their limit.

to its limit L when the three conditions of the theorem are satisfied. Starting from the origin O on a scale of real numbers (Fig. 11.12), we lay off the positive distance

$$s_1 = a_1.$$

To find the point corresponding to

$$s_2 = a_1 - a_2$$

we must back up a distance equal to a_2. Since $a_2 \leq a_1$, we do not back up any farther than O at most. Next we go forward a distance a_3 and mark the point corresponding to

$$s_3 = a_1 - a_2 + a_3.$$

Since $a_3 \leq a_2$, we go forward by an amount that is no greater than the previous backward step; that is, s_3 is less than or equal to s_1. We continue in this seesaw fashion, backing up or going forward as the signs in the series demand. But each forward or backward step is shorter than (or at most the same size as) the preceding step, because $a_{n+1} \leq a_n$. And since the nth term approaches zero as n increases, the size of step we take forward or backward gets smaller and smaller. We thus oscillate across the limit L, but the amplitude of oscillation continually decreases and approaches zero as its limit. The even-numbered partial sums $s_2, s_4, s_6, \ldots, s_{2m}$ continually increase toward L, while the odd-numbered sums $s_1, s_3, s_5, \ldots, s_{2m+1}$ continually decrease toward L. The limit L is between any two successive sums s_n and s_{n+1} and hence differs from s_n by an amount less than a_{n+1}.

It is because

$$|L - s_n| < a_{n+1} \qquad \text{for every } n \tag{8}$$

that we can make useful estimates of the sums of convergent alternating series.

THEOREM 2

The Alternating Series Estimation Theorem

If

$$\sum_{n=1}^{\infty} (-1)^{n+1} a_n$$

is an alternating series that satisfies the three conditions of Theorem 1, then

$$s_n = a_1 - a_2 + \cdots + (-1)^{n+1} a_n$$

approximates the sum L of the series with an error whose absolute value is less than a_{n+1}, the numerical value of the first unused term. Furthermore, the remainder, $L - s_n$, has the same sign as the first unused term.

We will leave the determination of the sign of the remainder as an exercise (see Problem 40).

EXAMPLE 5 Let us first try the estimation theorem on an alternating series whose sum we already know, namely, the geometric series

$$\sum_{n=0}^{\infty} (-1)^n \frac{1}{2^n} = 1 - \frac{1}{2} + \frac{1}{4} - \frac{1}{8} + \frac{1}{16} - \frac{1}{32} + \frac{1}{64} - \frac{1}{128} \;\Big|\; + \frac{1}{256} - \cdots.$$

Theorem 2 says that when we truncate the series after the eighth term, we throw away a total that is positive and less than $\frac{1}{256}$. A rapid calculation shows that the sum of the first eight terms is

$$0.6640625.$$

The sum of the series is

$$\frac{1}{1-(-\frac{1}{2})} = \frac{1}{\frac{3}{2}} = \frac{2}{3}. \tag{9}$$

The difference,

$$\frac{2}{3} - 0.6640625 = 0.0026041666\ldots,$$

is positive and less than

$$\frac{1}{256} = 0.00390625. \ \square$$

We shall show in Chapter 12 that a series for computing $\ln(1+x)$ when $|x| < 1$ is

$$\ln(1+x) = x - \frac{x^2}{2} + \frac{x^3}{3} - \cdots + (-1)^{n+1}\frac{x^n}{n} \cdots. \tag{10}$$

For $0 < x < 1$, this series satisfies all three conditions of Theorem 1, and we may use the estimation theorem to see how good an approximation of $\ln(1+x)$ we get from the first few terms of the series.

EXAMPLE 6 Calculate $\ln(1.1)$ with the approximation

$$\ln(1+x) \approx x - \frac{x^2}{2}, \tag{11}$$

and estimate the error involved. Is $x - (x^2/2)$ too large, or too small in this case?

Solution

$$\ln(1.1) \approx (0.1) - \frac{(0.1)^2}{2} = 0.095.$$

This approximation differs from the exact value of $\ln 1.1$ by less than

$$\frac{(0.1)^3}{3} = 0.000333\ldots.$$

Since the sign of this, the first unused term, is positive, the remainder is positive. That is, 0.095 underestimates $\ln(1.1)$. $\square$

EXAMPLE 7 How many terms of the series (10) do we need to use in order to be sure of calculating $\ln(1.2)$ with an error of less than 10^{-6}?

Solution

$$\ln(1.2) = (0.2) - \frac{(0.2)^2}{2} + \frac{(0.2)^3}{3} - \cdots$$

We find by trial that the eighth term

$$-\frac{(0.2)^8}{8} = -3.2 \times 10^{-7}$$

is the first term in the series whose absolute value is less than 10^{-6}. Therefore the sum of the first *seven* terms will give ln (1.2) with an error of less than 10^{-6}. The use of more terms would give an approximation that is even better, but seven terms are enough to guarantee the accuracy we wanted. Note also that we have not shown that six terms would *not* provide that accuracy. □

CAUTION ABOUT REARRANGEMENTS. If we rearrange an infinite number of terms of a conditionally convergent series, we can get results that are far different from the sum of the original series. We illustrate some of the things that can happen in the next example.

EXAMPLE 8 Consider the alternating harmonic series

$$\frac{1}{1} - \frac{1}{2} + \frac{1}{3} - \frac{1}{4} + \frac{1}{5} - \frac{1}{6} + \frac{1}{7} - \frac{1}{8} + \frac{1}{9} - \frac{1}{10} + \frac{1}{11} - \cdots.$$

Here, the series of terms $\Sigma\, 1/(2n-1)$ diverges to $+\infty$ and the series of terms $\Sigma -1/2n$ diverges to $-\infty$. We can always add enough positive terms (no matter how far out in the sequence of odd-numbered terms we begin) to get an arbitrarily large sum. Similarly, with the negative terms, no matter how far out we start, we can add enough consecutive even numbered terms to get a negative sum of arbitrarily large absolute value. If we wished to do so, we could start adding odd numbered terms until we had a sum equal to $+3$, say; then follow that with enough consecutive negative terms to make the new total less than -4. We could then add enough positive terms to make the total greater than $+5$ and follow with consecutive unused negative terms to make a new total less than -6, and so on. In this way, we could make the swings arbitrarily large in both the positive and negative directions.

Another possibility, with the same series, is to focus on a particular limit. Suppose we try to get sums that converge to 1. We start with the first term, $\frac{1}{1}$, and then subtract $\frac{1}{2}$. Next we add $\frac{1}{3}$ and $\frac{1}{5}$, which brings the total back to 1 or above. Then we subtract consecutive negative terms until the total is less than 1. Continue in this manner: when the sum is below 1, add positive terms until the total is 1 or more; then subtract (i.e., add negative) terms until the total is once more less than 1. This process can be continued indefinitely. Because both the odd-numbered terms and the even-numbered terms of the original series approach zero as $n \to \infty$, the amount by which our partial sums exceed 1 or fall below it approaches zero. So the new series converges to 1. The rearranged series starts like this:

$$\frac{1}{1} - \frac{1}{2} + \frac{1}{3} + \frac{1}{5} - \frac{1}{4} + \frac{1}{7} + \frac{1}{9} - \frac{1}{6} + \frac{1}{11} + \frac{1}{13} - \frac{1}{8} + \frac{1}{15} + \frac{1}{17} - \frac{1}{10}$$
$$+ \frac{1}{19} + \frac{1}{21} - \frac{1}{12} + \frac{1}{23} + \frac{1}{25} - \frac{1}{14} + \frac{1}{27} - \frac{1}{16} + \cdots. \quad \Box$$

The kind of behavior illustrated by this example is typical of what can happen with any conditionally convergent series. Moral: add the terms of such a series in the order given.

PROBLEMS

In Problems 1–10, determine which of the alternating series converge and which diverge.

1. $\sum_{n=1}^{\infty} (-1)^{n+1} \frac{1}{n^2}$
2. $\sum_{n=2}^{\infty} (-1)^{n+1} \frac{1}{\ln n}$
3. $\sum_{n=1}^{\infty} (-1)^{n+1}$
4. $\sum_{n=1}^{\infty} (-1)^{n+1} \frac{10^n}{n^{10}}$
5. $\sum_{n=1}^{\infty} (-1)^{n+1} \frac{\sqrt{n}+1}{n+1}$
6. $\sum_{n=1}^{\infty} (-1)^{n+1} \frac{\ln n}{n}$
7. $\sum_{n=1}^{\infty} (-1)^{n+1} \frac{1}{n^{3/2}}$
8. $\sum_{n=1}^{\infty} (-1)^{n+1} \frac{\ln n}{\ln n^2}$
9. $\sum_{n=1}^{\infty} (-1)^{n} \ln\left(1+\frac{1}{n}\right)$
10. $\sum_{n=1}^{\infty} (-1)^{n+1} \frac{3\sqrt{n+1}}{\sqrt{n}+1}$

In Problems 11–28, determine whether the series are absolutely convergent, conditionally convergent, or divergent.

11. $\sum_{n=1}^{\infty} (-1)^{n+1}(0.1)^n$
12. $\sum_{n=1}^{\infty} (-1)^{n+1} \frac{1}{\sqrt{n}}$
13. $\sum_{n=1}^{\infty} (-1)^{n+1} \frac{n}{n^3+1}$
14. $\sum_{n=1}^{\infty} \frac{n!}{2^n}$
15. $\sum_{n=1}^{\infty} (-1)^{n} \frac{1}{n+3}$
16. $\sum_{n=1}^{\infty} (-1)^{n} \frac{\sin n}{n^2}$
17. $\sum_{n=1}^{\infty} (-1)^{n+1} \frac{3+n}{5+n}$
18. $\sum_{n=2}^{\infty} (-1)^{n} \frac{1}{\ln n^3}$
19. $\sum_{n=1}^{\infty} (-1)^{n+1} \frac{1+n}{n^2}$
20. $\sum_{n=1}^{\infty} \frac{(-2)^{n+1}}{n+5^n}$
21. $\sum_{n=1}^{\infty} n^2(\frac{2}{3})^n$
22. $\sum_{n=1}^{\infty} (-1)^{n+1}(\sqrt[n]{10})$
23. $\sum_{n=1}^{\infty} (-1)^{n} \frac{\tan^{-1} n}{n^2+1}$
24. $\sum_{n=2}^{\infty} (-1)^{n+1} \frac{1}{n \ln n}$
25. $\sum_{n=1}^{\infty} \left(\frac{1}{n}-\frac{1}{2n}\right)$
26. $\sum_{n=1}^{\infty} (-1)^{n+1} \frac{(0.1)^n}{n}$
27. $\sum_{n=1}^{\infty} (-1)^{n+1}(\sqrt{n+1}-\sqrt{n})$
28. $\sum_{n=1}^{\infty} \frac{(-1)^{n+1}(n!)^2}{(2n)!}$

In Problems 29–32, estimate the magnitude of the error if the first four terms are used to approximate the series.

29. $\sum_{n=1}^{\infty} (-1)^{n+1} \frac{1}{n}$
30. $\sum_{n=1}^{\infty} (-1)^{n+1} \frac{1}{10^n}$
31. $\ln(1.01) = \sum_{n=1}^{\infty} (-1)^{n+1} \frac{(0.01)^n}{n}$
32. $\frac{1}{1+t} = \sum_{n=0}^{\infty} (-1)^n t^n, \quad 0<t<1$

Approximate the sums in Problems 33 and 34 to five decimal places (magnitude of the error less than 5×10^{-6}).

33. $\sum_{n=0}^{\infty} (-1)^n \frac{1}{(2n)!}$ (This is cos 1, the cosine of one radian.)
34. $\sum_{n=0}^{\infty} (-1)^n \frac{1}{n!}$ (This is 1/e.)
35. a) The series
$$\frac{1}{3}-\frac{1}{2}+\frac{1}{9}-\frac{1}{4}+\frac{1}{27}-\frac{1}{8}+\cdots+\frac{1}{3^n}-\frac{1}{2^n}+\cdots$$
does not meet one of the conditions of Theorem 1. Which one?
b) Find the sum of the series in (a).
36. The limit L of an alternating series that satisfies the conditions of Theorem 1 lies between the values of any two consecutive partial sums. This suggests using the average
$$\frac{s_n+s_{n+1}}{2} = s_n + \frac{1}{2}a_{n+1}$$
to estimate L. Compute
$$s_{20} + \frac{1}{2}\cdot\frac{1}{21}$$
as an approximation to the sum of the alternating harmonic series. The exact sum is $\ln 2 \approx 0.6931\ldots$.
37. Show that whenever an alternating series is approximated by one of its partial sums, if the three conditions of Leibniz's theorem are satisfied, then the remainder (sum of the unused terms) has the same sign as the first unused term. (*Hint:* Group the terms of the remainder in consecutive pairs.)
38. Prove the "zipper" theorem for sequences: If $\{a_n\}$ and $\{b_n\}$ both converge to L, then the sequence
$$a_1,\ b_1,\ a_2,\ b_2,\ \ldots,\ a_n,\ b_n,\ \ldots$$
also converges to L.

39. Show by example that $\Sigma_{n=1}^{\infty} a_n b_n$ may diverge even though $\Sigma_{n=1}^{\infty} a_n$ and $\Sigma_{n=1}^{\infty} b_n$ both converge.

40. Prove that the remainder, $L - s_n$, has the sign specified in Theorem 2.

41. (*Calculator*) In Example 8, suppose the goal is to arrange the terms to get a new series that converges to $-\frac{1}{2}$. Start the new arrangement with the first negative term, which is $-\frac{1}{2}$. Whenever you have a sum that is less than or equal to $-\frac{1}{2}$, start introducing positive terms, taken in order, until the new total is greater than $-\frac{1}{2}$. Then add negative terms until the total is less than or equal to $-\frac{1}{2}$ again. Continue this process until your partial sums have been above the target at least three times and finish at or below it. If s_n is the sum of the first n terms of your new series, plot the points (n, s_n) to illustrate how the sums are behaving.

***Toolkit* programs**

Sequences and Series

REVIEW QUESTIONS AND EXERCISES

1. Define "sequence," "series," "sequence of partial sums of a series."
2. Define "convergence" (a) of a sequence, (b) of an infinite series.
3. Which of the following statements are true, and which are false?
   a) If a sequence does not converge, then it diverges.
   b) If a sequence $\{n, f(n)\}$ does not converge, then $f(n)$ tends to infinity as n does.
   c) If a series does not converge, then its nth term does not approach zero as n tends to infinity.
   d) If the nth term of a series does not approach zero as n tends to infinity, then the series diverges.
   e) If a sequence $\{n, f(n)\}$ converges, then there is a number L such that $f(n)$ lies within 1 unit of L (i) for all values of n, (ii) for all but a finite number of values of n.
   f) If all partial sums of a series are less than some constant L, then the series converges.
   g) If a series converges, then its partial sums s_n are bounded (that is, $m \le s_n \le M$ for some constants m and M).
4. List three tests for convergence (or divergence) of an infinite series.
5. Under what circumstances do you know that a bounded sequence converges?
6. Define "absolute convergence" and "conditional convergence." Give examples of series that are (a) absolutely convergent, (b) conditionally convergent.
7. What test is usually used to decide whether a given alternating series converges? Give examples of convergent and divergent alternating series.

MISCELLANEOUS PROBLEMS

1. Find explicitly the nth partial sum of the series $\Sigma_{n=2}^{\infty} \ln(1 - 1/n^2)$, and thereby determine whether the series converges.
2. Evaluate $\Sigma_{k=2}^{\infty} 1/(k^2 - 1)$ by finding the nth partial sum and taking the limit as n becomes infinite.
3. Prove that the sequence $\{x_n\}$ and the series $\Sigma_{k=1}^{\infty}(x_{k+1} - x_k)$ both converge or both diverge.
4. In an attempt to find a root of the equation $x = f(x)$, a first approximation x_1 is estimated from the graphs of $y = x$ and $y = f(x)$. Then $x_2, x_3, \ldots, x_n, \ldots$ are computed successively from the formula $x_n = f(x_{n-1})$. If the points $x_1, x_2, \ldots, x_n, \ldots$ all lie on an interval $a \le x \le b$ on which $f(x)$ has a derivative such that $|f'(x)| < M < 1$, show that the sequence $\{x_n\}$ converges to a root of the given equation.
5. Assuming $|x| > 1$, show that
$$\frac{1}{1-x} = -\frac{1}{x} - \frac{1}{x^2} - \frac{1}{x^3} - \cdots.$$
6. Does the series $\Sigma_{n=1}^{\infty} \operatorname{sech} n$ converge? Why?
7. Does $\Sigma_{n=1}^{\infty} (-1)^n \tanh n$ converge? Why?

Establish the convergence or divergence of the series whose nth terms are given in Problems 8–19.

8. $\dfrac{1}{\ln(n+1)}$

9. $\dfrac{n}{2(n+1)(n+2)}$

10. $\dfrac{\sqrt{n+1} - \sqrt{n}}{\sqrt{n}}$

11. $\dfrac{1}{n(\ln n)^2}, \quad n \ge 2$

12. $\dfrac{1 + (-2)^{n-1}}{2^n}$

13. $\dfrac{n}{1000n^2 + 1}$

14. $e^n/n!$

15. $\dfrac{1}{n\sqrt{n^2+1}}$

16. $\dfrac{1}{n^{1+1/n}}$

17. $\dfrac{1\cdot 3\cdot 5\cdots(2n-1)}{2\cdot 4\cdot 6\cdots(2n)}$

18. $\dfrac{n^2}{n^3+1}$

19. $\dfrac{n+1}{n!}$

20. If the following series converges, find the sum

$$\sum_{n=1}^{\infty}\frac{1}{(n+1)(n+2)}.$$

21. a) Suppose $a_1, a_2, a_3, \ldots, a_n$ are positive numbers satisfying the following conditions:

i) $a_1 \geq a_2 \geq a_3 \geq \cdots$,

ii) the series $a_2 + a_4 + a_8 + a_{16} + \cdots$ diverges.

Show that the series

$$\frac{a_1}{1}+\frac{a_2}{2}+\frac{a_3}{3}+\cdots$$

diverges.

b) Use the result above to show that

$$\sum_{n=2}^{\infty}\frac{1}{n\ln n}$$

diverges.

22. Given $a_n \neq 1$, $a_n > 0$, $\Sigma\, a_n$ converges.

a) Show that $\Sigma\, a_n^2$ converges.

b) Does $\Sigma\, a_n/(1-a_n)$ converge? Explain.

23. Show that $\Sigma_{n=2}^{\infty} 1/[n(\ln n)^k]$ converges for $k > 1$.

12 Power Series

12.1 Power Series for Functions

The rational operations of arithmetic are addition, subtraction, multiplication, and division. Using only these simple operations, we can evaluate any rational function of x. But other functions, such as $\sqrt{x}$, ln x, cos x, and so on, cannot be evaluated so simply. These functions occur so frequently, however, that their values have been printed in mathematical tables, and many calculators and computers have been programmed to produce them on demand. One may wonder where the values in the tables came from. By and large, these numbers came from calculating partial sums of power series.

DEFINITION A *power series* is a series of the form

$$\sum_{n=0}^{\infty} a_n x^n = a_0 + a_1 x + a_2 x^2 + \cdots.$$

In this article we shall show how a power series can arise when we seek to approximate a function

$$y = f(x) \tag{1}$$

by a sequence of polynomials $f_n(x)$ of the form

$$f_n(x) = a_0 + a_1 x + a_2 x^2 + \cdots + a_n x^n. \tag{2}$$

We shall be interested, at least at first, in making the approximation for values of x near 0, because we want the term $a_n x^n$ to decrease as n increases. Hence we focus our attention on a portion of the curve $y = f(x)$ near the point $A(0, f(0))$, as shown in Fig. 12.1.

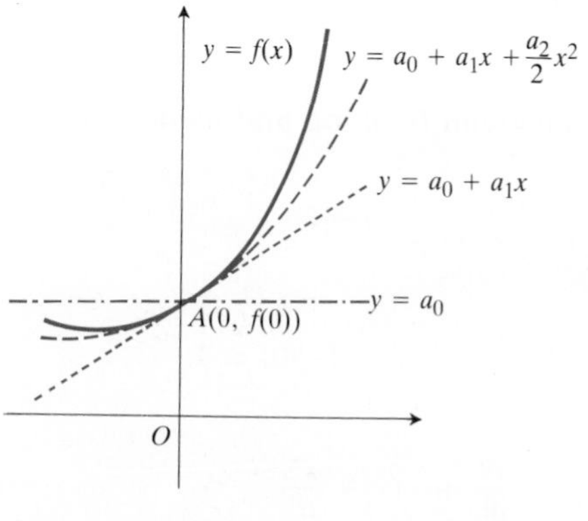

12.1 $f(x)$ is approximated near $x = 0$ by polynomials whose derivatives at $x = 0$ match the derivatives of f.

1. The graph of the polynomial $f_0(x) = a_0$ of degree zero will pass through $(0, f(0))$ if we take

$$a_0 = f(0).$$

2. The graph of the polynomial $f_1(x) = a_0 + a_1 x$ will pass through

$(0, f(0))$ and have the same slope as the given curve at that point if we choose

$$a_0 = f(0) \qquad \text{and} \qquad a_1 = f'(0).$$

3. The graph of the polynomial $f_2(x) = a_0 + a_1x + a_2x^2$ will pass through $(0, f(0))$ and have the same first and second derivative as the given curve at that point if

$$a_0 = f(0), \qquad a_1 = f'(0), \qquad \text{and } a_2 = \frac{f''(0)}{2}.$$

4. In general, the polynomial $f_n(x) = a_0 + a_1x + a_2x^2 + \cdots + a_nx^n$, which we choose to approximate $y = f(x)$ near $x = 0$, is the one whose graph passes through $(0, f(0))$ and whose first n derivatives match the derivatives of $f(x)$ at $x = 0$. To match the derivatives of f_n to those of f at $x = 0$, we merely have to choose the coefficients a_0 through a_n properly. To see how this may be done, we write down the polynomial and its derivatives as follows:

$$\begin{aligned} f_n(x) &= a_0 + a_1x + a_2x^2 + a_3x^3 + \cdots + a_nx^n \\ f_n'(x) &= a_1 + 2a_2x + 3a_3x^2 + \cdots + na_nx^{n-1} \\ f_n''(x) &= 2a_2 + 3\cdot 2a_3x + \cdots + n(n-1)a_nx^{n-2} \\ &\vdots \\ f_n^{(n)}(x) &= (n!)a_n. \end{aligned}$$

When we substitute 0 for x in the array above, we find that

$$a_0 = f(0), \qquad a_1 = f'(0), \qquad a_2 = \frac{f''(0)}{2!}, \qquad \ldots, \qquad a_n = \frac{f^{(n)}(0)}{n!}.$$

Thus,

$$f_n(x) = f(0) + f'(0)x + \frac{f''(0)}{2!}x^2 + \cdots + \frac{f^{(n)}(0)}{n!}x^n \tag{3}$$

is the polynomial we seek. Its graph passes through the point $(0, f(0))$, and its first n derivatives match the first n derivatives of $y = f(x)$ at $x = 0$. It is called the nth-degree *Taylor polynomial of f at x = 0.*

EXAMPLE 1 Find the Taylor polynomials $f_n(x)$ for the function $f(x) = e^x$ at $x = 0$.

Solution Expressed in terms of x, the given function and its derivatives are

$$f(x) = e^x, \qquad f'(x) = e^x, \qquad \ldots, \qquad f^{(n)}(x) = e^x,$$

so that

$$f(0) = e^0 = 1, \qquad f'(0) = 1, \qquad \ldots, \qquad f^{(n)}(0) = 1,$$

and

$$f_n(x) = 1 + x + \frac{x^2}{2!} + \frac{x^3}{3!} + \cdots + \frac{x^n}{n!}.$$

See Fig. 12.2. □

EXAMPLE 2 Find the Taylor polynomials $f_n(x)$ for $f(x) = \cos x$ at $x = 0$.

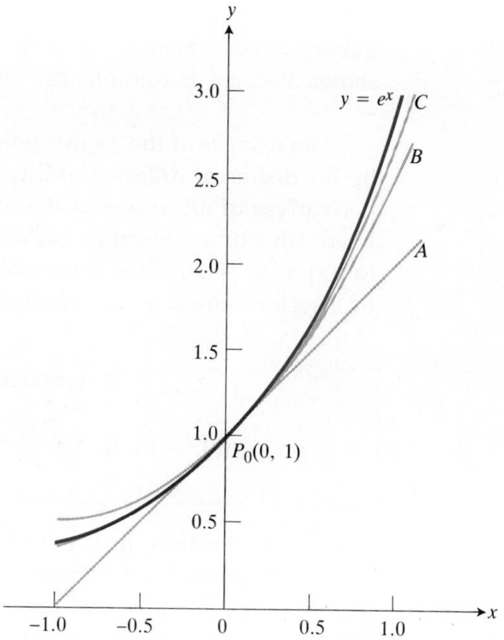

12.2 The graph of the function $y = e^x$, and graphs of three approximating polynomials, (*A*) a straight line, (*B*) a parabola, and (*C*) a cubic curve.

Solution The cosine and its derivatives are

$$\begin{aligned} f(x) &= \cos x, & f'(x) &= -\sin x, \\ f''(x) &= -\cos x, & f^{(3)}(x) &= \sin x, \\ &\vdots & &\vdots \\ f^{(2k)}(x) &= (-1)^k \cos x, & f^{(2k+1)}(x) &= (-1)^{k+1} \sin x. \end{aligned}$$

When $x = 0$, the cosines are 1 and the sines are 0, so that

$$f^{(2k)}(0) = (-1)^k, \qquad f^{(2k+1)}(0) = 0.$$

The Taylor polynomials have only even-powered terms, and for $n = 2k$ we have

$$f_{2k}(x) = 1 - \frac{x^2}{2!} + \frac{x^4}{4!} - \cdots + (-1)^k \frac{x^{2k}}{(2k)!}. \tag{4}$$

Figure 12.3 shows how well these polynomials can be expected to approx-

12.3 The polynomials

$$c_n(x) = \sum_{k=0}^{n} [(-1)^k x^{2k}/(2k)!]$$

converge to $\cos x$ as $n \to \infty$. (Adapted from Helen M. Kammerer, *American Mathematical Monthly,* 43(1936), 293–294.)

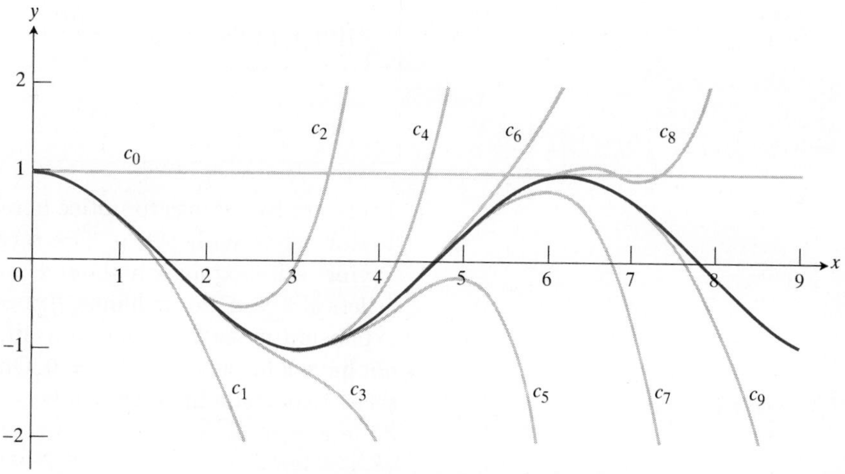

imate $y = \cos x$ near $x = 0$. Only the right-hand portions of the graphs are shown because the graphs are symmetric about the y-axis. □

The degrees of the Taylor polynomials of a given function are limited by the degree of differentiability of the function at $x = 0$. But if $f(x)$ has derivatives of all orders at the origin, it is natural to ask whether, for a fixed value of x, the values of these approximating polynomials converge to $f(x)$ as $n \to \infty$. Now, these polynomials are precisely the partial sums of a series known as the *Maclaurin series* for f:

The Maclaurin Series for f

$$f(0) + f'(0)x + \frac{f''(0)}{2!}x^2 + \cdots + \frac{f^{(n)}(0)}{n!}x^n + \cdots. \tag{5}$$

Thus, the question just posed is equivalent to asking whether the Maclaurin series for f converges to $f(x)$ as a sum. It certainly has the correct value, $f(0)$, at $x = 0$, but how far away from $x = 0$ may we go and still have convergence? And if the series does converge away from $x = 0$, does it still converge to $f(x)$? The graphs in Figs. 12.2 and 12.3 are encouraging, and the next few articles will confirm that we can normally expect a Maclaurin series to converge to its function in an interval about the origin. For many functions, this interval is the entire x-axis.

If, instead of approximating the values of f near zero, we are concerned with values of x near some other point a, we write our approximating polynomials in powers of $(x - a)$:

$$f_n(x) = a_0 + a_1(x - a) + a_2(x - a)^2 + \cdots + a_n(x - a)^n. \tag{6}$$

When we now determine the coefficients $a_0, a_1, \ldots, a_n$, so that the polynomial and its first n derivatives agree with the given function and its derivatives at $x = a$, we are led to a series that is called the *Taylor series expansion of f about* $x = a$, or simply the *Taylor series for f at* $x = a$.

The Taylor Series for f at $x = a$

$$f(a) + f'(a)(x - a) + \frac{f''(a)}{2!}(x - a)^2 + \cdots$$
$$+ \frac{f^{(n)}(a)}{n!}(x - a)^n + \cdots. \tag{7}$$

There are two things to notice here. The first is that Maclaurin series are Taylor series with $a = 0$. The second is that a function cannot have a Taylor series expansion about $x = a$ unless it has finite derivatives of all orders at $x = a$. For instance, $f(x) = \ln x$ does not have a Maclaurin series expansion, since the function itself, to say nothing of its derivatives, does not have a finite value at $x = 0$. On the other hand, it does have a Taylor series expansion in powers of $(x - 1)$, since $\ln x$ and all its derivatives are finite at $x = 1$.

Here are some examples of Taylor series.

EXAMPLE 3 From the formula derived for the Taylor polynomials of $\cos x$ in Example 2, it follows immediately that

$$\sum_{k=0}^{\infty} (-1)^k \frac{x^{2k}}{(2k)!} = 1 - \frac{x^2}{2!} + \frac{x^4}{4!} - \frac{x^6}{6!} + \cdots$$

is the Maclaurin series for $\cos x$. □

EXAMPLE 4 Find the Taylor series expansion of $\cos x$ about the point $a = 2\pi$.

Solution The values of $\cos x$ and its derivatives at $a = 2\pi$ are the same as their values at $a = 0$. Therefore,

$$f^{(2k)}(2\pi) = f^{(2k)}(0) = (-1)^k \quad \text{and} \quad f^{(2k+1)}(2\pi) = f^{(2k+1)}(0) = 0,$$

as in Example 2. The required series is

$$\sum_{k=0}^{\infty} (-1)^{2k} \frac{(x - 2\pi)^{2k}}{(2k)!} = 1 - \frac{(x - 2\pi)^2}{2!} + \frac{(x - 2\pi)^4}{4!} - \cdots. \ \square$$

REMARK. There is a convention about how formulas like

$$\frac{x^{2k}}{(2k)!} \quad \text{and} \quad \frac{(x - 2\pi)^{2k}}{(2k)!},$$

which arise in the power series of Examples 3 and 4, are to be evaluated when $k = 0$. Besides the usual agreement that $0! = 1$, we also assume that

$$\frac{x^0}{0!} = \frac{1}{1} = 1 \quad \text{and} \quad \frac{(x - 2\pi)^0}{0!} = \frac{1}{1} = 1,$$

even when $x = 0$ or 2π.

The Binomial Series

One of the most celebrated series of all times, the *binomial series*, is the Maclaurin series for the function $f(x) = (1 + x)^m$. Newton used it to estimate integrals (we will, in Article 12.5) and it can be used to give accurate estimates of roots. To derive the series, we first list the function and its derivatives:

$$\begin{aligned}
f(x) &= (1 + x)^m, \\
f'(x) &= m(1 + x)^{m-1}, \\
f''(x) &= m(m - 1)(1 + x)^{m-2}, \\
f'''(x) &= m(m - 1)(m - 2)(1 + x)^{m-3}, \\
&\vdots \\
f^{(k)}(x) &= m(m - 1)(m - 2) \cdots (m - k + 1)(1 + x)^{m-k}.
\end{aligned}$$

We then substitute the values of these at $x = 0$ in the basic Maclaurin series (5) to obtain

$$1 + mx + \frac{m(m-1)}{2!}x^2 + \cdots + \frac{m(m-1)(m-2)\cdots(m-k+1)}{k!}x^k + \cdots. \tag{8}$$

If m is an integer ≥ 0, the series terminates after $(m + 1)$ terms, because the coefficients from $k = m + 1$ on are 0. But when k is not an integer, the series is infinite. (For a proof that the series converges to $(1 + x)^m$ when $|x| < 1$, see Courant and John's *Introduction to Calculus and Analysis*, Wiley-Interscience, 1974.)

EXAMPLE 5 Use the binomial series to estimate $\sqrt{1.25}$ with an error of less than 0.001.

Solution We take $x = \frac{1}{4}$ and $m = \frac{1}{2}$ in (8) to obtain

$$\left(1 + \frac{1}{4}\right)^{1/2} = 1 + \frac{1}{2}\left(\frac{1}{4}\right) + \frac{(\frac{1}{2})(-\frac{1}{2})}{2!}\left(\frac{1}{4}\right)^2 + \frac{(\frac{1}{2})(-\frac{1}{2})(-\frac{3}{2})}{3!}\left(\frac{1}{4}\right)^3 + \cdots$$

$$= 1 + \frac{1}{8} - \frac{1}{128} + \frac{1}{1024} - \frac{1}{32768} + \cdots.$$

The series alternates after the first term, so that the approximation

$$\sqrt{1.25} \approx 1 + \frac{1}{8} - \frac{1}{128} \approx 1.117$$

is within $\frac{1}{1024}$ of the exact value and thus has the required accuracy. □

PROBLEMS

In Problems 1–9, use Eq. (3) to write the Taylor polynomials $f_3(x)$ and $f_4(x)$ for each of the following functions $f(x)$ at $x = 0$. In each case, your first step should be to complete a table like the one shown below.

n	$f^{(n)}(x)$	$f^{(n)}(0)$
0		
1		
2		
3		
4		

1. e^{-x} **2.** $\sin x$ **3.** $\cos x$

4. $\sin\left(x + \frac{\pi}{2}\right)$ **5.** $\sinh x$ **6.** $\cosh x$

7. $x^4 - 2x + 1$ **8.** $x^3 - 2x + 1$ **9.** $x^2 - 2x + 1$

In Problems 10–13, find the Maclaurin series for each function.

10. $\frac{1}{1 + x}$ **11.** x^2

12. $(1 + x)^2$ **13.** $(1 + x)^{3/2}$

14. Find the Maclaurin series for $f(x) = 1/(1 - x)$. Show that the series diverges when $|x| \geq 1$ and converges when $|x| < 1$.

In Problems 15–20, use Eq. (7) to write the Taylor series expansion of the given function about the given point a.

15. $f(x) = e^x$, $a = 10$ **16.** $f(x) = x^2$, $a = \frac{1}{2}$

17. $f(x) = \ln x$, $a = 1$ **18.** $f(x) = \sqrt{x}$, $a = 4$

19. $f(x) = \frac{1}{x}$, $a = -1$ **20.** $f(x) = \cos x$, $a = -\frac{\pi}{4}$

In Problems 21 and 22, write the sum of the first three terms of the Taylor series for the given function about the given point a.

21. $f(x) = \tan x$, $a = \frac{\pi}{4}$ **22.** $f(x) = \ln \cos x$, $a = \frac{\pi}{3}$

23. Use the binomial theorem to estimate $\sqrt{1.02}$ with an error of less than 0.001.

12.2

Taylor's Theorem with Remainder: Sines, Cosines, and e^x

In the previous article, we asked when a Taylor series for a function can be expected to converge to the function. In this article, we answer the question with a theorem named after the English mathematician Brook Taylor (1685–1731). In Article 3.10, we stated Taylor's Theorem and referred to an outline of the proof. We restate the theorem here and fill in some of the details of the proof.

THEOREM 1

Taylor's Theorem

If f and its first n derivatives $f', f'', \ldots, f^{(n)}$ are continuous on $[a, b]$, or on $[b, a]$, and $f^{(n)}$ is differentiable on (a, b), or on (b, a), then there exists a number c between a and b such that

$$f(b) = f(a) + f'(a)(b - a) + \frac{f''(a)}{2!}(b - a)^2 + \cdots + \frac{f^{(n)}(a)}{n!}(b - a)^n + \frac{f^{(n+1)}(c)}{(n + 1)!}(b - a)^{n+1}.$$

Proof. We assume that f satisfies the hypotheses of the theorem and define the Taylor polynomial about a of degree n:

$$f_n(x) = f(a) + f'(a)(x - a) + \frac{f''(a)}{2!}(x - a)^2 + \cdots + \frac{f^{(n)}(a)}{n!}(x - a)^n. \tag{1}$$

This polynomial and its first n derivatives match the function f and its first n derivatives at $x = a$. We do not disturb that matching by adding another term of the form $K(x - a)^{n+1}$, where K is any constant, because such a function and its first n derivatives are all equal to zero at $x = a$. Therefore, the new function

$$\phi_n(x) = f_n(x) + K(x - a)^{n+1} \tag{2}$$

and its first n derivatives still agree with f and its first n derivatives at $x = a$.

We now choose that particular value of K that makes the curve $y = \phi_n(x)$ agree with the original curve $y = f(x)$ at $x = b$. This can be done: we need only satisfy

$$f(b) = f_n(b) + K(b - a)^{n+1} \tag{3a}$$

or

$$K = \frac{f(b) - f_n(b)}{(b - a)^{n+1}}. \tag{3b}$$

With K defined by Eq. (3b), let $F(x) = f(x) - \phi_n(x)$, so that $F(x)$ measures the difference between the original function f and approximating function

ϕ_n, for each x in $[a, b]$, or in $[b, a]$ if $b < a$. To simplify the notation, we assume $a < b$, so that a is the left endpoint of all intervals mentioned. The same proof is valid if a is the right endpoint, instead of the left endpoint (for example, $[b, a]$, (b, a), $(c_1, a), \ldots, (c_n, a)$).

The remainder of the proof makes repeated use of Rolle's Theorem. First, because $F(a) = F(b) = 0$ and both F and F' are continuous on $[a, b]$, we know that

$$F'(c_1) = 0 \qquad \text{for some } c_1 \text{ in } (a, b).$$

Next, because $F'(a) = F'(c_1) = 0$ and both F' and F'' are continuous on $[a, c_1]$, we know that

$$F''(c_2) = 0 \qquad \text{for some } c_2 \text{ in } (a, c_1).$$

Rolle's Theorem, applied successively to $F'', F''', \ldots, F^{(n-1)}$ implies the existence of

$$\begin{array}{ll} c_3 \text{ in } (a, c_2) & \text{such that } F'''(c_3) = 0, \\ c_4 \text{ in } (a, c_3) & \text{such that } F^{(iv)}(c_4) = 0, \\ \vdots & \\ c_n \text{ in } (a, c_{n-1}) & \text{such that } F^{(n)}(c_n) = 0. \end{array}$$

Finally, because $F^{(n)}$ is continuous on $[a, c_n]$ and differentiable on (a, c_n) and $F^{(n)}(a) = F^{(n)}(c_n) = 0$, Rolle's Theorem implies that there is a number c_{n+1} in (a, c_n) such that

$$F^{(n+1)}(c_{n+1}) = 0. \tag{4}$$

When we differentiate

$$F(x) = f(x) - f_n(x) - K(x - a)^{n+1}$$

$n + 1$ times, we get

$$F^{(n+1)}(x) = f^{(n+1)}(x) - 0 - (n + 1)!\, K. \tag{5}$$

Eqs. (4) and (5) together lead to the result

$$K = \frac{f^{(n+1)}(c)}{(n + 1)!} \qquad \text{for some number } c = c_{n+1} \text{ in } (a, b). \tag{6}$$

Combining Eqs. (3b) and (6), we have

$$\frac{f(b) - f_n(b)}{(b - a)^{n+1}} = \frac{f^{(n+1)}(c)}{(n + 1)!}$$

or

$$f(b) = f_n(b) + \frac{f^{(n+1)}(c)}{(n + 1)!}(b - a)^{n+1} \qquad \text{for some } c \text{ between } a \text{ and } b. \blacksquare$$

COROLLARY 1

If f has derivatives of all orders in an open interval I containing a, then for each positive integer n and for each x in I,

$$f(x) = f(a) + f'(a)(x - a) + \frac{f''(a)}{2!}(x - a)^2 + \cdots + \frac{f^{(n)}(a)}{n!}(x - a)^n + R_n(x, a) \tag{7a}$$

where

$$R_n(x, a) = \frac{f^{(n+1)}(c)}{(n+1)!}(x-a)^{n+1} \qquad \text{for some } c \text{ between } a \text{ and } x. \tag{7b}$$

The corollary follows at once from Taylor's Theorem because the existence of derivatives of all orders in an interval I implies the continuity of those derivatives and we have merely replaced b by x in the final formula.

The function $R_n(x, a)$ is called the remainder of order n: it's the difference $f(x) - f_n(x)$ where $f_n(x)$ is the Taylor polynomial of degree n used to approximate $f(x)$ near $x = a$. This difference, also called the "error" in the approximation $f_n(x)$, can often be estimated by using Eq. (7b) as in the next example.

When $R_n(x, a) \to 0$ as $n \to \infty$, for all x in some interval around $x = a$, we say that the Taylor-series expansion for $f(x)$ converges to $f(x)$ on that interval and write

$$f(x) = \sum_{k=0}^{\infty} \frac{f^{(k)}(a)}{k!}(x-a)^k. \tag{8}$$

EXAMPLE 1 Show that the Taylor series of $f(x) = e^x$ about $a = 0$ converges to $f(x)$ for every real value of x.

Solution Let $f(x) = e^x$. This function and all its derivatives are continuous at every point, so Taylor's Theorem may be applied with any convenient value of a. We take $a = 0$, since the values of f and its derivatives are easy to compute there. Taylor's Theorem leads to

$$e^x = 1 + x + \frac{x^2}{2!} + \frac{x^3}{3!} + \cdots + \frac{x^n}{n!} + R_n(x, 0) \tag{9a}$$

where

$$R_n(x, 0) = \frac{e^c}{(n+1)!}x^{n+1} \qquad \text{for some } c \text{ between } 0 \text{ and } x. \tag{9b}$$

Because e^x is an increasing function of x, and c is between 0 and x, the value of e^c is between 1 and e^x. Therefore, if x is negative, so is c and $e^c < 1$; if x is positive, so is c and $e^c < e^x$. Thus we can write

$$R_n(x, 0) < \frac{|x|^{n+1}}{(n+1)!} \qquad \text{when} \quad x < 0, \tag{9c}$$

and

$$R_n(x, 0) < e^x\frac{x^{n+1}}{(n+1)!} \qquad \text{when} \quad x > 0. \tag{9d}$$

When $x = 0$, the first term of the series in (9a) is $1 = e^0$, so the "error" is zero. Finally, because

$$\lim_{n\to\infty} \frac{x^{n+1}}{(n+1)!} = 0 \qquad \text{for every } x,$$

it is also true that

$$\lim_{n\to\infty} R_n(x, 0) = 0 \qquad \text{for every value of } x:$$

$$e^x = \sum_{k=0}^{\infty} \frac{x^k}{k!} = 1 + x + \frac{x^2}{2!} + \cdots + \frac{x^k}{k!} + \cdots . \ \square \tag{10}$$

Estimating the Remainder

It is often possible to estimate $R_n(x, a)$ as we did in Example 1. This method of estimation is so convenient that we state it as a theorem for future reference.

THEOREM 2

The Remainder Estimation Theorem

If there are positive constants M and r such that $|f^{(n+1)}(t)| \leq Mr^{n+1}$ for all t between a and x, inclusive, then the remainder term $R_n(x, a)$ in Taylor's Theorem satisfies the inequality

$$|R_n(x, a)| \leq M\frac{r^{n+1}|x - a|^{n+1}}{(n + 1)!}.$$

Furthermore, if the conditions above hold and all the other conditions of Taylor's Theorem are satisfied by $f(x)$, then the series converges to $f(x)$.

We are now ready to look at some examples of how the Remainder Estimation Theorem and Taylor's Theorem can be used together to settle questions of convergence. As you will see, they can also be used to determine the accuracy with which a function is approximated by one of its Taylor polynomials.

EXAMPLE 2 The Maclaurin series for sin x converges to sin x for all x.

Expressed in terms of x, the function and its derivatives are

$$\begin{aligned} f(x) &= \sin x, & f'(x) &= \cos x, \\ f''(x) &= -\sin x, & f'''(x) &= -\cos x, \\ &\vdots & &\vdots \\ f^{(2k)}(x) &= (-1)^k \sin x, & f^{(2k+1)}(x) &= (-1)^k \cos x, \end{aligned}$$

so that

$$f^{(2k)}(0) = 0 \qquad \text{and} \qquad f^{(2k+1)}(0) = (-1)^k.$$

The series has only odd-powered terms and, for $n = 2k + 1$, Taylor's formula gives

$$\sin x = x - \frac{x^3}{3!} + \frac{x^5}{5!} - \cdots + \frac{(-1)^k x^{2k+1}}{(2k + 1)!} + R_{2k+1}(x, 0).$$

Now, since all the derivatives of sin x have absolute values less than or equal to 1, we can apply the Remainder Estimation Theorem with $M = 1$ and $r = 1$ to obtain

$$|R_{2k+1}(x, 0)| \leq 1 \cdot \frac{|x|^{2k+2}}{(2k + 2)!}.$$

Since $[|x|^{2k+2}/(2k+2)!] \to 0$ as $k \to \infty$, whatever the value of x,

$$R_{2k+1}(x, 0) \to 0,$$

and the Maclaurin series for sin x converges to sin x for every x:

$$\sin x = \sum_{k=0}^{\infty} \frac{(-1)^k x^{2k+1}}{(2k+1)!} = x - \frac{x^3}{3!} + \frac{x^5}{5!} - \frac{x^7}{7!} + \cdots. \square \qquad (11)$$

EXAMPLE 3 The Maclaurin series for cos x converges to cos x for every value of x.

We begin by adding the remainder term to the Taylor polynomial for cos x in Eq. (4) of the previous article, to obtain Taylor's formula for cos x with $n = 2k$:

$$\cos x = 1 - \frac{x^2}{2!} + \frac{x^4}{4!} - \cdots + (-1)^k \frac{x^{2k}}{(2k)!} + R_{2k}(x, 0).$$

Since the derivatives of the cosine have absolute value less than or equal to 1, we apply the Remainder Estimation Theorem with $M = 1$ and $r = 1$ to obtain

$$|R_{2k}(x, 0)| \le 1 \cdot \frac{|x|^{2k+1}}{(2k+1)!}.$$

For every value of x, $R_{2k} \to 0$ as $k \to \infty$. Therefore, the series converges to cos x for every value of x:

$$\cos x = \sum_{k=0}^{\infty} \frac{(-1)^k x^{2k}}{(2k)!} = 1 - \frac{x^2}{2!} + \frac{x^4}{4!} - \frac{x^6}{6!} + \cdots. \square \qquad (12)$$

EXAMPLE 4 Find the Maclaurin series for cos 2x and show that it converges to cos 2x for every value of x.

Solution The Maclaurin series for cos x converges to cos x for every value of x, and therefore converges for every value of 2x:

$$\cos 2x = \sum_{k=0}^{\infty} \frac{(-1)^k (2x)^{2k}}{(2k)!} = 1 - \frac{(2x)^2}{2!} + \frac{(2x)^4}{4!} - \frac{(2x)^6}{6!} + \cdots. \square$$

Taylor series can be added, subtracted, and multiplied by constants, just as other series can, and the results are once again Taylor series. The Taylor series for $f(x) + g(x)$ is the sum of the Taylor series for $f(x)$ and $g(x)$, because the nth derivative of $f(x) + g(x)$ is $f^{(n)}(x) + g^{(n)}(x)$, and so on. In the next example, we add the series for e^x and e^{-x} and divide by 2, to obtain the Taylor series for cosh x.

EXAMPLE 5 Find the Taylor series for cosh x.

Solution

$$e^x = 1 + x + \frac{x^2}{2!} + \frac{x^3}{3!} + \frac{x^4}{4!} + \frac{x^5}{5!} + \cdots,$$

$$e^{-x} = 1 - x + \frac{x^2}{2!} - \frac{x^3}{3!} + \frac{x^4}{4!} - \frac{x^5}{5!} + \cdots;$$

$$\cosh x = \frac{e^x + e^{-x}}{2} = 1 \qquad + \frac{x^2}{2!} \qquad + \frac{x^4}{4!} \qquad + \cdots = \sum_{k=0}^{\infty} \frac{x^{2k}}{(2k)!}. \square$$

EXAMPLE 6 The identity $e^{i\theta} = \cos\theta + i\sin\theta$. Up to this point we have not used imaginary numbers in our study of series. We recall that complex numbers occur in solving quadratic equations. The formula for the roots of the quadratic equation

$$ax^2 + bx + c = 0$$

is

$$x = \frac{-b \pm \sqrt{b^2 - 4ac}}{2a},$$

when a, b, and c are real numbers and $a \neq 0$. When the discriminant $D = b^2 - 4ac$ is negative, the two roots are complex numbers

$$u + iv \quad \text{and} \quad u - iv,$$

where

$$u = -\frac{b}{2a}, \qquad v = \frac{\sqrt{4ac - b^2}}{2a}.$$

We review these facts mainly to recall the symbol i:

$$i = \sqrt{-1},$$

and to remind ourselves that

$$i^2 = -1, \qquad i^3 = i^2 i = -i, \qquad i^4 = i^2 i^2 = 1, \qquad i^5 = i^4 i = i,$$

and so on.

With these facts in mind, we replace x by $i\theta$ in the Maclaurin series for e^x and simplify to get

$$\begin{aligned} e^{i\theta} &= 1 + \frac{i\theta}{1!} + \frac{i^2\theta^2}{2!} + \frac{i^3\theta^3}{3!} + \frac{i^4\theta^4}{4!} + \frac{i^5\theta^5}{5!} + \frac{i^6\theta^6}{6!} + \cdots \\ &= \left(1 - \frac{\theta^2}{2!} + \frac{\theta^4}{4!} - \frac{\theta^6}{6!} + \cdots\right) + i\left(\theta - \frac{\theta^3}{3!} + \frac{\theta^5}{5!} - \cdots\right) \\ &= \cos\theta + i\sin\theta. \end{aligned}$$

It would not be accurate to say that the calculations just completed have proved that

$$e^{i\theta} = \cos\theta + i\sin\theta. \tag{13}$$

Rather, we shall adopt the point of view that Eq. (13) is the *definition* of $e^{i\theta}$. This definition, which is standard, is motivated by the series expansions for $\cos\theta$, $\sin\theta$, and e^x with $x = i\theta$.

Once we have accepted Eq. (13) as the definition of $e^{i\theta}$, we quickly verify

The Law of Addition of Imaginary Exponents

If θ_1 and θ_2 are any real numbers then

$$e^{i\theta_1}e^{i\theta_2} = e^{i(\theta_1+\theta_2)}. \tag{14}$$

Proof. By definition,

$$e^{i\theta_1} = \cos\theta_1 + i\sin\theta_1, \qquad e^{i\theta_2} = \cos\theta_2 + i\sin\theta_2.$$

Multiplying and simplifying, we get

$$\begin{aligned} e^{i\theta_1}e^{i\theta_2} &= (\cos\theta_1\cos\theta_2 - \sin\theta_1\sin\theta_2) \\ &\quad + i(\sin\theta_1\cos\theta_2 + \cos\theta_1\sin\theta_2) \\ &= e^{i(\theta_1+\theta_2)}. \blacksquare \end{aligned}$$

Notice also that when $\theta = 0$, $i\theta = 0$ and Eq. (13) yields

$$e^0 = \cos 0 + i \sin 0 = 1,$$

which is consistent with $e^x = 1$ when $x = 0$.

If $\theta_1 = \theta$ and $\theta_2 = -\theta$, then $\theta_1 + \theta_2 = 0$, and Eq. (14) yields the result

$$e^{i\theta}e^{-i\theta} = e^0 = 1,$$

so that

$$e^{-i\theta} = 1/e^{i\theta}. \tag{15}$$

Thus the usual laws for the exponential function continue to apply to the function $e^{i\theta}$ defined by Eq. (13). □

Truncation Error

Here are some examples of how to use the Remainder Estimation Theorem to estimate truncation error.

EXAMPLE 7 Calculate e with an error of less than 10^{-6}.

Solution We can use the result of Example 1, Eq. (9a), with $x = 1$ to write

$$e = 1 + 1 + \frac{1}{2!} + \cdots + \frac{1}{n!} + R_n(1, 0)$$

with

$$R_n(1, 0) = e^c \frac{1}{(n+1)!} \qquad \text{for some } c \text{ between 0 and 1.}$$

For the purposes of this example, we do not assume that we already know that $e = 2.71828\ldots$, but we have earlier shown that $e < 3$. Hence, we are certain that

$$\frac{1}{(n+1)!} < R_n(1, 0) < \frac{3}{(n+1)!}$$

because $1 < e^c < 3$.

By trial we find that $1/9! > 10^{-6}$, while $3/10! < 10^{-6}$. Thus we should take $(n + 1)$ to be at least 10, or n to be at least 9. With an error of less than 10^{-6},

$$e = 1 + 1 + \frac{1}{2} + \frac{1}{3!} + \cdots + \frac{1}{9!} \approx 2.718282. \ \square$$

EXAMPLE 8 For what values of x can sin x be replaced by $x - (x^3/3!)$ with an error of magnitude no greater than 3×10^{-4}?

Solution Here we can take advantage of the fact that the Maclaurin series for sin x is an alternating series for every nonzero value of x. According to the Alternating Series Estimation Theorem in Article 11.7, the error in truncating

$$\sin x = x - \frac{x^3}{3!} + \frac{x^5}{5!} - \cdots$$

after $(x^3/3!)$ is no greater than

$$\left|\frac{x^5}{5!}\right| = \frac{|x|^5}{120}.$$

Therefore the error will be less than or equal to 3×10^{-4} if

$$\frac{|x|^5}{120} < 3 \times 10^{-4} \qquad \text{or} \qquad |x| < \sqrt[5]{360 \times 10^{-4}} \approx 0.514.$$

The Alternating Series Estimation Theorem tells us something that the Remainder Estimation Theorem does not: namely, that the estimate $x - (x^3/3!)$ for $\sin x$ is an underestimate when x is positive, because then $x^5/120$ is positive.

Figure 12.4 shows the graph of $\sin x$, along with the graphs of a number of its approximating Taylor polynomials. Note that the graph of $s_1 = x - (x^3/3!)$ is almost indistinguishable from the sine curve when $-1 \leq x \leq 1$. However, it crosses the x-axis at $\pm\sqrt{6} \approx \pm 2.45$, whereas the sine curve crosses the axis at $\pm\pi \approx \pm 3.14$.

One might wonder how the estimate given by the Remainder Estimation Theorem would compare with the one we just obtained from the Alternating Series Estimation Theorem. If we write

$$\sin x = x - \frac{x^3}{3!} + R_3,$$

then the Remainder Estimation Theorem gives

$$|R_3| \leq 1 \cdot \frac{|x|^4}{4!} = \frac{|x|^4}{24},$$

which is not very good. But, when we recognize that $x - x^3/3! = 0 + x + 0x^2 - x^3/3! + 0x^4$ is the Taylor polynomial of degree 4 as well as of

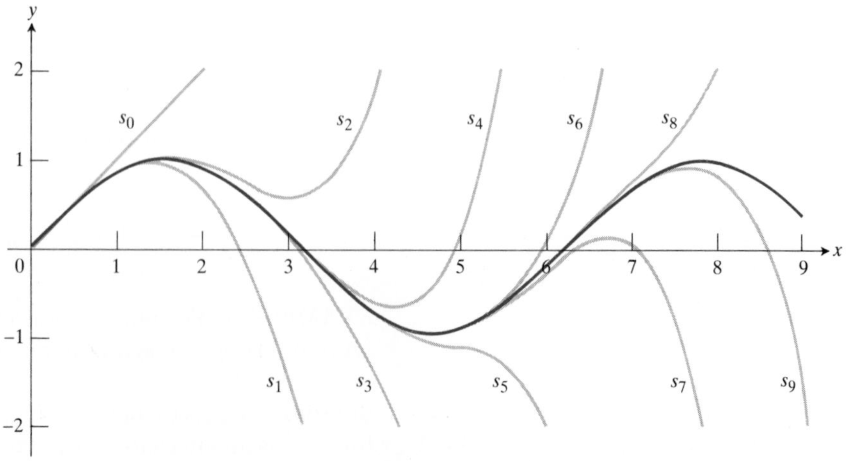

12.4 The polynomials

$$s_n(x) = \sum_{k=0}^{n} [(-1)^k x^{2k+1}/(2k+1)!]$$

converge to $\sin x$ as $n \to \infty$. (Adapted from Helen M. Kammerer. *American Mathematical Monthly,* 43(1936), 293–294.)

degree 3, then we have

$$\sin x = x - \frac{x^3}{3!} + 0 + R_4,$$

and the Remainder Estimation Theorem with $M = r = 1$ gives

$$|R_4| \le 1 \cdot \frac{|x|^5}{5!} = \frac{|x|^5}{120}.$$

This is what we had from the Alternating Series Estimation Theorem. □

In the preceding examples of application of the Remainder Estimation Theorem, we have been able to take $r = 1$. In the next example, $f(x) = \sin 2x$, a factor of 2 enters each time we differentiate, so that we have $r = 2$.

EXAMPLE 9 Let $f(x) = \sin 2x$. For what values of x is the approximation $\sin 2x \approx (2x) - (2x)^3/3! + (2x)^5/5!$ not in error by more than 5×10^{-6}?

Solution Because the Taylor polynomials of degree 5 and of degree 6 for $\sin 2x$ are identical, in that they differ only in a term that has a zero coefficient, we are certain that the error is no greater than $2^7|x|^7/7!$ (by comparison with Example 8). We therefore solve the inequality

$$\frac{|2x|^7}{7!} < 5 \times 10^{-6}.$$

The result is

$$|2x| < \sqrt[7]{7! \times 5 \times 10^{-6}} = \sqrt[7]{0.025200} \approx 0.59106$$

or

$$|x| < 0.29553 \text{ (radians)}. \ \square$$

PROBLEMS

In Problems 1–6, write the Maclaurin series for each function.

1. $e^{x/2}$

2. $\sin 3x$

3. $5 \cos \dfrac{x}{\pi}$

4. $\sinh x$

5. $\dfrac{x^2}{2} - 1 + \cos x$

6. $\cos^2 x = \dfrac{1 + \cos 2x}{2}$

7. Use series to verify that
a) $\cos(-x) = \cos x$
b) $\sin(-x) = -\sin x$

8. Show that

$$e^x = e^a\left[1 + (x - a) + \frac{(x - a)^2}{2!} + \cdots\right].$$

In Problems 9–11, write Taylor's formula (Eq. 7a), with $n = 2$ and $a = 0$, for the given function.

9. $\dfrac{1}{1 + x}$

10. $\ln(1 + x)$

11. $\sqrt{1 + x}$

12. Find the Taylor series for e^x at $a = 1$. Compare your series with the result in Problem 8.

13. For approximately what values of x can one replace $\sin x$ by $x - (x^3/6)$ with an error of magnitude no greater than 5×10^{-4}?

14. If $\cos x$ is replaced by $1 - (x^2/2)$ and $|x| < 0.5$, what estimate can be made of the error? Does $1 - (x^2/2)$ tend to be too large, or too small?

15. How close is the approximation $\sin x = x$ when $|x| < 10^{-3}$? For which of these values of x is $x < \sin x$?

16. The estimate $\sqrt{1 + x} = 1 + (x/2)$ is used when $|x|$ is small. Estimate the error when $|x| < 0.01$.

17. The approximation $e^x = 1 + x + (x^2/2)$ is used when x is small. Use the Remainder Estimation Theorem to estimate the error when $|x| < 0.1$.

18. When $x < 0$, the series for e^x is an alternating series. Use the Alternating Series Estimation Theorem to estimate the error that results from replacing e^x by $1 + x + (x^2/2)$ when $-0.1 < x < 0$. Compare with Problem 17.

19. Estimate the error in the approximation $\sinh x = x + (x^3/3!)$ when $|x| < 0.5$. (*Hint:* Use R_4, not R_3.)

20. When $0 \le h \le 0.01$, show that e^h may be replaced by $1 + h$ with an error of magnitude no greater than six-tenths of one percent of h. Use $e^{0.1} = 1.105$.

21. Let $f(x)$ and $g(x)$ have derivatives of all orders at $a = 0$. Show that the Maclaurin series for $f + g$ is

$$\sum_{n=0}^{\infty} \frac{f^{(n)}(0) + g^{(n)}(0)}{n!} x^n.$$

22. Each of the following sums is the value of an elementary function at some point. Find the function and the point.

a) $(0.1) - \dfrac{(0.1)^3}{3!} + \dfrac{(0.1)^5}{5!} - \cdots + \dfrac{(-1)^k(0.1)^{2k+1}}{(2k+1)!} + \cdots$

b) $1 - \dfrac{\pi^2}{4^2 \cdot 2!} + \dfrac{\pi^4}{4^4 \cdot 4!} - \cdots + \dfrac{(-1)^k(\pi)^{2k}}{4^{2k} \cdot (2k)!} + \cdots$

c) $1 + \dfrac{1}{2!} + \dfrac{1}{4!} + \cdots + \dfrac{1}{(2k)!} + \cdots$

23. Express each of the following in the form $u + iv$ with u and v real:

a) $e^{i\pi}$

b) $e^{i\pi/4}$

c) $e^{-i\pi/2}$

d) $e^{i\pi} \cdot e^{-i\pi/2}$

24. Using Eq. (13), show that

$$\cos\theta = \frac{e^{i\theta} + e^{-i\theta}}{2} \quad \text{and} \quad \sin\theta = \frac{e^{i\theta} - e^{-i\theta}}{2i}.$$

These are sometimes called Euler's identities.

25. Use the results of Problem 24 to show that

$$\cos^3\theta = \frac{1}{4}\cos 3\theta + \frac{3}{4}\cos\theta$$

and

$$\sin^3\theta = -\frac{1}{4}\sin 3\theta + \frac{3}{4}\sin\theta.$$

26. When a and b are real, we define $e^{(a+ib)x}$ to be $e^{ax}(\cos bx + i\sin bx)$. From this definition, show that

$$\frac{d}{dx}e^{(a+ib)x} = (a + ib)e^{(a+ib)x}.$$

27. Two complex numbers, $a + ib$ and $c + id$, are equal if and only if $a = c$ and $b = d$. Use this fact to evaluate

$$\int e^{ax}\cos bx\,dx \quad \text{and} \quad \int e^{ax}\sin bx\,dx$$

from

$$\int e^{(a+ib)x}\,dx = \frac{a - ib}{a^2 + b^2}e^{(a+ib)x} + C,$$

where

$$C = C_1 + iC_2$$

is a complex constant of integration.

***Toolkit* programs**

Taylor Series

12.3 Further Computations, Logarithms, Arctangents, and π

The Taylor-series expansion

$$f(x) = f(a) + f'(a)(x - a) + \frac{f''(a)}{2!}(x - a)^2 + \cdots + \frac{f^{(n)}(a)}{n!}(x - a)^n + R_n(x, a) \tag{1}$$

expresses the value of the function at x in terms of its value and the values of its derivatives at a, plus a remainder term, which we hope is so small that it may safely be omitted. In applying series to numerical computations, it is therefore *necessary* that a be chosen so that $f(a)$, $f'(a)$, $f''(a)$, . . . are known. In dealing with the trigonometric functions, for example, one might take $a = 0, \pm\pi/6, \pm\pi/4, \pm\pi/3, \pm\pi/2$, and so on. It is also clear that it is *desirable* to choose the value of a near to the value of x for which the function is to be computed, in order to make $(x - a)$ small, so that the terms of the series decrease rapidly as n increases.

EXAMPLE 1 What value of a might one choose in the Taylor series (1) to compute sin 35°?

Solution We could choose $a = 0$ and use the series

$$\sin x = x - \frac{x^3}{3!} + \frac{x^5}{5!} - \cdots + (-1)^n \frac{x^{2n+1}}{(2n+1)!} + 0 \cdot x^{2n+2} + R_{2n+2}(x, 0), \tag{2}$$

or we could choose $a = \pi/6$ (which corresponds to 30°) and use the series

$$\sin x = \sin\frac{\pi}{6} + \cos\frac{\pi}{6}\left(x - \frac{\pi}{6}\right) - \sin\frac{\pi}{6}\frac{(x - \pi/6)^2}{2!} - \cos\frac{\pi}{6}\frac{(x - \pi/6)^3}{3!} + \cdots + \sin\left(\frac{\pi}{6} + n\frac{\pi}{2}\right)\frac{(x - \pi/6)^n}{n!} + R_n\left(x, \frac{\pi}{6}\right).$$

The remainder in the series (2) satisfies the inequality

$$|R_{2n+2}(x, 0)| \le \frac{|x|^{2n+3}}{(2n+3)!}, \tag{3}$$

which tends to zero as n becomes infinite, no matter how large $|x|$ may be. We could therefore calculate sin 35° by placing

$$x = \frac{35\pi}{180} = 0.6108652$$

in the approximation

$$\sin x \approx x - \frac{x^3}{6} + \frac{x^5}{120} - \frac{x^7}{5040},$$

with an error of magnitude no greater than 3.3×10^{-8}, since

$$\left|R_8\left(\frac{35\pi}{180}, 0\right)\right| < \frac{(0.611)^9}{9!} < 3.3 \times 10^{-8}.$$

By using the series with $a = \pi/6$, we could obtain equal accuracy with a smaller exponent n, but at the expense of introducing $\cos \pi/6 = \sqrt{3}/2$ as one of the coefficients. In this series, with $a = \pi/6$, we would take

$$x = \frac{35\pi}{180},$$

but the quantity that appears raised to the various powers is

$$x - \frac{\pi}{6} = \frac{5\pi}{180} = 0.0872665,$$

which decreases rapidly when raised to high powers.

As a matter of fact, various trigonometric identities may be used, such as

$$\sin\left(\frac{\pi}{2} - x\right) = \cos x,$$

to facilitate the calculation of the sine or cosine of any angle with the Maclaurin series of the two functions. This method of finding the sine or cosine of an angle is used in computers. □

Computation of Logarithms

Natural logarithms may be computed from series. The starting point is the series for ln (1 + x) in powers of x:

$$\ln(1 + x) = x - \frac{x^2}{2} + \frac{x^3}{3} - \cdots + (-1)^{n-1}\frac{x^n}{n} + \cdots.$$

This series may be found directly from the Taylor-series expansion, Eq. (1), with $a = 0$. It may also be obtained by integrating the geometric series for $1/(1 + t)$ from $t = 0$ to $t = x$:

$$\int_0^x \frac{dt}{1+t} = \int_0^x (1 - t + t^2 - t^3 + \cdots)\,dt,$$

$$\ln(1 + t)\Big]_0^x = t - \frac{t^2}{2} + \frac{t^3}{3} - \frac{t^4}{4} + \cdots\Big]_0^x,$$

$$\ln(1 + x) = x - \frac{x^2}{2} + \frac{x^3}{3} - \frac{x^4}{4} + \cdots. \tag{4}$$

The expansion (4) is valid for $|x| < 1$, since then the remainder, $R_n(x, 0)$, approaches zero as $n \to \infty$, as we shall now see. The remainder is given by the integral of the remainder in the geometric series, that is,

$$R_n(x, 0) = \int_0^x \frac{(-1)^n t^n}{1+t}\,dt. \tag{5}$$

We now suppose that $|x| < 1$. For every t between 0 and x inclusive we have

$$|1 + t| \geq 1 - |x|$$

and

$$|(-1)^n t^n| = |t|^n,$$

so that

$$\left|\frac{(-1)^n t^n}{1+t}\right| \leq \frac{|t|^n}{1 - |x|}.$$

Therefore,

$$|R_n(x, 0)| \leq \int_0^{|x|} \frac{t^n}{1 - |x|}\,dt = \frac{1}{n+1}\cdot\frac{|x|^{n+1}}{1 - |x|}. \tag{6}$$

When $n \to \infty$, the right-hand side of the inequality (6) approaches zero, and so must the left-hand side. Thus (4) holds for $|x| < 1$.

Computation of π

Archimedes (287–212 B.C.) gave the approximation

$$3\tfrac{1}{7} > \pi > 3\tfrac{10}{71},$$

in the third century B.C. A French mathematician, Viéta (1540–1603), gave the formula

$$\frac{2}{\pi} = \sqrt{\tfrac{1}{2}} \times \sqrt{\left(\tfrac{1}{2} + \tfrac{1}{2}\sqrt{\tfrac{1}{2}}\right)} \times \sqrt{\left(\tfrac{1}{2} + \tfrac{1}{2}\sqrt{\left(\tfrac{1}{2} + \tfrac{1}{2}\sqrt{\tfrac{1}{2}}\right)}\right)} \times \cdots,$$

which Turnbull† calls "the first actual formula for the time-honoured number π." Other interesting formulas for π include‡

$$\frac{\pi}{4} = \cfrac{1}{1 + \cfrac{1^2}{2 + \cfrac{3^2}{2 + \cfrac{5^2}{2 + \cdots}}}},$$

which is credited to Lord Brouncker, an Irish peer;

$$\frac{\pi}{4} = \frac{2 \times 4 \times 4 \times 6 \times 6 \times 8 \times \cdots}{3 \times 3 \times 5 \times 5 \times 7 \times 7 \times \cdots},$$

discovered by the English mathematician Wallis; and

$$\frac{\pi}{4} = 1 - \frac{1}{3} + \frac{1}{5} - \frac{1}{7} + \cdots,$$

known as Leibniz's formula.

We now turn our attention to the series for $\tan^{-1} x$, since it leads to the Leibniz formula and others from which π has been computed to a great many decimal places.

Since

$$\tan^{-1} x = \int_0^x \frac{dt}{1 + t^2},$$

we integrate the geometric series, with remainder,

$$\frac{1}{1 + t^2} = 1 - t^2 + t^4 - t^6 + \cdots + (-1)^n t^{2n} + \frac{(-1)^{n+1} t^{2n+2}}{1 + t^2}. \tag{7}$$

Thus

$$\tan^{-1} x = x - \frac{x^3}{3} + \frac{x^5}{5} - \frac{x^7}{7} + \cdots + (-1)^n \frac{x^{2n+1}}{2n + 1} + R,$$

where

$$R = \int_0^x \frac{(-1)^{n+1} t^{2n+2}}{1 + t^2}\, dt.$$

The denominator of the integrand is greater than or equal to 1; hence

$$|R| \le \int_0^{|x|} t^{2n+2}\, dt = \frac{|x|^{2n+3}}{2n + 3}.$$

If $|x| \le 1$, the right side of this inequality approaches zero as $n \to \infty$.

† *World of Mathematics*, Vol. 1 (New York: Simon and Schuster, 1956), p. 121.
‡ *Ibid.*, p. 138.

Therefore R also approaches zero and we have

$$\tan^{-1} x = \sum_{n=0}^{\infty} \frac{(-1)^n x^{2n+1}}{2n+1},$$

or

$$\tan^{-1} x = x - \frac{x^3}{3} + \frac{x^5}{5} - \frac{x^7}{7} + \cdots, \quad |x| \le 1. \tag{8}$$

When we put $x = 1$, $\tan^{-1} 1 = \pi/4$, in Eq. (8) we get Leibniz's formula:

$$\frac{\pi}{4} = 1 - \frac{1}{3} + \frac{1}{5} - \frac{1}{7} + \frac{1}{9} - \cdots + \frac{(-1)^{n-1}}{2n-1} + \cdots.$$

Because this series converges very slowly, it is not used in approximating π to many decimal places. The series for $\tan^{-1}x$ converges most rapidly when x is near zero. For that reason, people who use the series for $\tan^{-1}x$ to compute π use various trigonometric identities.

For example, if

$$\alpha = \tan^{-1}\frac{1}{2} \quad \text{and} \quad \beta = \tan^{-1}\frac{1}{3},$$

then

$$\tan(\alpha + \beta) = \frac{\tan\alpha + \tan\beta}{1 - \tan\alpha\tan\beta} = \frac{\frac{1}{2} + \frac{1}{3}}{1 - \frac{1}{6}} = 1 = \tan\frac{\pi}{4}$$

and

$$\frac{\pi}{4} = \alpha + \beta = \tan^{-1}\frac{1}{2} + \tan^{-1}\frac{1}{3}. \tag{9}$$

Now Eq. (8) may be used with $x = \frac{1}{2}$ to evaluate $\tan^{-1}\frac{1}{2}$ and with $x = \frac{1}{3}$ to give $\tan^{-1}\frac{1}{3}$. The sum of these results, multiplied by 4, gives π.

In 1961, π was computed to more than 100,000 decimal places on an IBM 7090 computer. More recently, in 1973, Jean Guilloud and Martine Bouyer computed π to 1,000,000 decimal places on a CDC 7600 computer, by applying the arctangent series (8) to the formula

$$\pi = 48\tan^{-1}\frac{1}{18} + 32\tan^{-1}\frac{1}{57} - 20\tan^{-1}\frac{1}{239}. \tag{10}$$

They checked their work with the formula

$$\pi = 24\tan^{-1}\frac{1}{8} + 8\tan^{-1}\frac{1}{57} + 4\tan^{-1}\frac{1}{239}. \tag{11}$$

A number of current computations of π are being carried out with an algorithm discovered by Eugene Salamin (Problem 17). The algorithm produces sequences that converge to π even more rapidly than the sequence of partial sums of the arctangent series in Eqs. (10) and (11).†

REMARK. Two types of numerical error tend to occur in computing with series. The first is the *truncation error*, which is the remainder $R_n(x, a)$. This is the only error we have discussed so far.

†For a delightful account of attempts to compute, and even to *legislate*(!) the value of π, see Chapter 12 of David A. Smith's *Interface: Calculus and the Computer* (Boston, Massachusetts: Houghton Mifflin, 1976).

FREQUENTLY-USED MACLAURIN SERIES

$$\frac{1}{1-x} = 1 + x + x^2 + \cdots + x^n + \cdots = \sum_{n=0}^{\infty} x^n, \qquad |x| < 1$$

$$\frac{1}{1+x} = 1 - x + x^2 - \cdots + (-x)^n + \cdots = \sum_{n=0}^{\infty} (-1)^n x^n, \qquad |x| < 1$$

$$e^x = 1 + x + \frac{x^2}{2!} + \cdots + \frac{x^n}{n!} + \cdots = \sum_{n=0}^{\infty} \frac{x^n}{n!}, \qquad |x| < \infty$$

$$\sin x = x - \frac{x^3}{3!} + \frac{x^5}{5!} - \cdots + (-1)^n \frac{x^{2n+1}}{(2n+1)!} + \cdots = \sum_{n=0}^{\infty} \frac{(-1)^n x^{2n+1}}{(2n+1)!}, \qquad |x| < \infty$$

$$\cos x = 1 - \frac{x^2}{2!} + \frac{x^4}{4!} - \cdots + (-1)^n \frac{x^{2n}}{(2n)!} + \cdots = \sum_{n=0}^{\infty} \frac{(-1)^n x^{2n}}{(2n)!}, \qquad |x| < \infty$$

$$\ln(1+x) = x - \frac{x^2}{2} + \frac{x^3}{3} - \cdots + (-1)^{n-1} \frac{x^n}{n} + \cdots = \sum_{n=1}^{\infty} \frac{(-1)^{n-1} x^n}{n}, \qquad -1 < x \le 1$$

$$\ln \frac{1+x}{1-x} = 2 \tanh^{-1} x = 2\left(x + \frac{x^3}{3} + \frac{x^5}{5} + \cdots + \frac{x^{2n+1}}{2n+1} + \cdots\right) = 2 \sum_{n=0}^{\infty} \frac{x^{2n+1}}{2n+1}, \qquad |x| < 1$$

$$\tan^{-1} x = x - \frac{x^3}{3} + \frac{x^5}{5} - \cdots + (-1)^{n-1} \frac{x^{2n-1}}{2n-1} + \cdots = \sum_{n=1}^{\infty} \frac{(-1)^{n-1} x^{2n-1}}{2n-1}, \qquad |x| \le 1$$

BINOMIAL SERIES

$$\begin{aligned}(1+x)^m &= 1 + mx + \frac{m(m-1)\,x^2}{2!} + \frac{m(m-1)(m-2)\,x^3}{3!} + \cdots \\ &\quad + \frac{m(m-1)(m-2)\cdots(m-k+1)\,x^k}{k!} + \cdots \\ &= 1 + \sum_{k=1}^{\infty} \binom{m}{k} x^k, \qquad |x| < 1\end{aligned}$$

where

$$\binom{m}{1} = m,$$

$$\binom{m}{2} = \frac{m(m-1)}{2!},$$

$$\binom{m}{k} = \frac{m(m-1)\cdots(m-k+1)}{k!} \qquad \text{for } k \ge 3.$$

NOTE. It is customary to define $\binom{m}{0}$ to be 1 and to take $x^0 = 1$ (even in the usually excluded case where $x = 0$) in order to write the binomial series compactly as

$$(1+x)^m = \sum_{k=0}^{\infty} \binom{m}{k} x^k, \qquad |x| < 1.$$

If m is a *positive integer*, the series terminates at x^m and the result converges for all x.

The second is the *round-off error* that enters in calculating the sum of the finite number of terms

$$f(a) + f'(a)(x - a) + \cdots + \frac{f^{(n)}(a)(x - a)^n}{n!}$$

when we approximate each of these terms by a decimal number with only a finite number of decimal places. For example, taking 0.3333 in place of $\frac{1}{3}$ introduces a round-off error equal to $10^{-4}/3$. There are likely to be round-off errors associated with each term, some of these being positive and some negative. When the need for accuracy is paramount, it is important to control both the truncation error and the round-off errors. *Truncation* errors can be reduced by taking more terms of the series; *round-off* errors can be reduced by taking more decimal places.

PROBLEMS

In Problems 1–6, use a suitable series to calculate the indicated quantity to three decimal places. In each case, show that the remainder term does not exceed 5×10^{-4}.

1. cos 31° **2.** tan 46° **3.** sin 6.3

4. cos 69 **5.** ln 1.25 **6.** $\tan^{-1} 1.02$

7. Find the Maclaurin series for $\ln(1 + 2x)$. For what values of x does the series converge?

8. For what values of x can one replace $\ln(1 + x)$ by x with an error of magnitude no greater than one percent of the absolute value of x?

Use series to evaluate the integrals in Problems 9 and 10 to three decimals.

9. $\displaystyle\int_0^{0.1} \frac{\sin x}{x}\, dx$ **10.** $\displaystyle\int_0^{0.1} e^{-x^2}\, dx$

11. Show that the ordinate of the catenary $y = a \cosh x/a$ deviates from the ordinate of the parabola $x^2 = 2a(y - a)$ by less than $0.003|a|$ over the range $|x/a| \le \frac{1}{3}$.

12. a) Replace x by $-x$ in Eq. (4) to obtain a series for $\ln(1 - x)$. Combine this with the series for $\ln(1 + x)$ to show that

$$\ln \frac{1 + x}{1 - x} = 2\left(x + \frac{x^3}{3} + \frac{x^5}{5} + \cdots\right) \quad \text{for } |x| < 1.$$

b) For what value of x is $(1 + x)/(1 - x) = 2$? Use that value of x in the series of part (a) to estimate ln 2 to 3 decimals.

13. Find the sum of the series

$$\frac{1}{2} - \frac{1}{2}\left(\frac{1}{2}\right)^2 + \frac{1}{3}\left(\frac{1}{2}\right)^3 - \frac{1}{4}\left(\frac{1}{2}\right)^4 + \cdots.$$

14. How many terms of the series for $\tan^{-1} 1$ would you have to add for the Alternating Series Estimation Theorem to guarantee a calculation of $\pi/4$ to two decimals?

15. (*Calculator*) Equations (8) and (10) yield a series that converges to $\pi/4$ fairly rapidly. Estimate π to three decimal places with this series. In contrast, the convergence of $\sum_{n=1}^{\infty} (1/n^2)$ to $\pi^2/6$ is so slow that even fifty terms will not yield two-place accuracy.

16. (*Calculator*) a) Find π to two decimals with the formulas of Lord Brouncker and Wallis.
b) If your calculator is programmable, use Viéta's formula to calculate π to five decimal places.

17. (*Calculator*) A special case of Salamin's algorithm for estimating π begins with defining sequences $\{a_n\}$ and $\{b_n\}$ by the rules

$$a_0 = 1, \qquad b_0 = \frac{1}{\sqrt{2}},$$

$$a_{n+1} = \frac{(a_n + b_n)}{2}, \qquad b_{n+1} = \sqrt{a_n b_n},$$

Then the sequence $\{c_n\}$ defined for $n \ge 1$ by

$$c_n = \frac{4a_n b_n}{1 - \sum_{j=1}^{n} 2^{j+1}(a_j^2 - b_j^2)}$$

converges to π. Calculate c_3. (E. Salamin, "Computation of π using arithmetic-geometric mean," *Mathematics of Computation*, **30**, July, 1976, pp. 565–570.)

18. Show that the series in Eq. (8) for $\tan^{-1} x$ diverges for $|x| > 1$.

19. Show that

$$\int_0^x \frac{dt}{1 - t^2} = \int_0^x \left(1 + t^2 + t^4 + \cdots + t^{2n} + \frac{t^{2n+2}}{1 - t^2}\right) dt$$

or, in other words, that

$$\tanh^{-1} x = x + \frac{x^3}{3} + \frac{x^5}{5} + \cdots + \frac{x^{2n+1}}{2n + 1} + R,$$

where

$$R = \int_0^x \frac{t^{2n+2}}{1-t^2}\,dt.$$

20. Show that R in Problem 19 is no greater than

$$\frac{1}{1-x^2}\cdot\frac{|x|^{2n+3}}{2n+3}, \qquad \text{if } x^2 < 1.$$

21. a) Differentiate the identity

$$\frac{1}{1-x} = 1 + x + x^2 + \cdots + x^n + \frac{x^{n+1}}{1-x}$$

to obtain the expansion

$$\frac{1}{(1-x)^2} = 1 + 2x + 3x^2 + \cdots + nx^{n-1} + R.$$

b) Prove that, if $|x| < 1$, then $R \to 0$ as $n \to \infty$.

c) In one throw of two dice, the probability of getting a score of 7 is $p = \frac{1}{6}$. If the dice are thrown repeatedly, the probability that a 7 will appear for the first time at the nth throw is $q^{n-1}p$, where $q = 1 - p = \frac{5}{6}$. The expected number of throws until a 7 first appears is $\Sigma_{n=1}^{\infty} nq^{n-1}p$. Evaluate this series numerically.

d) In applying statistical quality control to an industrial operation, an engineer inspects items taken at random from the assembly line. Each item sampled is classified as "good' or "bad." If the probability of a good item is p and of a bad item is $q = 1 - p$, the probability that the first bad item found is the nth inspected is $p^{n-1}q$. The average number inspected up to and including the first bad item found is $\Sigma_{n=1}^{\infty} np^{n-1}q$. Evaluate this series, assuming $0 < p < 1$.

22. In probability theory, a random variable X may assume the values 1, 2, 3, . . . , with probabilities $p_1, p_2, p_3, \ldots$, where p_k is the probability that X is equal to k ($k = 1, 2, \ldots$). It is customary to assume $p_k \geq 0$ and $\Sigma_{k=1}^{\infty} p_k = 1$. The *expected value* of X denoted by E(X) is defined as $\Sigma_{k=1}^{\infty} kp_k$, provided this series converges. In each of the following cases, show that $\Sigma\, p_k = 1$ and find E(X), if it exists. (*Hint:* See Problem 21.)

a) $p_k = 2^{-k}$ b) $p_k = \dfrac{5^{k-1}}{6^k}$

c) $p_k = \dfrac{1}{k(k+1)} = \dfrac{1}{k} - \dfrac{1}{k+1}$

***Toolkit* programs**

Sequences and Series Taylor Series

12.4
Indeterminate Forms

In considering the ratio of two functions $f(x)$ and $g(x)$, we sometimes wish to know the value

$$\lim_{x\to a} \frac{f(x)}{g(x)} \tag{1}$$

at a point a where $f(x)$ and $g(x)$ are both zero. L'Hôpital's rule is often a help, but the differentiation involved can be time-consuming, especially if the rule has to be applied several times to reach a determinate form. In many instances, the limit in (1) can be calculated more quickly if the functions involved have power series expansions about $x = a$. In fact, the ease and reliability of the kind of calculation we are about to illustrate contributed to the early popularity of power series. The theoretical justification of the technique is too long to discuss here, but the formal manipulations are worth learning by themselves.

Suppose, then, that the functions f and g both have series expansions in powers of $x - a$,

$$f(x) = f(a) + f'(a)\cdot(x-a) + \frac{f''(a)}{2!}(x-a)^2 + \cdots, \tag{2a}$$

$$g(x) = g(a) + g'(a)\cdot(x-a) + \frac{g''(a)}{2!}(x-a)^2 + \cdots, \tag{2b}$$

that are known to us and that converge in some interval $|x - a| < \delta$. We then proceed to calculate the limit (1), provided the limit exists, in the manner shown by the following examples.

EXAMPLE 1 Evaluate $\lim_{x\to 1} [(\ln x)/(x - 1)]$.

Solution Let $f(x) = \ln x$, $g(x) = x - 1$. The Taylor series for $f(x)$, with $a = 1$, is found as follows:

$$\begin{aligned} f(x) &= \ln x, & f(1) &= \ln 1 = 0,\\ f'(x) &= 1/x, & f'(1) &= 1,\\ f''(x) &= -1/x^2, & f''(1) &= -1, \end{aligned}$$

so that

$$\ln x = 0 + (x - 1) - \frac{1}{2}(x - 1)^2 + \cdots.$$

Hence

$$\frac{\ln x}{x - 1} = 1 - \frac{1}{2}(x - 1) + \cdots$$

and

$$\lim_{x\to 1} \frac{\ln x}{x - 1} = \lim_{x\to 1} [1 - \frac{1}{2}(x - 1) + \cdots] = 1. \square$$

EXAMPLE 2 Evaluate $\lim_{x\to 0} [(\sin x - \tan x)/x^3]$.

Solution The Maclaurin series for $\sin x$ and $\tan x$, to terms in x^5, are

$$\sin x = x - \frac{x^3}{3!} + \frac{x^5}{5!} - \cdots,$$

$$\tan x = x + \frac{x^3}{3} + \frac{2x^5}{15} + \cdots.$$

Hence

$$\sin x - \tan x = -\frac{x^3}{2} - \frac{x^5}{8} - \cdots = x^3\left(-\frac{1}{2} - \frac{x^2}{8} - \cdots\right),$$

and

$$\lim_{x\to 0} \frac{\sin x - \tan x}{x^3} = \lim_{x\to 0} \left(-\frac{1}{2} - \frac{x^2}{8} - \cdots\right) = -\frac{1}{2}. \square$$

When we apply series to compute the limit $\lim_{x\to 0} (1/\sin x - 1/x)$ of Example 7, Article 3.9, we not only compute the limit successfully but also discover a nice approximation formula for $\csc x$.

EXAMPLE 3 Find

$$\lim_{x\to 0} \left(\frac{1}{\sin x} - \frac{1}{x}\right).$$

Solution

$$\frac{1}{\sin x}-\frac{1}{x}=\frac{x-\sin x}{x\sin x}=\frac{x-\left[x-\dfrac{x^3}{3!}+\dfrac{x^5}{5!}-\cdots\right]}{x\cdot\left[x-\dfrac{x^3}{3!}+\dfrac{x^5}{5!}-\cdots\right]}$$

$$=\frac{x^3\left[\dfrac{1}{3!}-\dfrac{x^2}{5!}+\cdots\right]}{x^2\left[1-\dfrac{x^2}{3!}+\cdots\right]}$$

$$=x\,\frac{\dfrac{1}{3!}-\dfrac{x^2}{5!}+\cdots}{1-\dfrac{x^2}{3!}+\cdots}.$$

Therefore,

$$\lim_{x\to 0}\left(\frac{1}{\sin x}-\frac{1}{x}\right)=\lim_{x\to 0}\left[x\,\frac{\dfrac{1}{3!}-\dfrac{x^2}{5!}+\cdots}{1-\dfrac{x^2}{3!}+\cdots}\right]=0.$$

In fact, from the series expressions above we can see that if $|x|$ is small, then

$$\frac{1}{\sin x}-\frac{1}{x}\approx x\cdot\frac{1}{3!}=\frac{x}{6}\qquad\text{or}\qquad \csc x\approx\frac{1}{x}+\frac{x}{6}.\ \square$$

PROBLEMS

Use series to evaluate the limits in Problems 1–20.

1. $\lim_{h\to 0}\dfrac{\sin h}{h}$

2. $\lim_{x\to 0}\dfrac{e^x-(1+x)}{x^2}$

3. $\lim_{t\to 0}\dfrac{1-\cos t-\frac{1}{2}t^2}{t^4}$

4. $\lim_{x\to\infty} x\sin\dfrac{1}{x}$

5. $\lim_{x\to 0}\dfrac{x^2}{1-\cosh x}$

6. $\lim_{h\to 0}\dfrac{(\sin h)/h-\cos h}{h^2}$

7. $\lim_{x\to 0}\dfrac{1-\cos x}{\sin x}$

8. $\lim_{x\to 0}\dfrac{\sin x}{e^x-1}$

9. $\lim_{z\to 0}\dfrac{\sin(z^2)-\sinh(x^2)}{z^6}$

10. $\lim_{t\to 0}\dfrac{\cos t-\cosh t}{t^2}$

11. $\lim_{x\to 0}\dfrac{\sin x-x+\dfrac{x^3}{6}}{x^5}$

12. $\lim_{x\to 0}\dfrac{e^x-e^{-x}-2x}{x-\sin x}$

13. $\lim_{x\to 0}\dfrac{x-\tan^{-1}x}{x^3}$

14. $\lim_{x\to 0}\dfrac{\tan x-\sin x}{x^3\cos x}$

15. $\lim_{x\to\infty} x^2(e^{-1/x^2}-1)$

16. $\lim_{x\to 0}\dfrac{\ln(1+x^2)}{1-\cos x}$

17. $\lim_{x\to 0}\dfrac{\tan 3x}{x}$

18. $\lim_{x\to 1}\dfrac{\ln x^2}{x-1}$

19. $\lim_{x\to\infty}\dfrac{x^{100}}{e^x}$

20. $\lim_{x\to 0}\left(\dfrac{1}{2-2\cos x}-\dfrac{1}{x^2}\right)$

21. a) Prove that $\int_0^x e^{t^2}\,dt\to+\infty$ as $x\to+\infty$.
    b) Find $\lim_{x\to\infty} x\int_0^x e^{t^2-x^2}\,dt$.

22. Find values of r and s such that
$$\lim_{x\to 0}(x^{-3}\sin 3x+rx^{-2}+s)=0.$$

23. (*Calculator*) The approximation for csc x in Example 3 leads to the approximation $\sin x\approx 6x/(6+x^2)$. Evaluate both sides of this approximation for $x=\pm1.0$, ±0.1, and ±0.01 radians. Try these values of x in the approximation $\sin x\approx x$. Which approximation appears to give better results?

***Toolkit* programs**

Taylor Series

12.5 Convergence of Power Series. Integration, Differentiation, Multiplication, and Division

We now know that some power series, like the Maclaurin series for sin x, cos x, and e^x, converge for all values of x, while others, like the series we derived for $\ln(1 + x)$ and $\tan^{-1} x$, converge only on finite intervals. But we learned all this by analyzing remainder formulas, and we have yet to face the question of how to investigate the convergence of a power series when there is no remainder formula to analyze. Moreover, all of the power series we have worked with have been Taylor series of functions for which we already had expressions in closed forms. What about other power series? Are they Taylor series, too, of functions otherwise unknown?

The first step in answering these questions is to note that a power series $\Sigma_{n=0}^{\infty} a_n x^n$ defines a function whenever it converges, namely, the function f whose value at each x is the number

$$f(x) = \sum_{n=0}^{\infty} a_n x^n. \tag{1}$$

We can then ask what kind of domain f has, how f is to be differentiated and integrated (if at all), whether f has a Taylor series, and, if it has, how its Taylor series is related to the defining series $\Sigma_{n=0}^{\infty} a_n x^n$.

The questions of what domain f has, and for what values the series (1) may be expected to converge, are answered by Theorem 1 and the discussion that follows it. We will prove Theorem 1, and then, after looking at examples, will proceed to Theorems 2 and 3, which answer the questions of whether f *can* be differentiated and integrated and *how* to do so when it can be. Theorem 3 also solves a problem that arose many chapters ago but that has remained unsolved until now: that of finding convenient expressions for evaluating integrals like

$$\int_0^1 \sin x^2\,dx \qquad \text{and} \qquad \int_0^{0.5} \sqrt{1 + x^4}\,dx,$$

which frequently arise in applications. Finally, we will see that, in the interior of its domain of definition, the function f does have a Maclaurin series, and that this is none other than the defining series $\Sigma_{n=0}^{\infty} a_n x^n$.

THEOREM 1

The Convergence Theorem for Power Series

If a power series

$$\sum_{n=0}^{\infty} a_n x^n = a_0 + a_1 x + a_2 x^2 + \cdots \tag{2}$$

converges for $x = c$ ($c \neq 0$), then it converges absolutely for all $|x| < |c|$. If the series diverges for $x = d$, then it diverges for all $|x| > |d|$.

Proof. Suppose the series

$$\sum_{n=0}^{\infty} a_n c^n \tag{3}$$

converges. Then

$$\lim_{n\to\infty} a_n c^n = 0.$$

Hence, there is an index N such that

$$|a_n c^n| < 1 \quad \text{for all} \quad n \geq N.$$

That is,

$$|a_n| < \frac{1}{|c|^n} \quad \text{for} \quad n \geq N. \tag{4}$$

Now take any x such that $|x| < |c|$ and consider

$$|a_0| + |a_1 x| + \cdots + |a_{N-1}x^{N-1}| + |a_N x^N| + |a_{N+1}x^{N+1}| + \cdots.$$

There is only a finite number of terms prior to $|a_N x^N|$ and their sum is finite. Starting with $|a_N x^N|$ and beyond, the terms are less than

$$\left|\frac{x}{c}\right|^N + \left|\frac{x}{c}\right|^{N+1} + \left|\frac{x}{c}\right|^{N+2} + \cdots \tag{5}$$

by virtue of the inequality (4). But the series in (5) is a geometric series with ratio $r = |x/c|$, which is less than one, since $|x| < |c|$. Hence the series (5) converges, so that the original series (3) converges absolutely. This proves the first half of the theorem.

The second half of the theorem involves nothing new. For if the series diverges at $x = d$ and converges at a value x_0 with $|x_0| > |d|$, we may take $c = x_0$ in the first half of the theorem and conclude that the series converges absolutely at d. But the series cannot both converge absolutely and diverge at one and the same time. Hence, if it diverges at d, it diverges for all $|x| > |d|$. ■

The significance of Theorem 1 is that a power series always behaves in exactly *one* of the following three ways (Fig. 12.5).

1. It converges at $x = 0$ and diverges everywhere else.
2. There is a positive number c such that the series diverges for $|x| > c$ but converges absolutely for $|x| < c$. It may or may not converge at either of the endpoints $x = c$ and $x = -c$.
3. It converges absolutely for every x.

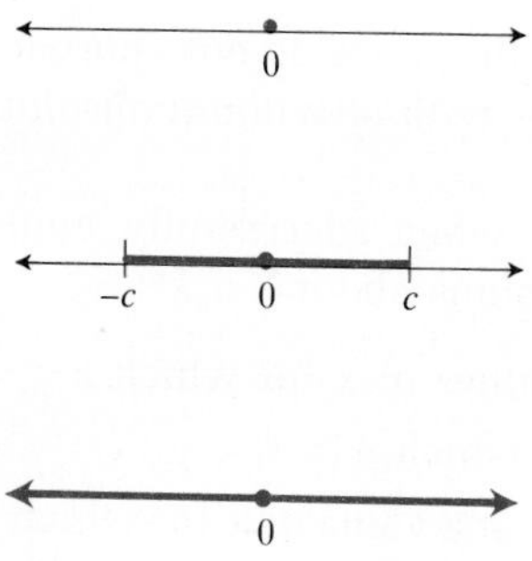

Figure 12.5

In Case 2, the set of points at which the series converges is a finite interval. We know from past examples that this interval may be open, half open, or closed, depending on the series in question. But no matter which kind of interval it is, c is called the *radius of convergence* of the series, and the convergence is absolute at every point in the interior of the interval. The interval is called the *interval of convergence*. If a power series converges absolutely for all values of x, we say that its radius of convergence is infinite. If it converges only at $x = 0$, we say that the radius of convergence is 0.

As examples of power series whose radii of convergence are infinite, we have the Taylor series of sin x, cos x, and e^x. These series converge for every value of $c = 2x$, and therefore converge absolutely for every value of x.

As examples of series whose radii of convergence are finite we have

Series	Interval of convergence
$\frac{1}{1-x} = 1 + x + x^2 + \cdots$	$-1 < x < 1$
$\ln(1+x) = x - \frac{x^2}{2} + \frac{x^3}{3} - \frac{x^4}{4} + \cdots$	$-1 < x \le 1$
$\tan^{-1} x = x - \frac{x^3}{3} + \frac{x^5}{5} - \cdots$	$-1 \le x \le 1$

The interval of convergence of a power series $\sum_{n=0}^{\infty} a_n x^n$ can often be found by applying the ratio test or the root test to the series of absolute values,

$$\sum_{n=0}^{\infty} |a_n x^n|.$$

Thus, if

$$\rho = \lim_{n\to\infty} \left| \frac{a_{n+1}x^{n+1}}{a_n x^n} \right| \quad \text{or if} \quad \rho = \lim_{n\to\infty} \sqrt[n]{|a_n x^n|},$$

then,

a) $\sum |a_n x^n|$ converges at all values of x for which $\rho < 1$,

b) $\sum |a_n x^n|$ diverges at all values of x for which $\rho > 1$,

c) $\sum |a_n x^n|$ may either converge or diverge at a value of x for which $\rho = 1$.

How do these three alternatives translate into statements about the series $\sum a_n x^n$? Case (a) says that $\sum a_n x^n$ converges absolutely at all values of x for which $\rho < 1$. Case (c) does not tell us anything more about the series $\sum a_n x^n$ than it does about the series $\sum |a_n x^n|$. Either series might converge or diverge at a value of x for which $\rho = 1$. In Case (b), we can actually conclude that $\sum a_n x^n$ diverges at all values of x for which $\rho > 1$. The argument goes like this: As you may recall from the discussions in Article 11.5, the fact that ρ is greater than 1 means that either

$$0 < |a_n x^n| < |a_{n+1}x^{n+1}| < |a_{n+2}x^{n+2}| < \cdots$$

or

$$\sqrt[n]{|a_n x^n|} > 1$$

for n sufficiently large. Thus the terms of the series do not approach 0 as n becomes infinite, and the series diverges *with or without absolute values*, by the nth-term test.

Therefore, the ratio and root tests, when successfully applied to $\sum |a_n x^n|$, lead us to the following conclusions about $\sum a_n x^n$:

A) $\sum a_n x^n$ converges absolutely for all values of x for which $\rho < 1$,

B) $\sum a_n x^n$ diverges at all values of x for which $\rho > 1$,

C) $\sum a_n x^n$ may either converge or diverge at a value of x for which $\rho = 1$.

In Case (C), another test is needed.

EXAMPLE 1 Find the interval of convergence of

$$\sum_{n=1}^{\infty} \frac{x^n}{n}. \tag{6}$$

Solution We apply the ratio test to the series of absolute values, and find

$$\rho = \lim_{n\to\infty} \left| \frac{x^{n+1}}{n+1} \cdot \frac{n}{x^n} \right| = |x|.$$

Therefore the original series converges absolutely if $|x| < 1$ and diverges if $|x| > 1$. When $x = +1$, the series becomes

$$1 + \frac{1}{2} + \frac{1}{3} + \frac{1}{4} + \cdots,$$

which diverges. When $x = -1$, the series becomes

$$-\left(1 - \frac{1}{2} + \frac{1}{3} - \frac{1}{4} + \cdots\right),$$

which converges, by Leibnitz's Theorem. Therefore the series (6) converges for $-1 \le x < 1$ and diverges for all other values of x. □

EXAMPLE 2 For what values of x does the series

$$\sum_{n=1}^{\infty} \frac{(2x-5)^n}{n^2}$$

converge?

Solution We treat the series as a power series in the variable $2x - 5$. An application of the root test to the series of absolute values yields

$$\rho = \lim_{n\to\infty} \sqrt[n]{\left| \frac{(2x-5)^n}{n^2} \right|} = \lim_{n\to\infty} \frac{|2x-5|}{\sqrt[n]{n^2}} = \frac{|2x-5|}{1} = |2x-5|.$$

The series converges absolutely for

$$|2x-5| < 1 \quad \text{or} \quad -1 < 2x - 5 < 1$$

or

$$4 < 2x < 6 \quad \text{or} \quad 2 < x < 3.$$

When $x = 2$, the series is $\sum_{n=1}^{\infty} [(-1)^n/n^2]$, which converges.

When $x = 3$, the series is $\sum_{n=1}^{\infty} [(1)^n/n^2]$, which converges. Therefore, the interval of convergence is $2 \le x \le 3$. □

Sometimes the comparison test does as well as any.

EXAMPLE 3 For what values of x does

$$\sum_{n=1}^{\infty} \frac{\cos^n x}{n!}$$

converge?

Solution For every value of x,

$$\left|\frac{\cos^n x}{n!}\right| \leq \frac{1}{n!}.$$

The series converges for every value of x. □

The next theorem says that a function defined by a power series has derivatives of all orders at every point in the interior of its interval of convergence. The derivatives can be obtained as power series by differentiating the terms of the original series. The first derivative is obtained by differentiating the terms of the original series once:

$$\frac{d}{dx}\sum_{n=0}^{\infty}(a_n x^n) = \sum_{n=0}^{\infty}\frac{d}{dx}(a_n x^n) = \sum_{n=0}^{\infty} n a_n x^{n-1}.$$

For the second derivative, the terms are differentiated again, and so on. We state the theorem without proof, and go directly to the examples.

THEOREM 2

The Term-by-Term Differentiation Theorem

If $f(x) = \sum_{n=0}^{\infty} a_n x^n$ has radius of convergence c, then,

1. $\sum_{n=0}^{\infty} n a_n x^{n-1}$ also has radius of convergence c,
2. $f(x)$ is differentiable on $(-c, c)$, and
3. $f'(x) = \sum_{n=0}^{\infty} n a_n x^{n-1}$ on $(-c, c)$.

Ostensibly, Theorem 2 mentions only f and f'. But because f' has the same radius of convergence that f has, the theorem applies equally well to f', saying that it has a derivative f'' on $(-c, c)$. This in turn implies that f'' is differentiable on $(-c, c)$, and so on. Thus, if $f(x) = \sum_{n=0}^{\infty} a_n x^n$ converges on $(-c, c)$, it has derivatives of all orders at every point of $(-c, c)$.

EXAMPLE 4 The relation $(d/dx)(\sin x) = \cos x$ is easily checked by differentiating the series for $\sin x$ term by term:

$$\sin x = x - \frac{x^3}{3!} + \frac{x^5}{5!} - \frac{x^7}{7!} + \cdots$$

$$\frac{d}{dx}(\sin x) = 1 - \frac{x^2}{2!} + \frac{x^4}{4!} - \frac{x^6}{6!} + \cdots = \cos x. \ \square$$

Convergence at one or both endpoints of the interval of convergence of a power series may be lost in the process of differentiation. That is why Theorem 2 mentions only the *open* interval $(-c, c)$.

EXAMPLE 5 The series $f(x) = \sum_{n=1}^{\infty}(x^n/n)$ of Example 1 converges for $-1 \leq x < 1$. The series of derivatives

$$f'(x) = \sum_{n=1}^{\infty} x^{n-1} = 1 + x + x^2 + x^3 + \cdots$$

is a geometric series that converges only for $-1 < x < 1$. The series diverges at the endpoint $x = -1$, as well as at the endpoint $x = 1$. □

Example 5 shows, however, that when the terms of a series are integrated, the resulting series may converge at an endpoint that was not a point of convergence before. The justification for term-by-term integration of a series is the following theorem, which we also state without proof.

THEOREM 3

The Term-by-Term Integration Theorem

If $f(x) = \sum_{n=0}^{\infty} a_n x^n$ has radius of convergence c, then,

1. $\displaystyle\sum_{n=0}^{\infty} \frac{a_n x^{n+1}}{n+1}$ also has radius of convergence c,

2. $\int f(x)\, dx$ exists for x in $(-c, c)$,

3. $\displaystyle\int f(x)\, dx = \sum_{n=0}^{\infty} \frac{a_n x^{n+1}}{n+1} + C$ on $(-c, c)$.

EXAMPLE 6 The series

$$\frac{1}{1+t} = 1 - t + t^2 - t^3 \cdots$$

converges on the open interval $-1 < t < 1$. Therefore,

$$\begin{aligned}\ln(1+x) &= \int_0^x \frac{1}{1+t}\, dt \\ &= t - \frac{t^2}{2} + \frac{t^3}{3} - \frac{t^4}{4} + \cdots \Big]_0^x, \\ &= x - \frac{x^2}{2} + \frac{x^3}{3} - \frac{x^4}{4} + \cdots, \qquad -1 < x < 1.\end{aligned}$$

As you know, the latter series also converges at $x = 1$, but that was not guaranteed by the theorem. □

EXAMPLE 7 By replacing t by t^2 in the series of Example 6, we obtain

$$\frac{1}{1+t^2} = 1 - t^2 + t^4 - t^6 + \cdots, \qquad -1 < t < 1.$$

Therefore

$$\begin{aligned}\tan^{-1} x &= \int_0^x \frac{1}{1+t^2}\, dt = t - \frac{t^3}{3} + \frac{t^5}{5} - \frac{t^7}{7} + \cdots \Big]_0^x \\ &= x - \frac{x^3}{3} + \frac{x^5}{5} - \frac{x^7}{7} + \cdots, \qquad -1 < x < 1.\end{aligned}$$

This is not as refined a result as the one we obtained in Article 12.3, where we were able to show that the interval of convergence was $-1 \le x \le 1$ by analyzing a remainder. But the result here is obtained more quickly. □

EXAMPLE 8 Express

$$\int \sin x^2 \, dx$$

as a power series.

Solution From the series for sin x we obtain

$$\sin x^2 = x^2 - \frac{x^6}{3!} + \frac{x^{10}}{5!} - \frac{x^{14}}{7!} + \cdots, \qquad -\infty < x < \infty.$$

Therefore,

$$\int \sin x^2 \, dx = C + \frac{x^3}{3} - \frac{x^7}{7 \cdot 3!} + \frac{x^{11}}{11 \cdot 5!} - \frac{x^{15}}{15 \cdot 7!} + \cdots, \qquad -\infty < x < \infty. \quad \square$$

EXAMPLE 9 Estimate $\int_0^1 \sin x^2 \, dx$ with an error of less than 0.001.

Solution From the indefinite integral in Example 8,

$$\int_0^1 \sin x^2 \, dx = \frac{1}{3} - \frac{1}{7 \cdot 3!} + \frac{1}{11 \cdot 5!} - \frac{1}{15 \cdot 7!} + \frac{1}{19 \cdot 9!} - \cdots.$$

The series alternates, and we find by trial that

$$\frac{1}{11 \cdot 5!} \approx 0.00076$$

is the first term to be numerically less than 0.001. The sum of the preceding two terms gives

$$\int_0^1 \sin x^2 \, dx \approx \frac{1}{3} - \frac{1}{42} \approx 0.310.$$

With two more terms we could estimate

$$\int_0^1 \sin x^2 \, dx \approx 0.310268$$

with an error of less than 10^{-6}; and with only one term beyond that we have

$$\int_0^1 \sin x^2 \, dx \approx \frac{1}{3} - \frac{1}{42} + \frac{1}{1320} - \frac{1}{75600} + \frac{1}{6894720} \approx 0.310268303,$$

with an error of less than 10^{-9}. To guarantee this accuracy with the error formula for the trapezoid rule would require using about 13,000 subintervals. □

EXAMPLE 10 Estimate $\int_0^{0.5} \sqrt{1 + x^4} \, dx$ with an error of less than 10^{-4}.

Solution The binomial expansion of $(1 + x^4)^{1/2}$ is

$$(1 + x^4)^{1/2} = 1 + \frac{1}{2}x^4 - \frac{1}{8}x^8 + \cdots,$$

a series whose terms alternate in sign after the second term. Therefore,

$$\begin{aligned}\int_0^{0.5} \sqrt{1+x^4}\,dx &= x + \frac{1}{2\cdot 5}x^5 - \frac{1}{8\cdot 9}x^9 + \cdots\Big]_0^{0.5}\\ &= 0.5 + 0.0031 - 0.00003 + \cdots\\ &\approx 0.5031,\end{aligned}$$

with an error of magnitude less than 0.00003. □

The Maclaurin Series for $\sum_{n=0}^{\infty} a_n x^n$

At the beginning of this article we asked whether a function

$$f(x) = \sum_{n=0}^{\infty} a_n x^n$$

defined by a convergent power series has a Taylor series. We can now answer that a function defined by a power series with a radius of convergence $c > 0$ has a Maclaurin series that converges to the function at every point of $(-c, c)$. Why? Because the Maclaurin series for the function $f(x) = \sum_{n=0}^{\infty} a_n x^n$ is the series $\sum_{n=0}^{\infty} a_n x^n$ itself. To see this, we differentiate

$$f(x) = a_0 + a_1 x + a_2 x^2 + \cdots + a_n x^n + \cdots$$

term by term and substitute $x = 0$ in each derivative $f^{(n)}(x)$. This produces

$$f^{(n)}(0) = n!\,a_n \qquad \text{or} \qquad a_n = \frac{f^{(n)}(0)}{n!}$$

for every n. Thus,

$$\begin{aligned}f(x) &= \sum_{n=0}^{\infty} a_n x^n\\ &= \sum_{n=0}^{\infty} \frac{f^{(n)}(0)}{n!} x^n, \qquad -c < x < c. \qquad (7)\end{aligned}$$

An immediate consequence of this is that series like

$$x \sin x = x^2 - \frac{x^4}{3!} + \frac{x^6}{5!} - \frac{x^8}{7!} + \cdots,$$

and

$$x^2 e^x = x^2 + x^3 + \frac{x^4}{2!} + \frac{x^5}{3!} + \cdots,$$

which are obtained by multiplying Maclaurin series by powers of x, as well as series obtained by integration and differentiation of power series, are themselves the Maclaurin series of the functions they represent.

Another consequence of (7) is that if two power series $\sum_{n=0}^{\infty} a_n x^n$ and $\sum_{n=0}^{\infty} b_n x^n$ are equal for all values of x in an open interval that contains the origin $x = 0$, then $a_n = b_n$ for every n. For if

$$\begin{aligned}f(x) &= \sum_{n=0}^{\infty} a_n x^n\\ &= \sum_{n=0}^{\infty} b_n x^n, \qquad -c < x < c,\end{aligned}$$

then a_n and b_n are both equal to $f^{(n)}(0)/n!$.

Multiplication of Power Series

We illustrate with an example.

EXAMPLE 11 Find the first five terms in the Maclaurin series expansion for $e^x \cos x$ by multiplying together the series for e^x and for $\cos x$.

Solution From Article 12.2, Eqs. (10) and (12), we have the series expansions

$$e^x = 1 + x + \frac{x^2}{2!} + \frac{x^3}{3!} + \frac{x^4}{4!} + \cdots,$$

$$\cos x = 1 - \frac{x^2}{2!} + \frac{x^4}{4!} - \cdots.$$

Obviously, if we need only terms involving x^n for $n \le 4$, we can truncate both series at their x^4-terms and multiply the resulting polynomials, discarding everything involving higher powers like $x^5, \ldots, x^8$. The result is

$$e^x \cos x = 1 + x + x^2\left(\frac{1}{2!} - \frac{1}{2!}\right) + x^3\left(\frac{1}{3!} - \frac{1}{2!}\right) + x^4\left(\frac{1}{4!} - \frac{1}{2!2!} + \frac{1}{4!}\right) + \cdots$$

$$= 1 + x - \frac{x^3}{3} - \frac{x^4}{6} + \cdots. \square \qquad (8)$$

We shall not prove it here, but using $\cos x = (e^{ix} + e^{-ix})/2$ we could establish the result

$$e^x \cos x = \frac{1}{2}\sum_{n=0}^{\infty} \frac{(1+i)^n + (1-i)^n}{n!} x^n = \sum_{n=0}^{\infty} \frac{(\sqrt{2})^n \cos(n\pi/4)}{n!} x^n, \qquad (9)$$

because

$$1 + i = \sqrt{2}\, e^{i\pi/4}, \qquad 1 - i = \sqrt{2}\, e^{-i\pi/4}.$$

REMARK. If $f(x) = u(x) + i\, v(x)$, where $u(x)$ and $v(x)$ are real valued functions of x, we call $u(x)$ the *real part* of $f(x)$ and write

$$u(x) = \operatorname{Re} f(x).$$

Similarly, $v(x)$ is called the *imaginary part* of $f(x)$ (that's right, even though v is itself real!) and write

$$v(x) = \operatorname{Im} f(x).$$

With this notation, and recalling that $e^{ix} = \cos x + i \sin x$, we see that

$$e^x \cos x = \operatorname{Re}(e^x e^{ix}) = \operatorname{Re}(e^{(1+i)x})$$

and

$$e^x \sin x = \operatorname{Im}(e^x e^{ix}) = \operatorname{Im}(e^{(1+i)x}).$$

From Eq. (9), it would be easy to show that the series for $e^x \cos x$ converges absolutely for all real values of x. This is also guaranteed by the following theorem, which we shall not prove.

THEOREM 4

The Series Multiplication Theorem for Power Series

If both $\Sigma\, a_n x^n$ and $\Sigma\, b_n x^n$ converge absolutely for $|x| < R$, and

$$\begin{aligned} c_n &= a_0 b_n + a_1 b_{n-1} + a_2 b_{n-2} + \cdots + a_{n-1} b_1 + a_n b_0 \\ &= \sum_{k=0}^{n} a_k b_{n-k}, \end{aligned} \tag{10a}$$

then the series $\Sigma\, c_n x^n$ also converges absolutely for $|x| < R$, and

$$\begin{aligned} &(a_0 + a_1 x + a_2 x^2 + \cdots) \cdot (b_0 + b_1 x + b_2 x^2 + \cdots) \\ &\qquad = c_0 + c_1 x + c_2 x^2 + \cdots. \end{aligned} \tag{10b}$$

Division of Series

Again, we illustrate with an example.

EXAMPLE 12 Find some of the terms in the Maclaurin series for tan x by dividing the series for sin x by the series for cos x.

Solution We proceed as in ordinary algebraic long division, keeping track of terms up to x^5 and disregarding all higher powers of x:

$$\begin{array}{rl} & x + \dfrac{x^3}{3} + \dfrac{2}{15}x^5 + \cdots = \tan x \\ \cos x = 1 - \dfrac{x^2}{2!} + \dfrac{x^4}{4!} - \cdots \Big) & x - \dfrac{x^3}{3!} + \dfrac{x^5}{5!} - \cdots = \sin x \\ & x - \dfrac{x^3}{2!} + \dfrac{x^5}{4!} \cdots \\ \hline & \dfrac{x^3}{3} - \dfrac{x^5}{30} \cdots \\ & \dfrac{x^3}{3} - \dfrac{x^5}{6} \cdots \\ \hline & \dfrac{2}{15}x^5 \cdots \\ & \dfrac{2}{15}x^5 \\ \hline \end{array}$$

To terms in x^5, we thus have

$$\tan x = \frac{\sin x}{\cos x} = x + \frac{x^3}{3} + \frac{2}{15}x^5 + \cdots. \tag{11}$$

We used these first few terms of the series for tan x in Article 12.4, Example 2. □

REMARK. Because $\cos(\pi/2) = 0$, we certainly cannot expect the Maclaurin series for tan x to converge outside the interval $|x| < \pi/2$. For the full Maclaurin series, which is not easy to obtain either by long division or by direct application of Taylor's formula, the reader is referred to page 75, Eq. 4.3.67, of the Handbook of Mathematical Functions, Applied Mathematics Series, of the U.S. Department of Commerce, National Bureau of Standards, AMS 55, edited by M. Abramowitz and I. A. Stegun, Dover Publications, Inc., New York.

There is another way to calculate the coefficients in the Maclaurin series for the quotient $f(x)/g(x)$, a way that is readily adaptable to machine computation. The facts are as follows:

1. If $f(x) = \Sigma\, a_n x^n$ for $|x| < R_1$, $g(x) = \Sigma\, b_n x^n$ for $|x| < R_2$, and $b_0 = g(0) \neq 0$, then $f(x)/g(x)$ has a power series representation $\Sigma\, c_n x^n$ on *some* interval $(-h, h)$.
2. Within that interval,

$$\begin{aligned}\Sigma\, a_n x^n &= f(x)\\ &= g(x) \cdot \frac{f(x)}{g(x)}\\ &= \Sigma\, b_n x^n \cdot \Sigma\, c_n x^n\\ &= \sum_n \left(\sum_{k=0}^{n} b_k c_{n-k} \right) x^n,\end{aligned}$$

and hence

$$a_n = \sum_{k=0}^{n} b_k c_{n-k}.$$

In other words,

$$\begin{aligned}a_0 &= b_0 c_0\\ a_1 &= b_0 c_1 + b_1 c_0, \qquad \text{etc.},\end{aligned}$$

so that the coefficients c_n can be found one after the other in this way:

$$c_0 = \frac{a_0}{b_0}, \qquad c_1 = \frac{a_1 - b_1 c_0}{b_0}, \tag{12a}$$

and, for all $n \geq 1$,

$$\begin{aligned}c_n &= \frac{a_n - (b_1 c_{n-1} + b_2 c_{n-2} + \cdots + b_n c_0)}{b_0}\\ &= \frac{a_n - \Sigma_{k=1}^{n} b_k c_{n-k}}{b_0}.\end{aligned} \tag{12b}$$

EXAMPLE 13 Let

$$f(x) = \sin x = x - \frac{x^3}{3!} + \frac{x^5}{5!} - \cdots$$

so that

$$a_0 = 0,\ a_1 = 1,\ a_2 = 0,\ a_3 = -\frac{1}{3!},\ a_4 = 0, \ldots$$

and let

$$g(x) = \cos x = 1 - \frac{x^2}{2!} + \frac{x^4}{4!} - \frac{x^6}{6!} + \cdots$$

so that

$$b_0 = 1,\ b_1 = 0,\ b_2 = -\frac{1}{2!},\ b_3 = 0,\ b_4 = \frac{1}{4!}, \text{ etc.}$$

Then

$$\tan x = \frac{\sin x}{\cos x} = c_0 + c_1 x + c_2 x^2 + \cdots$$

with

$$c_0 = \frac{a_0}{b_0} = 0, \; c_1 = \frac{a_1 - b_1 c_0}{b_0} = \frac{1 - 0}{1} = 1,$$

and so on. When we know the values of $c_0, c_1, \ldots, c_{n-1}$, the value of c_n is given by Eq. (12b) in terms of known coefficients. □

PROBLEMS

In Problems 1–20, find the interval of absolute convergence. If the interval is finite, determine whether the series converges at each endpoint.

1. $\sum_{n=0}^{\infty} x^n$
2. $\sum_{n=0}^{\infty} n^2 x^n$
3. $\sum_{n=1}^{\infty} \frac{n x^n}{2^n}$
4. $\sum_{n=0}^{\infty} \frac{(2x)^n}{n!}$
5. $\sum_{n=0}^{\infty} \frac{(-1)^n x^{2n+1}}{(2n+1)!}$
6. $\sum_{n=1}^{\infty} (-1)^{n-1} \frac{(x-1)^n}{n}$
7. $\sum_{n=0}^{\infty} \frac{n^2}{2^n} (x+2)^n$
8. $\sum_{n=0}^{\infty} \frac{x^{2n+1}}{2n+1}$
9. $\sum_{n=0}^{\infty} (-1)^n \frac{x^{2n+1}}{2n+1}$
10. $\sum_{n=1}^{\infty} \frac{(x-2)^n}{n^2}$
11. $\sum_{n=0}^{\infty} \frac{\cos nx}{2^n}$
12. $\sum_{n=1}^{\infty} \frac{2^n x^n}{n^5}$
13. $\sum_{n=0}^{\infty} \frac{x^n e^n}{n+1}$
14. $\sum_{n=1}^{\infty} \frac{(\cos x)^n}{n^n}$
15. $\sum_{n=0}^{\infty} n^n x^n$
16. $\sum_{n=0}^{\infty} \frac{(3x+6)^n}{n!}$
17. $\sum_{n=1}^{\infty} (-2)^n (n+1)(x-1)^n$
18. $\sum_{n=1}^{\infty} \frac{(-1)^{n+1}(x-2)^n}{n \cdot 2^n}$
19. $\sum_{n=0}^{\infty} \left(\frac{x^2-1}{2}\right)^n$
20. $\sum_{n=1}^{\infty} \frac{(x+3)^{n-1}}{n}$
21. Find the sum of the series in Problem 16.
22. When the series of Problem 19 converges, to what does it converge?
23. Use series to verify that

a) $\frac{d}{dx}(\cos x) = -\sin x$, b) $\int_0^x \cos t\, dt = \sin x$,

c) $y = e^x$ is a solution of the equation $y' = y$.

24. Obtain the Maclaurin series for $1/(1+x)^2$ from the series for $-1/(1+x)$.
25. Use the Maclaurin series for $1/(1-x^2)$ to obtain a series for $2x/(1-x^2)^2$.
26. Use the identity $\sin^2 x = (1 - \cos 2x)/2$ to obtain a series for $\sin^2 x$. Then differentiate this series to obtain a series for $2 \sin x \cos x$. Check that this is the series for $\sin 2x$.

(*Calculator*) In Problems 27–34, use series and a calculator to estimate each integral with an error of magnitude less than 0.001.

27. $\int_0^{0.2} \sin x^2\, dx$
28. $\int_0^{0.1} \tan^{-1} x\, dx$
29. $\int_0^{0.1} x^2 e^{-x^2}\, dx$
30. $\int_0^{0.1} \frac{\tan^{-1} x}{x}\, dx$
31. $\int_0^{0.4} \frac{1 - e^{-x}}{x}\, dx$
32. $\int_0^{0.1} \frac{\ln(1+x)}{x}\, dx$
33. $\int_0^{0.1} \frac{1}{\sqrt{1+x^4}}\, dx$
34. $\int_0^{0.25} \sqrt[3]{1+x^2}\, dx$
35. (*Calculator*) a) Obtain a power series for

$$\sinh^{-1} x = \int_0^x \frac{dt}{\sqrt{1+t^2}}.$$

b) Use the result of (a) to estimate $\sinh^{-1} 0.25$ to three decimal places.

36. (*Calculator*) Estimate $\int_0^1 \cos x^2\, dx$ with an error of less than one millionth.
37. Show by example that there are power series that converge only at $x = 0$.
38. Show by examples that the convergence of a series at an endpoint of its interval of convergence may be either conditional or absolute.
39. Let r be any positive number. Use Theorem 1 to show that if $\sum_{n=0}^{\infty} a_n x^n$ converges for $-r < x < r$, then it converges absolutely for $-r < x < r$.

40. Use the ratio test to show that the binomial series converges for $|x| < 1$. (This still does not show that the series converges to $(1 + x)^m$.)

41. Find terms through x^5 of the Maclaurin series for $e^x \sin x$ by appropriate multiplication. The series is the imaginary part of the series for

$$e^x \cdot e^{ix} = e^{(1+i)x}.$$

Use this fact to check your answer. For what values of x should the series for $e^x \sin x$ converge? Why?

42. Divide 1 by a sufficient number of terms of the Maclaurin series for cos x to obtain terms through x^4 in the Maclaurin series for sec x. Where do you think the resulting complete Maclaurin series should converge?

43. Integrate the first three nonzero terms of the Maclaurin series for tan t from 0 to x to get the first three nonzero terms in the Maclaurin series for ln sec x.

44. (Continuation) Another way to get some of the terms in the Maclaurin series for ln sec x is as follows: Let $\sec x = 1 + y$, so that $y = x^2/2 + 5x^4/24 + \cdots$ (from Problem 42). Then, for $|y| < 1$,

$$\ln \sec x = \ln (1 + y) = y - \frac{y^2}{2} + \cdots.$$

Neglecting powers of x higher than x^4, show that this also leads to

$$\ln \sec x = \frac{x^2}{2} + \frac{x^4}{12} + \cdots.$$

45. A circle of radius $r_1 = 1$ is inscribed in an equilateral triangle. A circle of radius r_2 passes through the vertices of that triangle and is inscribed in a square. A circle of radius r_3 passes through the vertices of that square and is inscribed in a regular pentagon. Continue in this fashion: a circle of radius r_{n-1} passes through the vertices of a regular polygon of n sides and is inscribed in a regular polygon of $n + 1$ sides.

a) Show that $r_2 = r_1 \sec (\pi/3)$, $r_3 = r_2 \sec (\pi/4)$, and, in general, that

$$r_n = r_{n-1} \sec \frac{\pi}{n+1}.$$

b) Next, show that

$$\ln r_n = \ln r_1 + \ln \sec \frac{\pi}{3} + \ln \sec \frac{\pi}{4} + \cdots + \ln \sec \frac{\pi}{n+1}.$$

c) Does r_n have a finite limit as n tends to infinity? (Suggestion: Compare the series $\sum_{n=3}^{\infty} \ln \sec (\pi/n)$ with the series $\sum_{n=3}^{\infty} 1/n^2$ using the limit comparison test.) You may wish to use the answer to Problem 44.

> ***Toolkit* programs**
> Taylor Series

REVIEW QUESTIONS AND EXERCISES

1. State Taylor's Theorem, with remainder.
2. It can be shown (though not very simply) that the function f defined by

$$f(x) = \begin{cases} 0 & \text{when } x = 0, \\ e^{-1/x^2} & \text{when } x \neq 0 \end{cases}$$

is everywhere continuous, together with its derivatives of all orders. At 0, the derivatives are all equal to 0.

a) Write the Taylor series expansion of f in powers of x.

b) What is the remainder $R_n(x, 0)$ for this function?

Does the Taylor series for f converge to $f(x)$ at some value of x different from zero? Give a reason for your answer.

3. If a Taylor series in powers of $x - a$ is to be used for the numerical evaluation of a function, what is necessary or desirable in the choice of a?
4. Describe a method that may be useful in finding $\lim_{x\to a} f(x)/g(x)$ if $f(a) = g(a) = 0$. Illustrate.
5. What tests may be used to find the interval of convergence of a power series? Do they also work at the endpoints of the interval? Illustrate with examples.

MISCELLANEOUS PROBLEMS

1. a) Find the expansion, in powers of x, of $x^2/(1 + x)$.

   b) Does the series expansion of $x^2/(1 + x)$ in powers of x converge when $x = 2$? (Give a brief reason.)
2. Obtain the Maclaurin series expansion for $\sin^{-1} x$ by integrating the series for $(1 - t^2)^{-1/2}$ from 0 to x. Find the intervals of convergence of these series.

3. Obtain the first four terms in the Maclaurin series for $e^{\sin x}$ by substituting the series for $y = \sin x$ in the series for e^y.

4. Assuming $|x| > 1$, obtain the expansions

$$\tan^{-1} x = \frac{\pi}{2} - \frac{1}{x} + \frac{1}{3x^3} - \frac{1}{5x^5} + \cdots, \qquad x > 1,$$

$$\tan^{-1} x = -\frac{\pi}{2} - \frac{1}{x} + \frac{1}{3x^3} - \frac{1}{5x^5} + \cdots, \qquad x < -1,$$

by integrating the series

$$\frac{1}{1+t^2} = \frac{1}{t^2} \cdot \frac{1}{1 + (1/t^2)} = \frac{1}{t^2} - \frac{1}{t^4} + \frac{1}{t^6} - \frac{1}{t^8} + \cdots$$

from $x(>1)$ to $+\infty$ or from $-\infty$ to $x\ (<-1)$.

5. a) Obtain the Maclaurin series, through the term in x^6, for $\ln(\cos x)$ by substituting the series for $y = 1 - \cos x$ in the series for $\ln(1 - y)$.
b) Use the result of part (a) to estimate

$$\int_0^{0.1} \ln(\cos x)\, dx$$

to five decimal places.

6. Compute $\int_0^1 [(\sin x)/x]\, dx$ to three decimal places.

7. Compute $\int_0^1 e^{-x^2}\, dx$ to three decimal places.

8. Expand the function $f(x) = \sqrt{1 + x^2}$ in powers of $(x - 1)$, obtaining three nonvanishing terms.

9. Expand the function $f(x) = 1/(1 - x)$ in powers of $(x - 2)$, and find the interval of convergence.

10. Expand $f(x) = 1/(x + 1)$ in powers of $(x - 3)$.

11. Expand $\cos x$ in powers of $(x - \pi/3)$.

12. Find the first three terms of the Taylor series expansion of the function $1/x$ about the point π.

13. Let f and g be functions satisfying the following conditions: (a) $f(0) = 1$, (b) $f'(x) = g(x)$, $g'(x) = f(x)$, (c) $g(0) = 0$. Estimate $f(1)$ to three decimal places.

14. Suppose $f(x) = \sum_{n=0}^{\infty} a_n x^n$. Prove that (a) if $f(x)$ is an even function, then $a_1 = a_3 = a_5 = \cdots = 0$; (b) if $f(x)$ is an odd function, then $a_0 = a_2 = a_4 = \cdots = 0$.

15. Find the first four terms (up to x^3) of the Maclaurin series of $f(x) = e^{(e^x)}$.

16. Estimate the error involved in using $x - x^2/2$ as an approximation to $\ln(1 + x)$ for values of x between 0 and 0.2, inclusive.

17. If $(1 + x)^{1/3}$ is replaced by $1 + x/3$ and $0 \le x \le \frac{1}{10}$, what estimate can be given for the error?

18. Use series to find

$$\lim_{x\to 0} \frac{\ln(1 - x) - \sin x}{1 - \cos^2 x}.$$

19. Find $\lim_{x\to 0} [(\sin x)/x]^{1/x^2}$.

20. Given a series of positive numbers a_n such that Σa_n converges. Does $\Sigma \ln(1 + a_n)$ converge? Explain.

In Problems 21–28, find the interval of convergence of each series. Test for convergence at the endpoints if the interval is finite.

21. $1 + \dfrac{x+2}{3 \cdot 1} + \dfrac{(x+2)^2}{3^2 \cdot 2} + \cdots + \dfrac{(x+2)^n}{3^n \cdot n} + \cdots$

22. $1 + \dfrac{(x-1)^2}{2!} + \dfrac{(x-1)^4}{4!} + \cdots + \dfrac{(x-1)^{2n-2}}{(2n-2)!} + \cdots$

23. $\displaystyle\sum_{n=1}^{\infty} \frac{x^n}{n^n}$

24. $\displaystyle\sum_{n=1}^{\infty} \frac{n!x^n}{n^n}$

25. $\displaystyle\sum_{n=0}^{\infty} \frac{n+1}{2n+1}\frac{(x-3)^n}{2^n}$

26. $\displaystyle\sum_{n=0}^{\infty} \frac{n+1}{2n+1}\frac{(x-2)^n}{3^n}$

27. $\displaystyle\sum_{n=1}^{\infty} \frac{(-1)^{n-1}(x-1)^n}{n^2}$

28. $\displaystyle\sum_{n=1}^{\infty} \frac{x^n}{\sqrt{n}}$

In Problems 29–31, determine *all* the values of x for which the series converge.

29. $\displaystyle\sum_{n=1}^{\infty} \frac{(x-2)^{3n}}{n!}$

30. $\displaystyle\sum_{n=1}^{\infty} \frac{2^n(\sin x)^n}{n^2}$

31. $\displaystyle\sum_{n=1}^{\infty} \frac{1}{n}\left(\frac{x-1}{x}\right)^n$

32. A function is defined by the power series

$$y = 1 + \frac{1}{6}x^3 + \frac{1}{180}x^6 + \cdots + \frac{1 \cdot 4 \cdot 7 \cdots (3n-2)}{(3n)!}x^{3n} + \cdots.$$

a) Find the interval of convergence of the series.
b) Show that there exist two constants a and b such that the function so defined satisfies a differential equation of the form $y'' = x^a y + b$.

33. If $a_n > 0$ and the series $\sum_{n=1}^{\infty} a_n$ converges, prove that $\sum_{n=1}^{\infty} a_n/(1 + a_n)$ converges.

34. If $1 > a_n > 0$ and $\sum_{n=1}^{\infty} a_n$ converges, prove that $\sum_{n=1}^{\infty} \ln(1 - a_n)$ converges. (*Hint:* First show that $|\ln(1 - a_n)| \le a_n/(1 - a_n)$.)

35. An infinite product, indicated by $\prod_{n=1}^{\infty} (1 + a_n)$, is said to converge if the series $\sum_{n=1}^{\infty} \ln(1 + a_n)$ converges. (The series is the natural logarithm of the product.) Prove that the product converges if every $a_n > -1$ and $\sum_{n=1}^{\infty} |a_n|$ converges. (*Hint:* Show that

$$|\ln(1 + a_n)| \le \frac{|a_n|}{1 - |a_n|} < 2|a_n|$$

when $|a_n| < \frac{1}{2}$.)

36. By multiplying the appropriate terms of the Maclaurin series for $\tan^{-1} x$ and $\ln(1 + x)$, find the terms through x^5 in the Maclaurin series for the product $(\tan^{-1} x) \cdot \ln(1 + x)$.

37. By appropriate division or multiplication of series, find the terms through x^5 in the Maclaurin series for $\tan^{-1} x/(1 - x)$.

38. When $f(0)$ and $g(0)$ are both zero, it may be possible to redefine the quotient $f(x)/g(x)$ as having the value $f'(0)/g'(0)$ provided $g'(0) \neq 0$. The result is equivalent to dividing both the numerator and the denominator by x in the series that represent f and g. For example, for all $x \neq 0$, we can write

$$\frac{x}{e^x - 1} = \frac{1}{1 + \dfrac{x}{2} + \dfrac{x^2}{6} + \dfrac{x^3}{24} + \cdots + \dfrac{x^n}{(n+1)!} + \cdots}.$$

By appropriate division or multiplication of series, find the coefficients c_0, c_1, and c_2 for the series representation $\Sigma\, c_n x^n$ of the function on the right.

13

Vectors

13.1

Vector Components and the Unit Vectors i and j.

Some physical quantities, like length and mass, are completely determined when their magnitudes are given in terms of specific units. Such quantities are called *scalars*. Other quantities, like forces and velocities, in which the direction as well as the magnitude is important, are called *vectors*. It is customary to represent a vector by a directed line segment whose direction represents the direction of the vector and whose length, in terms of some chosen unit, represents its magnitude.

We shall ordinarily deal with "free vectors," meaning that a vector is free to move about under parallel displacements. We say that two vectors are *equal* if they have the same direction and the same magnitude. See Fig. 13.1.

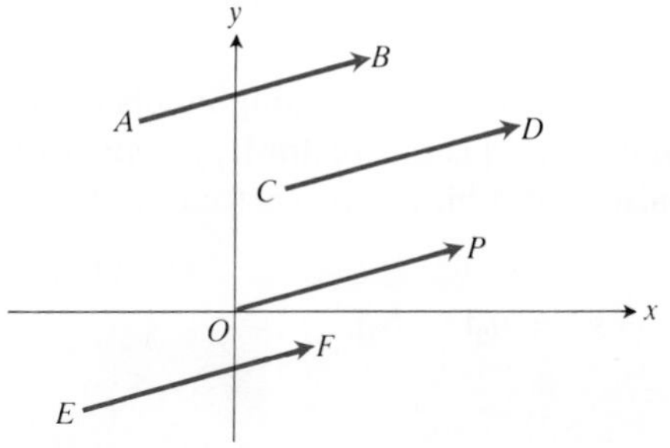

13.1 The arrows $\overrightarrow{AB}$, $\overrightarrow{CD}$, $\overrightarrow{OP}$, and $\overrightarrow{EF}$ shown here have the same direction and the same length. They all represent the same vector, and we write $\overrightarrow{AB} = \overrightarrow{CD} = \overrightarrow{OP} = \overrightarrow{EF}$.

Addition

Two vectors $\mathbf{v}_1$ and $\mathbf{v}_2$, may be added geometrically by drawing a representative of $\mathbf{v}_1$, say from A to B as in Fig. 13.2(a), and then a vector equal to $\mathbf{v}_2$ starting from the terminal point of $\mathbf{v}_1$.† In Fig. 13.2(a), $\mathbf{v}_2 = \overrightarrow{BC}$. The sum $\mathbf{v}_1 + \mathbf{v}_2$ is then the vector from the starting point A of $\mathbf{v}_1$ to the terminal point C of $\mathbf{v}_2$. That is, if

$$\mathbf{v}_1 = \overrightarrow{AB}, \quad \text{and} \quad \mathbf{v}_2 = \overrightarrow{BC},$$

then

$$\mathbf{v}_1 + \mathbf{v}_2 = \overrightarrow{AB} + \overrightarrow{BC} = \overrightarrow{AC}.$$

Whenever a vector $\mathbf{v}$ can be written as a sum

$$\mathbf{v} = \mathbf{v}_1 + \mathbf{v}_2,$$

the vectors $\mathbf{v}_1$ and $\mathbf{v}_2$ are called *components of* $\mathbf{v}$.

† In print, vectors are usually indicated by bold-faced Roman letters. In handwritten work it is customary to draw small arrows above letters that represent vectors, so: $\vec{v}$.

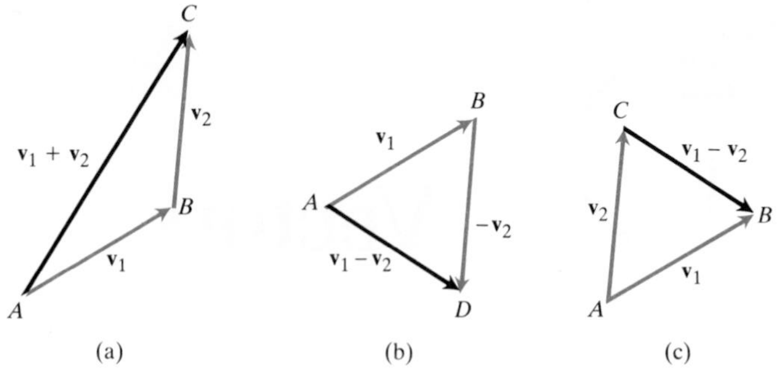

13.2 The sum (a) and difference (b and c) of two vectors $\mathbf{v}_1$ and $\mathbf{v}_2$.

The most satisfactory algebra of vectors is based on a representation of each vector in terms of components parallel to the axes of a cartesian coordinate system. This is accomplished by using the same unit of length on the two axes, with vectors of unit length along the axes used as *basic vectors* in terms of which the other vectors in the plane may be expressed.

Along the x-axis we choose the vector **i** from (0, 0) to (1, 0), as in Fig. 13.3. Along the y-axis we choose the vector **j** from (0, 0) to (0, 1). Then $a\mathbf{i}$, a being a scalar, represents a vector parallel to the x-axis, having magnitude $|a|$ and pointing to the right if a is positive, to the left if a is negative. Similarly, $b\mathbf{j}$ is a vector parallel to the y-axis and having the same direction as **j** if b is positive, or the opposite direction if b is negative. Figure 13.3 shows how the vector $\mathbf{v} = \overrightarrow{AC}$ is "resolved" into its x- and y- components as the sum

$$\mathbf{v} = a\mathbf{i} + b\mathbf{j}.$$

In this context it will also be convenient to call the numbers a and b *components* of **v**.

Components give us a way to define the equality of vectors algebraically:

DEFINITION

Equality of Vectors (Algebraic Definition)

$$a\mathbf{i} + b\mathbf{j} = a'\mathbf{i} + b'\mathbf{j} \Leftrightarrow a = a' \text{ and } b = b' \tag{1}$$

That is, two vectors are equal if and only if their corresponding components are equal. Thus, in Fig. 13.3, the vector $\overrightarrow{AB}$ and the vector $\overrightarrow{OP}$ from (0, 0) to $(a, 0)$ are both equal to $a\mathbf{i}$.

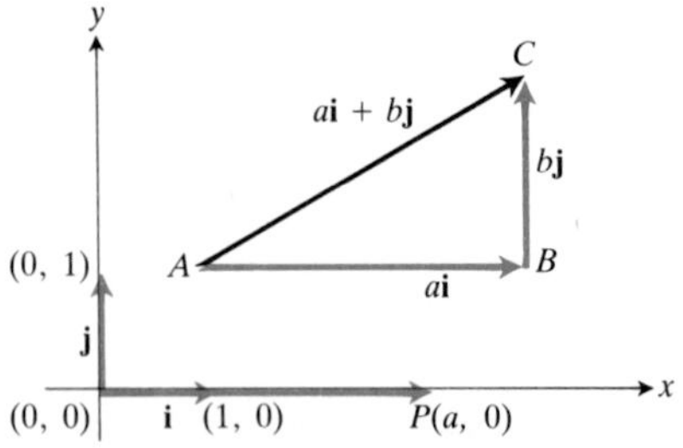

13.3 The vector $\overrightarrow{AC}$ expressed as a multiple of **i** plus a multiple of **j**.

Vectors may be added algebraically by adding their components in the following way. In Fig. 13.3, we see that $a\mathbf{i} + b\mathbf{j}$ is the vector hypotenuse of a right triangle whose vector sides are $a\mathbf{i}$ and $b\mathbf{j}$. If two vectors $\mathbf{v}_1$ and $\mathbf{v}_2$ are given in terms of components,

$$\mathbf{v}_1 = a_1\mathbf{i} + b_1\mathbf{j}, \qquad \mathbf{v}_2 = a_2\mathbf{i} + b_2\mathbf{j},$$

then

$$\mathbf{v}_1 + \mathbf{v}_2 = (a_1 + a_2)\mathbf{i} + (b_1 + b_2)\mathbf{j} \tag{2}$$

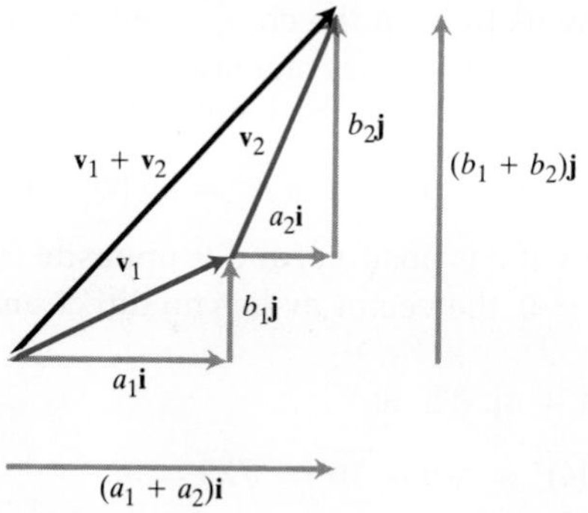

13.4 If $\mathbf{v}_1 = a_1\mathbf{i} + b_1\mathbf{j}$ and $\mathbf{v}_2 = a_2\mathbf{i} + b_2\mathbf{j}$, then $\mathbf{v}_1 + \mathbf{v}_2 = (a_1 + a_2)\mathbf{i} + (b_1 + b_2)\mathbf{j}$.

has x- and y-components obtained by adding the x- and y-components of $\mathbf{v}_1$ and $\mathbf{v}_2$. See Fig. 13.4.

EXAMPLE 1

$$(2\mathbf{i} - 4\mathbf{j}) + (5\mathbf{i} + 3\mathbf{j}) = (2 + 5)\mathbf{i} + (-4 + 3)\mathbf{j} = 7\mathbf{i} - \mathbf{j}. \ \square$$

Subtraction

The negative of a vector $\mathbf{v}$ is the vector $-\mathbf{v}$ that has the same length as $\mathbf{v}$ but points in the opposite direction. To subtract a vector $\mathbf{v}_2$ from a vector $\mathbf{v}_1$, we add $-\mathbf{v}_2$ to $\mathbf{v}_1$. This may be done geometrically by drawing $-\mathbf{v}_2$ from the tip of $\mathbf{v}_1$ and then drawing the vector from the initial point of $\mathbf{v}_1$ to the tip of $-\mathbf{v}_2$, as shown in Fig. 13.2(b), where

$$\overrightarrow{AD} = \overrightarrow{AB} + \overrightarrow{BD} = \mathbf{v}_1 + (-\mathbf{v}_2) = \mathbf{v}_1 - \mathbf{v}_2.$$

Another way to subtract $\mathbf{v}_2$ from $\mathbf{v}_1$ is to draw them both with a common initial point and then draw the vector from the tip of $\mathbf{v}_2$ to the tip of $\mathbf{v}_1$. This is illustrated in Fig. 13.2(c), where

$$\overrightarrow{CB} = \overrightarrow{CA} + \overrightarrow{AB} = -\mathbf{v}_2 + \mathbf{v}_1 = \mathbf{v}_1 - \mathbf{v}_2.$$

Thus, $\overrightarrow{CB}$ is the vector that when added to $\mathbf{v}_2$ gives $\mathbf{v}_1$:

$$\overrightarrow{CB} + \mathbf{v}_2 = (\mathbf{v}_1 - \mathbf{v}_2) + \mathbf{v}_2 = \mathbf{v}_1.$$

In terms of components, vector subtraction follows the simple algebraic law

$$\mathbf{v}_1 - \mathbf{v}_2 = (a_1 - a_2)\mathbf{i} + (b_1 - b_2)\mathbf{j}, \qquad (3)$$

which says that corresponding components are subtracted.

EXAMPLE 2

$$(6\mathbf{i} + 2\mathbf{j}) - (3\mathbf{i} - 5\mathbf{j}) = (6 - 3)\mathbf{i} + (2 - (-5))\mathbf{j} = 3\mathbf{i} + 7\mathbf{j}. \ \square$$

Length of a Vector

The length of the vector $\mathbf{v} = a\mathbf{i} + b\mathbf{j}$ is usually denoted by $|\mathbf{v}|$, which may be read "the magnitude of v." Figure 13.3 shows that $\mathbf{v}$ is the hypotenuse of a right triangle whose legs have lengths $|a|$ and $|b|$. Hence we may apply the theorem of Pythagoras to obtain

$$|\mathbf{v}| = |a\mathbf{i} + b\mathbf{j}| = \sqrt{a^2 + b^2}. \qquad (4)$$

EXAMPLE 3

$$|3\mathbf{i} - 5\mathbf{j}| = \sqrt{(3)^2 + (-5)^2} = \sqrt{9 + 25} = \sqrt{34}. \ \square$$

Multiplication by Scalars

The algebraic operation of multiplying a vector $\mathbf{v} = a\mathbf{i} + b\mathbf{j}$ by a scalar c is also simple, namely

$$c(a\mathbf{i} + b\mathbf{j}) = (ca)\mathbf{i} + (cb)\mathbf{j}. \qquad (5)$$

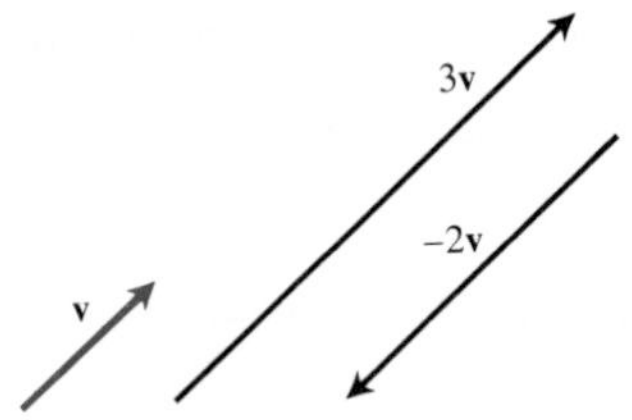

13.5 Scalar multiples of **v**.

Note how the unit vectors **i** and **j** allow us to keep the components separated from one another when we operate on vectors algebraically. Geometrically, $c\mathbf{v}$ is a vector whose length is $|c|$ times the length of **v**:

$$|c\mathbf{v}| = |(ca)\mathbf{i} + (cb)\mathbf{j}| = \sqrt{(ca)^2 + (cb)^2} = |c|\sqrt{a^2 + b^2} = |c|\,|\mathbf{v}|. \quad (6)$$

The direction of $c\mathbf{v}$ agrees with that of **v** if c is positive, and is opposite to that of **v** if c is negative (Fig. 13.5). If $c = 0$, the vector $c\mathbf{v}$ has no direction.

EXAMPLE 4 Let $c = 2$ and $\mathbf{v} = -3\mathbf{i} + 4\mathbf{j}$. Then,

$$|\mathbf{v}| = |-3\mathbf{i} + 4\mathbf{j}| = \sqrt{(-3)^2 + (4)^2} = \sqrt{9 + 16} = \sqrt{25} = 5,$$

and

$$|2\mathbf{v}| = |2(-3\mathbf{i} + 4\mathbf{j})| = |-6\mathbf{i} + 8\mathbf{j}| = \sqrt{(-6)^2 + (8)^2} = \sqrt{36 + 64}$$
$$= \sqrt{100} = 10 = 2|\mathbf{v}|.$$

If c had been -2 instead of 2, we would have found

$$|-2\mathbf{v}| = |-2(-3\mathbf{i} + 4\mathbf{j}) = |6\mathbf{i} - 8\mathbf{j}| = \sqrt{(6)^2 + (-8)^2} = \sqrt{100}$$
$$= 10 = |-2|\,|\mathbf{v}|. \ \square$$

Zero Vector

The vector

$$\mathbf{0} = 0\mathbf{i} + 0\mathbf{j} \quad (7)$$

is called the *zero vector*. It is the only vector whose length is zero, as we can see from the fact that

$$|a\mathbf{i} + b\mathbf{j}| = \sqrt{a^2 + b^2} = 0 \quad \Leftrightarrow \quad a = b = 0.$$

Unit Vector

Any vector **u** whose length is equal to the unit of length used along the coordinate axes is called a *unit vector*. If **u** is the unit vector obtained by rotating **i** through an angle θ in the positive direction, then (Fig. 13.6) **u** has a horizontal component

$$u_x = \cos\theta$$

and a vertical component

$$u_y = \sin\theta,$$

so that

$$\mathbf{u} = \mathbf{i}\cos\theta + \mathbf{j}\sin\theta. \quad (8)$$

If we allow the angle θ in Eq. (8) to vary from 0 to 2π, then the point P in Fig. 13.6 traces the unit circle $x^2 + y^2 = 1$ once in the counterclockwise direction. Every unit vector in the plane is given by Eq. (8) for some value of θ.

In physics it is common to denote unit vectors with small "hats," as in $\hat{u}$ (pronounced "u hat"). In hat notation, **i** and **j** become $\hat{\imath}$ and $\hat{\jmath}$.

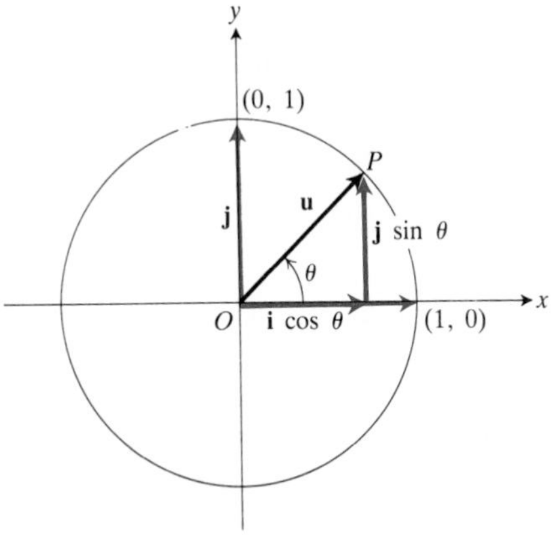

13.6 $\mathbf{u} = \mathbf{i}\cos\theta + \mathbf{j}\sin\theta$.

Direction

It is common in fields like classical electricity and magnetism, which use vectors a great deal, to define *direction* of a nonzero vector **A** to be the unit vector obtained by dividing **A** by its own length:

$$\text{Direction of } \mathbf{A} = \frac{\mathbf{A}}{|\mathbf{A}|}. \tag{9}$$

Instead of just saying that the unit vector $\mathbf{A}/|\mathbf{A}|$ *represents* the direction of **A**, we say that it *is* the direction of **A**.

To see that $\mathbf{A}/|\mathbf{A}|$ is indeed a unit vector, we can calculate its length directly:

$$\text{Length of } \frac{\mathbf{A}}{|\mathbf{A}|} = \left|\frac{\mathbf{A}}{|\mathbf{A}|}\right| = \left|\frac{1}{|\mathbf{A}|}\mathbf{A}\right| = \frac{1}{|\mathbf{A}|}|\mathbf{A}| = 1. \tag{10}$$

The zero vector **0** has no defined direction.

EXAMPLE 5 The direction of $\mathbf{A} = 3\mathbf{i} - 4\mathbf{j}$ is

$$\text{Direction of } \mathbf{A} = \frac{\mathbf{A}}{|\mathbf{A}|} = \frac{3\mathbf{i} - 4\mathbf{j}}{\sqrt{(3)^2 + (-4)^2}} = \frac{3\mathbf{i} - 4\mathbf{j}}{\sqrt{25}} = \frac{3}{5}\mathbf{i} - \frac{4}{5}\mathbf{j}.$$

To check the length, we can calculate

$$\left|\frac{3}{5}\mathbf{i} - \frac{4}{5}\mathbf{j}\right| = \sqrt{\left(\frac{3}{5}\right)^2 + \left(\frac{4}{3}\right)^2} = \sqrt{\frac{9}{25} + \frac{16}{25}} = \sqrt{\frac{25}{25}} = 1. \ \square$$

It follows from the definition of direction that two nonzero vectors **A** and **B** have the same direction if and only if

$$\frac{\mathbf{A}}{|\mathbf{A}|} = \frac{\mathbf{B}}{|\mathbf{B}|}, \tag{11}$$

or

$$\mathbf{A} = \frac{|\mathbf{A}|}{|\mathbf{B}|}\mathbf{B}. \tag{12}$$

Thus, if **A** and **B** have the same direction, then **A** is a positive scalar multiple of **B.** Conversely, if

$$\mathbf{A} = k\mathbf{B}, \qquad k > 0,$$

then

$$\frac{\mathbf{A}}{|\mathbf{A}|} = \frac{k\mathbf{B}}{|k\mathbf{B}|} = \frac{k}{|k|}\frac{\mathbf{B}}{|\mathbf{B}|} = \frac{k}{k}\frac{\mathbf{B}}{|\mathbf{B}|} = \frac{\mathbf{B}}{|\mathbf{B}|}. \tag{13}$$

Therefore, two nonzero vectors **A** and **B** point in the same direction if and only if **A** is a positive scalar multiple of **B.**

We say that two nonzero vectors **A** and **B** point in *opposite* directions if their directions are opposite in sign:

$$\frac{\mathbf{A}}{|\mathbf{A}|} = -\frac{\mathbf{B}}{|\mathbf{B}|} \tag{14}$$

From this it follows (Problem 22) that **A** and **B** have opposite directions if and only if **A** is a negative scalar multiple of **B.**

EXAMPLE 6

a) Same direction: $\mathbf{A} = 3\mathbf{i} - 4\mathbf{j}$ and $\mathbf{B} = \frac{3}{2}\mathbf{i} - 2\mathbf{j} = \frac{1}{2}\mathbf{A}$.

b) Opposite directions: $\mathbf{A} = 3\mathbf{i} - 4\mathbf{j}$ and $\mathbf{B} = -\mathbf{i} + \frac{4}{3}\mathbf{j} = -\frac{1}{3}\mathbf{A}$. □

Vectors Tangent and Normal to a Curve

Two vectors are said to be *parallel* if they are scalar multiples of each other, or the line segments representing them are parallel. Similarly, a vector is parallel to a line if the segments that represent the vector are parallel to the line. When we talk of a vector's being *tangent* or *normal* to a curve at a point, we mean that the vector is parallel to the line that is tangent or normal (perpendicular) to the curve at the point. The next example shows how such a vector may be found.

EXAMPLE 7 Find unit vectors tangent and normal to the curve $y = (x^3/2) + \frac{1}{2}$ at the point (1, 1).

Solution The slope of the line tangent to the curve at the point (1, 1) is

$$y' = \frac{3x^2}{2}\bigg|_{x=1} = \frac{3}{2}.$$

We find a unit vector with this slope. The vector $\mathbf{v} = 2\mathbf{i} + 3\mathbf{j}$ has slope $\frac{3}{2}$ (Fig. 13.7), as does every nonzero multiple of $\mathbf{v}$. To find a multiple of $\mathbf{v}$ that is a unit vector, we divide $\mathbf{v}$ by its length,

$$|\mathbf{v}| = \sqrt{2^2 + 3^2} = \sqrt{13}.$$

This produces the unit vector

$$\mathbf{u} = \frac{\mathbf{v}}{|\mathbf{v}|} = \frac{2}{\sqrt{13}}\mathbf{i} + \frac{3}{\sqrt{13}}\mathbf{j}.$$

The vector $\mathbf{u}$ is tangent to the curve at (1, 1) because it has the same direction as $\mathbf{v}$. Of course, the vector

$$-\mathbf{u} = -\frac{2}{\sqrt{13}}\mathbf{i} - \frac{3}{\sqrt{13}}\mathbf{j},$$

which points in the direction opposite to $\mathbf{u}$, is also tangent to the curve at (1, 1). Without some additional requirement, there is no reason to prefer one of these vectors to the other.

To find unit vectors normal to the curve at (1, 1), we look for unit vectors whose slopes are the negative reciprocal of the slope of $\mathbf{u}$. This is quickly done by interchanging the components of $\mathbf{u}$ and changing the sign of one of them. We obtain

$$\mathbf{n} = \frac{3}{\sqrt{13}}\mathbf{i} - \frac{2}{\sqrt{13}}\mathbf{j}, \quad \text{and} \quad -\mathbf{n} = -\frac{3}{\sqrt{13}}\mathbf{i} + \frac{2}{\sqrt{13}}\mathbf{j}.$$

Again, either one will do. The vectors have opposite directions, but both are normal to the curve at the point (1, 1). □

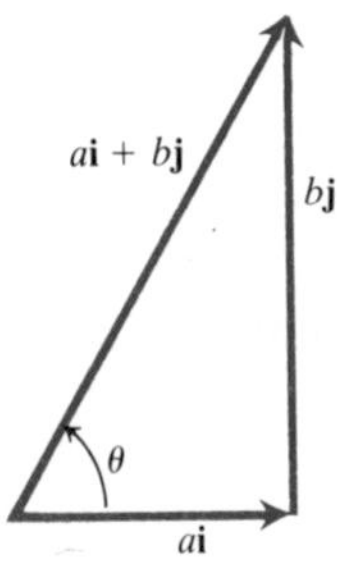

13.7 If $a \neq 0$, the vector $a\mathbf{i} + b\mathbf{j}$ has slope $b/a = \tan\theta$.

Vectors that are tangent and normal to curves in space will be discussed in the next chapter.

PROBLEMS

1. The three vectors **A**, **B**, and **C** shown below lie in a plane. Copy them on a sheet of paper.

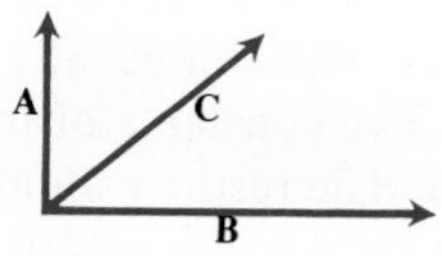

Then, by arranging the vectors in head-to-tail fashion, sketch
   a) **A** + **B**, to which then add **C**
   b) **A** + **C**, to which then add **B**
   c) **B** + **C**, to which then add **A**
   d) **A** − **B**, to which then add **C**
   e) **A** + **C**, from which then subtract **B**

In Problems 2–11 express each of the vectors in the form $a\mathbf{i} + b\mathbf{j}$. Indicate all quantities graphically.

2. $\overrightarrow{P_1P_2}$ if P_1 is the point (1, 3) and P_2 is the point (2, −1).
3. $\overrightarrow{OP_3}$ if O is the origin and P_3 is the midpoint of the vector $\overrightarrow{P_1P_2}$ joining $P_1(2, -1)$ and $P_2(-4, 3)$.
4. The vector from the point $A(2, 3)$ to the origin.
5. The sum of the vectors $\overrightarrow{AB}$ and $\overrightarrow{CD}$, given the four points $A(1, -1)$, $B(2, 0)$, $C(-1, 3)$, and $D(-2, 2)$.
6. A unit vector making an angle of 30° with the positive x-axis.
7. The unit vector obtained by rotating **j** through 120° in the clockwise direction.
8. A unit vector having the same direction as the vector $3\mathbf{i} - 4\mathbf{j}$.
9. A unit vector tangent to the curve $y = x^2$ at the point (2, 4).
10. A unit vector normal to the curve $y = x^2$ at the point $P(2, 4)$ and pointing from P toward the concave side of the curve (that is, an "inner" normal).
11. a) The unit vector tangent to the curve $y = x^2 + 2x$ at the point (1, 3) that points toward the x-axis.
    b) The unit vector normal to the curve at the point (1, 3) obtained by rotating the unit tangent vector found in part (a) 90° clockwise.
12. Show the unit vectors given by Eq. (8) for $\theta = 0, \pi/4, \pi/2, 2\pi/3, 5\pi/4$, and $5\pi/3$, together with a graph of the circle $x^2 + y^2 = 1$.

In Problems 13–18, find the length and direction of the vectors, and the angle that each makes with the positive x-axis.

13. $\mathbf{i} + \mathbf{j}$
14. $2\mathbf{i} - 3\mathbf{j}$
15. $\sqrt{3}\mathbf{i} + \mathbf{j}$
16. $-2\mathbf{i} + 3\mathbf{j}$
17. $5\mathbf{i} + 12\mathbf{j}$
18. $-5\mathbf{i} - 12\mathbf{j}$
19. The *speed* of a particle moving with velocity **v** is defined to be $|\mathbf{v}|$, the length (or magnitude) of **v**. Find the particle's speed if its velocity is $\mathbf{v} = -4\mathbf{i} + 2\mathbf{j}$.
20. Show that $\mathbf{A} = 3\mathbf{i} + 6\mathbf{j}$ and $\mathbf{B} = -\mathbf{i} - 2\mathbf{j}$ have opposite directions. Sketch them together.
21. Show that $\mathbf{C} = 3\mathbf{i} + 6\mathbf{j}$ and $\mathbf{D} = \frac{1}{2}\mathbf{i} + \mathbf{j}$ have the same direction.
22. Show that any two nonzero vectors **A** and **B** have opposite directions if and only if **A** is a negative scalar multiple of **B**. (*Hint:* See Eq. 14.)
23. Let A, B, C, D be the vertices, in order, of a quadrilateral. Let A', B', C', D' be the midpoints of the sides AB, BC, CD, and DA, in order. Prove that $A'B'C'D'$ is a parallelogram. (*Hint:* First show that $\overrightarrow{A'B'} = \overrightarrow{D'C'} = \frac{1}{2}\overrightarrow{AC}$.)
24. Show that the diagonals of a parallelogram bisect each other. (*Method:* Let A be one vertex and let M and N be the midpoints of the diagonals. Then show that $\overrightarrow{AM} = \overrightarrow{AN}$.)

13.2
Modeling Projectile Motion

In newtonian mechanics, the motion of a particle in a plane is usually described by a pair of differential equations

$$F_x = m\frac{dv_x}{dt}, \qquad F_y = m\frac{dv_y}{dt} \tag{1}$$

that express Newton's second law of motion

$$\mathbf{F} = m\mathbf{a} = m\frac{d\mathbf{v}}{dt} = m\frac{dv_x}{dt}\mathbf{i} + m\frac{dv_y}{dt}\mathbf{j} \tag{2}$$

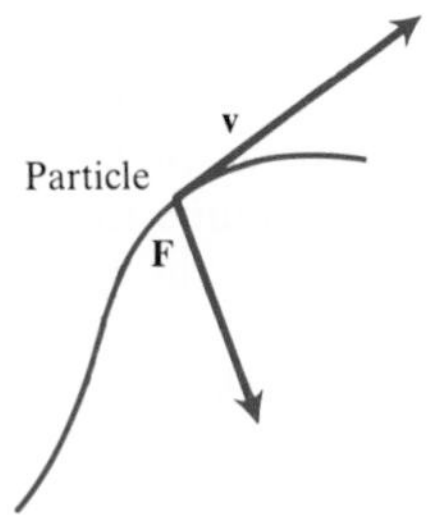

The force and velocity vectors of the motion of a particle in a plane might look like this at a particular time t.

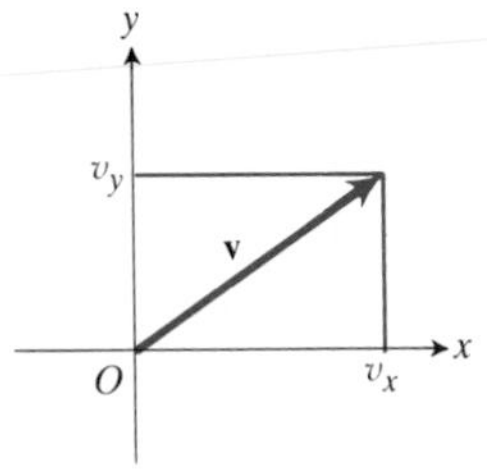

The components v_x and v_y of **v**.

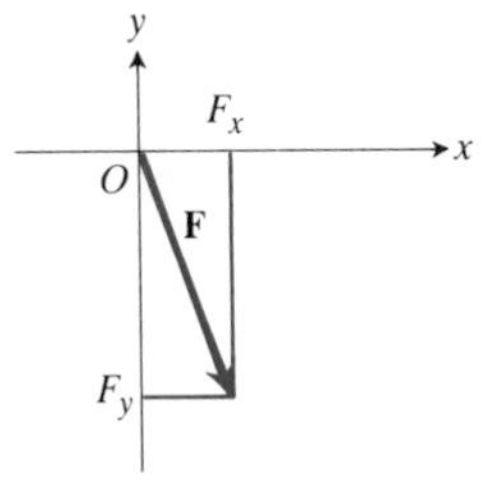

The components F_x and F_y of **F**.

13.8 Motion of a particle in a plane.

in parametric form. Here, **F** is a vector that represents a force acting on a particle of mass m at time t. The vector $\mathbf{v} = v_x\mathbf{i} + v_y\mathbf{j}$ is the velocity vector of the particle at time t (Fig. 13.8). The quantities F_x and F_y are the x- and y-components of **F**, while v_x and v_y are the components of **v**.

If we know the position and velocity of the particle at some given instant, then the position of the particle at any later time can usually be found by integrating Eqs. (1) with respect to time. The constants of integration are determined from given initial conditions. The result is another pair of parametric equations

$$x = f(t), \qquad y = g(t) \tag{3}$$

that give the coordinates x and y of the particle as functions of t.

The equations in (3) contain more information about the motion of the particle than the cartesian equation

$$y = F(x) \tag{4}$$

that we get from (3) by eliminating t. The parametric equations tell where the particle goes and *when* it gets to any given place, whereas the cartesian equation only tells the curve along which the particle travels. (Sometimes, too, a parametric representation of a curve is all that is possible; that is, a parameter cannot always be eliminated in practice.)

EXAMPLE 1 A projectile is fired with an initial velocity v_0 ft/s at an angle of elevation α. Assuming that gravity is the only force acting on the projectile, find parametric equations that give the coordinates of the projectile's position at any time t.

Solution We introduce coordinate axes with the origin at the point where the projectile begins its motion (Fig. 13.9). The distance traveled by the projectile over the ground is measured along the x-axis, and the height of the projectile above the ground is measured along the y-axis. At any time t, the projectile's position is given by a coordinate pair $x(t)$, $y(t)$, that we assume to be differentiable functions of t. If we measure distance in feet and time in seconds, with $t = 0$ at the instant the projectile is fired, then

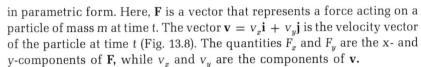

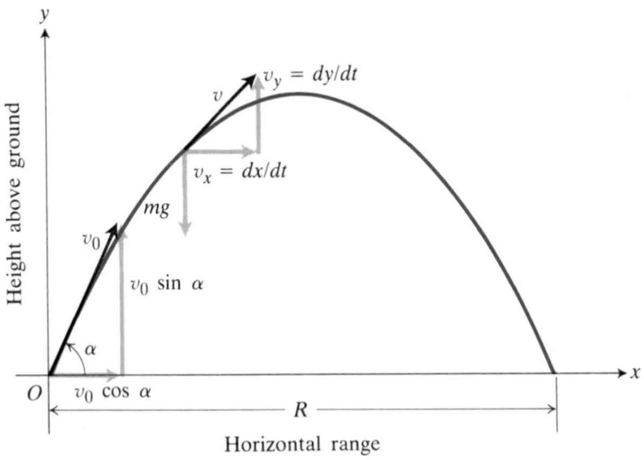

13.9 Ideal projectiles move along parabolas. The schematic figure here shows the lengths of the velocity vector and its components, first at time $t = 0$, and then later in the flight.

the initial conditions for the projectile's motion are

$$t = 0 \text{ sec}, \qquad x = 0 \text{ ft}, \qquad y = 0 \text{ ft},$$
$$v_x(0) = \left.\frac{dx}{dt}\right|_{t=0} = v_0 \cos\alpha \quad \text{ft/s}, \qquad v_y(0) = \left.\frac{dy}{dt}\right|_{t=0} = v_0 \sin\alpha \quad \text{ft/s}. \tag{5}$$

If the projectile is to travel only a few miles, and not go very high, it will cause no serious error to model the force of gravity with a constant vector **F** that points straight down. Its x- and y-components are

$$F_x = 0 \text{ lb}, \qquad F_y = -mg \text{ lb},$$

where m is the mass of the projectile and $g = 32$ ft/s is the acceleration of gravity. With these values for F_x and F_y, the equations in (1) become

$$0 = m\frac{d^2x}{dt^2}, \qquad -mg = m\frac{d^2y}{dt^2}. \tag{6}$$

To solve these equations for x and y, we integrate each one twice. This introduces four constants of integration, which may be evaluated by using the initial conditions (5). From the first equation in (6), we get

$$\frac{d^2x}{dt^2} = 0 \text{ ft/s}^2, \qquad \frac{dx}{dt} = c_1 \text{ ft/s}, \qquad x = c_1 t + c_2 \text{ ft}, \tag{7a}$$

and from the second equation in (6),

$$\frac{d^2y}{dt^2} = -g \text{ ft/s}^2, \qquad \frac{dy}{dt} = -gt + c_3 \text{ ft/s},$$
$$y = -\frac{1}{2}gt^2 + c_3 t + c_4 \text{ ft}. \tag{7b}$$

From the initial conditions, we find

$$c_1 = v_0 \cos\alpha \text{ ft/s}, \qquad c_2 = 0 \text{ ft}, \qquad c_3 = v_0 \sin\alpha \text{ ft/s}, \qquad c_4 = 0 \text{ ft}. \tag{7c}$$

The position of the projectile t seconds after firing is

$$x = (v_0 \cos\alpha)t \text{ ft}, \qquad y = -\frac{1}{2}gt^2 + (v_0 \sin\alpha)t \text{ ft}. \ \square \tag{8}$$

Height, Range, and Angle of Elevation

For a given angle of elevation α and a given muzzle velocity v_0, the position of the projectile in Example 1 at any time may be determined from the equations in (8). These equations may also be used to answer such questions as

1. How high does the projectile rise? When is it at its highest?
2. How far away does the projectile land, and how does the horizontal range R vary with the angle of elevation?
3. What angle of elevation gives the maximum range?

First, the projectile will reach its highest point when its y-(vertical) velocity component is zero, that is, when

$$\frac{dy}{dt} = -gt + v_0 \sin\alpha = 0 \text{ ft/s}$$

or

$$t = t_m = \frac{v_0 \sin \alpha}{g} \text{ sec.}$$

For this value of t, the value of y is

$$y_{\max} = -\frac{1}{2}g(t_m)^2 + (v_0 \sin \alpha)t_m = \frac{(v_0 \sin \alpha)^2}{2g} \text{ ft.}$$

Second, to find R we first find the time when the projectile strikes the ground. That is, we find the value of t for which $y = 0$. We then find the value of x for this value of t:

$$y = t\left(-\frac{1}{2}gt + v_0 \sin \alpha\right) = 0 \text{ ft}$$

when

$$t = 0 \quad \text{or} \quad t = \frac{2v_0 \sin \alpha}{g} = 2t_m \text{ sec.}$$

Since $t = 0$ is the instant when the projectile is fired, $t = 2t_m$ is the time when the projectile hits the ground. The corresponding value of x is

$$R = (v_0 \cos \alpha)(2t_m) = v_0 \cos \alpha \frac{2v_0 \sin \alpha}{g} = \frac{v_0^2}{g} \sin 2\alpha \text{ ft.}$$

Finally, the formula just given for R shows that the maximum range for a given muzzle velocity is obtained when $\sin 2\alpha = 1$, or $\alpha = 45°$.

A cartesian equation for the path of the projectile is readily obtained from (8). We need only substitute

$$t = \frac{x}{v_0 \cos \alpha}$$

from the first into the second equation of (8) to eliminate t and obtain

$$y = -\left(\frac{g}{2v_0^2 \cos^2 \alpha}\right)x^2 + (\tan \alpha)x. \tag{9}$$

Since this equation is linear in y and quadratic in x, it represents a *parabola*. Thus the path of a projectile (neglecting air resistance) is a parabola.

Differential equations of motion that take air resistance into account are usually too complicated for straightforward integration. The M.I.T. differential analyzers (early computers) were used to solve such equations during World War II to build up "range tables." In following moving targets, keeping track of *time* is of great importance, so that equations or tables that give x and y in terms of t are preferred over other forms.

Position Vectors

The vector $\mathbf{R}(t)$ from the origin to the position $(x(t), y(t))$ that the projectile in Example 1 has at time t is called the *position vector* of the projectile. It tells where the projectile is at time t. We can describe the motion in terms of this vector by writing the single vector equation

$$\mathbf{R}(t) = \mathbf{i}(v_0 \cos \alpha)t + \mathbf{j}\left(-\frac{1}{2}gt^2 + (v_0 \sin \alpha)t\right). \quad (10)$$

This formulation has the advantage that we can immediately write down the velocity and acceleration vectors of the motion by differentiating $\mathbf{R}(t)$ component by component with respect to t. We shall show how this is done in Chapter 14, where our study of vector functions and their derivatives will lead to Kepler's laws of planetary motion.

Other Motion

Not all motion can be considered to be projectile motion, and it is therefore a good idea to consider parametric equations in a less specific setting, as we do in the next example.

EXAMPLE 2 Find parametric equations for the plane curve traced by the point $P(x, y)$ if

$$\frac{dx}{dt} = 1 + x^2, \qquad \frac{dy}{dt} = \cos t, \qquad 0 \le t \le \frac{\pi}{4} \qquad (11)$$

and

$$x = 1, \qquad y = \sqrt{2} \qquad \text{when } t = \frac{\pi}{4}.$$

Solution The general rule for solving pairs of differential equations like these is to solve first for one coordinate in terms of t, then for the other. In this case we may begin with either equation, say with

$$\frac{dx}{dt} = 1 + x^2.$$

We separate the variables, as in Article 4.2, and integrate both sides of the resulting equation with respect to t to get

$$\frac{1}{1 + x^2}\frac{dx}{dt} = 1, \qquad \int \frac{1}{1 + x^2}\frac{dx}{dt}\,dt = \int dt, \qquad \tan^{-1} x = t + C.$$

We then determine C from the condition that $x = 1$ when $t = \pi/4$:

$$\tan^{-1} 1 = \frac{\pi}{4} + C, \qquad C = 0.$$

This gives $\tan^{-1} x = t$ or $x = \tan t$.

To express y in terms of t we integrate the second equation in (11):

$$\frac{dy}{dt} = \cos t, \qquad y = \sin t + C_1.$$

From the condition that $y = \sqrt{2}$ when $t = \pi/4$ we find

$$\sqrt{2} = \sin \frac{\pi}{4} + C_1, \qquad C_1 = \frac{\sqrt{2}}{2}.$$

Thus,

$$y = \sin t + \frac{\sqrt{2}}{2}.$$

The equations we seek are

$$x = \tan t, \quad \text{and} \quad y = \sin t + \frac{\sqrt{2}}{2}.$$

To check our solution (always a good idea) we differentiate both equations with respect to t,

$$\frac{dx}{dt} = \frac{d}{dt}(\tan t) = \sec^2 t = 1 + \tan^2 t = 1 + x^2,$$

$$\frac{dy}{dt} = \frac{d}{dt}\left(\sin t + \frac{\sqrt{2}}{2}\right) = \cos t,$$

to see that the original equations in (11) are satisfied. □

PROBLEMS

In Problems 1–4, the projectile is assumed to obey the laws of motion discussed above, in which air resistance is neglected.

1. Find two values of the angle of elevation that will enable a projectile to reach a target on the same level as the gun and 25,000 feet distance from it if the initial velocity is 1000 ft/s. Determine the times of flight corresponding to these two angles.

2. Show that doubling the initial velocity of a projectile multiplies both the maximum height and the range by a factor of four.

3. Show that a projectile attains three quarters of its maximum height in one half the time required to reach that maximum.

4. Suppose a target moving at the constant rate of a ft/s is level with and b ft away from a gun at the instant the gun is fired. If the target moves in a horizontal line directly away from the gun, show that the muzzle velocity v_0 and angle of elevation α must satisfy the equation.

$$v_0^2 \sin 2\alpha - 2av_0 \sin \alpha - bg = 0$$

if the projectile is to strike the target.

5. A human cannonball is to be fired with an initial speed of $v_0 = 80\sqrt{10/3}$ ft/s. The circus performer (of the right caliber, naturally) hopes to land on a special cushion located 200 ft down range. The circus is being held in a large room with a flat ceiling 75 ft high. Can the performer be fired to the cushion without striking the ceiling? If so, what should the cannon's angle of elevation be?

In Problems 6–10, find parametric equations for the curve traced by the point $P(x, y)$ if its coordinates satisfy the given differential equations and conditions. Sketch the curve. It may help to find a cartesian equation for the curve before you sketch.

6. $\dfrac{dx}{dt} = x, \quad \dfrac{dy}{dt} = y; \quad x = 1$, and $y = m$ (a constant) when $t = 0$

7. $\dfrac{dx}{dt} = x, \quad \dfrac{dy}{dt} = -y; \quad x > 0,\ y > 0$, and $x = 1,\ y = 2$ when $t = 0$

8. $\dfrac{dx}{dt} = x, \quad \dfrac{dy}{dt} = -x^2; \quad x > 0$, and $x = 1,\ y = -4$ when $t = 0$
(*Hint:* First solve for x in terms of t. Then use this result in finding y.)

9. $\dfrac{dx}{dt} = y, \quad \dfrac{dy}{dt} = y^2; \quad t < 1$, and $x = 0,\ y = 1$ when $t = 0$

10. $\dfrac{dx}{dt} = e^{-x}, \quad \dfrac{dy}{dt} = tx; \quad t > 0$, and $x = 0,\ y = 0$ when $t = 1$

11. a) Find parametric equations for the curve traced by the point $P(x, y)$ if

$$\frac{dx}{dt} = \sqrt{1 - x^2}, \qquad \frac{dy}{dt} = x,$$

and $x = 0,\ y = -1$ when $t = 0$.

b) Find a cartesian equation for the curve.

c) What are the coordinates of P when $t = \pi/2$? $t = \pi$?

12. Find parametric equations for the curve traced by the point $P(x, y)$ if

$$\frac{dx}{dt} = \sqrt{1 - x^2}, \qquad \frac{dy}{dt} = x^2,$$

and $x = 0,\ y = 1$ when $t = 0$. (*Hint:* Express x in terms of t before attempting to solve for y.)

***Toolkit* programs**	
Parametric Equations	Super * Grapher

13.3 Parametric Equations in Analytic Geometry

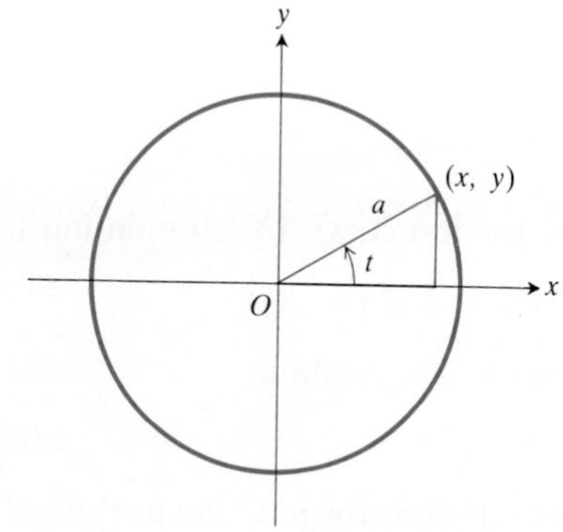

13.10 Circle: $x = a \cos t$, $y = a \sin t$.

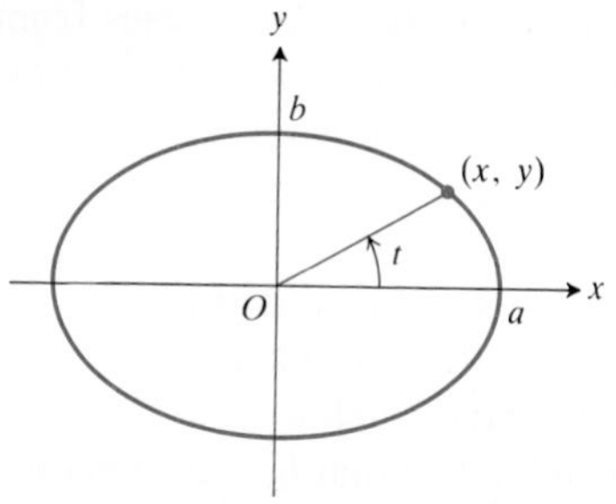

13.11 Ellipse: $x = a \cos t$, $y = b \sin t$.

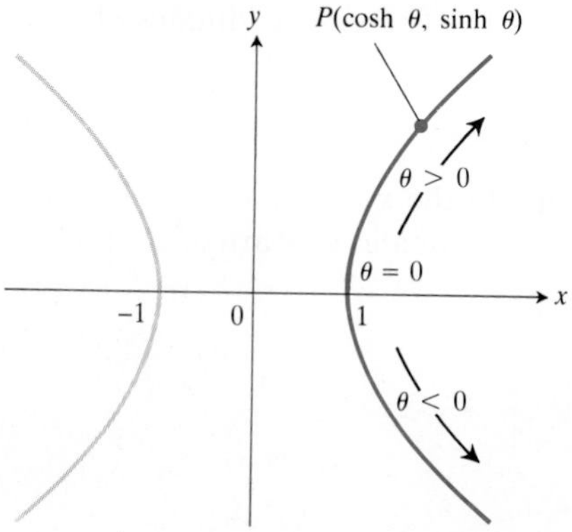

13.12 The parametric equations $x = \cosh \theta$, $y = \sinh \theta$, $-\infty < \theta < \infty$, give only the right branch of the hyperbola $x^2 - y^2 = 1$, because $\cosh \theta \geq 1$.

The solutions of differential equations of motion are not the only ways in which parametric equations arise.

EXAMPLE 1 *Parametric equations for the circle* $x^2 + y^2 = a^2$ (Fig. 13.10):

$$x = a \cos t, \qquad y = a \sin t, \qquad 0 \leq t \leq 2\pi. \;\square \tag{1}$$

EXAMPLE 2 *Parametric equations for the ellipse* $x^2/a^2 + y^2/b^2 = 1$ (Fig. 13.11):

$$x = a \cos t, \qquad y = b \sin t, \qquad 0 \leq t \leq 2\pi. \tag{2}$$

To see that these are parametric equations for the ellipse we first substitute $x = a \cos t$, $y = b \sin t$ into the cartesian equation:

$$\frac{x^2}{a^2} + \frac{y^2}{b^2} = \frac{a^2 \cos^2 t}{a^2} + \frac{b^2 \sin^2 t}{b^2} = \cos^2 t + \sin^2 t = 1. \tag{3}$$

This shows that the point $(a \cos t, b \sin t)$ lies on the ellipse.

We then observe that as t increases from 0 to 2π, x varies continuously from a to $-a$ to a, and y varies continuously from 0 to b, to 0, to $-b$, and back to 0. Thus, the point $(x, y) = (a \cos t, b \sin t)$ goes "once around the clock," tracing the entire ellipse. $\square$

Sometimes the parametric equations of a curve and the cartesian equation are not coextensive.

EXAMPLE 3 *One branch of an hyperbola.* Suppose the parametric equations of a curve are

$$x = \cosh \theta, \qquad y = \sinh \theta. \tag{4}$$

Then the hyperbolic identity

$$\cosh^2 \theta - \sinh^2 \theta = 1$$

enables us to eliminate θ and write

$$x^2 - y^2 = 1 \tag{5}$$

as a cartesian equation of the curve. Closer scrutiny, however, shows that Eq. (5) *includes too much,* for $x = \cosh \theta$ is always positive, so the parametric equations represent a curve lying wholly to the right of the y-axis, whereas the cartesian equation (5) represents both the right- and left-hand branches of the hyperbola (Fig. 13.12). The left-hand branch could be excluded by taking only positive values of x. That is,

$$x = \sqrt{1 + y^2} \tag{6}$$

does represent the curve given by (4). $\square$

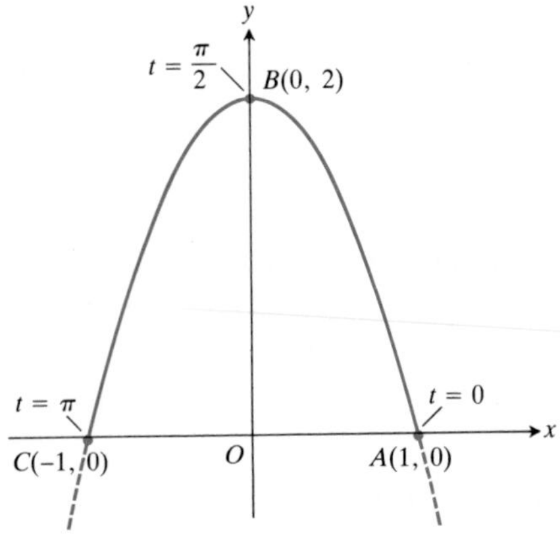

13.13 $x = \cos t,\ y = 1 - \cos 2t.$

EXAMPLE 4 *A parabolic arch.* Sketch the curve traced by the point $P(x, y)$ whose coordinates satisfy the equations

$$x = \cos t, \qquad y = 1 - \cos 2t. \tag{7}$$

Solution We find a cartesian equation for the curve by eliminating t:

$$y = 1 - \cos 2t = 1 - 2\cos^2 t + 1 = 2 - 2x^2.$$

Thus every point of the graph of (7) lies on the parabola

$$y = 2 - 2x^2. \tag{8}$$

The parametric equations in (7), however, describe only the portion of the parabola (Fig. 13.13) for which

$$-1 \le x = \cos t \le 1 \qquad \text{and} \qquad 0 \le y = 1 - \cos 2t \le 2.$$

From (7) we see that the point $P(x, y)$ starts at $A(1, 0)$ when $t = 0$. It then moves up and to the left as t increases, arriving at $B(0, 2)$ when $t = \pi/2$. It continues on to $C(-1, 0)$ as t increases to π. As t varies from π to 2π, the point retraces the arch CBA back to A. Since x and y are periodic, x with period 2π and y with period π, any further variation in t results in retracing a portion of the arch. □

EXAMPLE 5 *Trochoids and cycloids.* A wheel of radius a rolls along a horizontal straight line without slipping. Find the curve traced by a point P on a spoke of the wheel b units from its center. Such a curve is called a *trochoid* (one Greek word for wheel is *trochos*). When $b = a$, P is on the circumference and the curve is called a *cycloid*. This is like the path traveled by the head of a nail in a tire.

Solution In Fig. 13.14 we take the x-axis to be the line the wheel rolls along, with the y-axis through a low point of the trochoid. It is customary to use the angle ϕ through which CP has rotated as the parameter. Since the circle rolls without slipping, the distance OM that the wheel has moved horizontally is just equal to the circular arc $MN = a\phi$. (Roll the wheel back. Then N will fall at the origin O.) The xy-coordinates of C are therefore

$$h = a\phi, \qquad k = a. \tag{9}$$

We now introduce $x'y'$-axes parallel to the xy-axes and having their origin at C (Fig. 13.15). The xy- and $x'y'$-coordinates of P are related by the

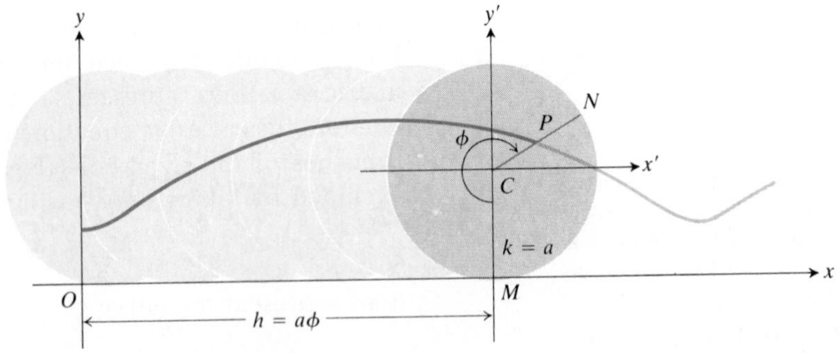

13.14 Trochoid: $x = a\phi - b\sin\phi$, $y = a - b\cos\phi$.

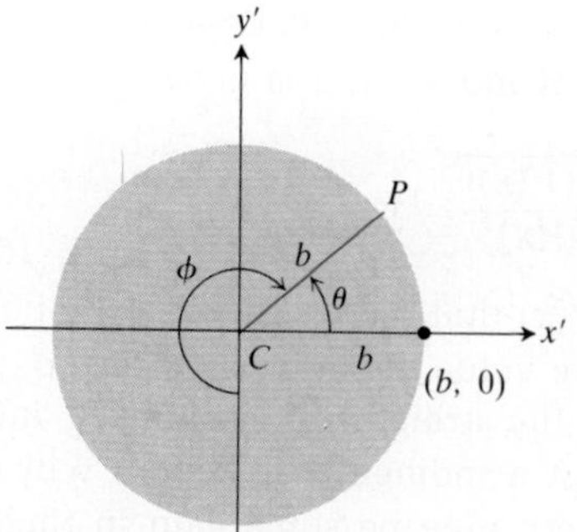

13.15 The $x'y'$-coordinates of P are $x' = b\cos\theta$, $y' = b\sin\theta$.

equations

$$x = h + x', \qquad y = k + y'. \tag{10}$$

From Fig. 13.15 we may immediately read

$$x' = b\cos\theta, \qquad y' = b\sin\theta,$$

or, since

$$\theta = \frac{3\pi}{2} - \phi,$$

$$x' = -b\sin\phi, \qquad y' = -b\cos\phi. \tag{11}$$

We substitute these results and Eqs. (10) into (9) and obtain

$$x = a\phi - b\sin\phi,$$
$$y = a - b\cos\phi \tag{12}$$

as parametric equations of the trochoid.

The cycloid (Fig. 13.16a),

$$x = a(\phi - \sin\phi), \qquad y = a(1 - \cos\phi), \tag{13}$$

obtained from (12) by taking $b = a$, is the most important special case. □

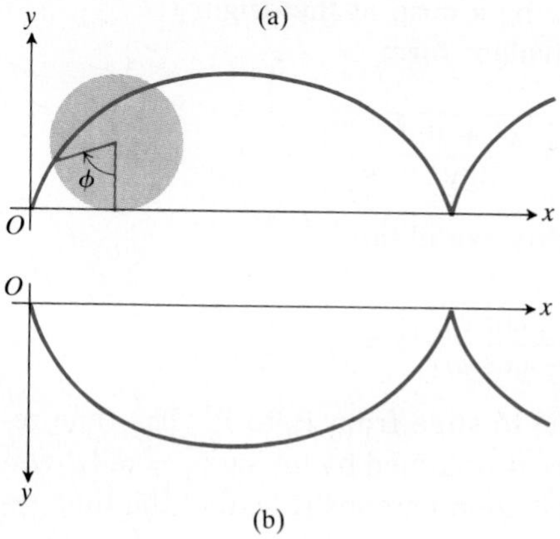

13.16 Cycloid: $x = a(\phi - \sin\phi)$, $y = a(1 - \cos\phi)$.

Brachistochrones and Tautochrones (Optional)

If we reflect both the cycloid and the y-axis across the x-axis, Eqs. (13) still apply, and the resulting curve (Fig. 13.16b) has several interesting properties, one of which we shall now discuss without proof. The proofs belong to a branch of mathematics known as the calculus of variations. Much of the fundamental theory of this subject is attributed to the Bernoulli brothers, John and James, who were friendly rivals and stimulated each other with mathematical problems in the form of challenges. One of these, the brachistochrone problem, was: Among all smooth curves joining two given points, to find that one along which a bead, subject only to the force of gravity, might slide *in the shortest time*.

The two points, labeled P_0 and P_1 in Fig. 13.17, may be taken to lie in a vertical plane at the origin and at (x_1, y_1), respectively. We can formulate the problem in mathematical terms as follows. The kinetic energy of the bead at the start is zero, since its velocity is zero. The work done by gravity in moving the particle from (0, 0) to any point (x, y) is mgy and this must be equal to the change in kinetic energy; that is,

$$mgy = \frac{1}{2}mv^2 - \frac{1}{2}m(0)^2.$$

Thus the velocity

$$v = ds/dt$$

that the particle has when it reaches $P(x, y)$ is

$$v = \sqrt{2gy}.$$

That is,

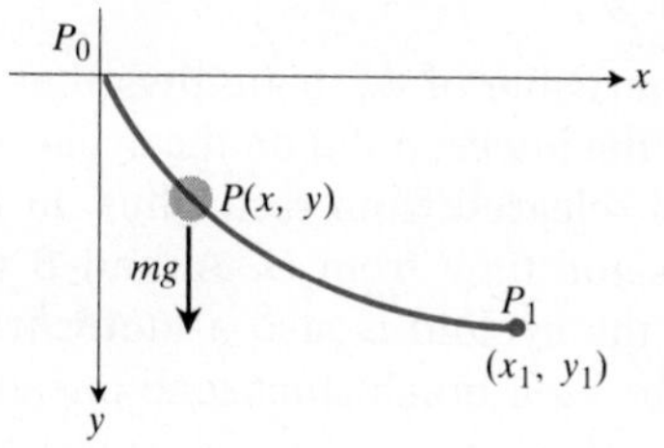

13.17 A bead sliding down a cycloid.

$$\frac{ds}{dt} = \sqrt{2gy} \qquad \text{or} \qquad dt = \frac{ds}{\sqrt{2gy}} = \frac{\sqrt{1 + \left(\frac{dy}{dx}\right)^2}\,dx}{\sqrt{2gy}}.$$

The time t_1 required for the bead to slide from P_0 to P_1 depends upon the particular curve $y = f(x)$ along which it moves and is given by

$$t_1 = \int_0^{x_1} \sqrt{\frac{1 + (f'(x))^2}{2gf(x)}}\, dx. \tag{14}$$

The problem is *to find the curve* $y = f(x)$ that passes through the points $P_0(0, 0)$ and $P_1(x_1, y_1)$ and minimizes the value of the integral in Eq. (14).

At first sight, one might guess that the straight line joining P_0 and P_1 would also yield the shortest time, but a moment's reflection will cast some doubt on this conjecture, for there may be some gain in time by having the particle start to fall vertically at first, thereby building up its velocity more quickly than if it were to slide along an inclined path. With this increased velocity, one may be able to afford to travel over a longer path and still reach P_1 in a shorter time. The solution of the problem is beyond the present book, but the brachistochrone curve is actually an arc of a cycloid through P_0 and P_1, having a cusp at the origin.

If we write Eq. (14) in the equivalent form

$$t_1 = \int \sqrt{\frac{dx^2 + dy^2}{2gy}}$$

and then substitute Eqs. (15) into this, we obtain

$$t_1 = \int_0^{\phi_1} \sqrt{\frac{a^2(2 - 2\cos\phi)}{2ga(1 - \cos\phi)}}\, d\phi = \phi_1 \sqrt{\frac{a}{g}}$$

as the time required for the particle to slide from P_0 to P_1. The time required to reach the bottom of the arc is obtained by taking $\phi_1 = \pi$. Now it is a remarkable fact, which we shall soon demonstrate, that the time required to slide along the cycloid from (0, 0) to the lowest point $(a\pi, 2a)$ is the same as the time required for the particle, starting from rest, to slide from *any intermediate point* of the arc, say (x_0, y_0), to $(a\pi, 2a)$. For the latter case, one has

$$v = \sqrt{2g(y - y_0)}$$

as the velocity at $P(x, y)$, and the time required is

$$\begin{aligned} T &= \int_{\phi_0}^{\pi} \sqrt{\frac{a^2(2 - 2\cos\phi)}{2ag(\cos\phi_0 - \cos\phi)}}\, d\phi = \sqrt{\frac{a}{g}} \int_{\phi_0}^{\pi} \sqrt{\frac{1 - \cos\phi}{\cos\phi_0 - \cos\phi}}\, d\phi \\ &= \sqrt{\frac{a}{g}} \int_{\phi_0}^{\pi} \sqrt{\frac{2\sin^2(\phi/2)}{[2\cos^2(\phi_0/2) - 1] - [2\cos^2(\phi/2) - 1]}}\, d\phi \\ &= 2\sqrt{\frac{a}{g}} \left[-\sin^{-1} \frac{\cos(\phi/2)}{\cos(\phi_0/2)} \right]_{\phi_0}^{\pi} = 2\sqrt{\frac{a}{g}}(-\sin^{-1} 0 + \sin^{-1} 1) = \pi \sqrt{\frac{a}{g}}. \end{aligned}$$

Since this answer is independent of the value of ϕ_0, it follows that the same length of time is required to reach the lowest point on the cycloid no matter where on the arc the particle is released from rest. Thus, in Fig. 13.18, three particles that start at the same time from O, A, and B will reach C simultaneously. In this sense, the cycloid is also a *tautochrone* (meaning "the same time") as well as being a brachistochrone (meaning "shortest time").

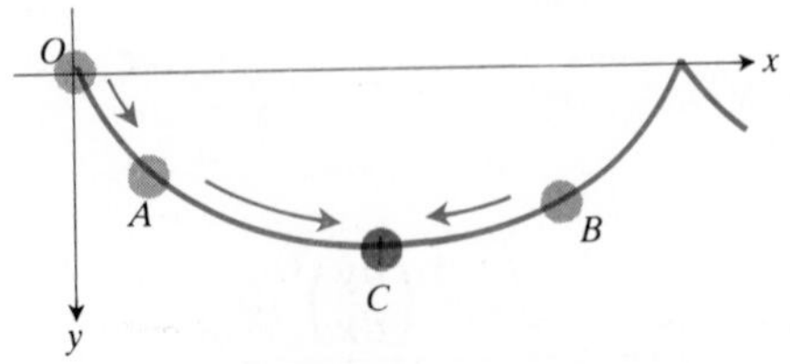

13.18 Beads released on the cycloid at O, A, and B will all take the same amount of time to reach C.

Standard Parametric Equations

For the circle $x^2 + y^2 = a^2$: $\begin{cases} x = a\cos t, \\ y = a\sin t, \end{cases} \quad 0 \le t \le 2\pi$

For the ellipse $\dfrac{x^2}{a^2} + \dfrac{y^2}{b^2} = 1$: $\begin{cases} x = a\cos t, \\ y = b\sin t, \end{cases} \quad 0 \le t \le 2\pi$

Cycloid: $x = a(\phi - \sin\phi), \quad y = a(1 - \cos\phi)$

PROBLEMS

In Problems 1–17, sketch the curve traced by the point $P(x, y)$ as the parameter t varies over the given domain. Also find a cartesian equation for each curve.

1. $x = \cos t, \quad y = \sin t, \quad 0 \le t \le 2\pi$

2. $x = \cos 2t, \quad y = \sin 2t, \quad 0 \le t \le \pi$

3. $x = 4\cos t, \quad y = 2\sin t, \quad 0 \le t \le 2\pi$

4. $x = 4\cos t, \quad y = 5\sin t, \quad 0 \le t \le 2\pi$

5. $x = \cos 2t, \quad y = \sin t, \quad 0 \le t \le 2\pi$

6. $x = \cos t, \quad y = \sin 2t, \quad 0 \le t \le 2\pi$

7. $x = \sec t, \quad y = \tan t, \quad -\pi/2 < t < \pi/2$

8. $x = \csc t, \quad y = \cot t, \quad 0 < t < \pi$

9. $x = t - \sin t, \quad y = 1 - \cos t, \quad 0 \le t \le 2\pi$

10. $x = 2 + 4\sin t, \quad y = 3 - 2\cos t, \quad 0 \le t \le 2\pi$

11. $x = t^3, \quad y = t^2, \quad -\infty < t < \infty$

12. $x = 2t + 3, \quad y = 4t^2 - 9, \quad -\infty < t < \infty$

13. $x = \cosh t, \quad y = 2\sinh t, \quad 0 \le t < \infty$

14. $x = 2 + 1/t, \quad y = 2 - t, \quad 0 < t < \infty$

15. $x = t + 1, \quad y = t^2 + 4, \quad 0 \le t < \infty$

16. $x = t^2 + t, \quad y = t^2 - t, \quad -\infty < t < \infty$

17. $x = 3 + 2\operatorname{sech} t, \quad y = 4 - 3\tanh t, \quad -\infty < t < \infty$

18. Find parametric equations of the semicircle

$$x^2 + y^2 = a^2, \qquad y > 0,$$

using as parameter the slope $t = dy/dx$ of the tangent to the curve at (x, y).

19. Find parametric equations of the semicircle

$$x^2 + y^2 = a^2, \qquad y > 0,$$

using as parameter the variable θ defined by the equation $x = a \tanh\theta$.

20. Find parametric equations of the circle

$$x^2 + y^2 = a^2,$$

using as parameter the arc length s measured counterclockwise from the point $(a, 0)$ to the point (x, y).

21. Find parametric equations of the catenary $y = a\cosh x/a$, using as parameter the length of arc s from the point $(0, a)$ to the point (x, y), with the sign of s taken to be the same as the sign of x.

22. If a string wound around a fixed circle is unwound while held taut in the plane of the circle, its end traces an *involute* of the circle (Fig. 13.19). Let the fixed circle be located with its center at the origin O and have radius a. Let the initial position of the tracing point P be $A(a, 0)$ and let the unwound portion of the string PT be tangent to the circle at T. Derive parametric equations of the involute, using the angle AOT as the parameter t.

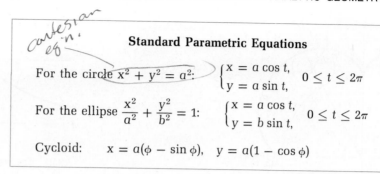

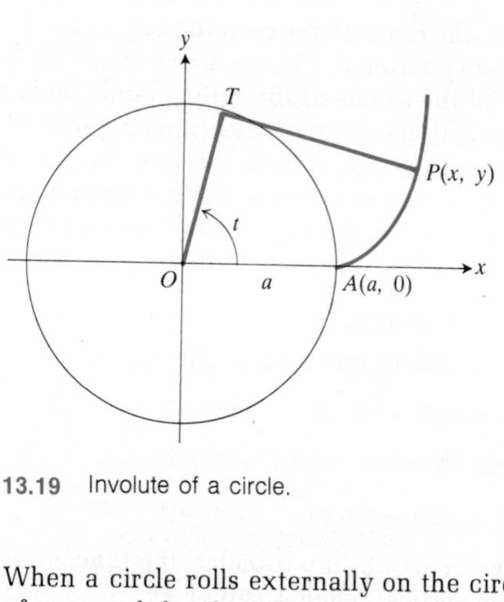

13.19 Involute of a circle.

23. When a circle rolls externally on the circumference of a second, fixed circle, any point P on the circumference of the rolling circle describes an *epicycloid* (Fig. 13.20). Let the fixed circle have its center at the origin O and have radius a. Let the radius of the rolling circle be b and let the initial position of the tracing point P be $A(a, 0)$. Determine parametric equations of the epicycloid, using as parameter the angle θ from the positive x-axis to the line of centers.

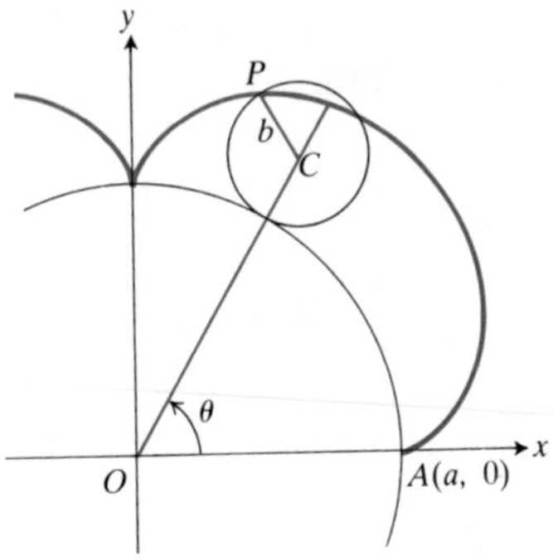

13.20 Epicycloid, with $b = a/4$.

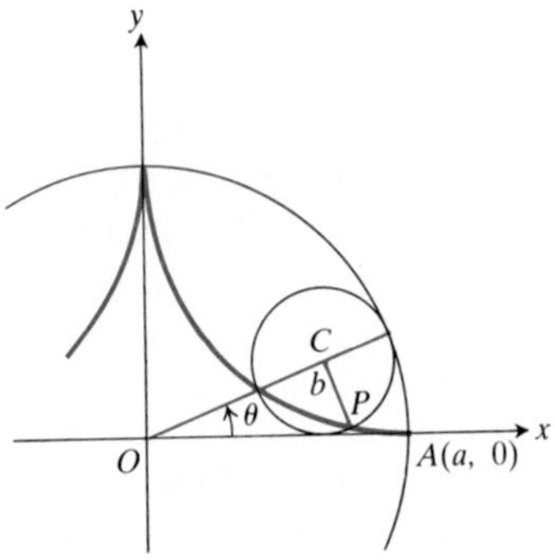

13.21 Hypocycloid, with $b = a/4$.

24. When a circle rolls on the inside of a fixed circle any point P on the circumference of the rolling circle describes a *hypocycloid*. Let the fixed circle be $x^2 + y^2 = a^2$, let the radius of the rolling circle be b, and let the initial position of the tracing point P be $A(a, 0)$. Use the angle θ from the positive x-axis to the line of centers as parameter and determine parametric equations of the hypocycloid. In particular, if $b = a/4$, as in Fig. 13.21, show that

$$x = a\cos^3\theta, \qquad y = a\sin^3\theta.$$

25. Find the length of one arch of the cycloid

$$x = a(\phi - \sin\phi), \qquad y = a(1 - \cos\phi).$$

26. Show that the slope of the cycloid

$$x = a(\phi - \sin\phi), \qquad y = a(1 - \cos\phi)$$

is $dy/dx = \cot\phi/2$. In particular, the tangent to the cycloid is vertical when ϕ is 0 or 2π.

27. Show that the slope of the trochoid

$$x = a\phi - b\sin\phi, \qquad y = a - b\cos\phi$$

is always finite if $b < a$.

28. The *witch of Maria Agnesi* is a bell-shaped curve that may be constructed as follows: Let C be a circle of radius a having its center at $(0, a)$ on the y-axis (Fig. 13.22). The variable line OA through the origin

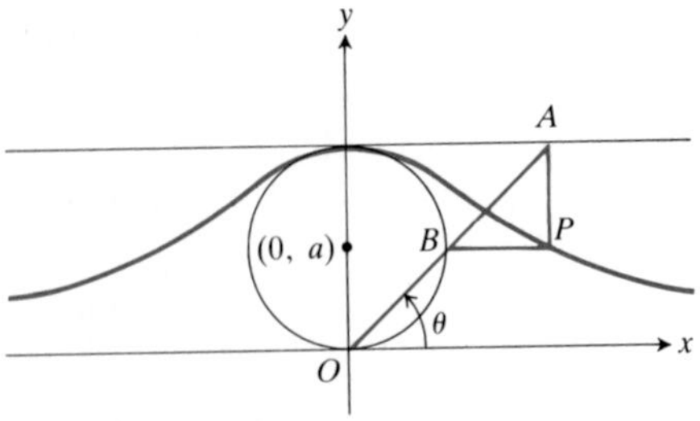

13.22 The witch of Maria Agnesi.

O intersects the line $y = 2a$ in the point A and intersects the circle in the point B. A point P on the witch is now located by taking the intersection of lines through A and B parallel to the y- and x-axes, respectively.

a) Find parametric equations of the witch, using as parameter the angle θ from the x-axis to the line OA.

b) Also find a cartesian equation for the witch.

Historical note: Maria Gaetana Agnesi (1718–1799), the daughter of a mathematics professor at the University of Bologna, wrote the first comprehensive calculus text. In four books, the text treated algebra and geometry, differential calculus, integral calculus, and differential equations. The text was translated into French and English, and it is a mistranslation that is responsible for our calling Agnesi's bell-shaped curve "the witch" today. This name, in fact, is found only in texts written in English. Agnesi's own name for the curve was "versiera," from the Latin verb *vertere,* to turn. The translator, a Cambridge scholar who had learned Italian expressly for the purpose of translating Agnesi's text, probably confused the Latin *versiera* with the Italian *avversiera,* "wife of the devil," carefully translating the latter as "the witch."

29. The following question appeared on a college entrance examination a few years ago.

Figure 13.23

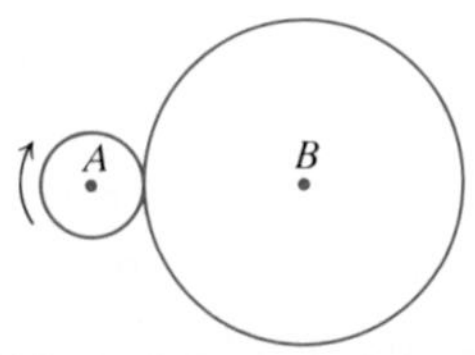

In the figure above, the radius of circle A is 1/3 the radius of circle B. Starting from the position shown in the figure, circle A rolls around circle B. At the end of how many revolutions of circle A will the center of circle A first reach its starting point?

(A) $\frac{3}{2}$ (B) 3 (C) 6 (D) $\frac{9}{2}$ (E) 9

None of the choices offered with the question was correct. What is the correct answer?

30. (*Calculator*) An automobile tire of radius 1 ft has a pebble stuck in the tread. Estimate to the nearest foot the length of the arched path traced by the pebble when the car goes one mile. Start by finding the ratio of the length of one arch of a cycloid to its base length.

31. The region bounded by the ellipse $x = 2\cos t$, $y = \sin t$, $0 \le t \le 2\pi$, is revolved about the y-axis. Find the volume swept out.

32. The equations $r = e^{2t}$, $\theta = 3t$, $0 \le t \le \pi/6$ give a curve in the polar coordinate plane.

a) Find the area bounded by the curve in the first quadrant.
b) Find the length of the curve.

33. Find parametric equations and a cartesian equation for the figure traced by the point $P(x, y)$ if

$$\frac{dx}{dt} = -2y, \qquad \frac{dy}{dt} = \cos t,$$

and $x = 3$, $y = 0$ when $t = 0$. Identify the figure.

***Toolkit* programs**	
Parametric Equations	Super * Grapher

13.4 Space Coordinates

The extension of coordinate geometry to three dimensions was begun in the seventeenth century, mainly by Descartes, Fermat, and Philippe de la Hire (1640–1718), a painter who later turned to geometry and astronomy. The full development, however, was the work of the eighteenth century. John Bernoulli, in a letter to Leibniz in 1715, introduced the three cartesian coordinate planes we use today, and the subsequent work of Euler, Lagrange, and Gaspard Monge (1746–1818) brought the analytic geometry of three dimensional space to the point we know today.

In this article, we look briefly at three coordinate systems for space. The first, cartesian coordinates, is the system we shall use the most. Cylindrical and spherical coordinates, however, will come in handy when we study integration in Chapter 16, because surfaces that have complicated cartesian equations sometimes have simpler equations in one of these other systems.

Cartesian Coordinates

Figure 13.24 shows a system of mutually orthogonal coordinate axes, Ox, Oy, and Oz. The system is called *right-handed* because a right-threaded screw pointing along Oz will advance when turned from Ox to Oy through an angle, say, of 90°. The cartesian coordinates of a point $P(x, y, z)$ in space may be read from the coordinate axes by passing planes through P perpendicular to each axis. The points on the x-axis have their y- and z-coordinates both zero. That is, they have coordinates of the form $(x, 0, 0)$. Points in a plane perpendicular to the z-axis, say, all have the same z-coordinate. Thus the points in the plane perpendicular to the z-axis and 5 units above the xy-plane all have coordinates of the form $(x, y, 5)$. We can write $z = 5$ as an equation for this plane. The three planes

$$x = 2, \qquad y = 3, \qquad z = 5$$

intersect in the point $P(2, 3, 5)$. The points of the yz-plane are obtained by setting $x = 0$. The three coordinate planes $x = 0$, $y = 0$, $z = 0$ divide the

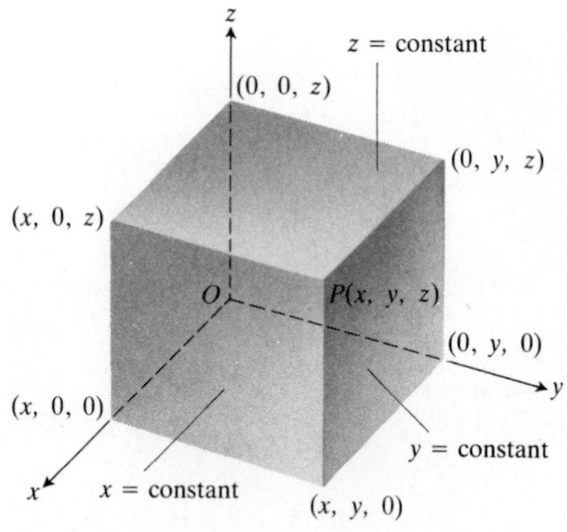

13.24 Right-handed cartesian coordinate system.

space into eight cells, called octants. The octant in which all three coordinates are positive is called the *first octant,* but there is no conventional numbering of the remaining seven octants.

EXAMPLE 1 Describe the set of points $P(x, y, z)$ whose cartesian coordinates satisfy the simultaneous equations

$$x^2 + y^2 = 4, \qquad z = 3.$$

Solution The points all lie in the horizontal plane $z = 3$, and in this plane they lie on the circle $x^2 + y^2 = 4$. Thus we may describe the set as the circle $x^2 + y^2 = 4$ in the plane $z = 3$. □

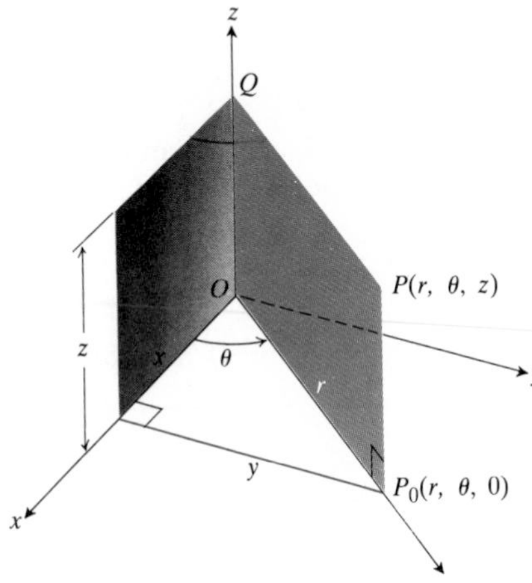

13.25 Cylindrical coordinates.

Cylindrical Coordinates

It is frequently convenient to use cylindrical coordinates (r, θ, z) to locate a point in space. These are just the polar coordinates (r, θ) used instead of (x, y) in the plane $z = 0$, coupled with the z-coordinate. Cylindrical and cartesian coordinates are therefore related by the following equations (Fig. 13.25):

Equations Relating Cartesian and Cylindrical Coordinates

$$x = r\cos\theta, \qquad r^2 = x^2 + y^2,$$
$$y = r\sin\theta, \qquad \tan\theta = y/x, \tag{1}$$
$$z = z.$$

The equation r = constant describes a circular cylinder of radius r whose axis is the z-axis, $r = 0$ being an equation for the z-axis itself. The equation θ = constant describes a plane containing the z-axis and making an angle θ with the positive x-axis (Fig. 13.26). (Some authors require the

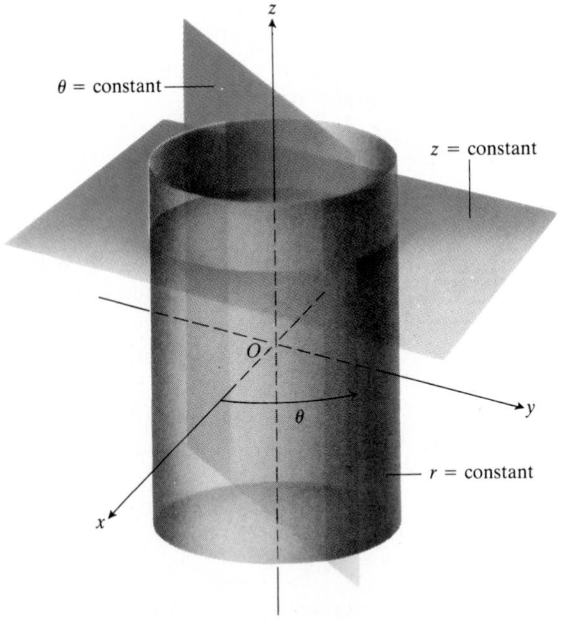

13.26 Some planes and cylinders have simple equations in cylindrical coordinates.

values of r in cylindrical coordinates to be nonnegative. In this case, the equation θ = constant describes a half-plane fanning out from the z-axis.)

Cylindrical coordinates are convenient when there is an axis of symmetry in a physical problem.

EXAMPLE 2 Describe the set of points $P(r, \theta, z)$ whose cylindrical coordinates satisfy the simultaneous equations

$$r = 2, \qquad \theta = \frac{\pi}{4}.$$

Solution These points make up the line of intersection of the cylinder $r = 2$ (cylinder of radius 2 about the z-axis) and the half-plane $\theta = \pi/4$, $r > 0$ (half-plane containing the z-axis and making an angle of $\pi/4$ radians with the positive x-axis). Thus we have the line that passes through the point $(2, \pi/4, 0)$ parallel to the z-axis. □

Spherical Coordinates

Spherical coordinates are useful when there is a center of symmetry that we can take as the origin. The *spherical coordinates* (ρ, ϕ, θ) are shown in Fig. 13.27.

The first coordinate $\rho = |OP|$ is the distance from the origin to the point P. It is never negative. The equation ρ = constant describes the surface of a sphere of radius ρ with center at O (Fig. 13.28).

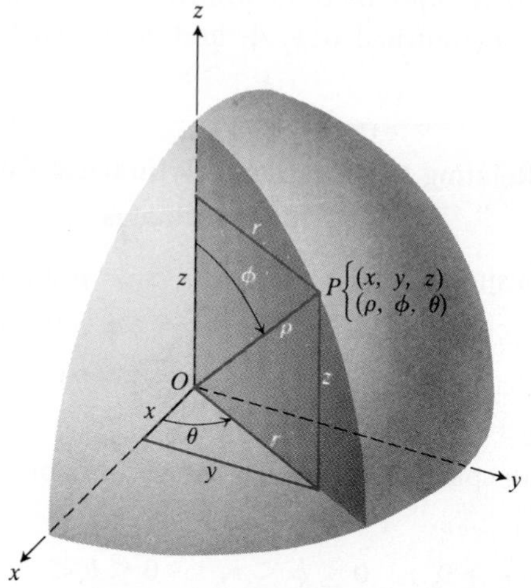

13.27 Spherical coordinates.

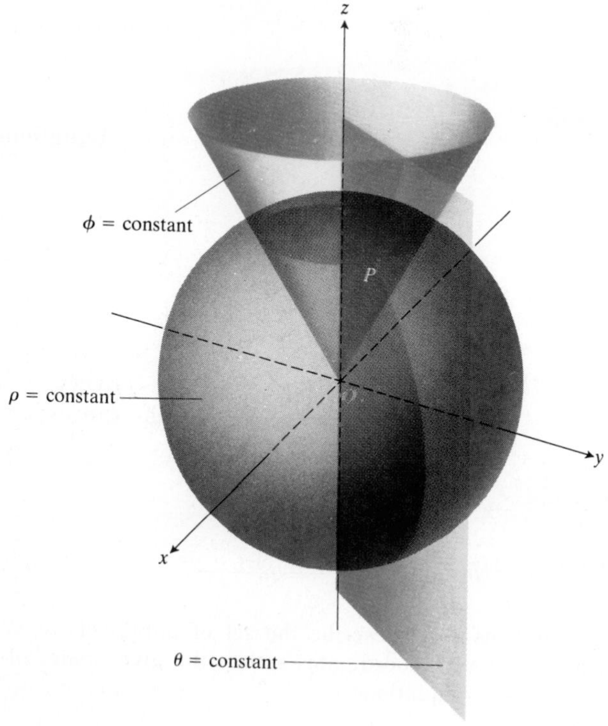

13.28 Spheres and cones whose centers are at the origin have simple equations in spherical coordinates.

The second spherical coordinate, ϕ, is the angle measured down from the z-axis to the line OP. The equation ϕ = constant describes a cone with vertex at O, axis Oz, and generating angle ϕ, provided we broaden our interpretation of the word "cone" to include the xy-plane for which $\phi = \pi/2$ and cones with generating angles greater than $\pi/2$.

The third spherical coordinate θ is the same as the angle θ in cylindrical coordinates, namely, the angle from the xz-plane to the plane through P and the z-axis. But, in contrast with cylindrical coordinates, the equation θ = constant in spherical coordinates defines a half-plane (Fig. 13.28) (because $\rho \geq 0$ and $0 \leq \phi \leq \pi$).

Incidentally, some books give spherical coordinates in the order (ρ, θ, ϕ) with the θ and ϕ reversed, and you should watch out for this when you read elsewhere.

EXAMPLE 3 Describe the set of points $P(\rho, \phi, \theta)$ whose spherical coordinates satisfy the simultaneous equations

$$\rho = 1, \qquad \phi = \frac{\pi}{3}.$$

Solution The equation $\rho = 1$ is an equation for the sphere of radius one unit centered at the origin. The equation $\phi = \pi/3$ is an equation for the cone that stands on its vertex at the origin and opens upward to make an angle of $\phi = \pi/3$ radians with the positive z-axis. The points in question make up the horizontal circle in which this cone intersects the sphere. □

From Fig. 13.27 we may read the following relationships between the cartesian (x, y, z), cylindrical (r, θ, z), and spherical (ρ, ϕ, θ) coordinate systems:

Equations Relating Cartesian and Cylindrical Coordinates to Spherical Coordinates

$$\begin{aligned} r &= \rho \sin \phi, & x &= r \cos \theta, & x &= \rho \sin \phi \cos \theta, \\ z &= \rho \cos \phi, & y &= r \sin \theta, & y &= \rho \sin \phi \sin \theta, \\ \theta &= \theta, & z &= z, & z &= \rho \cos \phi. \end{aligned} \tag{2}$$

Every point in space can be given spherical coordinates restricted to the ranges

$$\rho \geq 0, \qquad 0 \leq \phi \leq \pi, \qquad 0 \leq \theta < 2\pi. \tag{3}$$

PROBLEMS

In Problems 1–10, describe the set of points $P(x, y, z)$ whose cartesian coordinates satisfy the given pairs of simultaneous equations.

1. $x = 1, \quad y = 1$
2. $x = a, \quad y = b \qquad$ (a and b constant)
3. $y = x, \quad z = 5$
4. $x^2 + y^2 = 4, \quad z = 0$
5. $x^2 + y^2 = 4, \quad z = -2$
6. $x^2 + z^2 = 4, \quad y = 0$
7. $y^2 + z^2 = 1, \quad x = 0$

8. $z = 4y^2, \quad x = 0$

9. $z = y, \quad x = 0$

10. $y^2/a^2 + z^2/b^2 = 1, \quad x = 0$

The following table gives the coordinates of specific points in space in one of three coordinate systems. In Problems 11–19, find coordinates for each point in the other two systems. There may be more than one right answer, because points in cylindrical and spherical coordinates may have many coordinate triples.

	Cartesian (x, y, z)	Cylindrical (r, θ, z)	Spherical (ρ, ϕ, θ)
11.	$(1, 0, 0)$		
12.	$(0, 1, 0)$		
13.	$(0, 0, 1)$		
14.		$(1, 0, 0)$	
15.		$(\sqrt{2}, 0, 1)$	
16.		$(1, \pi/2, 1)$	
17.			$(\sqrt{3}, \pi/3, -\pi/2)$
18.			$(2\sqrt{2}, \pi/2, 3\pi/2)$
19.			$(\sqrt{2}, \pi, 3\pi/2)$

In Problems 20–23, describe the set of points $P(r, \theta, z)$ whose cylindrical coordinates satisfy the given pairs of simultaneous equations. Sketch.

20. $r = 2, \quad z = 3$

21. $\theta = \pi/6, \quad z = r$

22. $r = 3, \quad z = 2\theta$

23. $r = 2\theta, \quad z = 3\theta$

In Problems 24–28, describe the set of points $P(\rho, \phi, \theta)$ whose spherical coordinates satisfy the given pairs of simultaneous equations. Sketch.

24. $\rho = 5, \quad \theta = \pi/4$

25. $\rho = 5, \quad \phi = \pi/4$

26. $\theta = \pi/4, \quad \phi = \pi/4$

27. $\theta = \pi/2, \quad \rho = 4\cos\phi$

28. $\rho = 1, \quad \theta = \phi, \quad 0 \le \theta \le \pi/2$

In Problems 29–40, translate the equations from the given coordinate system (cartesian, cylindrical, or spherical) into equations in the other two systems.

29. $x^2 + y^2 + z^2 = 4$

30. $x^2 + y^2 + z^2 = 4z$

31. $z^2 = r^2$

32. $\rho = 6\cos\phi$

33. $x = y$

34. $\phi = 0$

35. $\theta = 0$

36. $z = -2$

37. $\rho = 1$

38. $\rho\cos\phi = 3$

39. $\rho\sin\phi\cos\theta = 2$

40. $x^2 + y^2 = 5$

Describe the sets in Problems 40–44.

40. $x \ge 0$

41. $3 \le \rho \le 5$

42. $r \ge 2, \quad \rho \le 5$

43. $0 \le \theta \le \pi/4, \quad 0 \le \phi \le \pi/4, \quad \rho \ge 0$

44. $4x^2 + 9y^2 \le 36$

13.5 Vectors and Distance in Space

Vectors in space are the three-dimensional analog of vectors in the plane and are subject to the same rules of addition, subtraction, and scalar multiplication that govern vectors in the plane. The vectors from the origin to the points whose cartesian coordinates are (1, 0, 0), (0, 1, 0), and (0, 0, 1) are the basic unit vectors. We denote them by **i**, **j**, and **k**, and write the vector from the origin O to the point $P(x, y, z)$ as

$$\mathbf{R} = \overrightarrow{OP} = \mathbf{i}x + \mathbf{j}y + \mathbf{k}z. \tag{1}$$

Vector Between Two Points

If $P_1(x_1, y_1, z_1)$ and $P_2(x_2, y_2, z_2)$ are two points in space (Fig. 13.29), then the vector from P_1 to P_2 is the vector sum

$$\overrightarrow{P_1P_2} = \overrightarrow{P_1O} + \overrightarrow{OP_2}.$$

Since

$$\overrightarrow{P_1O} = -\overrightarrow{OP_1},$$

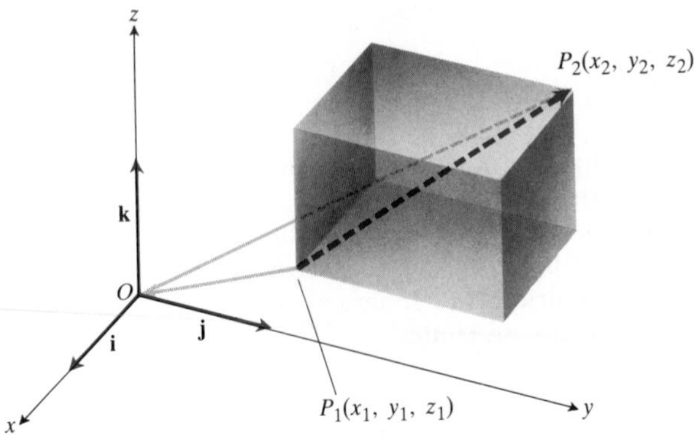

13.29 $\overrightarrow{P_1P_2} = \overrightarrow{P_1O} + \overrightarrow{OP_2}$.

this is the same as

$$\overrightarrow{P_1P_2} = \overrightarrow{OP_2} - \overrightarrow{OP_1}$$

or

$$\overrightarrow{P_1P_2} = \mathbf{i}(x_2 - x_1) + \mathbf{j}(y_2 - y_1) + \mathbf{k}(z_2 - z_1). \tag{2}$$

> The vector from $P_1(x_1, y_1, z_1)$ to $P_2(x_2, y_2, z_2)$ is
> $$\overrightarrow{P_1P_2} = (x_2 - x_1)\mathbf{i} + (y_2 - y_1)\mathbf{j} + (z_2 - z_1)\mathbf{k}.$$

The length of any vector

$$\mathbf{A} = a\mathbf{i} + b\mathbf{j} + c\mathbf{k}$$

is readily determined by applying the theorem of Pythagoras twice. In the right triangle ABC (Fig. 13.30),

$$|\overrightarrow{AC}| = |a\mathbf{i} + b\mathbf{j}| = \sqrt{a^2 + b^2},$$

and in the right triangle ACD,

$$|\overrightarrow{AD}| = \sqrt{|\overrightarrow{AC}|^2 + |\overrightarrow{CD}|^2} = \sqrt{(a^2 + b^2) + c^2}.$$

That is,

$$|a\mathbf{i} + b\mathbf{j} + c\mathbf{k}| = \sqrt{a^2 + b^2 + c^2}. \tag{3}$$

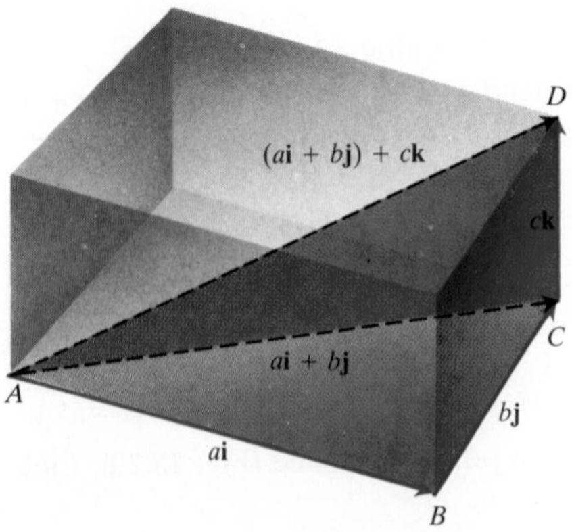

13.30 The length $|\overrightarrow{AD}|$ of the vector $\overrightarrow{AD}$ can be determined by applying the Pythagorean theorem to the right triangles ABC and ACD.

Distance

If we apply this result to the vector $\overrightarrow{P_1P_2}$ of Eq. (2), we obtain a formula for the distance between two points:

$$|\overrightarrow{P_1P_2}| = \sqrt{(x_2 - x_1)^2 + (y_2 - y_1)^2 + (z_2 - z_1)^2}. \tag{4}$$

EXAMPLE 1 The distance between

$$P_1(-2, 3, 0) \quad \text{and} \quad P_2(2, 1, 5)$$

is

$$|\overrightarrow{P_1P_2}| = \sqrt{(-2 - 2)^2 + (3 - 1)^2 + (0 - 5)^2} = \sqrt{16 + 4 + 25}$$
$$= \sqrt{45} = 3\sqrt{5}. \ \square$$

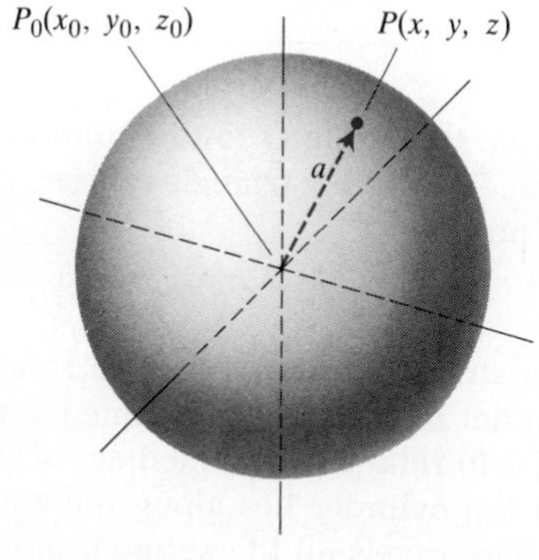

13.31 The equation of the sphere of radius a centered at the point (x_0, y_0, z_0) is $(x - x_0)^2 + (y - y_0)^2 + (z - z_0)^2 = a^2$.

Spheres

Equation (4) may be used to determine an equation for the sphere of radius a with center at $P_0(x_0, y_0, z_0)$ (Fig. 13.31). The point P is on the sphere if and only if

$$|\overrightarrow{P_0P}| = a,$$

or

$$(x - x_0)^2 + (y - y_0)^2 + (z - z_0)^2 = a^2. \tag{5}$$

EXAMPLE 2 Find the center and radius of the sphere

$$x^2 + y^2 + z^2 + 2x - 4y = 0.$$

Solution Complete the squares in the given equation to obtain

$$x^2 + 2x + 1 + y^2 - 4y + 4 + z^2 = 0 + 1 + 4$$
$$(x + 1)^2 + (y - 2)^2 + z^2 = 5.$$

Comparison with Eq. (5) shows that $x_0 = -1$, $y_0 = 2$, $z_0 = 0$, and $a = \sqrt{5}$. The center is $(-1, 2, 0)$ and the radius is $\sqrt{5}$. □

EXAMPLE 3 *Midpoints.* Find the coordinates of the midpoint M of the line segment that joins the points $P_1(x_1, y_1, z_1)$ and $P_2(x_2, y_2, z_2)$.

Solution Since

$$\overrightarrow{OM} = \overrightarrow{OP_1} + \frac{1}{2}(\overrightarrow{P_1P_2}) = \overrightarrow{OP_1} + \frac{1}{2}(\overrightarrow{OP_2} - \overrightarrow{OP_1}) = \frac{1}{2}(\overrightarrow{OP_1} + \overrightarrow{OP_2})$$
$$= \frac{x_1 + x_2}{2}\mathbf{i} + \frac{y_1 + y_2}{2}\mathbf{j} + \frac{z_1 + z_2}{2}\mathbf{k},$$

we see that the midpoint M is the point

$$M = \left(\frac{x_1 + x_2}{2}, \frac{y_1 + y_2}{2}, \frac{z_1 + z_2}{2}\right)$$

whose coordinates are obtained by averaging the coordinates of the points P_1 and P_2.

For instance, the midpoint of the segment joining $P_1(3, -2, 0)$ and $P_2(7, 4, 4)$ is

$$\left(\frac{3 + 7}{2}, \frac{-2 + 4}{2}, \frac{0 + 4}{2}\right) = (5, 1, 2). \ \square$$

Direction

We define the *direction* of a nonzero vector $\mathbf{A}$ in space to be the unit vector obtained by dividing $\mathbf{A}$ by its own length:

$$\text{Direction of } \mathbf{A} = \frac{\mathbf{A}}{|\mathbf{A}|}. \tag{6}$$

EXAMPLE 4 If $\mathbf{A} = 2\mathbf{i} - 3\mathbf{j} + 7\mathbf{k}$, then its length is $\sqrt{4 + 9 + 49} = \sqrt{62}$, and

$$\text{Direction of } (2\mathbf{i} - 3\mathbf{j} + 7\mathbf{k}) = \frac{2\mathbf{i} - 3\mathbf{j} + 7\mathbf{k}}{\sqrt{62}}.$$

In Problem 11 you will be asked to show that the vector found here really is a unit vector. □

PROBLEMS

Find the centers and radii of the spheres in Problems 1–4.

1. $x^2 + y^2 + z^2 + 4x - 4z = 0$
2. $2x^2 + 2y^2 + 2z^2 + x + y + z = 9$
3. $x^2 + y^2 + z^2 - 2az = 0$
4. $3x^2 + 3y^2 + 3z^2 + 2y - 2z = 9$
5. Find the distance between the point $P(x, y, z)$ and (a) the x-axis, (b) the y-axis, (c) the z-axis, (d) the xy-plane.
6. The distance from $P(x, y, z)$ to the origin is d_1 and the distance from P to $A(0, 0, 3)$ is d_2. Write an equation for the coordinates of P if

   a) $d_1 = 2d_2$, b) $d_1 + d_2 = 6$, c) $|d_1 - d_2| = 2$.

Find the lengths of the following vectors.

7. $2\mathbf{i} + \mathbf{j} - 2\mathbf{k}$
8. $3\mathbf{i} - 6\mathbf{j} + 2\mathbf{k}$
9. $\mathbf{i} + 4\mathbf{j} - 8\mathbf{k}$
10. $9\mathbf{i} - 2\mathbf{j} + 6\mathbf{k}$
11. Show that the vector $(2\mathbf{i} - 3\mathbf{j} + 7\mathbf{k})/\sqrt{62}$ found in Example 4 is a unit vector.
12. Find the directions of the following vectors:

    a) $4\mathbf{i} + 3\mathbf{j} + 5\mathbf{k}$ b) $\dfrac{\mathbf{i}}{\sqrt{3}} + \dfrac{\mathbf{j}}{\sqrt{3}} + \dfrac{\mathbf{k}}{\sqrt{3}}$

    c) $\dfrac{3}{5}\mathbf{i} + \dfrac{4}{5}\mathbf{k}$ d) $6\mathbf{i}$

    e) $-4\mathbf{j}$ f) $\mathbf{i} + \mathbf{j}$

    g) $\mathbf{i} + \mathbf{j} + \mathbf{k}$

13. Find the vector from the origin O to the point of intersection of the medians of the triangle whose vertices are the three points

    $$A(1, -1, 2), \quad B(2, 1, 3), \quad C(-1, 2, -1).$$

14. A bug is crawling straight up the side of a rotating right circular cylinder of radius 2 ft. At time $t = 0$, it is at the point $(2, 0, 0)$ relative to a fixed set of xyz-axes. The axis of the cylinder lies along the z-axis. Assume that the bug travels on the cylinder along a line parallel to the z-axis at the rate of c ft/s, and that the cylinder rotates (counterclockwise as viewed from above) at the rate of b radians/second. If $P(x, y, z)$ is the bug's position at the end of t seconds, show that

    $$\overrightarrow{OP} = \mathbf{i}(2 \cos bt) + \mathbf{j}(2 \sin bt) + \mathbf{k}(ct).$$

15. Let $ABCD$ be a general (not necessarily planar) quadrilateral in space. Show that the two segments joining the midpoints of opposite sides of $ABCD$ bisect each other (*Hint:* Show that the segments have the same midpoint.)
16. Use similar triangles to find the coordinates of the point Q that divides the segment from $P_1(x_1, y_1, z_1)$ to $P_2(x_2, y_2, z_2)$ into two lengths whose ratio is $p/q = r$ (that is, (distance P_1Q)/(distance QP_2) $= p/q = r$).
17. Show that $|c\mathbf{v}| = |c||\mathbf{v}|$ for any real number c and any vector $\mathbf{v} = \overrightarrow{P_1P_2}$.

13.6 The Scalar Product of Two Vectors

Definition

In mechanics, the work done by a constant force $\mathbf{F}$ when the point of application of $\mathbf{F}$ undergoes a displacement $\overrightarrow{PQ}$ (Fig. 13.32) is defined to be

$$\text{Work} = (|\mathbf{F}| \cos \theta)|\overrightarrow{PQ}| = |\mathbf{F}||\overrightarrow{PQ}| \cos \theta. \tag{1}$$

For instance, if the magnitude of $\mathbf{F}$ is $|\mathbf{F}| = 40$ newtons (about 9 pounds), $|\overrightarrow{PQ}| = 3$ m, and $\theta = 60°$ so that $\cos \theta = \frac{1}{2}$, then the work done is

$$\text{Work} = (40)(3)\left(\frac{1}{2}\right) = 60 \text{ newton meters.}$$

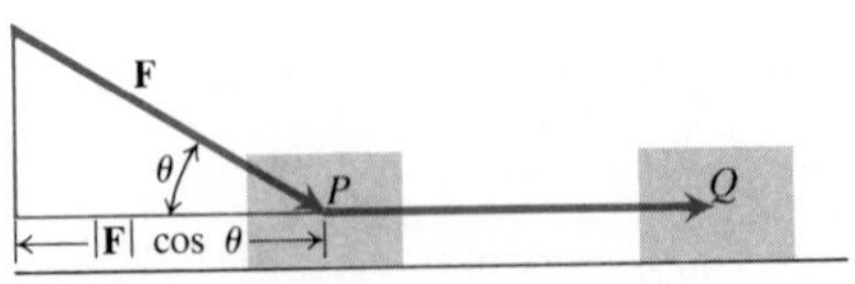

13.32 The work done by **F** during a displacement $\overrightarrow{PQ}$ is $(|\mathbf{F}| \cos \theta)|\overrightarrow{PQ}|$.

In mathematics, we call the quantity

$$|\mathbf{F}||\overrightarrow{PQ}| \cos \theta$$

the *scalar product* of $\mathbf{F}$ and $\overrightarrow{PQ}$.

DEFINITION

Scalar Product

The *scalar product* $\mathbf{A} \cdot \mathbf{B}$ ("$\mathbf{A}$ dot $\mathbf{B}$") of two vectors $\mathbf{A}$ and $\mathbf{B}$ is the number

$$\mathbf{A} \cdot \mathbf{B} = |\mathbf{A}||\mathbf{B}| \cos \theta, \tag{2}$$

where θ measures the smaller angle determined by $\mathbf{A}$ and $\mathbf{B}$ when their initial points coincide (as in Fig. 13.33).

In words, the scalar product of $\mathbf{A}$ and $\mathbf{B}$ is the length of $\mathbf{A}$ times the length of $\mathbf{B}$ times the cosine of the angle between them. The scalar product is a scalar, not a vector. It is sometimes called the *dot product,* because of the dot in the notation $\mathbf{A} \cdot \mathbf{B}$. In more advanced settings in mathematics, it is also called the *inner product* of $\mathbf{A}$ and $\mathbf{B}$.

Geometric and Algebraic Properties

If the dot product is negative, then $\cos \theta$ is negative and the angle between $\mathbf{A}$ and $\mathbf{B}$ is greater than 90°.

From Eq. (2) we can see that interchanging the two factors $\mathbf{A}$ and $\mathbf{B}$ does not change the dot product. That is,

$$\mathbf{A} \cdot \mathbf{B} = \mathbf{B} \cdot \mathbf{A} \tag{3}$$

Scalar multiplication is commutative.

If c is any number, then

$$(c\mathbf{A}) \cdot \mathbf{B} = c(\mathbf{A} \cdot \mathbf{B}) = \mathbf{A} \cdot (c\mathbf{B}) \tag{4}$$

(Problem 43).

The angle a vector $\mathbf{A}$ makes with itself is $\theta = 0$, and $\cos 0 = 1$. Therefore,

$$\mathbf{A} \cdot \mathbf{A} = |\mathbf{A}||\mathbf{A}|(1) = |\mathbf{A}|^2 \quad \text{or} \quad |\mathbf{A}| = \sqrt{\mathbf{A} \cdot \mathbf{A}} \tag{5}$$

This gives a convenient way to calculate a vector's length, as we shall see when we work Example 1.

The vector we get by projecting $\mathbf{B}$ onto the line through $\mathbf{A}$ is called the *vector projection of* $\mathbf{B}$ *onto* $\mathbf{A}$. We denote it by $\text{proj}_{\mathbf{A}}\,\mathbf{B}$ (Fig. 13.33).

The *component of* $\mathbf{B}$ *in the direction of* $\mathbf{A}$ is a number that is plus or minus the length of the vector projection of $\mathbf{B}$ onto $\mathbf{A}$. The sign is plus if $\text{proj}_{\mathbf{A}}\,\mathbf{B}$ has the same direction as $+\mathbf{A}$, and is minus if it has the same

13.33 Vector projections of **B** onto **A**. In (a), the component of **B** in the direction of **A** is the length of the vector projection. In (b), it is *minus* the length of the vector projection.

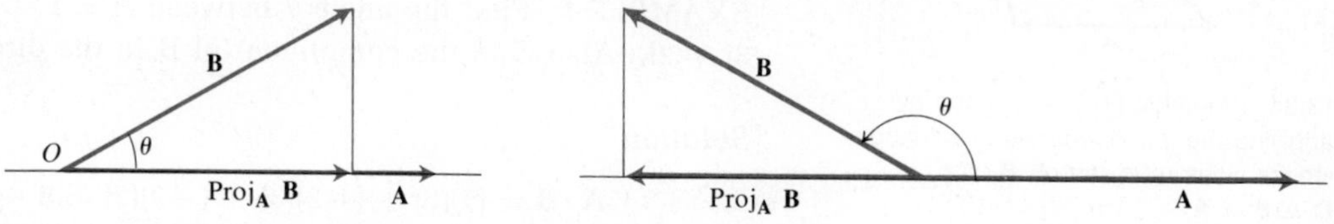

direction as $-\mathbf{A}$. In either case, the component of $\mathbf{B}$ in the direction of $\mathbf{A}$ is equal to $|\mathbf{B}| \cos \theta$ (see Fig. 13.33 again).

The dot product gives a convenient way to calculate the component of $\mathbf{B}$ in the direction of $\mathbf{A}$. We solve Eq. (2) for $|\mathbf{B}| \cos \theta$ to get

$$\mathbf{B}\text{-component in } \mathbf{A}\text{-direction} = |\mathbf{B}| \cos \theta = \frac{\mathbf{A} \cdot \mathbf{B}}{|\mathbf{A}|} = \mathbf{B} \cdot \frac{\mathbf{A}}{|\mathbf{A}|}. \tag{6}$$

Thus the component of $\mathbf{B}$ in the direction of $\mathbf{A}$ is the dot product of $\mathbf{B}$ with the direction of $\mathbf{A}$.

Multiplying both sides of Eq. (6) by $|\mathbf{A}|$ leads to a geometric interpretation of $\mathbf{A} \cdot \mathbf{B}$:

$$\begin{aligned} \mathbf{A} \cdot \mathbf{B} &= |\mathbf{A}|(|\mathbf{B}| \cos \theta) \\ &= (\text{length of } \mathbf{A}) \text{ times } (\mathbf{B}\text{-component in } \mathbf{A}\text{-direction}). \end{aligned} \tag{7}$$

Of course we may interchange the roles of $|\mathbf{A}|$ and $|\mathbf{B}|$ and write the dot product in the alternative form of

$$\begin{aligned} \mathbf{A} \cdot \mathbf{B} &= |\mathbf{B}|(|\mathbf{A}| \cos \theta) \\ &= (\text{length of } \mathbf{B}) \text{ times } (\mathbf{A}\text{-component in } \mathbf{B}\text{-direction}). \end{aligned} \tag{8}$$

Calculation

To calculate $\mathbf{A} \cdot \mathbf{B}$ from the components of $\mathbf{A}$ and $\mathbf{B}$, we let

$$\mathbf{A} = a_1\mathbf{i} + a_2\mathbf{j} + a_3\mathbf{k}, \qquad \mathbf{B} = b_1\mathbf{i} + b_2\mathbf{j} + b_3\mathbf{k},$$

and

$$\mathbf{C} = \mathbf{B} - \mathbf{A} = (b_1 - a_1)\mathbf{i} + (b_2 - a_2)\mathbf{j} + (b_3 - a_3)\mathbf{k}.$$

Then we apply the law of cosines to a triangle whose sides represent the vectors $\mathbf{A}$, $\mathbf{B}$, and $\mathbf{C}$ (Fig. 13.34) and obtain

$$\begin{aligned} |\mathbf{C}|^2 &= |\mathbf{A}|^2 + |\mathbf{B}|^2 - 2|\mathbf{A}||\mathbf{B}| \cos \theta, \\ |\mathbf{A}||\mathbf{B}| \cos \theta &= \frac{|\mathbf{A}|^2 + |\mathbf{B}|^2 - |\mathbf{C}|^2}{2}. \end{aligned} \tag{9}$$

The left side of this equation is $\mathbf{A} \cdot \mathbf{B}$, and we may calculate all terms on the right side of (9) by applying Eq. (3) of Article 13.5 to find the lengths of $\mathbf{A}$, $\mathbf{B}$, and $\mathbf{C}$. The result of this algebra is the formula

$$\mathbf{A} \cdot \mathbf{B} = a_1b_1 + a_2b_2 + a_3b_3. \tag{10}$$

Thus, to find the scalar product of two given vectors we multiply their *corresponding* $\mathbf{i}$, $\mathbf{j}$, and $\mathbf{k}$ components together and add the results. In particular, from Eq. (5) we have

$$|\mathbf{A}| = \sqrt{\mathbf{A} \cdot \mathbf{A}} = \sqrt{a_1^2 + a_2^2 + a_3^2}. \tag{11}$$

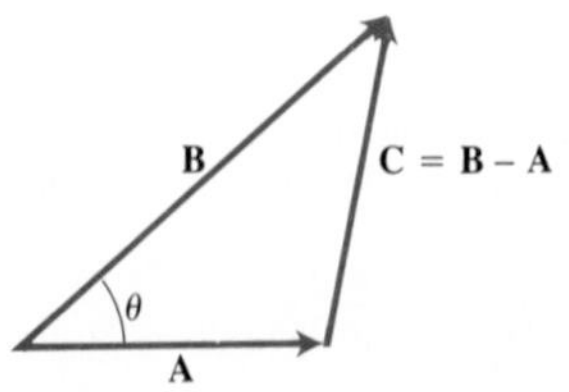

13.34 Equation (10) is obtained by applying the law of cosines to a triangle whose sides represent **A**, **B**, and **C** = **B** − **A**.

EXAMPLE 1 Find the angle θ between $\mathbf{A} = \mathbf{i} - 2\mathbf{j} - 2\mathbf{k}$ and $\mathbf{B} = 6\mathbf{i} + 3\mathbf{j} + 2\mathbf{k}$. Also, find the component of $\mathbf{B}$ in the direction of $\mathbf{A}$.

Solution

$$\mathbf{A} \cdot \mathbf{B} = (1)(6) + (-2)(3) + (-2)(2) = 6 - 6 - 4 = -4$$

from Eq. (10), while

$$\mathbf{A}\cdot\mathbf{B} = |\mathbf{A}||\mathbf{B}|\cos\theta$$

from Eq. (2). Since $|\mathbf{A}| = \sqrt{1+4+4} = 3$ and $|\mathbf{B}| = \sqrt{36+9+4} = 7$, we have

$$\cos\theta = \frac{\mathbf{A}\cdot\mathbf{B}}{|\mathbf{A}||\mathbf{B}|} = \frac{-4}{21}, \qquad \text{or} \qquad \theta = \cos^{-1}\frac{-4}{21} \approx 101°.$$

The component of **B** in the direction of **A** is

$$\mathbf{B}\cdot\frac{\mathbf{A}}{|\mathbf{A}|} = -\frac{4}{3}.$$

This is the negative of the length of the vector projection of **B** onto **A**. □

From Eq. (10), it is readily seen that if

$$\mathbf{C} = c_1\mathbf{i} + c_2\mathbf{j} + c_3\mathbf{k}$$

is any third vector, then

$$\begin{aligned}\mathbf{A}\cdot(\mathbf{B}+\mathbf{C}) &= a_1(b_1+c_1) + a_2(b_2+c_2) + a_3(b_3+c_3)\\ &= (a_1b_1 + a_2b_2 + a_3b_3) + (a_1c_1 + a_2c_2 + a_3c_3)\\ &= \mathbf{A}\cdot\mathbf{B} + \mathbf{A}\cdot\mathbf{C}.\end{aligned}$$

Hence scalar multiplication obeys the *distributive* law:

$$\mathbf{A}\cdot(\mathbf{B}+\mathbf{C}) = \mathbf{A}\cdot\mathbf{B} + \mathbf{A}\cdot\mathbf{C}. \tag{12}$$

If we combine this with the commutative law, Eq. (3), it is also evident that

$$(\mathbf{A}+\mathbf{B})\cdot\mathbf{C} = \mathbf{A}\cdot\mathbf{C} + \mathbf{B}\cdot\mathbf{C}. \tag{13}$$

Equations (12) and (13) together permit us to multiply sums of vectors by the familiar laws of algebra. For example,

$$(\mathbf{A}+\mathbf{B})\cdot(\mathbf{C}+\mathbf{D}) = \mathbf{A}\cdot\mathbf{C} + \mathbf{A}\cdot\mathbf{D} + \mathbf{B}\cdot\mathbf{C} + \mathbf{B}\cdot\mathbf{D}. \tag{14}$$

Orthogonal Vectors

It is clear from Eq. (2) that the dot product of two vectors is zero when the vectors are perpendicular, since $\cos 90° = 0$. Conversely, if $\mathbf{A}\cdot\mathbf{B} = 0$ then one of the vectors is zero or else the vectors are perpendicular. The zero vector has no definite direction, and we can adopt the convention that it is perpendicular to any vector. Then we can say that $\mathbf{A}\cdot\mathbf{B} = 0$ if and only if the vectors **A** and **B** are perpendicular. Perpendicular vectors are also said to be *orthogonal*.

The zero vector $\mathbf{0} = 0\mathbf{i} + 0\mathbf{j} + 0\mathbf{k}$ is orthogonal to *every* vector, because its dot product with every vector is zero.

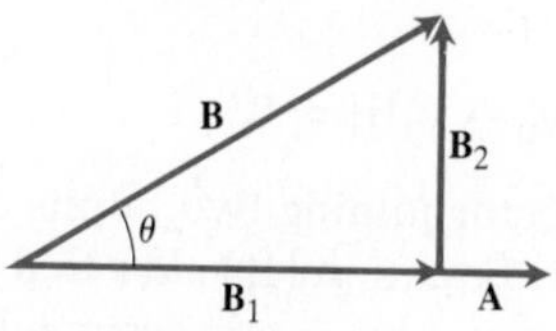

13.35 The vector **B** as the sum of vectors parallel and perpendicular to **A**.

EXAMPLE 2 Let **A** and **B** be nonzero vectors. Express the vector **B** as the sum of a vector $\mathbf{B}_1$ parallel to **A** and a vector $\mathbf{B}_2$ perpendicular to **A** (Fig. 13.35).

Solution We wish to write

$$\mathbf{B} = \mathbf{B}_1 + \mathbf{B}_2 \quad \text{with} \quad \mathbf{B}_1 = c\mathbf{A} \quad \text{and} \quad \mathbf{B}_2\cdot\mathbf{A} = 0. \tag{15}$$

To determine an appropriate value for the number c, we have

$$\mathbf{B}_2 = \mathbf{B} - \mathbf{B}_1 = \mathbf{B} - c\mathbf{A} \tag{16}$$

and

$$0 = \mathbf{B}_2 \cdot \mathbf{A} = (\mathbf{B} - c\mathbf{A}) \cdot \mathbf{A} = \mathbf{B} \cdot \mathbf{A} - c(\mathbf{A} \cdot \mathbf{A})$$

$$c(\mathbf{A} \cdot \mathbf{A}) = \mathbf{B} \cdot \mathbf{A}$$

$$c = \frac{\mathbf{B} \cdot \mathbf{A}}{\mathbf{A} \cdot \mathbf{A}}. \tag{17}$$

From (17) and (15) we now get the equations for calculating $\mathbf{B}_1$ and $\mathbf{B}_2$:

$$\mathbf{B}_1 = \frac{\mathbf{B} \cdot \mathbf{A}}{\mathbf{A} \cdot \mathbf{A}}\mathbf{A}, \qquad \mathbf{B}_2 = \mathbf{B} - \mathbf{B}_1. \tag{18}$$

For example, if

$$\mathbf{B} = 2\mathbf{i} + \mathbf{j} - 3\mathbf{k} \quad \text{and} \quad \mathbf{A} = 3\mathbf{i} - \mathbf{j},$$

then

$$\frac{\mathbf{B} \cdot \mathbf{A}}{\mathbf{A} \cdot \mathbf{A}} = \frac{6 - 1}{9 + 1} = \frac{1}{2},$$

and

$$\mathbf{B}_1 = \frac{1}{2}\mathbf{A} = \frac{3}{2}\mathbf{i} - \frac{1}{2}\mathbf{j}$$

is parallel to **A**, while

$$\mathbf{B}_2 = \mathbf{B} - \mathbf{B}_1 = \frac{1}{2}\mathbf{i} + \frac{3}{2}\mathbf{j} - 3\mathbf{k}$$

is perpendicular to **A**. □

As Fig. 13.35 suggests, the vector $\mathbf{B}_1$ in Eq. (18) is the vector projection of **B** onto **A** (Problem 44).

EXAMPLE 3 Show that the vector $\mathbf{N} = a\mathbf{i} + b\mathbf{j}$ is perpendicular to the line $ax + by = c$ in the xy-plane (Fig. 13.36).

Solution Let $P_1(x_1, y_1)$ and $P_2(x_2, y_2)$ be any two points on the line; that is,

$$ax_1 + by_1 = c, \qquad ax_2 + by_2 = c.$$

By subtraction, we eliminate c and obtain

$$a(x_2 - x_1) + b(y_2 - y_1) = 0,$$

or

$$(a\mathbf{i} + b\mathbf{j}) \cdot [(x_2 - x_1)\mathbf{i} + (y_2 - y_1)\mathbf{j}] = 0. \tag{19}$$

Now $(x_2 - x_1)\mathbf{i} + (y_2 - y_1)\mathbf{j} = \overrightarrow{P_1P_2}$ is a vector joining two points on the line, while $\mathbf{N} = a\mathbf{i} + b\mathbf{j}$ is the given vector. Equation (19) says that either $\mathbf{N} = \mathbf{0}$, or $\overrightarrow{P_1P_2} = \mathbf{0}$, or else $\mathbf{N} \perp \overrightarrow{P_1P_2}$. But $ax + by = c$ is assumed to be an honest equation of a straight line, so that a and b are not both zero and

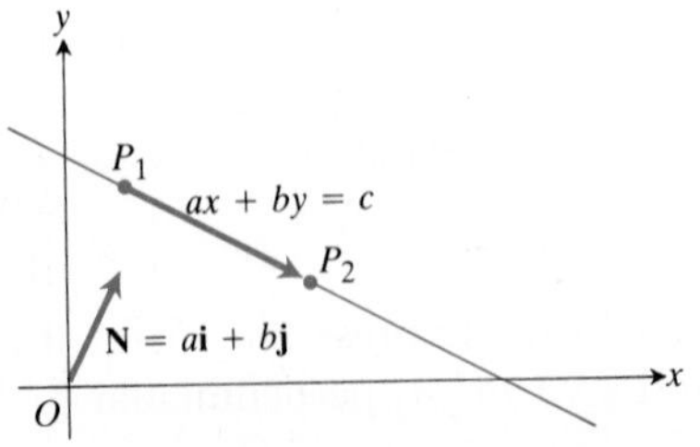

13.36 The vector $\mathbf{N} = a\mathbf{i} + b\mathbf{j}$ is normal to the line $ax + by = c$. Example 3 explains why.

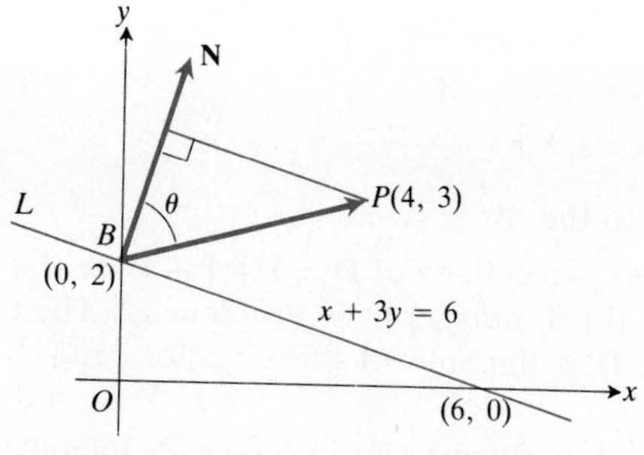

13.37 The distance from P to line L is the length of the vector projection of $\overrightarrow{BP}$ onto **N**.

$\mathbf{N} \neq \mathbf{0}$. Furthermore, we may surely choose P_2 different from P_1 on the line to make $\overrightarrow{P_1P_2} \neq \mathbf{0}$. Hence $\mathbf{N} \perp \overrightarrow{P_1P_2}$.

For example, $\mathbf{N} = 2\mathbf{i} - 3\mathbf{j}$ is normal to the line $2x - 3y = 5$. □

EXAMPLE 4 Using vector methods, find the distance d between the point $P(4, 3)$ and the line L: $x + 3y = 6$ (Fig. 13.37).

Solution The line cuts the y-axis at $B(0, 2)$. At B, draw the vector

$$\mathbf{N} = \mathbf{i} + 3\mathbf{j}$$

normal to L (see Example 3). Then the distance between P and L is the component of $\overrightarrow{BP}$ in the direction of **N**. Since

$$\overrightarrow{BP} = (4 - 0)\mathbf{i} + (3 - 2)\mathbf{j} = 4\mathbf{i} + \mathbf{j},$$

we have

$$d = |\text{proj}_{\mathbf{N}}\, \overrightarrow{BP}| = \left| \overrightarrow{BP} \cdot \frac{\mathbf{N}}{|\mathbf{N}|} \right| = \frac{4 + 3}{\sqrt{10}} = \frac{7\sqrt{10}}{10}. \quad \square$$

Unit Vectors and Direction Cosines

Suppose that **u** is a unit vector that makes angles α, β, and γ with the coordinate axes when we represent it by an arrow with its initial point at the origin. Figure 13.38 shows the case in which the three angles are acute. Then

$$\begin{aligned} \mathbf{u} \cdot \mathbf{i} &= |\mathbf{u}||\mathbf{i}| \cos \alpha = \cos \alpha, \\ \mathbf{u} \cdot \mathbf{j} &= |\mathbf{u}||\mathbf{j}| \cos \beta = \cos \beta, \\ \mathbf{u} \cdot \mathbf{k} &= |\mathbf{u}||\mathbf{k}| \cos \gamma = \cos \gamma. \end{aligned} \tag{20}$$

If we write

$$\mathbf{u} = u_1\mathbf{i} + u_2\mathbf{j} + u_3\mathbf{k},$$

then we also see that

$$\mathbf{u} \cdot \mathbf{i} = u_1, \qquad \mathbf{u} \cdot \mathbf{j} = u_2, \qquad \mathbf{u} \cdot \mathbf{k} = u_3. \tag{21}$$

Thus,

$$u_1 = \cos \alpha, \qquad u_2 = \cos \beta, \qquad u_3 = \cos \gamma. \tag{22}$$

That is, the components of any unit vector **u** are the cosines of the angles it makes with the coordinate axes:

$$\mathbf{u} = u_1\mathbf{i} + u_2\mathbf{j} + u_3\mathbf{k} = \mathbf{i} \cos \alpha + \mathbf{j} \cos \beta + \mathbf{k} \cos \gamma. \tag{23}$$

If **A** is any nonzero vector, then **A** and its direction $\mathbf{u} = \mathbf{A}/|\mathbf{A}|$ make the same angles with the coordinate axes. To find the cosines of these angles we need only calculate the components of the unit vector

$$\frac{\mathbf{A}}{|\mathbf{A}|} = \mathbf{i} \cos \alpha + \mathbf{j} \cos \beta + \mathbf{k} \cos \gamma.$$

These cosines are called the *direction cosines* of **A**. That is, the direction cosines of **A** are the components of the direction of **A**.

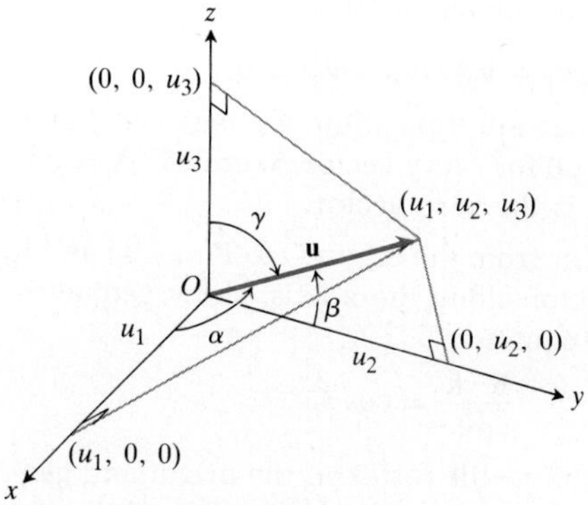

13.38 If $\mathbf{u} = u_1\mathbf{i} + u_2\mathbf{j} + u_3\mathbf{k}$ is a unit vector, then $u_1 = \cos \alpha$, $u_2 = \cos \beta$, $u_3 = \cos \gamma$ from the three right triangles shown here.

PROBLEMS

In Problems 1–10, calculate $\mathbf{A} \cdot \mathbf{B}$, $|\mathbf{A}|$, and $|\mathbf{B}|$, and find the cosine of the angle between **A** and **B**.

1. $\mathbf{A} = 3\mathbf{i} + 2\mathbf{j}$, $\mathbf{B} = 5\mathbf{j} + \mathbf{k}$
2. $\mathbf{A} = \mathbf{i}$, $\mathbf{B} = 5\mathbf{j} - 3\mathbf{k}$
3. $\mathbf{A} = 3\mathbf{i} - 2\mathbf{j} - \mathbf{k}$, $\mathbf{B} = -2\mathbf{j}$
4. $\mathbf{A} = -2\mathbf{i} + 7\mathbf{j}$, $\mathbf{B} = \mathbf{k}$
5. $\mathbf{A} = 5\mathbf{j} - 3\mathbf{k}$, $\mathbf{B} = \mathbf{i} + \mathbf{j} + \mathbf{k}$
6. $\mathbf{A} = \frac{1}{\sqrt{2}}\mathbf{i} + \frac{1}{\sqrt{3}}\mathbf{j} + \frac{1}{\sqrt{6}}\mathbf{k}$, $\mathbf{B} = \frac{1}{\sqrt{2}}\mathbf{j} - \mathbf{k}$
7. $\mathbf{A} = -\mathbf{i} + \mathbf{j}$, $\mathbf{B} = \sqrt{2}\mathbf{i} + \sqrt{3}\mathbf{j} + 2\mathbf{k}$
8. $\mathbf{A} = \mathbf{i} + \mathbf{k}$, $\mathbf{B} = \mathbf{i} + \mathbf{j} + \mathbf{k}$
9. $\mathbf{A} = 2\mathbf{i} - 4\mathbf{j} + \sqrt{5}\mathbf{k}$, $\mathbf{B} = -2\mathbf{i} + 4\mathbf{j} - \sqrt{5}\mathbf{k}$
10. $\mathbf{A} = -5\mathbf{i} + \mathbf{j}$, $\mathbf{B} = 2\mathbf{i} + \sqrt{17}\mathbf{j} + 10\mathbf{k}$
11. Find the interior angles of the triangle ABC whose vertices are $A(-1, 0, 2)$, $B(2, 1, -1)$, and $C(1, -2, 2)$.
12. Find the angle between $\mathbf{A} = 2\mathbf{i} + 2\mathbf{j} + \mathbf{k}$ and $\mathbf{B} = 2\mathbf{i} + 10\mathbf{j} - 11\mathbf{k}$.
13. Find the angle between the diagonal of a cube and the diagonal of one of its faces. (*Hint:* Use a cube three of whose edges represent the vectors **i**, **j**, and **k**.)
14. Find the angle between the diagonal of a cube and one of the edges it meets at a vertex.
15. Find (a) the vector projection of $\mathbf{B} = \mathbf{i} + 3\mathbf{j} + 4\mathbf{k}$ onto the vector $\mathbf{A} = 10\mathbf{i} + 11\mathbf{j} - 2\mathbf{k}$ and (b) the component of **B** in the direction of **A**.
16. Find the component of $\mathbf{A} = 2\mathbf{i} + 2\mathbf{j} + \mathbf{k}$ in the direction of $\mathbf{B} = 2\mathbf{i} + 10\mathbf{j} - 11\mathbf{k}$.
17. Find the point $A(a, a, 0)$ at the foot of the perpendicular from the point $B(2, 4, -3)$ to the line $y = x$ in the xy-plane.
18. How many lines through the origin make angles of 60° with both the y- and z-axes? What angles do these lines make with the positive x-axis?
19. Write the vector $\mathbf{B} = 3\mathbf{j} + 4\mathbf{k}$ as the sum of a vector $\mathbf{B}_1$ parallel to $\mathbf{A} = \mathbf{i} + \mathbf{j}$ and a vector $\mathbf{B}_2$ perpendicular to **A**.
20. Repeat Problem 19 for $\mathbf{B} = \mathbf{i} + \mathbf{j} + \mathbf{k}$.
21. Find the distance in the xy-plane between the line $x + 3y = 6$ and
    a) the point (2, 8), b) the origin.
22. Show that each of the vectors

$$\mathbf{A} = \frac{1}{\sqrt{3}}(\mathbf{i} - \mathbf{j} + \mathbf{k}),$$

$$\mathbf{B} = \frac{1}{\sqrt{2}}(\mathbf{j} + \mathbf{k}),$$

$$\mathbf{C} = \frac{1}{\sqrt{6}}(-2\mathbf{i} - \mathbf{j} + \mathbf{k})$$

is orthogonal to the other two.

23. Find the vector projections of $\mathbf{D} = \mathbf{i} + \mathbf{j} + \mathbf{k}$ on the vectors (a) **A**, (b) **B**, and (c) **C** of Problem 22. Then (d) show that **D** is the sum of these vector projections.

In Problems 24–32, find the direction cosines of the vector **A** in the following problems.

24. Problem 1
25. Problem 2
26. Problem 3
27. Problem 4
28. Problem 5
29. Problem 6
30. Problem 7
31. Problem 8
32. Problem 9
33. Find the work done by a force $\mathbf{F} = -5\mathbf{k}$ (magnitude 5 newtons) as its point of application moves from the origin to the point (1, 1, 1).
34. A locomotive exerted a constant force of 60,000 newtons on a freight train while drawing it for 1 km along a straight track. How much work did the locomotive do?
35. How much work is performed in sliding a crate 20 m along a loading dock by pushing on it with a constant force of 200 newtons at an angle of 30° from the horizontal?
36. Suppose it is known that $\mathbf{A} \cdot \mathbf{B}_1 = \mathbf{A} \cdot \mathbf{B}_2$, and **A** is not zero, but nothing more is known about the vectors $\mathbf{B}_1$ and $\mathbf{B}_2$. Is it permissible to cancel **A** from both sides of the equation? Give a reason for your answer.
37. In Fig. 13.2 it looks as if $\mathbf{v}_1 + \mathbf{v}_2$ and $\mathbf{v}_1 - \mathbf{v}_2$ are orthogonal. Is this mere coincidence, or are there circumstances under which we may expect the sum of two vectors to be perpendicular to the difference of the same two vectors? Find out by expanding the left-hand side of the equation

$$(\mathbf{v}_1 + \mathbf{v}_2) \cdot (\mathbf{v}_1 - \mathbf{v}_2) = 0.$$

38. Show that scalar multiplication is *positive definite;* that is, $\mathbf{A} \cdot \mathbf{A} \geq 0$ for every vector **A**, and $\mathbf{A} \cdot \mathbf{A} = 0$ if and only if **A** is the zero vector.
39. If **R** is the vector from the origin O to $P(x, y, z)$ and **k** is the unit vector along the z-axis, show geometrically that the equation

$$\frac{\mathbf{R} \cdot \mathbf{k}}{|\mathbf{R}|} = \cos 45°$$

represents a cone with vertex at the origin and generating angle of 45°. Express the equation in cartesian form.

40. If $a = |\mathbf{A}|$ and $b = |\mathbf{B}|$, show that the vector

$$\mathbf{C} = a\mathbf{B} + b\mathbf{A}$$

bisects the angle between **A** and **B**.

41. With the same notation as in Problem 40, show that the vectors $a\mathbf{B} + b\mathbf{A}$ and $\mathbf{A}b - \mathbf{B}a$ are perpendicular.

42. Using vector methods, show that the distance d between the point (x_1, y_1) and the line $ax + by = c$ is

$$d = \frac{|ax_1 + by_1 - c|}{\sqrt{a^2 + b^2}}.$$

43. Show that if c is any number, and $\mathbf{A}$ and $\mathbf{B}$ any vectors, then

$$(c\mathbf{A})\cdot\mathbf{B} = c(\mathbf{A}\cdot\mathbf{B}) = \mathbf{A}\cdot(c\mathbf{B}).$$

44. Show that the vector $\mathbf{B}_1$ in Example 2 is the vector projection of $\mathbf{B}$ onto $\mathbf{A}$.

45. Let α, β, and γ be the angles that a line through the origin makes with the positive x-, y-, and z-axes. Show that $\cos^2\alpha + \cos^2\beta + \cos^2\gamma = 1$.

13.7 The Vector Product of Two Vectors in Space

Cross Products

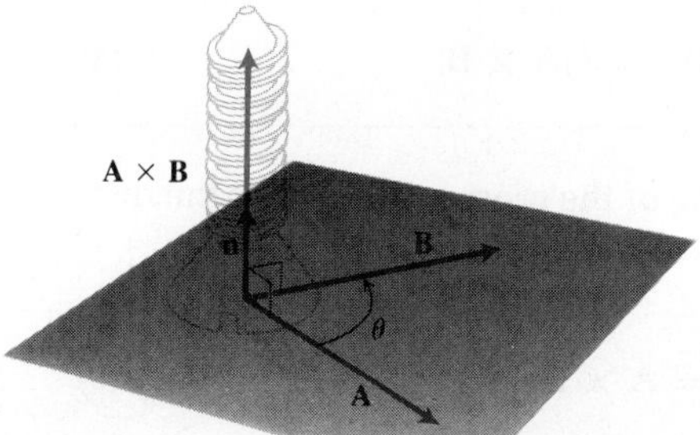

13.39 $\mathbf{A} \times \mathbf{B}$ and $\mathbf{B} \times \mathbf{A}$ have the same magnitude but point in opposite directions from the plane of **A** and **B**.

Two nonzero vectors $\mathbf{A}$ and $\mathbf{B}$ in space may be moved without changing their magnitude or direction so that their initial points coincide. Suppose that this has been done and let θ be the angle from $\mathbf{A}$ to $\mathbf{B}$, with $0 \le \theta \le \pi$. Unless $\mathbf{A}$ and $\mathbf{B}$ are parallel, they now determine a plane. Let $\mathbf{n}$ be a unit vector perpendicular to this plane and pointing in the direction a right-threaded screw advances when its head is rotated from $\mathbf{A}$ to $\mathbf{B}$ through the angle θ. The *vector product*, or *cross product*, of $\mathbf{A}$ and $\mathbf{B}$, in that order, is then defined by the equation (Fig. 13.39)

$$\mathbf{A} \times \mathbf{B} = \mathbf{n}|\mathbf{A}||\mathbf{B}| \sin\theta. \tag{1}$$

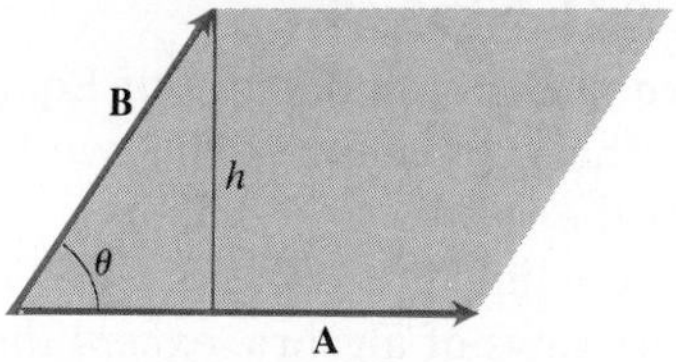

13.40 The area of the parallelogram is $|\mathbf{A} \times \mathbf{B}|$.

Like the definition of the scalar product of two vectors, given in Article 13.6, the definition of the vector product given here is coordinate-free. We emphasize, however, that the vector product $\mathbf{A} \times \mathbf{B}$ is a *vector*, while the scalar product $\mathbf{A}\cdot\mathbf{B}$ is a *scalar*. Cross products play an important role in the study of electricity and magnetism.

If $\mathbf{A}$ and $\mathbf{B}$ are parallel, then $\theta = 0$ or $180°$ and $\sin\theta = 0$, so that $\mathbf{A} \times \mathbf{B} = \mathbf{0}$. In this case, the direction of $\mathbf{n}$ is not determined, but this is immaterial, since the zero vector has no specific direction. In all other cases, however, $\mathbf{n}$ is determined and the cross product is a vector having the same direction as $\mathbf{n}$ and having magnitude equal to the area, $|\mathbf{A}||\mathbf{B}| \sin\theta$, of the parallelogram determined by the vectors $\mathbf{A}$ and $\mathbf{B}$ (Fig. 13.40).

If the order of the factors $\mathbf{A}$ and $\mathbf{B}$ is reversed in the construction of the cross product, the direction of the unit vector perpendicular to their plane is reversed (Fig. 13.41). The right-handed screw that turns through θ from $\mathbf{B}$ to $\mathbf{A}$ points the other way. The original unit vector $\mathbf{n}$ is now replaced by $-\mathbf{n}$, with the result that

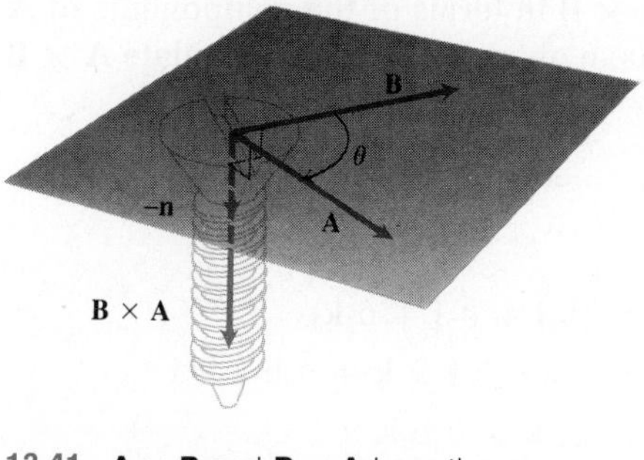

13.41 $\mathbf{A} \times \mathbf{B}$ and $\mathbf{B} \times \mathbf{A}$ have the same magnitude but point in opposite directions from the plane of **A** and **B**.

$$\mathbf{B} \times \mathbf{A} = -\mathbf{A} \times \mathbf{B}. \tag{2}$$

Thus, cross-product multiplication is not commutative. Reversing the order of the factors changes the sign of the product.

When the definition of the cross product is applied to the unit vectors, **i**, **j**, and **k**, one readily finds that

$$\begin{aligned} \mathbf{i} \times \mathbf{j} &= -\mathbf{j} \times \mathbf{i} = \mathbf{k}, \\ \mathbf{j} \times \mathbf{k} &= -\mathbf{k} \times \mathbf{j} = \mathbf{i}, \\ \mathbf{k} \times \mathbf{i} &= -\mathbf{i} \times \mathbf{k} = \mathbf{j}, \end{aligned} \tag{3}$$

while

$$\mathbf{i} \times \mathbf{i} = \mathbf{j} \times \mathbf{j} = \mathbf{k} \times \mathbf{k} = \mathbf{0}.$$

The Associative and Distributive Laws

The associative law

$$(r\mathbf{A}) \times (s\mathbf{B}) = (rs)\mathbf{A} \times \mathbf{B}, \tag{4}$$

follows from the geometric meaning of the cross product, but the distributive law

$$\mathbf{A} \times (\mathbf{B} + \mathbf{C}) = \mathbf{A} \times \mathbf{B} + \mathbf{A} \times \mathbf{C} \tag{5}$$

is not so easy to establish. We shall assume it here and leave its proof to Appendix 10.

The companion law

$$(\mathbf{B} + \mathbf{C}) \times \mathbf{A} = \mathbf{B} \times \mathbf{A} + \mathbf{C} \times \mathbf{A} \tag{6}$$

now follows at once from Eq. (5) if we multiply both sides of Eq. (5) by minus one and take account of the fact that interchanging the two factors in a cross product changes the sign of the result.

From Eqs. (4), (5), and (6), we may conclude that cross-product multiplication of vectors follows the ordinary laws of algebra, *except that the order of the factors is not reversible.*

The Determinant Formula

Our next objective is to express $\mathbf{A} \times \mathbf{B}$ in terms of the components of **A** and **B**. If we apply the results we have obtained so far to calculate $\mathbf{A} \times \mathbf{B}$ with

$$\mathbf{A} = a_1\mathbf{i} + a_2\mathbf{j} + a_3\mathbf{k}, \qquad \mathbf{B} = b_1\mathbf{i} + b_2\mathbf{j} + b_3\mathbf{k},$$

we obtain

$$\begin{aligned} \mathbf{A} \times \mathbf{B} &= (a_1\mathbf{i} + a_2\mathbf{j} + a_3\mathbf{k}) \times (b_1\mathbf{i} + b_2\mathbf{j} + b_3\mathbf{k}) \\ &= a_1b_1\mathbf{i} \times \mathbf{i} + a_1b_2\mathbf{i} \times \mathbf{j} + a_1b_3\mathbf{i} \times \mathbf{k} + a_2b_1\mathbf{j} \times \mathbf{i} \\ &\quad + a_2b_2\mathbf{j} \times \mathbf{j} + a_2b_3\mathbf{j} \times \mathbf{k} + a_3b_1\mathbf{k} \times \mathbf{i} \\ &\quad + a_3b_2\mathbf{k} \times \mathbf{j} + a_3b_3\mathbf{k} \times \mathbf{k} \\ &= \mathbf{i}(a_2b_3 - a_3b_2) + \mathbf{j}(a_3b_1 - a_1b_3) + \mathbf{k}(a_1b_2 - a_2b_1), \end{aligned} \tag{7}$$

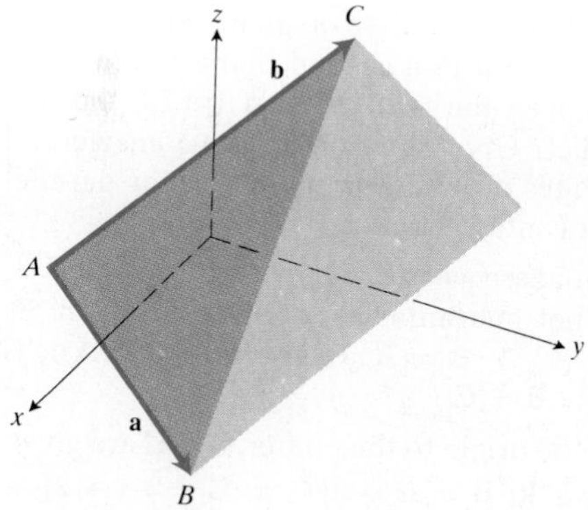

13.42 The area of ΔABC is half of $|\mathbf{a} \times \mathbf{b}|$.

where Eqs. (3) have been used to evaluate the products $\mathbf{i} \times \mathbf{i} = \mathbf{0}$, $\mathbf{i} \times \mathbf{j} = \mathbf{k}$, etc. The terms on the right-hand side of Eq. (7) are the same as the terms in the expansion of the third-order determinant below, so that the cross product may conveniently be calculated from the equation

$$\mathbf{A} \times \mathbf{B} = \begin{vmatrix} \mathbf{i} & \mathbf{j} & \mathbf{k} \\ a_1 & a_2 & a_3 \\ b_1 & b_2 & b_3 \end{vmatrix}. \tag{8}$$

(Determinants are reviewed in Appendix 1.)

EXAMPLE 1 Find the area of the triangle whose vertices are $A(1, -1, 0)$, $B(2, 1, -1)$, and $C(-1, 1, 2)$ (Fig. 13.42).

Solution Two sides of the given triangle are represented by the vectors

$$\mathbf{a} = \overrightarrow{AB} = (2-1)\mathbf{i} + (1+1)\mathbf{j} + (-1-0)\mathbf{k} = \mathbf{i} + 2\mathbf{j} - \mathbf{k},$$
$$\mathbf{b} = \overrightarrow{AC} = (-1-1)\mathbf{i} + (1+1)\mathbf{j} + (2-0)\mathbf{k} = -2\mathbf{i} + 2\mathbf{j} + 2\mathbf{k}.$$

The area of the triangle is one half the area of the parallelogram represented by these vectors. The area of the parallelogram is the magnitude of the vector

$$\mathbf{c} = \mathbf{a} \times \mathbf{b} = \begin{vmatrix} \mathbf{i} & \mathbf{j} & \mathbf{k} \\ 1 & 2 & -1 \\ -2 & 2 & 2 \end{vmatrix}$$

$$= \mathbf{i}\begin{vmatrix} 2 & -1 \\ 2 & 2 \end{vmatrix} - \mathbf{j}\begin{vmatrix} 1 & -1 \\ -2 & 2 \end{vmatrix} + \mathbf{k}\begin{vmatrix} 1 & 2 \\ -2 & 2 \end{vmatrix} = 6\mathbf{i} + 6\mathbf{k},$$

which is $|\mathbf{c}| = \sqrt{6^2 + 6^2} = 6\sqrt{2}$. Therefore, the area of the triangle is $\frac{1}{2}|\mathbf{a} \times \mathbf{b}| = 3\sqrt{2}$. □

EXAMPLE 2 Find a unit vector perpendicular to both $\mathbf{A} = 2\mathbf{i} + \mathbf{j} - \mathbf{k}$ and $\mathbf{B} = \mathbf{i} - \mathbf{j} + 2\mathbf{k}$.

Solution The vector $\mathbf{N} = \mathbf{A} \times \mathbf{B}$ is perpendicular to both $\mathbf{A}$ and $\mathbf{B}$. We divide $\mathbf{N}$ by $|\mathbf{N}|$ to obtain a unit vector $\mathbf{u}$ that has the same direction as $\mathbf{N}$:

$$\mathbf{u} = \frac{\mathbf{N}}{|\mathbf{N}|} = \frac{\mathbf{A} \times \mathbf{B}}{|\mathbf{A} \times \mathbf{B}|} = \frac{\mathbf{i} - 5\mathbf{j} - 3\mathbf{k}}{\sqrt{1^2 + (-5)^2 + (-3)^2}} = \frac{\mathbf{i} - 5\mathbf{j} - 3\mathbf{k}}{\sqrt{35}}.$$

Either $\mathbf{u}$ or its negative will do. □

PROBLEMS

1. Find $\mathbf{A} \times \mathbf{B}$ if $\mathbf{A} = 2\mathbf{i} - 2\mathbf{j} - \mathbf{k}$, $\mathbf{B} = \mathbf{i} + \mathbf{j} + \mathbf{k}$.

2. Find the direction of $\mathbf{A} \times \mathbf{B}$ if
a) $\mathbf{A} = \mathbf{i}$, $\mathbf{B} = \mathbf{j}$
b) $\mathbf{A} = \mathbf{i} + \mathbf{k}$, $\mathbf{B} = \mathbf{j}$
c) $\mathbf{A} = \mathbf{i} + \mathbf{k}$, $\mathbf{B} = \mathbf{j} + \mathbf{k}$
d) $\mathbf{A} = \sqrt{5}\mathbf{i} - 2\mathbf{j} - 3\mathbf{k}$, $\mathbf{B} = 2\mathbf{i} + 2\mathbf{j} + \mathbf{k}$

3. Find a vector $\mathbf{N}$ perpendicular to the plane determined by the points $A(1, -1, 2)$, $B(2, 0, -1)$, and $C(0, 2, 1)$.

4. Find a vector that is perpendicular to both of the vectors $\mathbf{A} = \mathbf{i} + \mathbf{j} + \mathbf{k}$ and $\mathbf{B} = \mathbf{i} + \mathbf{j}$.

5. Find the distance between the origin and the plane

ABC of Problem 3 by projecting $\overrightarrow{OA}$ onto the normal vector **N**.

6. Find the area of the triangle ABC in Problem 3.
7. If $\overrightarrow{AB} \times \overrightarrow{AC} = 2\mathbf{i} - 4\mathbf{j} + 4\mathbf{k}$, find the area of triangle ABC.
8. a) Find a vector normal to the plane through the three points

$$A(0, 2, 1), \quad B(2, 1, 2), \quad C(1, 1, 3).$$

   b) Find the area of the triangle formed by the three points in part (a).
9. If $\mathbf{A} = 2\mathbf{i} - \mathbf{j}$, $\mathbf{B} = \mathbf{i} + 3\mathbf{j} - 2\mathbf{k}$, find $\mathbf{A} \times \mathbf{B}$. Then calculate $(\mathbf{A} \times \mathbf{B}) \cdot \mathbf{A}$ and $(\mathbf{A} \times \mathbf{B}) \cdot \mathbf{B}$.
10. Is $(\mathbf{A} \times \mathbf{B}) \cdot \mathbf{A}$ always zero? Explain. What about $(\mathbf{A} \times \mathbf{B}) \cdot \mathbf{B}$?
11. Let $\mathbf{A} = 5\mathbf{i} - \mathbf{j} + \mathbf{k}$, $\mathbf{B} = \mathbf{j} - 5\mathbf{k}$, $\mathbf{C} = -15\mathbf{i} + 3\mathbf{j} - 3\mathbf{k}$. Which pairs of vectors are (a) perpendicular? (b) parallel?
12. The vector **A** is 4 units long and its direction is **i**. The vector **B** is 6 units long, and its direction is **k**.
    a) What is the direction of $\mathbf{A} \times \mathbf{B}$? of $\mathbf{B} \times \mathbf{A}$?
    b) What is the magnitude of $\mathbf{A} \times \mathbf{B}$? of $\mathbf{B} \times \mathbf{A}$?
13. $\mathbf{A} = 3\mathbf{i} + \mathbf{j} - \mathbf{k}$ is normal to a plane M_1 and $\mathbf{B} = 2\mathbf{i} - \mathbf{j} + \mathbf{k}$ is normal to a second plane M_2. (a) Find the angle between the two normals. (b) Do the two planes intersect? Give a reason for your answer. (c) If the two planes do intersect, find a vector parallel to their line of intersection.
14. Let **A** be a nonzero vector. Show that (a) $\mathbf{A} \times \mathbf{B} = \mathbf{A} \times \mathbf{C}$ does not guarantee $\mathbf{B} = \mathbf{C}$ (see Problem 36, Article 13.6); (b) $\mathbf{A} \cdot \mathbf{B} = \mathbf{A} \cdot \mathbf{C}$ and $\mathbf{A} \times \mathbf{B} = \mathbf{A} \times \mathbf{C}$ together imply $\mathbf{B} = \mathbf{C}$.
15. Vectors from the origin to the points A, B, C are given by $\mathbf{A} = \mathbf{i} - \mathbf{j} + \mathbf{k}$, $\mathbf{B} = 2\mathbf{i} + 3\mathbf{j} - \mathbf{k}$, $\mathbf{C} = -\mathbf{i} + 2\mathbf{j} + 2\mathbf{k}$. Find all points $P(x, y, z)$ that satisfy the following requirements: $\overrightarrow{OP}$ is a unit vector perpendicular to **C** and P lies in the plane determined by **A** and **B**.
16. Find the distance between the line L_1 through the two points $A(1, 0, -1)$, $B(-1, 1, 0)$ and the line L_2 through the two points $C(3, 1, -1)$, $D(4, 5, -2)$. (The distance is to be measured along a line perpendicular to L_1 and L_2. First find a vector **N** perpendicular to both lines, and then project $\overrightarrow{AC}$ onto **N**.)
17. Repeat Problem 16 for the points $A(4, 0, 2)$, $B(2, 4, 1)$, $C(1, 3, 2)$, $D(2, 2, 4)$.

13.8
Equations of Lines, Line Segments, and Planes

Lines

Suppose L is a line in space that passes through a point $P_0(x_0, y_0, z_0)$ and is parallel to a given nonzero vector

$$\mathbf{v} = A\mathbf{i} + B\mathbf{j} + C\mathbf{k}.$$

Then L is the set of all points $P(x, y, z)$ for which the vector $\overrightarrow{P_0P}$ is parallel to **v** (Fig. 13.43).

That is, P is on the line L if and only if there is a scalar t such that

$$\overrightarrow{P_0P} = t\mathbf{v} \tag{1a}$$

or

$$(x - x_0)\mathbf{i} + (y - y_0)\mathbf{j} + (z - z_0)\mathbf{k} = t(A\mathbf{i} + B\mathbf{j} + C\mathbf{k}). \tag{1b}$$

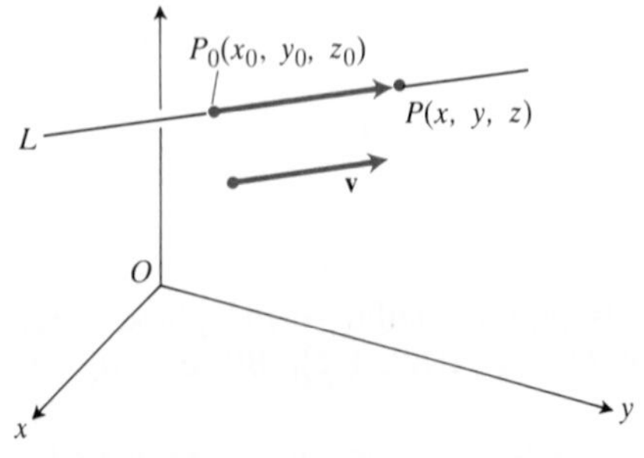

13.43 P is on the line through P_0 parallel to **v** if and only if $\overrightarrow{P_0P}$ is a scalar multiple of **v**.

When we match the components in Eq. (1b) we have

$$x - x_0 = tA, \qquad y - y_0 = tB, \qquad z - z_0 = tC \qquad \text{(2a)}$$

or

$$x = x_0 + tA, \qquad y = y_0 + tB, \qquad z = z_0 + tC. \qquad \text{(2b)}$$

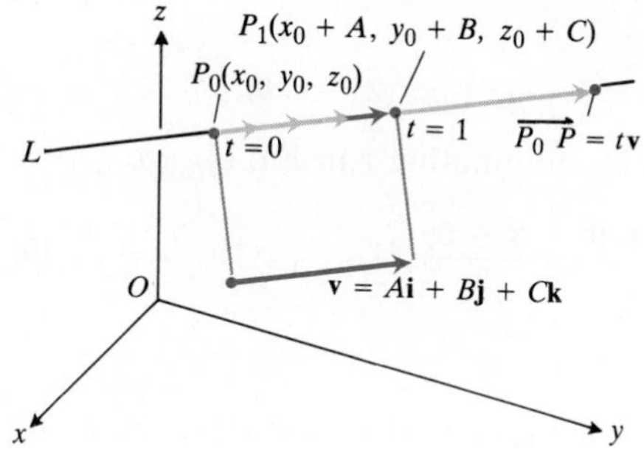

13.44 As t increases from 0 to 1, the point P moves from $P_0(x_0, y_0, z_0)$ to $P_1(x_0 + A, y_0 + B, z_0 + C)$.

The equations in (2a) and (2b) are *parametric equations* for the line, as opposed to Eqs. (1a) and (1b), which are *vector equations* for the line. If we let t vary between $-\infty$ and $+\infty$, the point $P(x, y, z)$ given by Eq. (2b) will traverse the entire line L through P_0.

If we let t vary through the closed interval $[a, b]$ from $t = a$ to $t = b$, the point $P(x, y, z)$ given by Eq. (2b) will trace out the segment from the point where $t = a$ to the point where $t = b$.

For example, when t varies from 0 to 1, the point

$$P(x, y, z) = P(x_0 + tA, x_0 + tB, x_0 + tC)$$

in Fig. 13.44 moves from $P_0(x_0, y_0, z_0)$ to $P_1(x_0 + A, y_0 + B, z_0 + C)$.

We may eliminate t from the equations in (2) to obtain the following *cartesian equations* for the line:

$$\frac{x - x_0}{A} = \frac{y - y_0}{B} = \frac{z - z_0}{C}, \qquad \text{(3)}$$

provided A, B, and C are different from zero.

EXAMPLE 1 The line through $P(2, -9, 5)$ parallel to $\mathbf{v} = 3\mathbf{i} - \mathbf{j} + 4\mathbf{k}$ is

$$\frac{x - 2}{3} = \frac{y + 9}{-1} = \frac{z - 5}{4}.$$

Remember: To say that these are cartesian equations for the line is to say that (x, y, z) lies on the line if and only if these equations hold. □

If any one of the constants A, B, or C is zero in Eqs. (3), the corresponding numerator is also zero. This follows at once from the parametric equations (2), which show, for example, that

$$x - x_0 = tA \qquad \text{and} \qquad A = 0$$

together imply that

$$x - x_0 = 0.$$

Thus, when one of the denominators in (3) is zero, we interpret the equations to say that the corresponding numerator is zero. With this interpretation, Eqs. (3) may always be used.

EXAMPLE 2 Find parametric and cartesian equations for the line through $P(2, -9, 5)$ parallel to $\mathbf{v} = 2\mathbf{j} + 3\mathbf{k}$.

Solution The parametric equations given by Eq. (2b) are

$$x = 2 + 0 \cdot t, \qquad y = -9 + 2t, \qquad z = 5 + 3t$$

or

$$x = 2, \qquad y = -9 + 2t, \qquad z = 5 + 3t. \tag{4}$$

As t increases from $t = 0$ to $t = 1$, the point (x, y, z) in (4) traces the segment from

$$t = 0: \qquad (2, -9 + 2 \cdot 0, 5 + 3 \cdot 0) = (2, -9, 5)$$

to

$$t = 1: \qquad (2, -9 + 2 \cdot 1, 5 + 3 \cdot 1) = (2, -7, 8).$$

The cartesian equations found by eliminating t in Eq. (4) are

$$x = 2, \qquad \frac{y + 9}{2} = \frac{z - 5}{3}. \; \square \tag{5}$$

Planes

To obtain an equation for a *plane*, we suppose that a point $P_0(x_0, y_0, z_0)$ on the plane and a nonzero vector

$$\mathbf{N} = A\mathbf{i} + B\mathbf{j} + C\mathbf{k} \tag{6}$$

perpendicular to the plane are given (Fig. 13.45). Then the point $P(x, y, z)$ will lie in the plane if and only if the vector $\overrightarrow{P_0P}$ is perpendicular to $\mathbf{N}$; that is, if and only if

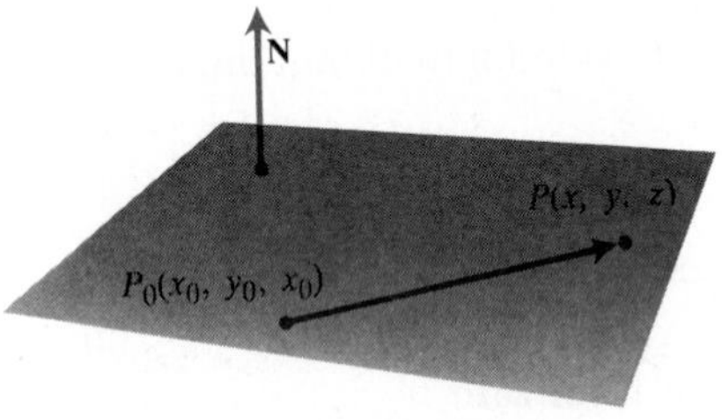

13.45 P lies in the plane through P_0 perpendicular to $\mathbf{N}$ if and only if $\overrightarrow{P_0P} \cdot \mathbf{N} = 0$.

$$\mathbf{N} \cdot \overrightarrow{P_0P} = 0$$

or

$$A(x - x_0) + B(y - y_0) + C(z - z_0) = 0. \tag{7}$$

This equation may also be put in the form

$$Ax + By + Cz = Ax_0 + By_0 + Cz_0 \tag{8}$$

or

$$Ax + By + Cz = D, \qquad \text{where} \qquad D = Ax_0 + By_0 + Cz_0. \tag{9}$$

Conversely, if we start from any linear equation such as (9), we may find a point $P_0(x_0, y_0, z_0)$ whose coordinates do satisfy it; that is, such that

$$Ax_0 + By_0 + Cz_0 = D.$$

Then, by subtraction, we may put the given equation (9) into the form of Eq. (7) and factor it into the dot product

$$\mathbf{N} \cdot \overrightarrow{P_0P} = 0,$$

with $\mathbf{N}$ as in Eq. (6). This says that the constant vector $\mathbf{N}$ is perpendicular to the vector $\overrightarrow{P_0P}$ for every pair of points P_0 and P whose coordinates satisfy the equation. Hence the set of points $P(x, y, z)$ whose coordinates satisfy such a linear equation is a plane and the vector $A\mathbf{i} + B\mathbf{j} + C\mathbf{k}$, with the same coefficients that x, y, and z have in the given equation, is normal to the plane.

Equation (7) is called a *vector equation* for the plane through P, perpendicular to **N**. Equation (9) is a *cartesian equation* for the plane.

EXAMPLE 3 Find a cartesian equation for the plane through $P_0(-3, 0, 7)$ perpendicular to the vector $\mathbf{N} = 5\mathbf{i} + 2\mathbf{j} - \mathbf{k}$.

Solution From Eq. (9),

$$D = 5(-3) + 2(0) - 1(7) = -15 - 7 = -22,$$

and the corresponding cartesian equation for this line is

$$5x + 2x - y = -22. \square$$

EXAMPLE 4 Find the distance d between the point $P(2, -3, 4)$ and the plane $x + 2y + 2z = 13$.

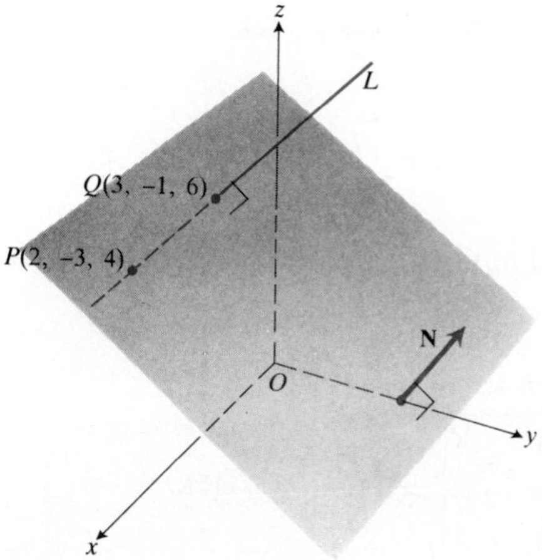

13.46 The distance between P and the plane is the distance between P and Q.

Solution 1 Carry out the following steps:

STEP 1. Find a line L through P normal to the plane.

STEP 2. Find the coordinates of the point Q in which the line meets the plane.

STEP 3. Compute the distance between P and Q.

The vector $\mathbf{N} = \mathbf{i} + 2\mathbf{j} + 2\mathbf{k}$ is normal to the given plane, and the line

$$x = 2 + t, \qquad y = -3 + 2t, \qquad z = 4 + 2t \tag{10}$$

goes through P, and, being parallel to **N**, is normal to the plane. The point (x, y, z) from (10) will lie in the plane

$$x + 2y + 2z = 13$$

if

$$\begin{aligned}(2 + t) + 2(-3 + 2t) + 2(4 + 2t) &= 13,\\ 9t + 4 &= 13,\\ t &= 1.\end{aligned}$$

Thus, the point Q in which the line meets the plane can be obtained by setting $t = 1$ in Eq. (11):

$$Q(x, y, z) = (2 + 1, -3 + 2, 4 + 2) = (3, -1, 6).$$

The distance between $P(2, -3, 4)$ and $Q(3, -1, 6)$ (Fig. 13.46) is

$$\sqrt{(2-3)^2 + (-3+1)^2 + (4-6)^2} = \sqrt{1 + 4 + 4} = \sqrt{9} = 3.$$

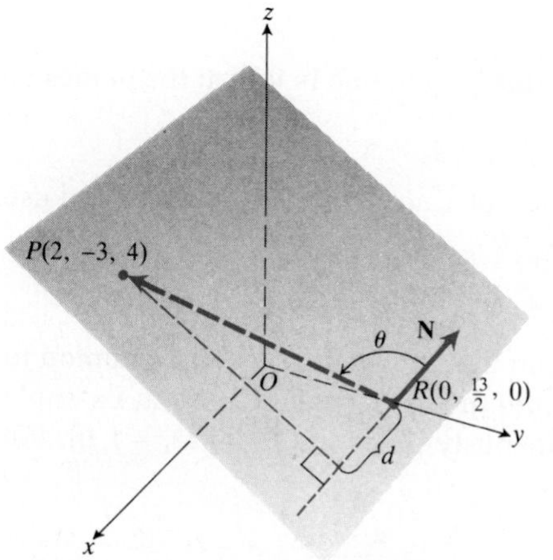

13.47 d is the length of the $\text{proj}_{\mathbf{N}} \overrightarrow{RP}$.

Solution 2 Let R be any point in the plane, and find the component of $\overrightarrow{RP}$ in the direction of **N**. This will be plus d or minus d, the sign depending on the direction of the vector projection of $\overrightarrow{RP}$ onto **N**. Figure 13.47 shows that the component is negative in this case, but we do not need to know this to find d.

Since R can be any point in the given plane, we might as well choose R to be a point whose coordinates are simple, say the point $R(0, \frac{13}{2}, 0)$ where the plane meets the y-axis. Then,

$$\overrightarrow{RP} = (2 - 0)\mathbf{i} + (-3 - \tfrac{13}{2})\mathbf{j} + (4 - 0)\mathbf{k} = 2\mathbf{i} - \tfrac{19}{2}\mathbf{j} + 4\mathbf{k}.$$

The component of $\overrightarrow{RP}$ in the direction of $\mathbf{N} = \mathbf{i} + 2\mathbf{j} + 2\mathbf{k}$ is

$$\overrightarrow{RP} \cdot \frac{\mathbf{N}}{|\mathbf{N}|} = \frac{\mathbf{N} \cdot \overrightarrow{RP}}{|\mathbf{N}|} = \frac{2 - 19 + 8}{\sqrt{(1)^2 + (2)^2 + (2)^2}} = \frac{-9}{\sqrt{9}} = -3.$$

Therefore, $d = 3$. □

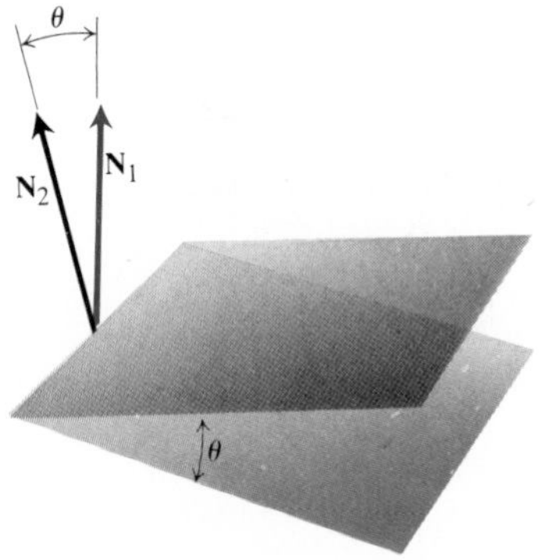

13.48 The angle between two planes can be obtained from their normals.

EXAMPLE 5 Find the angle between the two planes $3x - 6y - 2z = 7$ and $2x + y - 2z = 5$.

Solution Clearly the angle between two planes (Fig. 13.48) is the same as the angle between their normals. (Actually there are two angles in each case, namely θ and $180° - \theta$.) From the equations of the planes we may read off their normal vectors:

$$\mathbf{N}_1 = 3\mathbf{i} - 6\mathbf{j} - 2\mathbf{k}, \qquad \mathbf{N}_2 = 2\mathbf{i} + \mathbf{j} - 2\mathbf{k}.$$

Then

$$\cos\theta = \frac{\mathbf{N}_1 \cdot \mathbf{N}_2}{|\mathbf{N}_1||\mathbf{N}_2|} = \frac{4}{21}, \qquad \theta = \cos^{-1}\left(\frac{4}{21}\right) \approx 79°. \ \square$$

EXAMPLE 6 Find a vector parallel to the line of intersection of the two planes in Example 5.

Solution The requirements are met by the vector

$$\mathbf{v} = \mathbf{N}_1 \times \mathbf{N}_2 = \begin{vmatrix} \mathbf{i} & \mathbf{j} & \mathbf{k} \\ 3 & -6 & -2 \\ 2 & 1 & -2 \end{vmatrix} = 14\mathbf{i} + 2\mathbf{j} + 15\mathbf{k}.$$

The vector $\mathbf{v}$ is perpendicular to both of the normals $\mathbf{N}_1$ and $\mathbf{N}_2$, and is therefore parallel to both planes. □

Intersecting Planes

EXAMPLE 7 Find cartesian equations for the line in which the planes of Example 5 intersect.

Solution We find a vector parallel to the line, a point on the line, and use Eq. (3). Example 6 gives

$$\mathbf{v} = 14\mathbf{i} + 2\mathbf{j} + 15\mathbf{k}$$

parallel to the line. To find a point on the line we find a point common to the two planes. Substituting $z = 0$ into the plane equations from Example 5 and solving for x and y simultaneously gives the point $(3, -1, 0)$. We therefore obtain

$$\frac{(x - 3)}{14} = \frac{y - (-1)}{2} = \frac{z - 0}{15}$$

or

$$\frac{x-3}{14} = \frac{y+1}{2} = \frac{z}{15}$$

as cartesian equations for the line. □

EXAMPLE 8 Find an equation of the plane that passes through the two points $P_1(1, 0, -1)$ and $P_2(-1, 2, 1)$ and is parallel to the line of intersection of the planes $3x + y - 2z = 6$ and $4x - y + 3z = 0$.

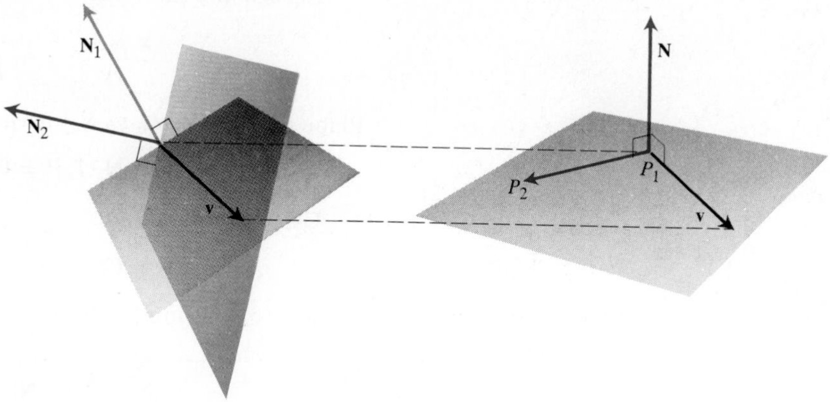

13.49 Constructing a plane through P_1 and P_2 that is parallel to the line of intersection of two other planes.

Solution The coordinates of either one of the points P_1 or P_2 will do for the x_0, y_0, and z_0 in Eq. (9). What remains, then, is to find a vector **N** normal to the plane in question to furnish the coefficients A, B, and C of Eq. (9) (Fig. 13.49).

The line of intersection of the two given planes is parallel to the vector

$$\mathbf{v} = \mathbf{N}_1 \times \mathbf{N}_2 = \begin{vmatrix} \mathbf{i} & \mathbf{j} & \mathbf{k} \\ 3 & 1 & -2 \\ 4 & -1 & 3 \end{vmatrix} = \mathbf{i} - 17\mathbf{j} - 7\mathbf{k},$$

where $\mathbf{N}_1$ and $\mathbf{N}_2$ are normals to the two given planes. The vector

$$\overrightarrow{P_1P_2} = -2\mathbf{i} + 2\mathbf{j} + 2\mathbf{k}$$

is to lie in the required plane. Now we may slide **v** parallel to itself until it also lies in the required plane (since the plane is to be parallel to **v**). Hence we may take

$$\mathbf{N} = \overrightarrow{P_1P_2} \times \mathbf{v} = 20\mathbf{i} - 12\mathbf{j} + 32\mathbf{k}$$

as a vector normal to the plane. Actually,

$$\frac{1}{4}\mathbf{N} = 5\mathbf{i} - 3\mathbf{j} + 8\mathbf{k}$$

serves just as well. From this normal vector, we may substitute

$$A = 5, \qquad B = -3, \qquad C = 8$$

in Eq. (9), together with $x_0 = 1$, $y_0 = 0$, $z_0 = -1$, since the point $(1, 0, -1)$ is to lie in the plane. The required plane is therefore

$$5x - 3y + 8z = -3. \ \square$$

Equations for Lines and Planes in Space

Line through $P_0(x_0, y_0, z_0)$ parallel to $\mathbf{v} = A\mathbf{i} + B\mathbf{j} + C\mathbf{k}$:

Vector equation: $\overrightarrow{P_0P} = t\mathbf{v}$

Parametric equations:

$$x = x_0 + tA, \qquad y = y_0 + tB, \qquad z = z_0 + tC$$

Cartesian equations:

$$\frac{x - x_0}{A} = \frac{y - y_0}{B} = \frac{z - z_0}{C}$$

Plane through $P_0(x_0, y_0, z_0)$ perpendicular to $\mathbf{N} = A\mathbf{i} + B\mathbf{j} + C\mathbf{k}$:

Vector equation: $\mathbf{N} \cdot \overrightarrow{P_0P} = 0$

Cartesian equation:

$$Ax + By + Cz = D, \qquad D = Ax_0 + By_0 + Cz_0$$

Distance from point P to a plane:

$$\text{Distance} = \left|\overrightarrow{PR} \cdot \frac{\mathbf{N}}{|\mathbf{N}|}\right| = \left|\frac{\mathbf{N} \cdot \overrightarrow{PR}}{|\mathbf{N}|}\right| \qquad \begin{pmatrix}\mathbf{N} \text{ any normal to the plane,} \\ R \text{ any point in the plane}\end{pmatrix}$$

PROBLEMS

1. Find the coordinates of the point P in which the line

$$\frac{x - 1}{2} = \frac{y + 1}{-1} = \frac{z}{3}$$

intersects the plane $3x + 2y - z = 5$.

2. Find parametric and cartesian equations of the line joining the points $A(1, 2, -1)$ and $B(-1, 0, 1)$.

3. Find parametric equations for the line through the points $(1, 2, 0)$ and $(1, 1, -1)$.

4. Find equations for the line through $P(2, 4, 5)$ perpendicular to the plane $3x + 7y - 5z = 21$.

5. Find the point of intersection of the lines

$$\frac{x}{1} = \frac{y}{3} = \frac{z}{2} \quad \text{and} \quad \frac{x - 2}{1} = \frac{y - 3}{4} = \frac{z - 4}{2}.$$

6. Find an equation for the plane through the point $P_0(0, 2, -1)$ perpendicular to $\mathbf{N} = 3\mathbf{i} - 2\mathbf{j} - \mathbf{k}$.

7. Find an equation for the plane through $(1, -1, 3)$ parallel to the plane $3x + y + z = 7$.

8. Find an equation for the plane through the points $(1, 1, -1)$, $(2, 0, 2)$, and $(0, -2, 1)$.

9. Find an equation for the plane through the points $(2, 4, 5)$, $(1, 5, 7)$, and $(-1, 6, 8)$.

10. Find an equation for the plane through the points $(1, 0, 0)$, $(0, 1, 0)$, and $(0, 0, 1)$.

11. Find an equation for the plane through $P_0(2, 4, 5)$ perpendicular to the line

$$\frac{x - 5}{1} = \frac{y - 1}{3} = \frac{z}{4}.$$

12. Find an equation for the plane through the origin that contains the line

$$\frac{x - 1}{2} = \frac{y + 1}{3} = \frac{z}{4}.$$

13. Find a plane through $A(1, -2, 1)$ perpendicular to the vector from the origin to A.

14. Find a plane through $P_0(2, 1, -1)$ perpendicular to the line of intersection of the planes $2x + y - z = 3$, $x + 2y + z = 2$.

15. Find a plane through the points $P_0(1, 2, 3)$ and $P_2(3, 2, 1)$ perpendicular to the plane $4x - y + 2z = 7$.

16. Find the distance from the point $(1, 1, 1)$ to the plane $2x - 3y + 7z = 11$.

17. a) Find an equation for the plane through $P_0(4, 2, 1)$ perpendicular to $\mathbf{N} = 6\mathbf{i} - 2\mathbf{j} + 3\mathbf{k}$.
b) Find the distance from the origin to the plane in part (a).
c) Find the distance from the point $(6, 6, 8)$ to the plane in part (a).
d) Find equations for the line through $P_0(4, 2, -1)$ perpendicular to the plane in part (a).

18. Find the angle between the planes
a) $x + y + z = 5$, $x + 2y + z = 7$,
b) $x + y = 1$, $x + z = 1$,
c) $5x + y - z = 10$, $x - 2y + 3 = -1$.

19. Find the line of intersection of the planes
a) $x + y + z = 1$, $x + y = 2$,
b) $3x - 6y - 2z = 3$, $2x + y - 2z = 2$,
c) $x - 2y + 4z = 2$, $x + y - 2z = 5$.

20. The line L through the origin is normal to the plane $2x - y - z = 4$. Find the point in which L meets the plane $x + y - 2z = 2$.

21. a) What is meant by the angle between a line and a plane?
b) Find the acute angle between the line

$$x = -1 + 2t, \qquad y = 3t, \qquad z = 3 + 6t$$

and the plane $10x + 2y - 11z = 3$.

22. Let $P_i(x_i, y_i, z_i)$, $i = 1, 2, 3$, be three points. What set is described by the equation

$$\begin{vmatrix} x & y & z & 1 \\ x_1 & y_1 & z_1 & 1 \\ x_2 & y_2 & z_2 & 1 \\ x_3 & y_3 & z_3 & 1 \end{vmatrix} = 0?$$

23. Find the distance between the origin and the line

$$\frac{x-5}{3} = \frac{y-5}{4} = \frac{z+3}{-5}.$$

24. The equation $\mathbf{N} \cdot \overrightarrow{P_0P} = 0$ represents a plane through P_0 perpendicular to $\mathbf{N}$. What set does the inequality $\mathbf{N} \cdot \overrightarrow{P_0P} > 0$ represent? Give a reason for your answer.

25. The unit vector $\mathbf{u}$ makes angles α, β, γ, respectively, with the positive x-, y-, z-axes. Find the plane normal to $\mathbf{u}$ through $P_0(x_0, y_0, z_0)$.

26. Show that the planes obtained by substituting different values for the constant D in the equation

$$2x + 3y - 6z = D$$

are parallel. Find the distance between two of these planes, one corresponding, say, to $D = D_1$ and the other to $D = D_2$.

27. Prove that the line

$$x = 1 + 2t, \qquad y = -1 + 3t, \qquad z = 2 + 4t$$

is parallel to the plane $x - 2y + z = 6$.

28. a) Prove that three points A, B, C are collinear if and only if $\overrightarrow{AC} \times \overrightarrow{AB} = \mathbf{0}$.
b) Are the points $A(1, 2, -3)$, $B(3, 1, 0)$, $C(-3, 4, -9)$ collinear?

29. Prove that four points A, B, C, D are coplanar if and only if $\overrightarrow{AD} \cdot (\overrightarrow{AB} \times \overrightarrow{BC}) = 0$.

30. Show that the line of intersection of the planes

$$x + 2y - 2z = 5 \qquad \text{and} \qquad 5x - 2y - z = 0$$

is parallel to the line

$$x = -3 + 2t, \qquad y = 3t, \qquad z = 1 + 4t.$$

Find the plane determined by these two lines.

31. Show that the lines

$$\frac{x+1}{3} = \frac{y-6}{1} = \frac{z-3}{2}$$

and

$$\frac{x-6}{2} = \frac{y-11}{2} = \frac{z-3}{-1}$$

intersect. Find the plane determined by these two lines.

32. Show, by vector methods, that the distance from the point $P_1(x_1, y_1, z_1)$ to the plane $Ax + By + Cz - D = 0$ is

$$\frac{|Ax_1 + By_1 + Cz_1 - D|}{\sqrt{A^2 + B^2 + C^2}}.$$

***Toolkit* programs**

3D Grapher

13.9

Products of Three Vectors or More

Products of three vectors or more often arise in physical and engineering problems. For example, the electromotive force $d\mathcal{E}$ induced in an element of a conducting wire $d\mathbf{l}$ moving with velocity $\mathbf{v}$ through a magnetic field $\mathbf{B}$ is given by $d\mathcal{E} = (\mathbf{v} \times \mathbf{B}) \cdot d\mathbf{l}$.† Here the factor in parentheses is a vector, and the result of forming the scalar product of this vector and $d\mathbf{l}$ is a scalar. It is a real economy in thinking to represent the result in the compact vector form that removes the necessity of carrying factors such as the sine of the angle between $\mathbf{B}$ and $\mathbf{v}$ and the cosine of the angle between the normal to their plane and the vector $d\mathbf{l}$. These are all taken into account by the given product of the three vectors.

Triple Scalar Product

The product $(\mathbf{A} \times \mathbf{B}) \cdot \mathbf{C}$, called the *triple scalar product*, has the following geometrical significance. The vector $\mathbf{N} = \mathbf{A} \times \mathbf{B}$ is normal to the base of the parallelepiped determined by the vectors **A**, **B**, and **C** in Fig. 13.50. The magnitude of **N** equals the area of the base determined by **A** and **B**. Thus

$$(\mathbf{A} \times \mathbf{B}) \cdot \mathbf{C} = \mathbf{N} \cdot \mathbf{C} = |\mathbf{N}|\,|\mathbf{C}| \cos \theta$$

is, except perhaps for sign, the *volume of a box* of edges **A**, **B**, and **C**, since

$$|\mathbf{N}| = |\mathbf{A} \times \mathbf{B}| = \text{area of base}$$

and

$$|\mathbf{C}| \cos \theta = \pm h = \pm \text{altitude of box.}$$

If **C** and $\mathbf{A} \times \mathbf{B}$ lie on the same side of the plane determined by **A** and **B**, the triple scalar product will be positive. But if the vectors **A**, **B**, and **C** form a left-handed system, then $(\mathbf{A} \times \mathbf{B}) \cdot \mathbf{C}$ is negative. By successively considering the plane of **B** and **C**, then the plane of **C** and **A**, as the base of the parallelepiped, we can readily see that

$$(\mathbf{A} \times \mathbf{B}) \cdot \mathbf{C} = (\mathbf{B} \times \mathbf{C}) \cdot \mathbf{A} = (\mathbf{C} \times \mathbf{A}) \cdot \mathbf{B}. \tag{1}$$

Since the dot product is commutative, we also have

$$(\mathbf{B} \times \mathbf{C}) \cdot \mathbf{A} = \mathbf{A} \cdot (\mathbf{B} \times \mathbf{C}),$$

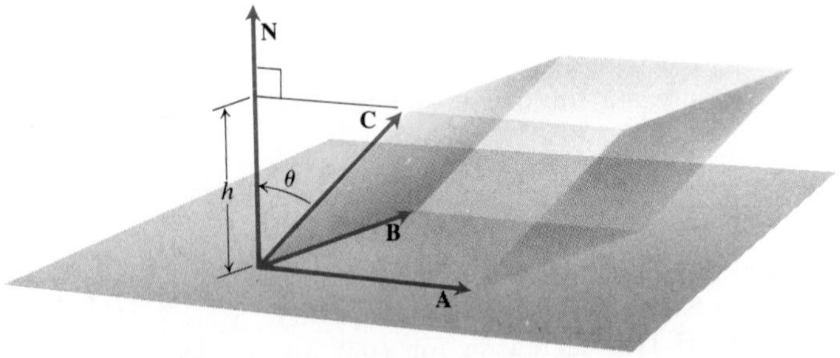

13.50 Except perhaps for sign, the number $(\mathbf{A} \times \mathbf{B}) \cdot \mathbf{C}$ is the volume of the parallelepiped (parallelogram-sided box) shown here.

† See F. W. Sears, M. W. Zemansky, H. D. Young, *University Physics*, 6th ed. (Reading, Mass.: Addison-Wesley, 1982), Chapter 33.

so that Eq. (1) gives the result

$$(\mathbf{A} \times \mathbf{B}) \cdot \mathbf{C} = \mathbf{A} \cdot (\mathbf{B} \times \mathbf{C}). \tag{2}$$

Equation (2) says that the dot and the cross may be interchanged in the triple scalar product, provided only that the multiplications are performed in a way that "makes sense." Thus $(\mathbf{A} \cdot \mathbf{B}) \times \mathbf{C}$ is excluded on the ground that $(\mathbf{A} \cdot \mathbf{B})$ is a scalar and we never "cross" a scalar and a vector.

The triple scalar product in Eq. (2) is conveniently expressed in determinant form as follows:

$$\begin{aligned}
\mathbf{A} \cdot (\mathbf{B} \times \mathbf{C}) &= A \cdot \left[\begin{vmatrix} b_2 & b_3 \\ c_2 & c_3 \end{vmatrix} \mathbf{i} - \begin{vmatrix} b_1 & b_3 \\ c_1 & c_3 \end{vmatrix} \mathbf{j} + \begin{vmatrix} b_1 & b_2 \\ c_1 & c_2 \end{vmatrix} \mathbf{k} \right] \\
&= a_1 \begin{vmatrix} b_2 & b_3 \\ c_2 & c_3 \end{vmatrix} - a_2 \begin{vmatrix} b_1 & b_3 \\ c_1 & c_3 \end{vmatrix} + a_3 \begin{vmatrix} b_1 & b_2 \\ c_1 & c_2 \end{vmatrix} \qquad (3) \\
&= \begin{vmatrix} a_1 & a_2 & a_3 \\ b_1 & b_2 & b_3 \\ c_1 & c_2 & c_3 \end{vmatrix}.
\end{aligned}$$

$$\mathbf{A} \cdot (\mathbf{B} \times \mathbf{C}) = \begin{vmatrix} a_1 & a_2 & a_3 \\ b_1 & b_2 & b_3 \\ c_1 & c_2 & c_3 \end{vmatrix}.$$

EXAMPLE 1 See Fig. 13.51. Let

$$\mathbf{A} = \overrightarrow{PQ}, \qquad \mathbf{B} = \overrightarrow{PS}$$
$$\mathbf{A}' = \overrightarrow{P'Q'}, \qquad \mathbf{B}' = \overrightarrow{P'S'}$$

be sides of parallelograms $PQRS$ and $P'Q'R'S'$ that are related in such a way that PP', QQ', and SS' are parallel to one another and to the unit vector $\mathbf{n}$. Show that

$$(\mathbf{A} \times \mathbf{B}) \cdot \mathbf{n} = (\mathbf{A}' \times \mathbf{B}') \cdot \mathbf{n} \tag{4}$$

and discuss the geometrical meaning of this identity.

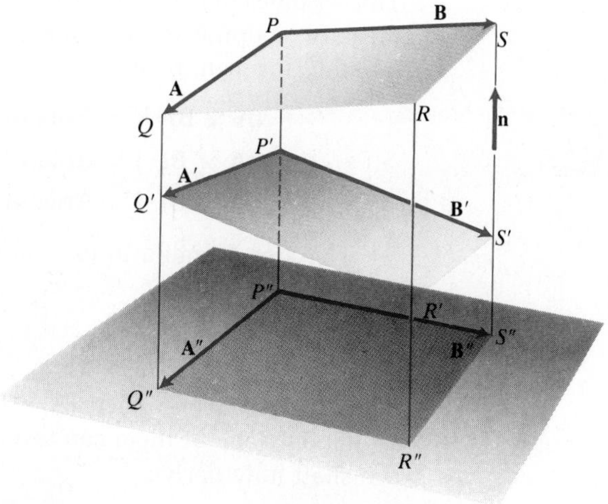

13.51 If $|\mathbf{n}| = 1$, then $(\mathbf{A} \times \mathbf{B}) \cdot \mathbf{n}$ is the area of the projection of the parallelogram determined by $\mathbf{A}$ and $\mathbf{B}$ on a plane perpendicular to $\mathbf{n}$.

Verification of Eq. (4) From the way the parallelograms are related, it follows that

$$\begin{aligned}\mathbf{A} = \overrightarrow{PQ} &= \overrightarrow{PP'} + \overrightarrow{P'Q'} + \overrightarrow{Q'Q} \\ &= \overrightarrow{P'Q'} + (\overrightarrow{PP'} - \overrightarrow{QQ'}) \\ &= \mathbf{A}' + s\mathbf{n}\end{aligned}$$

for some scalar s, since both $\overrightarrow{PP'}$ and QQ' are parallel to **n.** Similarly,

$$\mathbf{B} = \mathbf{B}' + t\mathbf{n}$$

for some scalar t. Hence

$$\begin{aligned}\mathbf{A} \times \mathbf{B} &= (\mathbf{A}' + s\mathbf{n}) \times (\mathbf{B}' + t\mathbf{n}) \\ &= \mathbf{A}' \times \mathbf{B}' + t(\mathbf{A}' \times \mathbf{n}) + s(\mathbf{n} \times \mathbf{B}') + st(\mathbf{n} \times \mathbf{n}). \qquad (5)\end{aligned}$$

But $\mathbf{n} \times \mathbf{n} = \mathbf{0}$, while $\mathbf{A}' \times \mathbf{n}$ and $\mathbf{n} \times \mathbf{B}'$ are both perpendicular to **n.** Therefore when we dot both sides of (5) with **n** we get Eq. (4).

Geometrical meaning of Eq. (4) The result (4) says that when the parallelograms $PQRS$ and $P'Q'R'S'$ are any two plane sections of a prism with sides parallel to **n,** then the box determined by **A, B,** and **n** has the same volume as the box determined by **A**′, **B**′, and **n.** Thus, in particular, we may replace the right-hand side of (4) by $(\mathbf{A}'' \times \mathbf{B}'') \cdot \mathbf{n}$, where $\mathbf{A}''$ and $\mathbf{B}''$ are sides of a *right* section $P''Q''R''S''$ as in Fig. 13.51. Then $\mathbf{A}'' \times \mathbf{B}''$ is parallel to **n,** and

$$\mathbf{A}'' \times \mathbf{B}'' = (\text{Area right section}) \text{ times } \mathbf{n}$$

and

$$(\mathbf{A}'' \times \mathbf{B}'') \cdot \mathbf{n} = \text{Area right section.}$$

Therefore, by Eq. (4), we have the following interpretation:

> $(\mathbf{A} \times \mathbf{B}) \cdot \mathbf{n}$ is the area of the orthogonal projection of the parallelogram determined by **A** and **B** onto a plane whose unit normal is **n.** (6)

This assumes that $\mathbf{A} \times \mathbf{B}$ and **n** lie on the same side of the plane $PQRS$. If they are on opposite sides, take the absolute value to get the area. Except possibly for sign, then,

$$(\mathbf{A} \times \mathbf{B}) \cdot \mathbf{k} = \text{Area of projection in the } xy\text{-plane}, \qquad (7a)$$

$$(\mathbf{A} \times \mathbf{B}) \cdot \mathbf{j} = \text{Area of projection in the } xz\text{-plane}, \qquad (7b)$$

$$(\mathbf{A} \times \mathbf{B}) \cdot \mathbf{i} = \text{Area of projection in the } yz\text{-plane}. \qquad (7c)$$

The area formula in (6) will be used in calculating surface areas in Article 16.9. □

Triple Vector Product

The triple vector products $(\mathbf{A} \times \mathbf{B}) \times \mathbf{C}$ and $\mathbf{A} \times (\mathbf{B} \times \mathbf{C})$ are usually not equal, but each of them can be evaluated rather simply by formulas that we shall now derive.

We start by showing that the vector product $(\mathbf{A} \times \mathbf{B}) \times \mathbf{C}$ is given by

$$(\mathbf{A} \times \mathbf{B}) \times \mathbf{C} = (\mathbf{A} \cdot \mathbf{C})\mathbf{B} - (\mathbf{B} \cdot \mathbf{C})\mathbf{A}. \tag{8}$$

In this formula, $\mathbf{A} \cdot \mathbf{C}$ is a scalar multiplying $\mathbf{B}$, and $\mathbf{B} \cdot \mathbf{C}$ is a scalar multiplying $\mathbf{A}$.

CASE 1. If one of the vectors is the zero vector, Eq. (8) is true because both sides of it are zero.

CASE 2. If none of the vectors is zero, but if $\mathbf{B} = s\mathbf{A}$ for some scalar s, then both sides of Eq. (8) are zero again.

CASE 3. Suppose that none of the vectors is zero and that $\mathbf{A}$ and $\mathbf{B}$ are not parallel. The vector on the left in Eq. (8) is parallel to the plane determined by $\mathbf{A}$ and $\mathbf{B}$, so that it is possible to find scalars m and n such that

$$(\mathbf{A} \times \mathbf{B}) \times \mathbf{C} = m\mathbf{A} + n\mathbf{B}. \tag{9}$$

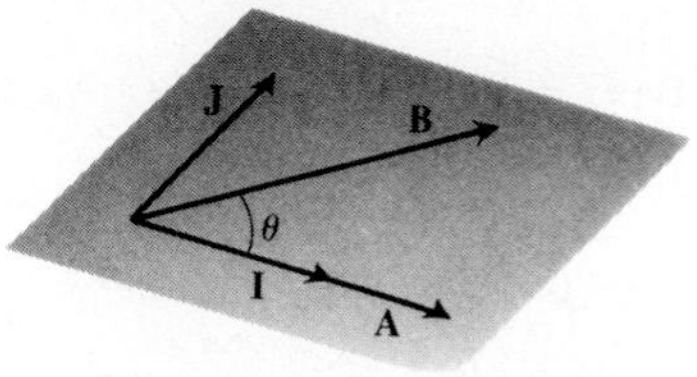

13.52 Orthogonal vectors **I** and **J** in the plane of **A** and **B**.

To calculate m and n, we introduce orthogonal unit vectors $\mathbf{I}$ and $\mathbf{J}$ in the plane of $\mathbf{A}$ and $\mathbf{B}$ with $\mathbf{I} = \mathbf{A}/|\mathbf{A}|$ (Fig. 13.52). We also introduce a third unit vector $\mathbf{K} = \mathbf{I} \times \mathbf{J}$, and write all our vectors in terms of these unit vectors $\mathbf{I}$, $\mathbf{J}$, and $\mathbf{K}$:

$$\begin{aligned} \mathbf{A} &= a_1\mathbf{I}, \\ \mathbf{B} &= b_1\mathbf{I} + b_2\mathbf{J}, \\ \mathbf{C} &= c_1\mathbf{I} + c_2\mathbf{J} + c_3\mathbf{K}. \end{aligned} \tag{10}$$

Then

$$\mathbf{A} \times \mathbf{B} = a_1 b_2 \mathbf{K}$$

and

$$(\mathbf{A} \times \mathbf{B}) \times \mathbf{C} = a_1 b_2 c_1 \mathbf{J} - a_1 b_2 c_2 \mathbf{I}. \tag{11}$$

Comparing this with the right-hand side of Eq. (9), we have

$$m(a_1\mathbf{I}) + n(b_1\mathbf{I} + b_2\mathbf{J}) = a_1 b_2 c_1 \mathbf{J} - a_1 b_2 c_2 \mathbf{I}.$$

This is equivalent to the pair of scalar equations

$$\begin{aligned} ma_1 + nb_1 &= -a_1 b_2 c_2, \\ nb_2 &= a_1 b_2 c_1. \end{aligned}$$

If b_2 were equal to zero, $\mathbf{A}$ and $\mathbf{B}$ would be parallel, contrary to hypothesis. Hence b_2 is not zero and we may solve the last equation for n. We find

$$n = a_1 c_1 = \mathbf{A} \cdot \mathbf{C}.$$

Then, by substitution,

$$\begin{aligned} ma_1 &= -nb_1 - a_1 b_2 c_2 \\ &= -a_1 c_1 b_1 - a_1 b_2 c_2, \end{aligned}$$

and since $|\mathbf{A}| = a_1 \neq 0$, we may divide by a_1 and have

$$m = -(b_1 c_1 + b_2 c_2) = -(\mathbf{B} \cdot \mathbf{C}).$$

When these values are substituted for m and n in Eq. (9), we obtain the result given in Eq. (8).

The identity

$$(\mathbf{B} \times \mathbf{C}) \times \mathbf{A} = (\mathbf{B} \cdot \mathbf{A})\mathbf{C} - (\mathbf{C} \cdot \mathbf{A})\mathbf{B} \tag{12a}$$

follows from Eq. (8) by a simple interchange of the letters **A, B,** and **C.** If we now interchange the factors $\mathbf{B} \times \mathbf{C}$ and **A** we must change the sign on the right-hand side of the equation. This gives the following identity, which is a companion of Eq. (8):

$$\mathbf{A} \times (\mathbf{B} \times \mathbf{C}) = (\mathbf{A} \cdot \mathbf{C})\mathbf{B} - (\mathbf{A} \cdot \mathbf{B})\mathbf{C}. \tag{12b}$$

EXAMPLE 2 Verify Eq. (8) for the vectors

$$\mathbf{A} = \mathbf{i} - \mathbf{j} + 2\mathbf{k}, \qquad \mathbf{B} = 2\mathbf{i} + \mathbf{j} + \mathbf{k}, \qquad \mathbf{C} = \mathbf{i} + 2\mathbf{j} - \mathbf{k}.$$

Solution Since

$$\mathbf{A} \cdot \mathbf{C} = -3, \qquad \mathbf{B} \cdot \mathbf{C} = 3,$$

the right-hand side of Eq. (8) is

$$(\mathbf{A} \cdot \mathbf{C})\mathbf{B} - (\mathbf{B} \cdot \mathbf{C})\mathbf{A} = -3\mathbf{B} - 3\mathbf{A} = -3(3\mathbf{i} + 3\mathbf{k}) = -9\mathbf{i} - 9\mathbf{k}.$$

To calculate the left-hand side of Eq. (8) we have

$$\mathbf{A} \times \mathbf{B} = \begin{vmatrix} \mathbf{i} & \mathbf{j} & \mathbf{k} \\ 1 & -1 & 2 \\ 2 & 1 & 1 \end{vmatrix} = -3\mathbf{i} + 3\mathbf{j} + 3\mathbf{k},$$

so that

$$(\mathbf{A} \times \mathbf{B}) \times \mathbf{C} = \begin{vmatrix} \mathbf{i} & \mathbf{j} & \mathbf{k} \\ -3 & 3 & 3 \\ 1 & 2 & -1 \end{vmatrix} = -9\mathbf{i} - 9\mathbf{k}. \ \square$$

EXAMPLE 3 Use Eqs. (8) and (12b) to express

$$(\mathbf{A} \times \mathbf{B}) \times (\mathbf{C} \times \mathbf{D})$$

in terms of scalar multiplication and cross products involving no more than two factors.

Solution Write, for convenience,

$$\mathbf{C} \times \mathbf{D} = \mathbf{V}.$$

Then use Eq. (8) to evaluate

$$(\mathbf{A} \times \mathbf{B}) \times \mathbf{V} = (\mathbf{A} \cdot \mathbf{V})\mathbf{B} - (\mathbf{B} \cdot \mathbf{V})\mathbf{A}$$

or

$$(\mathbf{A} \times \mathbf{B}) \times (\mathbf{C} \times \mathbf{D}) = (\mathbf{A} \cdot \mathbf{C} \times \mathbf{D})\mathbf{B} - (\mathbf{B} \cdot \mathbf{C} \times \mathbf{D})\mathbf{A}.$$

The result, as written, expresses the answer as a scalar times **B** minus a scalar times **A.** One could also represent the answer as a scalar times **C** minus a scalar times **D.** Geometrically, the vector is parallel to the line of intersection of the **A, B**-plane and the **C, D**-plane. □

PROBLEMS

In Problems 1–3, take

$$\mathbf{A} = 4\mathbf{i} - 8\mathbf{j} + \mathbf{k},$$
$$\mathbf{B} = 2\mathbf{i} + \mathbf{j} - 2\mathbf{k},$$
$$\mathbf{C} = 3\mathbf{i} - 4\mathbf{j} + 12\mathbf{k}.$$

1. Find $(\mathbf{A} \cdot \mathbf{B})\mathbf{C}$ and $\mathbf{A}(\mathbf{B} \cdot \mathbf{C})$.

2. Find the volume of the box having **A, B, C** as three co-terminous edges.

3. a) Find $\mathbf{A} \times \mathbf{B}$ and use the result to find $(\mathbf{A} \times \mathbf{B}) \times \mathbf{C}$.
b) Find $(\mathbf{A} \times \mathbf{B}) \times \mathbf{C}$ by another method.

4. Find the volume of the parallelepiped whose edges are determined by the vectors $\mathbf{A} = 3\mathbf{j}$, $\mathbf{B} = -2\mathbf{i} + \mathbf{j}$, and $\mathbf{C} = -\mathbf{i} + \mathbf{j} + \mathbf{k}$.

In Problems 5–7, find the volume of the tetrahedron whose vertices are the given points.

5. $(0, 0, 0)$, $(1, -1, 1)$, $(2, 1, -2)$, $(-1, 2, -1)$

6. $(0, 0, 0)$, $(2, 1, 0)$, $(2, -1, 1)$, $(1, 0, 2)$

7. $(1, 0, 3)$, $(2, 1, 1)$, $(0, 0, 2)$, $(3, 4, 1)$

8. Which of the following are *not always* true?
a) $\mathbf{A} \times \mathbf{B} = \mathbf{B} \times \mathbf{A}$
b) $\mathbf{A} \times (\mathbf{B} + \mathbf{C}) = \mathbf{A} \times \mathbf{B} + \mathbf{A} \times \mathbf{C}$
c) $\mathbf{A} \times (-\mathbf{A}) = \mathbf{0}$
d) $\mathbf{A} \cdot (\mathbf{B} \times \mathbf{C}) = (\mathbf{A} \cdot \mathbf{B}) \times \mathbf{C}$
e) $(\mathbf{A} \times \mathbf{B}) \cdot \mathbf{A} = \mathbf{B} \cdot (\mathbf{A} \times \mathbf{B})$
f) $(\mathbf{A} \times \mathbf{A}) \cdot \mathbf{A} = 0$
g) $(\mathbf{A} \times \mathbf{B}) \cdot \mathbf{C} = (\mathbf{B} \times \mathbf{C}) \cdot \mathbf{A}$
h) $(\mathbf{A} \times \mathbf{B}) \times \mathbf{C} = \mathbf{A} \times (\mathbf{B} \times \mathbf{C})$

9. Suppose that

$$\mathbf{A} \cdot \mathbf{A} = 4, \quad \mathbf{B} \cdot \mathbf{B} = 4, \quad \mathbf{A} \cdot \mathbf{B} = 0,$$
$$(\mathbf{A} \times \mathbf{B}) \times \mathbf{C} = \mathbf{0}, \quad (\mathbf{A} \times \mathbf{B}) \cdot \mathbf{C} = 8.$$

Find each of the following:
a) $\mathbf{A} \cdot \mathbf{C}$, b) $|\mathbf{C}|$, c) $\mathbf{B} \times \mathbf{C}$.

(*Hint:* Picture the vectors, and think geometrically. Use basic, coordinate-free definitions. Avoid long calculations.)

10. Prove that any vector **A** satisfies the identity

$$\mathbf{A} = \tfrac{1}{2}[\mathbf{i} \times (\mathbf{A} \times \mathbf{i}) + \mathbf{j} \times (\mathbf{A} \times \mathbf{j}) + \mathbf{k} \times (\mathbf{A} \times \mathbf{k})].$$

11. Express the product $\mathbf{R} = (\mathbf{A} \times \mathbf{B}) \times (\mathbf{C} \times \mathbf{D})$ in the form $a\mathbf{C} + b\mathbf{D}$ with scalars a and b.

12. Use Eq. (3) to show that
a) $\mathbf{A} \cdot (\mathbf{C} \times \mathbf{B}) = -\mathbf{A} \cdot (\mathbf{B} \times \mathbf{C})$,
b) $\mathbf{A} \cdot (\mathbf{A} \times \mathbf{B}) = 0$,
c) $(\mathbf{A} + \mathbf{D}) \cdot (\mathbf{B} \times \mathbf{C}) = \mathbf{A} \cdot (\mathbf{B} \times \mathbf{C}) + \mathbf{D} \cdot (\mathbf{B} \times \mathbf{C})$.
Interpret the results geometrically.

13. Explain the statement in the text that $(\mathbf{A} \times \mathbf{B}) \times \mathbf{C}$ is parallel to the plane determined by **A** and **B.** Illustrate with a sketch.

14. Explain the statement, at the end of Example 3, that $(\mathbf{A} \times \mathbf{B}) \times (\mathbf{C} \times \mathbf{D})$ is parallel to the line of intersection of the **A, B**-plane and the **C, D**-plane. Illustrate with a sketch.

15. Find a line in the plane of $P_0(0, 0, 0)$, $P_1(2, 2, 0)$, $P_2(0, 1, -2)$, and perpendicular to the line

$$\frac{x+1}{3} = \frac{y-1}{2} = 2z.$$

16. Let $P(1, 2, -1)$, $Q(3, -1, 4)$, and $R(2, 6, 2)$ be three vertices of a parallelogram $PQRS$.
a) Find the coordinates of S.
b) Find the area of $PQRS$.
c) Find the area of the projection of $PQRS$ in the xy-plane; in the yz-plane; in the xz-plane.

17. Show that the area of a parallelogram in space is the square root of the sum of the squares of the areas of its projections on any three mutually orthogonal planes.

13.10 Cylinders

In this article and the next, we shall consider some extensions of analytic geometry to space. We begin with the notion of a surface.

The set of points $P(x, y, z)$ that satisfy an equation

$$F(x, y, z) = 0 \tag{1}$$

may be interpreted in a broad sense as being a surface. The simplest

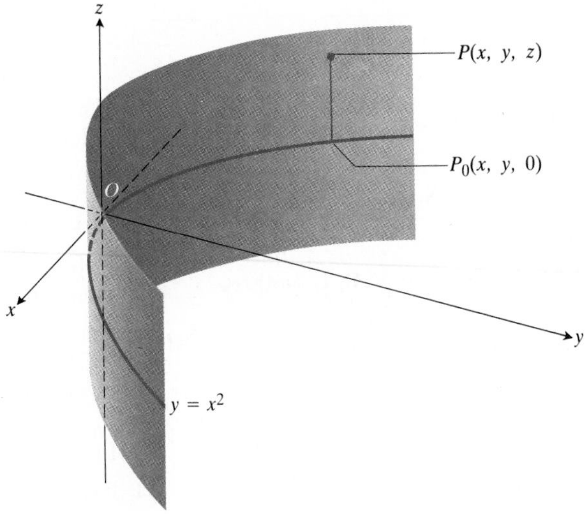

13.53 Parabolic cylinder.

examples of surfaces are planes, which have equations of the form $Ax + By + Cz - D = 0$. Almost as simple as planes are the surfaces called *cylinders*.

In general, a cylinder is a surface that is generated by moving a straight line along a given curve while holding the line parallel to a given fixed line.

EXAMPLE 1 The *parabolic cylinder* of Fig. 13.53 is generated by a line parallel to the z-axis that moves along the curve $y = x^2$ in the xy-plane. If a point $P_0(x, y, 0)$ lies on the parabola, then every point $P(x, y, z)$ with the same x- and y-coordinates lies on the line through P_0 parallel to the z-axis, and hence belongs to the surface. Conversely, if $P(x, y, z)$ lies on the surface, its projection $P_0(x, y, 0)$ on the xy-plane lies on the parabola $y = x^2$, so that its coordinates satisfy the equation $y = x^2$. Regardless of the value of z, the points of the surface are the points whose coordinates satisfy this equation. Thus, the equation $y = x^2$ is an equation for the cylinder as well as for the generating parabola. The cross sections of the cylinder perpendicular to the z-axis are parabolas, too, all of them congruent to the parabola in the xy-plane. □

In general, any curve

$$f(x, y) = 0 \tag{2}$$

in the xy-plane defines a cylinder in space whose equation is also $f(x, y) = 0$, and which is made up of the points of the lines through the curve that are parallel to the z-axis. The lines are sometimes called *elements* of the cylinder.

The discussion above can be carried through for cylinders with elements parallel to the other coordinate axes, and the result is summarized by saying that *an equation in cartesian coordinates, from which one coordinate variable is missing, represents a cylinder with elements parallel to the axis associated with the missing variable.*

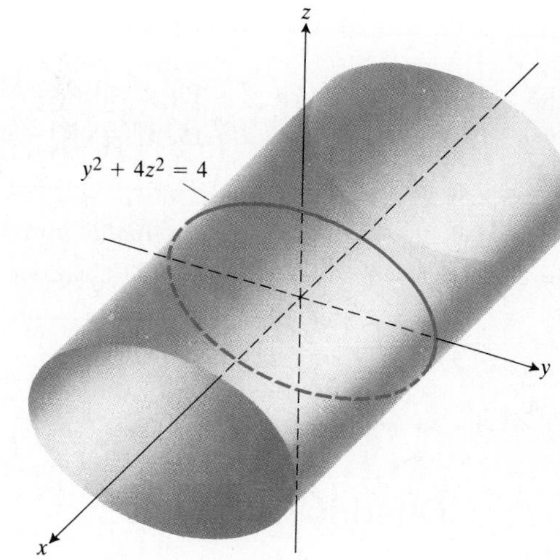

13.54 Elliptic cylinder.

EXAMPLE 2 The surface

$$y^2 + 4z^2 = 4$$

is an *elliptic cylinder* with elements parallel to the x-axis. It extends indefinitely in both the negative and positive directions along the x-axis, which is the axis of the cylinder, since it passes through the centers of the elliptical cross sections of the cylinder (Fig. 13.54). □

EXAMPLE 3 The surface

$$r^2 = 2a \cos 2\theta$$

in cylindrical coordinates is a cylinder with elements parallel to the z-axis. Each section perpendicular to the z-axis is a lemniscate. The cylinder extends indefinitely in both the positive and negative directions along the z-axis (Fig. 13.55). □

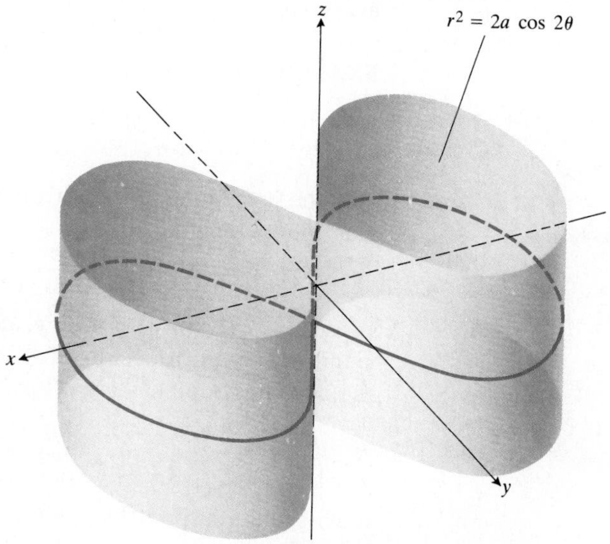

13.55 A cylinder whose cross sections are lemniscates.

PROBLEMS

Describe and sketch each of the following surfaces [(r, θ, z) are cylindrical coordinates].

1. $x^2 + z^2 = 1$
2. $z = x^2$
3. $x = -y^2$
4. $4x^2 + y^2 = 4$
5. $z = -y$
6. $y^2 - x^2 = 1$
7. $x^2 - z^2 = 1$
8. $z^2 - y^2 = 1$
9. $r = 4$
10. $r = \sin\theta$
11. $r = \cos\theta$
12. $r = 1 + \cos\theta$
13. $x^2 + y^2 = a^2$
14. $y^2 + z^2 - 4z = 0$
15. $x^2 + 4z^2 - 4z = 0$

Toolkit programs

3D Grapher

13.11 Quadric Surfaces

A surface whose equation is a quadratic in the variables x, y, and z is called a *quadric* surface. We indicate briefly how some of the simpler ones may be recognized from their equations.

The *sphere*

$$(x - x_0)^2 + (y - y_0)^2 + (z - z_0)^2 = a^2 \tag{1}$$

with center at (x_0, y_0, z_0) and radius a has already been mentioned in Article 13.5. Likewise, the various *cylinders*

$$Ax^2 + Bxy + Cy^2 + Dx + Ey + F = 0 \tag{2}$$

with elements parallel to the z-axis, and others with elements parallel to the other coordinate axes, are familiar and will not be further discussed. In the examples that follow, we choose coordinate axes that yield simple forms of the equations. For example, we take the origin to be at the center of the ellipsoid in Example 1 below. If the center were at (x_0, y_0, z_0) instead, the equation would simply have $x - x_0$, $y - y_0$, and $z - z_0$, in place of x, y, z, respectively. We take a, b, and c to be positive constants in every case.

EXAMPLE 1 *The ellipsoid* (Fig. 13.56)

$$\frac{x^2}{a^2} + \frac{y^2}{b^2} + \frac{z^2}{c^2} = 1 \tag{3}$$

cuts the coordinate axes at $(\pm a, 0, 0)$, $(0, \pm b, 0)$, and $(0, 0, \pm c)$. It lies inside the rectangular box

$$|x| \le a, \qquad |y| \le b, \qquad |z| \le c.$$

Since only even powers of x, y, and z occur in the equation, this surface is symmetric with respect to each coordinate plane. The sections cut out by the coordinate planes are ellipses. For example,

$$\frac{x^2}{a^2} + \frac{y^2}{b^2} = 1 \qquad \text{when} \quad z = 0.$$

Each section cut out by a plane

$$z = z_1, \qquad |z_1| < c,$$

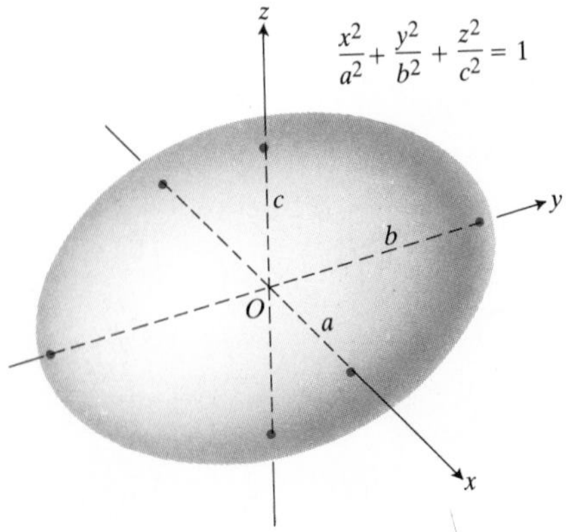

13.56 Ellipsoid.

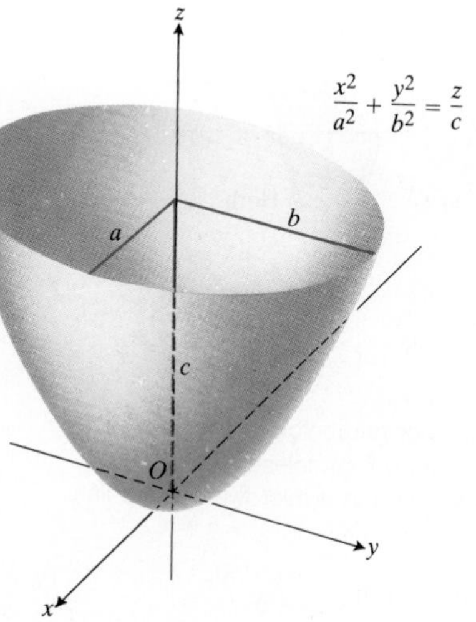

13.57 Elliptic paraboloid.

is an ellipse

$$\frac{x^2}{a^2[1-(z_1^2/c^2)]}+\frac{y^2}{b^2[1-(z_1^2/c^2)]}=1$$

with center on the z-axis and having semiaxes

$$\frac{a}{c}\sqrt{c^2-z_1^2} \quad \text{and} \quad \frac{b}{c}\sqrt{c^2-z_1^2}.$$

The surface is sketched in Fig. 13.56. When two of the three semiaxes a, b, and c are equal, the surface is an ellipsoid of revolution, and when all three are equal, it is a sphere. □

EXAMPLE 2 The elliptic paraboloid (Fig. 13.57)

$$\frac{x^2}{a^2}+\frac{y^2}{b^2}=\frac{z}{c} \tag{4}$$

is symmetric with respect to the planes $x = 0$ and $y = 0$. The only intercept on the axes is at the origin. Since the left-hand side of the equation is nonnegative, the surface is limited to the region $z \geq 0$. That is, away from the origin it lies above the xy-plane. The section cut out from the surface by the yz-plane is

$$x = 0, \qquad y^2 = \frac{b^2}{c}z,$$

which is a parabola with vertex at the origin and opening upward. Similarly, one finds that when $y = 0$,

$$x^2 = \frac{a^2}{c}z,$$

which is also such a parabola. When $z = 0$, the cut reduces to the single point (0, 0, 0). Each plane $z = z_1 > 0$ perpendicular to the z-axis cuts the surface in an ellipse of semiaxes

$$a\sqrt{z_1/c} \quad \text{and} \quad b\sqrt{z_1/c}.$$

These semiaxes increase in magnitude as z_1 increases. The paraboloid extends indefinitely upward. □

EXAMPLE 3 Circular paraboloid, or paraboloid of revolution:

$$\frac{x^2}{a^2}+\frac{y^2}{a^2}=\frac{z}{c}. \tag{5a}$$

The equation is obtained by taking $b = a$ in Eq. (4) for the elliptic paraboloid. The cross sections of the surface by planes perpendicular to the z-axis are circles centered on the z-axis. The cross sections by planes containing the z-axis are congruent parabolas with a common focus at the point $(0, 0, a^2/4c)$. In cylindrical coordinates, (5a) becomes

$$\frac{r^2}{a^2}=\frac{z}{c}. \tag{5b}$$

Shapes cut from circular paraboloids are used for antennas in radio telescopes, satellite trackers, and microwave radio links (Fig. 13.58). □

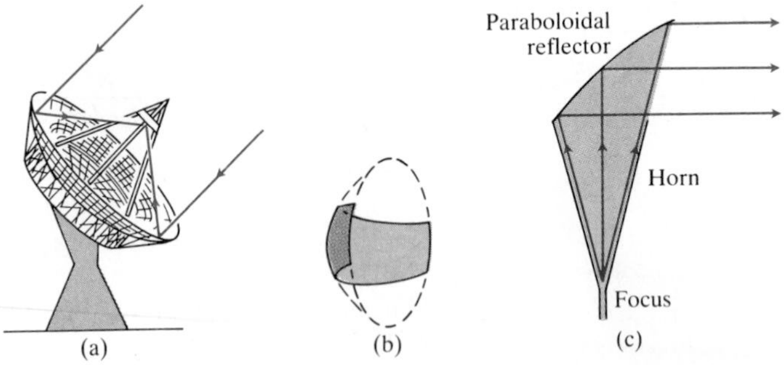

13.58 Many antennas are shaped like pieces of paraboloids of revolution. (a) Radio telescopes use the same principles as optical telescopes. (b) A "rectangular-cut" radar reflector. (c) Horn antenna in a microwave radio link.

EXAMPLE 4 *The elliptic cone* (Fig. 13.59)

$$\frac{x^2}{a^2} + \frac{y^2}{b^2} = \frac{z^2}{c^2} \tag{6}$$

is symmetric with respect to all three coordinate planes. The plane $z = 0$ cuts the surface in the single point (0, 0, 0). The plane $x = 0$ cuts it in the two intersecting straight lines

$$x = 0, \qquad \frac{y}{b} = \pm\frac{z}{c}, \tag{7}$$

and when

$$y = 0, \qquad \frac{x}{a} = \pm\frac{z}{c}. \tag{8}$$

The section cut out by a plane $z = z_1 > 0$ is an ellipse with center on the z-axis and vertices lying on the straight lines (7) and (8). In fact, the whole surface is generated by a straight line L that passes through the origin and a point Q on the ellipse

$$z = c, \qquad \frac{x^2}{a^2} + \frac{y^2}{b^2} = 1.$$

As the point Q traces out the ellipse, the infinite line L generates the surface, which is a cone with elliptic cross sections. To see why, suppose that $Q(x_1, y_1, z_1)$ is a point on the surface and t is any scalar. Then the vector from O to the point $P(tx_1, ty_1, tz_1)$ is simply t times $\overrightarrow{OQ}$, so that as t varies from $-\infty$ to $+\infty$ the point P traces out the infinite line L. But since Q is assumed to be on the surface, the equation

$$\frac{x_1^2}{a^2} + \frac{y_1^2}{b^2} = \frac{z_1^2}{c^2}$$

is satisfied. Multiplying both sides of this equation by t^2 shows that the point $P(tx_1, ty_1, tz_1)$ is also on the surface. This establishes the validity of the remark that the surface is a cone generated by the line L through O and the point Q on the ellipse.

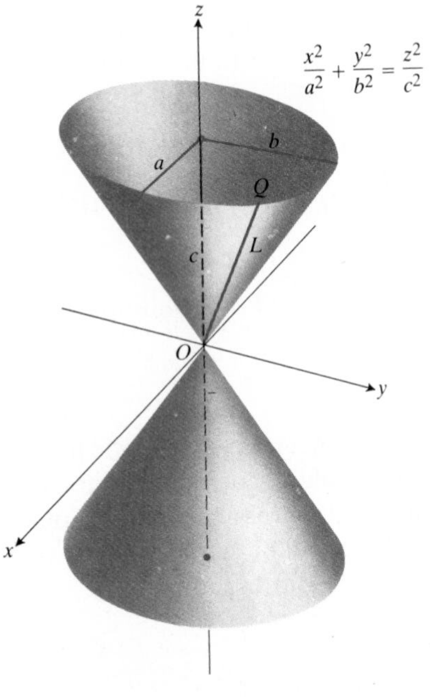

13.59 Elliptic cone.

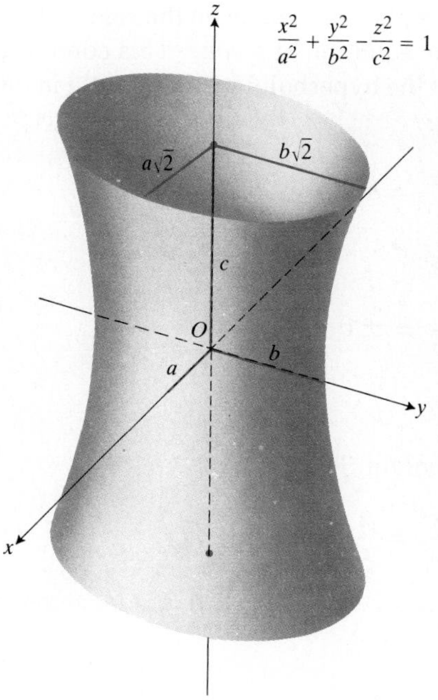

13.60 Hyperboloid of one sheet.

In case $a = b$, the cone is a right circular cone and its equation in cylindrical coordinates is simply

$$\frac{r}{a} = \frac{z}{c}. \ \square \tag{9}$$

EXAMPLE 5 The hyperboloid of one sheet (Fig. 13.60)

$$\frac{x^2}{a^2} + \frac{y^2}{b^2} - \frac{z^2}{c^2} = 1 \tag{10}$$

is symmetric with respect to each of the three coordinate planes. The sections cut out by the coordinate planes are

$$\begin{aligned} x = 0: &\quad \text{the hyperbola} \quad \frac{y^2}{b^2} - \frac{z^2}{c^2} = 1, \\ y = 0: &\quad \text{the hyperbola} \quad \frac{x^2}{a^2} - \frac{z^2}{c^2} = 1, \\ z = 0: &\quad \text{the ellipse} \quad \frac{x^2}{a^2} + \frac{y^2}{b^2} = 1. \end{aligned} \tag{11}$$

The plane $z = z_1$ cuts the surface in an ellipse with center on the z-axis and vertices on the hyperbolas in (11). The surface is connected, meaning that it is possible to travel from any point on it to any other point on it without leaving the surface. For this reason, it is said to have *one* sheet, in contrast to the next example, which consists of *two* sheets.

In the special case where $a = b$, the surface is a hyperboloid of revolution with equation given in cylindrical coordinates by

$$\frac{r^2}{a^2} - \frac{z^2}{c^2} = 1. \ \square \tag{12}$$

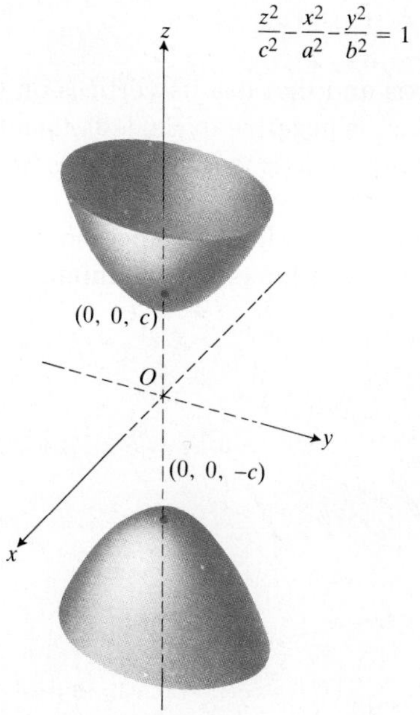

13.61 Hyperboloid of two sheets.

EXAMPLE 6 The hyperboloid of two sheets (Fig. 13.61)

$$\frac{z^2}{c^2} - \frac{x^2}{a^2} - \frac{y^2}{b^2} = 1 \tag{13}$$

is symmetric with respect to the three coordinate planes. The plane $z = 0$ does not intersect the surface; in fact, one must have

$$|z| \geq c$$

for real values of x and y in Eq. (13). The hyperbolic sections

$$\begin{aligned} x = 0: &\quad \frac{z^2}{c^2} - \frac{y^2}{b^2} = 1, \\ y = 0: &\quad \frac{z^2}{c^2} - \frac{x^2}{a^2} = 1 \end{aligned}$$

have their vertices and foci on the z-axis. The surface is separated into two portions, one above the plane $z = c$ and the other below the plane $z = -c$. This accounts for its name.

Equations (10) and (13) differ in the number of negative terms that each contains on the left side when the right side is $+1$. The number of negative signs is the same as the number of sheets of the hyperboloid. If

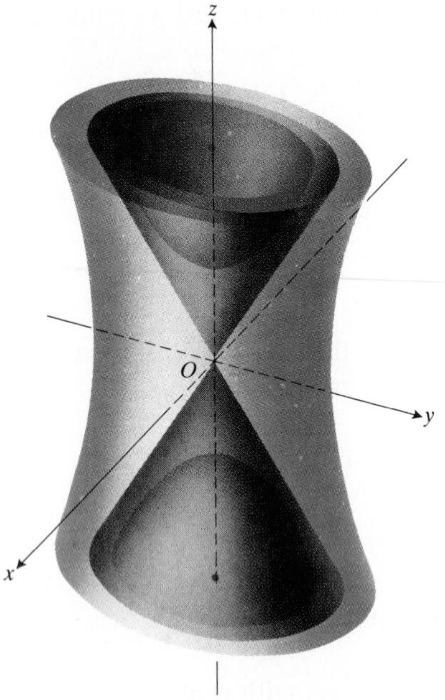

13.62 Cone asymptotic to hyperboloid of one sheet and hyperboloid of two sheets.

we compare with Eq. (6), we see that replacing the one on the right side of either Eq. (10) or (13) by zero gives the equation of a cone. This cone (Fig. 13.62) is, in fact, asymptotic to both of the hyperboloids (10) and (13) in the same way that the lines

$$\frac{x^2}{a^2} - \frac{y^2}{b^2} = 0$$

are asymptotic to the two hyperbolas

$$\frac{x^2}{a^2} - \frac{y^2}{b^2} = \pm 1$$

in the xy-plane. □

EXAMPLE 7 *The hyperbolic paraboloid* (Fig. 13.63)

$$\frac{y^2}{b^2} - \frac{x^2}{a^2} = \frac{z}{c} \tag{14}$$

has symmetry with respect to the planes $x = 0$ and $y = 0$. The sections in these planes are

$$x = 0: \quad y^2 = b^2\frac{z}{c}, \tag{15a}$$

$$y = 0: \quad x^2 = -a^2\frac{z}{c}, \tag{15b}$$

which are parabolas. In the plane $x = 0$, the parabola opens upward and has vertex at the origin. The parabola in the plane $y = 0$ has the same vertex, but it opens downward.

If we cut the surface by a plane $z = z_1 > 0$, the section is a hyperbola,

$$\frac{y^2}{b^2} - \frac{x^2}{a^2} = \frac{z_1}{c}, \tag{16}$$

whose focal axis is parallel to the y-axis and that has its vertices on the parabola in (15a). If, on the other hand, z_1 is negative in Eq. (16), then the focal axis of the hyperbola is parallel to the x-axis, and its vertices lie on the parabola in (15b).

Near the origin the surface is shaped like a saddle. To a person traveling along the surface in the yz-plane, the origin looks like a minimum. To

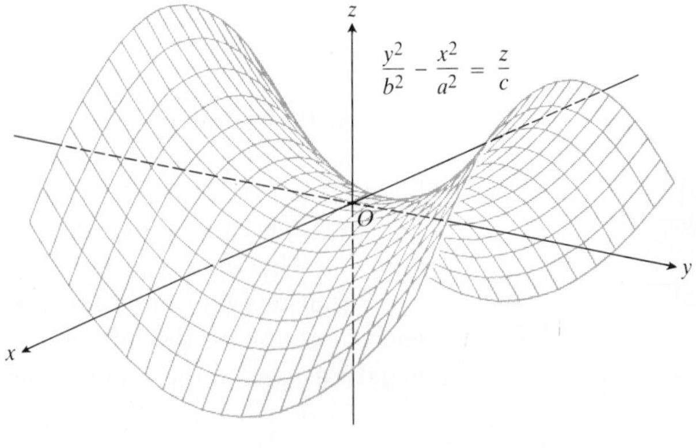

13.63 Hyperbolic paraboloid.

a person traveling in the xz-plane, on the other hand, the origin looks like a maximum. Such a point is called a *minimax* or *saddle point* of a surface (Fig. 13.63). We shall discuss maximum and minimum points on surfaces in Chapter 15.

If $a = b$ in Eq. (14), the surface is not a surface of revolution, but it is possible to express the equation in the alternative form

$$\frac{2x'y'}{a^2} = \frac{z}{c} \tag{17}$$

if we refer it to $x'y'$-axes obtained by rotating the xy-axes through 45°. □

PROBLEMS

In Problems 1–24, describe and sketch each of the surfaces [(r, θ, z) are cylindrical coordinates]. Complete the square, when necessary, to put the equation into one of the standard forms shown in the examples.

1. $x^2 + y^2 = z - 1$

2. $x^2 + y^2 + z^2 + 4x - 6y = 3$

3. $x^2 + 4y^2 + z^2 = 4$

4. $x^2 + 4y^2 + z^2 - 8y = 0$

5. $4x^2 + 4y^2 + 4z^2 - 8y = 0$

6. $x^2 - y^2 + z^2 + 4x - 6y = 9$

7. $x^2 - y^2 - z^2 + 4x - 6y = 9$

8. $z^2 = 4x$

9. $z^2 = 4xy$

10. $z = 4xy$

11. $z = r^2$

12. $z = r$

13. $z^2 = r$

14. $z^2 = x^2 + 4y^2$

15. $z^2 = x^2 - 4y^2$

16. $z^2 = 4y^2 - x^2$

17. $z^2 = x^2 + 4y^2 - 2x + 8y + 4z$

18. $z^2 = x^2 + 4y^2 - 2x + 8y + 4z + 1$

19. $x^2 + 4z^2 = 4$

20. $x = y^2 + 4z^2 + 1$

21. $z = r \cos \theta$

22. $z = r \sin \theta$

23. $z = \sin \theta \ (0 \le \theta \le \pi/2)$

24. $z = \cosh \theta \ (0 \le \theta \le \pi/2)$

25. a) Express the area, $A(z_1)$, of the cross section cut from the ellipsoid

$$\frac{x^2}{a^2} + \frac{y^2}{b^2} + \frac{z^2}{c^2} = 1$$

by the plane $z = z_1$, as a function of z_1. (The area of an ellipse of semiaxes A and B is πAB.)

b) By integration, find the volume of the ellipsoid of part (a). Consider slices made by planes perpendicular to the z-axis. Does your answer give the correct volume of a sphere in case $a = b = c$?

26. By integration, prove that the volume of the segment of the elliptic paraboloid

$$\frac{x^2}{a^2} + \frac{y^2}{b^2} = \frac{z}{c}$$

cut off by the plane $z = h$ is equal to one half the area of its base times its altitude.

27. Given the hyperboloid of one sheet of Eq. (10).

a) By integration, find the volume between the plane $z = 0$ and the plane $z = h$, enclosed by the hyperboloid.

b) Express your answer to part (a) in terms of the altitude h and the areas A_0 and A_h of the plane ends of the segment of the hyperboloid.

c) Verify that the volume of part (a) is also given exactly by the prismoid formula

$$V = h(A_0 + 4A_m + A_h)/6,$$

where A_0 and A_h are the areas of the plane ends of the segment of the hyperboloid and A_m is the area of its midsection cut out by the plane $z = h/2$.

28. If the hyperbolic paraboloid

$$\frac{y^2}{b^2} - \frac{x^2}{a^2} = \frac{z}{c}$$

is cut by the plane $y = y_1$, the resulting curve is a parabola. Find its vertex and focus.

29. What is the nature, in general, of a surface whose equation in spherical coordinates has the form $\rho = F(\phi)$? Give reasons for your answer.

Describe and sketch the following surfaces, which are special cases of Problem 29.

30. $\rho = a \cos \phi$

31. $\rho = a(1 + \cos \phi)$

32. A surface has the equation $z^2 = 3(x^2 + y^2)$ in rectangular coordinates. Find equations for the surface in (a) cylindrical and (b) spherical coordinates. Identify and sketch the surface.

***Toolkit* programs**

3D Grapher

REVIEW QUESTIONS AND EXERCISES

1. When are two vectors equal?
2. How are two vectors added? Subtracted?
3. If a vector is multiplied by a scalar, how is the result related to the original vector? In your discussion include all possible values of the scalar: positive, negative, and zero.
4. In a single diagram, show the cartesian, cylindrical, and spherical coordinates of an arbitrary point P, and write the expressions for each set of coordinates in terms of the other two kinds.
5. What set in space is described by
   a) $x = \text{constant}$, b) $r = \text{constant}$,
   c) $\theta = \text{constant}$, d) $\rho = \text{constant}$,
   e) $\phi = \text{constant}$, f) $ax + by + cz = d$,
   g) $ax^2 + by^2 + cz^2 = d$?
6. What is the length of the vector $a\mathbf{i} + b\mathbf{j} + c\mathbf{k}$? On what theorem of plane geometry does this result depend?
7. Define *scalar product* of two vectors. Which algebraic laws (commutative, associative, distributive) are satisfied by the operations of addition and scalar multiplication of vectors? Which of these laws is (are) not satisfied? Explain. When is the scalar product equal to zero?
8. Suppose that $\mathbf{i}, \mathbf{j}, \mathbf{k}$ is one set of mutually orthogonal unit vectors and that $\mathbf{i}', \mathbf{j}', \mathbf{k}'$ is another set of such vectors. Suppose that all the scalar products of a unit vector from one set with a unit vector from the other set are known. Let
$$\mathbf{A} = a\mathbf{i} + b\mathbf{j} + c\mathbf{k} = a'\mathbf{i}' + b'\mathbf{j}' + c'\mathbf{k}'$$
and express a, b, c in terms of a', b', c'; and conversely. (Expressions involve $\mathbf{i}\cdot\mathbf{i}'$, $\mathbf{i}\cdot\mathbf{j}'$, $\mathbf{i}\cdot\mathbf{k}'$, and so forth.)
9. List four applications of the scalar product.
10. Define *vector product* of two vectors. Which algebraic laws (commutative, associative, distributive) are satisfied by the vector product operation (combined with addition), and which are not? Explain. When is the vector product equal to zero?
11. Derive the formula for expressing the vector product of two vectors as a determinant. What is the effect of interchanging the order of the two vectors and the corresponding rows of the determinant?
12. How may vector and scalar products be used to find the equation of a plane through three given points?
13. With the book closed, develop equations for a line
    a) through two given points,
    b) through one point and parallel to a given line.
14. With the book closed, develop the equation of a plane
    a) through a given point and normal to a given vector,
    b) through one point and parallel to a given plane,
    c) through a point and perpendicular to each of two given planes.
15. What is the geometrical interpretation of
$$\mathbf{A}\cdot(\mathbf{B}\times\mathbf{C})?$$
When is this triple scalar product equal to zero?
16. Given a parallelogram $PQRS$ in space, how could you find a vector normal to its plane and with length equal to its area?
17. What set in space is described by an equation of the form
    a) $f(x, y) = 0$, b) $f(z, r) = 0$,
    c) $z = f(\theta)$, $0 \le \theta \le 2\pi$?
18. Define *quadric surface*. Name and sketch six different quadric surfaces and indicate their equations.

MISCELLANEOUS PROBLEMS

In Problems 1 through 10, find parametric equations of the path traced by the point $P(x, y)$ for the data given.

1. $\dfrac{dx}{dt} = x^2$, $\dfrac{dy}{dt} = x$; $t = 0$, $x = 1$, $y = 1$
2. $\dfrac{dx}{dt} = \cos^2 x$, $\dfrac{dy}{dt} = x$; $t = 0$, $x = \dfrac{\pi}{4}$, $y = 0$
3. $\dfrac{dx}{dt} = e^t$, $\dfrac{dy}{dt} = xe^x$; $t = 0$, $x = 1$, $y = 0$
4. $\dfrac{dx}{dt} = 6\sin 2t$, $\dfrac{dy}{dt} = 4\cos 2t$; $t = 0$, $x = 0$, $y = 4$
5. $\dfrac{dx}{dt} = 1 - \cos t$, $\dfrac{dy}{dt} = \sin t$; $t = 0$, $x = 0$, $y = 0$
6. $\dfrac{dx}{dt} = \sqrt{1 + y}$, $\dfrac{dy}{dt} = y$; $t = 0$, $x = 0$, $y = 1$
7. $\dfrac{dx}{dt} = \operatorname{sech} x$, $\dfrac{dy}{dt} = x$; $t = 0$, $x = 0$, $y = 0$
8. $\dfrac{dx}{dt} = \cosh\dfrac{t}{2}$, $\dfrac{dy}{dt} = x$; $t = 0$, $x = 2$, $y = 0$
9. $\dfrac{dx}{dt} = y$, $\dfrac{dy}{dt} = -x$; $t = 0$, $x = 0$, $y = 4$

10. $\dfrac{d^2x}{dt^2} = -\dfrac{dx}{dt}, \quad \dfrac{dy}{dt} = x; \quad t = 0, x = 1, y = 1, \dfrac{dx}{dt} = 1$

11. A particle is projected with velocity v at an angle α to the horizontal from a point that is at the foot of a hill inclined at an angle ϕ to the horizontal, where

$$0 < \phi < \alpha < (\pi/2).$$

Show that it reaches the ground at a distance

$$\frac{2v^2 \cos \alpha}{g \cos^2 \phi} \sin (\alpha - \phi)$$

measured up the face of the hill. Hence show that the greatest range achieved for a given v is when $\alpha = (\phi/2) + (\pi/4)$.

12. A wheel of radius 4 in. rolls along the x-axis with angular velocity 2 rad/s. Find the curve described by a point on a spoke and 2 in. from the center of the wheel if it starts from the point (0, 2) at time $t = 0$.

13. OA is the diameter of a circle of radius a. AN is tangent to the circle at A. A line through O making angle θ with diameter OA intersects the circle at M and tangent line at N. On ON a point P is located so that $OP = MN$. Taking O as origin, OA along the y-axis, and angle θ as parameter, find parametric equations of the locus described by P.

14. Let a line AB be the x-axis of a system of rectangular coordinates. Let the point C be the point (0, 1). Let the line DE through C intersect AB at F. Let P and P' be the points on DE such that $PF = P'F = a$. Express the coordinates of P and P' in terms of the angle $\theta = \angle CFB$.

15. For the curve

$$x = a(t - \sin t), \qquad y = a(1 - \cos t),$$

find the following quantities.

a) The area bounded by the x-axis and one loop of the curve
b) The length of one loop
c) The area of the surface of revolution obtained by rotating one loop about the x-axis
d) The coordinates of the centroid of the area in (a).

16. In Fig. 13.64, D is the midpoint of side AB and E is one third of the way between C and B. *Using vectors,* prove that F is midpoint of the line CD.

Figure 13.64

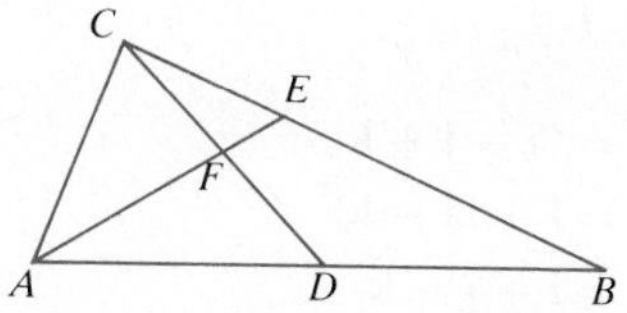

17. The vectors $2\mathbf{i} + 3\mathbf{j}$, $4\mathbf{i} + \mathbf{j}$, and $5\mathbf{i} + y\mathbf{j}$ have their initial points at the origin. Find the value of y so that the vectors terminate on one straight line.

18. **A** and **B** are vectors from the origin to the two points A and B. The point P is determined by the vector $\overrightarrow{OP} = x\mathbf{A} + y\mathbf{B}$, where x and y are positive numbers whose sum is equal to one. Prove that P lies on the line segment AB.

19. Using vector methods, prove that the segment joining the midpoints of two sides of a triangle is parallel to, and half the length of, the third side.

20. Let ABC be a triangle and let M be the midpoint of AB. Let P be the point on CM that is two thirds of the way from C to M. Let O be any point in space.

a) Show that

$$\overrightarrow{OP} = \left(\frac{\overrightarrow{OA} + \overrightarrow{OB} + \overrightarrow{OC}}{3}\right).$$

b) Show how the result in part (a) leads to the conclusion that the medians of a triangle meet in a point.

21. A, B, C are the vertices of a triangle and a, b, c are the midpoints of the opposite sides. Show that

$$\overrightarrow{Aa} + \overrightarrow{Bb} + \overrightarrow{Cc} = \mathbf{0}.$$

Interpret the result geometrically.

22. Vectors are drawn from the center of a regular polygon to its vertices. Show that their sum is zero.

23. Let **A**, **B**, **C** be vectors from a common point O to points A, B, C.

a) If A, B, C are collinear, show that three constants x, y, z (not all zero) exist such that

$$x + y + z = 0 \quad \text{and} \quad x\mathbf{A} + y\mathbf{B} + z\mathbf{C} = \mathbf{0}.$$

b) Conversely, if three constants x, y, z (not all zero) exist such that

$$x + y + z = 0 \qquad \text{and} \qquad x\mathbf{A} + y\mathbf{B} + z\mathbf{C} = \mathbf{0},$$

show that A, B, C are collinear.

24. Find the vector projection of **B** onto **A** if

$$\mathbf{A} = 3\mathbf{i} - \mathbf{j} + \mathbf{k}, \qquad \mathbf{B} = 2\mathbf{i} + \mathbf{j} - 2\mathbf{k}.$$

25. Find the cosine of the angle between the line

$$\frac{1 - x}{4} = \frac{y}{3} = -\frac{z}{5}$$

and the vector $\mathbf{i} + \mathbf{j}$.

26. Given two noncollinear vectors **A** and **B**. Given also that **A** can be expressed in the form $\mathbf{A} = \mathbf{C} + \mathbf{D}$, where **C** is a vector parallel to **B**, and **D** is a vector perpendicular to **B**. Express **C** and **D** in terms of **A** and **B**.

27. The curve whose vector equation is

$$\mathbf{r} = (t^4 + 2t^2 + 1)\mathbf{i} + (1 + 4t - t^4)\mathbf{j}$$

intersects the line $x + y = 0$, $z = 0$. Find the cosine of the angle which the acceleration vector makes with the radius vector at the point of intersection.

28. *Using vectors*, prove that for any four numbers a, b, c, d we have the inequality

$$(a^2 + b^2)(c^2 + d^2) \geq (ac + bd)^2.$$

(*Hint*: Consider $\mathbf{A} = a\mathbf{i} + b\mathbf{j}$ and $\mathbf{B} = c\mathbf{i} + d\mathbf{j}$.)

29. Find a vector parallel to the plane $2x - y - z = 4$ and perpendicular to the vector $\mathbf{i} + \mathbf{j} + \mathbf{k}$.

30. Find a vector which is normal to the plane determined by the points

$$A(1, 0, -1), \quad B(2, -1, 1), \quad C(-1, 1, 2).$$

31. Given vectors

$$\mathbf{A} = 2\mathbf{i} - \mathbf{j} + \mathbf{k},$$
$$\mathbf{B} = \mathbf{i} + 2\mathbf{j} - \mathbf{k},$$
$$\mathbf{C} = \mathbf{i} + \mathbf{j} - 2\mathbf{k},$$

find a *unit* vector in the plane of $\mathbf{B}$ and $\mathbf{C}$ that is $\perp \mathbf{A}$.

32. By forming the cross product of two appropriate vectors, derive the trigonometric identity

$$\sin(\alpha - \beta) = \sin\alpha\cos\beta - \cos\alpha\sin\beta.$$

33. Find a vector of *length two* parallel to the line

$$x + 2y + z - 1 = 0, \qquad x - y + 2z + 7 = 0.$$

34. Given a tetrahedron with vertices O, A, B, C. A vector is constructed normal to each face, pointing outward, and having length equal to the area of the face. Using cross products, prove that the sum of these four outward normals is the zero vector.

35. What angle does the line of intersection of the two planes

$$2x + y - z = 0, \qquad x + y + 2z = 0$$

make with the x-axis?

36. Let $\mathbf{A}$ and $\mathbf{C}$ be given vectors in space, with $\mathbf{A} \neq \mathbf{0}$ and $\mathbf{A} \cdot \mathbf{C} = 0$, and let d be a given scalar. Find a vector $\mathbf{B}$ that satisfies both equations $\mathbf{A} \times \mathbf{B} = \mathbf{C}$ and $\mathbf{A} \cdot \mathbf{B} = d$ simultaneously. The answer should be given as a formula involving $\mathbf{A}$, $\mathbf{C}$, and d.

37. Given any two vectors

$$\mathbf{A} = a_1\mathbf{i} + a_2\mathbf{j}, \qquad \mathbf{B} = b_1\mathbf{i} + b_2\mathbf{j}$$

in the plane, define a new vector, $\mathbf{A} \otimes \mathbf{B}$, called their "circle product," as follows:

$$\mathbf{A} \otimes \mathbf{B} = (a_1b_1 - a_2b_2)\mathbf{i} + (a_1b_2 + a_2b_1)\mathbf{j}.$$

This product satisfies the following algebraic laws.

a) $\mathbf{A} \otimes \mathbf{B} = \mathbf{B} \otimes \mathbf{A}$
b) $\mathbf{A} \otimes (\mathbf{B} \otimes \mathbf{C}) = (\mathbf{A} \otimes \mathbf{B}) \otimes \mathbf{C}$
c) $\mathbf{A} \otimes (\mathbf{B} + \mathbf{C}) = (\mathbf{A} \otimes \mathbf{B}) + (\mathbf{A} \otimes \mathbf{C})$
d) $|\mathbf{A} \otimes \mathbf{B}| = |\mathbf{A}|\,|\mathbf{B}|$

Prove (a) and (d).

38. Find the equations of the straight line that passes through the point (1, 2, 3) and makes an angle of 30° with the positive x-axis and an angle of 60° with the positive y-axis.

39. The line L, whose equations are

$$x - 2z - 3 = 0, \qquad y - 2z = 0,$$

intersects the plane

$$x + 3y - z + 4 = 0.$$

Find the point of intersection P and find the equation of that line in this plane that passes through P and is perpendicular to L.

40. Find the distance between the point (2, 2, 3) and the plane $2x + 3y + 5z = 0$.

41. Given the two parallel planes

$$Ax + By + Cz = D_1, \qquad Ax + By + Cz = D_2,$$

show that the distance between them is given by the formula

$$\frac{|D_1 - D_2|}{|A\mathbf{i} + B\mathbf{j} + C\mathbf{k}|}.$$

42. Consider the straight line through the point (3, 2, 1) and perpendicular to the plane

$$2x - y + 2z + 2 = 0.$$

Compute the coordinates of the point of intersection of that line and that plane.

43. Find the distance between the point (2, 2, 0) and the line

$$x + y = 0, \qquad y - z = 1.$$

44. Find an equation of the plane parallel to the plane $2x - y + 2z + 4 = 0$ if the point $(3, 2, -1)$ is equidistant from both planes.

45. Given the four points

$$A(-2, 0, -3), \qquad B(1, -2, 1),$$
$$C(-2, -\tfrac{13}{5}, \tfrac{26}{5}), \qquad D(\tfrac{16}{5}, -\tfrac{13}{5}, 0).$$

a) Find the equation of the plane through AB that is parallel to CD.
b) Compute the shortest distance between the lines AB and CD.

46. The three vectors

$$\mathbf{A} = 3\mathbf{i} - \mathbf{j} + \mathbf{k},$$
$$\mathbf{B} = \mathbf{i} + 2\mathbf{j} - \mathbf{k},$$
$$\mathbf{C} = \mathbf{i} + \mathbf{j} + \mathbf{k}$$

are all drawn from the origin. Find an equation for the plane through their endpoints.

47. Show that the plane through the three points

$$(x_1, y_1, z_1), \qquad (x_2, y_2, z_2), \qquad (x_3, y_3, z_3)$$

is given by

$$\begin{vmatrix} x_1 - x & y_1 - y & z_1 - z \\ x_2 - x & y_2 - y & z_2 - z \\ x_3 - x & y_3 - y & z_3 - z \end{vmatrix} = 0.$$

48. Given the two straight lines

$$x = a_1t + b_1, \qquad y = a_2t + b_2, \qquad z = a_3t + b_3,$$
$$x = c_1\tau + d_1, \qquad y = c_2\tau + d_2, \qquad z = c_3\tau + d_3,$$

where t and τ are parameters. Show that the necessary and sufficient condition that the two lines either intersect or are parallel is

$$\begin{vmatrix} a_1 & c_1 & b_1 - d_1 \\ a_2 & c_2 & b_2 - d_2 \\ a_3 & c_3 & b_3 - d_3 \end{vmatrix} = 0.$$

49. Given the vectors

$$\mathbf{A} = \mathbf{i} + \mathbf{j} - \mathbf{k},$$
$$\mathbf{B} = 2\mathbf{i} + \mathbf{j} + \mathbf{k},$$
$$\mathbf{C} = -\mathbf{i} - 2\mathbf{j} + 3\mathbf{k},$$

evaluate

a) $\mathbf{A} \cdot (\mathbf{B} \times \mathbf{C})$, b) $\mathbf{A} \times (\mathbf{B} \times \mathbf{C})$.

50. Given four points

$$A\,(1, 1, 1), \qquad B\,(0, 0, 2),$$
$$C\,(0, 3, 0), \qquad D\,(4, 0, 0),$$

find the volume of the tetrahedron with vertices at A, B, C, D, and find the angle between the edges AB and AC.

51. Prove or disprove the formula

$$\mathbf{A} \times [\mathbf{A} \times (\mathbf{A} \times \mathbf{B})] \cdot \mathbf{C} = -|\mathbf{A}|^2\mathbf{A} \cdot \mathbf{B} \times \mathbf{C}.$$

52. If the four vectors **A, B, C, D** are coplanar, show that

$$(\mathbf{A} \times \mathbf{B}) \times (\mathbf{C} \times \mathbf{D}) = \mathbf{0}.$$

53. Prove the following identities, in which **i, j, k** are three mutually perpendicular unit vectors, and **A, B, C** are any vectors.

a) $\mathbf{A} \times (\mathbf{B} \times \mathbf{C}) + \mathbf{B} \times (\mathbf{C} \times \mathbf{A}) + \mathbf{C} \times (\mathbf{A} \times \mathbf{B}) = \mathbf{0}$.

b) $\mathbf{A} \times \mathbf{B} = [\mathbf{A} \cdot (\mathbf{B} \times \mathbf{i})]\mathbf{i} + [\mathbf{A} \cdot (\mathbf{B} \times \mathbf{j})]\mathbf{j} + [\mathbf{A} \cdot (\mathbf{B} \times \mathbf{k})]\mathbf{k}$.

54. Show that

$$(\mathbf{a} \times \mathbf{b}) \cdot (\mathbf{c} \times \mathbf{d}) = \begin{vmatrix} \mathbf{a} \cdot \mathbf{c} & \mathbf{b} \cdot \mathbf{c} \\ \mathbf{a} \cdot \mathbf{d} & \mathbf{b} \cdot \mathbf{d} \end{vmatrix}.$$

55. Sketch the surfaces

a) $(x - 1)^2 + 4(y^2 + z^2) = 16$,

b) $z = r^2$ (cylindrical coordinates),

c) $\rho = a \sin \phi$ (spherical coordinates).

56. Find an equation for the set of points in space whose distance from the point $(2, -1, 3)$ is twice their distance from the xy-plane. Name the surface and find its center of symmetry.

57. Find an equation of the sphere that has the two planes

$$x + y + z - 3 = 0, \qquad x + y + z - 9 = 0$$

as tangent planes, if the two planes

$$2x - y = 0, \qquad 3x - z = 0$$

pass through the center of the sphere.

58. The two cylinders $z^3 - x = 0$ and $x^2 - y = 0$ intersect in a curve C. Find an equation of a cylinder parallel to the x-axis which passes through C. This cylinder traces out a curve C' in the yz-plane. Rotate C' about an y-axis and obtain the equation for the surface so generated.

14 Vector Functions and Their Derivatives

14.1 Derivatives of Vector Functions

Vector Functions

Let **i**, **j**, and **k** be the unit vectors along the x-, y-, and z-axes in space. Then a function

$$\mathbf{F}(t) = \mathbf{i}x(t) + \mathbf{j}y(t) + \mathbf{k}z(t), \tag{1}$$

where $x(t)$, $y(t)$, and $z(t)$ are real-valued functions of the real variable t, is a vector-valued function or *vector function* of t. We shall use such functions primarily to describe the motion of a particle in space or in the plane. When the motion is in the plane, we shall ordinarily assume that coordinates are chosen to make the plane of motion the xy-plane. This allows us to consider motion in the xy-plane as a special case of motion in space, the special condition being that the z-coordinate of the particle is zero.

EXAMPLE 1 The position of a particle in the xy-plane at time t is given by

$$\mathbf{R}(t) = x\mathbf{i} + y\mathbf{j} = e^t\mathbf{i} + te^t\mathbf{j}$$

(Fig. 14.1). Where is the particle at time $t = 0$, and how fast is it moving and in what direction?

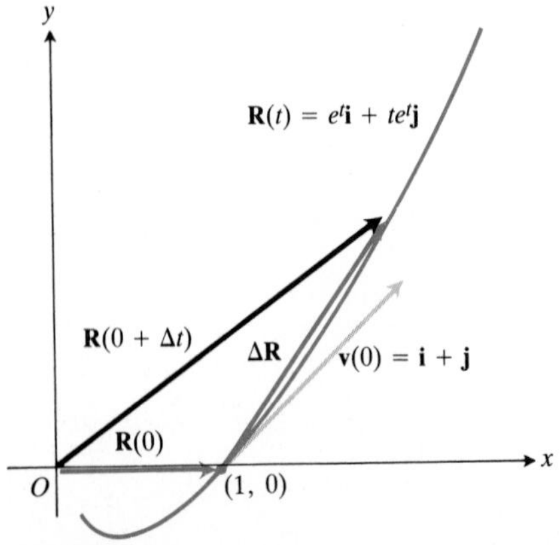

14.1 At $t = 0$, the position and velocity vectors are $\mathbf{R}(0) = \mathbf{i}$ and $\mathbf{v}(0) = \mathbf{i} + \mathbf{j}$.

Solution At $t = 0$, we have $x = 1$ and $y = 0$, so

$$\mathbf{R}(0) = \mathbf{i}.$$

This is the vector from the origin to the position $P(0, 1)$ of the particle at time $t = 0$. Next, how can we find the speed and direction of motion? If you have studied particle mechanics from a vector viewpoint, then you probably know the answer. If not, think where the particle might be a short time Δt past $t = 0$. Both x and y will have increased somewhat, and **R** will have changed by the amount

$$\Delta\mathbf{R} = \mathbf{i}\,\Delta x + \mathbf{j}\,\Delta y.$$

The particle's speed will therefore be approximately

$$\frac{|\Delta \mathbf{R}|}{\Delta t} = \left|\frac{\Delta \mathbf{R}}{\Delta t}\right|.$$

If we let $\Delta t \to 0$, the right-hand side of

$$\frac{\Delta \mathbf{R}}{\Delta t} = \mathbf{i}\frac{\Delta x}{\Delta t} + \mathbf{j}\frac{\Delta y}{\Delta t} \tag{2}$$

will approach the vector

$$\mathbf{v} = \mathbf{i}\frac{dx}{dt} + \mathbf{j}\frac{dy}{dt} = \mathbf{i}e^t + \mathbf{j}(te^t + e^t),$$

whose value at $t = 0$ is

$$\mathbf{v}(0) = \mathbf{i} + \mathbf{j}.$$

The left-hand side of Eq. (2) therefore has the limit

$$\lim_{\Delta t \to 0} \frac{\Delta \mathbf{R}}{\Delta t} = \mathbf{i} + \mathbf{j},$$

and it is natural to call this limit the velocity of the motion at $t = 0$. Accordingly, the particle's speed and direction at $t = 0$ are

$$\text{Speed} = |\mathbf{i} + \mathbf{j}| = \sqrt{2},$$

$$\text{Direction} = \frac{\mathbf{i} + \mathbf{j}}{\sqrt{2}}. \ \square$$

We now pause to define more precisely the notions of limit and derivative for vector functions. We shall return to velocity in the next article as one of several vector derivatives that are important in studying motion.

Limits

Roughly speaking, we want to say that the limit of the vector $\mathbf{F}(t)$ as $t \to a$ is the vector $\mathbf{L}$ if $\mathbf{F}(t)$ moves into the "position" occupied by $\mathbf{L}$ as $t \to a$. In the limit, the length and direction of $\mathbf{F}$ should match the length and direction of $\mathbf{L}$. If we assume that $\mathbf{L}$ and $\mathbf{F}(t)$ are located so that their initial points coincide, then we may picture the convergence of $\mathbf{F}(t)$ to $\mathbf{L}$ as in Fig. 14.2. This picture shows that for $\mathbf{F}(t)$ to approach $\mathbf{L}$, the difference vector $\mathbf{F}(t) - \mathbf{L}$ must shrink to zero, which is equivalent to saying that its length $|\mathbf{F}(t) - \mathbf{L}|$ goes to zero as $t \to a$. This length is a real-valued function of the single variable t, and we know what it means for such a function to have zero as a limit as $t \to a$. Thus we may define limits of vector-valued functions in terms of the familiar limits of real-valued functions in the following way.

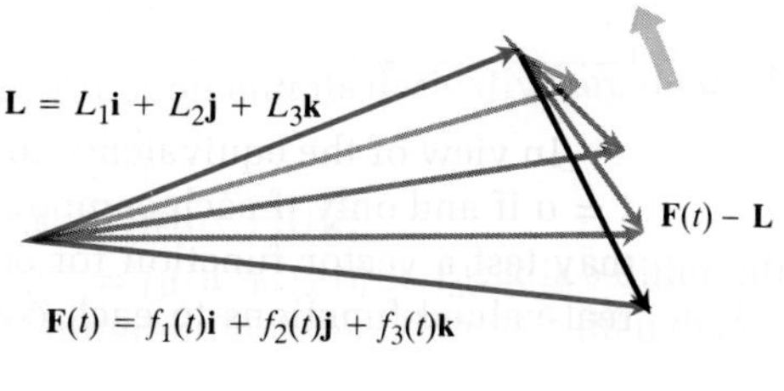

14.2 We say that $\lim_{t\to a} \mathbf{F}(t) = \mathbf{L}$ if $\lim_{t\to a} |\mathbf{F}(t) - \mathbf{L}| = 0$.

ble function of s and

$$\frac{d\mathbf{F}}{ds} = \mathbf{F}'(g(s))g'(s). \tag{8}$$

We often write Eq. (8) in the abbreviated form

$$\frac{d\mathbf{F}}{ds} = \frac{d\mathbf{F}}{dt}\frac{dt}{ds}. \tag{9}$$

The Chain Rule given for vector functions by Eq. (9) is an immediate consequence of the Chain Rule for scalar functions that applies to the components f_1, f_2, and f_3. According to this rule, f_1, f_2, and f_3 are differentiable functions of s and

$$\frac{df_1}{ds} = \frac{df_1}{dt}\frac{dt}{ds}, \qquad \frac{df_2}{ds} = \frac{df_2}{dt}\frac{dt}{ds}, \qquad \frac{df_3}{ds} = \frac{df_3}{dt}\frac{dt}{ds}.$$

Therefore,

$$\begin{aligned}\frac{d\mathbf{F}}{ds} &= \left(\frac{df_1}{ds}\right)\mathbf{i} + \left(\frac{df_2}{ds}\right)\mathbf{j} + \left(\frac{df_3}{ds}\right)\mathbf{k}\\ &= \left(\frac{df_1}{dt}\frac{dt}{ds}\right)\mathbf{i} + \left(\frac{df_2}{dt}\frac{dt}{ds}\right)\mathbf{j} + \left(\frac{df_3}{dt}\frac{dt}{ds}\right)\mathbf{k}\\ &= \left(\frac{df_1}{dt}\mathbf{i} + \frac{df_2}{dt}\mathbf{j} + \frac{df_3}{dt}\mathbf{k}\right)\frac{dt}{ds}\\ &= \left(\frac{d\mathbf{F}}{dt}\right)\frac{dt}{ds}.\end{aligned}$$

EXAMPLE 5 Express $d\mathbf{F}/ds$ in terms of s if $\mathbf{F}(t) = \mathbf{i} + \sin(t+1)\mathbf{j} + e^{t+1}\mathbf{k}$ and $t = g(s) = s^2 - 1$.

Solution From the Chain Rule we have

$$\frac{d\mathbf{F}}{ds} = \frac{d\mathbf{F}}{dt}\frac{dt}{ds} = (\cos(t+1)\mathbf{j} + e^{t+1}\mathbf{k})(2s) = 2s\cos(s^2)\mathbf{j} + 2se^{s^2}\mathbf{k}.$$

This is what we would have obtained had we first substituted $t = g(s) = s^2 - 1$ in the formula for $\mathbf{F}(t)$ and then differentiated with respect to s:

$$\mathbf{F}(g(s)) = \mathbf{i} + \sin(s^2)\mathbf{j} + e^{s^2}\mathbf{k},$$

$$\frac{d}{ds}\mathbf{F}(g(s)) = 2s\cos(s^2)\mathbf{j} + 2se^{s^2}\mathbf{k}. \quad \square$$

PROBLEMS

1. Find $\lim_{t\to 0}\mathbf{F}(t)$ if $\mathbf{F}(t) = e^t\mathbf{i} + te^t\mathbf{j}$.

2. Find $\lim_{t\to 1}\mathbf{F}(t)$ if $\mathbf{F}(t) = (t-1)\mathbf{i} + 3\mathbf{j} + 7t\mathbf{k}$.

3. Find $\lim_{t\to 0}\mathbf{F}(t)$ if $\mathbf{F}(t) = (e^t\sin t)\mathbf{i} + (e^t\cos t)\mathbf{j} - e^t\mathbf{k}$.

4. Find $\lim_{t\to -3}\mathbf{F}(t)$ if

$$\mathbf{F}(t) = \frac{t^2+4t+3}{t+3}\mathbf{j} + \ln(4+t)\mathbf{k}.$$

5. Find $\lim_{t\to 0}\mathbf{F}(t)$ if

$$\mathbf{F}(t) = \frac{t}{\sin t}\mathbf{i} + \frac{1-\cos t}{t}\mathbf{j} + \mathbf{k}.$$

6. Find $\lim_{t\to 0}\mathbf{F}(t)$ if

$$\mathbf{F}(t) = \frac{t}{\cos t}\mathbf{i} + t\left(1 + \frac{1}{t}\right)\mathbf{j} + (2 + \sin t)\mathbf{k}.$$

At what values of t are the vector functions in Problems 7–12 continuous?

7. $\mathbf{F}(t) = (\cos t)\mathbf{i} + (\sin t)\mathbf{j} + \mathbf{k}$

8. $\mathbf{F}(t) = (t^2 - 1)\mathbf{i} + 3\mathbf{k}$

9. $\mathbf{F}(t) = e^t\mathbf{i} + \cos\frac{1}{t+1}\mathbf{j} + \ln|1+t|\mathbf{k}$

10. $\mathbf{F}(t) = \frac{1}{t-1}\mathbf{i} + \frac{1}{t-2}\mathbf{j} + \frac{1}{t-3}\mathbf{k}$

11. $\mathbf{F}(t) = (\tan^{-1} t)\mathbf{i} + t^2\mathbf{k}$

12. $\mathbf{F}(t) = \sqrt{1-t^2}\,\mathbf{i} + 3t\mathbf{j} - 7\mathbf{k}$

Find the derivatives of the vector functions in Problems 13–18, and give the domain of each derivative.

13. $\mathbf{F}(t) = e^{2t}\mathbf{i} + te^{-t}\mathbf{j}$

14. $\mathbf{F}(t) = (\ln\sqrt{1+t})\mathbf{i} + \sqrt{1-t^2}\,\mathbf{j}$

15. $\mathbf{F}(t) = \mathbf{i} + 3\mathbf{j} - \mathbf{k}$

16. $\mathbf{F}(t) = \mathbf{i}\sin^{-1} 2t + \mathbf{j}\tan^{-1} 3t + \mathbf{k}(1/t)$

17. $\mathbf{F}(x) = \mathbf{i}\sec^{-1} 3x + \mathbf{j}\cosh 2x + \mathbf{k}\tanh 4x$

18. $\mathbf{F}(t) = \mathbf{i}\left(\frac{2t-1}{2t+1}\right) + \mathbf{j}\ln(1-4t^2)$

19. Find $\mathbf{F}'(s)$ if $\mathbf{F}(t) = (1/\sqrt{t})\mathbf{i} + (\sin t)\mathbf{j} + \ln(1+t^2)\mathbf{k}$ and $t = \sqrt{s-1}$.

20. Find $\mathbf{F}'(t)$ if $\mathbf{F}(x) = \mathbf{i}xe^x + \mathbf{j}\ln 3x$ and

a) $x = \ln t$, b) $x = e^t$.

21. Find $\mathbf{F}'(x)$ if

$$\mathbf{F}(u) = \mathbf{i}\cos\sqrt{u} + \mathbf{j}\tan^{-1} u - \mathbf{k}(1/(1+\sqrt{u})),$$

$u = x^2 + 2x + 1$, and $x + 1 > 0$.

22. Find $\mathbf{F}'(x)$ if

$$\mathbf{F}(u) = \mathbf{i}\cos\sqrt{u} + \mathbf{j}\tan^{-1} u - \mathbf{k}(1/(1+\sqrt{u})),$$

$u = x^2 + 2x + 1$, and $x + 1 > 0$.

23. Find $\mathbf{F}'(s)$ if

$$\mathbf{F}(t) = -\mathbf{i}\sin t + \mathbf{j}\cos t + \mathbf{k}$$

and $dt/ds = \sqrt{2}$.

24. Find $\mathbf{F}'(s)$ as a function of t if

$$\mathbf{F}(t) = (\cos t + t\sin t)\mathbf{i} + (\sin t - t\cos t)\mathbf{j}$$

and $dt/ds = t$.

25. Find $\mathbf{F}'(t)$ if $\mathbf{F}(\theta) = \mathbf{i}\sin 2\theta + \mathbf{j}\cos 2\theta$ and $d\theta/dt = 2t$.

***Toolkit* programs**	
Parametric Equations	Super * Grapher

14.2 Tangent Vectors, Velocity, and Acceleration

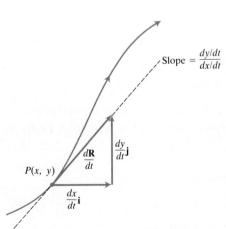

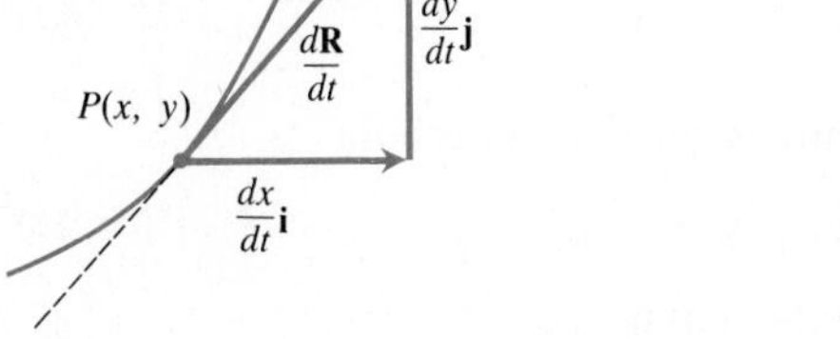

14.3 At each point where $dx/dt \neq 0$ on the curve $\mathbf{R} = x(t)\mathbf{i} + y(t)\mathbf{j}$, the slope of $d\mathbf{R}/dt$ is the slope of the tangent to the curve.

When we model the motion of a particle along a curve in the plane, we want the particle's velocity vector $\mathbf{v}$ to be tangent to the curve and to point in the direction of motion. We also want its length $|\mathbf{v}|$ to be the particle's speed $|ds/dt|$ (where s is arc length measured along the curve from some preselected base point P_0). The remarkable fact is that if

$$\mathbf{R} = x(t)\mathbf{i} + y(t)\mathbf{j}$$

gives the particle's position on the curve at time t, then the vector

$$\frac{d\mathbf{R}}{dt} = \frac{dx}{dt}\mathbf{i} + \frac{dy}{dt}\mathbf{j}$$

has these properties. The slope of the curve at any point $P(x, y)$ where the tangent is not vertical is

$$\frac{dy}{dx} = \frac{dy/dt}{dx/dt} = \text{the slope of } \frac{d\mathbf{R}}{dt}$$

(see Fig. 14.3). Furthermore, if s is arc length measured along the curve from the point $P_0(x(t_0), y(t_0))$, then the value of s at any value of t is given by

$$s(t) = \begin{cases} \displaystyle\int_{t_0}^{t} \sqrt{x'(\tau)^2 + y'(\tau)^2}\, d\tau & \text{if } t \geq t_0, \\ \\ \displaystyle\int_{t}^{t_0} \sqrt{x'(\tau)^2 + y'(\tau)^2}\, d\tau & \text{if } t < t_0, \end{cases} \tag{1}$$

as we saw in Article 5.6. The Second Fundamental Theorem of Calculus (Article 4.7) applied to these integrals gives

$$\left|\frac{ds}{dt}\right| = \sqrt{x'(t)^2 + y'(t)^2} = \left|\frac{d\mathbf{R}}{dt}\right|. \tag{2}$$

Because $d\mathbf{R}/dt$ has the properties we want a velocity vector to have, we model velocity in mathematics by defining it to be the vector

$$\text{Velocity:} \qquad \mathbf{v} = \frac{d\mathbf{R}}{dt}. \tag{3}$$

At a point where $dx/dt = 0$ and $dy/dt \neq 0$, $d\mathbf{R}/dt = \mathbf{j}(dy/dt)$ is still the appropriate vector. For example, $\mathbf{R}(t) = (\cos t)\mathbf{i} + (\sin t)\mathbf{j}$ describes motion in a counterclockwise direction on a unit circle with center at the origin. Here,

$$\frac{d\mathbf{R}}{dt} = -(\sin t)\mathbf{i} + (\cos t)\mathbf{j}$$

points vertically upward at $t = 0$.

If both dx/dt and dy/dt are zero, then the speed is zero and a vector of zero length has no direction. This happens, for example, at each cusp on the four-cusped hypocycloid

$$x = \cos^3 t, \qquad y = \sin^3 t.$$

The acceleration of the motion is defined to be

$$\text{Acceleration:} \qquad \mathbf{a} = \frac{d\mathbf{v}}{dt} = \frac{d^2\mathbf{R}}{dt^2} = \frac{d^2x}{dt^2}\mathbf{i} + \frac{d^2y}{dt^2}\mathbf{j}. \tag{4}$$

For a particle of constant mass m moving under the action of an applied force $\mathbf{F}$, Newton's second law of motion states that

$$\mathbf{F} = m\mathbf{a}. \tag{5}$$

Since one ordinarily visualizes the force vector as being *applied at P*, it is customary to adopt the same viewpoint about the acceleration vector $\mathbf{a}$ (Fig. 14.4).

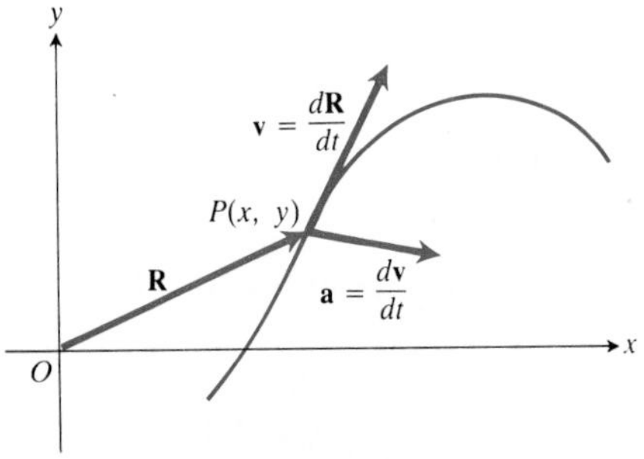

14.4 Typical position, velocity, and acceleration vectors of a particle moving in the plane.

EXAMPLE 1 A particle $P(x, y)$ moves on the hyperbola

$$x = r\cosh \omega t, \qquad y = r \sinh \omega t, \tag{6}$$

where r and ω are positive constants. Then

$$\mathbf{R} = \mathbf{i}(r\cosh \omega t) + \mathbf{j}(r \sinh \omega t),$$

$$\mathbf{v} = \frac{d\mathbf{R}}{dt} = \mathbf{i}(\omega r \sinh \omega t) + \mathbf{j}(\omega r \cosh \omega t),$$

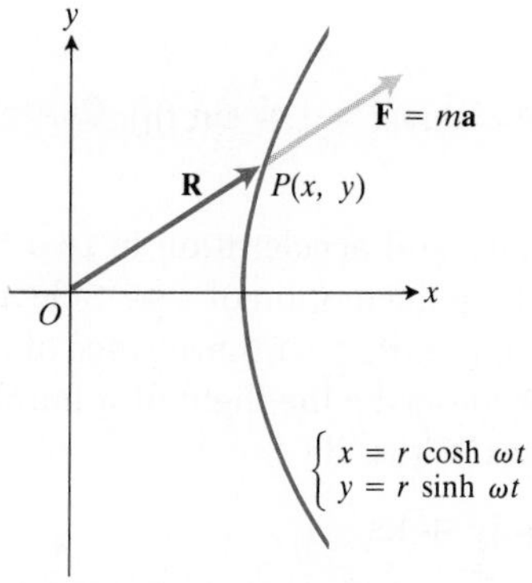

14.5 The force acting on the particle of Example 1 points directly away from O at all times.

and

$$\mathbf{a} = \mathbf{i}(\omega^2 r \cosh \omega t) + \mathbf{j}(\omega^2 r \sinh \omega t) = \omega^2 \mathbf{R}.$$

This means that the force $\mathbf{F} = m\mathbf{a} = m\omega^2\mathbf{R}$ has a magnitude $m\omega^2|\mathbf{R}| = m\omega^2|\overrightarrow{OP}|$ which is directly proportional to the distance OP, and that its direction is the same as the direction of $\mathbf{R}$. Thus the force is directed away from O (Fig. 14.5). □

The next example illustrates how we obtain the path of motion by integrating Eq. (5) when the force $\mathbf{F}$ is a given function of time, and the initial position and initial velocity of the particle are given. In general, the force $\mathbf{F}$ may depend upon the position of P as well as upon the time, and the problem of integrating the differential equations so obtained is usually discussed in textbooks on that subject.

EXAMPLE 2 The force acting on a particle P of mass m in the plane is given as a function of time t by

$$\mathbf{F} = \mathbf{i}\cos t + \mathbf{j}\sin t.$$

If the particle starts at the point (2, 0) with the initial velocity $v_0\mathbf{j}$ (Fig. 14.6), find its position at time t.

Solution If we denote the position vector by

$$\mathbf{R}(t) = \mathbf{i}x(t) + \mathbf{j}y(t),$$

we may restate the problem as follows: Find $\mathbf{R}(t)$ if

$$\mathbf{F} = m\frac{d^2\mathbf{R}}{dt^2} = \mathbf{i}\cos t + \mathbf{j}\sin t \tag{7}$$

and

$$\mathbf{R}(0) = 2\mathbf{i}, \qquad \mathbf{v}(0) = v_0\mathbf{j}. \tag{8}$$

Since $\mathbf{v} = d\mathbf{R}/dt$, we may rewrite (7) in the form

$$m\frac{d\mathbf{v}}{dt} = \mathbf{i}\cos t + \mathbf{j}\sin t.$$

Integrating both sides of this equation with respect to t gives

$$m\mathbf{v} = \mathbf{i}\sin t - \mathbf{j}\cos t + \mathbf{C}_1, \tag{9}$$

where the constant of integration $\mathbf{C}_1$ is a *vector*. The value of $\mathbf{C}_1$ may be found by using the initial condition $\mathbf{v}(0) = v_0\mathbf{j}$ in Eq. (9) with $t = 0$:

$$mv_0\mathbf{j} = -\mathbf{j} + \mathbf{C}_1, \qquad \mathbf{C}_1 = (mv_0 + 1)\mathbf{j}.$$

Thus, Eq. (9) gives

$$m\mathbf{v} = \mathbf{i}\sin t + (mv_0 + 1 - \cos t)\mathbf{j}.$$

Another integration gives

$$m\mathbf{R} = -\mathbf{i}\cos t + \mathbf{j}(mv_0 t + t - \sin t) + \mathbf{C}_2. \tag{10}$$

The initial condition $\mathbf{R}(0) = 2\mathbf{i}$ enables us to evaluate $\mathbf{C}_2$:

$$2m\mathbf{i} = -\mathbf{i} + \mathbf{C}_2, \qquad \mathbf{C}_2 = (2m + 1)\mathbf{i}.$$

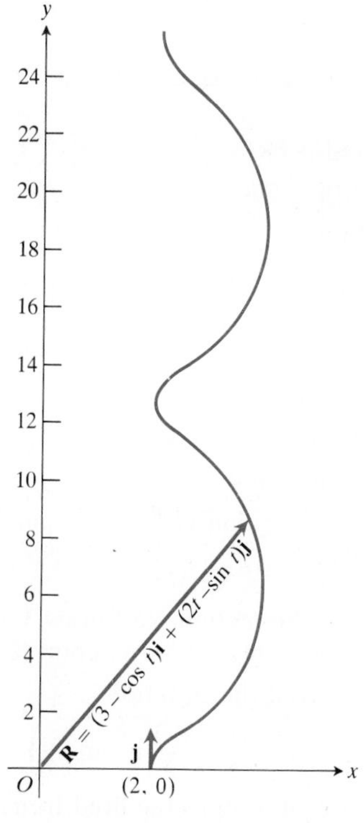

14.6 When $m = v_0 = 1$, the particle P in Example 2 follows the path shown here: $\mathbf{R} = (3 - \cos t)\mathbf{i} + (2t - \sin t)\mathbf{j}$.

Therefore,

$$\mathbf{R}(t) = \frac{1}{m}[(2m + 1 - \cos t)\mathbf{i} + (mv_0t + t - \sin t)\mathbf{j}]. \ \square$$

The foregoing equations for velocity and acceleration in two dimensions would be appropriate for describing the motion of a particle moving on a flat surface, such as a water bug skimming over the surface of a pond or a hockey puck sliding on ice. But to describe the flight of a bumblebee or a rocket, we need three coordinates. Thus, if

$$\mathbf{R}(t) = \mathbf{i}x + \mathbf{j}y + \mathbf{k}z, \tag{11}$$

where x, y, z are functions of t that are twice-differentiable, then the velocity of $P(x, y, z)$ is defined to be

$$\mathbf{v} = \frac{d\mathbf{R}}{dt} = \mathbf{i}\frac{dx}{dt} + \mathbf{j}\frac{dy}{dt} + \mathbf{k}\frac{dz}{dt}, \tag{12}$$

and the acceleration is defined to be

$$\mathbf{a} = \mathbf{i}\frac{d^2x}{dt^2} + \mathbf{j}\frac{d^2y}{dt^2} + \mathbf{k}\frac{d^2z}{dt^2}. \tag{13}$$

As in the plane, the length or magnitude of the velocity vector is the *speed* with which the object moves along its path:

$$\text{speed} = |\mathbf{v}|. \tag{14}$$

We shall see why this is so in the next article.

PROBLEMS

In Problems 1–12, $\mathbf{R}$ is the position vector of a moving particle at time t. Find the velocity and acceleration at time t. Then find the velocity, acceleration, and speed at the particular instant given.

1. $\mathbf{R} = (a \cos \omega t)\mathbf{i} + (a \sin \omega t)\mathbf{j}$, a and ω being positive constants: $t = \pi/(3\omega)$.
2. $\mathbf{R} = (2 \cos t)\mathbf{i} + (3 \sin t)\mathbf{j}$, $t = \pi/4$
3. $\mathbf{R} = (t + 1)\mathbf{i} + (t^2 - 1)\mathbf{j}$, $t = 2$
4. $\mathbf{R} = (\cos 2t)\mathbf{i} + (2 \sin t)\mathbf{j}$, $t = 0$
5. $\mathbf{R} = e^t\mathbf{i} + e^{-2t}\mathbf{j}$, $t = \ln 3$
6. $\mathbf{R} = (\sec t)\mathbf{i} + (\tan t)\mathbf{j}$, $t = \pi/6$
7. $\mathbf{R} = (\cosh 3t)\mathbf{i} + (2 \sinh t)\mathbf{j}$, $t = 0$
8. $\mathbf{R} = [\ln (t + 1)]\mathbf{i} + t^2\mathbf{j}$, $t = 1$
9. $\mathbf{R} = (e^{-t})\mathbf{i} + (2 \cos 3t)\mathbf{j} + (2 \sin 3t)\mathbf{k}$, $t = 0$
10. $\mathbf{R} = (1 + t)\mathbf{i} + (t^2/\sqrt{2})\mathbf{j} + (t^3/3)\mathbf{k}$, $t = 1$
11. $\mathbf{R} = (4 \cos t)\mathbf{i} - (3 \sin t)\mathbf{j} + 2t\mathbf{k}$, $t = \pi/3$
12. $\mathbf{R} = (\sqrt{2}/(t + 1))\mathbf{i} + (t^3/3)\mathbf{j} + (\sqrt{2} \ln t)\mathbf{k}$, $t = 1$

In Problems 13–15, $\mathbf{R}$ is the position vector of a particle in space for time $t > 0$. Find the time or times at which the velocity and acceleration vectors are perpendicular.

13. $\mathbf{R} = (2t^3 + 3)\mathbf{i} + (\ln t)\mathbf{j} + 3\mathbf{k}$
14. $\mathbf{R} = (t^4 + 3)\mathbf{i} + (4 \ln t)\mathbf{j} - 5t\mathbf{k}$
15. $\mathbf{R} = (\sin t)\mathbf{i} + (\cos t)\mathbf{j} + t\mathbf{k}$

In Problems 16–18, $\mathbf{R}$ is the position vector of a particle in space. Find the angle between the velocity and acceleration vectors in each case.

16. $\mathbf{R} = (e^t)\mathbf{i} + (e^t \sin t)\mathbf{j} + (e^t \cos t)\mathbf{k}$
17. $\mathbf{R} = (\tan t)\mathbf{i} + (\sinh 2t)\mathbf{j} + (\operatorname{sech} 3t)\mathbf{k}$
18. $\mathbf{R} = (\ln (t^2 + 1))\mathbf{i} + (\tan^{-1} t)\mathbf{j} + (\sqrt{t^2 + 1})\mathbf{k}$
19. Let $\mathbf{R} = t\mathbf{i} + t^2\mathbf{j}$ be the position vector of a particle moving in the xy-plane.
    a) Find $\mathbf{v}$ and $\mathbf{a}$.
    b) Sketch the path of the motion, and draw $\mathbf{v}$ and $\mathbf{a}$ for $t = 2$ (as vectors starting at the point (2, 4)).
20. In Article 13.2 we derived the equation
$$\mathbf{R} = \mathbf{i}(v_0 \cos \alpha)t + \mathbf{j}\left(-\frac{1}{2}gt^2 + (v_0 \sin \alpha)t\right)$$
for the position vector of a projectile fired from the origin into the first quadrant with an initial speed v_0 at an angle α with the x-axis. Find the velocity and acceleration.

21. A Howitzer at the point (1726, 0) in the xy-plane fires a practice round whose trajectory $\mathbf{R} = x(t)\mathbf{i} + y(t)\mathbf{j}$ is given by

$$\mathbf{R} = (256t + 1726)\mathbf{i} + (-4.9t^2 + 153t)\mathbf{j}, \qquad t \geq 0,$$

with t in seconds, and x and y in meters. Down range, 9900 m from the Howitzer, is a hill that rises 1100 m above the x-axis. Assuming no other obstructions, will the round clear the top of the hill? If so, by how many meters?

22. The acceleration of a particle in the plane as a function of time t is $\mathbf{a}(t) = -3t\mathbf{i}$. When $t = 0$, the particle's velocity and position are $\mathbf{v}(0) = 2\mathbf{j}$ and $\mathbf{R}(0) = 4\mathbf{i}$. Find the velocity $\mathbf{v}$ and position $\mathbf{R}$ as functions of time.

23. The acceleration of a particle in space as a function of time t is $\mathbf{a}(t) = 3t\mathbf{i} + 4\mathbf{j} + \mathbf{k}$. When $t = 0$, the particle's velocity and position are $\mathbf{v}(0) = 4\mathbf{i}$ and $\mathbf{R}(0) = 5\mathbf{j}$. Find the velocity $\mathbf{v}$ and position $\mathbf{R}$ as functions of time.

24. The force acting on a particle of mass m in the plane is given as a function of time t by the equation $\mathbf{F} = (1 + t)^{-1/2}\mathbf{i} - e^{-t}\mathbf{j}$. When $t = 0$, the particle's position and velocity are $\mathbf{R}(0) = (1/3)\mathbf{i} - \mathbf{j}$ and $\mathbf{v}(0) = -\mathbf{i} + 0.5\mathbf{j}$. Find the particle's velocity and position as functions of t.

25. If the projectile encounters a resistance proportional to the velocity, the force is

$$\mathbf{F} = -mg\mathbf{j} - k\frac{d\mathbf{R}}{dt}.$$

Show that one integration of $\mathbf{F} = md^2\mathbf{R}/dt^2$ leads to the differential equation

$$\frac{d\mathbf{R}}{dt} + \frac{k}{m}\mathbf{R} = \mathbf{v}_0 - gt\mathbf{j}.$$

(To solve this equation, one can multiply both sides of the equation by $e^{(k/m)t}$. Then the left side is the derivative of the product $\mathbf{R}e^{(k/m)t}$ and both sides can be integrated.)

26. The plane $z = 2x + 3y$ intersects the cylinder $x^2 + y^2 = 9$ in an ellipse.

a) Express the position of a point $P(x, y, z)$ on this ellipse as a vector function $\mathbf{R} = \overrightarrow{OP} = \mathbf{R}(\theta)$, where θ is a measure of the dihedral angle between the xz-plane and the plane containing the z-axis and P.

b) Using the equations of part (a), find the velocity and acceleration of P, assuming that $d\theta/dt = \omega$ is constant.

***Toolkit* programs**	
Parametric Equations	Super * Grapher

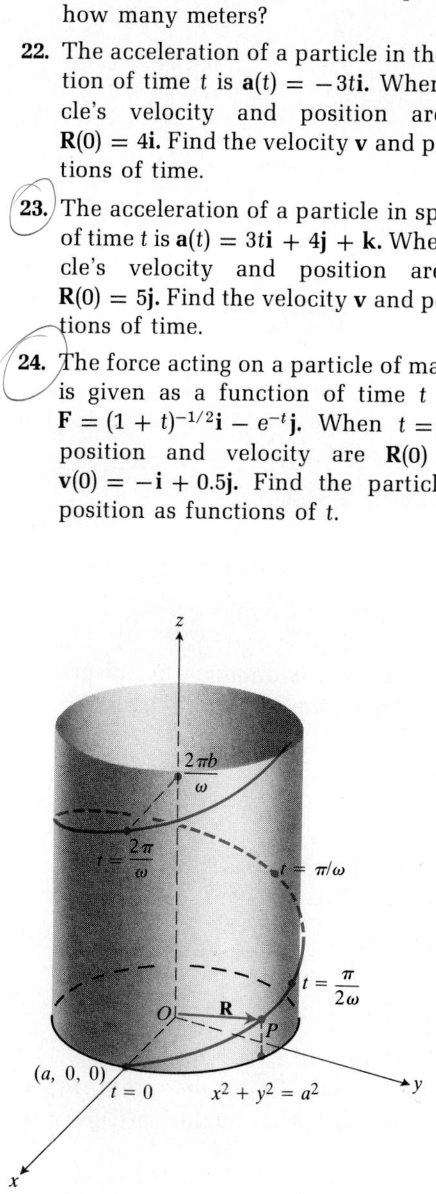

14.7 The helix traced by $\mathbf{R} = (a\cos\omega t)\mathbf{i} + (a\sin\omega t)\mathbf{j} + (bt)\mathbf{k}$, $b > 0$, spirals up from the xy-plane as t increases from zero.

14.3 Arc Length for Space Curves. The Unit Tangent Vector T

Space Curves

Everything we have done so far for two-dimensional motion in the plane can be extended to three-dimensional motion in space. To this end, let $P(x, y, z)$ be a point whose position in space is given by the vector

$$\mathbf{R}(t) = x(t)\mathbf{i} + y(t)\mathbf{j} + z(t)\mathbf{k},$$

where x, y, and z are differentiable functions of t. As t varies continuously, P traces a curve in space.

EXAMPLE 1 The equation

$$\mathbf{R} = (a\cos\omega t)\mathbf{i} + (a\sin\omega t)\mathbf{j} + (bt)\mathbf{k},$$

where a, b, and ω are positive constants, represents a circular helix (Fig. 14.7). The projection of the point $P(x, y, z)$ onto the xy-plane moves around the circle $x^2 + y^2 = a^2$, $z = 0$, as t varies, while the distance $|bt|$ between P and the xy-plane changes steadily with t. □

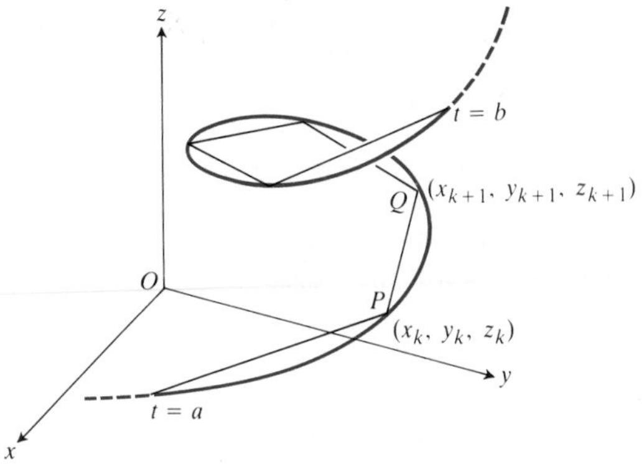

14.8 The length of the arc from $t = a$ to $t = b$ is a limit of lengths of polygonal paths.

Arc Length

As Fig. 14.8 suggests, we define the length of a space curve traced by a moving particle as a limit of lengths

$$\sum_a^b \sqrt{(x_{k+1} - x_k)^2 + (y_{k+1} - y_k)^2 + (z_{k+1} - z_k)^2} \tag{1}$$

of approximating polygonal paths in much the same way that we defined the length of a plane curve in Article 5.6. The resulting formula for the length of a space curve

$$\mathbf{R}(t) = x(t)\mathbf{i} + y(t)\mathbf{j} + z(t)\mathbf{k}$$

from $t = a$ to $t = b$ is

$$L_a^b = \int_a^b \sqrt{x'(t)^2 + y'(t)^2 + z'(t)^2}\, dt. \tag{2}$$

Thus the length of curve from $t = a$ to $t = b$ is obtained by integrating $|\mathbf{v}| = |d\mathbf{R}/dt|$ from a to b. Dropping the z-term in Eq. (2) gives the arc length formula for curves in the plane.

DEFINITION

Arc Length

The length of the curve $\mathbf{R}(t)$ from $t = a$ to $t = b$ is the integral of the length of the velocity vector from $t = a$ to $t = b$:

$$L_a^b = \int_a^b |\mathbf{v}|\, dt. \tag{3}$$

If we choose a base point $P_0(t_0)$ on the curve, as shown in Fig. 14.9, then the length $s(t)$ along the curve from P_0 to the point $P(t) = (x(t), y(t), z(t))$ for any $t \geq t_0$ is a function of t whose values are given by the integral

$$s(t) = \int_{t_0}^t \sqrt{x'(\tau)^2 + y'(\tau)^2 + z'(\tau)^2}\, d\tau. \tag{4}$$

If the derivatives beneath the radical are continuous, then the Second Fundamental Theorem of Calculus tells us that s is a differentiable func-

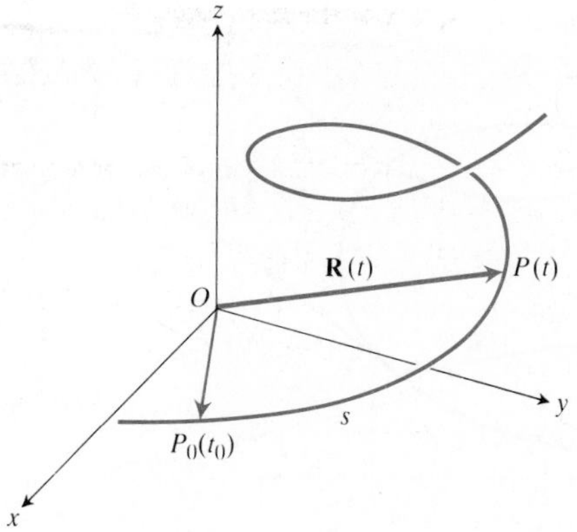

14.9 The distance s along the curve from $P_0(t_0)$ to any point $P(t)$, $t \geq t_0$, is

$$s = \int_{t_0}^{t} |\mathbf{v}|\, d\tau.$$

tion of t whose derivative is

$$\frac{ds}{dt} = \sqrt{x'(t)^2 + y'(t)^2 + z'(t)^2} = \left|\frac{d\mathbf{R}}{dt}\right| = |\mathbf{v}|. \tag{5}$$

Once again we find that the length of the particle's velocity vector is the particle's speed along the curve:

$$\text{Speed} = \left|\frac{ds}{dt}\right| = |\mathbf{v}|. \tag{6}$$

EXAMPLE 2 Find the length of one turn of the helix

$$\mathbf{R} = (\cos t)\mathbf{i} + (\sin t)\mathbf{j} + t\mathbf{k}.$$

Solution This is the helix shown in Fig. 14.7, with $a = b = \omega = 1$. Since $\cos t$ and $\sin t$ have period 2π, the helix makes one full turn as t runs from 0 to 2π. According to Eq. (3), therefore, the length we seek is

$$\int_0^{2\pi} |\mathbf{v}|\, dt = \int_0^{2\pi} \sqrt{(-\sin t)^2 + (\cos t)^2 + (1)^2}\, dt = \int_0^{2\pi} \sqrt{2}\, dt = 2\pi\sqrt{2}.$$

This is $\sqrt{2}$ times the length of the unit circle in the xy-plane over which the helix stands. (Also, see Problem 21.) □

The Unit Tangent Vector T

Suppose that $\mathbf{R}(t)$ is a curve in the plane or in space whose component functions are differentiable, and that $s(t)$ gives arc length along the curve measured from some preselected base point P_0. Then we know from Eq. (4) and the Second Fundamental Theorem that $s(t)$ is a differentiable function of t. If, further, $ds/dt = |\mathbf{v}|$ is never zero (as we shall assume from now on), then $s(t)$ is one-to-one and has an inverse that gives t as a differentiable function of s whose derivative is

$$\frac{dt}{ds} = \frac{1}{|\mathbf{v}|}. \tag{7}$$

(See Articles 2.10 and 3.8.) This makes $\mathbf{R}(t)$ a differentiable function of s, whose derivative can be calculated from the Chain Rule in Article 14.1 to be

$$\frac{d\mathbf{R}}{ds} = \frac{d\mathbf{R}}{dt}\frac{dt}{ds} = \mathbf{v}\frac{dt}{ds}. \tag{8}$$

From this equation we see that

$$\left|\frac{d\mathbf{R}}{ds}\right| = |\mathbf{v}|\left|\frac{dt}{ds}\right| = |\mathbf{v}|\frac{1}{|\mathbf{v}|} = 1. \tag{9}$$

Thus the vector $d\mathbf{R}/ds$ is *always* a unit vector. Furthermore, it points in the direction of $\mathbf{v}$ because it is a positive scalar multiple $dt/ds = 1/|\mathbf{v}|$ of $\mathbf{v}$. We call $d\mathbf{R}/ds$ the unit tangent vector of the curve and denote it by $\mathbf{T}$. See Figs. 14.10 and 14.11.

From a geometric point of view, $\mathbf{T}$ arises in the following way: As $Q \to P$ and $\Delta s \to 0$, the direction of the secant line PQ approaches the direction of the tangent to the curve at P, while the ratio of chord PQ to

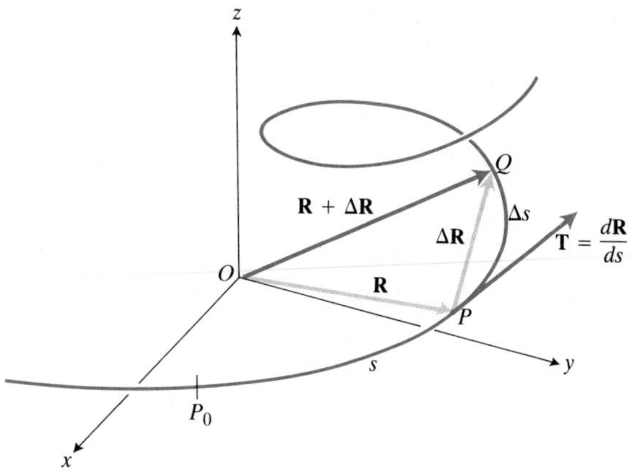

14.10 If **R** is differentiable, then $\mathbf{T} = d\mathbf{R}/ds = \lim_{\Delta s \to 0} \Delta\mathbf{R}/\Delta s$ is a unit vector tangent to the curve traced out by P.

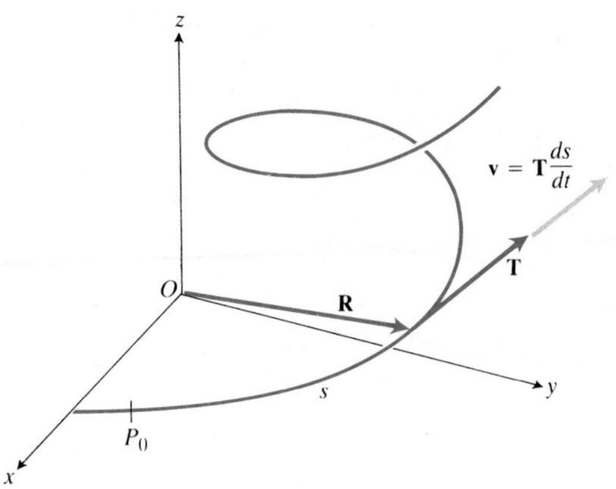

14.11 Divide **v** by ds/dt to get **T**.

arc PQ approaches one (for a "smooth" curve). Therefore the limit of $\Delta\mathbf{R}/\Delta s$ is a unit vector tangent to the curve at P and pointing in the direction in which arc length increases along the curve.

DEFINITION

Unit Tangent Vector

The unit tangent vector to the curve $\mathbf{R}(t)$ is

$$\mathbf{T} = \frac{d\mathbf{R}}{ds}. \tag{10}$$

This definition assumes that $|\mathbf{v}| = |d\mathbf{R}/dt|$ is never zero; we shall require this of all our parameterized curves.

While $\mathbf{T}$ is defined to be $d\mathbf{R}/ds$, the natural parameterization of a motion in many cases is likely to be time and not arc length. The best way to find $\mathbf{T}$ in such cases is not from the equation $\mathbf{T} = d\mathbf{R}/ds$ but from the equation

$$\mathbf{T} = \frac{\mathbf{v}}{|\mathbf{v}|}. \tag{11}$$

This equation follows from observing that since $\mathbf{T}$ is a unit vector in the direction of $\mathbf{v}$, it must be $\mathbf{v}/|\mathbf{v}|$.

EXAMPLE 3 Find $\mathbf{T}$ for the helix of Example 1.

Solution The velocity vector is

$$\mathbf{v} = \mathbf{i}(-a\omega \sin \omega t) + \mathbf{j}(a\omega \cos \omega t) + \mathbf{k}(b),$$

whose length is

$$|\mathbf{v}| = \sqrt{a^2\omega^2 \sin^2 \omega t + a^2\omega^2 \cos^2 \omega t + b^2} = \sqrt{a^2\omega^2 + b^2}.$$

Therefore,

$$\mathbf{T} = \frac{\mathbf{v}}{|\mathbf{v}|} = \frac{a\omega(-\mathbf{i} \sin \omega t + \mathbf{j} \cos \omega t) + b\mathbf{k}}{\sqrt{a^2\omega^2 + b^2}}. \quad \square$$

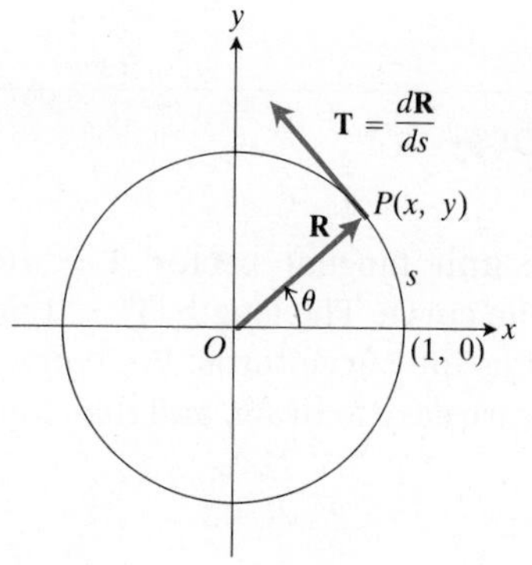

14.12 $\mathbf{R} = \mathbf{i}\cos\theta + \mathbf{j}\sin\theta$.

EXAMPLE 4 Find **T** for the motion

$$\mathbf{R} = (\cos t + t\sin t)\mathbf{i} + (\sin t - t\cos t)\mathbf{j}, \qquad t \geq 0.$$

Solution

$$\mathbf{v} = \frac{d\mathbf{R}}{dt} = (-\sin t + t\cos t + \sin t)\mathbf{i} + (\cos t + t\sin t - \cos t)\mathbf{j}$$

$$= (t\cos t)\mathbf{i} + (t\sin t)\mathbf{j},$$

$$|\mathbf{v}| = \sqrt{t^2\cos^2 t + t^2\sin^2 t} = |t| = t,$$

$$\mathbf{T} = \frac{\mathbf{v}}{|\mathbf{v}|} = (\cos t)\mathbf{i} + (\sin t)\mathbf{j}. \ \square$$

EXAMPLE 5 For the counterclockwise motion

$$\mathbf{R} = \mathbf{i}\cos\theta + \mathbf{j}\sin\theta$$

around the unit circle, we find that

$$\mathbf{v} = (-\sin\theta)\mathbf{i} + (\cos\theta)\mathbf{j}$$

is already a unit vector, so that $\mathbf{v} = \mathbf{T}$. In fact, **T** is **R** rotated 90° counterclockwise (Fig. 14.12). □

PROBLEMS

Each of Problems 1–8 gives the position vector **R** of a particle moving in the plane. Find **T** in each case.

1. $\mathbf{R} = 2\mathbf{i}\cos t + 2\mathbf{j}\sin t$
2. $\mathbf{R} = e^t\mathbf{i} + t^2\mathbf{j}$
3. $\mathbf{R} = (\cos^3 t)\mathbf{i} + (\sin^3 t)\mathbf{j}$
4. $\mathbf{R} = \mathbf{i}x + \mathbf{j}x^2$
5. $\mathbf{R} = (\cos 2t)\mathbf{i} + (2\cos t)\mathbf{j}$
6. $\mathbf{R} = \dfrac{t^3}{3}\mathbf{i} + \dfrac{t^2}{2}\mathbf{j}$
7. $\mathbf{R} = (e^t\cos t)\mathbf{i} + (e^t\sin t)\mathbf{j}$
8. $\mathbf{R} = \cosh t\mathbf{i} + t\mathbf{j}$
9. Find the length of the curve $\mathbf{R} = t\mathbf{i} + (2/3)t^{3/2}\mathbf{j}$ from $t = 0$ to $t = 8$.
10. Find the length of the curve

$$\mathbf{R} = (t\sin t + \cos t)\mathbf{i} + (t\cos t - \sin t)\mathbf{j}$$

from $t = 0$ to $t = 2$.

11. Find **T** at $t = 0$ if $\mathbf{R} = (2t^3 + 3t^2)\mathbf{i} + (t^2 + 2t)\mathbf{j}$

In Problems 12–15, find **T**, the speed $|ds/dt| = |\mathbf{v}|$, and the length of the curve from $t = 0$ to $t = \pi$.

12. $\mathbf{R} = (6\sin 2t)\mathbf{i} + (6\cos 2t)\mathbf{j} + (5t)\mathbf{k}$
13. $\mathbf{R} = (e^t\cos t)\mathbf{i} + (e^t\sin t)\mathbf{j} + e^t\mathbf{k}$
14. $\mathbf{R} = (3\cosh 2t)\mathbf{i} + (3\sinh 2t)\mathbf{j} + (6t)\mathbf{k}$
15. $\mathbf{R} = (3t\cos t)\mathbf{i} + (3t\sin t)\mathbf{j} + (4t)\mathbf{k}$
16. Find **T** when $t = 1$ if $\mathbf{R} = t\mathbf{i} + t^2\mathbf{j} - \dfrac{1}{t}\mathbf{k}$.
17. Find **v**, **T**, the speed $ds/dt = |\mathbf{v}|$, and **a**, all at $t = 0$, if

$$\mathbf{R} = 3t^2\mathbf{i} + (6t + 7)\mathbf{j} + 2t^4\mathbf{k}.$$

18. Find the length of the curve $\mathbf{R} = \mathbf{i} + t\mathbf{j} + t^2\mathbf{k}$ from the point (1, 0, 0) to the point (1, 1, 1).
19. Find the length of the curve

$$\mathbf{R} = (\cos t)\mathbf{i} + (\sin t)\mathbf{j} + t^2\mathbf{k}, \qquad 0 \leq t \leq 1.$$

20. Find the length of the curve

$$\mathbf{R} = t\mathbf{i} + t\mathbf{j} + (4 - t^2)\mathbf{k}$$

from the point (0, 0, 4) to the point (1, 1, 3).

21. The length $2\pi\sqrt{2}$ of one turn of the helix in Example 2 is also the length of the diagonal of a square 2π units on a side. Show how to obtain this square by cutting away and flattening a portion of the cylinder around which the helix winds.

***Tookit* programs**

Parametric Equations   Super * Grapher

14.4 Curvature and Normal Vectors

In this article we study the way the unit tangent vector $\mathbf{T} = d\mathbf{R}/ds$ changes as we move along a differentiable curve. The length $|\mathbf{T}| = 1$ never changes, but the direction of $\mathbf{T}$ changes as the curve turns. We work first with curves in the plane, where pictures are easy to draw, and then look at a helix in space.

Motion in a Plane

We measure the rate at which $\mathbf{T}$ turns as it moves along the curve by measuring the change in ϕ, the *direction angle* or *slope angle* that $\mathbf{T}$ makes with $\mathbf{i}$ (Fig. 14.13). At each point, the absolute value of $d\phi/ds$, measured in radians per unit of length along the curve, is called the *curvature* of the curve. The usual notation for the curvature is the Greek letter κ (kappa).

$$\text{Curvature:} \qquad \kappa = \left|\frac{d\phi}{ds}\right| \tag{1}$$

If the curve is the graph of a function $y = f(x)$, then at any point P on the curve

$$\frac{dy}{dx} = \tan\phi,$$

both being the slope of the curve at P. To calculate κ, we solve this equation for ϕ,

$$\phi = \tan^{-1}\frac{dy}{dx}, \tag{2}$$

and apply the Chain Rule in the form

$$\frac{d\phi}{ds} = \frac{d\phi}{dx}\frac{dx}{ds}. \tag{3}$$

Since

$$\frac{d\phi}{dx} = \frac{d^2y/dx^2}{1 + (dy/dx)^2}, \tag{4}$$

from Eq. (2), and

$$\frac{dx}{ds} = \frac{1}{ds/dx} = \frac{1}{\pm\sqrt{1 + (dy/dx)^2}}, \tag{5}$$

from Article 5.6, Eq. (3) gives

$$\kappa = \left|\frac{d\phi}{ds}\right| = \left|\frac{d\phi}{dx}\frac{dx}{ds}\right| = \frac{|d^2y/dx^2|}{[1 + (dy/dx)^2]^{3/2}}. \tag{6a}$$

All the derivatives in (6a) must exist for the formula to make sense.

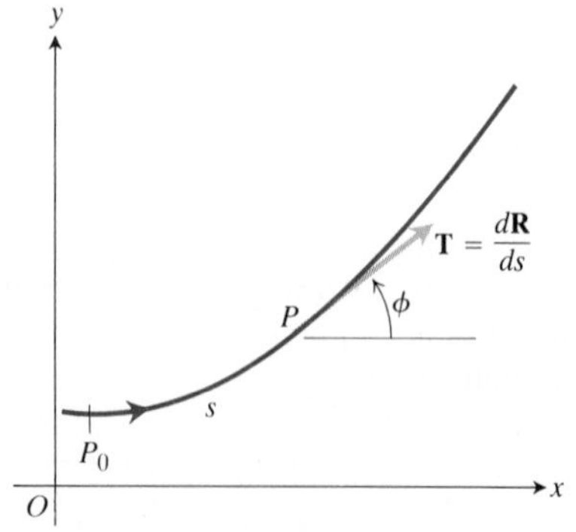

14.13 The value of $|d\phi/ds|$ at P is called the curvature of the curve at P.

If the curve we are studying is the graph of a function $x = g(y)$, the formula for curvature is

$$\kappa = \frac{|d^2x/dy^2|}{[1 + (dx/dy)^2]^{3/2}}. \tag{6b}$$

This is just (6a) with x and y interchanged.

If the curve is given parametrically by

$$x = f(t), \qquad y = g(t),$$

then

$$\kappa = \frac{|\dot{x}\ddot{y} - \dot{y}\ddot{x}|}{[\dot{x}^2 + \dot{y}^2]^{3/2}}. \tag{6c}$$

This formula is derived in Problem 27. The dot notation for derivatives was frequently used by Newton, and it is a standard notation in applications that involve time derivatives.

As we shall see in Article 14.5, the curvature formulas 6(a)–6(c) are special cases of the vector curvature formula

$$\kappa = \frac{|\mathbf{v} \times \mathbf{a}|}{|\mathbf{v}|^3}. \tag{7}$$

In this formula, **v** and **a** are the velocity and acceleration vectors of the curve. If you want to remember only one curvature formula, then remember Eq. (7). The others come in handy now and then, but in practice the vector equation is the most useful one because it works in space as well as in the plane. (Also see Problem 27.)

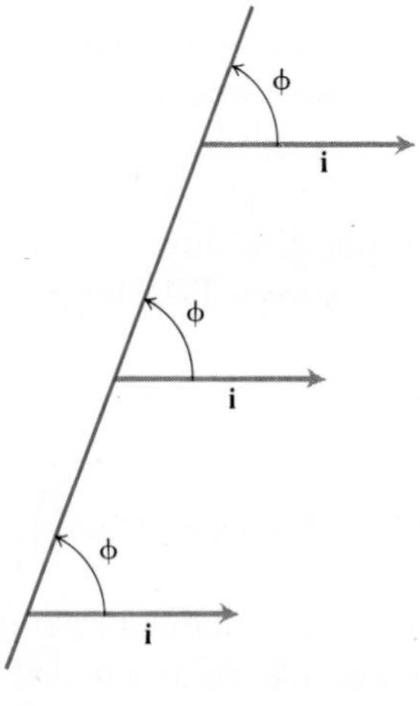

14.14 Along a line, the angle ϕ stays the same from point to point.

EXAMPLE 1 Show that the curvature of a straight line is 0.

Solution On a straight line, ϕ has a constant value (Fig. 14.14). Therefore, $d\phi/ds = 0$ and $\kappa = |d\phi/ds| = 0$. □

EXAMPLE 2 Show that the curvature of a circle of radius a is $1/a$.

Solution 1 Parameterize the circle with the equations $x = a \cos t$, $y = a \sin t$, and use Eq. (6c):

$$\dot{x} = -a \sin t, \qquad \dot{y} = a \cos t,$$
$$\ddot{x} = -a \cos t, \qquad \ddot{y} = -a \sin t,$$

and by (6c)

$$\kappa = \frac{|(-a \sin t)(-a \sin t) - (a \cos t)(-a \cos t)|}{[a^2 \sin^2 t + a^2 \cos^2 t]^{3/2}} = \frac{a^2}{a^3} = \frac{1}{a}.$$

Solution 2 Use Eq. (7) with

$$\mathbf{R}(t) = (a \cos t)\mathbf{i} + (a \sin t)\mathbf{j} + (0)\mathbf{k},$$
$$\mathbf{v}(t) = (-a \sin t)\mathbf{i} + (a \cos t)\mathbf{j},$$
$$\mathbf{a}(t) = (-a \cos t)\mathbf{i} - (a \sin t)\mathbf{j}.$$

Then

$$\mathbf{v} \times \mathbf{a} = \begin{vmatrix} \mathbf{i} & \mathbf{j} & \mathbf{k} \\ -a \sin t & a \cos t & 0 \\ -a \cos t & -a \sin t & 0 \end{vmatrix} = (a^2 \sin^2 t + a^2 \cos^2 t)\mathbf{k} = a^2\mathbf{k},$$

$$|\mathbf{v}|^3 = \left[\sqrt{(-a \sin t)^2 + (a \cos t)^2}\right]^3 = a^3,$$

and

$$\frac{|\mathbf{v} \times \mathbf{a}|}{|\mathbf{v}|^3} = \frac{|a^2\mathbf{k}|}{a^3} = \frac{a^2|\mathbf{k}|}{a^3} = \frac{1}{a}. \quad \square$$

Circle of Curvature and Radius of Curvature

The *circle of curvature* or *osculating circle* at a point P on a plane curve where $\kappa \neq 0$ is the circle in the plane of the curve that

1. is tangent to the curve at P (has the same tangent that the curve does),
2. has the same curvature the curve does at P, and
3. lies toward the concave or inner side of the curve (as in Fig. 14.15).

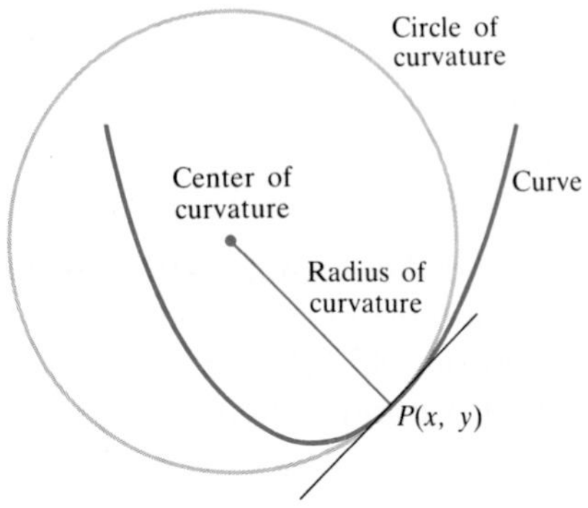

14.15 The osculating circle or circle of curvature at $P(x, y)$ lies toward the inner side of the curve.

The *radius of curvature* of the curve at P is the radius of the circle of curvature, which, according to Example 2, is

$$\text{Radius of curvature} = \rho = \frac{1}{\kappa}. \tag{8}$$

To calculate ρ, we calculate κ and take its reciprocal.

The *center of curvature* of the curve at P is the center of the circle of curvature.

If we write a cartesian equation for the circle of curvature at a point P on the curve $y = f(x)$, and differentiate the circle equation implicitly to find dy/dx and d^2y/dx^2, the values of these derivatives will agree with the values of f' and f'' at P. (We shall not prove this, but see Problem 25.)

Unit Normal Vector

We can now calculate the change $d\phi/ds$ in the direction angle of $\mathbf{T}$, but how about the change in $\mathbf{T}$ itself? We can express $\mathbf{T}$ in terms of ϕ as

$$\mathbf{T} = \mathbf{i} \cos \phi + \mathbf{j} \sin \phi. \tag{9}$$

Therefore,

$$\frac{d\mathbf{T}}{d\phi} = -\mathbf{i} \sin \phi + \mathbf{j} \cos \phi = \mathbf{i} \cos\left(\phi + \frac{\pi}{2}\right) + \mathbf{j} \sin\left(\phi + \frac{\pi}{2}\right), \tag{10}$$

from which it follows that $d\mathbf{T}/d\phi$ is the unit vector obtained by rotating $\mathbf{T}$ counterclockwise through $\pi/2$ radians. Thus, $d\mathbf{T}/d\phi$ is normal to the curve at all times.

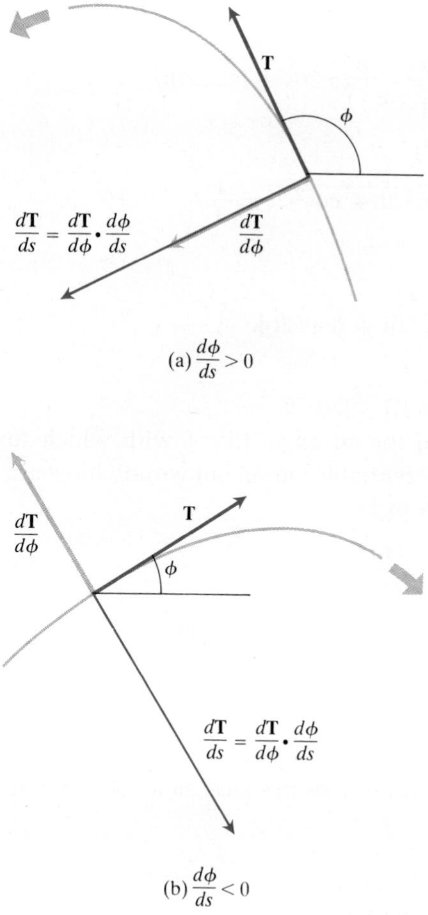

14.16 The vector $\frac{d\mathbf{T}}{ds}$ always points inward.

From the Chain Rule, we have

$$\frac{d\mathbf{T}}{ds} = \frac{d\mathbf{T}}{d\phi}\frac{d\phi}{ds}, \tag{11}$$

a vector whose magnitude is

$$\left|\frac{d\mathbf{T}}{ds}\right| = \left|\frac{d\mathbf{T}}{d\phi}\right| \left|\frac{d\phi}{ds}\right| = 1 \cdot \kappa = \kappa. \tag{12}$$

The vector $d\mathbf{T}/ds$ is normal to the curve because it is a scalar multiple of $d\mathbf{T}/d\phi$. It points in the direction of $d\mathbf{T}/d\phi$ when $d\phi/ds > 0$ and in the opposite direction when $d\phi/ds < 0$. As we can see in Fig. 14.16, this means that $d\mathbf{T}/ds$ points inward toward the concave side of the curve at all times.

The unit vector $\mathbf{N}$ obtained by dividing $d\mathbf{T}/ds$ by its length κ is called the *principal unit normal vector* to the curve. Like $d\mathbf{T}/ds$, the vector $\mathbf{N}$ always points toward the concave side of the curve.

Principal Unit Normal Vector

$$\mathbf{N} = \frac{d\mathbf{T}/ds}{|d\mathbf{T}/ds|} = \frac{1}{\kappa}\frac{d\mathbf{T}}{ds} \qquad (\kappa \neq 0). \tag{13}$$

At a point P_0 where the curvature is zero,

$$\frac{d\mathbf{T}}{ds} = \mathbf{N}\kappa = 0$$

and there is no (unique) principal normal. All vectors in the plane through P_0 and perpendicular to $\mathbf{T}$ are normal vectors and none of them is determined as $\mathbf{N}$. This is the case, for example, at every point on a straight line.

To calculate $\mathbf{N}$ it is often useful to apply the formula

$$\mathbf{N} = \frac{1}{\kappa}\frac{d\mathbf{T}/dt}{ds/dt} = \frac{1}{\kappa}\frac{d\mathbf{T}/dt}{|\mathbf{v}|}, \tag{14}$$

obtained from the right-hand side of Eq. (13) by the Chain Rule.

EXAMPLE 3 Find $\mathbf{T}$, κ, and $\mathbf{N}$ for the curve

$$\mathbf{R} = -(3\sin 2t)\mathbf{i} - (3\cos 2t)\mathbf{j}.$$

Solution We first find $\mathbf{T}$:

$$\mathbf{v} = -(6\cos 2t)\mathbf{i} + (6\sin 2t)\mathbf{j},$$

$$|\mathbf{v}| = \sqrt{36\cos^2 2t + 36\sin^2 2t} = 6,$$

$$\mathbf{T} = \frac{\mathbf{v}}{|\mathbf{v}|} = -(\cos 2t)\mathbf{i} + (\sin 2t)\mathbf{j}.$$

From this we find

$$\frac{d\mathbf{T}}{dt} = (2\sin 2t)\mathbf{i} + (2\cos 2t)\mathbf{j}$$

and

$$\frac{d\mathbf{T}}{ds} = \frac{d\mathbf{T}/dt}{ds/dt} = \frac{d\mathbf{T}/dt}{|\mathbf{v}|} = \frac{1}{3}((\sin 2t)\mathbf{i} + (\cos 2t)\mathbf{j}).$$

The curvature is

$$\kappa = \left|\frac{d\mathbf{T}}{ds}\right| = \frac{1}{3}\sqrt{\sin^2 2t + \cos^2 2t} = \frac{1}{3},$$

and

$$\mathbf{N} = \frac{1}{\kappa}\frac{d\mathbf{T}}{ds} = (\sin 2t)\mathbf{i} + (\cos 2t)\mathbf{j}. \quad \square$$

Curvature and **N** for Curves in Space

In space there is no natural choice for an angle like ϕ with which to measure the change in **T** along a differentiable curve. But we still have arc length and can define the curvature to be

$$\kappa = \left|\frac{d\mathbf{T}}{ds}\right|, \tag{15}$$

as it worked out to be in Eq. (12) for curves in the plane.

The formula

$$\kappa = \frac{|\mathbf{v} \times \mathbf{a}|}{|\mathbf{v}|^3} \tag{16}$$

from Eq. (7) also gives the curvature for curves in space, as we shall see in the next article.

EXAMPLE 4 Find the curvature of the helix

$$\mathbf{R}(t) = (a \cos \omega t)\mathbf{i} + (a \sin \omega t)\mathbf{j} + (bt)\mathbf{k}$$

of Article 14.3, Example 3.

Solution In Example 3, Article 14.3, we found

$$\mathbf{T} = \frac{a\omega(-\mathbf{i}\sin\omega t + \mathbf{j}\cos\omega t) + b\mathbf{k}}{\sqrt{a^2\omega^2 + b^2}} \tag{17}$$

and

$$\frac{ds}{dt} = \sqrt{a^2\omega^2 + b^2}.$$

Hence

$$\frac{d\mathbf{T}}{ds} = \frac{d\mathbf{T}/dt}{ds/dt} = \frac{-a\omega^2}{a^2\omega^2 + b^2}(\mathbf{i}\cos\omega t + \mathbf{j}\sin\omega t), \tag{18}$$

and

$$\kappa = \left|\frac{d\mathbf{T}}{ds}\right| = \frac{a\omega^2}{a^2\omega^2 + b^2}. \tag{19}$$

Two limiting cases of Eq. (19) are worth checking. First, if $b = 0$, then $z = 0$, and the helix reduces to a circle of radius a in the xy-plane, while Eq. (19) reduces to $\kappa = 1/a$ as it should. Second, if $a = 0$, then $x = y = 0$

and $z = bt$. This tells us that the point moves along the z-axis. Again Eq. (19) gives the correct curvature, namely $\kappa = 0$. In the general case, the curvature of a circular helix is constant and less than $1/a$, the curvature of the circle that is the cross section of the cylinder around which the helix winds (Fig. 14.7). □

To define the principal unit normal vector **N** in space, we use the same definition we use for plane curves:

$$\mathbf{N} = \frac{d\mathbf{T}/ds}{|d\mathbf{T}/ds|} = \frac{1}{\kappa}\frac{d\mathbf{T}}{ds}. \tag{20}$$

This equation certainly defines **N** as a unit vector, and it remains only to show that $\mathbf{T} \cdot \mathbf{N} = 0$ to know that **N** is normal to the curve. This is an immediate consequence of a fact we shall establish in the next article, and we shall defer the proof until then. In the meantime, we check that $\mathbf{T} \cdot \mathbf{N} = 0$ for the helix in Example 4.

EXAMPLE 5 Show that $\mathbf{T} \cdot \mathbf{N} = 0$ for the helix in Example 4.

Solution We have

$$\mathbf{N} = \frac{d\mathbf{T}/ds}{|d\mathbf{T}/ds|} = -\mathbf{i}\cos\omega t - \mathbf{j}\sin\omega t.$$

Note that **N** always points inward toward the z-axis. The **k** component of **N** is zero, so that when we calculate $\mathbf{T} \cdot \mathbf{N}$ using the formula for **T** in Eq. (17) we have

$$\mathbf{T} \cdot \mathbf{N} = \frac{a\omega}{\sqrt{a^2\omega^2 + b^2}}(\sin\omega t\cos\omega t - \cos\omega t\sin\omega t) = 0. \ \square$$

The Binormal Vector B

We can define a vector **B** perpendicular to both **T** and **N** by the formula

$$\mathbf{B} = \mathbf{T} \times \mathbf{N}. \tag{21}$$

This vector is called the *binormal vector* of the curve. The three vectors **T**, **N**, and **B** (in this order) define a moving right-handed *coordinate frame* or *reference frame* of mutually orthogonal vectors (Fig. 14.17). They play a

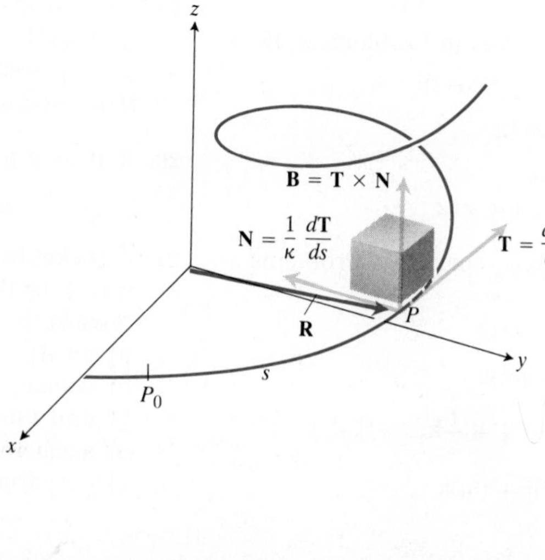

14.17 **T**, **N**, and **B** form a right-handed coordinate frame.

significant role in the geometry of space curves (see any current text on *differential geometry*). Moving reference frames are important in physics and in the determination of the flight paths of space vehicles.

Curvature and Vector Formulas

In the plane:

$$\kappa = \frac{|d^2y/dx^2|}{[1 + (dy/dx)^2]^{3/2}} \quad (\text{if } y = f(x))$$

$$\kappa = \frac{|d^2x/dy^2|}{[1 + (dx/dy)^2]^{3/2}} \quad (\text{if } x = g(y))$$

$$\kappa = \frac{|\dot{x}\ddot{y} - \dot{y}\ddot{x}|}{[\dot{x}^2 + \dot{y}^2]^{3/2}} \quad (\text{if } x = f(t),\ y = g(t))$$

In plane and space:

$$\kappa = \left|\frac{d\mathbf{T}}{ds}\right| = \frac{|\mathbf{v} \times \mathbf{a}|}{|\mathbf{v}|^3}$$

Radius of curvature: $\rho = 1/\kappa \quad (\text{if } \kappa \neq 0)$

Unit tangent vector: $\mathbf{T} = \dfrac{\mathbf{v}}{|\mathbf{v}|}$

Principal normal vector: $\mathbf{N} = \dfrac{d\mathbf{T}/ds}{|d\mathbf{T}/ds|} = \dfrac{1}{\kappa}\dfrac{d\mathbf{T}}{ds} \quad (\kappa \neq 0)$

Binormal vector: $\mathbf{B} = \mathbf{T} \times \mathbf{N}$

PROBLEMS

Find the curvature of the plane curves in Problems 1–8.

1. $y = \ln(\cos x)$
2. $y = e^{2x}$
3. $x = \ln \sec y$
4. $x = \frac{1}{3}(y^2 + 2)^{3/2}$
5. $x = \dfrac{y^4}{4} + \dfrac{1}{8y^2}$
6. $y = a \cosh(x/a)$
7. $x = 2t + 3, \quad y = 5 - t^2$
8. $x = t^3/3, \quad y = t^2/2$

Find **T**, κ, and **N** for the plane curves in Problems 9–12.

9. $\mathbf{R} = (\cos t + t \sin t)\mathbf{i} + (\sin t - t \cos t)\mathbf{j}$
10. $\mathbf{R} = a(t - \sin t)\mathbf{i} + a(1 - \cos t)\mathbf{j}$
11. $\mathbf{R} = (e^t \cos t)\mathbf{i} + (e^t \sin t)\mathbf{j}$
12. $\mathbf{R} = (\cos^3 t)\mathbf{i} + (\sin^3 t)\mathbf{j}, \quad 0 < t < \pi/2$

Find **T**, κ, **N**, and **B** for the space curves in Problems 13–16.

13. $\mathbf{R} = (6 \sin 2t)\mathbf{i} + (6 \cos 2t)\mathbf{j} + (5t)\mathbf{k}$
14. $\mathbf{R} = (e^t \cos t)\mathbf{i} + (e^t \sin t)\mathbf{j} + (e^t)\mathbf{k}$
15. $\mathbf{R} = \dfrac{1}{3}(1 + t)^{3/2}\mathbf{i} + \dfrac{1}{3}(1 - t)^{3/2}\mathbf{j} + \dfrac{t}{2}\mathbf{k}, \quad -1 < t < 1$
16. $\mathbf{R} = (3 \cosh 2t)\mathbf{i} + (3 \sinh 2t)\mathbf{j} + (6t)\mathbf{k}$

17. Let $\mathbf{R} = (\cos t + t \sin t)\mathbf{i} + (\sin t - t \cos t)\mathbf{j} + 3\mathbf{k}$ describe a space curve for $t \geq 0$. Find **T**, κ, **N**, and **B**. Also, find equations for the line tangent to the curve at the point $(\pi/2, 1, 3)$.

18. Let $\mathbf{R} = (3 \sin t)\mathbf{i} + (3 \cos t)\mathbf{j} + (4t)\mathbf{k}$. Find **T**, κ, **N**, and **B**. Also, find equations for the line tangent to the curve at the point $(0, 3, 8\pi)$.

19. Let $\mathbf{R} = (2 \cos t)\mathbf{i} + (2 \sin t)\mathbf{j} + t\mathbf{k}$. Find **T**, κ, **N**, and **B**. Also, find the arc length s as a function of t. Assume $s = 0$ when $t = 0$ and that s increases with t. (*Hint:* Remember that $ds/dt = |\mathbf{v}|$.)

20. Repeat Problem 19 for
$$\mathbf{R} = (e^t \cos t)\mathbf{i} + (e^t \sin t)\mathbf{j} + 2\mathbf{k}.$$

21. A rocket leaves the point $(1, -2, 3)$ at time $t = 0$ and travels with constant speed 1 unit in a straight line toward the point $(3, 0, 0)$. Find, as functions of t, the
    a) position vector **R**,
    b) velocity **v**,
    c) unit tangent vector **T**,
    d) acceleration **a**,
    e) curvature κ.

22. Find the radius of curvature of the plane curve $y = \ln(\cos x)$ as a function of x.

23. Find an equation for the circle of curvature of the plane curve

$$\mathbf{R} = (2 \ln t)\mathbf{i} - (t + (1/t))\mathbf{j}.$$

24. Find an equation for the circle of curvature of the curve $y = \sin x$ at the point $(\pi/2, 1)$.

25. Find the equation of the osculating circle associated with the curve $y = e^x$ at the point (0, 1). By calculating dy/dx and d^2y/dx^2 at the point (0, 1) from the equation of this circle, verify that these derivatives have the same values there as do the corresponding derivatives for the curve $y = e^x$. Sketch the curve and the osculating circle.

26. The figure below shows arc length s on the circle $x^2 + y^2 = a^2$, measured counterclockwise from the positive x-axis to a point P. It also shows the angle ϕ that the tangent to the circle at P makes with the x-axis. Use the equations $s = a\theta$, $\phi = \theta + \pi/2$ to calculate the circle's curvature directly from the definition $\kappa = |d\phi/ds|$.

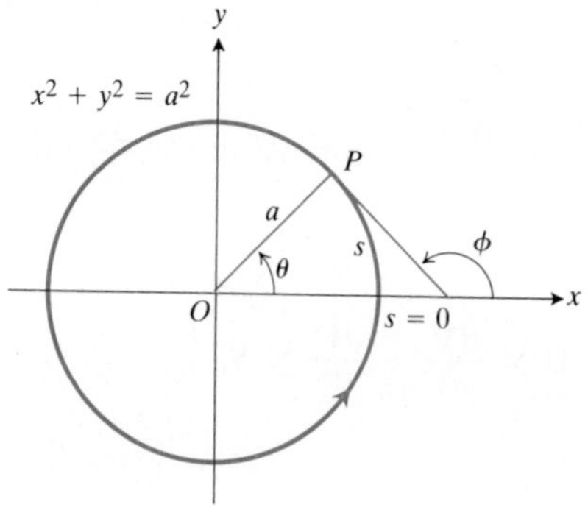

27. The curve $\mathbf{R}(t) = x(t)\mathbf{i} + y(t)\mathbf{j}$ in the plane may be thought of as a curve $\mathbf{R}(t) = x(t)\mathbf{i} + y(t)\mathbf{j} + 0 \cdot \mathbf{k}$ in space whose **k** component is zero. Show that for such a curve the formula $\kappa = |\mathbf{v} \times \mathbf{a}|/|\mathbf{v}|^3$ reduces to

$$\kappa = |\dot{x}\ddot{y} - \dot{y}\ddot{x}|/[\dot{x}^2 + \dot{y}^2]^{3/2}.$$

28. Show that when x and y are considered as functions of arc length s, the unit vectors **T** and **N** may be expressed as follows:

$$\mathbf{T} = \mathbf{i}\frac{dx}{ds} + \mathbf{j}\frac{dy}{ds}, \qquad \mathbf{N} = -\mathbf{i}\frac{dy}{ds} + \mathbf{j}\frac{dx}{ds},$$

where $dx/ds = \cos\phi$, $dy/ds = \sin\phi$, and ϕ is the angle from the positive x-axis to the tangent line.

29. Let $\mathbf{R} = \overrightarrow{OP}$ be the vector from the origin to a moving point P. Let **T** and **N** be the unit tangent and principal normal vectors, respectively, for the curve described by P. Express the velocity and acceleration vectors $d\mathbf{R}/dt$ and $d^2\mathbf{R}/dt^2$ in terms of their **T**- and **N**-components.

***Toolkit* programs**

Parametric Equations Super * Grapher

14.5

Derivatives of Vector Products. The Tangential and Normal Components of **v** and **a**

If $\mathbf{U}(t)$, $\mathbf{V}(t)$ are differentiable vector functions of t, then the products $\mathbf{U} \cdot \mathbf{V}$ and $\mathbf{U} \times \mathbf{V}$ are differentiable, and their derivatives may be calculated by applying rules like the one we know for products of scalar functions:

$$\frac{d}{dt}(\mathbf{U} \cdot \mathbf{V}) = \frac{d\mathbf{U}}{dt} \cdot \mathbf{V} + \mathbf{U} \cdot \frac{d\mathbf{V}}{dt}, \tag{1}$$

$$\frac{d}{dt}(\mathbf{U} \times \mathbf{V}) = \frac{d\mathbf{U}}{dt} \times \mathbf{V} + \mathbf{U} \times \frac{d\mathbf{V}}{dt}. \tag{2}$$

One way to verify Eqs. (1) and (2) would be to write

$$\begin{aligned} \mathbf{U} &= \mathbf{i}f_1(t) + \mathbf{j}f_2(t) + \mathbf{k}f_3(t), \\ \mathbf{V} &= \mathbf{i}g_1(t) + \mathbf{j}g_2(t) + \mathbf{k}g_3(t), \end{aligned} \tag{3}$$

with the component f's and g's being differentiable functions of t. We would then substitute these expressions for **U** and **V** into Eqs. (1) and (2), multiply out, and apply the ordinary rules for differentiating products of scalar functions. However, instead of appealing to the component-wise verification of the identities in Eqs. (1) and (2), it is instructive to establish these equations by the Δ-process. For example, let

$$\mathbf{W} = \mathbf{U} \times \mathbf{V},$$

where t has some specific value. Then give t an increment Δt and denote the new values of the vectors by $\mathbf{U} + \Delta\mathbf{U}$, etc., so that

$$\begin{aligned}\mathbf{W} + \Delta\mathbf{W} &= (\mathbf{U} + \Delta\mathbf{U}) \times (\mathbf{V} + \Delta\mathbf{V}) \\ &= \mathbf{U} \times \mathbf{V} + \mathbf{U} \times \Delta\mathbf{V} + \Delta\mathbf{U} \times \mathbf{V} + \Delta\mathbf{U} \times \Delta\mathbf{V},\end{aligned}$$

and

$$\frac{\Delta\mathbf{W}}{\Delta t} = \mathbf{U} \times \frac{\Delta\mathbf{V}}{\Delta t} + \frac{\Delta\mathbf{U}}{\Delta t} \times \mathbf{V} + \frac{\Delta\mathbf{U}}{\Delta t} \times \Delta\mathbf{V}.$$

Now take limits as $\Delta t \to 0$, noting that

$$\lim \frac{\Delta\mathbf{W}}{\Delta t} = \frac{d\mathbf{W}}{dt}, \qquad \lim \frac{\Delta\mathbf{U}}{\Delta t} = \frac{d\mathbf{U}}{dt}, \qquad \lim \Delta\mathbf{V} = \lim \frac{\Delta\mathbf{V}}{\Delta t} \cdot \lim \Delta t = 0,$$

so that

$$\frac{d\mathbf{W}}{dt} = \mathbf{U} \times \frac{d\mathbf{V}}{dt} + \frac{d\mathbf{U}}{dt} \times \mathbf{V},$$

which is equivalent to Eq. (2).

EXAMPLE 1 The formula for the derivative of the triple scalar product leads to an interesting identity for the derivative of a determinant of order three. Let

$$\begin{aligned}\mathbf{U} &= u_1\mathbf{i} + u_2\mathbf{j} + u_3\mathbf{k}, \\ \mathbf{V} &= v_1\mathbf{i} + v_2\mathbf{j} + v_3\mathbf{k}, \\ \mathbf{W} &= w_1\mathbf{i} + w_2\mathbf{j} + w_3\mathbf{k},\end{aligned} \tag{4}$$

where the components are differentiable functions of a scalar t. Then the identity

$$\frac{d}{dt}(\mathbf{U} \cdot \mathbf{V} \times \mathbf{W}) = \frac{d\mathbf{U}}{dt} \cdot \mathbf{V} \times \mathbf{W} + \mathbf{U} \cdot \frac{d\mathbf{V}}{dt} \times \mathbf{W} + \mathbf{U} \cdot \mathbf{V} \times \frac{d\mathbf{W}}{dt} \tag{5}$$

(Problem 21) is equivalent to

$$\frac{d}{dt}\begin{vmatrix} u_1 & u_2 & u_3 \\ v_1 & v_2 & v_3 \\ w_1 & w_2 & w_3 \end{vmatrix} = \begin{vmatrix} \frac{du_1}{dt} & \frac{du_2}{dt} & \frac{du_3}{dt} \\ v_1 & v_2 & v_3 \\ w_1 & w_2 & w_3 \end{vmatrix} + \begin{vmatrix} u_1 & u_2 & u_3 \\ \frac{dv_1}{dt} & \frac{dv_2}{dt} & \frac{dv_3}{dt} \\ w_1 & w_2 & w_3 \end{vmatrix} + \begin{vmatrix} u_1 & u_2 & u_3 \\ v_1 & v_2 & v_3 \\ \frac{dw_1}{dt} & \frac{dw_2}{dt} & \frac{dw_3}{dt} \end{vmatrix} \tag{6}$$

This says that the derivative of a determinant of order three is the sum of three determinants obtained from the original determinant by differentiating one row at a time. The result may be extended to determinants of any order n. □

Derivatives of Vectors of Constant Length

An important geometrical result is obtained by differentiating the identity

$$\mathbf{V}\cdot\mathbf{V} = |\mathbf{V}|^2 \tag{7a}$$

when $\mathbf{V}$ is a vector of constant magnitude. For then $|\mathbf{V}|^2$ is a constant, so its derivative is zero, and one has

$$\mathbf{V}\cdot\frac{d\mathbf{V}}{dt} + \frac{d\mathbf{V}}{dt}\cdot\mathbf{V} = 0$$

or, since the scalar product is commutative,

$$2\mathbf{V}\cdot\frac{d\mathbf{V}}{dt} = 0 \quad \text{or} \quad \mathbf{V}\cdot\frac{d\mathbf{V}}{dt} = 0. \tag{7b}$$

This means that either $\mathbf{V}$ is zero, $d\mathbf{V}/dt$ is zero (and hence $\mathbf{V}$ is constant in direction as well as magnitude), or else that $d\mathbf{V}/dt$ is perpendicular to $\mathbf{V}$.

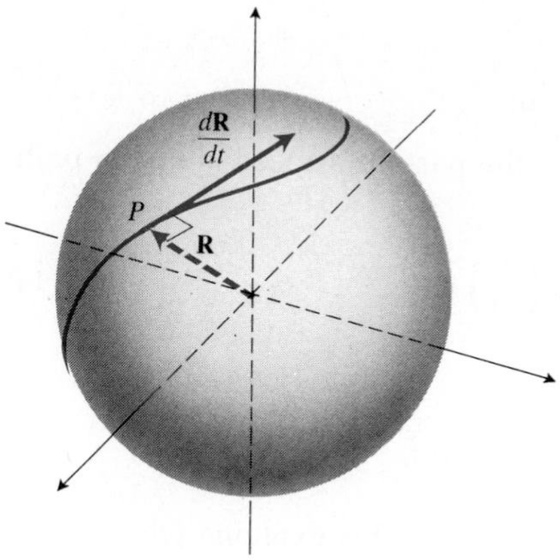

14.18 The velocity vector of a particle P that moves on the surface of a sphere is tangent to the sphere.

EXAMPLE 2 Suppose that a point P moves about on the surface of a sphere. Then the magnitude of the vector $\mathbf{R}$ from the center to P is a constant equal to the radius of the sphere. Therefore, the velocity vector $d\mathbf{R}/dt$ is always perpendicular to $\mathbf{R}$ (Fig. 14.18). □

N · T Is Zero

We can now show that the principal normal vector

$$\mathbf{N} = \frac{1}{\kappa}\frac{d\mathbf{T}}{ds} \tag{8}$$

defined in Article 14.4 is orthogonal to the unit tangent vector $\mathbf{T}$. The vector $\mathbf{T}$ has a constant length of $|\mathbf{T}| = 1$. We may therefore apply Eq. (7b) with $\mathbf{V} = \mathbf{T}$ and $t = s$ to conclude that

$$\mathbf{T}\cdot\frac{d\mathbf{T}}{ds} = 0 \quad \text{or} \quad \mathbf{T}\cdot\mathbf{N} = 0. \tag{9}$$

The Tangential and Normal Components of a

In mechanics it is often useful to describe the acceleration of a moving object in terms of its components tangent and normal to the path of motion; that is, in terms of its components in the directions of $\mathbf{T}$ and $\mathbf{N}$. This may be done for curves in either space or plane in the following way.

From the Chain Rule we see that

$$\mathbf{v} = \frac{d\mathbf{R}}{dt} = \frac{d\mathbf{R}}{ds}\frac{ds}{dt} = \mathbf{T}\,\frac{ds}{dt}, \tag{10}$$

$$\mathbf{a} = \frac{d\mathbf{v}}{dt} = \frac{d}{dt}\left(\mathbf{T}\,\frac{ds}{dt}\right) = \mathbf{T}\,\frac{d^2s}{dt^2} + \frac{d\mathbf{T}}{dt}\frac{ds}{dt}. \tag{11}$$

By the Chain Rule and Eq. (8) we also have

$$\frac{d\mathbf{T}}{dt} = \frac{d\mathbf{T}}{ds}\frac{ds}{dt} = \mathbf{N}\kappa\frac{ds}{dt},$$

so that

$$\mathbf{a} = \mathbf{T}\frac{d^2s}{dt^2} + \mathbf{N}\kappa\left(\frac{ds}{dt}\right)^2 = a_T\mathbf{T} + a_N\mathbf{N}. \tag{12}$$

Equation (12) expresses **a** in terms of its tangential and normal components. The *tangential component*,

$$a_T = \frac{d^2s}{dt^2} \tag{13}$$

is simply the derivative of $v = ds/dt$, the particle's speed along its path. The *normal component*

$$a_N = \kappa\left(\frac{ds}{dt}\right)^2 \tag{14}$$

can be written as

$$a_N = \kappa\left(\frac{ds}{dt}\right)^2 = \kappa v^2,$$

the curvature times the square of the speed. This explains why a large normal force must be supplied by friction between the tires and the roadway to hold an automobile on a level road when it makes a sharp turn (large κ) or a moderate turn at high speed (large v^2).

If the particle moves in a circle with constant speed $v = ds/dt$, then $d^2s/dt^2 = 0$, and the only acceleration is the normal acceleration

$$\mathbf{N}\kappa\left(\frac{ds}{dt}\right)^2 = \mathbf{N}\kappa v^2 = \mathbf{N}\frac{v^2}{\rho},$$

which is toward the center of the circle. If the speed around the circle is not constant, then **a** has a nonzero tangential component as well (Fig. 14.19).

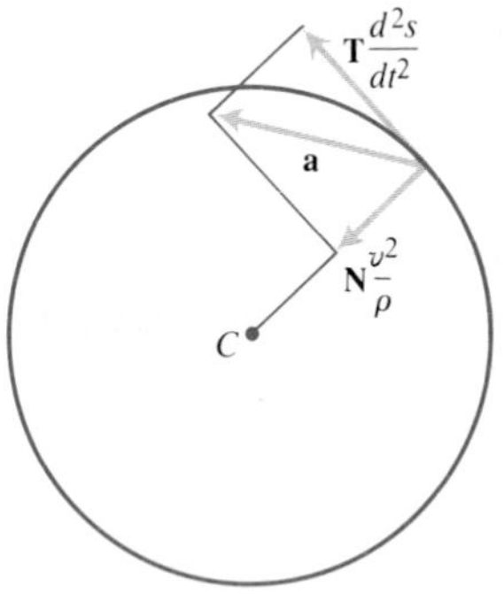

14.19 Tangential and normal components of an acceleration vector.

To calculate a_T we can use Eq. (13) directly, but instead of using Eq. (14) to calculate a_N it is often better to use the equation

$$a_N = \sqrt{|\mathbf{a}|^2 - a_T^2} \tag{15}$$

(which comes from solving the equation

$$|\mathbf{a}|^2 = a_T^2 + a_N^2 \tag{16}$$

for a_N). With Eq. (15) we can find a_N without having to find κ first.

EXAMPLE 3 The position of a moving particle at time $t \geq 0$ is

$$\mathbf{R} = (\cos t + t\sin t)\mathbf{i} + (\sin t - t\cos t)\mathbf{j}.$$

Find **v**, **a**, $|ds/dt|$, a_T, and a_N. Then express **a** in terms of a_T and a_N.

Solution We differentiate **R** with respect to t to find

$$\begin{aligned}\mathbf{v} &= (-\sin t + \sin t + t\cos t)\mathbf{i} + (\cos t - \cos t + t\sin t)\mathbf{j}\\ &= (t\cos t)\mathbf{i} + (t\sin t)\mathbf{j}\end{aligned}$$

and

$$\mathbf{a} = \frac{d\mathbf{v}}{dt} = (\cos t - t \sin t)\mathbf{i} + (\sin t + t \cos t)\mathbf{j}.$$

The speed is

$$\left|\frac{ds}{dt}\right| = |\mathbf{v}| = \sqrt{(t \cos t)^2 + (t \sin t)^2} = t.$$

From Eq. (13),

$$a_T = \frac{d^2s}{dt^2} = \frac{d}{dt}\left(\frac{ds}{dt}\right) = \frac{d}{dt}(t) = 1.$$

From Eq. (15),

$$a_N = \sqrt{|\mathbf{a}|^2 - a_T^2} = \sqrt{(\cos t - t \sin t)^2 + (\sin t + t \cos t)^2 - 1} = t.$$

Finally,

$$\mathbf{a} = a_T\mathbf{T} + a_N\mathbf{N} = (1)\mathbf{T} + t\mathbf{N} = \mathbf{T} + t\mathbf{N}.$$

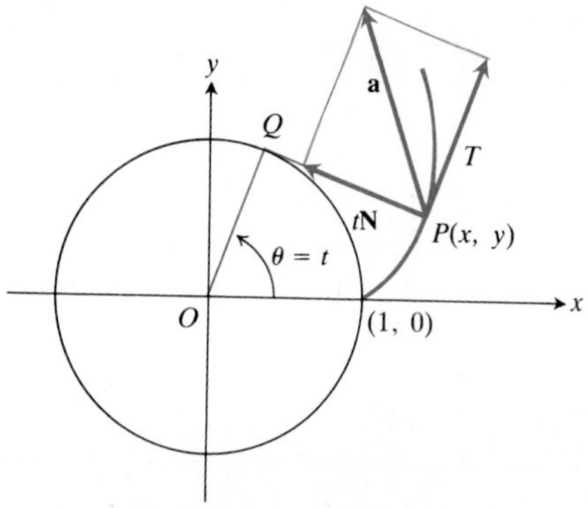

14.20 Motion on the involute of a circle when θ increases steadily with time.

Here the tangential acceleration has constant magnitude and the normal acceleration starts with zero magnitude at $t = 0$ and increases with time. The equations of motion are the same as the parametric equations for the *involute* of a circle of unit radius. This is the path of the endpoint P of a string that is held taut as it is unwound from the circle. To get the parameterization, the origin is taken as the center of the circle (Fig. 14.20), with (1, 0) being the position where P starts. The angle t is measured from the positively directed x-axis counterclockwise to the ray from O to the point of tangency Q.

Incidentally, we can find the radius of curvature from the equation $a_N = v^2/\rho$, since $v^2 = |\mathbf{v}|^2 = t^2$ and $a_N = t$ and hence

$$\rho = \frac{v^2}{a_N} = \frac{t^2}{t} = t. \quad \square$$

The Vector Formula for Curvature

Equations (10) and (12) for **v** and **a** can be used to derive the vector curvature formula

$$\kappa = \frac{|\mathbf{v} \times \mathbf{a}|}{|\mathbf{v}|^3} \tag{17}$$

that we introduced in Article 14.4.

First we compute $\mathbf{v} \times \mathbf{a}$:

$$\mathbf{v} \times \mathbf{a} = \mathbf{T}\frac{ds}{dt} \times \left[\mathbf{T}\frac{d^2s}{dt^2} + \mathbf{N}\kappa\left(\frac{ds}{dt}\right)^2\right] = \mathbf{T} \times \mathbf{N}\kappa\left(\frac{ds}{dt}\right)^3 \tag{18}$$

because we can apply the distributive law for the cross product, and $\mathbf{T} \times \mathbf{T} = \mathbf{0}$. Moreover, $\mathbf{T} \times \mathbf{N}$ is the unit binormal vector **B**. Therefore

$$\mathbf{v} \times \mathbf{a} = \mathbf{B}\kappa\left(\frac{ds}{dt}\right)^3. \tag{19}$$

Since **B** is a *unit* vector, the magnitude of $\mathbf{v} \times \mathbf{a}$ is

$$|\mathbf{v} \times \mathbf{a}| = \kappa\left|\frac{ds}{dt}\right|^3 = \kappa|\mathbf{v}|^3.$$

Finally, if $|\mathbf{v}| \neq 0$, we get (by division)

$$\kappa = \frac{|\mathbf{v} \times \mathbf{a}|}{|\mathbf{v}|^3},$$

which is Eq. (17).

Useful Formulas for Curves in Plane and Space

Position vector in the plane: $\mathbf{R} = x(t)\mathbf{i} + y(t)\mathbf{j}$

Position vector in space: $\mathbf{R} = x(t)\mathbf{i} + y(t)\mathbf{j} + z(t)\mathbf{k}$

Velocity: $\mathbf{v} = \dfrac{d\mathbf{R}}{dt}$

Acceleration: $\mathbf{a} = \dfrac{d\mathbf{v}}{dt} = \dfrac{d^2\mathbf{R}}{dt^2}$

$$\mathbf{T} = \frac{\mathbf{v}}{|\mathbf{v}|} \text{ if } |\mathbf{v}| \neq 0, \qquad \mathbf{v} = \mathbf{T}\frac{ds}{dt}, \qquad \left|\frac{ds}{dt}\right| = |\mathbf{v}| = \text{speed}$$

$$\kappa = \frac{|\mathbf{v} \times \mathbf{a}|}{|\mathbf{v}|^3} \text{ if } |\mathbf{v}| \neq 0, \qquad \kappa\mathbf{N} = \frac{d\mathbf{T}}{ds}$$

$$a_{\mathrm{T}} = \frac{d^2s}{dt^2} = \frac{d}{dt}|\mathbf{v}|, \qquad a_{\mathrm{N}} = \sqrt{|\mathbf{a}|^2 - a_{\mathrm{T}}^2}$$

$$\mathbf{a} = a_{\mathrm{T}}\mathbf{T} + a_{\mathrm{N}}\mathbf{N}$$

PROBLEMS

In Problems 1–7, find $\mathbf{v}$, $\mathbf{a}$, the speed $|ds/dt|$, a_{T}, and a_{N}. Then express $\mathbf{a}$ in terms of a_{T} and a_{N} (*without* finding $\mathbf{T}$ and $\mathbf{N}$).

1. $\mathbf{R} = (2t + 3)\mathbf{i} + (t^2 - 1)\mathbf{j}$
2. $\mathbf{R} = \ln(t^2 + 1)\mathbf{i} + (t - 2\tan^{-1} t)\mathbf{j}$
3. $\mathbf{R} = (2\cos t)\mathbf{i} + (2\sin t)\mathbf{j}$
4. $\mathbf{R} = (a\cos \omega t)\mathbf{i} + (a \sin \omega t)\mathbf{j}$
5. $\mathbf{R} = (e^t \cos t)\mathbf{i} + (e^t \sin t)\mathbf{j}$
6. $\mathbf{R} = (\cosh 2t)\mathbf{i} + (\sinh 2t)\mathbf{j}$
7. a) $\mathbf{R} = (\cos t)\mathbf{i} + (\sin t)\mathbf{j} + (bt)\mathbf{k}$.
   b) Calculate the curvature of the helix in part (a). What effect does increasing $|b|$ have on the curvature? (This explains mathematically why stretching a spring tends to straighten it.)

In Problems 8–14, find $\mathbf{v}$, $\mathbf{a}$, the speed $|ds/dt|$, a_{T}, and a_{N} *at the given value of t*.

8. $\mathbf{R} = (1 - \sin t)\mathbf{i} + (1 - \sqrt{2}\cos t)\mathbf{j}$, at $t = \pi/6$
9. $\mathbf{R} = (t + 1)\mathbf{i} + 2t\mathbf{j} + t^2\mathbf{k}$, at $t = 1$
10. $\mathbf{R} = (t\cos t)\mathbf{i} + (t \sin t)\mathbf{j} + t\mathbf{k}$, at $t = 0$
11. $\mathbf{R} = t^2\mathbf{i} + (t + \frac{1}{3}t^3)\mathbf{j} + (t - \frac{1}{3}t^3)\mathbf{k}$, at $t = 0$
12. $\mathbf{R} = (e^t \cos t)\mathbf{i} + (e^t \sin t)\mathbf{j} + \sqrt{2}e^t\mathbf{k}$, at $t = 0$
13. $\mathbf{R} = (2 + 3t + 3t^2)\mathbf{i} + (4t + 4t^2)\mathbf{j} - (6\cos t)\mathbf{k}$, at $t = 0$
14. $\mathbf{R} = (2 + t)\mathbf{i} + (t + 2t^2)\mathbf{j} + (1 + t^2)\mathbf{k}$, at $t = 0$
15. For the curve in Problem 14, find (a) the curvature at $t = 0$, (b) equations for the line tangent to the curve at $t = 0$, and (c) an equation for the plane through the point (2, 0, 1) that contains $v(0)$ and $a(0)$.
16. Deduce from Eq. (12) that a particle will move in a straight line if the normal component of acceleration is identically zero.
17. If a particle moves in a curve with constant speed, show that the force is always directed along the normal.
18. If the force acting on a particle is at all times perpendicular to the direction of motion, show that the speed remains constant.
19. Show that the radius of curvature of a plane curve $\mathbf{R} = x(t)\mathbf{i} + y(t)\mathbf{j}$ is given by

$$\rho = \frac{(\dot{x}^2 + \dot{y}^2)}{\sqrt{\ddot{x}^2 + \ddot{y}^2 - \ddot{s}^2}},$$

where the dots mean derivatives with respect to t ($\dot{x} = dx/dt$, $\ddot{x} = d^2x/dt^2$, and so on) and

$$\ddot{s} = \frac{d}{dt}\sqrt{\dot{x}^2 + \dot{y}^2}.$$

20. Derive Eq. (1) by the Δ-process.

21. Apply Eqs. (1) and (2) to $\mathbf{U} \cdot \mathbf{V}_1$ with $\mathbf{V}_1 = \mathbf{V} \times \mathbf{W}$ and thereby derive Eq. (5) for

$$\frac{d}{dt}[\mathbf{U} \cdot (\mathbf{V} \times \mathbf{W})].$$

22. If $\mathbf{F}(t) = \mathbf{i}f(t) + \mathbf{j}g(t) + \mathbf{k}h(t)$, where f, g, and h are functions of t that have derivatives of orders one, two, and three, show that

$$\frac{d}{dt}\left[\mathbf{F} \cdot \left(\frac{d\mathbf{F}}{dt} \times \frac{d^2\mathbf{F}}{dt^2}\right)\right] = \mathbf{F} \cdot \left(\frac{d\mathbf{F}}{dt} \times \frac{d^3\mathbf{F}}{dt^3}\right).$$

Explain why the answer contains just this one term rather than the three terms that one might expect.

23. With the book closed, derive vector expressions for the velocity and acceleration in terms of tangential and normal components. Check your derivations with those given in the text.

14.6 Planetary Motion and Satellites

In this article we show that the motion of a planet or particle under the influence of a central force field takes place in a plane. We then derive Kepler's second law (equal areas in equal times) and formulate Kepler's first and third laws.

Polar and Cylindrical Coordinates

If a particle P moves on a plane curve whose equation is given in polar coordinates, it is convenient to express the velocity and acceleration vectors in terms of still a third set of unit vectors. We introduce unit vectors

$$\mathbf{u}_r = \mathbf{i}\cos\theta + \mathbf{j}\sin\theta, \qquad \mathbf{u}_\theta = -\mathbf{i}\sin\theta + \mathbf{j}\cos\theta, \tag{1}$$

which point respectively along the radius vector $\overrightarrow{OP}$, and at right angles to $\overrightarrow{OP}$ and in the direction of increasing θ, as shown in Fig. 14.21. Then we find from (1) that

$$\frac{d\mathbf{u}_r}{d\theta} = -\mathbf{i}\sin\theta + \mathbf{j}\cos\theta = \mathbf{u}_\theta, \qquad \frac{d\mathbf{u}_\theta}{d\theta} = -\mathbf{i}\cos\theta - \mathbf{j}\sin\theta = -\mathbf{u}_r. \tag{2}$$

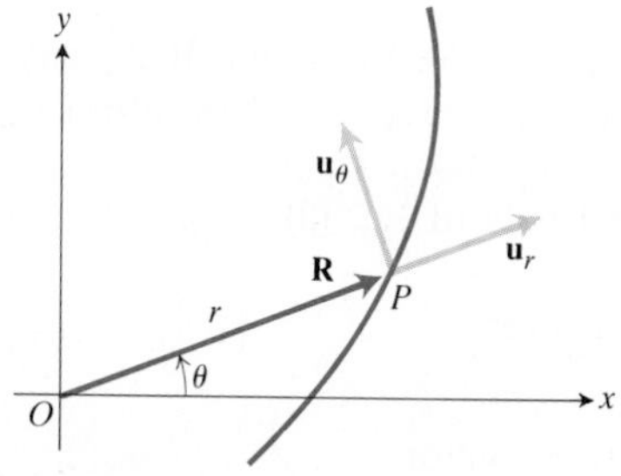

14.21 The unit vectors $\mathbf{u}_r$ and $\mathbf{u}_\theta$.

This says that the result of differentiating either one of the unit vectors $\mathbf{u}_r$ and $\mathbf{u}_\theta$ with respect to θ is equivalent to rotating that vector through 90° in the positive (counterclockwise) direction.

Since the vectors $\mathbf{R} = \overrightarrow{OP}$ and $r\mathbf{u}_r$ have the same direction, and the length of $\mathbf{R}$ is the absolute value of the polar coordinate r of $P(r, \theta)$, we have

$$\mathbf{R} = r\mathbf{u}_r. \tag{3}$$

To obtain the velocity, we must differentiate this with respect to t, remembering that both r and $\mathbf{u}_r$ may be variables. From (2) and the Chain Rule we get

$$\begin{aligned} \frac{d\mathbf{u}_r}{dt} &= \frac{d\mathbf{u}_r}{d\theta}\frac{d\theta}{dt} = \mathbf{u}_\theta \frac{d\theta}{dt}, \\ \frac{d\mathbf{u}_\theta}{dt} &= \frac{d\mathbf{u}_\theta}{d\theta}\frac{d\theta}{dt} = -\mathbf{u}_r \frac{d\theta}{dt}. \end{aligned} \tag{4}$$

Hence

$$\mathbf{v} = \frac{d\mathbf{R}}{dt} = \mathbf{u}_r \frac{dr}{dt} + r\frac{d\mathbf{u}_r}{dt}$$

becomes

$$\mathbf{v} = \mathbf{u}_r \frac{dr}{dt} + \mathbf{u}_\theta r \frac{d\theta}{dt}. \tag{5}$$

Of course this velocity vector is tangent to the curve at P and has magnitude

$$|\mathbf{v}| = \sqrt{(dr/dt)^2 + r^2(d\theta/dt)^2} = |ds/dt|.$$

In fact, if the three sides of the "differential triangle" of sides dr, $r\,d\theta$, and ds are all divided by dt, the result will be a similar triangle having sides dr/dt, $r\,d\theta/dt$, and ds/dt, which illustrates the vector equation

$$\mathbf{v} = \mathbf{T}\frac{ds}{dt} = \mathbf{u}_r \frac{dr}{dt} + \mathbf{u}_\theta r \frac{d\theta}{dt}.$$

(See Fig. 14.22.)

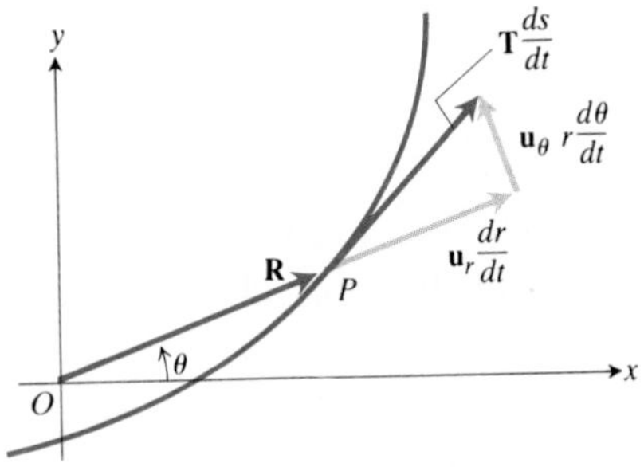

14.22 $\mathbf{u}_r \frac{dr}{dt} + \mathbf{u}_\theta r \frac{d\theta}{dt}$ is the velocity vector.

The acceleration vector is found by differentiating the velocity vector in (5) as follows:

$$\mathbf{a} = \frac{d\mathbf{v}}{dt} = \left(\mathbf{u}_r \frac{d^2r}{dt^2} + \frac{dr}{dt}\frac{d\mathbf{u}_r}{dt}\right) + \left(\mathbf{u}_\theta r \frac{d^2\theta}{dt^2} + \mathbf{u}_\theta \frac{dr}{dt}\frac{d\theta}{dt} + \frac{d\mathbf{u}_\theta}{dt} r \frac{d\theta}{dt}\right).$$

When Eqs. (4) are used to evaluate the derivatives of $\mathbf{u}_r$ and $\mathbf{u}_\theta$ and the components are separated, the result becomes

$$\mathbf{a} = \mathbf{u}_r\left[\frac{d^2r}{dt^2} - r\left(\frac{d\theta}{dt}\right)^2\right] + \mathbf{u}_\theta\left[r\frac{d^2\theta}{dt^2} + 2\frac{dr}{dt}\frac{d\theta}{dt}\right]. \tag{6}$$

Equations (5) and (6) apply to motion in the xy-plane. It is easy to modify them to apply to motion in space. First, we need to add a term

$\mathbf{k}z$ to the right-hand side of Eq. (3),

$\mathbf{k}\frac{dz}{dt}$ to the velocity vector,

$\mathbf{k}\frac{d^2z}{dt^2}$ to the acceleration vector.

Thus we get

$$\mathbf{R} = r\mathbf{u}_r + \mathbf{k}z, \tag{7a}$$

$$\mathbf{v} = \mathbf{u}_r \frac{dr}{dt} + \mathbf{u}_\theta r \frac{d\theta}{dt} + \mathbf{k}\frac{dz}{dt}, \tag{7b}$$

$$\mathbf{a} = \mathbf{u}_r\left[\frac{d^2r}{dt^2} - r\left(\frac{d\theta}{dt}\right)^2\right] + \mathbf{u}_\theta\left[r\frac{d^2\theta}{dt^2} + 2\frac{dr}{dt}\frac{d\theta}{dt}\right] + \mathbf{k}\frac{d^2z}{dt^2}. \tag{7c}$$

Equations (7) are particularly useful in connection with cylindrical coordinates. The three vectors $\mathbf{u}_r$, $\mathbf{u}_\theta$, and $\mathbf{k}$ are mutually orthogonal unit vectors that form a right-handed frame:

$$\begin{aligned} \mathbf{u}_r \times \mathbf{u}_\theta &= \mathbf{k}, \\ \mathbf{k} \times \mathbf{u}_r &= \mathbf{u}_\theta, \\ \mathbf{u}_\theta \times \mathbf{k} &= \mathbf{u}_r. \end{aligned} \tag{8}$$

Planets Move in Planes

If we assume that the motion obeys Newton's second law, $\mathbf{F} = m\mathbf{a}$, and that the only force acting on the planet is a gravitational attraction directed toward a fixed point O, of magnitude inversely proportional to the square of the distance of the planet from O, Kepler's laws may be deduced by integrating differential equations of motion. We get these equations from the physical assumptions

$$\mathbf{F} = m\mathbf{a} = m\frac{d^2\mathbf{R}}{dt^2} = -\frac{GmM}{|\mathbf{R}|^2}\frac{\mathbf{R}}{|\mathbf{R}|}, \tag{9}$$

where G is the gravitational constant, m the mass of the moving particle (or planet), and M the mass of the attracting object (or sun) at O. Equation (9) tells us that the acceleration vector $d^2\mathbf{R}/dt^2$ is proportional to the position vector $\mathbf{R}$. This is typical of a *central force field* (see Fig. 14.23), and it leads to the result

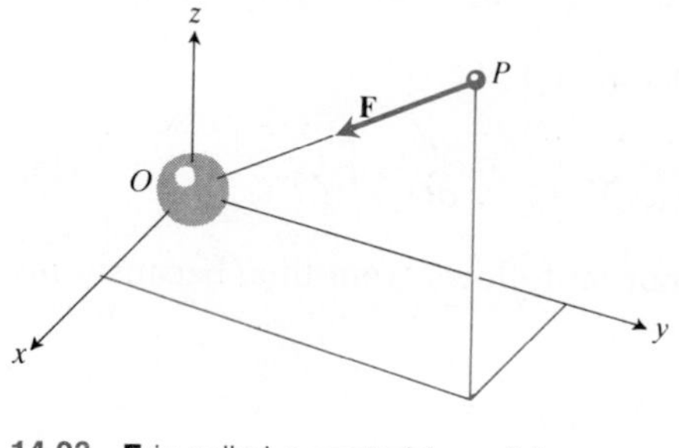

14.23 **F** is called a *central force* if it points toward a fixed point (here the origin) no matter how P moves.

$$\mathbf{R} \times \frac{d^2\mathbf{R}}{dt^2} = \mathbf{0}, \tag{10}$$

because $\mathbf{A} \times \mathbf{B}$ is always equal to the zero vector if $\mathbf{B}$ is a scalar times $\mathbf{A}$. Now we can deduce that the left-hand side of Eq. (10) is the derivative of the cross product of $\mathbf{R}$ and $d\mathbf{R}/dt$. First we have

$$\frac{d}{dt}\left(\mathbf{R} \times \frac{d\mathbf{R}}{dt}\right) = \frac{d\mathbf{R}}{dt} \times \frac{d\mathbf{R}}{dt} + \mathbf{R} \times \frac{d^2\mathbf{R}}{dt^2} = \mathbf{0} + \mathbf{R} \times \frac{d^2\mathbf{R}}{dt^2},$$

because the cross product of any vector with itself is the zero vector. When the derivative of a vector is zero, each of its components (relative to a fixed coordinate system) is a constant, so the vector is a constant vector. Thus, Eq. (10) implies that

$$\mathbf{R} \times \frac{d\mathbf{R}}{dt} = \mathbf{C}. \tag{11}$$

From the geometric interpretation of the cross product, we conclude from Eq. (11) that both $\mathbf{R}$ and $\mathbf{v} = d\mathbf{R}/dt$ lie in a plane perpendicular to the fixed vector $\mathbf{C}$. We choose a reference frame so that this plane, which includes the path of the particle, is the same as the xy-plane, and so that the unit vector in the direction of $\mathbf{C}$ is $\mathbf{k}$. We introduce polar coordinates in this plane, choosing as initial line $\theta = 0$, the direction of $\mathbf{R}$ when $|\mathbf{R}|$ is a *minimum*. In planetary motion, this corresponds to *perihelion* position of the planet (the position nearest to the sun). If we also measure the time t from the instant of passage through perihelion, we get the following initial values: when $t = 0$,

$$\begin{aligned} r &= r_0 = \text{minimum value of } r \Rightarrow \left(\frac{dr}{dt}\right)_0 = 0, \\ \theta &= 0, \\ |\mathbf{v}| &= v_0 = \left(r\frac{d\theta}{dt}\right)_0. \end{aligned} \tag{12}$$

Kepler's Second Law

Because the motion takes place in a plane, we can use Eqs. (3), (5), and (6) for $\mathbf{R}$, $\mathbf{v}$, and $\mathbf{a}$. Also, since $\mathbf{R} \times \mathbf{v}$ is constant (Eq. 11), we can evaluate that constant by forming the cross product at time $t = 0$:

$$\mathbf{C} = (\mathbf{R}_0) \times (\mathbf{v}_0) = \mathbf{k}\left(r^2\frac{d\theta}{dt}\right)_0 = \mathbf{k}(r_0 v_0). \tag{13}$$

To get Eq. (13), cross the right-hand side of Eq. (3) and the right-hand side of Eq. (5), and recall that

$$\mathbf{u}_r \times \mathbf{u}_r = \mathbf{0} \quad \text{and} \quad \mathbf{u}_r \times \mathbf{u}_\theta = \mathbf{k}.$$

If we substitute this result for $\mathbf{C}$ in Eq. (11), we get

$$\mathbf{k}\left(r^2\frac{d\theta}{dt}\right) = \mathbf{k}(r_0 v_0), \qquad \text{or} \qquad r^2\frac{d\theta}{dt} = r_0 v_0. \tag{14}$$

Kepler's second law of planetary motion follows from this, because the element of area in polar coordinates is

$$dA = \frac{1}{2}r^2\,d\theta$$

and Eq. (14) implies that

$$\frac{dA}{dt} = \frac{1}{2}r^2\frac{d\theta}{dt} = \frac{1}{2}r_0 v_0 = \text{constant}. \tag{15}$$

This is the result: *The radius vector sweeps over area at a constant rate in a central force field* (Fig. 14.24).

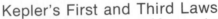

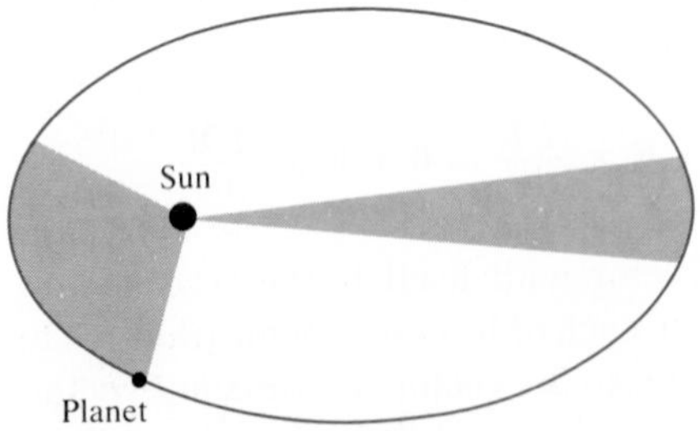

14.24 The line joining a planet and its sun sweeps over equal areas in equal times.

Kepler's First and Third Laws

In deriving Kepler's second law we introduced the constants that appear in Kepler's other two laws, so we can at least state these laws explicitly even though we shall not derive them here.

Table 14.1 Numerical Data

Gravitational constant:	$G = 6.673 \times 10^{-8} \dfrac{\text{cm}^3}{\text{gm} \cdot \text{s}^2}$
Sun's mass:	2×10^{33} gm
Earth's mass:	5.976×10^{27} gm
One year = 365.256 days = 3.156×10^7 seconds	
One mile = 1.609×10^5 cm	

Kepler's first law says that the path of a planet's motion, which we already know lies in a plane, is actually a conic section whose equation is

$$r = \frac{(1/h)}{1 + e\cos\theta}, \tag{16}$$

with

$$\frac{1}{h} = \frac{r_0^2 v_0^2}{GM} \quad \text{and} \quad e = \frac{1}{r_0 h} - 1 = \frac{r_0 v_0^2}{GM} - 1. \tag{17}$$

In the coordinate system we have been using, Eq. (16) represents a conic section of eccentricity e with one focus at the origin (sun's center). For given values of r_0, M, and G, Eq. (17) shows that the eccentricity depends on the v_0, velocity at perihelion. For a circular orbit, $e = 0$ and $v_0 = \sqrt{GM/r_0}$. For $0 < e < 1$, the orbit is an ellipse with one focus at O; for $e = 1$, the orbit is a parabola; and for $e > 1$, the orbit is a hyperbola.

Kepler's third law says that when the orbit is an ellipse, the time T required to go around once (the *period*) is given by the equation

$$\frac{T^2}{a^3} = \frac{4\pi^2}{GM}, \tag{18}$$

where a is the length of the ellipse's semimajor axis. Thus, for every planet in a given solar system, the ratio T^2/a^3 has the constant value $4\pi^2/GM$, where M is the sun's mass and G is Newton's gravitational constant. If a is measured in centimeters, M in grams, and G in $\text{cm}^3/\text{gm} \cdot \text{s}^2$, then Eq. (18) calculates T in seconds.

PROBLEMS

In Problems 1–5, express $\mathbf{v}$ and $\mathbf{a}$ in terms of $\mathbf{u}_r$ and $\mathbf{u}_\theta$.

1. $r = a(1 - \cos\theta)$ and $\dfrac{d\theta}{dt} = 3$

2. $r = a\sin 2\theta$ and $\dfrac{d\theta}{dt} = 2t$

3. $r = e^{a\theta}$ and $\dfrac{d\theta}{dt} = 2$

4. $r = a(1 + \sin t)$ and $\theta = 1 - e^{-t}$

5. $r = 2\cos 4t$ and $\theta = 2t$

6. Let $r = \sin\theta$ and $\theta = \pi t$. Express $\mathbf{v}$ and $\mathbf{a}$ in terms of $\mathbf{u}_r$ and $\mathbf{u}_\theta$ when $t = 1/4$.

7. Let $\mathbf{R} = (t\cos t)\,\mathbf{i} + (t\sin t)\,\mathbf{j} + t\mathbf{k}$, $t \geq 0$. Express $\mathbf{v}$ and $\mathbf{a}$ (a) in the frame $\mathbf{i}, \mathbf{j}, \mathbf{k}$; (b) in the frame $\mathbf{u}_r, \mathbf{u}_\theta, \mathbf{k}$.

8. Let $r = a(1 + \sin t)$, $\theta = 1 - e^{-t}$, $z = \cos t$. Express $\mathbf{v}$ and $\mathbf{a}$ in terms of $\mathbf{u}_r, \mathbf{u}_\theta, \mathbf{k}$ at $t = 0$.

9. If a particle moves in an ellipse whose polar equation is $r = c/(1 - e\cos\theta)$ and the force is directed toward the origin, show that the magnitude of the force is proportional to $1/r^2$.

10. Since the orbit of the Vanguard satellite had major axis $2a = 10{,}784$ mi (approximately), Eq. (18), with M equal to Earth's mass, should give the period. Compute it.

11. In May 1965, the U.S.S.R. launched Proton I, weighing 26,900 lb (at launch), with a perigee of 118 miles, an apogee of 390 miles, and a period of 92 minutes. Using the period $T = 5520$ s and the relevant data for the mass of the earth and the gravitational constant G, compute the semimajor axis a of the orbit from Eq. (18). Use 1 lb = 2204.62 gm.

12. Read the article *Orbit* in the Encyclopaedia Britannica. What dates does it give for Kepler's announcements of his first two laws? Of his third law?

13. Read the article *Space Exploration* in the most recent Yearbook of the Encyclopaedia Britannica (or other source). Report the perigee, apogee, and orbital period for at least one earth satellite as given in the article you read.

14. In the Encyclopaedia Britannica (or elsewhere), read an article on the scientific work of Kepler. In what way was Kepler's work dependent on earlier work of Tycho Brahe? On contemporary work of Galileo?

15. Without introducing coordinates, give a geometric argument for the validity of the equation

$$\frac{dA}{dt} = \frac{1}{2}\left|\mathbf{R} \times \frac{d\mathbf{R}}{dt}\right|,$$

where $\mathbf{R}$ is the position vector of a particle moving in a plane curve, and dA/dt is the rate at which that vector sweeps out area.

16. For what values of v_0 is the orbit of Eq. (16) a parabola? A circle? An ellipse? A hyperbola?

17. Assuming that the earth's distance from the sun at perihelion is approximately 93,000,000 miles, and that the eccentricity of the earth's orbit about the sun is 0.0167, compute the velocity of the earth in its orbit at perihelion. (Use Eq. 17.)

REVIEW QUESTIONS AND EXERCISES

1. Define the derivative of a vector function.

2. Develop formulas for the derivatives, with respect to θ, of the unit vectors $\mathbf{u}_r$ and $\mathbf{u}_\theta$.

3. Develop vector formulas for velocity and acceleration of a particle moving in a plane curve:
   a) in terms of cartesian coordinates,
   b) in terms of polar coordinates,
   c) in terms of distance traveled along the curve and unit vectors tangent and normal to the curve.

4. a) Define curvature of a plane curve.
   b) Define radius of curvature.
   c) Define center of curvature.
   d) Define osculating circle.

5. Develop a formula for the curvature of a curve whose parametric equations are $x = f(t)$, $y = g(t)$.

6. In what way does the curvature of a curve affect the acceleration of a particle moving along the curve? In particular, discuss the case of constant-speed motion along a curve.

7. State and derive Kepler's second law concerning motion in a central force field.

8. If a vector $\mathbf{V}$ is a differentiable function of t and $|\mathbf{V}|$ = constant, what do you know about $d\mathbf{V}/dt$?

9. Define arc length and curvature of a space curve.

10. Explain how to $\mathbf{T}$, $\mathbf{N}$, and $\mathbf{B}$ for a space curve.

11. Express the vector $\mathbf{R} = \overrightarrow{OP}$ in terms of cylindrical coordinates and the unit vectors $\mathbf{u}_r$, $\mathbf{u}_\theta$, and $\mathbf{k}$.

12. Derive formulas for the velocity $\mathbf{v} = d\mathbf{R}/dt$ and acceleration $\mathbf{a} = d\mathbf{v}/dt$ in terms of cylindrical coordinates and the unit vectors $\mathbf{u}_r$, $\mathbf{u}_\theta$, and $\mathbf{k}$.

MISCELLANEOUS PROBLEMS

1. A particle moves in the xy-plane according to the time law

$$x = 1/\sqrt{1 + t^2}, \quad y = t/\sqrt{1 + t^2}.$$

   a) Compute the velocity vector and acceleration vector when $t = 1$.
   b) At what time is the speed of the particle a maximum?

2. A circular wheel with unit radius rolls along the x-axis uniformly, rotating one half-turn per second (Fig. 14.25). The position of a point P on the circumference is given by the formula

$$\overrightarrow{OP} = \mathbf{R} = \mathbf{i}(\pi t - \sin \pi t) + \mathbf{j}(1 - \cos \pi t).$$

   a) Determine the velocity (*vector*) $\mathbf{v}$ and the acceleration (*vector*) $\mathbf{a}$ at time t.
   b) Determine the slopes (as functions of t) of the two straight lines PC and PQ joining P to the center C of the wheel and to the point Q that is topmost at the instant.

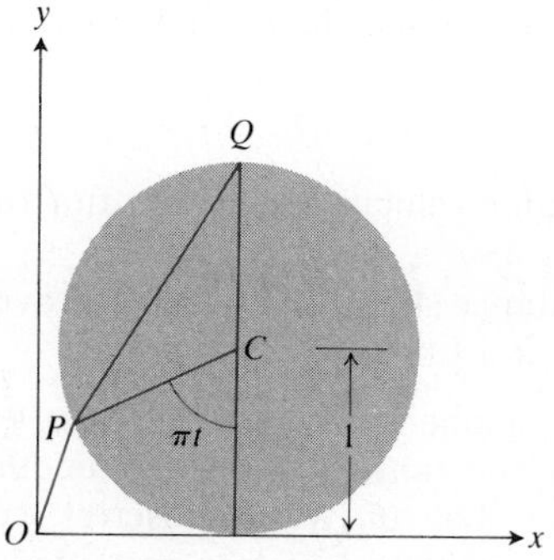

Figure 14.25

c) Show that the directions of the vectors **v** and **a** can be expressed in terms of the straight lines described in part (b).

3. The motion of a particle in the xy-plane is given by

$$\mathbf{R} = \mathbf{i}at \cos t + \mathbf{j}at \sin t.$$

Find the speed, and the tangential and normal components of the acceleration.

4. A particle moves in such a manner that the derivative of the position vector is always perpendicular to the position vector. Show that the particle moves on a circle with center at the origin.

5. The position of a point at time t is given by the formulas $x = e^t \cos t$, $y = e^t \sin t$.

a) Show that $\mathbf{a} = 2\mathbf{v} - 2\mathbf{r}$.

b) Show that the angle between the radius vector **r** and the acceleration vector **a** is constant, and find this angle.

6. Given the instantaneous velocity $\mathbf{v} = a\mathbf{i} + b\mathbf{j}$ and acceleration $\mathbf{a} = c\mathbf{i} + d\mathbf{j}$ of a particle at a point P on its path of motion, determine the curvature of the path at P.

7. Find the parametric equations, in terms of the parameter θ, of the locus of the center of curvature of the cycloid

$$x = a(\theta - \sin\theta), \quad y = a(1 - \cos\theta).$$

8. Find the point on the curve $y = e^x$ for which the radius of curvature is a minimum.

9. a) Given a closed curve having the property that every line parallel to the x-axis or the y-axis has at most two points in common with the curve. Let

$$x = x(t), \quad y = y(t), \quad \alpha \le t \le \beta,$$

be equations of the curve. Prove that if dx/dt and dy/dt are continuous, then the area bounded by the curve is

$$\frac{1}{2}\left|\int_\alpha^\beta \left[x(t)\frac{dy}{dt} - y(t)\frac{dx}{dt}\right] dt\right|.$$

b) Use the result of part (a) to find the area inside the ellipse

$$x = a\cos\phi, \quad y = b\sin\phi, \quad 0 \le \phi \le 2\pi.$$

What does the answer become when $a = b$?

10. For the curve defined by the equations

$$x = \int_0^\theta \cos\left(\frac{1}{2}\pi t^2\right) dt, \quad y = \int_0^\theta \sin\left(\frac{1}{2}\pi t^2\right) dt,$$

calculate the curvature κ as a function of the length of arc s, where s is measured from $(0, 0)$.

11. The curve for which the length of the tangent intercepted between the point of contact and the y-axis is always equal to 1 is called the *tractrix*. Find its equation. Show that the radius of curvature at each point of the curve is inversely proportional to the length of the normal intercepted between the point on the curve and the y-axis. Calculate the length of arc of the tractrix, and find the parametric equations in terms of the length of arc.

12. Let $x = x(t)$, $y = y(t)$ be a closed curve. A constant length p is measured off along the normal to the curve. The extremity of this segment describes a curve that is called a *parallel curve* to the original curve. Find the length of arc, the radius of curvature, and the area enclosed by the parallel curve.

13. Given the curve represented by the parametric equations

$$x = 32t, \quad y = 16t^2 - 4.$$

a) Calculate the radius of curvature of the curve at the point where $t = 3$.

b) Find the length of the curve between the points where $t = 0$ and $t = 1$.

14. Find the velocity, acceleration, and speed of a particle whose position at time t is

$$x = 3\sin t, \quad y = 2\cos t.$$

Also find the tangential and normal components of the acceleration.

15. The position of a particle at time t is given by the equations

$$x = 1 + \cos 2t, \quad y = \sin 2t.$$

Find

a) the normal and tangential components of acceleration at time t,

b) the radius of curvature of the path,

c) the equation of the path in polar coordinates, using the x-axis as the line $\theta = 0$ and the y-axis as the line $\theta = \pi/2$.

16. A particle moves so that its position at time t has the polar coordinates $r = t$, $\theta = t$. Find the velocity **v**, the acceleration **a**, and the curvature κ at any time t.

17. Find an expression for the curvature of the curve whose equation in polar coordinates is $r = f(\theta)$.

18. a) Find the equation in *polar coordinates* of the curve

$$x = e^{2t} \cos t, \quad y = e^{2t} \sin t.$$

b) Find the length of this curve from $t = 0$ to $t = 2\pi$.

19. Express the velocity vector in terms of $\mathbf{u}_r$ and $\mathbf{u}_\theta$ for a point moving in the xy-plane according to the law

$$\mathbf{r} = (t + 1)\mathbf{i} + (t - 1)\mathbf{j}.$$

20. The polar coordinates of a particle at time t are

$$r = e^{\omega t} + e^{-\omega t}, \quad \theta = t,$$

where ω is a constant. Find the acceleration vector when $t = 0$.

21. A slender rod, passing through the fixed point O, is rotating about O in a plane at the constant rate of 3 rad/min. An insect is crawling along the rod toward O at the constant rate of 1 in/min. Use polar coordinates in the plane, with point O as the origin, and assume that the insect starts at the point $r = 2$, $\theta = 0$.
 a) Find, in polar form, the vector velocity and vector acceleration of the insect when it is halfway to the origin.
 b) What will be the length of the path, in the plane, that the insect has traveled when it reaches the origin?

22. A smooth ball rolls inside a long hollow tube while the tube rotates with constant angular velocity ω about an axis perpendicular to the axis of the tube. Assuming no friction between the ball and the sides of the tube, show that the distance r from the axis of rotation to the ball satisfies the differential equation

$$d^2r/dt^2 - \omega^2 r = 0.$$

If at time $t = 0$ the ball is at rest (relative to the tube) at $r = a > 0$, find r as a function of t.

23. A particle P slides without friction along a coil spring having the form of a right circular helix. If the positive z-axis is taken downward, the cylindrical coordinates of P at time t are $r = a$, $z = b\theta$, where a and b are positive constants. If the particle starts at $r = a$, $\theta = 0$ with zero velocity and falls under gravity, the law of conservation of energy then tells us that its speed after it has fallen a vertical distance z is $\sqrt{2gz}$.
 a) Find the angular velocity $d\theta/dt$ when $\theta = 2\pi$.
 b) Express θ and z as functions of the time t.
 c) Determine the tangential and normal components of the velocity $d\mathbf{R}/dt$ and acceleration $d^2\mathbf{R}/dt^2$ as functions of t. Is there any component of acceleration in the direction of the binormal **B**?

24. Suppose the curve in Problem 23 is replaced by the conical helix

$$r = a\theta, \quad z = b\theta.$$

 a) Express the angular velocity $d\theta/dt$ as a function of θ.
 b) Express the distance that the particle travels along this helix as a function of θ.

25. Hold two of the three spherical coordinates ρ, ϕ, θ of point P in Fig. 13.27 constant while letting the other coordinate increase. Let **u**, with subscript corresponding to the coordinate which is permitted to vary, denote the unit vector that points in the direction in which P starts to move under these conditions.
 a) Express the three unit vectors $\mathbf{u}_\rho$, $\mathbf{u}_\phi$, $\mathbf{u}_\theta$ that are obtained in this manner, in terms of ρ, ϕ, θ, and the unit vectors **i, j, k.**
 b) Show that $\mathbf{u}_\rho \cdot \mathbf{u}_\phi = 0$.
 c) Show that $\mathbf{u}_\theta = \mathbf{u}_\rho \times \mathbf{u}_\phi$.
 d) Do the vectors $\mathbf{u}_\rho$, $\mathbf{u}_\phi$, $\mathbf{u}_\theta$ form a system of mutually orthogonal vectors? Is the system, in the order given, a right-handed or a left-handed system?

26. If the spherical coordinates ρ, ϕ, θ of a moving point P are differentiable functions of the time t, and $\mathbf{R} = \overrightarrow{OP}$ is the vector from the origin to P, express **R** and $d\mathbf{R}/dt$ in terms of ρ, ϕ, θ and their derivatives and the unit vectors $\mathbf{u}_\rho$, $\mathbf{u}_\phi$, $\mathbf{u}_\theta$ of Problem 25.

27. Express $ds^2 = dx^2 + dy^2 + dz^2$ in terms of (a) cylindrical coordinates r, θ, z, and (b) spherical coordinates ρ, ϕ, θ (see Problem 26). Interpret your results geometrically in terms of the sides and a diagonal of a rectangular box. Sketch.

28. Using the results of Problem 27, find the lengths of the following curves between $\theta = 0$ and $\theta = \ln 8$.
 a) $z = r = ae^\theta$
 b) $\phi = \pi/6$, $\rho = 2e^\theta$

29. Determine parametric equations giving x, y, z in terms of the parameter θ for the curve of intersection of the sphere $\rho = a$ and the plane $y + z = 0$, and find its length.

30. In Article 14.6, we found the velocity vector of a particle moving in a plane to be

$$\mathbf{v} = \frac{d\mathbf{R}}{dt} = \mathbf{i}\frac{dx}{dt} + \mathbf{j}\frac{dy}{dt} = \mathbf{u}_r\frac{dr}{dt} + \mathbf{u}_\theta\frac{r\,d\theta}{dt}.$$

 a) Express dx/dt and dy/dt in terms of dr/dt and $r\,d\theta/dt$ by computing $\mathbf{v} \cdot \mathbf{i}$ and $\mathbf{v} \cdot \mathbf{j}$.
 b) Express dr/dt and $r\,d\theta/dt$ in terms of dx/dt and dy/dt by computing $\mathbf{v} \cdot \mathbf{u}_r$ and $\mathbf{v} \cdot \mathbf{u}_\theta$.

31. The line through OA, A being the point (1, 1, 1), is the axis of rotation of a rigid body that is rotating with a constant angular speed of 6 rad/s. The rotation ap-

pears clockwise when we look toward the origin from A. Find the velocity vector of the point of the body that is at the position (1, 3, 2).

32. Consider the space curve whose parametric equations are

$$x = t, \quad y = t, \quad z = \frac{2}{3}t^{3/2}.$$

Compute the equation of the plane that passes through the point $(1, 1, \frac{2}{3})$ of this curve, and is perpendicular to the tangent of this curve at the same point.

33. A curve is given by the parametric equations

$$x = e^t \sin 2t, \quad y = e^t \cos 2t, \quad z = 2e^t.$$

Let P_0 be the point where $t = 0$. Determine

a) the direction cosines of the tangent, principal normal, and binormal at P_0,

b) the curvature at P_0.

34. The *normal plane* to a space curve at any point P of the curve is defined as the plane through P that is perpendicular to the tangent vector. The *osculating plane* at P is the plane containing the tangent and the principal normal. Given the space curve whose vector equation is

$$\mathbf{r}(t) = t\mathbf{i} + t^2\mathbf{j} + t^3\mathbf{k},$$

find

a) the equation of the normal plane at (1, 1, 1),

b) the equation of the osculating plane at (1, 1, 1).

35. Given the curve whose vector is

$$\mathbf{r}(t) = (3t - t^3)\mathbf{i} + 3t^2\mathbf{j} + (3t + t^3)\mathbf{k},$$

compute the curvature.

36. Show that the length of the arc described by the endpoint of

$$\mathbf{R} = (3 \cos t)\mathbf{i} + (3 \sin t)\mathbf{j} + t^2\mathbf{k},$$

as t varies from 0 to 2, is $5 + \frac{9}{4} \ln 3$.

37. The curve whose vector equation is

$$\mathbf{r}(t) = (2\sqrt{t} \cos t)\mathbf{i} + (3\sqrt{t} \sin t)\mathbf{j} + \sqrt{1 - t}\,\mathbf{k}, \quad 0 \le t \le 1,$$

lies on a quadric surface. Find the equation of this surface and describe it.

15 Partial Derivatives

15.1 Functions of Two or More Variables

In many applications, the values of the functions under study are determined by the values of more than one independent variable. The function may be as simple as the rule $V = \pi r^2 h$ for calculating the volume of a circular cylinder from its radius and height. Or it might be as complicated as the rule

$$w(x, t) = \cos(1.7 \times 10^{-2}t - 0.2x)e^{-0.2x},$$

which calculates the temperature x feet below the surface of the earth during the tth day of the year as a fraction of the average surface temperature on that day (Example 5 below).

Like functions of a single variable, these functions have *domains* (the sets of allowable input pairs, triples, or whatever, of real numbers) and *ranges* (their sets of output values).

DEFINITION

> Suppose D is a collection of n-tuples of real numbers
>
> $$(x_1, x_2, \ldots, x_n).$$
>
> A *function* f with domain D is a rule that assigns a number
>
> $$w = f(x_1, x_2, \ldots, x_n)$$
>
> to each n-tuple in D. The function's range is the set of w-values the function takes on. The symbol w is called the *dependent variable* of f, and f is said to be a function of the n *independent variables* $x_1, x_2, \ldots, x_n$.

In the function $V = \pi r^2 h$, the dependent variable is V. The independent variables are r and h.

Functions given by formulas are evaluated in the usual way by substituting values of the independent variables and calculating the corresponding values of the dependent variable. Thus the value of the function $w = \sqrt{x^2 + y^2 + z^2}$ at the point (3, 0, 4) is $\sqrt{(3)^2 + (0)^2 + (4)^2} = \sqrt{25} = 5$.

The Convention About Domains

In defining functions of more than one variable we observe the same convention that we do for functions of a single variable (Article 1.6). We never let the independent variables take on values that require division by zero. Also, we require the outputs to be real numbers unless we specifically say otherwise. Points that give imaginary numbers as outputs are to be excluded from the domain. Thus, if $f(x, y) = 1/(x^2 - y^2)$, we require $x^2 \neq y^2$, and if $f(x, y) = \sqrt{y - x^2}$, we require $y \geq x^2$.

Except for these restrictions, the domains of functions unless otherwise stated are assumed to be the largest possible sets for which their defining rules assign real numbers. These sets are sometimes called the "natural" domains of the functions.

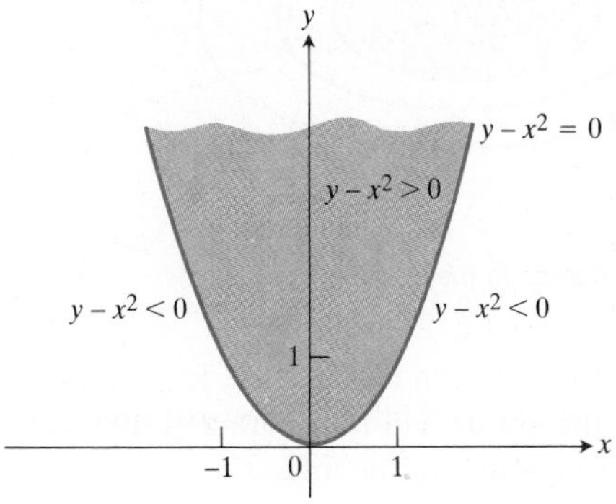

15.1 The domain of $f(x, y) = \sqrt{y - x^2}$ consists of the shaded region and its bounding parabola $y = x^2$. Elsewhere in the plane, $y - x^2 < 0$ and f is not defined.

EXAMPLE 1 Sketch the domain of

$$f(x, y) = \sqrt{y - x^2}.$$

What is the function's range?

Solution According to our convention, the domain D is the set of all number pairs (x, y) in the plane for which $\sqrt{y - x^2}$ is real (Fig. 15.1). This means the points for which

$$y - x^2 \geq 0 \qquad \text{or} \qquad y \geq x^2.$$

Therefore, D is the set of points that lie above and on the parabola $y = x^2$. The range of f is the set of all nonnegative numbers, $z \geq 0$. □

EXAMPLE 2 What are the domain and range of

$$f(x, y) = \frac{xy}{x^2 - y^2}?$$

Solution By our convention, x and y may take on any values except the ones for which $x^2 - y^2 = 0$, or $x^2 = y^2$. The domain therefore consists of all points in the plane except the points on the lines $y = x$ and $y = -x$. The range of f is the entire set of real numbers $-\infty < z < \infty$, as we may see by letting the point $P(x, y)$ traverse the hyperbola $x^2 - y^2 = 1$. On the right-hand branch of the hyperbola, $x = \sqrt{1 + y^2}$ and the values of f are given by the formula

$$f(x, y) = \frac{(\sqrt{1 + y^2})\, y}{1} = y\sqrt{1 + y^2}. \; \square$$

Graphs, Level Curves, Contours

There are two standard ways to picture a function $z = f(x, y)$. One is to draw a number of its *level curves*, the curves in the domain along which f has some constant value $f(x, y) = c$. The other is to draw the graph of f as a *surface* $z = f(x, y)$ in space, the surface being the set of points $(x, y, f(x, y))$. In the next example we do both.

EXAMPLE 3 Graph

$$z = f(x, y) = 100 - x^2 - y^2,$$

and plot the level curves $f(x, y) = 0$, 51, and 75.

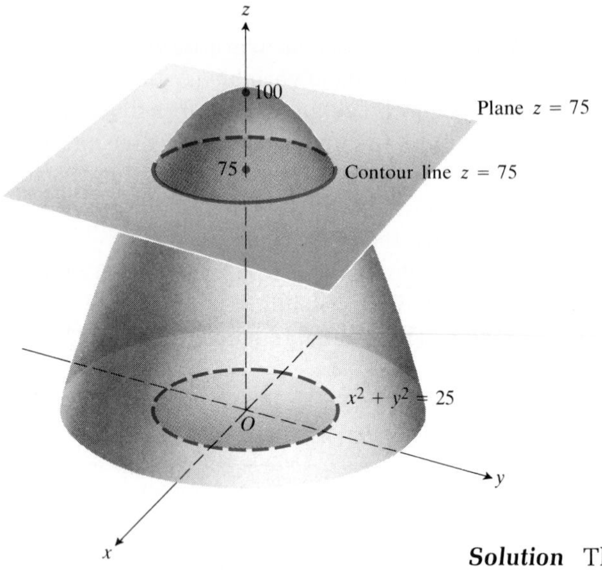

15.2 The contour line $z = 75$ on the surface $z = f(x, y) = 100 - x^2 - y^2$.

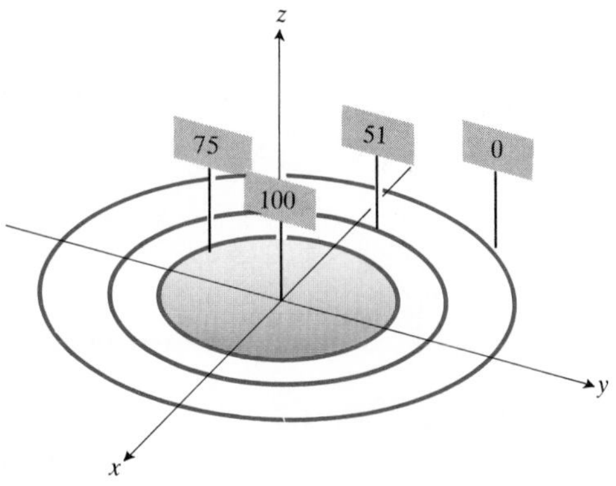

15.3 Level curves in the domain of f.

Solution The graph of f is a paraboloid, shown in Fig. 15.2. Its domain is the entire xy-plane, and its range is the set of real numbers $z \leq 100$. (The figure does not show the graph for negative values of z.)

The level curve $f(x, y) = 0$ is the set of points in the xy-plane at which

$$f(x, y) = 100 - x^2 - y^2 = 0, \qquad \text{or} \qquad x^2 + y^2 = 100,$$

the circle of radius 10 centered at the origin (Fig. 15.3). Similarly, the level curves $f(x, y) = 51$ and $f(x, y) = 75$ are the circles

$$f(x, y) = 100 - x^2 - y^2 = 75 \qquad \text{or} \qquad x^2 + y^2 = 25,$$
$$f(x, y) = 100 - x^2 - y^2 = 51 \qquad \text{or} \qquad x^2 + y^2 = 49. \ \square$$

Contour Lines

The points on a surface $z = f(x, y)$ that represent a fixed function value $f(x, y) = c$ make a *contour line* on the surface. These points all have the same distance $|c|$ above (or below) the xy-plane. In other words, the contour line is the curve in which the surface is cut by the plane $z = c$. Figure 15.2 shows the contour line $z = f(x, y) = 75$ lying directly above the circle $x^2 + y^2 = 25$, which is the level curve $f(x, y) = 75$ in the xy-plane.

Because of the close association of contour lines with level curves there is no firm agreement about which word to use for which kind of curve. On most maps, for example, curves that represent constant elevation are called contours, not level curves. The convention in this book is that level curves lie in the domain of f (in the plane $z = 0$) while contour lines lie on the surface defined by f (in the appropriate plane $z = c$).

Level Surfaces

The points (x, y, z) in space at which a function $w = f(x, y, z)$ of three independent variables takes on a constant value $f(x, y, z) = c$ usually make up a surface called a *level surface* of the function.

EXAMPLE 4 Describe the level surfaces of the function

$$w = \sqrt{x^2 + y^2 + z^2}.$$

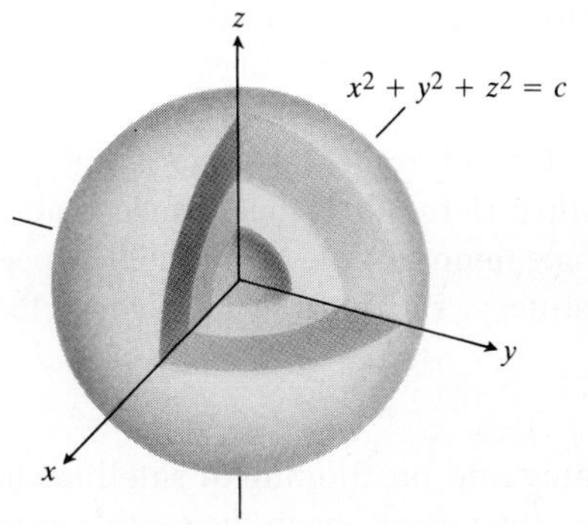

15.4 The level surfaces of $w = \sqrt{x^2 + y^2 + z^2}$ are concentric spheres.

Solution The value of function w can be interpreted as the distance from the origin to the point (x, y, z) in rectangular coordinates. Each level surface

$$\sqrt{x^2 + y^2 + z^2} = a, \qquad a > 0$$

is a sphere of radius a centered at the origin. Figure 15.4 shows a cutaway view of three of these spheres. The "level surface" $\sqrt{x^2 + y^2 + z^2} = 0$ consists of the origin alone.

It is important to keep in mind that we are *not* graphing the function $w = \sqrt{x^2 + y^2 + z^2}$ here. The graph of the function lies in a four-variable space. Instead, we are looking at surfaces in the function's domain. These surfaces show how the function values change as we move about in the domain. If we stay on the sphere of radius a centered at the origin, the function maintains the constant value a. If we move from sphere to sphere, the function values change. They increase as we move away from the origin, and decrease as we move toward it. Thus, the way the function values change as we move away from any given point depends on the direction we take. This dependence of change on direction is an important idea, and we shall return to it in Article 15.6. □

Three-dimensional graphing programs now available for computers make it possible to graph many functions of two variables with only a few keystrokes. We can often get information about a function from one of these graphs more quickly than we can from a formula.

EXAMPLE 5 Figure 15.5 shows a computer-generated graph of the function

$$w(x, t) = \cos(1.7 \times 10^{-2}t - 0.2x)e^{-0.2x} \qquad (t \text{ in days, } x \text{ in feet}).$$

The graph shows how the temperature beneath the earth's surface varies with time. The variation is given as a fraction of the variation at the

15.5 This computer-generated graph of $w(x, t) = \cos(1.7 \times 10^{-2}t - 0.2x)e^{-0.2x}$ shows the seasonal variation of the temperature below ground as a fraction of surface temperature. $\Delta x = 0.375$ ft, $\Delta t = 15.625$ days. At $x = 15$ ft the variation is only 5% of the variation at the surface. At $x = 30$ ft the variation is less than 0.25% of the surface variation. (Adapted by John G. Aspinall from art provided by Norton Starr for G. C. Berresford's "Differential Equations and Root Cellars," *The UMAP Journal,* Volume 2, Number 3, (1981), pp. 53–75.)

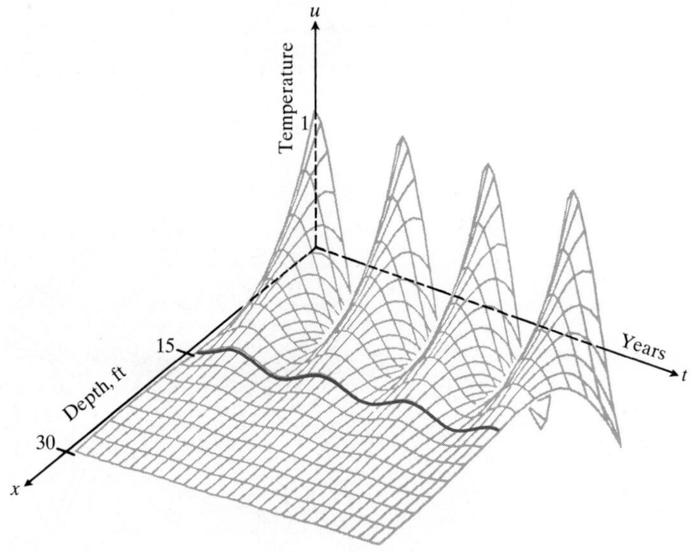

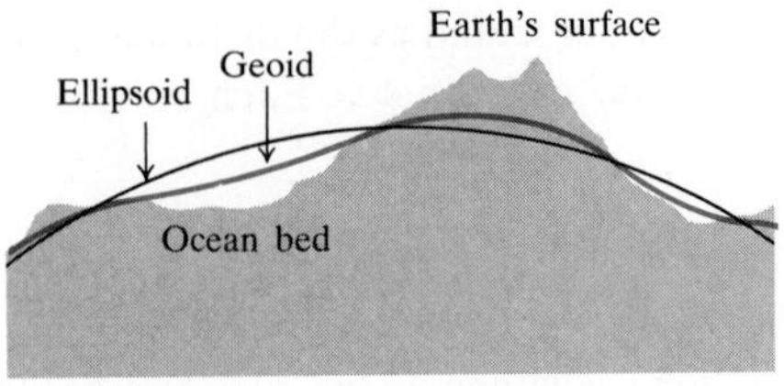

15.6 A profile of the earth, geoid, and ellipsoid.

surface. At a depth of 15 ft, the variation (vertical amplitude in the figure) is about 5 percent of the surface variation. At 30 ft, there is almost no variation in temperature throughout the year.

Another thing the graph shows that is not immediately apparent from the equation for f is that the temperature 15 ft below the surface is about half a year out of phase with the surface temperature. When the temperature is lowest on the surface (late January, say) it is at its highest 15 ft below. The seasons are reversed. □

EXAMPLE 6 Along with the launching and positioning of satellites has come an increased interest in mapping the variations in the earth's gravitational field. The strength of the field may vary significantly from place to place on the surface of the earth. In fact, changes in the performance of pendulum clocks taken by voyagers from Europe to other continents are said to have provided early supporting evidence of Newton's law of gravitation. If one takes the strength of the field at mean sea level as a standard, then the points in the earth's gravitational field that have the same gravitational potential as the standard value constitute a potato-shaped surface that geophysicists call the *geoid*. The geoid differs from the surface of the earth itself in most places, and it also differs from the ellipsoid that is normally used to approximate the surface of the earth (Fig. 15.6). The height of the geoid above or below the ellipsoid is called the *geoidal height*. Geoidal heights are counted as positive when the geoid rises above the ellipsoid, and negative when it dips below the ellipsoid. Figure 15.7 shows contours of geoidal height on a Mercator map of the earth. These contours are the level curves of the geoidal height function. □

15.7 A contour map of geoidal height measured in meters above (+) and below (−) an ideal earth shaped like an ellipsoid.

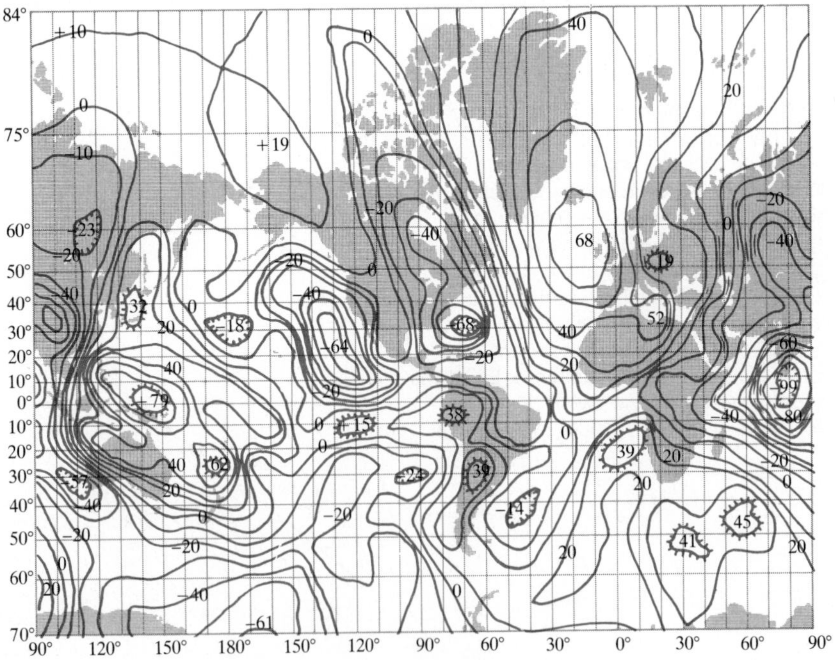

PROBLEMS

In Problems 1–12, give the domain and range, and describe the level curves.

1. $f(x, y) = 1/xy$
2. $f(x, y) = \sqrt{x + y}$
3. $f(x, y) = 1/\sqrt{x + y}$
4. $f(x, y) = \ln (x + y)$
5. $f(x, y) = \ln (x^2 + y^2)$
6. $f(x, y) = \sqrt{4 - x^2 - y^2}$
7. $f(x, y) = \sqrt{y/x}$
8. $f(x, y) = y/x^2$
9. $f(x, y) = \tan^{-1} (y/x)$
10. $f(x, y) = e^{x-y}$
11. $f(x, y) = \cos xy$
12. $f(x, y) = \sinh (x^2 + y^2)$.

In Problems 13–18, find the domain and range.

13. $f(x, y) = \dfrac{xy}{y^2 - x^2}$
14. $f(x, y, z) = \sqrt{1 - (x^2 + y^2 + z^2)}$
15. $f(x, y, z) = \sqrt{z}$
16. $f(x, y, z) = 5$
17. $f(x, y, z) = \dfrac{x^2}{x^2 + y^2 + z^2}$
18. $f(x, y, z) = \dfrac{1}{1 - (x^2 + y^2 + z^2)}$

Describe the level surfaces of the functions in Problems 19–32. You may wish to refer to Articles 13.8, 13.10, and 13.11.

19. $f(x, y, z) = x^2 + y^2 + z^2$
20. $f(x, y, z) = \ln (x^2 + y^2 + z^2)$
21. $f(x, y, z) = 1/(x^2 + y^2 + z^2)$
22. $f(x, y, z) = (x^2/25) + (y^2/16) + (z^2/9)$
23. $f(x, y, z) = x$
24. $f(x, y, z) = x + y$
25. $f(x, y, z) = x + y + z$
26. $f(x, y, z) = x^2 + y^2$
27. $f(x, y, z) = y^2 + z^2$
28. $f(x, y, z) = x^2 + z^2$
29. $f(x, y, z) = x^2 + y^2 - z$
30. $f(x, y, z) = \sqrt{x^2 + y^2 - z}$
31. $f(x, y, z) = x^2 - y$
32. $f(x, y, z) = \sin y$.

In Problems 33–38, show two ways of representing the function $z = f(x, y)$: (a) by sketching the surface $z = f(x, y)$ in space, and (b) by drawing a family of level curves in the plane.

33. $f(x, y) = x$
34. $f(x, y) = y$
35. $f(x, y) = x^2 + y^2$
36. $f(x, y) = x^2 - y^2$
37. $f(x, y) = \sqrt{y - x^2}$ (Example 1)
38. $f(x, y) = x \sin y$
39. Sketch the level curves of
$$f(x, y) = -(x - 1)^2 - y^2 + 1$$
for $f(x, y) = 1, 0, -3, -8$.
40. Sketch the level curves of $f(x, y) = 2x^2 - y + 1$ for $f(x, y) = 0, 1, 2$.
41. Find an equation for the level curve of $f(x, y) = 16 - x^2 - y^2$ that passes through the point $(2\sqrt{2}, \sqrt{2})$.
42. Repeat Problem 41 for $f(x, y) = \sqrt{x^2 - 1}$ and $(1, 0)$.
43. Sketch in the xy-plane the domain of the function $z = \sqrt{\sin (\pi xy)}$, $xy \geq 0$, and shade the points that lie in the domain.
44. Find the maximum value of the function $w = xyz$ on the line $x = 20 - t$, $y = t$, $z = 20$. At what point or points on the line does the maximum occur? (*Hint:* Along the line, w can be expressed as a function of the single variable t.)

***Toolkit* programs**
3D Grapher

15.2 Limits and Continuity

If the values of a function $z = f(x, y)$ can be made as close as we like to a fixed number L by taking the point (x, y) close to the point (x_0, y_0), but not equal to (x_0, y_0), we say that L is the limit of f as (x, y) approaches (x_0, y_0).

In symbols, we write

$$\lim_{(x,\, y)\to(x_0,\, y_0)} f(x, y) = L, \tag{1}$$

and we say "the limit of f as (x, y) approaches (x_0, y_0) equals L." This is like the limit of a function of one variable, except that there are two independent variables involved instead of one.

For (x, y) to be "close" to (x_0, y_0) means that the cartesian distance $\sqrt{(x - x_0)^2 + (y - y_0)^2}$ is small in some sense. Since

$$|x - x_0| = \sqrt{(x - x_0)^2} \leq \sqrt{(x - x_0)^2 + (y - y_0)^2} \tag{2}$$

and

$$|y - y_0| = \sqrt{(y - y_0)^2} \leq \sqrt{(x - x_0)^2 + (y - y_0)^2}, \tag{3}$$

the inequality

$$\sqrt{(x - x_0)^2 + (y - y_0)^2} < \delta$$

implies

$$|x - x_0| < \delta \qquad \text{and} \qquad |y - y_0| < \delta.$$

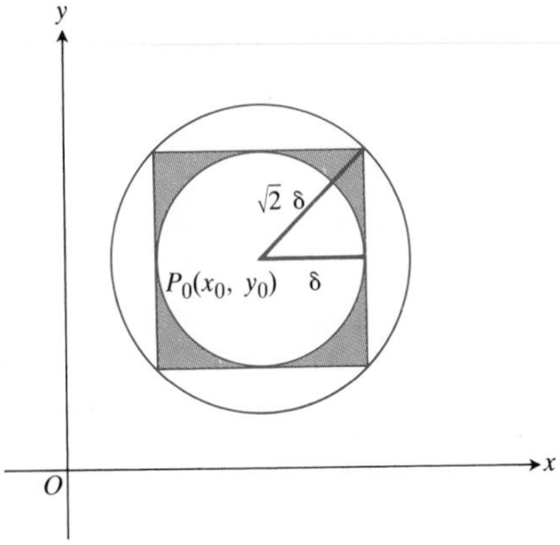

15.8 The open square $|x - x_0| < \delta$, $|y - y_0| < \delta$ lies inside the open disc $\sqrt{(x - x_0)^2 + (y - y_0)^2} < \sqrt{2}\delta$ and contains the open disc $\sqrt{(x - x_0)^2 + (y - y_0)^2} < \delta$.

Conversely, if for some $\delta > 0$ both $|x - x_0| < \delta$ and $|y - y_0| < \delta$, then

$$\sqrt{(x - x_0)^2 + (y - y_0)^2} < \sqrt{\delta^2 + \delta^2} = \sqrt{2}\delta,$$

which is small if δ is small. Thus, in calculating limits we may think either in terms of the distance in the plane or in terms of differences in coordinates. See Fig. 15.8. We therefore have two equivalent definitions of limit.

Two Equivalent Definitions of Limit

The limit of $f(x, y)$ as $(x, y) \to (x_0, y_0)$ is the number L if for any $\varepsilon > 0$ there exists a $\delta > 0$ such that for all points (x, y) either

1. $0 < \sqrt{(x - x_0)^2 + (y - y_0)^2} < \delta \Rightarrow |f(x, y) - L| < \varepsilon$,

or

2. $0 < |x - x_0| < \delta$ and $0 < |y - y_0| < \delta \Rightarrow |f(x, y) - L| < \varepsilon$.

As we did with functions of a single variable in Article 1.10, we may use the definition of limit to prove directly that

$$\begin{aligned} \lim_{(x,\, y)\to(x_0,\, y_0)} x &= x_0, \\ \lim_{(x,\, y)\to(x_0,\, y_0)} y &= y_0, \\ \lim_{(x,\, y)\to(x_0,\, y_0)} k &= k. \end{aligned} \tag{4}$$

The proofs are similar and we shall not repeat them here.

There is also a theorem like Theorem 1 of Article 1.10 that says that the limit of a sum $f(x, y) + g(x, y)$ is the sum of the limits of f and g (when they exist), with similar results for differences, products, constant multiples, and quotients.

THEOREM 1

If

$$\lim_{(x, y)\to(x_0, y_0)} f(x, y) = L_1 \qquad \text{and} \qquad \lim_{(x, y)\to(x_0, y_0)} g(x, y) = L_2,$$

then

i) $\lim[f(x, y) + g(x, y)] = L_1 + L_2$,

ii) $\lim[f(x, y) - g(x, y)] = L_1 - L_2$,

iii) $\lim[f(x, y)g(x, y)] = L_1 L_2$,

iv) $\lim[kf(x, y)] = kL_1$ (any number k),

v) $\lim \dfrac{f(x, y)}{g(x, y)} = \dfrac{L_1}{L_2}$ if $L_2 \neq 0$.

The limits are all to be taken as $(x, y) \to (x_0, y_0)$.

When we apply Theorem 1 to the functions and limits in (4) we obtain the useful result that the limits of polynomials and rational functions like

$$x^2 + y^2, \qquad \frac{x - xy + 3}{x^2y + 5xy - y^3}$$

as $(x, y) \to (x_0, y_0)$ may be calculated by evaluating the functions at the point (x_0, y_0). The only requirement is that the functions be defined there. In other words, polynomials and rational functions in two variables are *continuous* wherever they are defined, as we shall see in a moment.

EXAMPLE 1

a) $\displaystyle\lim_{(x, y)\to(3, -4)} (x^2 + y^2) = (3)^2 + (-4)^2 = 9 + 16 = 25.$

b) $\displaystyle\lim_{(x, y)\to(0, 1)} \frac{x - xy + 3}{x^2y + 5xy - y^3} = \frac{0 - 0(1) + 3}{(0)^2(1) + 5(0)(1) - (1)^3} = -3.$ □

Continuity

A function $f(x, y)$ is said to be *continuous* at the point (x_0, y_0) if

1. f is defined at (x_0, y_0),
2. $\displaystyle\lim_{(x, y)\to(x_0, y_0)} f(x, y)$ exists, and (5)
3. $\displaystyle\lim_{(x, y)\to(x_0, y_0)} f(x, y) = f(x_0, y_0)$.

It is a consequence of Theorem 1, although we shall not prove it, that if $f(x, y)$ and $g(x, y)$ are both continuous at a point, then their sum $f(x, y) + g(x, y)$ is continuous at that point. Similar results hold for differences, products, and multiples of continuous functions. Also, the quotient of two continuous functions is continuous wherever it is defined. It follows that polynomials and rational functions in two variables are continuous at any point at which they are defined.

If $z = f(x, y)$ is a continuous function of x and y, and $w = g(z)$ is a continuous function of z, then the composite

$$w = g(f(x, y))$$

is continuous. Thus, composites like

$$e^{x-y}, \qquad \cos\frac{xy}{x^2 + 1}, \qquad \ln(1 + x^2y^2)$$

are continuous at every point (x, y).

As with functions of a single variable, the general rule is that composites of continuous functions are continuous. The only requirement is that each function be continuous where it is applied.

EXAMPLE 2 Given $\varepsilon > 0$, how close to $(0, 0)$ should we take (x, y) to make $|f(x, y) - f(0, 0)| < \varepsilon$ if

$$f(x, y) = \frac{x + y}{x^2 + y^2 + 1}?$$

Solution Since $f(0, 0) = 0$,

$$|f(x, y) - f(0, 0)| = \left|\frac{x + y}{x^2 + y^2 + 1}\right|,$$

and we are being asked how close to $(0, 0)$ we should take (x, y) to make

$$\left|\frac{x + y}{x^2 + y^2 + 1}\right| < \varepsilon.$$

Since the denominator of the fraction is never less than one, we have

$$\begin{aligned}\left|\frac{x + y}{x^2 + y^2 + 1}\right| &\le |x + y| \le |x| + |y| \\ &\le \sqrt{x^2 + y^2} + \sqrt{x^2 + y^2} = 2\sqrt{x^2 + y^2},\end{aligned}$$

which will be less than ε if

$$\sqrt{x^2 + y^2} < \frac{\varepsilon}{2}.$$

Therefore, the original ε inequality will hold if the distance from (x, y) to $(0, 0)$ is less than $\varepsilon/2$. That is, we may choose our δ in the definition of limit to be the number $\varepsilon/2$. □

EXAMPLE 3 Show that the function

$$f(x, y) = \begin{cases} \dfrac{xy}{x^2 + y^2}, & (x, y) \ne (0, 0) \\ 0, & (x, y) = (0, 0) \end{cases}$$

is continuous at every point except the origin. See Fig. 15.9.

Solution The function f is continuous at any point $(x, y) \ne (0, 0)$ because its values are then given by a rational function of x and y.

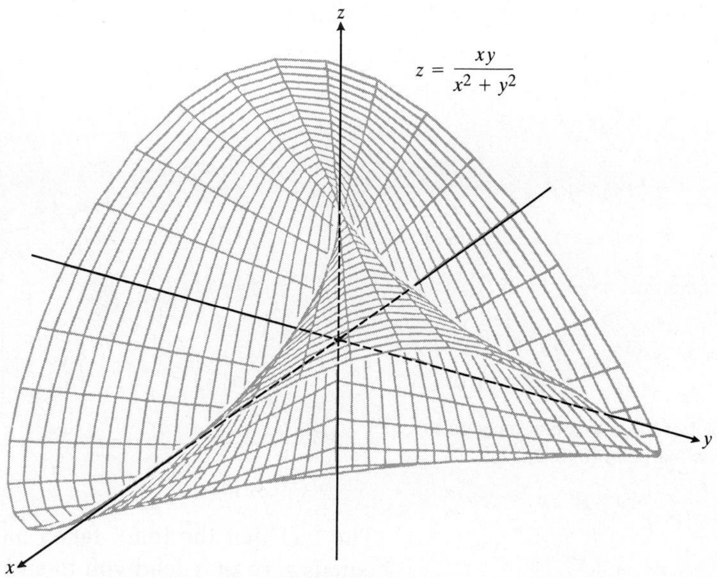

15.9 The graph of

$$f(x, y) = \begin{cases} \dfrac{xy}{x^2 + y^2}, & (x, y) \neq (0, 0) \\ 0, & (x, y) = (0, 0). \end{cases}$$

The only point of the z-axis that belongs to this surface is the point (0, 0, 0).

At (0, 0) the value of f is defined, but f, we claim, has no limit as $(x, y) \to (0, 0)$. The reason is that different paths of approach to the origin may give different limits, as we shall now see.

If $(x, y) \to (0, 0)$ along the line $y = mx$, but $(x, y) \neq (0, 0)$, then

$$f(x, y) = \frac{xy}{x^2 + y^2} = \frac{x(mx)}{x^2 + m^2x^2} = \frac{m}{1 + m^2},$$

$$\lim_{\substack{(x, y)\to(0, 0) \\ \text{along } y=mx}} f(x, y) = \frac{m}{1 + m^2}.$$

This limit changes as m changes. In fact, the substitution $m = \tan\theta$ produces

$$f(x, y) = \frac{m}{1 + m^2} = \frac{\tan\theta}{\sec^2\theta} = \frac{1}{2}\sin 2\theta,$$

which shows that the limit varies from $-\frac{1}{2}$ to $\frac{1}{2}$ depending on the angle of approach. There is no single number L that we may call the limit of f as $(x, y) \to (0, 0)$. Therefore, the limit fails to exist and the function cannot be continuous at (0, 0). □

EXAMPLE 4 Examine the limits of

$$f(x, y) = \frac{x^2y}{x^4 + y^2}, \qquad (x, y) \neq (0, 0)$$

as $(x, y) \to (0, 0)$ along the line $y = mx$ and along the parabola $y = x^2$. Does f have a limit as $(x, y) \to (0, 0)$?

Solution (Fig. 15.10) Along the line $y = mx$

$$f(x, y) = f(x, mx) = \frac{x^2(mx)}{x^4 + m^2x^2} = \frac{mx}{x^2 + m^2}.$$

Therefore,

$$\lim_{\substack{(x, y)\to(0, 0) \\ \text{along } y=mx}} f(x, y) = \lim_{x\to 0} \frac{mx}{x^2 + m^2} = 0.$$

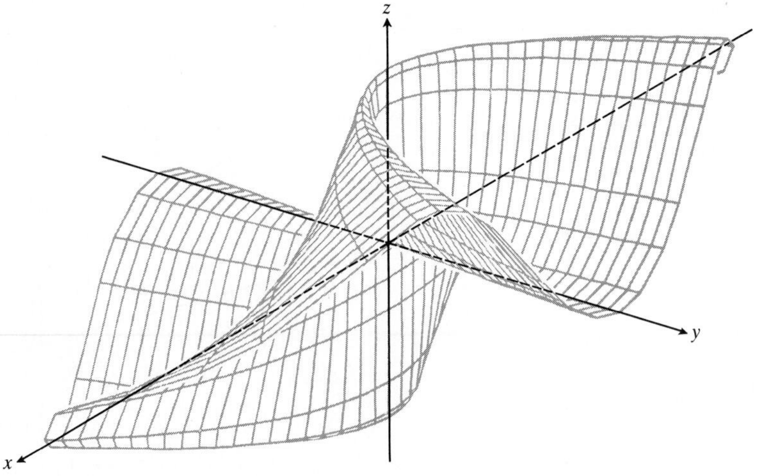

15.10 The graph of

$$f(x, y) = \begin{cases} \dfrac{x^2y}{x^4 + y^2}, & (x, y) \neq (0, 0) \\ 0, & (x, y) = (0, 0). \end{cases}$$

The only point of the z-axis that belongs to this surface is the point (0, 0, 0).

The fact that the limit, taken along all *linear* paths to (0, 0), exists and equals zero may lead you to suspect that $\lim_{(x, y)\to(0, 0)} f(x, y)$ exists. However, along the parabola $y = x^2$,

$$f(x, y) = f(x, x^2) = \frac{x^2(x^2)}{x^4 + x^4} = \frac{1}{2}.$$

Therefore,

$$\lim_{\substack{(x, y)\to(0, 0) \\ \text{along } y=x^2}} f(x, y) = \frac{1}{2}.$$

For f to have a limit as $(x, y) \to (0, 0)$, the limits along all paths of approach would have to agree. Since we have reached different limits on different paths, we conclude that f has no limit as $(x, y) \to (0, 0)$. □

Functions of Three Variables or More

The definitions of limit and continuity for functions of two variables and the conclusions about limits and continuity for sums, products, quotients, and composites all extend to functions of three variables or more. Thus, functions like

$$\ln \sqrt{x^2 + y^2 + z^2} \qquad \text{and} \qquad \frac{e^{x-y+z}}{z^2 + \cos \sqrt{xy}}$$

are continuous at every point at which they are defined, and limits like

$$\lim_{P\to(1, 0, 1)} \frac{e^{x-y+z}}{z^2 + \cos(\sqrt{xy})} = \frac{e^{1-0+1}}{(1)^2 + \cos(0)} = \frac{e^2}{2},$$

where P denotes the point (x, y, z), may be evaluated by direct substitution.

Continuous Functions Defined in Closed, Bounded Regions

As we know, a function of a single variable that is continuous throughout a closed, bounded interval $[a, b]$ takes on an absolute maximum value and an absolute minimum value at least once in $[a, b]$. The same is true of a function $z = f(x, y)$ that is continuous on a closed, bounded region R in the plane (like a line segment, a disc, or a filled-in triangle, square, or rectangle). It takes on an absolute maximum value at some point in R, and an absolute minimum value at some point in R.

Similar results are true for functions of three or more variables. A continuous function $w = f(x, y, z)$, for example, must take on absolute maximum and minimum values on any closed, bounded region (solid ball or cube, spherical shell, rectangular plate) on which it is defined.

Techniques for finding these maximum and minimum values will be the subject of Articles 15.9 and 15.10.

PROBLEMS

Find the limits (if they exist) in Problems 1–18.

1. $\displaystyle\lim_{(x,y)\to(0,0)} \frac{3x^2 - y^2 + 5}{x^2 + y^2 + 2}$

2. $\displaystyle\lim_{(x,y)\to(1,1)} \ln|1 + x^2y^2|$

3. $\displaystyle\lim_{(x,y)\to(0,\ln 2)} e^{x-y}$

4. $\displaystyle\lim_{(x,y)\to(0,4)} \frac{x}{\sqrt{y}}$

5. $\displaystyle\lim_{P\to(1,3,4)} \sqrt{x^2 + y^2 + z^2 - 1}$

6. $\displaystyle\lim_{P\to(1,2,6)} \left(\frac{1}{x} + \frac{1}{y} + \frac{1}{z}\right)$

7. $\displaystyle\lim_{(x,y)\to(0,0)} \frac{e^y \sin x}{x}$

8. $\displaystyle\lim_{(x,y)\to(0,\pi/2)} \sec x \tan y$

9. $\displaystyle\lim_{(x,y)\to(0,0)} \tan^{-1}\frac{1}{\sqrt{x^2 + y^2}}$

10. $\displaystyle\lim_{(x,y)\to(0,0)} \cos\frac{x^2 + y^2}{x + y + 1}$

11. $\displaystyle\lim_{(x,y)\to(1,1)} \cos\sqrt[3]{|xy| - 1}$

12. $\displaystyle\lim_{(x,y)\to(1,0)} \frac{x \sin y}{x^2 + 1}$

13. $\displaystyle\lim_{(x,y)\to(0,0)} y \sin(1/x)$

14. $\displaystyle\lim_{(x,y)\to(0,2)} \left(\frac{\cos x - 1}{x^2}\right)\left(\frac{y - 2}{y^2 - 4}\right)$

15. $\displaystyle\lim_{(x,y)\to(1,1)} \frac{x^2 - 2xy + y^2}{x - y}$

16. $\displaystyle\lim_{(x,y)\to(-2,2)} \frac{xy + y - 2x - 2}{x + 1}$

17. $\displaystyle\lim_{(x,y)\to(1,1)} \frac{x^2 - y^2}{x - y}$

18. $\displaystyle\lim_{(x,y)\to(1,1)} \frac{x^3y^3 - 1}{xy - 1}$

By considering different lines of approach, show that the functions in Problems 19–24 have no limit as $(x, y) \to (0, 0)$.

19. $\dfrac{x + y}{x - y}$

20. $\dfrac{x - y}{x + y}$

21. $\dfrac{x^2 - y^2}{x^2 + y^2}$

22. $\dfrac{x}{\sqrt{x^2 + y^2}}$

23. $\dfrac{x^2 + y^2}{xy}$

24. $\dfrac{xy}{|xy|}$

In Problems 25–28, find how close to the origin one should take the point (x, y) to make $|f(x, y) - f(0, 0)| < 0.01$.

25. $f(x, y) = x^2 + y^2$

26. $f(x, y) = y/(x^2 + 1)$

27. $f(x, y) = (x + y)/(x^2 + 1)$

28. $f(x, y) = (x + y)/(2 + \cos x)$

In Problems 29–32, find how close to the origin one should take the point (x, y, z) to make

$$|f(x, y, z) - f(0, 0, 0)| < \varepsilon$$

for the given ε.

29. $f(x, y, z) = x^2 + y^2 + z^2, \quad \varepsilon = 0.01$

30. $f(x, y, z) = xyz, \quad \varepsilon = 0.008$

31. $f(x, y, z) = \dfrac{x + y + z}{x^2 + y^2 + z^2 + 1}, \quad \varepsilon = 0.015$

32. $f(x, y, z) = \tan^2 x + \tan^2 y + \tan^2 z, \quad \varepsilon = 0.03$

33. Is $f(x, y, z) = \sqrt{x^2 + y^2 + z^2}$ continuous at $(0, 0, 0)$? Explain.

34. Show that the function $f(x, y) = x^2/(x^2 + y)$ has no limit as $(x, y) \to (0, 0)$ by examining the limits of f as $(x, y) \to (0, 0)$ along the parabola $y = kx^2$ for selected values of k.

15.3 Partial Derivatives

Partial derivatives are the derivatives we get when we hold constant all but one of the independent variables in a function and differentiate with respect to that one. In this article, we show how partial derivatives arise geometrically and how they may be calculated by applying the rules we already know for differentiating functions of a single variable. We begin our discussion with functions of two independent variables.

If (x_0, y_0) is a point in the domain of a function $z = f(x, y)$, the plane $y = y_0$ will cut the surface $z = f(x, y)$ in the curve $z = f(x, y_0)$, as shown in Fig. 15.11. This curve is the graph of the function $z = f(x, y_0)$ in the plane $y = y_0$. The vertical coordinate in this plane is z, the height above (below) the xy-plane. The horizontal coordinate is x.

The derivative of $z = f(x, y_0)$ with respect to x at $x = x_0$ is defined in the usual way as the limit

$$\left.\frac{d}{dx} f(x, y_0)\right|_{\text{at } x = x_0} = \lim_{\Delta x \to 0} \frac{f(x_0 + \Delta x, y_0) - f(x_0, y_0)}{\Delta x}, \tag{1}$$

provided that the limit exists. The limit is called the *partial derivative* of *f with respect to* x at the point (x_0, y_0). The *slope* of the curve $z = f(x, y_0)$ at $x = x_0$ is defined to be the value of the partial derivative with respect to x there. The *tangent* to the curve $z = f(x, y_0)$ at $x = x_0$ is defined to be the line in the plane $y = y_0$ that passes through the point $(x_0, y_0, f(x_0, y_0))$ with this slope.

The usual notations for the partial derivative of $z = f(x, y)$ with respect to x at (x_0, y_0) are as follows:

$\frac{\partial f}{\partial x}(x_0, y_0)$ or $f_x(x_0, y_0)$ "Partial derivative of f with respect to x at (x_0, y_0)" or "f sub x at (x_0, y_0)." Convenient for stressing the point (x_0, y_0).

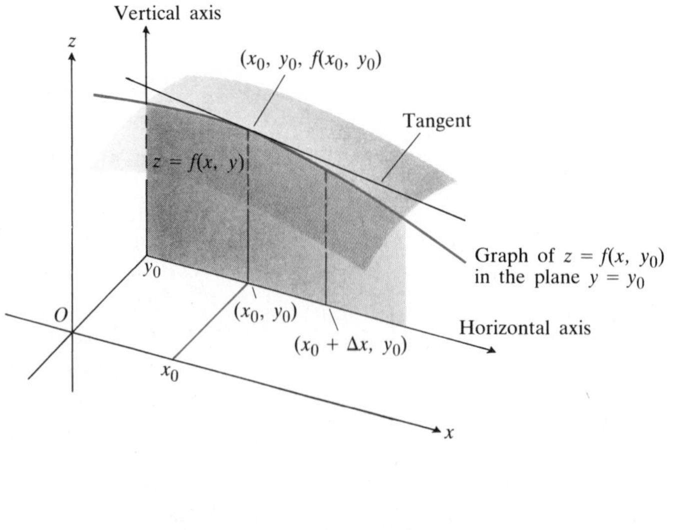

15.11 The intersection of the plane $y = y_0$ with the surface $z = f(x, y)$, viewed from a point above the third quadrant of the xy-plane.

$\left.\dfrac{\partial z}{\partial x}\right\|_{(x_0, y_0)}$	"Partial derivative of z with respect to x at (x_0, y_0)." Common in science and engineering when you are dealing with variables and do not mention the function explicitly.
$f_x, \dfrac{\partial f}{\partial x}, z_x$, or $\dfrac{\partial z}{\partial x}$	"Partial derivative of f (or z) with respect to x." Convenient when you regard the partial derivative as a function in its own right.

As Eq. (1) shows, $\partial f/\partial x$ is the ordinary derivative of f with respect to x with y held fixed. Thus, to calculate $\partial f/\partial x$ we may differentiate f with respect to x in the usual way while treating y as a constant.

EXAMPLE 1 Find $\partial f/\partial x$ if

$$f(x, y) = 100 - x^2 - y^2.$$

Solution We regard y as a constant and calculate

$$\frac{\partial f}{\partial x} = \frac{d}{dx}(100 - x^2 - y^2) = 0 - 2x - 0 = -2x. \square$$

The definition of the partial derivative $\partial f/\partial y$ of $z = f(x, y)$ with respect to y at (x_0, y_0) is like the definition of $\partial f/\partial x$. We regard $z = f(x_0, y)$ as a function of the single independent variable y (Fig. 15.12) and define

$$\frac{\partial f}{\partial y}(x_0, y_0) = \left.\frac{d}{dy}f(x_0, y)\right|_{y=y_0} = \lim_{\Delta y \to 0} \frac{f(x_0, y_0 + \Delta y) - f(x_0, y_0)}{\Delta y}, \quad (2)$$

provided the limit exists. This limit is also taken to be the slope of the curve $z = f(x_0, y)$ in the plane $x = x_0$. The tangent to the curve is the line in the plane $x = x_0$ that passes through $(x_0, y_0, f(x_0, y_0))$ with this slope.

We now have two tangent lines associated with $z = f(x, y)$ at (x_0, y_0). Figure 15.13 shows them together. If $\partial f/\partial x$ and $\partial f/\partial y$ are both continuous at (x_0, y_0) the plane determined by these two lines will be tangent to the

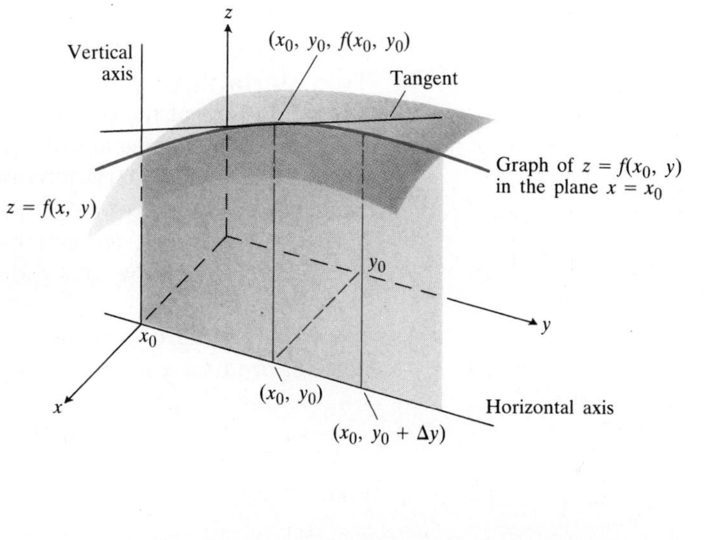

15.12 The intersection of the plane $x = x_0$ with the surface $z = f(x, y)$, viewed from the first quadrant of the xy-plane.

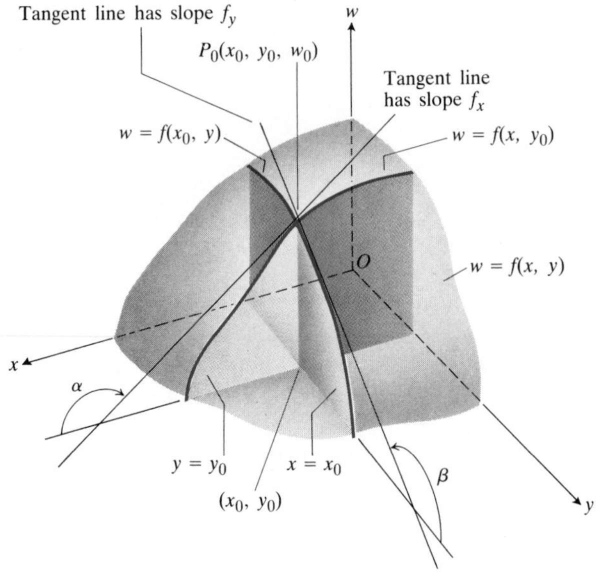

15.13 The plane $y = y_0$ cuts the surface $w = f(x, y)$ in the curve $w = f(x, y_0)$. At each x, the slope of this curve is $f_x(x, y_0)$. Similarly, the plane $x = x_0$ cuts the surface in a curve whose slope is $f_y(x_0, y)$.

surface $z = f(x, y)$ at (x_0, y_0). Tangency is a more complicated phenomenon for surfaces than it is for curves, however, and we shall have to wait until Article 15.6 to deal with it.

EXAMPLE 2 Find $\partial f/\partial y$ if

$$f(x, y) = e^x \ln (x^2 + y^2 + 1).$$

Solution Treating x, e^x, and x^2 as constants we apply the chain rule for functions of a single variable (in this case y) to find

$$\begin{aligned}\frac{\partial f}{\partial y} &= \frac{\partial}{\partial y}(e^x \ln (x^2 + y^2 + 1)) = e^x \cdot \frac{\partial}{\partial x} \ln (x^2 + y^2 + 1)\\ &= e^x \cdot \frac{1}{x^2 + y^2 + 1} \cdot \frac{\partial}{\partial y}(x^2 + y^2 + 1) = \frac{e^x}{x^2 + y^2 + 1} \cdot 2y\\ &= \frac{2ye^x}{x^2 + y^2 + 1}. \quad \square\end{aligned}$$

Functions with More Than Two Variables

The definitions of the partial derivatives of functions of more than two independent variables are like the definitions for functions of two variables. They are ordinary derivatives with respect to one variable, taken while all the other independent variables are regarded as constants. Thus, if $w = f(x, y, z, u, v)$, we may have as many as five partial derivatives, w_x, w_y, w_z, w_u, and w_v.

EXAMPLE 3 Three resistors of resistances R_1, R_2, and R_3 connected in parallel produce a resistance R given by

$$\frac{1}{R} = \frac{1}{R_1} + \frac{1}{R_2} + \frac{1}{R_3}.$$

Find $\partial R/\partial R_2$.

Solution Treat R_1 and R_3 as constants and differentiate both sides of the equation with respect to R_2. Then

$$-\frac{1}{R^2}\frac{\partial R}{\partial R_2} = -\frac{1}{R_2^2}, \qquad \text{or} \qquad \frac{\partial R}{\partial R_2} = \left(\frac{R}{R_2}\right)^2. \ \square$$

EXAMPLE 4 Find

$$\frac{\partial}{\partial z}(xy)^{\sin z}.$$

Solution We treat xy as a constant and apply the single variable rule

$$\frac{d}{dz}(a)^u = a^u \ln a \frac{du}{dz}$$

with $a = xy$ and $u = \sin z$. The result is

$$\frac{\partial}{\partial z}(xy)^{\sin z} = (xy)^{\sin z} \ln (xy) \frac{d}{dz} \sin z = (xy)^{\sin z} \ln (xy) \cos z. \ \square$$

If $z = f(x, y)$ is continuous we can expect any change in z created by small changes in x and y to be small. If, in addition, f has continuous partial derivatives, we may estimate Δz by the approximation

$$\Delta z \approx f_x(x_0, y_0)\,\Delta x + f_y(x_0, y_0)\,\Delta y. \tag{3}$$

The theorem below establishes this fact and gives some information about the error involved. We call it the increment theorem for functions of two variables. As we shall see, it provides a basis for much of the later work in this chapter. Applications of the approximation in Eq. (3), and a useful estimate of the error involved, will be presented in Article 15.8.

THEOREM 2

Increment Theorem for Functions of Two Variables

Suppose that $z = f(x, y)$ is continuous and has partial derivatives throughout a region

$$R: \quad |x - x_0| < h, \qquad |y - y_0| < k$$

in the xy-plane. Suppose also that Δx and Δy are small enough for the point $(x_0 + \Delta x, y_0 + \Delta y)$ to lie in R. If f_x and f_y are continuous at (x_0, y_0), then the increment

$$\Delta z = f(x_0 + \Delta x, y_0 + \Delta y) - f(x_0, y_0) \tag{4}$$

can be written as

$$\Delta z = f_x(x_0, y_0)\,\Delta x + f_y(x_0, y_0)\,\Delta y + \varepsilon_1\,\Delta x + \varepsilon_2\,\Delta y, \tag{5}$$

where

$$\varepsilon_1, \varepsilon_2 \to 0 \qquad \text{as} \qquad \Delta x, \Delta y \to 0.$$

Theorem 2 is analagous to the result in Article 2.5 that the increment in a differentiable function $y = g(x)$ of a single variable that takes place when x changes from x_0 to $x_0 + \Delta x$ can be written as

$$\Delta y = g'(x_0)\,\Delta x + \varepsilon \cdot \Delta x, \tag{6}$$

where $\varepsilon \to 0$ as $\Delta x \to 0$. A proof of Theorem 2 may be found in Appendix 4.

Our first application of the increment theorem will be to develop chain rules for differentiating functions of two variables in the next article.

Functions of Three or More Variables

Analogous results hold for functions of any finite number of independent variables. For a function of three variables

$$w = f(x, y, z),$$

that is continuous and has partial derivatives f_x, f_y, f_z at and in some neighborhood of the point (x_0, y_0, z_0), and whose derivatives are continuous at the point, we have

$$\begin{aligned}\Delta w &= f(x_0 + \Delta x, y_0 + \Delta y, z_0 + \Delta z) - f(x_0, y_0, z_0)\\ &= f_x\,\Delta x + f_y\,\Delta y + f_z\,\Delta z + \varepsilon_1\,\Delta x + \varepsilon_2\,\Delta y + \varepsilon_3\,\Delta z, \end{aligned} \tag{7}$$

where

$$\varepsilon_1, \varepsilon_2, \varepsilon_3 \to 0 \quad \text{when} \quad \Delta x, \Delta y, \text{and } \Delta z \to 0.$$

The partial derivatives f_x, f_y, f_z in this formula are to be evaluated at the point (x_0, y_0, z_0).

PROBLEMS

Find $\partial f/\partial x$ and $\partial f/\partial y$ in Problems 1–24.

1. $f(x, y) = x + y$

2. $f(x, y) = x^2 + y^2$

3. $f(x, y) = ye^x$

4. $f(x, y) = e^x \cos y$

5. $f(x, y) = e^x \sin y$

6. $f(x, y) = \tan^{-1}(y/x)$

7. $f(x, y) = \ln\sqrt{x^2 + y^2}$

8. $f(x, y) = \cosh(y/x)$

9. $f(x, y) = x$

10. $f(x, y) = y$

11. $f(x, y) = 4$

12. $f(x, y) = \sqrt{9 - x^2 - y^2}$

13. $f(x, y) = x^2 - xy + y^2$

14. $f(x, y) = x/(x^2 + y^2)$

15. $f(x, y) = 1/(x + y)$

16. $f(x, y) = (x + 2)(y + 3)$

17. $f(x, y) = (x + 2)/(y + 3)$

18. $f(x, y) = e^{x \ln y}$

19. $f(x, y) = (x + y)/(xy - 1)$

20. $f(x, y) = 5xy - 7x^2 - y^2 + 3x - 6y + 2$

21. $f(x, y) = \sec(x + y)$

22. $f(x, y) = \tanh(2x + 5y)$

23. $f(x, y) = (xy)^e$

24. $f(x, y) = \ln|\sec xy + \tan xy|$

In Problems 25–36, find the partial derivatives of the given function with respect to each variable.

25. $f(x, y, z, w) = x^2e^{2y+3z}\cos(4w)$

26. $f(x, y, z) = z\sin^{-1}(y/x)$

27. $f(u, v, w) = \dfrac{u^2 - v^2}{v^2 + w^2}$

28. $f(r, \theta, z) = \dfrac{r(2 - \cos 2\theta)}{r^2 + z^2}$

29. $f(x, y, u, v) = \dfrac{x^2 + y^2}{u^2 + v^2}$

30. $f(x, y, r, s) = \sin 2x \cosh 3r + \sinh 3y \cos 4s$

31. $f(x, y, z) = xy + yz + zx$

32. $f(u, v, w) = (u^2 + v^2 + w^2)^{-1/2}$

33. $f(x, y, z) = 1 + y^2 + 2z^2$

34. $f(x, y, z) = (xy)^z$

35. $f(P, Q, R) = PQR$

36. $f(u, v, w, x) = \ln(uvwx)$

In Problems 37 and 38, A, B, C are the angles of a triangle and a, b, c are the respective opposite sides (Fig. 15.14).

37. Express A (explicitly or implicitly) as a function of a, b, c and calculate $\partial A/\partial a$ and $\partial A/\partial b$.

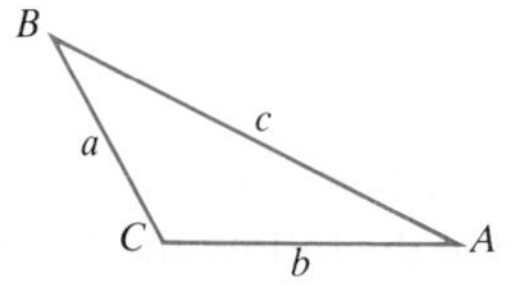

Figure 15.14

38. Express a (explicitly or implicitly) as a function of A, b, B and calculate $\partial a/\partial A$ and $\partial a/\partial B$.

In Problems 39–44, express the spherical coordinates ρ, ϕ, θ as functions of the cartesian coordinates x, y, z and calculate:

39. $\partial\rho/\partial x$ **40.** $\partial\phi/\partial z$ **41.** $\partial\theta/\partial y$

42. $\partial\theta/\partial z$ **43.** $\partial\phi/\partial x$ **44.** $\partial\theta/\partial x$

In Problems 45–47, let $\mathbf{R} = \mathbf{i}x + \mathbf{j}y + \mathbf{k}z$ be the vector from the origin to (x, y, z). Express x, y, z as functions of the spherical coordinates ρ, ϕ, θ, and calculate:

45. $\partial\mathbf{R}/\partial\rho$ **46.** $\partial\mathbf{R}/\partial\phi$ **47.** $\partial\mathbf{R}/\partial\theta$

48. Express the answers to Problems 45–47 in terms of the unit vectors $\mathbf{u}_\rho$, $\mathbf{u}_\phi$, $\mathbf{u}_\theta$ discussed in Miscellaneous Problem 25, Chapter 14.

15.4 Chain Rules

Chain Rules for Functions Defined Along Paths

When we are interested in the temperature $f(x, y, z)$ at points on a path

$$x = x(t), \qquad y = y(t), \qquad z = z(t)$$

in space (or in the pressure or density at points along a path through a gas or fluid) we may think of f as a function of the single variable t. At each t the temperature at the point $(x(t), y(t), z(t))$ on the path is the value of the composite $f(x(t), y(t), z(t))$. If we then wish to know the rate at which f changes with respect to t as we travel along the path, we have only to differentiate this composite with respect to t (provided, of course, df/dt exists).

Many times df/dt can be found by substituting $x(t)$, $y(t)$, and $z(t)$ into the formula for f and differentiating directly with respect to t. But often we work with functions whose formulas are complicated or for which formulas are not readily available. To find their derivatives we need chain rules that express them in terms of the derivatives of the functions being composed. These rules and their application are the subject of this article. As we shall see, they are like the chain rule from Chapter 2, with an additional term for each additional variable.

In Chapter 2, when we had a differentiable function $y = f(x)$ of a single variable x that was itself a differentiable function of t, we found that y was a differentiable function of t with

$$\frac{dy}{dt} = \frac{dy}{dx}\frac{dx}{dt}. \tag{1}$$

If we write $x = g(t)$, then $y = f(x) = f(g(t))$ and Eq. (1) takes the form

$$f'(t) = f'(x)g'(t)$$

or

$$f'(t) = f'(g(t))g'(t). \tag{2}$$

This shows how the derivatives in Eq. (1) are to be evaluated.

EXAMPLE 2 Express dw/dt as a function of t if $w = xy + z$ and $x = \cos t$, $y = \sin t$, $z = t$. (This derivative shows how w varies along a helix in its domain.)

Solution Starting with Eq. (7b) we have

$$\begin{aligned}\frac{dw}{dt} &= \frac{\partial w}{\partial x}\frac{dx}{dt} + \frac{\partial w}{\partial y}\frac{dy}{dt} + \frac{\partial w}{\partial z}\frac{dz}{dt} \\ &= y(-\sin t) + x(\cos t) + 1(1) \\ &= -\sin^2 t + \cos^2 t + 1 \\ &= 1 + \cos 2t. \ \square\end{aligned}$$

If the parameter t for a curve

$$x = x(t), \qquad y = y(t), \qquad z = z(t)$$

in the domain of a function $f(x, y, z)$ happens to be arc length, then the derivative df/dt may be interpreted as the rate at which f changes with respect to distance along the curve. We shall explore this extremely important interpretation in Article 15.6.

Chain Rules for Functions Defined on Surfaces

No essential complication is introduced by considering a surface (instead of a curve) passing through the domain of a function $w = f(x, y, z)$. For example, we might be interested in how the temperature $w = f(x, y, z)$ varies over the surface of some sphere in space. If the points of the sphere are given in terms of their latitude r and longitude s, then we can think of the temperature on the sphere as a function of these two independent variables. More generally, if $w = f(x, y, z)$, and if

$$x = x(r, s), \qquad y = y(r, s), \qquad z = z(r, s) \tag{8}$$

are any functions whatsoever, then the composite

$$w = f(x(r, s), y(r, s), z(r, s)), \tag{9}$$

when defined, is a function of r and s. If x, y, z, and f also have continuous partial derivatives, then the partial derivatives of w with respect to r and s exist and are given by the following equations:

$$\frac{\partial w}{\partial r} = \frac{\partial f}{\partial x}\frac{\partial x}{\partial r} + \frac{\partial f}{\partial y}\frac{\partial y}{\partial r} + \frac{\partial f}{\partial z}\frac{\partial z}{\partial r}, \tag{10a}$$

$$\frac{\partial w}{\partial s} = \frac{\partial f}{\partial x}\frac{\partial x}{\partial s} + \frac{\partial f}{\partial y}\frac{\partial y}{\partial s} + \frac{\partial f}{\partial z}\frac{\partial z}{\partial s}. \tag{10b}$$

Equation (10a) can be derived from Eq. (7b) by holding s fixed and taking $r = t$. Similarly, Eq. (10b) can be obtained by holding r fixed and taking $s = t$. Or we can go back to increments in each case and apply Eq. (7) of Article 15.3.

The tree diagrams for Eqs. (10a) and (10b) are shown in Fig. 15.17.

Chain Rules

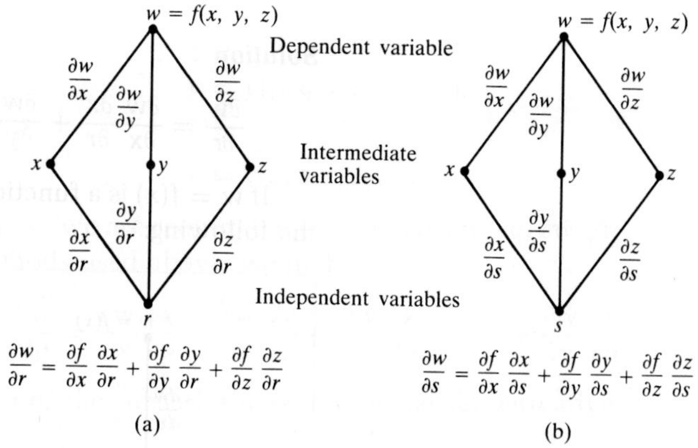

Figure 15.17

EXAMPLE 3 Find $\partial w/\partial r$ and $\partial w/\partial s$ as functions of r and s if $w = f(x, y, z) = x + 2y + z^2$, and $x = r/s$, $y = r^2 + e^s$, $z = 2r$.

Solution From Eq. (10a),

$$\frac{\partial w}{\partial r} = \frac{\partial f}{\partial x}\frac{\partial x}{\partial r} + \frac{\partial f}{\partial y}\frac{\partial y}{\partial r} + \frac{\partial f}{\partial z}\frac{\partial z}{\partial r} = 1\left(\frac{1}{s}\right) + 2(2r) + 2z(2) = \frac{1}{s} + 12r.$$

Similarly,

$$\frac{\partial w}{\partial s} = \frac{\partial f}{\partial x}\frac{\partial x}{\partial s} + \frac{\partial f}{\partial y}\frac{\partial y}{\partial s} + \frac{\partial f}{\partial z}\frac{\partial z}{\partial s} = 1\left(-\frac{r}{s^2}\right) + 2(e^s) + 2z(0)$$

$$= -\frac{r}{s^2} + 2e^s. \square$$

The chain rules in Eqs. (10a) and (10b) both have useful special cases, depending on the number of variables w has. For instance, if $w = f(x, y)$ is a function of only two variables, x and y, then Eq. (10a) reduces to the following:

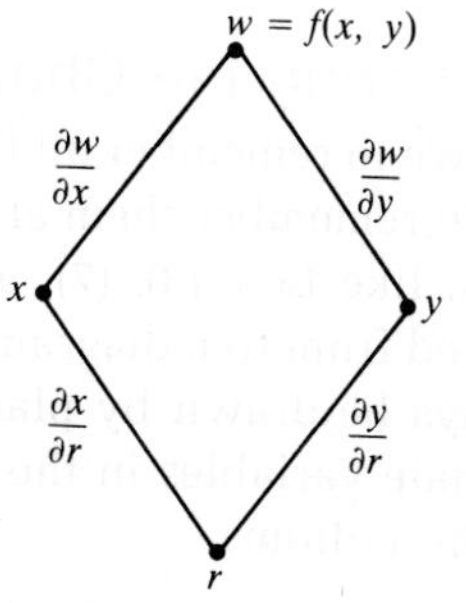

$$\frac{\partial w}{\partial r} = \frac{\partial f}{\partial x}\frac{\partial x}{\partial r} + \frac{\partial f}{\partial y}\frac{\partial y}{\partial r}. \qquad (11)$$

Figure 15.18

Equation (11) is Eq. (10a) without the z-term.

9. a) $\dfrac{\partial w}{\partial r}$

b) $\dfrac{\partial w}{\partial s}$; if $w = f(x, y, z, t)$, $x = x(r, s)$, $y = y(r, s)$,

$z = z(r, s)$, $t = t(r, s)$

10. a) $\dfrac{\partial w}{\partial u}$

b) $\dfrac{\partial w}{\partial v}$; if $w = f(x, y)$, $x = x(u)$, $y = y(v)$

11. Find $\partial z/\partial v$ if $z = x^2 + 2xy$, $x = u \cos v$, $y = u \sin v$. (Answer in x and y.)

12. Find $\partial w/\partial x$ if $w = uv + \ln v$, $u = x + y^2$, $v = e^x \cos y$. (Answer in u and v.)

13. Find $\partial z/\partial u$ if $z = e^x + xy^2$, $x = u + v$, $y = e^{u+v}$. (Answer in x and y.)

14. Find $\partial w/\partial u$ if $w = x^2 + y^2$, $x = u - v$, $y = ve^{2u}$. (Answer in x and y.)

15. Find $\partial z/\partial u$ when $u = 0$, $v = 1$ if $z = \sin xy + x \sin y$, $x = u^2 + v^2$, $y = uv$.

16. Find $\partial w/\partial v$ when $u = 0$, $v = 0$ if $w = (x^2 + y - 2)^4 + (x - y + 2)^3$, $x = u - 2v + 1$, $y = 2u + v - 2$.

17. If $w = \sqrt{x^2 + y^2 + z^2}$, $x = e^r \cos s$, $y = e^r \sin s$, $z = e^s$, find $\partial w/\partial r$ and $\partial w/\partial s$ by the chain rule, and check your answer by using a different method.

18. If $w = \ln (x^2 + y^2 + 2z)$, $x = r + s$, $y = r - s$, $z = 2rs$, find $\partial w/\partial r$ and $\partial w/\partial s$ by the chain rule, and check your answer by using a different method.

19. Find $\partial w/\partial x$ at the point $(x, y, z) = (1, 1, 1)$ if $w = \cos uv$, $u = xyz$, $v = \pi/(4(x^2 + y^2))$.

20. The dimensions of a rectangular solid at a given instant t_0 are $L(t_0) = 13$ cm, $W(t_0) = 9$ cm, and $H(t_0) = 5$ cm. If L and H are increasing at 2 cm/s and W is decreasing at 4 cm/s, find the rates of change of the volume V and surface area S at time t_0. Are V and S increasing, decreasing, or neither?

21. Let $z = f(t)$, $t = (x + y)/xy$. Show that

$$x^2 \frac{\partial z}{\partial x} = y^2 \frac{\partial z}{\partial y}.$$

22. Show that $w = e^{2x+2ax} + \tan^{-1}(y + ax)$ is a solution of the partial differential equation from Example 5,

$$\frac{\partial w}{\partial x} - a\frac{\partial w}{\partial y} = 0.$$

23. If a and b are constants and $w = (ax + by)^3 + \tanh (ax + by) + \cos (ax + by)$, show that

$$a\frac{\partial w}{\partial y} = b\frac{\partial w}{\partial x}.$$

24. If a and b are constants and $w = f(ax + by)$ is a differentiable function of $u = ax + by$, show that

$$a\frac{\partial w}{\partial y} = b\frac{\partial w}{\partial x}.$$

(*Hint*: Apply the chain rule with u as the only independent variable in the first set of variables.)

25. If $w = f[xy/(x^2 + y^2)]$ is a differentiable function of $u = xy/(x^2 + y^2)$, show that

$$x(\partial w/\partial x) + y(\partial w/\partial y) = 0.$$

(See the hint for Problem 24.)

26. If $w = f(x + y, x - y)$ has continuous partial derivatives with respect to $u = x + y$, $v = x - y$, show that

$$\frac{\partial w}{\partial x}\frac{\partial w}{\partial y} = \left(\frac{\partial f}{\partial u}\right)^2 - \left(\frac{\partial f}{\partial v}\right)^2.$$

27. If we substitute polar coordinates $x = r \cos \theta$ and $y = r \sin \theta$ in a function $w = f(x, y)$ that has continuous partial derivatives, show that

$$\frac{\partial w}{\partial r} = f_x \cos \theta + f_y \sin \theta,$$

and

$$\frac{1}{r}\frac{\partial w}{\partial \theta} = -f_x \sin \theta + f_y \cos \theta.$$

28. Using determinants, solve the equations given in Problem 27 for f_x and f_y in terms of $(\partial w/\partial r)$ and $(\partial w/\partial \theta)$.

29. In connection with Problem 27, show that

$$\left(\frac{\partial w}{\partial r}\right)^2 + \frac{1}{r^2}\left(\frac{\partial w}{\partial \theta}\right)^2 = f_x^2 + f_y^2.$$

30. Find $\partial F/\partial x$ if $F(x, y) = \int_0^{x^2 y} e^{t^2} dt$.

15.5 Nonindependent Variables

In finding partial derivatives of functions like $w = f(x, y)$ so far we have assumed x and y to be independent. But in many applications this is not the case. For example, the internal energy U of a gas may be expressed in terms of its pressure p, volume v, and temperature T:

$$U = f(p, v, T).$$

If the gas is an ideal gas, however, then p, v, and T obey the ideal gas law

$$pv = nRT \qquad (n,\ R \text{ constant}),$$

and are therefore not independent. Finding partial derivatives in situations like these can be complicated. But it is better to face the complication now than to meet it for the first time while you are also trying to learn physics, engineering, economics, or whatever.

EXAMPLE 1 Find $\partial w/\partial x$ if

$$w = x^2 + y^2 + z^2 \qquad \text{and} \qquad z = x^2 + y^2.$$

1. *Solution with* x *and* y *as independent variables:* With $z = x^2 + y^2$ we have

$$w = x^2 + y^2 + z^2 = x^2 + y^2 + (x^2 + y^2)^2$$
$$= x^2 + y^2 + x^4 + 2x^2y^2 + y^4,$$
$$\frac{\partial w}{\partial x} = 2x + 4x^3 + 4xy^2. \tag{1}$$

2. *Solution with* x *and* z *as independent variables:* With $y^2 = z - x^2$, we have

$$w = x^2 + y^2 + z^2 = x^2 + (z - x^2) + z^2 = z + z^2,$$
$$\frac{\partial w}{\partial x} = 0. \tag{2}$$

The answers in the two solutions in this example are genuinely different. We cannot change one to the other by using the relation $z = x^2 + y^2$. Which is the real $\partial w/\partial x$?

There is no single right answer to this question because the original problem is not properly stated. When the variables are not all independent (as is the case with x, y, and z when $z = x^2 + y^2$) the meaning of an expression like $\partial w/\partial x$ depends on which of the variables are assumed to be the independent ones.

To see why, let us interpret the answers in (1) and (2) geometrically. The function $w = x^2 + y^2 + z^2$ measures the square of the distance from the point (x, y, z) to the origin. The condition $z = x^2 + y^2$ says that the point (x, y, z) lies on the surface of the paraboloid of revolution shown in Fig. 15.20. What does it mean to calculate $\partial w/\partial x$ at a point $P(x, y, z)$ that can move only on this surface? What is the value of $\partial w/\partial x$ when the coordinates of P are, say, $(1, 0, 1)$?

If we take x and y to be independent, then we find $\partial w/\partial x$ by holding y fixed (at $y = 0$ in this case) and letting x vary. This means P moves along the parabola $z = x^2$ in the xz-plane. As P moves on this parabola, w, which is the square of the distance from P to the origin, changes. We calculate $\partial w/\partial x$ in this case (our first solution above) to be

$$\frac{\partial w}{\partial x} = 2x + 4x^3 + 4xy^2.$$

At the point $P(1, 0, 1)$ the value of this derivative is

$$\frac{\partial w}{\partial x} = 2 + 4 + 0 = 6.$$

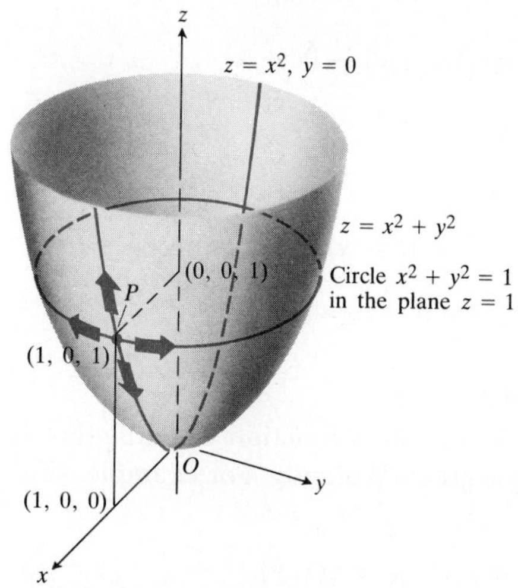

15.20 (a) As x changes, with $y = 0$, P moves up or down the surface on the parabola $z = x^2$ in the xz-plane. (b) As x changes, with $z = 1$, P moves on the circle $x^2 + y^2 = 1$, $z = 1$.

If we take x and z to be independent, then we find $\partial w/\partial x$ by holding z fixed while x varies. Since the z-coordinate of P is 1, varying x moves P along a circle in the plane $z = 1$. As P moves along this circle, its distance from the origin remains constant, and w, being the square of this distance, does not change. That is,

$$\partial w/\partial x = 0,$$

as we found in our second solution.

In short, the value of $\partial w/\partial x$ changes according to which variables we take to be independent. We are measuring rates of change in w as P moves along different paths on the surface $z = x^2 + y^2$. □

To show what variables are assumed to be independent in calculating a derivative, we can use the following notation:

$$\left(\frac{\partial w}{\partial x}\right)_y \qquad \partial w/\partial x \text{ with } x \text{ and } y \text{ independent,} \tag{3}$$

$$\left(\frac{\partial w}{\partial x}\right)_z \qquad \partial w/\partial x \text{ with } x \text{ and } z \text{ independent,} \tag{4}$$

$$\left(\frac{\partial f}{\partial y}\right)_{x,t} \qquad \partial f/\partial y \text{ with } y,\ x, \text{ and } t \text{ independent.} \tag{5}$$

EXAMPLE 2 If

$$w = x^2 + y - z + \sin t \qquad \text{and} \qquad x + y = t,$$

find

a) $\left(\frac{\partial w}{\partial x}\right)_{y,z}$ b) $\left(\frac{\partial w}{\partial x}\right)_{t,z}$.

Solution a) With x, y, z independent, we have

$$t = x + y, \qquad w = x^2 + y - z + \sin(x + y),$$

$$\left(\frac{\partial w}{\partial x}\right)_{y,z} = 2x + 0 - 0 + \cos(x + y)\frac{\partial}{\partial x}(x + y)$$

$$= 2x + \cos(x + y).$$

b) With x, t, z independent, we have

$$y = t - x, \qquad w = x^2 + (t - x) - z + \sin t,$$

$$\left(\frac{\partial w}{\partial x}\right)_{t,z} = 2x - 1 - 0 + 0 = 2x - 1. \ \square$$

Functions Not Given Explicitly

In applications we often have to differentiate a function when neither it nor the relation among its variables is given explicitly. For example, suppose that

$$w = f(x, y, z) \qquad \text{and} \qquad z = g(x, y), \tag{6}$$

and that we wish to express

$$\left(\frac{\partial w}{\partial x}\right)_y$$

in terms of the derivatives of f and g. Because $z = g(x, y)$, we may regard all three of x, y, and z as functions of the two independent variables x and y. We may then apply Eq. (10a) of Article 15.4,

$$\frac{\partial w}{\partial r} = \frac{\partial f}{\partial x}\frac{\partial x}{\partial r} + \frac{\partial f}{\partial y}\frac{\partial y}{\partial r} + \frac{\partial f}{\partial z}\frac{\partial z}{\partial r},$$

with $r = x$ to find

$$\left(\frac{\partial w}{\partial x}\right)_y = \frac{\partial f}{\partial x}\frac{\partial x}{\partial x} + \frac{\partial f}{\partial y}\frac{\partial y}{\partial x} + \frac{\partial f}{\partial z}\frac{\partial z}{\partial x}, \tag{7}$$

Equals 1   Equals 0 because y is independent of x

$$\left(\frac{\partial w}{\partial x}\right)_y = \frac{\partial f}{\partial x} + \frac{\partial f}{\partial z}\frac{\partial z}{\partial x}. \tag{8}$$

The partial derivatives of $w = f(x, y, z)$ on the right-hand side of Eqs. (7) and (8) are to be calculated formally, as if all three of x, y, and z were independent variables.

EXAMPLE 3 Verify Eq. (8) for

$$w = f(x, y, z) = x^2 + y^2 + z^2, \qquad z = g(x, y) = x^2 + y^2$$

by calculating $\left(\frac{\partial w}{\partial x}\right)_y$.

Solution These are the functions of Example 1, so we should get the answer in the first solution there,

$$\left(\frac{\partial w}{\partial x}\right)_y = 2x + 4x^3 + 4xy^2. \tag{9}$$

To see that this is the case, we first calculate

$$\frac{\partial f}{\partial x} = 2x, \qquad \frac{\partial f}{\partial z} = 2z, \qquad \frac{\partial z}{\partial x} = 2x$$

formally (without regard to dependence). We then evaluate the right-hand side of Eq. (8). This gives

$$\frac{\partial f}{dx} + \frac{\partial f}{\partial z}\frac{\partial z}{\partial x} = 2x + 2z \cdot 2x = 2x + 4(x^2 + y^2)x = 2x + 4x^3 + 4xy^2,$$

in agreement with Eq. (9). □

EXAMPLE 4 Suppose that the equation $f(x, y, z) = 0$ determines z as a differentiable function of the independent variables x and y and that $\partial f/\partial z \neq 0$. Show that

$$\left(\frac{\partial z}{\partial x}\right)_y = -\frac{\partial f/\partial x}{\partial f/\partial z}.$$

Solution We regard all three of x, y, and z to be functions of the two independent variables x and y and calculate

$$\frac{\partial}{\partial x} f(x, y, z) = \frac{\partial}{\partial x}(0),$$

$$\frac{\partial f}{\partial x}\frac{\partial x}{\partial x} + \frac{\partial f}{\partial y}\frac{\partial y}{\partial x} + \frac{\partial f}{\partial z}\frac{\partial z}{\partial x} = 0,$$

$$\frac{\partial f}{\partial x}\cdot 1 + \frac{\partial f}{\partial y}\cdot 0 + \frac{\partial f}{\partial z}\frac{\partial z}{\partial x} = 0,$$

$$\frac{\partial f}{\partial x} + \frac{\partial f}{\partial z}\frac{\partial z}{\partial x} = 0,$$

$$\frac{\partial z}{\partial x} = -\frac{\partial f/\partial x}{\partial f/\partial z}. \quad \square$$

PROBLEMS

1. If $w = x^2 + y^2 + z^2$ and $z = x^2 + y^2$, find

a) $\left(\frac{\partial w}{\partial y}\right)_z$, b) $\left(\frac{\partial w}{\partial z}\right)_x$, c) $\left(\frac{\partial w}{\partial z}\right)_y$.

2. If $w = x^2 + y - z + \sin t$ and $x + y = t$, find

a) $\left(\frac{\partial w}{\partial y}\right)_{x,t}$, b) $\left(\frac{\partial w}{\partial y}\right)_{z,t}$, c) $\left(\frac{\partial w}{\partial z}\right)_{x,y}$,

d) $\left(\frac{\partial w}{\partial z}\right)_{y,t}$, e) $\left(\frac{\partial w}{\partial t}\right)_{x,z}$, f) $\left(\frac{\partial w}{\partial t}\right)_{y,z}$.

3. Let $U = f(p, v, T)$ be the internal energy of a gas that obeys the ideal gas law $pv = nRT$ (n and R constant). Find

a) $\left(\frac{\partial U}{\partial p}\right)_v$, b) $\left(\frac{\partial U}{\partial T}\right)_v$.

4. Let $z = f(u, v)$ where $u = x$, $v = \ln(x^2 + 1)$. Express $\partial z/\partial x$ in terms of x, f_x, and f_y.

5. Let $w = f(r, \theta)$, where $r = \sqrt{x^2 + y^2}$ and $\theta = \tan^{-1}(y/x)$. Find $\partial w/\partial x$ and $\partial w/\partial y$ and express the answers in terms of r and θ.

6. Let $z = f(u, v)$, where $u = ax + by$ and $v = ax - by$. Find formulas for $\partial z/\partial x$ and $\partial z/\partial y$.

7. If $z = x + f(u)$, where $u = xy$, show that

$$x\frac{\partial z}{\partial x} - y\frac{\partial z}{\partial y} = x.$$

8. Suppose that the equation $g(x, y, z) = 0$ determines z as a differentiable function of the independent variables x and y and that $g_z \neq 0$. Show that

$$\left(\frac{\partial z}{\partial y}\right)_x = -\frac{\partial g/\partial y}{\partial g/\partial z}.$$

9. Establish the fact, widely used in hydrodynamics, that if $f(x, y, z) = 0$, then

$$\left(\frac{\partial x}{\partial y}\right)_z\left(\frac{\partial y}{\partial z}\right)_x\left(\frac{\partial z}{\partial x}\right)_y = -1.$$

(*Hint:* Express all the derivatives in terms of the formal partial derivatives $\partial f/\partial x$, $\partial f/\partial y$, and $\partial f/\partial z$.)

10. Let $w = f(r, s, t)$, with $r = x - y$, $s = y - z$, $t = z - x$. Show that

$$w_x + w_y + w_z = 0.$$

15.6

Gradients, Directional Derivatives, and Tangent Planes

We know from Article 15.4 that if the partial derivatives of f are continuous, then the rate at which the values of f change with respect to t along a differentiable curve

$$x = x(t), \qquad y = y(t), \qquad z = z(t)$$

in its domain may be calculated from the formula

$$\frac{df}{dt} = \frac{\partial f}{\partial x}\frac{dx}{dt} + \frac{\partial f}{\partial y}\frac{dy}{dt} + \frac{\partial f}{\partial z}\frac{dz}{dt}. \tag{1}$$

To use this formula at any particular point

$$P_0(x_0, y_0, z_0) = P_0(x(t_0), y(t_0), z(t_0)),$$

we evaluate the partial derivatives of f at P_0 and the derivatives of $x(t)$, $y(t)$, and $z(t)$ at t_0.

The derivative df/dt in Eq. (1) is interpreted as the rate of change of f at P_0 with respect to increasing t and therefore depends, among other things, on the direction of motion along the curve. This observation is particularly important when the curve is a straight line through P_0 and the parameter t is arc length (distance along the line) measured from P_0 in the direction of a given unit vector **u.** For then df/dt is the rate of change of f with respect to distance in its domain in the direction of **u.** Thus, by varying **u,** we can find the rates at which f changes with respect to distance as we move through P_0 in different directions. These "directional derivatives" are extremely useful in science and engineering (as well as in mathematics), and one of the goals of this article is to develop a simple formula for calculating them.

Calculating Directional Derivatives

Let

$$\mathbf{u} = u_1\mathbf{i} + u_2\mathbf{j} + u_3\mathbf{k}$$

be a unit vector, and let L be the line through a point P_0 whose equations are

$$x = x_0 + tu_1, \qquad y = y_0 + tu_2, \qquad z = z_0 + tu_3. \tag{2}$$

Then, motion along L in the direction of increasing t is motion in the direction of **u.** Also, $|t|$ measures distance along the line from $P_0(x_0, y_0, z_0)$, as we can see from the following calculation:

$$\begin{aligned}
&\text{Distance from } P_0 \text{ to any point } P(x, y, z) \text{ on this line}\\
&\quad = \sqrt{(x - x_0)^2 + (y - y_0)^2 + (z - z_0)^2}\\
&\quad = \sqrt{(tu_1)^2 + (tu_2)^2 + (tu_3)^2}\\
&\quad = |t|\sqrt{u_1^2 + u_2^2 + u_3^2}\\
&\quad = |t| \cdot 1 \qquad (\mathbf{u} \text{ is a unit vector})\\
&\quad = |t|.
\end{aligned}$$

Substituting the derivatives

$$\frac{dx}{dt} = \frac{d}{dt}(x_0 + tu_1) = u_1,$$

$$\frac{dy}{dt} = \frac{d}{dt}(y_0 + tu_2) = u_2,$$

$$\frac{dz}{dt} = \frac{d}{dt}(z_0 + tu_3) = u_3,$$

in Eq. (1) gives

$$\frac{df}{dt} = \frac{\partial f}{\partial x}u_1 + \frac{\partial f}{\partial y}u_2 + \frac{\partial f}{\partial z}u_3. \tag{3}$$

The expression on the right in this equation is the dot product of **u** and the vector

$$\nabla f = \frac{\partial f}{\partial x}\mathbf{i} + \frac{\partial f}{\partial y}\mathbf{j} + \frac{\partial f}{\partial z}\mathbf{k}.$$

This vector is called the *gradient vector* of f at the point P_0. It is customary to picture it as a vector in the domain of f. Its components are calculated by evaluating the three partial derivatives of f at (x_0, y_0, z_0). The derivative on the left-hand side of Eq. (3) is called "the derivative of f at the point P_0 in the direction of **u**." It is often denoted by

$$(D_{\mathbf{u}}f)_{P_0}.$$

DEFINITION

Definitions of Directional Derivative and Gradient

1. If the partial derivatives of $f(x, y, z)$ are defined at $P_0(x_0, y_0, z_0)$, then the *gradient* of f at P_0 is the vector

$$\nabla f = \frac{\partial f}{\partial x}\mathbf{i} + \frac{\partial f}{\partial y}\mathbf{j} + \frac{\partial f}{\partial z}\mathbf{k} \tag{4}$$

obtained by evaluating the partial derivatives of f at P_0.

2. If $f(x, y, z)$ has continuous partial derivatives at $P_0(x_0, y_0, z_0)$, and **u** is a unit vector, then the *derivative of f at P_0 in the direction of* **u** is the number

$$(D_{\mathbf{u}}f)_{P_0} = (\nabla f)_{P_0} \cdot \mathbf{u}, \tag{5}$$

which is the scalar product of **u** and the gradient of f at P_0.

Another notation in use for the gradient of f is

$$\text{grad } f,$$

read the way it is written. The symbol ∇f may be read "grad f" as well as "gradient of f" or "del f."

Another common notation for the directional derivative is

$$\left(\frac{df}{ds}\right)_{\mathbf{u},P_0}.$$

EXAMPLE 1 Find the derivative of

$$f(x, y, z) = x^3 - xy^2 - z$$

at $P_0(1, 1, 0)$ in the direction of the vector $\mathbf{A} = 2\mathbf{i} - 3\mathbf{j} + 6\mathbf{k}$.

Solution The direction of **A** is obtained by dividing **A** by its length:

$$|\mathbf{A}| = \sqrt{(2)^2 + (-3)^2 + (6)^2} = \sqrt{49} = 7,$$

$$\mathbf{u} = \frac{\mathbf{A}}{|\mathbf{A}|} = \frac{2}{7}\mathbf{i} - \frac{3}{7}\mathbf{j} + \frac{6}{7}\mathbf{k}.$$

The partial derivatives of f at P_0 are

$$\begin{aligned} f_x &= 3x^2 - y^2|_{(1,1,0)} = 2, \\ f_y &= -2xy|_{(1,1,0)} = -2, \\ f_z &= -1|_{(1,1,0)} = -1. \end{aligned}$$

The gradient of f at P_0 is

$$\nabla f|_{(1,1,0)} = 2\mathbf{i} - 2\mathbf{j} - \mathbf{k}.$$

The derivative of f at P_0 in the direction $\mathbf{A}$ is therefore

$$\begin{aligned} (D_{\mathbf{u}}f)|_{(1,1,0)} &= \nabla f|_{(1,1,0)} \cdot \mathbf{u} \\ &= (2\mathbf{i} - 2\mathbf{j} - \mathbf{k}) \cdot \left(\frac{2}{7}\mathbf{i} - \frac{3}{7}\mathbf{j} + \frac{6}{7}\mathbf{k}\right) \\ &= \frac{4}{7} + \frac{6}{7} - \frac{6}{7} \\ &= \frac{4}{7}. \ \square \end{aligned}$$

EXAMPLE 2 Estimate how much

$$f(x, y, z) = xe^y + yz$$

will change if the point $P(x, y, z)$ is moved from $P_0(2, 0, 0)$ straight toward $P_1(4, 1, -2)$ a distance of $\Delta s = 0.1$ units.

Solution We first find the derivative of f at P_0 in the direction of the vector

$$\overrightarrow{P_0P_1} = 2\mathbf{i} + \mathbf{j} - 2\mathbf{k}.$$

The direction of this vector is

$$\mathbf{u} = \frac{\overrightarrow{P_0P_1}}{|\overrightarrow{P_0P_1}|} = \frac{\overrightarrow{P_0P_1}}{3} = \frac{2}{3}\mathbf{i} + \frac{1}{3}\mathbf{j} - \frac{2}{3}\mathbf{k}.$$

The gradient of f at P_0 is

$$\begin{aligned} \nabla f|_{(2,0,0)} &= (e^y\mathbf{i} + (xe^y + z)\mathbf{j} + y\mathbf{k})|_{(2,0,0)} \\ &= \mathbf{i} + 2\mathbf{j}. \end{aligned}$$

Therefore,

$$(D_{\mathbf{u}}f)_{P_0} = (\mathbf{i} + 2\mathbf{j}) \cdot \left(\frac{2}{3}\mathbf{i} + \frac{1}{3}\mathbf{j} - \frac{2}{3}\mathbf{k}\right) = \frac{2}{3} + \frac{2}{3} = \frac{4}{3}.$$

The change Δf in f that results from moving $\Delta s = 0.1$ units away from P_0 in the direction of $\mathbf{u}$ is approximately the derivative of f in this direction times Δs:

$$\Delta f \approx (D_{\mathbf{u}}f)\,\Delta s = \left(\frac{4}{3}\right)(0.1) \approx 0.13. \ \square$$

If we write the directional derivative in the form

$$D_{\mathbf{u}}f = \nabla f \cdot \mathbf{u} = |\nabla f||\mathbf{u}| \cos\theta = |\nabla f| \cos\theta, \tag{6}$$

the following facts come to light.

Properties of the Directional Derivative

$$D_{\mathbf{u}}f = \nabla f \cdot \mathbf{u} = |\nabla f| \cos \theta$$

1. The directional derivative has its largest positive value when $\cos \theta = 1$, or when $\mathbf{u}$ is the direction of the gradient. That is, f increases most rapidly in its domain in the direction of ∇f. The derivative in this direction is

$$D_{\mathbf{u}}f = |\nabla f| \cos (0) = |\nabla f|.$$

2. Similarly, f decreases most rapidly in the direction of $-\nabla f$. The derivative in this direction is

$$D_{\mathbf{u}}f = |\nabla f| \cos (\pi) = -|\nabla f|.$$

3. $D_{-\mathbf{u}}f = \nabla f \cdot (-\mathbf{u}) = -\nabla f \cdot \mathbf{u} = -D_{\mathbf{u}}f.$ (7)
4. The relationships of the partial derivatives of f to the directional derivative are

$$D_{\mathbf{i}}f = \nabla f \cdot \mathbf{i} = f_x, \qquad D_{\mathbf{j}}f = \nabla f \cdot \mathbf{j} = f_y, \qquad D_{\mathbf{k}}f = \nabla f \cdot \mathbf{k} = f_z.$$

Thus,

f_x = derivative of f in the $\mathbf{i}$ direction,
f_y = derivative of f in the $\mathbf{j}$ direction,
f_z = derivative of f in the $\mathbf{k}$ direction.

Combining these results with Eq. (7) gives

$$D_{-\mathbf{i}}f = -f_x, \qquad D_{-\mathbf{j}}f = -f_y, \qquad D_{-\mathbf{k}}f = -f_z.$$

5. Any direction $\mathbf{u}$ normal (perpendicular) to the gradient is a direction of zero change in f because

$$D_{\mathbf{u}}f = |\nabla f| \cos (\pi/2) = |\nabla f| \cdot 0 = 0.$$

For functions of two variables, we get results much like the ones for functions of three variables. The two-variable formulas are obtained by dropping the z-terms from the three-variable formulas. Thus, for a function $f(x, y)$, and a unit vector $\mathbf{u} = u_1\mathbf{i} + u_2\mathbf{j}$, we have the following formulas:

$$\nabla f = \frac{\partial f}{\partial x}\mathbf{i} + \frac{\partial f}{\partial y}\mathbf{j}, \tag{8}$$

$$D_{\mathbf{u}}f = \nabla f \cdot \mathbf{u} = \frac{\partial f}{\partial x}u_1 + \frac{\partial f}{\partial y}u_2. \tag{9}$$

EXAMPLE 3 a) Find the derivative of

$$f(x, y) = 100 - x^2 - y^2$$

at the point $P_0(3, 4)$ in the direction of the unit vector $\mathbf{u} = u_1\mathbf{i} + u_2\mathbf{j}$. (b) In what direction in its domain (the xy-plane) is f increasing most rapidly at P_0? What is the derivative of f in this direction? (c) Identify the directions in which the derivative of f is zero.

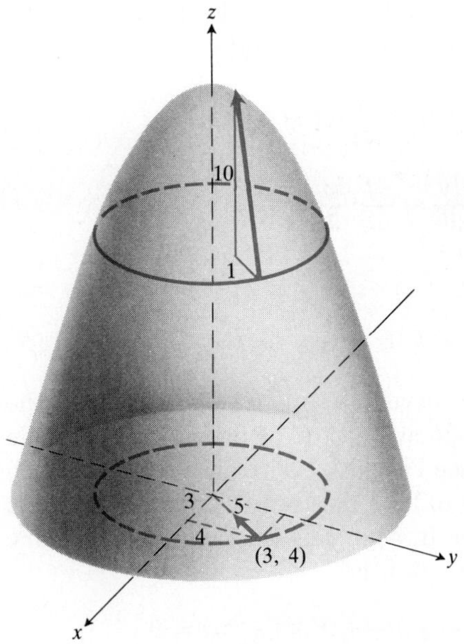

15.21 At (3, 4) the change in $f(x, y) = 100 - x^2 - y^2$ is greatest in the direction toward the origin. This corresponds to the direction of steepest ascent on the surface $z = 100 - x^2 - y^2$.

Solution a) We have

$$f(x, y) = 100 - x^2 - y^2,$$
$$f_x(3, 4) = -2x\,|_{(3,4)} = -6,$$
$$f_y(3, 4) = -2y\,|_{(3,4)} = -8,$$

and

$$D_{\mathbf{u}} f = \nabla f \cdot \mathbf{u} = (-6\mathbf{i} - 8\mathbf{j}) \cdot \mathbf{u} = -6u_1 - 8u_2.$$

b) The function increases most rapidly in the direction of the gradient. Since

$$|\nabla f| = \sqrt{(-6)^2 + (-8)^2} = \sqrt{36 + 64} = 10,$$

this direction is

$$\frac{\nabla f}{|\nabla f|} = -\frac{6}{10}\mathbf{i} - \frac{8}{10}\mathbf{j} = -\frac{3}{5}\mathbf{i} - \frac{4}{5}\mathbf{j}.$$

See Fig. 15.21. The derivative in this direction is $|\nabla f| = 10$.

c) The derivative of f is zero in the directions perpendicular to ∇f. We can obtain one of these directions by interchanging the components of

$$\frac{\nabla f}{|\nabla f|} = -\frac{3}{5}\mathbf{i} - \frac{4}{5}\mathbf{j}$$

and changing the sign of the new first component (see Example 7, Article 13.1). The result is

$$\mathbf{n} = \frac{4}{5}\mathbf{i} - \frac{3}{5}\mathbf{j}.$$

As a check, we can calculate $\nabla f \cdot \mathbf{n}$ to see that it is zero:

$$\nabla f \cdot \mathbf{n} = (-6\mathbf{i} - 8\mathbf{j})\left(\frac{4}{5}\mathbf{i} - \frac{3}{5}\mathbf{j}\right)$$
$$= -\frac{24}{5} + \frac{24}{5} = 0.$$

The other direction of zero change in f in the xy-plane is

$$-\mathbf{n} = -\frac{4}{5}\mathbf{i} + \frac{3}{5}\mathbf{j}. \ \square$$

The Geometry of Gradients and Level Surfaces

Suppose that $f(x, y, z)$ has continuous partial derivatives, that the surface

$$S: \quad f(x, y, z) = c$$

is one of the level surfaces of f, and that

$$x = x(t), \qquad y = y(t), \qquad z = z(t)$$

is a differentiable curve on S through a point P_0 on S. Then

$$f(x(t), y(t), z(t)) = c$$

for every value of t. If we take the derivative of both sides of this equation with respect to t and apply the chain rule on the left-hand side, we get

$$\frac{d}{dt}f(x(t), y(t), z(t)) = \frac{d}{dt}(c) = 0,$$

$$\frac{\partial f}{\partial x}\frac{dx}{dt} + \frac{\partial f}{\partial y}\frac{dy}{dt} + \frac{\partial f}{\partial z}\frac{dz}{dt} = 0,$$

or

$$\nabla f \cdot \mathbf{v} = 0, \tag{10}$$

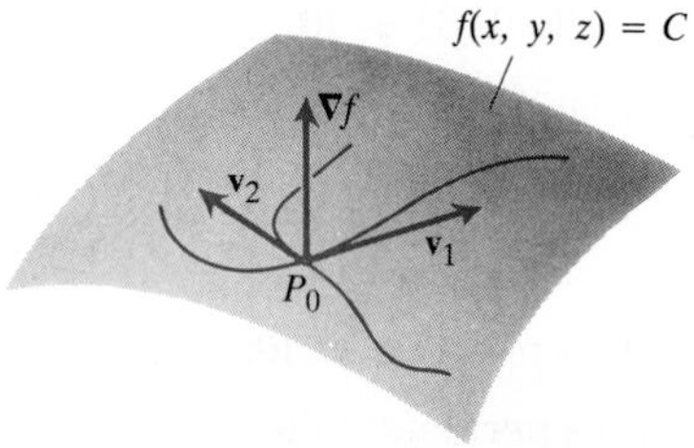

15.22 ∇f is perpendicular to the velocity vector of every differentiable curve in the surface through P_0. The velocity vectors at P_0 therefore lie in a common plane.

where $\mathbf{v}$ is the velocity vector of the curve. We conclude from this that the gradient is perpendicular to the velocity vector of every differentiable curve on S that passes through P_0 (see Fig. 15.22). Therefore, ∇f is perpendicular to the tangent line of every differentiable curve on S that passes through P_0. These lines therefore lie in a single plane, namely the plane through P_0 normal to ∇f, whose equation is

$$f_x(P_0)(x - x_0) + f_y(P_0)(y - y_0) + f_z(P_0)(z - z_0) = 0.$$

We define this plane to be the plane tangent to S at P_0. The line through P_0 perpendicular to this plane and parallel to ∇f at P_0 is called the normal line to the surface at P_0.

Tangent Plane and Normal Line to a Level Surface $f(x, y, z) = c$

If $f(x, y, z)$ has continuous partial derivatives at a point $P_0(x_0, y_0, z_0)$ on the level surface

$$S: \quad f(x, y, z) = c,$$

the *tangent plane* to S at P_0 is the plane through P_0 normal to ∇f at P_0. Its equation is

$$f_x(P_0)(x - x_0) + f_y(P_0)(y - y_0) + f_z(P_0)(z - z_0) = 0. \tag{11}$$

The partial derivatives in this equation are to be evaluated at P_0.

The *normal line* to S at P_0 is the line perpendicular to the tangent plane and parallel to ∇f at P_0, given by the equations

$$\begin{aligned} x &= x_0 + f_x(P_0)t, \\ y &= y_0 + f_y(P_0)t, \\ z &= z_0 + f_z(P_0)t. \end{aligned} \tag{12}$$

If none of the partial derivatives of f is zero at P_0, the normal line is also given by the equations

$$\frac{x - x_0}{f_x(P_0)} = \frac{y - y_0}{f_y(P_0)} = \frac{z - z_0}{f_z(P_0)}. \tag{13}$$

EXAMPLE 4 Find the tangent plane and normal line to the surface

$$x^2 + xyz - z^3 = 1$$

at the point $P_0(1, 1, 1)$.

Solution We first find the gradient vector at P_0:

$$\begin{aligned} f_x(1, 1, 1) &= (2x + yz)|_{(1,1,1)} = 3, \\ f_y(1, 1, 1) &= xz|_{(1,1,1)} = 1, \\ f_z(1, 1, 1) &= (xy - 3z^2)|_{(1,1,1)} = -2, \\ \nabla f|_{(1,1,1)} &= 3\mathbf{i} + \mathbf{j} - 2\mathbf{k}. \end{aligned}$$

From Eq. (11) the plane tangent to the surface at $P_0(1, 1, 1)$ is

$$3(x - 1) + (y - 1) - 2(z - 1) = 0,$$

or

$$3x + y - 2z = 2.$$

From Eq. (12), the normal to the surface at $P_0(1, 1, 1)$ is the line

$$x = 1 + 3t, \qquad y = 1 + t, \qquad z = 1 - 2t. \ \square$$

EXAMPLE 5 The surfaces

$$f(x, y, z) = x^2 + y^2 - z^2 = 1, \qquad g(x, y, z) = x + y + z = 5$$

intersect in a curve C. Find the line tangent to C at the point $P_0(1, 2, 2)$.

Solution The tangent line is normal to both ∇f and ∇g at P_0. Therefore,

$$\mathbf{v} = \nabla f \times \nabla g$$

will be a vector parallel to the line. We use the components of $\mathbf{v}$ and the coordinates of P_0 to write the equations for the line.

We have

$$\begin{aligned} \nabla f|_{(1,2,2)} &= (2x\mathbf{i} + 2y\mathbf{j} - 2z\mathbf{k})|_{(1,2,2)} \\ &= 2\mathbf{i} + 4\mathbf{j} - 4\mathbf{k}, \\ \nabla g &= \mathbf{i} + \mathbf{j} + \mathbf{k}. \end{aligned}$$

Therefore,

$$\mathbf{v} = \nabla f \times \nabla g = \begin{vmatrix} \mathbf{i} & \mathbf{j} & \mathbf{k} \\ 2 & 4 & -4 \\ 1 & 1 & 1 \end{vmatrix} = 8\mathbf{i} - 6\mathbf{j} - 2\mathbf{k}.$$

The line is

$$x = 1 + 8t, \qquad y = 2 - 6t, \qquad z = 2 - 2t. \ \square$$

Just as the gradient of $f(x, y, z)$ is perpendicular to the level surfaces of f, the gradient of a function $g(x, y)$ of two variables is perpendicular to the function's level curves (Problems 81 and 82).

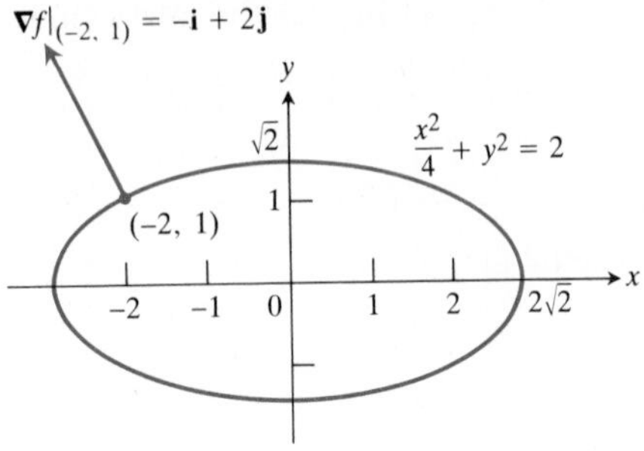

15.23 The level curve and gradient of $f(x, y) = (x^2/4) + y^2$ at $P_0\ (-2, 1)$.

EXAMPLE 6 Sketch the level curve of

$$f(x, y) = \frac{x^2}{4} + y^2$$

that passes through the point $P_0(-2, 1)$. Find a vector in the plane of the curve and normal to it at P_0, and include it in the sketch.

Solution The value of f at P_0 is

$$\frac{(-2)^2}{4} + (1)^2 = 2.$$

See Fig. 15.23. Therefore the level curve of f through P_0 is the ellipse

$$\frac{x^2}{4} + y^2 = 2.$$

For a vector normal to the ellipse at $(-2, 1)$ we may take

$$\nabla f\Big|_{(-2,\,1)} = \left(\frac{x}{2}\mathbf{i} + 2y\mathbf{j}\right)\Big|_{(-2,\,1)} = -\mathbf{i} + 2\mathbf{j}. \ \square$$

The Tangent Plane for a Surface $z = f(x, y)$

The equation for a surface $z = f(x, y)$ can be written in the form

$$\underbrace{f(x, y) - z}_{F(x,\,y,\,z)} = 0. \qquad (14)$$

From this we see that the surface is also the level surface

$$F(x, y, z) = 0$$

of the function $w = F(x, y, z)$. Thus if F has continuous partial derivatives, we may define the tangent plane to the surface $z = f(x, y)$ at a point $P_0(x_0, y_0, z_0)$ on the surface to be the plane perpendicular to ∇F at P_0. The equation for this plane is

$$F_x(P_0)(x - x_0) + F_y(P_0)(y - y_0) + F_z(P_0)(z - z_0) = 0.$$

But now

$$\begin{aligned} F_x(P_0) &= \frac{\partial}{\partial x}(f(x, y) - z)\big|_{(x_0,\,y_0,\,z_0)} = f_x(x_0, y_0), \\ F_y(P_0) &= \frac{\partial}{\partial y}(f(x, y) - z)\big|_{(x_0,\,y_0,\,z_0)} = f_y(x_0, y_0), \\ F_z(P_0) &= \frac{\partial}{\partial z}(f(x, y) - z)\big|_{(x_0,\,y_0,\,z_0)} = -1, \end{aligned} \qquad (15)$$

and the equation of the tangent plane reduces to

$$f_x(x_0, y_0)(x - x_0) + f_y(x_0, y_0)(y - y_0) - (z - z_0) = 0.$$

We can see from Eqs. (15) that F_x, F_y, and F_z will be continuous at P_0 if f_x and f_y are continuous at (x_0, y_0).

Tangent Plane and Normal Line for a Surface $z = f(x, y)$

If $P_0(x_0, y_0, z_0)$ is a point on the surface $z = f(x, y)$ and f_x and f_y are continuous at (x_0, y_0), then the tangent plane to the surface at P_0 is the plane

$$f_x(x_0, y_0)(x - x_0) + f_y(x_0, y_0)(y - y_0) - (z - z_0) = 0. \quad (16)$$

The normal line to the surface at P_0 is the line

$$x = x_0 + tf_x(x_0, y_0),\quad y = y_0 + tf_y(x_0, y_0),\quad z = z_0 - t, \quad (17)$$

or

$$\frac{x - x_0}{f_x(x_0, y_0)} = \frac{y - y_0}{f_y(x_0, y_0)} = \frac{z - z_0}{-1} \quad (18)$$

if $f_x(x_0, y_0) \neq 0$ and $f_y(x_0, y_0) \neq 0$.

EXAMPLE 7 Find equations for the tangent plane and normal line to the surface

$$z = f(x, y) = 9 - x^2 - y^2$$

at the point $P_0(1, 2, 4)$. See Fig. 15.24.

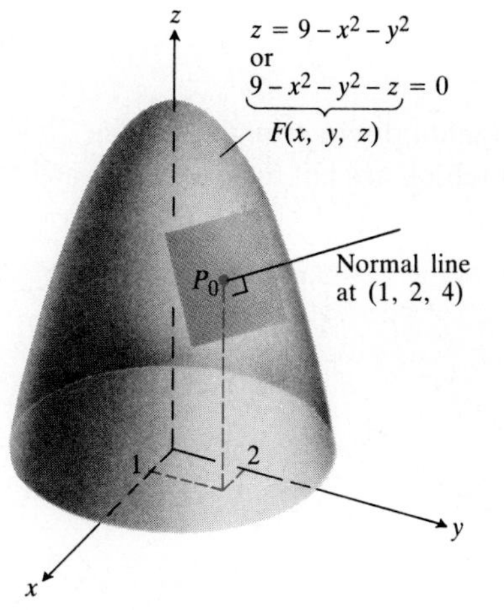

15.24 The tangent plane and normal line to the surface $z = 9 - x^2 - y^2$ at $P_0(1, 2, 4)$.

Solution 1 With

$$f_x(1, 2) = -2x|_{(1,2)} = -2, \qquad f_y(1, 2) = -2y|_{(1,2)} = -4,$$

Equations (16) and (17) give

Tangent plane: $(-2)(x - 1) + (-4)(y - 2) - (z - 4) = 0$ or $2x + 4y + z = 14$,

Normal line: $x = 1 - 2t, \quad y = 2 - 4t, \quad z = 4 - t.$

Solution 2 If we rewrite the equation

$$z = 9 - x^2 - y^2$$

in the form

$$9 - x^2 - y^2 - z = 0,$$

we see that the surface $z = f(x, y)$ is the same as the level surface

$$F(x, y, z) = 0$$

of the function

$$F(x, y, z) = 9 - x^2 - y^2 - z.$$

The tangent plane we seek is therefore in the plane normal to ∇F at $P_0(1, 2, 5)$. With

$$F_x = -2x|_{(1, 2, 4)} = -2,$$
$$F_y = -2y|_{(1, 2, 4)} = -4,$$
$$F_z = -1|_{(1, 2, 4)} = -1,$$

Equations (11) and (12) give

Tangent plane: $(-2)(x - 1) + (-4)(y - 2) + (-1)(z - 4) = 0$

or $2x + 4y + z = 14,$

Normal line: $x = 1 - 2t, \quad y = 2 - 4t, \quad z = 4 - t,$

in agreement with Solution 1. □

The Continuity Requirement for Partial Derivatives

Why do we require functions defining surfaces to have continuous first partial derivatives before we define tangent planes? The answer is that surfaces defined by functions with discontinuous partial derivatives may be too uneven to have satisfactory tangent planes. Here are two examples of such surfaces.

EXAMPLE 8 The "ridge" surface

$$z = \frac{1}{2}(||x| - |y|| - |x| - |y|) \tag{19}$$

shown in Fig. 15.25 consists of two upside-down troughs whose "backbones" lie along the x- and y-axes and which are cut to fit together as they come in to the origin.

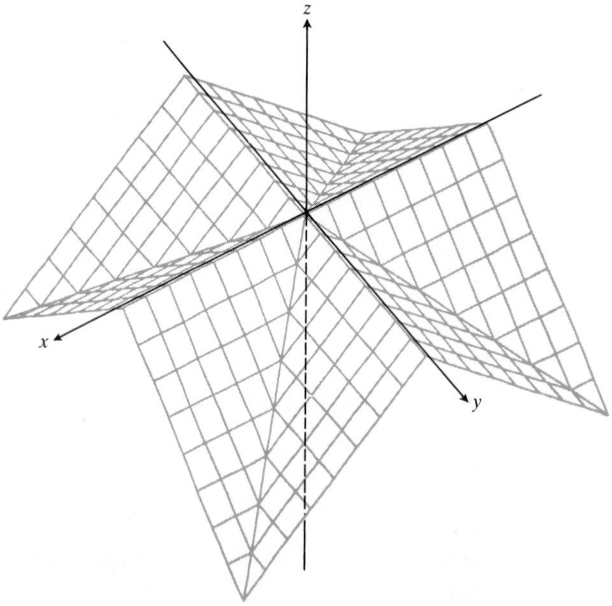

15.25 The "ridge" surface

$z = \frac{1}{2}(||x| - |y|| - |x| - |y|)$

viewed from the point (10, 15, 20).

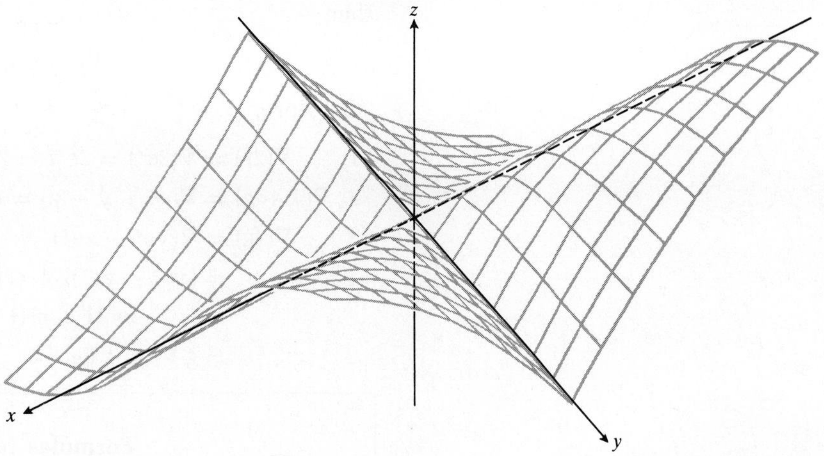

15.26 The "twisted butterfly" surface

$$z = \begin{cases} \dfrac{2|x|y}{\sqrt{x^2 + y^2}}, & (x, y) \neq (0, 0) \\ 0, & (x, y) = (0,0) \end{cases}$$

viewed from the point (10, 15, 10).

There is no unique plane tangent to the surface at the origin or at points where the two troughs meet. □

EXAMPLE 9 The "twisted butterfly" surface

$$z = \begin{cases} \dfrac{2|x|y}{\sqrt{x^2 + y^2}}, & (x, y) \neq (0, 0) \\ 0, & (x, y) = (0, 0) \end{cases} \tag{20}$$

shown in Fig. 15.26 can be generated by pivoting the x-axis at the origin and swinging it around like a wobbly compass needle. As the axis swings from $\theta = -\pi/2$ to $\theta = \pi/2$ (using cylindrical coordinates) the point r units from the origin rises, or falls, according to the law

$$z = r \sin 2\theta. \tag{21}$$

For every position the pivoted x-axis takes, there is a line in the surface through the origin perpendicular to it. But the plane determined by these two lines is not tangent to the surface. Otherwise, every such plane would have an equal claim to that honor. □

Algebraic Properties of the Gradient Vector

If $f(x, y, z)$ and $g(x, y, z)$ have partial derivatives, then the constant multiple kf, the sum $f + g$, and the product fg have gradient vectors, and

$$\begin{aligned} &1. \quad \nabla(kf) = k\nabla f \quad \text{(any number } k\text{)}, \\ &2. \quad \nabla(f + g) = \nabla f + \nabla g, \\ &3. \quad \nabla(fg) = f\,\nabla g + g\,\nabla f. \end{aligned} \tag{22}$$

The proofs are left as exercises.

EXAMPLE 10 If

$$f(x, y, z) = e^x, \qquad g(x, y, z) = y - z,$$

then

$$\nabla f = e^x\mathbf{i}, \qquad \nabla g = \mathbf{j} - \mathbf{k}.$$

Therefore,

1. $\nabla(2f) = \nabla(2e^x) = 2e^x\mathbf{i} = 2\,\nabla f,$
2. $\nabla(f + g) = \nabla(e^x + y - z) = e^x\mathbf{i} + \mathbf{j} - \mathbf{k} = \nabla f + \nabla g,$
3. $$\begin{aligned}\nabla(fg) &= \nabla(ye^x - ze^x)\\ &= (ye^x - ze^x)\mathbf{i} + e^x\mathbf{j} - e^x\mathbf{k}\\ &= (y - z)e^x\mathbf{i} + e^x(\mathbf{j} - \mathbf{k})\\ &= g\,\nabla f + f\,\nabla g.\ \square\end{aligned}$$

Formulas for Functions with Continuous First Partial Derivatives

Gradient vector:

$\nabla f = f_x\mathbf{i} + f_y\mathbf{j}$ (two-dimensional)

$\nabla f = f_x\mathbf{i} + f_y\mathbf{j} + f_z\mathbf{k}$ (three-dimensional)

$\nabla f(x, y)$ is normal to the level curves of f.

$\nabla f(x, y, z)$ is normal to the level surfaces of f.

Directional derivative: The derivative of f in the direction $\mathbf{u}$ at the point P_0 is

$$(D_{\mathbf{u}}f)_{P_0} = (\nabla f)_{P_0} \cdot \mathbf{u} = (\text{grad } f)_{P_0} \cdot \mathbf{u}$$

Tangent plane and normal line to the surface $f(x, y, z) = 0$ at the point $P_0(x_0, y_0, z_0)$:

Tangent plane:
$f_x(P_0)(x - x_0) + f_y(P_0)(y - y_0) + f_z(P_0)(z - z_0) = 0$

Normal line:
$x = x_0 + f_x(P_0)t, \qquad y = y_0 + f_y(P_0)t, \qquad z = z_0 + f_z(P_0)t$

Tangent plane and normal line to a surface $z = f(x, y)$ at the point $P_0(x_0, y_0, f(x_0, y_0))$:

Tangent plane:
$f_x(x_0, y_0)(x - x_0) + f_y(x_0, y_0)(y - y_0) - (z - z_0) = 0$

Normal line:
$x = x_0 + tf_x(x_0, y_0), \qquad y = y_0 + tf_y(x_0, y_0), \qquad z = z_0 - t$

PROBLEMS

In Problems 1–7, find the electric intensity vector $\mathbf{E} = -\nabla V$ for each potential function V at the given point.

1. $V = x^2 + y^2 - 2z^2$, $(1, 1, 1)$

2. $V = 2z^3 - 3(x^2 + y^2)z$, $(1, 1, 1)$

3. $V = e^{-2y} \cos 2x$, $(\pi/4, 0, 0)$

4. $V = \ln \sqrt{x^2 + y^2}$, $(3, 4, 0)$

5. $V = (x^2 + y^2 + z^2)^{-(1/2)}$, $(1, 2, -2)$

6. $V = e^{3x+4y} \cos 5z$, $(0, 0, \pi/6)$

7. $V = \cos 3x \cos 4y \sinh 5z$, $(0, \pi/4, 0)$

In Problems 8–18, find the derivative of f at the point P_0 in the direction of the vector **A**.

8. $f = x^2 + y^2$, $P_0(1, 0)$, $\mathbf{A} = \mathbf{i} - \mathbf{j}$

9. $f = e^x \sin \pi y$, $P_0(0, 1)$, $\mathbf{A} = 4\mathbf{i} + 4\mathbf{j}$

10. $f = \cos xy$, $P_0(2, \pi/4)$, $\mathbf{A} = 4\mathbf{i} - \mathbf{j}$

11. $f = \dfrac{x - y^2}{x}$, $P_0(1, 1)$, $\mathbf{A} = 12\mathbf{i} + 5\mathbf{j}$

12. $f = x^2 + 2xy - 3y^2$, $P_0(1/2, 1/2)$, $\mathbf{A} = \sqrt{3}\,\mathbf{i} + \mathbf{j}$

13. $f = x \tan^{-1}(y/x)$, $P_0(1, 1)$, $\mathbf{A} = 2\mathbf{i} - \mathbf{j}$

14. $f = xy + yz + zx$, $P_0(1, -1, 2)$, $\mathbf{A} = 10\mathbf{i} + 11\mathbf{j} - 2\mathbf{k}$

15. $f = x^2 + 2y^2 + 3z^2$, $P_0(1, 1, 1)$, $\mathbf{A} = \mathbf{i} + \mathbf{j} + \mathbf{k}$

16. $f = \ln \sqrt{x^2 + y^2 + z^2}$, $P_0(3, 4, 12)$, $\mathbf{A} = 3\mathbf{i} + 6\mathbf{j} - 2\mathbf{k}$

17. $f = e^x \cos yz$, $P_0(0, 0, 0)$, $\mathbf{A} = 2\mathbf{i} + \mathbf{j} - 2\mathbf{k}$

18. $f = \cos xy + e^{yz} + \ln zx$,
$P_0(1, 0, 1/2)$, $\mathbf{A} = \mathbf{i} + 2\mathbf{j} + 2\mathbf{k}$

In Problems 19–21, find (a) the direction in which f increases most rapidly at P_0, and (b) the rate at which f changes in that direction.

19. $f(x, y) = x^2 + \cos xy$, $P_0(1, \pi/2)$

20. $f(x, y, z) = e^{xy} + z^2$, $P_0(0, 2, 3)$

21. $f(x, y, z) = (x + y - 2)^2 + (3x - y - 6)^2$, $P_0(1, 1, 0)$

In Problems 22–24, find (a) the direction in which f decreases most rapidly at P_0, and (b) the rate at which f changes in that direction.

22. $f(x, y) = x^2 + xy + y^2$, $P_0(-1, 1)$

23. $f(x, y, z) = (x + y)^2 + (y + z)^2 + (z + x)^2$,
$P_0(2, -1, 2)$

24. $f(x, y, z) = z \ln (x^2 + y^2)$, $P_0(1, 1, 1)$

25. Use the directional derivative to estimate how much $f(x, y) = \cos \pi xy + xy^2$ will change if the point $P(x, y)$ is moved from $P(-1, -1)$ a distance of $\Delta s = 0.1$ unit along the vector $\mathbf{A} = \mathbf{i} + \mathbf{j}$.

26. By about how much will

$$f(x, y, z) = \ln \sqrt{x^2 + y^2 + z^2}$$

change if the point $P(x, y, z)$ is moved from $P(3, 4, 12)$ a distance of $\Delta s = 0.1$ unit along the vector $3\mathbf{i} + 6\mathbf{j} - 2\mathbf{k}$?

27. By about how much will

$$f(x, y, z) = e^x \cos yz$$

change as the point $P(x, y, z)$ moves from the origin at a distance of $\Delta s = 0.1$ unit in the direction of $\mathbf{A} = 2\mathbf{i} + \mathbf{j} - 2\mathbf{k}$?

28. By about how much will

$$f(x, y, z) = x + x \cos z - y \sin z + y$$

change if the point $P(x, y, z)$ moves from $P_0(2, -1, 0)$ a distance of $\Delta s = 0.2$ units toward the point $P_1(0, 1, 2)$?

29. In which two directions is the derivative of $f(x, y) = xy + y^2$ at the point $P_0(2, 5)$ equal to zero?

30. In which two directions is the derivative of $f(x, y) = (x^2 - y^2)/(x^2 + y^2)$ at $P_0(1, 1)$ equal to zero?

In Problems 31–35, sketch the level curve of $f(x, y)$ through the point P_0. Find a vector normal to the curve at P_0 and include it in your sketch.

31. $f(x, y) = x^2 - y^2$, $P_0(2, -1)$

32. $f(x, y) = x^2/3 + 3y^2/4$, $P_0(2, 2\sqrt{5/3})$

33. $f(x, y) = x^2 - y$, $P_0(\sqrt{2}, 1)$

34. $f(x, y) = \sqrt{x^2 - y}$, $P_0(1, 0)$

35. $f(x, y) = 12 + 2x^2 - y^2$, $P_0(1, 2)$

In Problems 36–43, find equations for (a) the tangent plane and (b) the normal line to the given level surface at the point P_0.

36. $x^2 + y^2 + z^2 = 3$, $P_0(1, 1, 1)$

37. $x^2 + y^2 - z^2 = 18$, $P_0(3, 5, -4)$

38. $z^2 - x^2 - y^2 = 0$, $P_0(3, 4, -5)$

39. $2z - x^2 = 0$, $P_0(2, 0, 2)$

40. $z - \ln (x^2 + y^2) = 0$, $P_0(1, 0, 0)$

41. $(x + y)^2 + z^2 = 25$, $P_0(1, 2, 4)$

42. $x^2 + 2xy - y^2 + z^2 = 7$, $P_0(1, -1, 3)$

43. $\cos \pi x - x^2y + e^{xz} + yz = 4$, $P_0(0, 1, 2)$

In Problems 44–57, find (a) the tangent plane and (b) the normal line to the given surface at the given point.

44. $z = x^2 + y^2$, $(3, 4, 25)$

45. $z = \sqrt{9 - x^2 - y^2}$, $(1, -2, 2)$

46. $z = x^2 - xy - y^2$, $(1, 1, -1)$

47. $z = \tan^{-1} \dfrac{y}{x}$, $(1, 1, \pi/4)$

48. $z = x/\sqrt{x^2 + y^2}$, $(3, -4, \frac{3}{5})$

49. $z = 9x^2 + y^2$, $(0, 0, 0)$

50. $z = \cos (\pi x/2)$, $(1, 0, 0)$

51. $z = 1 - x - y$, $(0, 1, 0)$

52. $z = (x + y)/(xy - 1)$, $(1, 2, 3)$

53. $z = x^2 + y^2 - 2xy + 3y - x + 4$, $(2, -3, 18)$

54. $x = 2z^2 - zy + y^3$, $(-4, -2, 1)$
(as a surface over the yz-plane)

55. $x = 1 - y^2 - z^2$, $(0, 0, 1)$

56. $y = 4 - x^2 - 4z^2$, $(0, 0, 1)$
(as a surface over the xz-plane)

57. $y = \sin x$, $(0, 0, 0)$

In Problems 58–65, sketch the portion of the given level surface that lies in the first octant (x, y, and z nonnegative). Then find a vector perpendicular to the surface at the given point P_0. Include the vector in your sketch.

58. $x^2 + y^2 + z^2 = 9, \quad P_0(1, 2, 2)$

59. $x + y^2 + z = 2, \quad P_0(\frac{1}{2}, 1, \frac{1}{2})$

60. $2x + y^2 + z = 4, \quad P_0(1, 1, 1)$

61. $x^2 + y^2 + z = 4, \quad P_0(0, 1, 3)$

62. $2x + 3y + 6z = 18, \quad P_0(3, 2, 1)$

63. $z - \sqrt{9 - x^2 - y^2} = 0, \quad P_0(1, 2, 2)$

64. $1/\sqrt{x^2 + y^2 + z^2} = 1, \quad P_0(\frac{1}{2}, \frac{1}{2}, \sqrt{2}/2)$

65. $x^2 + 2y^2 - z = 2, \quad P_0(1, 1, 1)$

66. The derivative of $f(x, y)$ at $P_0(1, 2)$ in the direction of the vector $\mathbf{i} + \mathbf{j}$ is $2\sqrt{2}$, and in the direction of the vector $-2\mathbf{j}$ is -3. What is the derivative of f in the direction of the vector $-\mathbf{i} - 2\mathbf{j}$?

67. The derivative of $f(x, y, z)$ at a given point P is greatest in the direction of the vector $\mathbf{A} = \mathbf{i} + \mathbf{j} - \mathbf{k}$. In this direction the value of the derivative is $2\sqrt{3}$. (a) Find the gradient vector of f at P. (b) Find the derivative of f at P in the direction of the vector $\mathbf{i} + \mathbf{j}$.

68. a) Find equations for all lines through the origin normal to the surface $xy + z = 2$.
b) Find the points in which these lines meet the surface.

69. Find the points on the surface

$$(y + z)^2 + (z - x)^2 = 16$$

where the normal is parallel to the yz-plane.

70. Find the points on the surface

$$xy + yz + zx - x - z^2 = 0$$

where the tangent plane is parallel to the xy-plane.

71. A curve is said to be *tangent to a surface at a point P* if the line tangent to the curve at P lies in the plane that is tangent to the surface at P.
a) Show that the curve $x = \ln t$, $y = t \ln t$, $z = t$ is tangent to the surface

$$xz^2 - yz + \cos xy = 1$$

at $P(0, 0, 1)$.
b) Show that the curve $x = (t^3/4) - 2$, $y = (4/t) - 3$, $z = \cos(t - 2)$ is tangent to the surface

$$x^3 + y^3 + z^3 - xyz = 0$$

at $t = 2$.

72. Find the derivative of $f(x, y, z) = xyz$ in the direction of the velocity vector of the helix

$$R(t) = (\cos 3t)\mathbf{i} + (\sin 3t)\mathbf{j} + 3t\mathbf{k}$$

at time $t = \pi/3$.

73. For the function $f(x, y) = x^2y + 2y^2x$, at the point $P_0(1, 3)$, find
a) the direction of greatest increase in f,
b) the derivative of f in the direction of greatest increase in f,
c) the direction of greatest decrease in f,
d) the directions in which the derivative of f is zero,
e) an equation for the plane tangent to the surface $z = f(x, y)$ at (1, 3, 21).

74. If $z = f(x, y)$ has continuous partial derivatives at the point $P_0(x_0, y_0)$, which of the following statements are true?
a) If $\mathbf{u}$ is a unit vector, then the derivative of f in the direction of $\mathbf{u}$ is $(f_x(x_0, y_0)\mathbf{i} + f_y(x_0, y_0)\mathbf{j}) \cdot \mathbf{u}$.
b) The derivative of f in the direction of $\mathbf{u}$ is a vector.
c) The directional derivative of f at P_0 has its greatest value in the direction of ∇f.
d) At (x_0, y_0), ∇f is normal to the level curve $f(x, y) = f(x_0, y_0)$.

In Problems 75–80, find equations for the line tangent to the curve of intersection of the two surfaces at the given point.

75. Surfaces: $x^2 + y^2 = 4, \quad z = x^2 + y^2$
Point: $(\sqrt{2}, \sqrt{2}, 4)$

76. Surfaces: $xyz = 1, \quad x^2 + 2y^2 + 3z^2 = 6$
Point: (1, 1, 1)

77. Surfaces: $x^2 + 2y + 2z = 4, \quad y = 1$
Point: $(1, 1, \frac{1}{2})$

78. Surfaces: $x + y^2 + z = 2, \quad y = 1$
Point: $(\frac{1}{2}, 1, \frac{1}{2})$

79. Surfaces: $x + y^2 + 2z = 4, \quad x = 1$
Point: (1, 1, 1)

80. Surfaces: $x^3 + 3x^2y^2 + y^3 + 4xy - z^2 = 0$,
$x^2 + y^2 + z^2 = 11$
Point: (1, 1, 3)

81. Suppose that $g(x, y)$ has continuous partial derivatives and that g has the constant value c on the differentiable curve $x = x(t)$, $y = y(t)$. That is,

$$g(x(t), y(t)) = c$$

for all values of t. Differentiate both sides of this equation with respect to t to show that ∇g is normal to the tangent vector at every point of the curve.

82. *Tangent lines to level curves.* A quick way to find the tangent line to a level curve $f(x, y) = c$ at a point $P_0(x_0, y_0)$ on the curve is as follows: Since ∇f is normal to the tangent line, the points $P(x, y)$ on the line satisfy the equation $\nabla f \cdot \overrightarrow{P_0P} = 0$, or

$$f_x(x_0, y_0)(x - x_0) + f_y(x_0, y_0)(y - y_0) = 0.$$

Use this equation to find the lines tangent to the following curves at the given points.

a) $x^2 + y^2 = 4, \quad P_0(\sqrt{2}, \sqrt{2})$
b) $x^2 + xy + y^2 = 7, \quad P_0(1, 2)$
(This was Example 6 in Article 2.4.)
c) $x^5 + 4xy^3 - 3y^5 = 2, \quad P_0(1, 1)$

83. Show that the curve

$$R = -t\mathbf{i} + \sqrt{t}\,\mathbf{j} + (\ln t)\mathbf{k}$$

intersects the surface

$$z = \ln\left(\frac{y - 2x^2 - y^2}{4}\right)$$

at a right angle when $t = 1$ (i.e., that the curve's velocity vector is normal to the surface's tangent plane).

84. Suppose cylindrical coordinates r, θ, z are introduced into a function $w = f(x, y, z)$ to yield $w = F(r, \theta, z)$. Show that the gradient may be expressed in terms of cylindrical coordinates and the unit vectors $\mathbf{u}_r$, $\mathbf{u}_\theta$, $\mathbf{k}$ as follows:

$$\nabla w = \mathbf{u}_r \frac{\partial w}{\partial r} + \frac{1}{r}\mathbf{u}_\theta \frac{\partial w}{\partial \theta} + \mathbf{k}\frac{\partial w}{\partial z}.$$

(*Hint:* The component of ∇w in the direction of $\mathbf{u}_r$ is equal to the directional derivative in that direction. But this is precisely $\partial w/\partial r$. Reason similarly for the components of ∇w in the directions of $\mathbf{u}_\theta$ and $\mathbf{k}$.)

85. Verify the result given in Problem 84 by transforming the given expression on the right-hand side of the equation into **i**, **j**, **k** components and replacing the cylindrical coordinates r, θ by cartesian coordinates x, y, and making use of the chain rule for partial derivatives.

86. Express the gradient in terms of spherical coordinates and the appropriate unit vectors $\mathbf{u}_\rho$, $\mathbf{u}_\phi$, $\mathbf{u}_\theta$. Use a geometrical argument to determine the component of ∇w in each of these directions. (See the hint for Problem 84.)

87. Verify the answer obtained in Problem 86 by transforming the expression you obtained in spherical coordinates back into cartesian coordinates. Make use of a chain rule.

88. In Fig. 15.13 let

$$\mathbf{R} = \mathbf{i}x + \mathbf{j}y + \mathbf{k}f(x, y)$$

be the vector from the origin to (x, y, w). What can you say about the direction of the vector (a) $\partial\mathbf{R}/\partial x$, (b) $\partial\mathbf{R}/\partial y$? (c) Calculate the vector product

$$\mathbf{v} = \left(\frac{\partial \mathbf{R}}{\partial x}\right) \times \left(\frac{\partial \mathbf{R}}{\partial y}\right).$$

What can you say about the direction of this vector **v** with respect to the surface $w = f(x, y)$?

89. Verify the gradient formulas in Eq. (22).

15.7 Higher Order Derivatives. Partial Differential Equations from Physics

This article is devoted mainly to second order partial derivatives, which are denoted by

$$\frac{\partial^2 f}{\partial x^2}, \quad \frac{\partial^2 y}{\partial y^2}, \quad \frac{\partial^2 f}{\partial x\,\partial y}, \quad \frac{\partial^2 f}{\partial y\,\partial x}$$

or

$$f_{xx}, \quad f_{yy}, \quad f_{yx}, \quad f_{xy},$$

and defined by the equations

$$\frac{\partial^2 f}{\partial x^2} = \frac{\partial}{\partial x}\left(\frac{\partial f}{\partial x}\right), \qquad \frac{\partial^2 f}{\partial x\,\partial y} = \frac{\partial}{\partial x}\left(\frac{\partial f}{\partial y}\right),$$

and so on. Note the order in which the derivatives are taken:

$\dfrac{\partial^2 f}{\partial x\,\partial y}$ Differentiate first with respect to y, then with respect to x,

f_{yx} Means the same thing.

Second order partial derivatives appear in equations that express important physical laws for wave motion, heat flow, and gravitational potential. (A gravitational potential function measures the work done in moving a unit of mass against a gravitational field from a central reference point, say the Sun, to a new position, say Mars.) Second order partial derivatives are also used to test for maxima and minima of functions of two variables, as we shall see in the next article.

EXAMPLE 1 If

$$f(x, y) = x \cos y + ye^x,$$

then

$$\frac{\partial f}{\partial x} = \cos y + ye^x,$$

$$\frac{\partial}{\partial y}\left(\frac{\partial f}{\partial x}\right) = -\sin y + e^x = \frac{\partial^2 f}{\partial y\, \partial x},$$

$$\frac{\partial}{\partial x}\left(\frac{\partial f}{\partial x}\right) = ye^x = \frac{\partial^2 f}{\partial x^2},$$

$$\frac{\partial}{\partial x}\left(\frac{\partial^2 f}{\partial x^2}\right) = ye^x = \frac{\partial^3 f}{\partial x^3},$$

$$\frac{\partial}{\partial y}\left(\frac{\partial^2 f}{\partial x^2}\right) = e^x = \frac{\partial^3 f}{\partial y\, \partial x^2},$$

while

$$\frac{\partial f}{\partial y} = -x \sin y + e^x,$$

$$\frac{\partial}{\partial x}\left(\frac{\partial f}{\partial y}\right) = -\sin y + e^x = \frac{\partial^2 f}{\partial x\, \partial y},$$

$$\frac{\partial}{\partial y}\left(\frac{\partial f}{\partial y}\right) = -x \cos y = \frac{\partial^2 f}{\partial y^2},$$

$$\frac{\partial}{\partial x}\left(\frac{\partial^2 f}{\partial x\, \partial y}\right) = e^x = \frac{\partial^3 f}{\partial x^2\, \partial y}. \quad \square$$

You may have noticed in Example 1 that the "mixed" second order partial derivatives

$$\frac{\partial^2 f}{\partial y\, \partial x} \quad \text{and} \quad \frac{\partial^2 f}{\partial x\, \partial y}$$

were equal. This was not a mere coincidence. Whenever $f, f_x, f_y, f_{xy}, f_{yx}$ are all continuous at a point, the mixed partial derivatives f_{xy} and f_{yx} will be equal at that point.

THEOREM 4

The Mixed-Derivative Theorem

If $f(x, y)$ and its partial derivatives f_x, f_y, f_{xy}, f_{yx} are defined in a region containing a point (a, b) and are all continuous at (a, b), then

$$f_{xy}(a, b) = f_{yx}(a, b). \tag{1}$$

A proof of Theorem 4 is given in Appendix 4.

EXAMPLE 2 Find z_{xy} if

$$z = f(u, v), \qquad u = x^2 - y^2, \qquad v = 2xy,$$

and f and its partial derivatives are all continuous.

Solution From the chain rule,

$$z_x = \frac{\partial}{\partial x} f(u, v) = f_u u_x + f_v v_x = f_u 2x + f_v 2y = 2xf_u + 2yf_v.$$

Therefore,

$$\begin{aligned} z_{xy} &= \frac{\partial}{\partial y}(2xf_u + 2yf_v) \\ &= 2x\frac{\partial}{\partial y}(f_u) + f_u\frac{\partial}{\partial y}(2x) + 2y\frac{\partial}{\partial y}(f_v) + f_v\frac{\partial}{\partial y}(2y) \\ &= 2x\frac{\partial}{\partial y}(f_u) + 2y\frac{\partial}{\partial y}(f_v) + 2f_v. \end{aligned} \tag{2}$$

We use the chain rule again to calculate the two remaining derivatives:

$$\frac{\partial}{\partial y}(f_u) = f_{uu}\frac{\partial u}{\partial y} + f_{uv}\frac{\partial v}{\partial y} = -2yf_{uu} + 2xf_{uv},$$

$$\frac{\partial}{\partial y}(f_v) = f_{vu}\frac{\partial u}{\partial y} + f_{vv}\frac{\partial v}{\partial y} = -2yf_{vu} + 2xf_{vv}.$$

Thus we may continue from Eq. (2) to get

$$\begin{aligned} z_{xy} &= 2x(-2yf_{uu} + 2xf_{uv}) + 2y(-2yf_{vu} + 2xf_{vv}) + 2f_v \\ &= -4xyf_{uu} + 4x^2f_{uv} - 4y^2f_{vu} + 4xyf_{vv} + 2f_v \qquad (3) \\ &= 4xy(f_{vv} - f_{uu}) + 4(x^2 - y^2)f_{uv} + 2f_v, \qquad (4) \end{aligned}$$

or, if we want the answer entirely in terms of u and v,

$$z_{xy} = 2v(f_{vv} - f_{uu}) + 4uf_{uv} + 2f_v.$$

We used the equality $f_{uv} = f_{vu}$ to get from (3) to (4). □

EXAMPLE 3 Find d^2w/dt^2 if

$$w = f(x, y), \qquad x = e^t, \qquad y = 2t - 1,$$

and w and its partial derivatives are all continuous.

Solution We find dw/dt, and then differentiate it with respect to t. From the chain rule,

$$\frac{dw}{dt} = f_x\frac{dx}{dt} + f_y\frac{dy}{dt} = f_x \cdot e^t + f_y \cdot 2 = e^tf_x + 2f_y.$$

Therefore,

$$\begin{aligned} \frac{d^2w}{dt^2} &= \frac{d}{dt}(e^tf_x) + \frac{d}{dt}(2f_y) \\ &= e^tf_x + e^t\frac{d}{dt}(f_x) + 2\frac{d}{dt}(f_y). \end{aligned} \tag{5}$$

We use the chain rule again to calculate the two remaining derivatives:

$$\frac{d}{dt}(f_x) = f_{xx}\frac{dx}{dt} + f_{xy}\frac{dy}{dt} = f_{xx}e^t + f_{xy}\cdot 2 = e^t f_{xx} + 2f_{xy}.$$

$$\frac{d}{dt}(f_y) = f_{yx}\frac{dx}{dt} + f_{yy}\frac{dy}{dt} = f_{yx}e^t + f_{yy}\cdot 2 = e^t f_{yx} + 2f_{yy}.$$

Substituting these in Eq. (5) gives

$$\begin{aligned}\frac{d^2w}{dt^2} &= e^t f_x + e^t(e^t f_{xx} + 2f_{xy}) + 2(e^t f_{yx} + 2f_{yy})\\ &= e^t f_x + e^{2t} f_{xx} + 2e^t f_{xy} + 2e^t f_{yx} + 4f_{yy}\\ &= e^t f_x + e^{2t} f_{xx} + 4e^t f_{xy} + 4f_{yy}. \ \square\end{aligned}$$

Partial Derivatives of Still Higher Order

If we refer once more to Example 1 we note not only that

$$\frac{\partial^2 f}{\partial x\,\partial y} = \frac{\partial^2 f}{\partial y\,\partial x}$$

but also that

$$\frac{\partial^3 f}{\partial x^2\,\partial y} = \frac{\partial^3 f}{\partial y\,\partial x^2}.$$

This equality may be derived from Theorem 4 as follows:

$$\begin{aligned}\frac{\partial^3 f}{\partial x^2\,\partial y} &= \frac{\partial}{\partial x}\left(\frac{\partial^2 f}{\partial x\,\partial y}\right) = \frac{\partial}{\partial x}\left(\frac{\partial^2 f}{\partial y\,\partial x}\right)\\ &= \frac{\partial}{\partial x}\left(\frac{\partial}{\partial y} f_x\right) = \frac{\partial}{\partial y}\left(\frac{\partial}{\partial x} f_x\right)\\ &= \frac{\partial}{\partial y}\left(\frac{\partial^2 f}{\partial x^2}\right) = \frac{\partial^3 f}{\partial y\,\partial x^2}.\end{aligned}$$

In fact, if all the partial derivatives that appear are continuous, the notation

$$\frac{\partial^{m+n} f}{\partial x^m\,\partial y^n}$$

may be used to denote the result of differentiating the function $f(x, y)$ m times with respect to x and n times with respect to y. The final result will be the same no matter what the order of differentiation is.

Being able to control the order of differentiation can work to our advantage, as we see in the next example.

EXAMPLE 4 To calculate

$$\frac{\partial^5}{\partial x^2\,\partial y^3}(x \sin y + e^y) \tag{6}$$

we can differentiate with respect to x first to show without any further work that the final result is zero:

$$\frac{\partial^5}{\partial x^2\,\partial y^3}(x \sin y + e^y) = \frac{\partial^3}{\partial y^3}\frac{\partial^2}{\partial x^2}(x \sin y + e^y)$$

$$= \frac{\partial^3}{\partial y^3}\frac{\partial}{\partial x}(\sin y) = \frac{\partial^3}{\partial y^3}(0) = 0. \ \square$$

The One-dimensional Heat Equation (= Diffusion Equation = Telegraph Equation)

If $w(x, t)$ represents the temperature at position x at time t in a uniform conducting rod with perfectly insulated sides (no heat flow through the sides—see Fig. 15.27), then the partial derivatives w_{xx} and w_t satisfy a differential equation of the following form:

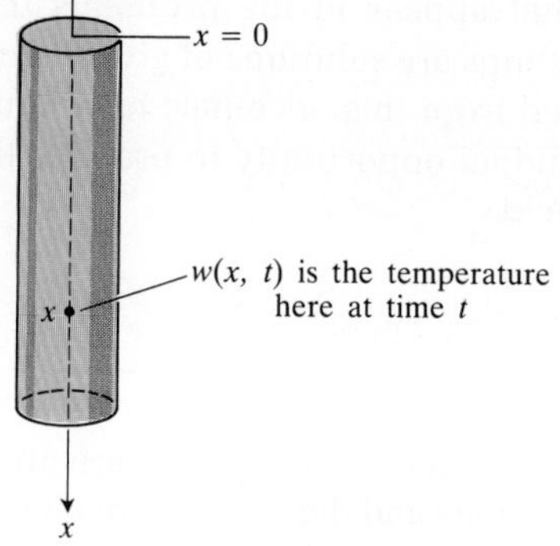

15.27 The temperature distribution in a heat-conducting rod satisfies the equation

$$w_{xx} = \frac{1}{c^2}w_t.$$

$$w_{xx} = \frac{1}{c^2}w_t. \tag{7}$$

This equation is called the *one-dimensional heat equation*. The value of c^2, which is determined by the material from which the rod is made, has been determined experimentally for a broad range of materials, and for a given application one finds the appropriate value in a table.

Equation (7) can be used effectively to model the seasonal temperature variation beneath the earth's surface, $w(x, t)$ then being the temperature on the tth day of the year x ft below the surface. (One thinks of a vertical core of the earth's surface as an insulated rod.) For dry soil,

$$c^2 = 0.19 \text{ ft}^2/\text{day},$$

and Eq. (7) becomes

$$w_{xx} = \frac{1}{0.19}w_t, \qquad \text{or} \qquad w_t = 0.19w_{xx}. \tag{8}$$

The solution that matches the seasonal temperature variation at the earth's surface is

$$w(x, t) = \cos(1.7 \times 10^{-2}t - 0.2x)e^{-0.2x}. \tag{9}$$

This can be verified by calculating w_{xx} and w_t, and seeing that $w_t = 0.19w_{xx}$. The calculation is tedious, however, and we shall omit it. The solution in (9) was graphed in Fig. 15.5.

In chemistry and biochemistry, the heat equation is known as the *diffusion equation*. In this context, $w(x, t)$ represents the concentration of a dissolved substance, a salt, for instance, diffusing along a tube filled with liquid. The value of $w(x, t)$ is the concentration at point x at time t. In other applications, $w(x, t)$ represents the diffusion of a gas down a long, thin pipe.

In electrical engineering, the heat equation appears in the forms

$$v_{xx} = RCv_t \tag{10}$$

and

$$i_{xx} = RCi_t, \tag{11}$$

which are known as the *telegraph equations*. These equations describe the voltage v and the flow of current i in a coaxial cable, or in any other cable in which leakage and inductance are negligible. The functions and constants in these equations are

$$\begin{aligned} v(x, t) &= \text{voltage at point } x \text{ at time } t, \\ R &= \text{resistance per unit length}, \\ C &= \text{capacitance to ground per unit of cable length}, \\ i(x, t) &= \text{current at point } x \text{ at time } t. \end{aligned}$$

Partial differential equations are generally hard to solve, in part because their solutions can take so many different forms. You will not be asked to solve any of the equations that appear in the problems that follow, but only to verify that given functions are solutions of given equations. There are still benefits to be gained from this: a chance to become acquainted with important equations, and an opportunity to practice the chain rule in another professional context.

PROBLEMS

In Problems 1–6, find $\partial^2 f/\partial x^2$, $\partial^2 f/\partial y^2$, and $\partial^2 f/\partial y\, \partial x$.

1. $f(x, y) = \ln(x^2 + y^2)$

2. $f(x, y) = e^x \ln(3 - y^2)$

3. $f(x, y) = x^2 y + \cos y + y \sin x$

4. $f(x, y) = (x - y)/xy$

5. $f(x, y, z) = xyz$

6. $f(x, y, z) = xy + yz + zx$

In Problems 7–10, verify that $w_{xy} = w_{yx}$.

7. $w = \ln(2x + 3y)$

8. $w = \tan^{-1}(y/x)$

9. $w = xy^2 + x^2y^3 + x^3y^4$

10. $w = e^x \sinh y + \cos(2x - 3y)$

11. Which order of differentiation will calculate f_{xy} faster: x first, or y first? Try to answer without writing anything down.

a) $f(x, y) = x \sin y + e^y$

b) $f(x, y) = 1/x$

c) $f(x, y) = y + (x/y)$

d) $f(x, y) = y + x^2y + 4y^3 - \ln(y^2 + 1)$

e) $f(x, y) = x^2 + 5xy + \sin x + 7e^x$

f) $f(x, y) = x \ln xy$

12. The derivative $\partial^5 f/\partial x^2\, \partial y^3$ is zero for each of the following functions. To show this as quickly as possible, which variable should one differentiate with respect to first: x, or y? Try to answer without writing anything down.

a) $f(x, y) = y^2x^4e^x + 2$

b) $f(x, y) = y + y^2[\ln(x^2 + 5) + \sin x - 7x^3]$

c) $f(x, y) = xe^y + x \sin y - x \cos y$

d) $f(x, y) = x \int_1^y e^{t^2/2}\, dt$

In Problems 13–18, assume that the necessary derivatives exist, and that the functions and derivatives are all continuous.

13. Find w_{xy} if $w = f(u, v)$, $u = x + y$, and $v = xy$. Express the answer in terms of u, v, f_u, and f_v.

14. Find z_{xx} if $z = f(u, v)$, $u = x^2 - y^2$, and $v = 2xy$. Express the answer in terms of x, y, f_u, and f_v.

15. Find u_{ss} if $u = f(x, y)$, $x = r^2 + s^2$ and $y = 2rs$. Express the answer in terms of r, s, f_x, and f_y.

16. Find $\partial^2 w/\partial x\, \partial y$ if $w = f(u, v)$, $u = x + y$, and $v = y^2$. Express the answer in terms of x, y, f_u, and f_v.

17. Find d^2w/dt^2 if $w = f(x, y)$, $x = \sin t$, and $y = t^2$.

18. Let $w = f(u)$, where $u = xg(y)$. Show that $w_{xx} = f''(xg(y)) \cdot g^2(y)$.

19. Find the value of $\partial^2 w/\partial \theta^2$ at $r = 2$, $\theta = \pi/2$ if $w = f(x, y)$, $x = r\cos\theta$, $y = r\sin\theta$, and $f_x = f_y = f_{xx} = f_{yy} = 1$ when $r = 2$, $\theta = \pi/2$. (Be sure to work out all the derivatives before substituting numerical values.)

Laplace equations

The *three-dimensional Laplace equation*

$$\frac{\partial^2 f}{\partial x^2} + \frac{\partial^2 f}{\partial y^2} + \frac{\partial^2 f}{\partial z^2} = 0 \tag{12}$$

is satisfied by steady-state heat distributions $T = f(x, y, z)$ in space, by gravitational potentials, and by electrostatic potentials. The *two-dimensional Laplace equation*,

$$\frac{\partial^2 f}{\partial x^2} + \frac{\partial^2 f}{\partial y^2} = 0, \tag{13}$$

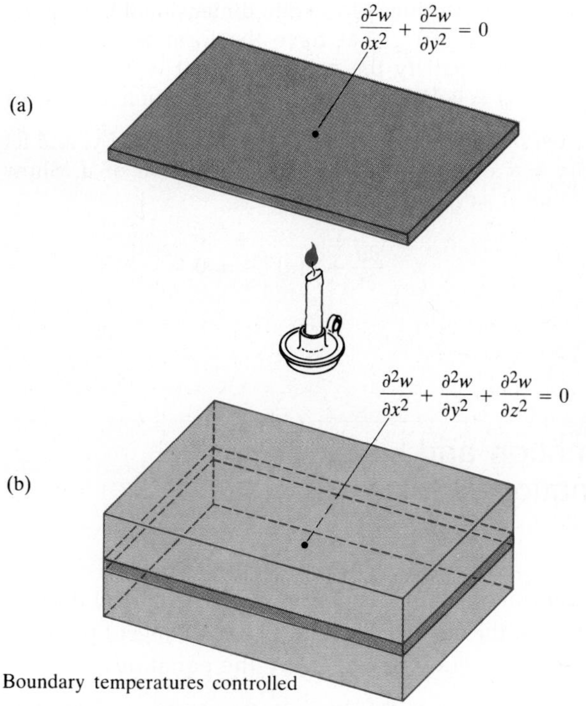

15.28 Steady-state heat distributions in planes and solids satisfy Laplace equations. The plane (a) may be treated as a thin slice of the solid (b) perpendicular to the z-axis.

obtained by dropping the $\partial^2 f/\partial z^2$ term from (12), describes potentials and steady-state heat distributions in a plane. See Fig. 15.28.

Show that the functions in Problems 20–27 satisfy a Laplace equation.

20. $f = x^2 + y^2 - 2z^2$

21. $f = 2z^3 - 3(x^2 + y^2)z$

22. $f = e^{-2y} \cos 2x$

23. $f = \ln \sqrt{x^2 + y^2}$

24. $f = (x^2 + y^2 + z^2)^{-1/2}$

25. $f = e^{3x+4y} \cos 5z$

26. $f = \tan^{-1}(y/x)$

27. $f = \cos 3x \cos 4y \sinh 5z$

28. Show that if $w = f(u, v)$ satisfies the Laplace equation $f_{uu} + f_{vv} = 0$, and if $u = (x^2 - y^2)/2$, $v = xy$, then $w_{xx} + w_{yy} = 0$.

29. Show that $u = f(x - iy) + g(x + iy)$ is a solution of the equation

$$\frac{\partial^2 u}{\partial x^2} + \frac{\partial^2 u}{\partial y^2} = 0,$$

if all the necessary partial derivatives exist. (Here, $i = \sqrt{-1}$.)

30. For what values of n does

$$f(x, y, z) = (x^2 + y^2 + z^2)^n$$

satisfy the three-dimensional Laplace equation (12)?

31. Find all solutions of the two-dimensional Laplace equation (13) of the form
a) $f(x, y) = ax^2 + bxy + c^2$,
b) $f(x, y) = ax^3 + bx^2y + cxy^2 + dy^3$.

The wave equation

If we stand on an ocean shore and take a snapshot of the waves, the picture shows a regular pattern of peaks and valleys at an instant in time (Fig. 15.29). We see periodic vertical motion in space, with respect to distance. If we stand in the water we can feel the rise and fall of the water as the waves go by. We see periodic vertical motion in time. In physics, this beautiful symmetry is expressed by the *one-dimensional wave equation*

$$\frac{\partial^2 w}{\partial t^2} = c^2 \frac{\partial^2 w}{\partial x^2}, \tag{14}$$

where w is the wave height, x is the distance variable, t is the time variable, and c is the velocity with which the waves are propagated.

In our example, x is the distance across the ocean's surface, but in other applications x might be the distance along a vibrating string, distance through air (sound waves), or distance through space (light waves). The number c varies with the medium and type of wave.

Show that the functions in Problems 32–37 are solutions of the wave equation, Eq. (14).

32. $w = \sin(x + ct)$

33. $w = \cos(2x + 2ct)$

34. $w = \sin(x + ct) + \cos(2x + 2ct)$

35. $w = \ln(2x + 2ct)$

36. $w = \tan(2x - 2ct)$

37. $w = 5\cos(3x + 3ct) - 7\sinh(4x - 4ct)$

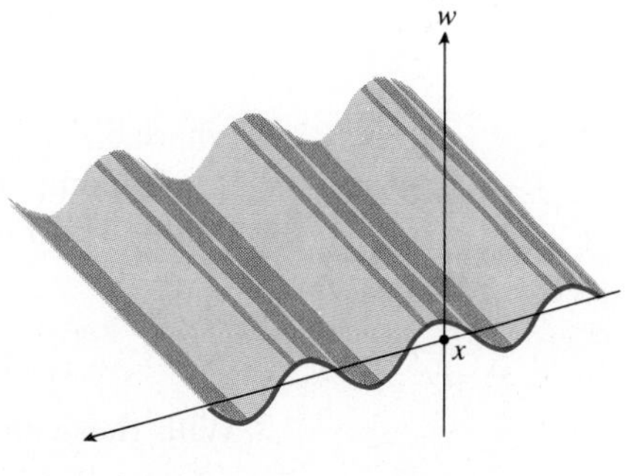

15.29 Waves in water at an instant in time. As time passes,

$$\frac{\partial^2 w}{\partial t^2} = c^2 \frac{\partial^2 w}{\partial x^2}.$$

38. If c is a constant and $w = f(u) + g(v)$, where $u = x + ct$ and $v = x - ct$, show that

$$\frac{\partial^2 w}{\partial t^2} = c^2 \frac{\partial^2 w}{\partial x^2} = c^2(f''(u) + g''(v)),$$

assuming that all the necessary derivatives exist.

The heat equation

39. Find all solutions of the one-dimensional heat equation $w_t = c^2 w_{xx}$ of the form $w = e^{rt} \sin \pi x$, r a constant.

40. Find all solutions of the one-dimensional heat equation $w_t = c^2 w_{xx}$ that have the form $w = e^{rt} \sin kx$ and that satisfy the conditions that $w(0, t) = 0$ and $w(L, t) = 0$. What happens to the solution as $t \to \infty$?

41. Let $u = f(y)$ be a differentiable function of y, and let $y = x - tg(u)$, where g is any function of u. Show that u satisfies the equation

$$\frac{\partial u}{\partial t} + g(u)\frac{\partial u}{\partial x} = 0.$$

15.8 Linear Approximation and Increment Estimation

Linearization

If a function $f(x, y)$ is continuous and has continuous first partial derivatives at a point (x_0, y_0), then the surface $z = f(x, y)$ has a tangent plane at the point $P_0(x_0, y_0, z_0) = (x_0, y_0, f(x_0, y_0))$ given by the equation

$$z = T(x, y) = f(x_0, y_0) + f_x(x_0, y_0)(x - x_0) + f_y(x_0, y_0)(y - y_0). \quad (1)$$

Since the surface and plane will be close together near P_0, we can use the values of T to approximate the values of f (Fig. 15.30). The function $T(x, y)$ is called the *linearization* or *linear approximation* of f at (x_0, y_0).

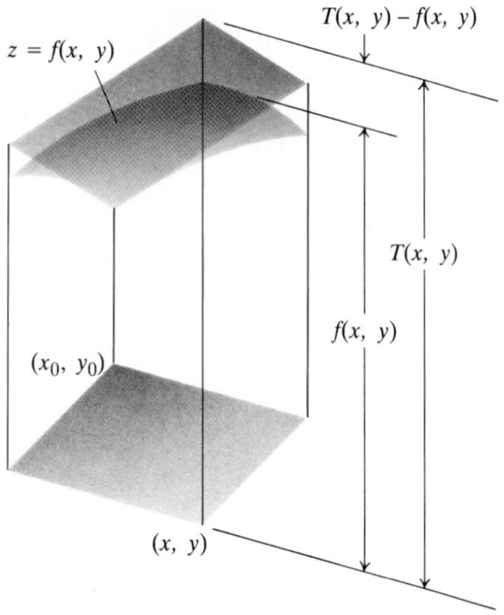

15.30 The tangent plane and surface are close together near the point of tangency.

The approximation

$$f(x, y) \approx T(x, y) \quad (2)$$

plays the same role for functions of two variables that the tangent line approximation plays for functions of a single variable. It provides a useful, simple way to estimate the values of functions that arise in science, engineering, and mathematics.

How good is the approximation in (2)? From the increment theorem for functions of two variables (Theorem 2, Article 15.3), we know that if $(x_0 + \Delta x, y_0 + \Delta y)$ is a point near (x_0, y_0), then

$$f(x_0 + \Delta x, y_0 + \Delta y) = f(x_0, y_0) + f_x(x_0, y_0)\,\Delta x + f_y(x_0, y_0)\,\Delta y + \varepsilon_1\,\Delta x + \varepsilon_2\,\Delta y, \quad (3)$$

where $\varepsilon_1, \varepsilon_2 \to 0$ as $\Delta x, \Delta y \to 0$.

To compare the values of $T(x, y)$ with those of $f(x, y)$ we change the notation in Eq. (3), writing

$$\Delta x = (x - x_0), \qquad \Delta y = (y - y_0)$$
$$x = x_0 + \Delta x, \qquad y = y_0 + \Delta y.$$

With these changes, Eq. (3) becomes

$$f(x, y) = \underbrace{f(x_0, y_0) + f_x(x_0, y_0)(x - x_0) + f_y(x_0, y_0)(y - y_0)}_{\text{This part is } T(x, y).} + \varepsilon_1(x - x_0) + \varepsilon_2(y - y_0). \quad (4)$$

Thus,

$$\begin{aligned} |f(x, y) - T(x, y)| &= |\varepsilon_1(x - x_0) + \varepsilon_2(y - y_0)| \\ &\leq |\varepsilon_1|\,|x - x_0| + |\varepsilon_2|\,|y - y_0|. \end{aligned} \tag{5}$$

This inequality, combined with the fact that $\varepsilon_1, \varepsilon_2 \to 0$ as $\Delta x, \Delta y \to 0$, assures us that the approximation $f(x, y) \approx T(x, y)$ will be close in some rectangle

$$R: |x - x_0| \leq \delta_1, \qquad |y - y_0| \leq \delta_2,$$

centered at (x_0, y_0).

To see how close the approximation is, we need the second order partial derivatives of f. If f_{xx}, f_{yy}, f_{xy} are continuous throughout R, then their absolute values are all less than or equal to

$$B = \text{max on } R \text{ of } |f_{xx}|, |f_{yy}|, |f_{xy}|, \tag{6}$$

and it turns out (for reasons explained in Article 15.9) that

$$|f(x, y) - T(x, y)| \leq \tfrac{1}{2}B(|x - x_0| + |y - y_0|)^2 \tag{7}$$

throughout R. (In practice the exact value of B is hard to find, and we replace it with any reasonable upper bound.) If we set $x - x_0 = \Delta x$, $y - y_0 = \Delta y$, Eq. (7) takes the form

$$|f(x, y) - T(x, y)| \leq \tfrac{1}{2}B(|\Delta x| + |\Delta y|)^2. \tag{8}$$

The following statements summarize the discussion so far:

Linear Approximation of $f(x, y)$ Near (x_0, y_0)

Suppose $f(x, y)$ and its first and second order partial derivatives are continuous throughout a rectangle R

$$R: |x - x_0| \leq \delta_1, \qquad |y - y_0| \leq \delta_2,$$

centered at (x_0, y_0), and that $|f_{xx}|$, $|f_{yy}|$, and $|f_{xy}|$ are less than or equal to some number B throughout R. Then throughout R

$$f(x, y) \approx \underbrace{f(x_0, y_0) + f_x(x_0, y_0)(x - x_0) + f_y(x_0, y_0)(y - y_0)}_{T(x, y)} \tag{9}$$

with an error E that is bounded by the inequality

$$|E| = |f(x, y) - T(x, y)| \leq \tfrac{1}{2}B(|x - x_0| + |y - y_0|)^2. \tag{10}$$

EXAMPLE 1 Find the linearization $T(x, y)$ of

$$f(x, y) = x^2 - xy + \tfrac{1}{2}y^2 + 3$$

at the point (3, 2). Then use Eq. (10) to give an upper bound for the error in the approximation $f(x, y) \approx T(x, y)$ over the rectangle

$$R: |x - 3| \leq 0.1, \qquad |y - 2| \leq 0.1.$$

Express the possible error as a percent of the value of f at (3, 2), the center of the rectangle.

Solution

$$f(3, 2) = 8,$$
$$f_x(3, 2) = 2x - y|_{(3,2)} = 4,$$
$$f_y(3, 2) = -x + y|_{(3,2)} = -1.$$

Therefore,

$$\begin{aligned} T(x, y) &= f(3, 2) + f_x(3, 2)(x - 3) + f_y(3, 2)(y - 2) \\ &= 8 + 4(x - 3) - (y - 2) \\ &= 4x - y - 2. \end{aligned}$$

To get an upper bound for the error in the approximation

$$x^2 - xy + \tfrac{1}{2}y^2 + 3 \approx 4x - y - 2 \tag{11}$$

over the rectangle R, we calculate

$$|f_{xx}| = |2| = 2,$$
$$|f_{yy}| = |1| = 1,$$
$$|f_{xy}| = |-1| = 1.$$

The largest of these is $B = 2$, and Eq. (10) gives

$$|E| = |f(x, y) - T(x, y)| \le \tfrac{1}{2}(2)(|x - 3| + |y - 2|)^2$$

throughout R. Since

$$|x - 3| \le 0.1 \qquad \text{and} \qquad |y - 2| \le 0.1,$$

on R, we have

$$|E| \le (0.1 + 0.1)^2 = 0.04.$$

Since $f(3, 2) = 8$, the estimate in (11) will be in error by no more than

$$\frac{0.04}{8} = 0.005 = 0.5\%$$

of the value of f at $(3, 2)$ as long as (x, y) lies in R. □

EXAMPLE 2 Show that

$$f(x, y) = \frac{1}{1 + x - y} \approx 1 - x + y \tag{12}$$

for $(x, y) \approx (0, 0)$.

Solution The function f is continuous for $|x - y| < 1$, as are the derivatives

$$f_x(x, y) = \frac{-1}{(1 + x - y)^2}, \qquad f_y(x, y) = \frac{1}{(1 + x - y)^2}. \tag{13}$$

At (0, 0) Eq. (9) takes the form

$$f(x, y) \approx f(0, 0) + xf_x(0, 0) + yf_y(0, 0),$$

or

$$f(x, y) = \frac{1}{1 + x - y} \approx 1 + x(-1) + y(1) = 1 - x + y,$$

which is Eq. (12). □

EXAMPLE 3 Find the linear approximation of

$$f(x, y) = \frac{1}{1 + x - y}$$

near the point (2, 1).

Solution Using Eq. (9) and the formulas for f_x and f_y from Example 2 gives

$$f(x, y) \approx f(2, 1) + f_x(2, 1)(x - 2) + f_y(2, 1)(y - 1),$$

$$\frac{1}{1 + x - y} \approx \frac{1}{2} + \left(-\frac{1}{4}\right)(x - 2) + \left(\frac{1}{4}\right)(y - 1) = \frac{3}{4} - \frac{x}{4} + \frac{y}{4}. \; \square$$

Increment Estimation

To the extent that a function $f(x, y)$ is approximated by its linearization $T(x, y)$, changes in f are approximated by changes in T. To be precise, suppose we start at a point $P_0(x_0, y_0)$ and move to a nearby point $Q(x, y) = (x_0 + \Delta x, y_0 + \Delta y)$. Then

$$\begin{aligned} T(x, y) &= f(x_0, y_0) + f_x(x_0, y_0)(x - x_0) + f_y(x_0, y_0)(y - y_0) \\ &= f(x_0, y_0) + f_x(x_0, y_0)\,\Delta x + f_y(x_0, y_0)\,\Delta y, \end{aligned}$$

and the change in the value of T resulting from the move from P to Q is

$$\begin{aligned} \Delta T &= T(x, y) - T(x_0, y_0) \\ &= T(x, y) - f(x_0, y_0) = f_x(x_0, y_0)\,\Delta x + f_y(x_0, y_0)\,\Delta y. \end{aligned} \tag{14}$$

See Fig. 15.31.

The change in the value of f that results from the move from P to Q is

$$\Delta f = f(x, y) - f(x_0, y_0),$$

which, according to Eq. (9), can be approximated as

$$\begin{aligned} \Delta f = f(x, y) - f(x_0, y_0) &\approx f_x(x_0, y_0)(x - x_0) + f_y(x_0, y_0)(y - y_0) \\ &\approx \underbrace{f_x(x_0, y_0)\,\Delta x + f_y(x_0, y_0)\,\Delta y}_{\Delta T}. \end{aligned}$$

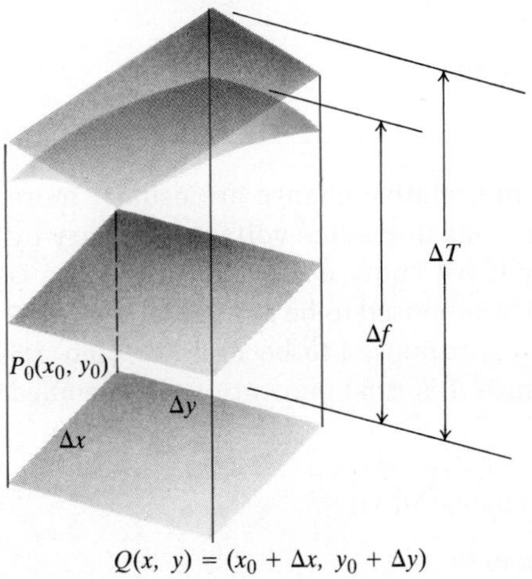

15.31 $\Delta T = T(x, y) - f(x_0, y_0)$.

The right-hand side of this approximation is the formula we derived for ΔT in Eq. (14).

Increment Estimation Formula for $f(x, y)$

$$\Delta f = f(x, y) - f(x_0, y_0) \approx \underbrace{f_x(x_0, y_0)\,\Delta x + f_y(x_0, y_0)\,\Delta y}_{\Delta T} \tag{15}$$

The increment approximation in (15) gives a way to see how *sensitive* a function $f(x, y)$ is to small changes in x and y near the point (x_0, y_0). This is illustrated in the following example.

EXAMPLE 4 How sensitive is the volume

$$V = \pi r^2 h$$

of a right circular cylinder to small changes in its radius and height near the point $(r_0, h_0) = (1, 5)$?

Solution We use the increment approximation in Eq. (15) to obtain

$$\begin{aligned}\Delta V &\approx V_r(r_0, h_0)\,\Delta r + V_h(r_0, h_0)\,\Delta h \\ &\approx V_r(1, 5)\,\Delta r + V_h(1, 5)\,\Delta h \\ &\approx 2\pi rh|_{(1,5)}\,\Delta r + \pi r^2|_{(1,5)}\,\Delta h = 10\pi\,\Delta r + \pi\,\Delta h,\end{aligned}$$

or

$$\Delta V \approx 10\pi\,\Delta r + \pi\,\Delta h. \tag{16}$$

This shows that a one-unit change in r will change V by about 10π units. A one-unit change in h will change V by about π units. Therefore, the volume of a cylinder with radius $r = 1$ and height $h = 5$ is nearly ten times as sensitive to small changes in r as it is to small changes in h.

In contrast, if the values of r and h are reversed, so that $r = 5$ and $h = 1$, then

$$\Delta V \approx 2\pi rh|_{(5,1)}\,\Delta r + \pi r^2|_{(5,1)}\,\Delta h = 10\pi r\,\Delta r + 25\pi\,\Delta h.$$

The volume is now more sensitive to small changes in h than it is to small changes in r. Thus, the sensitivity to change depends not only on the increments but also on the relative sizes of r and h (Fig. 15.32). □

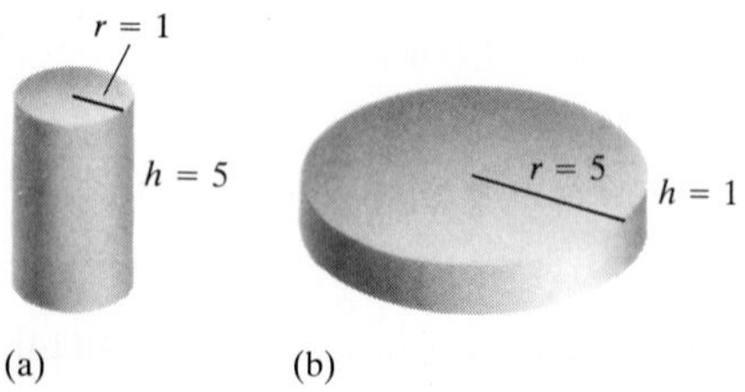

15.32 The volume of cylinder (a) is more sensitive to a small change in r than it is to an equally small change in h. The volume of cylinder (b) is more sensitive to small changes in h than it is to small changes in r.

There is a general rule to be learned from Example 4: A function is most sensitive to small changes in the variables that give the largest partial derivatives.

If the value of a function $z = f(x, y)$ changes from z_0 to $z_0 + \Delta z$, the change Δz may be looked at from three points of view:

Absolute change:	Δz	
Relative change:	$\dfrac{\Delta z}{z_0}$	(17)
Percentage change:	$\dfrac{\Delta z}{z_0} \times 100$	

In practice, percentage change and relative change are usually more important than absolute change. Knowing that a line voltage may vary by ±5 volts is useful information only if we know what the line voltage is supposed to be in the first place. If it is supposed to be 220,000 volts, then 5 volts more or less won't matter. If it is supposed to be 10 volts, then ±5 volts is likely to be significant because it is ±50 percent of the specified value.

EXAMPLE 5 How is the relative change in

$$V = \pi r^2 h$$

related to relative change in r and h? How are the percentage changes related?

Solution Starting with

$$\Delta V \approx V_r(r, h)\,\Delta r + V_h(r, h)\,\Delta h = 2\pi rh\,\Delta r + \pi r^2\,\Delta h,$$

we divide both sides by $V = \pi r^2 h$ to obtain

$$\frac{\Delta V}{V} \approx \frac{2\pi rh}{\pi r^2 h}\,\Delta r + \frac{\pi r^2}{\pi r^2 h}\,\Delta h$$

or

$$\frac{\Delta V}{V} \approx 2\frac{\Delta r}{r} + \frac{\Delta h}{h}. \tag{18}$$

The relative change in V is the relative change in h plus twice the relative change in r. Multiplying through by 100 gives

$$\frac{\Delta V}{V} \times 100 \approx 2\frac{\Delta r}{r} \times 100 + \frac{\Delta h}{h} \times 100, \tag{19}$$

which shows that the percentage change in V is about equal to the percentage change in h plus twice the percentage change in r.

A 3 percent increase in r and a 2 percent decrease in h give

$$\frac{r}{\Delta r} \times 100 = 3, \qquad \frac{h}{\Delta h} \times 100 = -2,$$

respectively. The resulting change in V is an increase of about four percent:

$$\frac{\Delta V}{V} \times 100 \approx 2(3) + (-2) = 6 - 2 = 4\%. \ \square$$

EXAMPLE 6 If r is measured with an accuracy of ± 2 percent and h with an accuracy of ± 0.5 percent, about how accurately can we calculate V from the formula

$$V = \pi r^2 h?$$

Solution From Eq. (18) in Example 5 we have

$$\frac{\Delta V}{V} \approx 2\frac{\Delta r}{r} + \frac{\Delta h}{h}.$$

We are told that

$$\Delta r = \pm 2\frac{r}{100}, \qquad \Delta h = \pm 0.5\frac{h}{100} = \pm\frac{h}{200},$$

so that

$$|\Delta r| \le \frac{r}{50}, \qquad |\Delta h| \le \frac{h}{200}.$$

Therefore,

$$\left|\frac{\Delta V}{V}\right| \approx \left|2\frac{\Delta r}{r} + \frac{\Delta h}{h}\right| \le \frac{2}{r}|\Delta r| + \frac{1}{h}|\Delta h|$$

$$\le \frac{2}{r}\frac{r}{50} + \frac{1}{h}\frac{h}{200} = \frac{1}{25} + \frac{1}{200} = 0.045.$$

The maximum percentage change or error resulting from the possible errors in measuring r and h will therefore be about 4.5 percent:

$$\left|\frac{\Delta V}{V}\right| \times 100 \approx 0.045 \times 100 = 4.5\%. \quad \square$$

How accurately do we have to measure r and h to have a reasonable chance of calculating $V = \pi r^2 h$ with an error of less than 2 percent? Questions like this are hard to answer for functions of more than one variable because there is usually no single right answer. Since

$$\frac{\Delta V}{V} \approx 2\frac{\Delta r}{r} + \frac{\Delta h}{h},$$

we see that $\Delta V/V$ is controlled by $\Delta r/r$ and $\Delta h/h$ *together*. If we can measure h with great accuracy we might come out all right even if we are sloppy in measuring r. On the other hand, our measurement of h may have such a large possible Δh that the resulting estimate of $\Delta V/V$ would be too crude to be useful even if Δr were zero.

What we do in such cases is look for a "reasonable" square about the measured values (r_0, h_0) in which V will not vary by more than the allowable amount from $V_0 = \pi r_0^2 h_0$. The next example shows how this may be done.

EXAMPLE 7 Find a reasonable square about the point $(r_0, h_0) = (5, 12)$ in which the value of $V = \pi r^2 h$ will not vary by more than ± 0.1.

Solution We start with the approximation

$$\begin{aligned} \Delta V &\approx 2\pi r_0 h_0\, \Delta r + \pi r_0^2\, \Delta h \\ &\approx 2\pi(5)(12)\, \Delta r + \pi(5)^2\, \Delta h = 120\pi\, \Delta r + 25\pi\, \Delta h. \end{aligned}$$

Since we are looking for a *square* about the point (5, 12), we may take $\Delta h = \Delta r$ and

$$\Delta V \approx 120\pi\, \Delta r + 25\pi\, \Delta r = 145\pi\, \Delta r.$$

To get

$$|\Delta V| \approx 145\pi\, |\Delta r| \le 0.1,$$

we take

$$|\Delta r| \le \frac{0.1}{145\pi} \approx 2.1 \times 10^{-4} \qquad \text{(rounding down).}$$

With $\Delta h = \Delta r$, the square we seek about (5, 12) is given by

$$|r - 5| \le 2.1 \times 10^{-4}, \qquad |h - 12| \le 2.1 \times 10^{-4}.$$

As long as (r, h) lies in this square, we may expect $|\Delta V| \le 0.1$. $\square$

Functions of More Than Two Variables

Analogous results, which we shall state without proof, hold for functions of any finite number of variables

For a function $w = f(x, y, z)$ of three independent variables that is continuous and has partial derivatives f_x, f_y, f_z at and in some neighborhood of the point (x_0, y_0, z_0), and whose derivatives are continuous at the point, we have

$$\begin{aligned}\Delta w &= f(x_0 + \Delta x, y_0 + \Delta y, z_0 + \Delta z) - f(x_0, y_0, z_0)\\ &= f_x\,\Delta x + f_y\,\Delta y + f_z\,\Delta z + \varepsilon_1\,\Delta x + \varepsilon_2\,\Delta y + \varepsilon_3\,\Delta z,\end{aligned} \tag{20}$$

where

$$\varepsilon_1, \varepsilon_2, \varepsilon_3 \to 0 \qquad \text{when} \qquad \Delta x, \Delta y, \text{ and } \Delta z \to 0.$$

The partial derivatives f_x, f_y, f_z in this formula are to be evaluated at the point (x_0, y_0, z_0).

Equation (20) leads to the following linear approximation and increment estimation formulas for small Δx, Δy, and Δz.

Linear Approximation and Increment Estimation Formulas for $w = f(x, y, z)$

1. *Linear approximation:*

$$f(x, y, z) \approx T(x, y, z), \tag{21}$$

where

$$\begin{aligned}T(x, y, z) = f(x_0, y_0, z_0) &+ f_x(x_0, y_0, z_0)(x - x_0)\\ &+ f_y(x_0, y_0, z_0)(y - y_0)\\ &+ f_z(x_0, y_0, z_0)(z - z_0)\end{aligned} \tag{22}$$

2. *Error bound:* Suppose that $f(x, y, z)$ and its first and second order partial derivatives are continuous throughout a three-dimensional rectangular region

$$R: |x - x_0| \le \delta_1, \qquad |y - y_0| \le \delta_2, \qquad |z - z_0| \le \delta_3,$$

centered at (x_0, y_0, z_0), and that $|f_{xx}|, |f_{yy}|, |f_{zz}|, |f_{xy}|, |f_{xz}|$, and $|f_{yz}|$ are all less than or equal to some number B throughout R. Then throughout R the error E in the approximation (21) is bounded by the inequality

$$\begin{aligned}|E| &= |f(x, y, z) - T(x, y, z)|\\ &\le \frac{1}{2}B(|x - x_0| + |y - y_0| + |z - z_0|)^2.\end{aligned} \tag{23}$$

3. *Increment estimate:*

$$\Delta f \approx \underbrace{f_x(x_0, y_0, z_0)\,\Delta x + f_y(x_0, y_0, z_0)\,\Delta y + f_z(x_0, y_0, z_0)\,\Delta z}_{\Delta T} \tag{24}$$

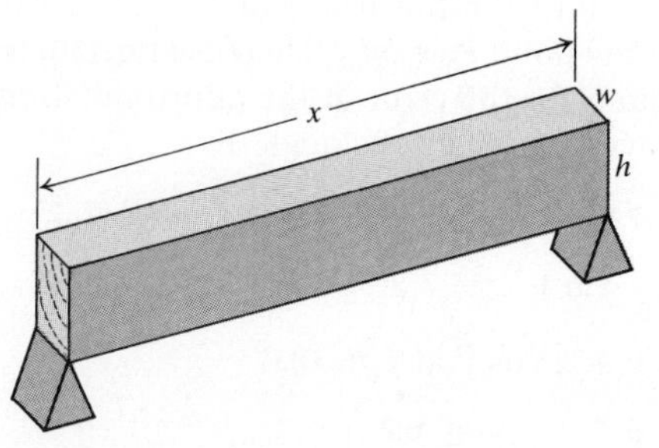

15.33 Beam supported at ends.

EXAMPLE 8 *Deflection of loaded beams.* A rectangular beam that is supported at its two ends (Fig. 15.33) will sag in the middle when subjected to a uniform load. The amount S of sag, called the *deflection* of the beam, may be estimated from the formula

$$S = C\frac{px^4}{wh^3} \quad \text{(m)},$$

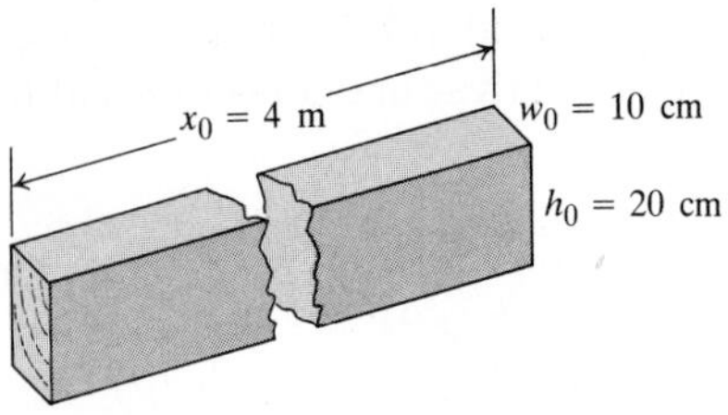

15.34 The dimensions of the beam in Example 8.

where

p = the load (kg/meter of beam length),
x = the length between supports (m),
w = the width of the beam (m),
h = the height of the beam (m),
C = a constant that depends on the units of measurement and on the material from which the beam is made.

When

$$\Delta S \approx S_p\,\Delta p + S_x\,\Delta x + S_w\,\Delta w + S_h\,\Delta h$$

is written out for a particular set of values p_0, x_0, w_0, h_0, and simplified, the resulting approximation is

$$\Delta S \approx S_0\left[\frac{\Delta p}{p_0} + \frac{4\,\Delta x}{x_0} - \frac{\Delta w}{w_0} - \frac{3\,\Delta h}{h_0}\right], \tag{25}$$

where $S_0 = S(p_0, x_0, h_0, w_0) = Cp_0x_0^4/w_0h_0^3$.
At $p_0 = 100$ kg/m, $x_0 = 4$ m, $w_0 = 0.1$ m, and $h_0 = 0.2$ m,

$$\Delta S \approx S_0\left[\frac{\Delta p}{100} + \Delta x - 10\,\Delta w - 15\,\Delta h\right]. \tag{26}$$

(See Fig. 15.34.) □

Conclusions about this beam from Eq. (26) Since Δp and Δx appear with positive coefficients in Eq. (26), increases in p and in x will increase the sag. But Δw and Δh appear with negative coefficients, so that increases in w and h will *decrease* the sag (make the beam stiffer). The sag is not very sensitive to small changes in load, because the coefficient of Δp is 1/100. The coefficient of Δh is a negative number of greater magnitude than the coefficient of Δw. Therefore, making the beam $\Delta h = 1$ cm higher will decrease the sag more than making the beam $\Delta w = 1$ cm wider.

PROBLEMS

In Problems 1–6, find the linearization $T(x, y)$ of the function $f(x, y)$ at each point.

1. $f(x, y) = x^2 + y^2 + 1$ at (a) $(0, 0)$, (b) $(1, 1)$.

2. $f(x, y) = x^3y^4$ at (a) $(1, 1)$, (b) $(0, 0)$.

3. $f(x, y) = e^x \cos y$ at (a) $(0, 0)$, (b) $(0, \pi/2)$.

4. $f(x, y) = (x + y + 2)^2$ at (a) $(0, 0)$, (b) $(1, 2)$.

5. $f(x, y) = 3x - 4y + 5$ at (a) $(0, 0)$, (b) $(1, 1)$.

6. $f(x, y) = e^{2y-z}$ at (a) $(0, 0)$, (b) $(1, 2)$.

In Problems 7–10, find the linearization $T(x, y, z)$ of the function $f(x, y, z)$ at each point.

7. $f(x, y, z) = \sqrt{x^2 + y^2 + z^2}$ at
(a) $(1, 0, 0)$, (b) $(1, 1, 0)$, (c) $(1, 1, 1)$.

8. $f(x, y, z) = (\sin xy)/z$ at
(a) $(\pi/2, 1, 1)$, (b) $(2, 0, 1)$.

9. $f(x, y, z) = e^x \cos(y + z)$ at
(a) $(0, 0, 0)$, (b) $(0, \pi/4, \pi/4)$.

10. $f(x, y, z) = \tan^{-1}(xyz)$ at
(a) $(1, 0, 0)$, (b) $(1, 1, 0)$, (c) $(1, 1, 1)$.

In Problems 11–15, find the linearization $T(x, y)$ of the function $f(x, y)$ at the point $P_0(x_0, y_0)$. Then use Eq. (10) to give an upper bound for the error in the approximation $f(x, y) \approx T(x, y)$ over the given rectangle R.

11. $f(x, y) = \frac{1}{2}x^2 + xy + \frac{1}{4}y^2 + 3x - 3y + 4$ at $P_0(2, 2)$.

R: $|x - 2| \le 0.1$, $|y - 2| \le 0.1$

12. $f(x, y) \times 1 + y + x \cos y$ at $P_0(0, 0)$.

R: $|x| \le 0.2$, $|y| \le 0.2$

(Use $|\cos y| \le 1$ and $|\sin y| \le 1$ in estimating E.)

13. $f(x, y) = \ln x + \ln y$ at $P_0(1, 1)$.

$R: |x - 1| \le 0.2, \quad |y - 1| \le 0.2$

14. $f(x, y) = e^x \cos y$ at $P_0(0, 0)$.

$R: |x| \le 0.1, \quad |y| \le 0.1$

(Use $e^x \le 3$ and $|\cos y| \le 1$ in estimating E.)

15. $f(x, y) = xy^2 + y \cos (x - 1)$ at $P_0(1, 2)$.

$R: |x - 1| \le 0.1, \quad |y - 2| \le 0.1$

16. Estimate how much simultaneous errors of 2 percent in a, b, and c may affect the product abc.

17. Find the linearization $T(x, y, z)$ of the function $f(x, y, z)$ at the point $P_0(x_0, y_0, z_0)$. Then use Eq. (23) to give a bound for the error in the approximation $f(x, y, z) \approx T(x, y, z)$ over the given rectangular region R.

a) $f(x, y, z) = x^2 + xy + yz + (1/4)z^2$ at $P_0(1, 1, 2)$.
$R: |x - 1| \le 0.01, |y - 1| \le 0.01, |z - 2| \le 0.08$

b) $f(x, y, z) = xy + 2yz - 3xz$ at $P_0(1, 1, 0)$.
$R: |x - 1| \le 0.01, |y - 1| \le 0.01, |z| \le 0.01$

c) $f(x, y, z) = \sqrt{2} \cos x \sin (y + z)$ at $P_0(0, 0, \pi/4)$.
$R: |x| \le 0.01, |y| \le 0.01, |z - \pi/4| \le 0.01$.

18. What relationship must hold between the r and h in Example 4 if the volume of the cylinder is to be equally sensitive to small changes in the two variables?

19. The beam of Example 8 is tipped on its side, so that $h = 0.1$ m and $w = 0.2$ m.

a) What is the approximation for ΔS now?

b) Compare the sensitivity of the beam to a small change in height with its sensitivity to a change of the same amount in width.

20. Estimate the amount of material in a hollow rectangular box whose inside measurements are 5 ft long, 3 ft wide, 2 ft deep, if the box is made of lumber that is $\frac{1}{2}$ in. thick and the box has no top.

21. The area of a triangle is $A = \frac{1}{2}ab \sin C$, where a and b are two sides of the triangle and C is the included angle. In surveying a particular triangular plot of land, a and b are measured to be 150 ft and 200 ft, respectively, and C is read to be 60°. By how much (approximately) is the computed area in error if a and b are in error by $\frac{1}{2}$ ft each and C is in error by 2°?

22. Suppose that $u = xe^y + y \sin z$, and that x, y, and z can be measured with maximum possible errors of ± 0.2, ± 0.6, and $\pm \pi/180$, respectively. Estimate the maximum possible error in calculating u from the measured values $x = 2$, $y = \ln 3$, $z = \pi/2$.

23. Suppose T is to be found from the formula $T = x \cosh y$, where x and y are found to be 2 and $\ln (2)$ with maximum possible errors of $|\Delta x| = 0.1$ and $|\Delta y| = 0.02$. Estimate the maximum possible error in the computed value of T.

24. About how accurately may $V = \pi r^2 h$ be calculated from measurements of r and h that are in error by 1%?

25. If $r = 5$ cm and $h = 12$ cm to the nearest millimeter, what should we expect the maximum percentage error in calculating $V = \pi r^2 h$ to be?

26. To estimate the volume of a cylinder of radius about 2 m and height about 3 m, about how accurately should the radius and height be measured so that the error in the volume estimate will not exceed 0.1 m^3? Assume that the possible error Δr in measuring r is equal to the possible error Δh in measuring h.

27. Give a reasonable square centered at (1, 1) over which the value of $f(x, y) = x^3 y^4$ will not vary by more than ± 0.1.

28. When an x ohm and y ohm resistor are in parallel, the resistance R they produce may be calculated from the formula $1/R = 1/x + 1/y$. By about what percentage will R change if x increases from 20 to 20.1 ohms and y decreases from 25 to 24.9 ohms?

29. a) If $x = 3 \pm 0.01$, $y = 4 \pm 0.01$, with what accuracy can the polar coordinates r and θ of the point (x, y) be calculated from the formulas $r^2 = x^2 + y^2$, $\theta = \tan^{-1} (y/x)$?

b) At the point $(x, y) = (3, 4)$, are r and θ more sensitive to changes in x, or to changes in y? Draw a figure showing x, y, r, θ and confirm your results by appealing to the geometry of the figure.

30. A function $w = f(x, y)$ is said to be *differentiable* at $P(a, b)$ if there are constants M and N (possibly depending on f and P) such that

$$\Delta w = M\,\Delta x + N\,\Delta y + \alpha[|\Delta x| + |\Delta y|]$$

and $\alpha \to 0$ as $|\Delta x| + |\Delta y| \to 0$. Here $\Delta w = f(a + \Delta x, b + \Delta y) - f(a, b)$. Prove that if f is differentiable at (a, b), then $M = f_x(a, b)$ and $N = f_y(a, b)$.

31. If the function $w = f(x, y)$ is differentiable at $P(a, b)$ (see Problem 30), prove it is continuous there.

32. To which of the variables K, M, and h is the function $Q = \sqrt{2KM/h}$ most sensitive near the point $(K_0, M_0, h_0) = (2, 20, 0.05)$? least sensitive?

33. a) Around the point (1, 0) is $f(x, y) = x^2(y + 1)$ more sensitive to changes in x, or to changes in y?

b) What ratio of Δx to Δy will make ΔT in Eq. (15) equal to zero at (1, 0)?

34. If $|a|$ is much greater than $|b|$, $|c|$, and $|d|$, to which entry is the value of the determinant

$$\begin{vmatrix} a & b \\ c & d \end{vmatrix}$$

most sensitive?

35. *Method of steepest descent.* Suppose it is desired to find a solution of the equation $f(x, y, z) = 0$. Let

$P_0(x_0, y_0, z_0)$ be a first guess, and suppose $f(x_0, y_0, z_0) = f_0$ is not zero. Let $(\nabla f)_0$ be the gradient vector normal to the surface $f(x, y, z) = f_0$ at P_0. If f_0 is positive, we want to decrease the value of f. The gradient points in the direction of most rapid increase, its negative in the direction of "steepest descent." We therefore take as next approximation

$$x_1 = x_0 - hf_x(x_0, y_0, z_0),$$
$$y_1 = y_0 - hf_y(x_0, y_0, z_0),$$
$$z_1 = z_0 - hf_z(x_0, y_0, z_0).$$

What value of h corresponds to making $\Delta T = -f_0$? What change is suggested if f_0 is negative? (The method could be applied to the problem of solving the simultaneous equations

$$x - y + z = 3,$$
$$x^2 + y^2 + z^2 = 20,$$
$$xyz = 8,$$

by writing

$$f(x, y, z) = (x - y + z - 3)^2 + (x^2 + y^2 + z^2 - 20)^2 + (xyz - 8)^2.)$$

15.9 Maxima, Minima, and Saddle Points

As we can see in Figs. 15.35 and 15.36, functions of two variables can have relative and absolute maximum and minimum values just the way functions of a single variable can. The way we go about finding these extreme values is much the same, except that we now have more derivatives to work with because we have more variables.

Testing for Extreme Values

1. *Look for critical points and boundary points.*

As we know, a continuous function $z = f(x, y)$ takes on an absolute maximum value and an absolute minimum value on any closed, bounded region R on which it is defined. Moreover (as we shall show later), these and any other relative maxima and minima can occur only at

i) boundary points of R,

ii) interior points of R where $f_x = f_y = 0$, or points where f_x or f_y fail to exist. (We call these the *critical points* of f.)

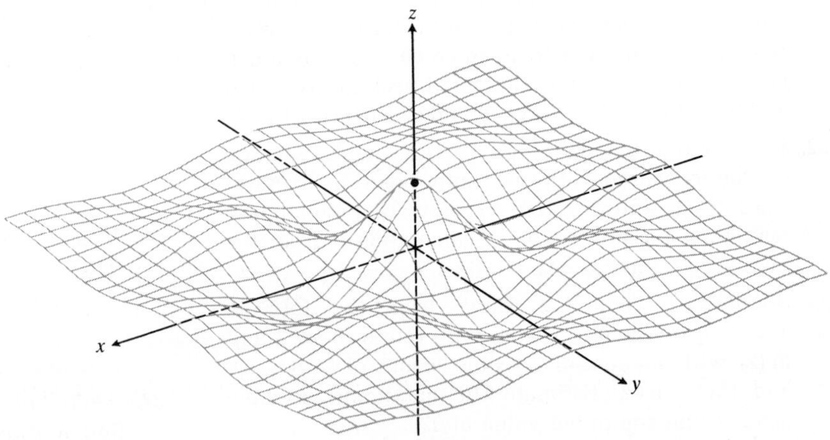

15.35 The function $z = (\cos x)(\cos y)e^{-\sqrt{x^2+y^2}}$ has many relative maxima and minima.

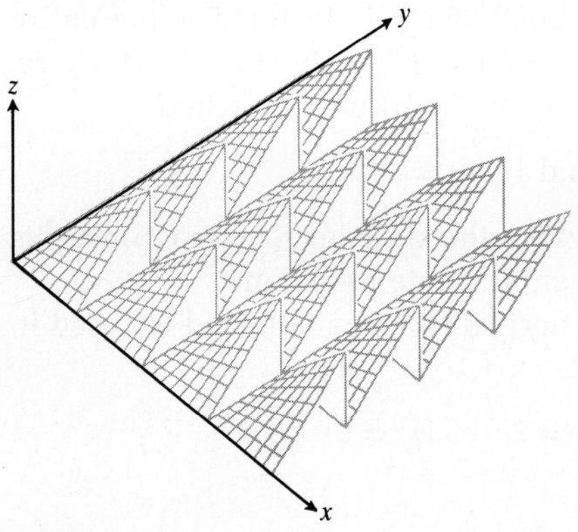

15.36 The surface $z = (x - [x])(y - [y])$ for $x \geq 0$, $y \geq 0$, where $[x]$ is the greatest integer function.

Usually there are only a few such points, so we can evaluate f at all of them and choose the largest and smallest values. No further tests are required if all we wish to find is the absolute maximum and minimum.

2. *If there are no boundary points, look for critical points.*

A function $z = f(x, y)$ defined on a region R without boundary points (for instance, the interior of a disc, triangle, or rectangle, a quadrant minus the axes, or the entire plane) may have no relative maxima or minima on R. However, if it does they must occur at points of R where

$$f_x = f_y = 0.$$

or where one or both of these derivatives fails to exist (as with $z = \sqrt{x^2 + y^2}$ at $(0, 0)$).

3. *A second derivative test may be applied at interior points where $f_x = f_y = 0$ and the first and second order partial derivatives of f are continuous.*

The fact that $f_x = f_y = 0$ at an interior point (a, b) does not guarantee that f will have an extreme value there. There is, however, a second derivative test that may help to identify the behavior of f at (a, b). It goes like this:

If $f_x(a, b) = f_y(a, b) = 0$, then

i) f has a *relative maximum* at (a, b) if $f_{xx} < 0$ and $f_{xx}f_{yy} - f_{xy}^2 > 0$ at (a, b).
ii) f has a *relative minimum* at (a, b) if $f_{xx} > 0$ and $f_{xx}f_{yy} - f_{xy}^2 > 0$ at (a, b).
iii) f has a *saddle point* at (a, b) if $f_{xx}f_{yy} - f_{xy}^2 < 0$ at (a, b).
iv) The test is *inconclusive* at (a, b) if $f_{xx}f_{yy} - f_{xy}^2 = 0$ at (a, b). We must find some other way to determine the behavior of f at (a, b).

The expression $f_{xx}f_{yy} - f_{xy}^2$ is called the *discriminant* of f. It is sometimes easier to remember in the determinant form

$$f_{xx}f_{yy} - f_{xy}^2 = \begin{vmatrix} f_{xx} & f_{xy} \\ f_{yx} & f_{yy} \end{vmatrix}.$$

We shall now look at examples that show these tests at work. After that, we shall show why the condition $f_x = f_y = 0$ is a necessary condition for having an extreme value at an interior point of the domain of a differentiable function and look into the mathematics behind the second derivative test.

The Tests at Work

In the first example we look at the function $f(x, y) = x^2 + y^2$, whose behavior we already know from looking at the formula: Its value is zero at the origin and increases steadily as (x, y) moves away from the origin. The point of Example 1 is to show how the tests above reveal this.

EXAMPLE 1 Find the extreme value of

$$f(x, y) = x^2 + y^2.$$

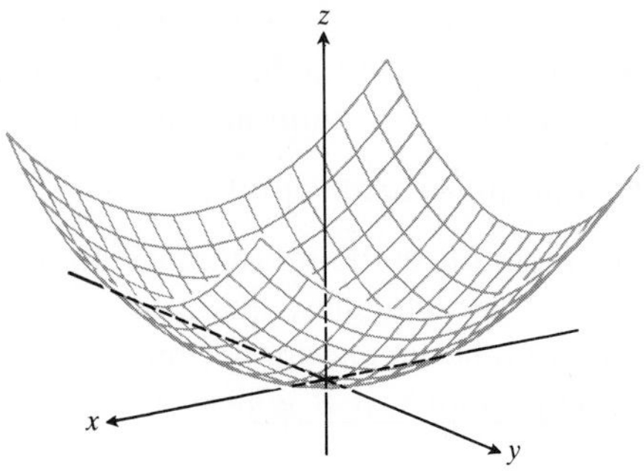

15.37 The function $f(x, y) = x^2 + y^2$ has an absolute minimum at the origin.

Solution The domain of f has no boundary points, for it is the entire plane (Fig. 15.37). The derivatives $f_x = 2x$ and $f_y = 2y$ exist everywhere. Therefore, relative maxima and minima can occur only where

$$f_x = 2x = 0 \qquad \text{and} \qquad f_y = 2y = 0.$$

The only possibility is the origin, where the value of f is zero. Since f is never negative, we see that zero is an absolute minimum.

We have not needed the second derivative test at all. Had we used it, we would have found

$$f_{xx} = 2, \qquad f_{yy} = 2, \qquad f_{xy} = 0$$

and

$$f_{xx}f_{yy} - f_{xy}^2 = (2)(2) - (0)^2 = 4 > 0,$$

identifying (0, 0) as a relative minimum. This in itself does not identify (0, 0) as an absolute minimum. It takes more information to do that. □

EXAMPLE 2 Find the extreme values of the function

$$f(x, y) = xy - x^2 - y^2 - 2x - 2y + 4.$$

Solution The function is defined and differentiable for all x and y and therefore has extreme values only at the points where f_x and f_y are simultaneously zero. This leads to

$$f_x = y - 2x - 2 = 0, \qquad f_y = x - 2y - 2 = 0,$$

or

$$x = y = -2.$$

Therefore, the point $(-2, -2)$ is the only point where f may take on an extreme value. To see if it does so, we calculate

$$f_{xx} = -2, \qquad f_{yy} = -2, \qquad f_{xy} = 1.$$

The discriminant of f at $(a, b) = (-2, -2)$ is

$$f_{xx}f_{yy} - f_{xy}^2 = (-2)(-2) - (1)^2 = 4 - 1 = 3.$$

The combination

$$f_{xx} < 0 \qquad \text{and} \qquad f_{xx}f_{xy} - f_{xy}^2 > 0$$

tells us that f has a relative maximum at $(-2, -2)$. The value of f at this point is $f(-2, -2) = 8$. □

EXAMPLE 3 Find the extreme values of

$$f(x, y) = xy.$$

Solution Since the function is differentiable everywhere and its domain has no boundary points (Fig. 15.38), the function can assume extreme values only where

$$f_x = y = 0 \qquad \text{and} \qquad f_y = x = 0.$$

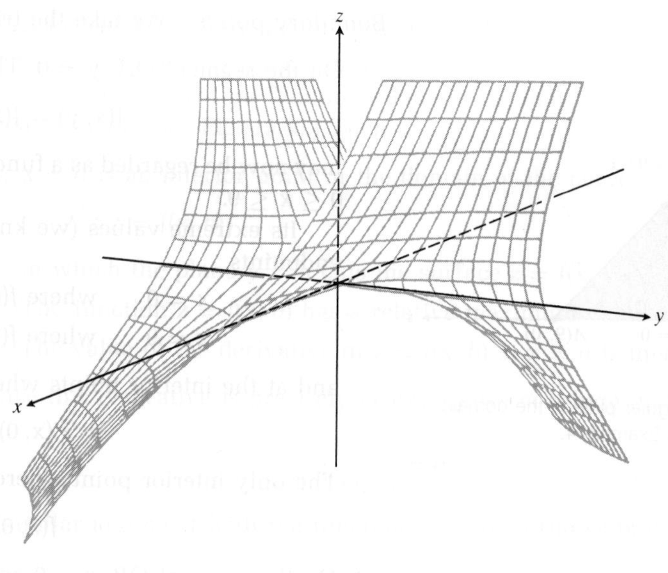

15.38 The surface $z = xy$ has a saddle point at the origin.

Thus, the origin is the only point where f might have an extreme value. To see what happens there, we calculate

$$f_{xx} = 0, \qquad f_{yy} = 0, \qquad f_{xy} = 1.$$

The discriminant,

$$f_{xx}f_{yy} - f_{xy}^2 = -1,$$

is negative. Therefore the function has a saddle point at (0, 0). We conclude that $f(x, y) = xy$ assumes no extreme values at all.

If we restrict the domain of f to the disc $x^2 + y^2 \leq 1$, then the maximum value of f is $+\frac{1}{2}$ and the minimum is $-\frac{1}{2}$. (Change to polar coordinates: $xy = r^2 \sin\theta \cos\theta = \frac{1}{2}r^2 \sin 2\theta$.) □

EXAMPLE 4 Find the absolute maximum and minimum values of

$$f(x, y) = 2 + 2x + 2y - x^2 - y^2$$

on the triangular plate in the first quadrant bounded by the lines $x = 0$, $y = 0$, $y = 9 - x$.

Solution The only places where f can assume these values are points on the boundary of the triangle (Fig. 15.39) and points inside it where $f_x = f_y = 0$.

Interior points. For these we have

$$f_x = 2 - 2x = 0,$$
$$f_y = 2 - 2y = 0,$$

or

$$(x, y) = (1, 1),$$

where

$$f(1, 1) = 4.$$

PROBLEMS

Test the following surfaces for maxima, minima, and saddle points. Find the function values at these points.

1. $z = x^2 + xy + y^2 + 3x - 3y + 4$
2. $z = x^2 + 3xy + 3y^2 - 6x + 3y - 6$
3. $z = 5xy - 7x^2 + 3x - 6y + 2$
4. $z = 2xy - 5x^2 - 2y^2 + 4x + 4y - 4$
5. $z = x^2 + xy + 3x + 2y + 5$
6. $z = y^2 + xy - 2x - 2y + 2$
7. $z = 2xy - 5x^2 - 2y^2 + 4x - 4$
8. $z = 2xy - x^2 - 2y^2 + 3x + 4$
9. $z = x^2 + xy + y^2 + 3y + 3$
10. $z = 3x^2 + 6xy + 7y^2 - 2x + 4y$
11. $z = 2x^2 + 3xy + 4y^2 - 5x + 2y$
12. $z = 4x^2 - 6xy + 5y^2 - 20x + 26y$
13. $z = x^2 - 4xy + y^2 + 5x - 2y$
14. $z = x^2 + y^2 - 2x + 4y + 6$
15. $z = x^2 - y^2 - 2x + 4y + 6$
16. $z = x^2 - 2xy + 2y^2 - 2x + 2y + 1$
17. $z = x^2 + 2xy$
18. $z = 3 + 2x + 2y - 2x^2 - 2xy - y^2$
19. $z = x^2 + xy + y^2 + x - 4y + 5$
20. $z = x^2 - xy + y^2 + 2x + 2y - 4$
21. $z = 3x^2 - xy + 2y^2 - 8x + 9y + 10$
22. $z = 5x^2 - 4xy + 2y^2 + 4x - 4y + 10$
23. $z = x^3 - y^3 - 2xy + 6$
24. $z = x^3 + y^3 + 3x^2 - 3y^2 - 8$
25. $z = 6x^2 - 2x^3 + 3y^2 + 6xy$
26. $z = 9x^3 + y^3/3 - 4xy$
27. $z = x^3 + y^3 - 3xy + 15$
28. $z = x^3 + 3xy + y^3$
29. $z = 4xy - x^4 - y^4$
30. $z = x \sin y$

In Problems 31–38, find the absolute maxima and minima of the functions on the given domains.

31. $f(x, y) = 2x^2 - 4x + y^2 - 4y + 1$ on the closed triangular plate bounded by the lines $x = 0$, $y = 2$, $y = 2x$ in the first quadrant.
32. $D(x, y) = x^2 - xy + y^2 + 1$ on the closed triangular plate in the first quadrant bounded by the lines $x = 0$, $y = 4$, $y = x$.
33. $f(x, y) = x^2 + y^2$ on the closed triangular plate bounded by the lines $x = 0$, $y = 0$, $y + 2x = 2$ in the first quadrant.
34. $T(x, y) = x^2 + xy + y^2 - 6x$ on the rectangular plate $0 \leq x \leq 5$, $-3 \leq y \leq 3$.
35. $T(x, y) = x^2 + xy + y^2 - 6x + 2$ on the rectangular plate $0 \leq x \leq 5$, $-3 \leq y \leq 0$.
36. $f(x, y) = 48xy - 32x^3 - 24y^2$ on the rectangular plate $0 \leq x \leq 1$, $0 \leq y \leq 1$.
37. $f(x, y) = (x^2 - 4x) \cos y$ over the region $1 \leq x \leq 3$, $-\pi/4 \leq y \leq \pi/4$.
38. $f(x, y) = x - (x - \frac{1}{4})(y + \frac{1}{2})$ on the closed triangular region bounded by the lines $x = 0$, $y = 0$, $x + y = 1$ in the first quadrant.
39. A flat circular plate has the shape of the region $x^2 + y^2 \leq 1$. The plate, including the boundary where $x^2 + y^2 = 1$, is heated so that the temperature at any point (x, y) is
$$T(x, y) = x^2 + 2y^2 - x.$$
Find the hottest and coldest points on the plate, and the temperature at each of these points.
40. Find the critical point of
$$f(x, y) = xy + 2x - \ln x^2y$$
in the open first quadrant $(x > 0, y > 0)$ and show that f takes on a minimum there.
41. Find all relative and absolute maxima and minima of
$$f(x, y) = -x^2 - y^2 + 2x + 2y + 2$$
on the closed first quadrant $x \geq 0$, $y \geq 0$.
42. Determine the maxima, minima, and saddle points of $f(x, y)$, if any, given that
 a) $f_x = 2x - 4y$, $f_y = 2y - 4x$,
 b) $f_x = 2(x - 1)$, $f_y = 2(y - 2)$,
 c) $f_x = 9x^2 - 9$, $f_y = 2y + 4$.
43. Sketch the surface
$$z = \sqrt{x^2 + y^2}$$
over the region R: $|x| \leq 1$, $|y| \leq 1$. Find the high and low points of the surface over R. Discuss the existence, and the values, of $\partial z/\partial x$ and $\partial z/\partial y$ at these points.
44. The discriminant $f_{xx}f_{yy} - f_{xy}^2$ of each of the following functions is zero at the origin. Determine whether the function has a maximum, a minimum, or neither at the origin by imagining in each case what the surface $z = f(x, y)$ looks like.
 a) $f(x, y) = x^2y^2$
 b) $f(x, y) = 1 - x^2y^2$
 c) $f(x, y) = xy^2$
 d) $f(x, y) = x^3y^2$
 e) $f(x, y) = x^3y^3$
 f) $f(x, y) = x^4y^4$

45. *The error bound for linear approximations of functions of three variables* (Article 15.8, Eq. (23)).

a) Suppose that $f(x, y, z)$ and its first and second order partial derivatives are continuous throughout a region R whose interior contains the point $P(a, b, d)$. Suppose also that the increments h, k, and m are small enough for the line segment PS joining P to the point $S(a + h, b + k, d + m)$ to lie in the interior of R. Parameterize PS by the equations

$$x = a + th,$$
$$y = b + tk,$$
$$z = d + tm, \qquad 0 \le t \le 1.$$

Apply Taylor's theorem to the function

$$F(t) = f(a + th, b + tk, d + tm)$$

to show that

$$\begin{aligned} f(a + h, b + k, d + m) &= f(a, b, c) + hf_x(a, b, c) + kf_y(a, b, c) + mf_z(a, b, c) \\ &+ \frac{1}{2}\Big| h^2 f_{xx} + k^2 f_{yy} + m^2 f_{zz} + 2hkf_{xy} \\ &+ 2hmf_{xz} + 2kmf_{yz} \Big|_{(a+ch,\, b+ck,\, d+cm)} \end{aligned} \tag{6'}$$

for some c between 0 and 1. This is the three-variable version of Eq. (6) of the present article.

b) Use Eq. (6′) from part (a) to derive the inequality in Article 15.8, Eq. (23).

***Toolkit* programs**
3D Grapher

15.10 Lagrange Multipliers

Constrained Maxima and Minima

As we saw in Article 15.9, it is sometimes necessary to find the extreme values of a function $f(x, y)$ when its domain is subject to some kind of constraint, for example, that it be a particular triangular plate or quadrant of the plane. But, as Fig. 15.42 suggests, functions may be subject to other kinds of constraints as well. In this article, we explore an important method of finding the maxima and minima of constrained functions—the method of *Lagrange multipliers.* We set the stage with two examples, and then discuss the method in general terms and look at more examples.

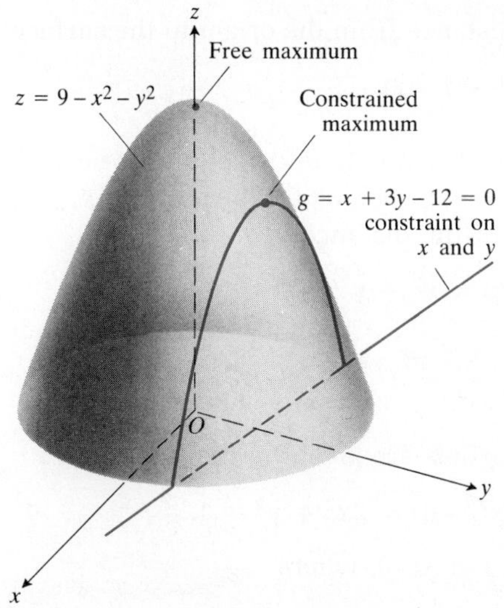

15.42 $f(x, y) = 9 - x^2 - y^2$, subject to the constraint $g(x, y) = x + 3y - 12 = 0$.

EXAMPLE 1 Find the point $P(x, y, z)$ on the plane

$$2x + y - z - 5 = 0$$

that lies closest to the origin.

Solution The problem asks us to find the minimum value of the function

$$|\overrightarrow{OP}| = \sqrt{(x - 0)^2 + (y - 0)^2 + (z - 0)^2} = \sqrt{x^2 + y^2 + z^2}$$

subject to the constraint that

$$2x + y - z - 5 = 0.$$

Since $|\overrightarrow{OP}|$ has a minimum value wherever the function

$$f(x, y, z) = x^2 + y^2 + z^2$$

has a minimum value, we may solve the problem by finding the minimum value of $f(x, y, z)$ subject to the constraint

$$z = 2x + y - 5.$$

This leads to finding the points (x, y) at which the function

$$h(x, y) = f(x, y, 2x + y - 5) = x^2 + y^2 + (2x + y - 5)^2$$

has its minimum value or values. Since the domain of h is the entire xy-plane, the first derivative test of the preceding article tells us that any minima h might have must occur at points where

$$h_x = h_y = 0.$$

This leads to

$$10x + 4y = 20,$$
$$4x + 4y = 10,$$

and the solution.

$$x = \frac{5}{3}, \qquad y = \frac{5}{6}.$$

We may apply either geometric arguments or the second derivative test to show that these values minimize h. The z-coordinate of the corresponding point on the plane $z = 2x + y - 5$ is

$$z = 2\left(\frac{5}{3}\right) + \frac{5}{6} - 5 = -\frac{5}{6}.$$

Therefore, the point we seek is

$$\text{Closest point: } P\left(\frac{5}{3}, \frac{5}{6}, -\frac{5}{6}\right). \quad \square$$

Attempts to solve a constrained maximum or minimum problem by substitution, as we might call the method of Example 1, do not always go smoothly, as the next example shows. This is one of the reasons for learning the new method of this article, which does not require us to decide in advance which of the constrained variables to regard as independent.

EXAMPLE 2 Find the minimum distance from the origin to the surface

$$x^2 - z^2 - 1 = 0.$$

Solution 1 We begin by looking for the points $P(x, y, z)$ on the surface that are closest to the origin. To find them, we look for the points $P(x, y, z)$ in space that minimize the square of the distance

$$|\overrightarrow{OP}|^2 = f(x, y, z) = x^2 + y^2 + z^2$$

subject to the constraint that

$$x^2 - z^2 - 1 = 0 \qquad \text{or} \qquad z^2 = x^2 - 1.$$

The values of f at points on this surface are given by the function

$$h(x, y) = x^2 + y^2 + (x^2 - 1) = 2x^2 + y^2 - 1.$$

The only extreme value of h occurs at (0, 0), where

$$h_x = 4x = 0 \qquad \text{and} \qquad h_y = 2y = 0.$$

But here we run into trouble because $z^2 = x^2 - 1 = -1$ means that z is imaginary when $x = 0$. A point $P(x, y, z)$ can lie on the surface $z^2 = x^2 - 1$ only if $|x| \geq 1$. What went wrong?

What happened was that the first derivative test found (as it should have) the point *in the domain of h* where h has a minimum value, but we are looking for points *on the surface* where h has a minimum value.

One way to go from here is to substitute $x^2 = z^2 + 1$ into the formula $f(x, y, z) = x^2 + y^2 + z^2$ and look for points where

$$k(y, z) = (z^2 + 1) + y^2 + z^2 = 1 + y^2 + 2z^2$$

takes on its smallest values. These occur where

$$k_y = 2y = 0 \quad \text{and} \quad k_z = 4z = 0,$$

or where

$$y = z = 0.$$

This leads to

$$x^2 = z^2 + 1 = 1, \qquad x = \pm 1.$$

The resulting points, $(\pm 1, 0, 0)$, are now on the surface. Moreover, it is clear from the inequality

$$k(y, z) = 1 + y^2 + 2z^2 \geq 1$$

that the points $(\pm 1, 0, 0)$ give a minimum of k. The minimum distance from the origin to the surface is therefore 1 unit. The surface is shown in Fig. 15.43.

Solution 2 Another way to find the minimum distance from the origin to the surface is to start with a small sphere centered at the origin and let it expand like a soap bubble until it just touches the surface. At the two points of contact, the surface and sphere have the same tangent planes and normal lines. Therefore, if the surface and sphere are represented as level surfaces

$$g(x, y, z) = x^2 - z^2 - 1 = 0,$$
$$f(x, y, z) = x^2 + y^2 + z^2 - 1 = 0$$

of the functions g and f, then the gradients ∇g and ∇f will be parallel at the two points of contact. In other words, at each point of contact we will be able to find a scalar λ such that

$$\nabla f = \lambda \nabla g,$$

or

$$2x\mathbf{i} + 2y\mathbf{j} + 2z\mathbf{k} = \lambda(2x\mathbf{i} - 2z\mathbf{k}).$$

Thus, the coordinates of either point of tangency will have to satisfy the three scalar equations

$$2x = \lambda 2x, \tag{1a}$$
$$2y = \lambda \cdot 0 \tag{1b}$$
$$2z = -2z\lambda, \tag{1c}$$

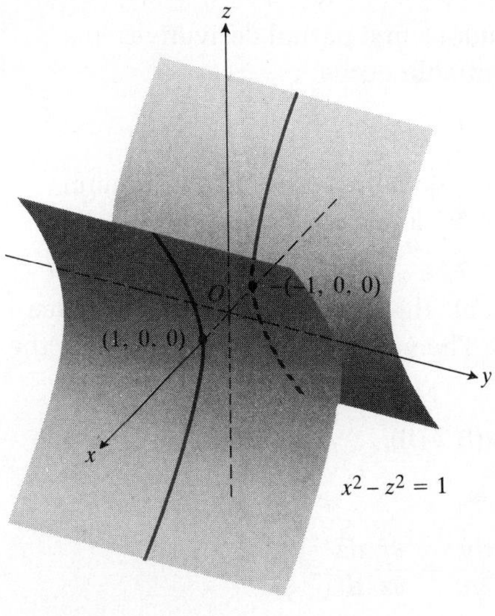

15.43 The surface $x^2 - z^2 = 1$.

or

$$x(2 - 2\lambda) = 0, \quad (2a)$$

$$y = 0, \quad (2b)$$

$$z(2 + 2\lambda) = 0. \quad (2c)$$

For what values of λ will a point $P(x, y, z)$ whose coordinates satisfy Eqs. (2a)–(2c) lie on the surface $x^2 - z^2 - 1 = 0$? Since every point on the surface has $x \neq 0$, we may divide both sides of Eq. (2a) by x to obtain

$$2 - 2\lambda = 0, \qquad \lambda = 1.$$

For $\lambda = 1$, Eq. (2c) gives

$$z(2 + 2 \cdot 1) = 0, \qquad z = 0.$$

Since $y = 0$ also (from Eq. 2b) the points we seek on the surface all have coordinates of the form

$$(x, 0, 0),$$

which means that $x = \pm 1$. As in Solution 1, we find that the points on the surface closest to the origin are $(\pm 1, 0, 0)$, and the minimum distance is one unit. □

The Method of Lagrange Multipliers

In Solution 2 above we solved the problem presented in Example 2 by the *method of Lagrange multipliers*. In general terms, the method says that the extreme values of a function $f(x, y, z)$ whose variables are subject to a constraint of the form $g(x, y, z) = 0$ are to be found on the surface $g = 0$ at the points where

$$\nabla f = \lambda \nabla g$$

for some scalar λ (called a *Lagrange multiplier*).

To explore the method further and see why it works, we first make the following observation, which we state as a theorem.

THEOREM 6

> Suppose that $f(x, y, z)$ has continuous first partial derivatives in a region that contains the differentiable curve
>
> $$C: \quad \mathbf{R}(t) = x(t)\mathbf{i} + y(t)\mathbf{j} + z(t)\mathbf{k}.$$
>
> If P_0 is a point on C where f has a local maximum or minimum relative to its values on C, then ∇f is perpendicular to C at P_0.

Proof. We show that $\nabla f \cdot \mathbf{v} = 0$ at the points in question, where $\mathbf{v} = d\mathbf{R}/dt$ is the tangent vector to C. The values of f on C are given by the composite function

$$f(x(t), y(t), z(t)),$$

whose derivative with respect to t is

$$\frac{df}{dt} = \frac{\partial f}{\partial x}\frac{dx}{dt} + \frac{\partial f}{\partial y}\frac{dy}{dt} + \frac{\partial f}{\partial z}\frac{dz}{dt} = \nabla f \cdot \mathbf{v}.$$

At any point P_0 where f has a local maximum or minimum relative to its values on the curve, $df/dt = 0$, so that

$$\nabla f \cdot \mathbf{v} = 0. \blacksquare$$

By dropping the z-terms in the statement and proof of Theorem 6 we obtain a similar result for functions of two variables: The points on a differentiable curve

$$\mathbf{R}(t) = x(t)\mathbf{i} + y(t)\mathbf{j}$$

where a function $f(x, y)$ takes on its local maxima and minima relative to its values on the curve are the points where $\nabla f \cdot d\mathbf{R}/dt = 0$.

Theorem 6 is the key to why the method of Lagrange multipliers works, as we shall now see. Suppose that $f(x, y, z)$ and $g(x, y, z)$ have continuous first partial derivatives and that P_0 is a point on the surface $g(x, y, z) = 0$ where f has a local maximum or minimum value relative to its other values on the surface. Then f takes on a local maximum or minimum at P_0 relative to its values on every differentiable curve through P_0 on the surface $g(x, y, z) = 0$. Therefore, ∇f is perpendicular to the velocity vector of every such differentiable curve through P_0. But so is ∇g (because ∇g is perpendicular to the level surface $g = 0$, as we saw in Article 15.6). Therefore, at P_0, ∇f is some scalar multiple λ of ∇g.

The argument for the two-variable case is left as an exercise (Problem 36).

The Method of Lagrange Multipliers

Suppose that $f(x, y, z)$ and $g(x, y, z)$ have continuous partial derivatives. To find the local maximum and minimum values of f subject to the constraint $g(x, y, z) = 0$, find the values of x, y, z, and λ that satisfy the equations

$$\nabla f = \lambda \nabla g \quad \text{and} \quad g(x, y, z) = 0 \tag{3}$$

simultaneously.

Comment on notation. Some books describe the method of Lagrange multipliers without vector notation in the following equivalent way:

To maximize or minimize a function $f(x, y, z)$ subject to the constraint $g(x, y, z) = 0$, construct the auxiliary function

$$H(x, y, z, \lambda) = f(x, y, z) - \lambda g(x, y, z). \tag{4}$$

Then find the values of x, y, z, and λ for which the partial derivatives of H are all zero:

$$H_x = 0, \quad H_y = 0, \quad H_z = 0, \quad H_\lambda = 0. \tag{5}$$

These requirements are equivalent to the requirements in (3), as we can see by calculating

$$\begin{aligned} H_x &= f_x - \lambda g_x = 0 &&\text{or}\quad f_x = \lambda g_x, \\ H_y &= f_y - \lambda g_y = 0 &&\text{or}\quad f_y = \lambda g_y, \\ H_z &= f_z - \lambda g_z = 0 &&\text{or}\quad f_z = \lambda g_z, \\ H_\lambda &= -g(x, y, z) = 0 &&\text{or}\quad g(x, y, z) = 0. \end{aligned} \tag{6}$$

The first three equations give $\nabla f = \lambda \nabla g$, and the last is $g(x, y, z) = 0$.

EXAMPLE 3 Find the greatest and smallest values that the function

$$f(x, y) = xy$$

takes on the ellipse

$$\frac{x^2}{8} + \frac{y^2}{2} = 1.$$

Solution We are asked to find the extreme values of $f(x, y) = xy$ subject to the constraint

$$g(x, y) = \frac{x^2}{8} + \frac{y^2}{2} - 1 = 0.$$

To do so, we first find the values of x, y, and λ for which

$$\nabla f = \lambda \nabla g \qquad \text{and} \qquad g(x, y) = 0.$$

The gradient equation gives

$$y\mathbf{i} + x\mathbf{j} = \frac{\lambda}{4}x\mathbf{i} + \lambda y\mathbf{j},$$

from which we find

$$y = \frac{\lambda}{4}x, \qquad x = \lambda y,$$

and

$$y = \frac{\lambda}{4}(\lambda y) = \frac{\lambda^2}{4}y, \quad \text{so that} \quad \lambda = \pm 2 \quad \text{or} \quad y = 0.$$

We now consider two cases.

CASE 1. If $y = 0$, then $x = y = 0$, but the point (0, 0) is not on the ellipse. Hence, $y \neq 0$.

CASE 2. If $y \neq 0$, then

$$x = \pm 2y.$$

Substituting this in the equation $g(x, y) = 0$ gives

$$\frac{(+2y)^2}{8} + \frac{y^2}{2} = 1$$

$$4y^2 + 4y^2 = 8$$

$$y = \pm 1.$$

The function $f(x, y) = xy$ therefore takes on its extreme values on the ellipse at the four points $(\pm 2, 1)$, $(\pm 2, -1)$, and the extreme values are $xy = 2$ and $xy = -2$. □

Geometric interpretation of the solution (see Fig. 15.44). The level curves of the function $f(x, y) = xy$ are the hyperbolas $xy = c$. The farther they are from the origin, the larger the absolute value of f. We want to find the extreme values of $f(x, y)$, given that the point (x, y) also lies on the ellipse $x^2 + 4y^2 = 8$. Which hyperbolas intersecting the ellipse are farthest from

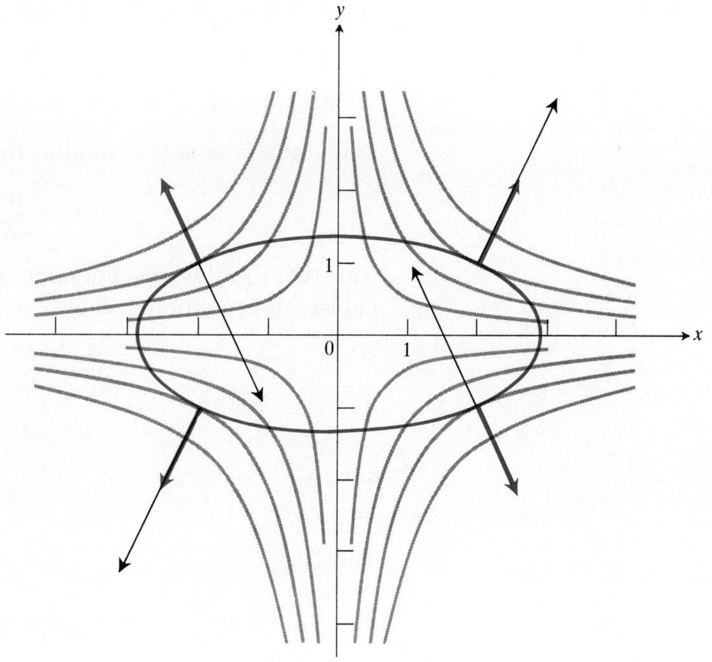

15.44 When subjected to the constraint $g(x, y) = x^2/8 + y^2/2 - 1 = 0$, the function $f(x, y) = xy$ takes on extreme values at the four points $(\pm 2, \pm 1)$.

the origin? The hyperbolas that just graze the ellipse, the ones that are tangent to it. At these points,

a) the normal to the hyperbola is normal to the ellipse;
b) the gradient

$$\nabla f = y\mathbf{i} + x\mathbf{j}$$

is a multiple ($\lambda = \pm 2$) of the gradient

$$\nabla g = \frac{x}{4}\mathbf{i} + y\mathbf{j}.$$

For example, at the point (2, 1),

$$\nabla f = \mathbf{i} + 2\mathbf{j} \quad \text{and} \quad \nabla g = \tfrac{1}{2}\mathbf{i} + \mathbf{j},$$

so that $\nabla f = 2\nabla g$. At the point $(-2, 1)$,

$$\nabla f = \mathbf{i} - 2\mathbf{j} \quad \text{and} \quad \nabla g = -\tfrac{1}{2}\mathbf{i} + \mathbf{j},$$

so that $\nabla f = -2\nabla g$.

EXAMPLE 4 Let a and b be positive constants. Find the maximum and minimum values of the function

$$ax + by$$

subject to the constraint

$$x^2 + y^2 = 1.$$

Solution We model this as a Lagrange multiplier problem with

$$f(x, y) = ax + by, \qquad g(x, y) = x^2 + y^2 - 1$$

and find the values of x, y, and λ that satisfy the equations

$$\nabla f = \lambda \nabla g: \quad a\mathbf{i} + b\mathbf{j} = 2x\lambda\,\mathbf{i} + 2y\lambda\,\mathbf{j},$$
$$g(x, y) = 0: \quad x^2 + y^2 - 1 = 0.$$

The gradient equation implies that $\lambda \neq 0$, and gives

$$x = \frac{a}{2\lambda}, \qquad y = \frac{b}{2\lambda}.$$

Note that x and y have the same sign because $a > 0$ and $b > 0$. With these values, the equation $g(x, y) = 0$ gives

$$\frac{a^2}{4\lambda^2} + \frac{b^2}{4\lambda^2} = 1$$
$$a^2 + b^2 = 4\lambda^2$$
$$\lambda = \pm\frac{\sqrt{a^2 + b^2}}{2}.$$

Thus,

$$x = \frac{a}{2\lambda} = \pm\frac{a}{\sqrt{a^2 + b^2}}, \qquad y = \frac{b}{2\lambda} = \pm\frac{b}{\sqrt{a^2 + b^2}},$$

and $f(x, y) = ax + by$ has its extreme values at the two points

$$(x, y) = \pm\left(\frac{a}{\sqrt{a^2 + b^2}}, \frac{b}{\sqrt{a^2 + b^2}}\right).$$

By calculating the values of f at these points we see that its maximum value on the circle $x^2 + y^2 = 1$ is

$$\frac{a^2}{\sqrt{a^2 + b^2}} + \frac{b^2}{\sqrt{a^2 + b^2}} = \sqrt{a^2 + b^2}, \tag{7a}$$

and its minimum value is

$$-\frac{a^2}{\sqrt{a^2 + b^2}} - \frac{b^2}{\sqrt{a^2 + b^2}} = -\sqrt{a^2 + b^2}. \; \square \tag{7b}$$

Lagrange Multipliers with Two Constraints

If there are two constraints on the variables of $f(x, y, z)$, say

$$g(x, y, z) = 0 \quad \text{and} \quad h(x, y, z) = 0,$$

we may find the constrained relative maxima and minima of f by introducing two Lagrange multipliers λ and μ. That is, we locate the points $P(x, y, z)$ where f takes on its constrained extreme values by finding the values of x, y, z, λ, and μ that satisfy the equations

$$\nabla f = \lambda \nabla g + \mu \nabla h, \tag{8a}$$
$$g(x, y, z) = 0, \tag{8b}$$
$$h(x, y, z) = 0 \tag{8c}$$

simultaneously. The functions f, g, and h are required to have continuous first partial derivatives.

The equations in (8) have a nice geometric interpretation. The surfaces $g = 0$ and $h = 0$ (usually) intersect in a differentiable curve, say C (Fig.

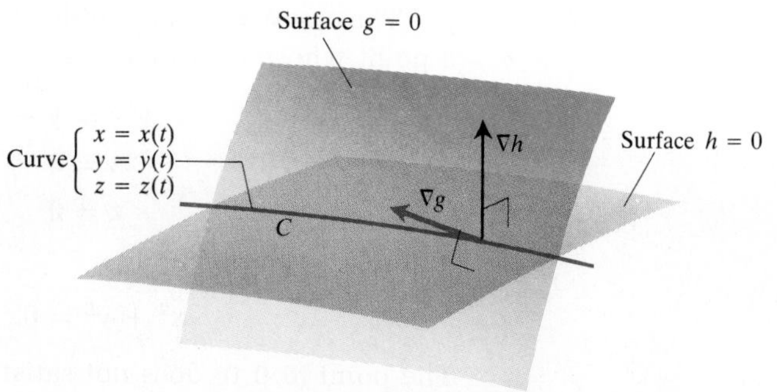

15.45 Vectors ∇g and ∇h are in a plane perpendicular to curve C because ∇g is normal to the surface $g = 0$, and ∇h is normal to the surface $h = 0$.

15.45), and along this curve we seek the points where f has local maximum and minimum values relative to its other values on the curve. These are the points where ∇f is normal to C, as we saw in Theorem 6. But ∇g and ∇h are also normal to C at these points because C lies in the surfaces $g = 0$ and $h = 0$. Therefore ∇f lies in the plane determined by ∇g and ∇h, which means that

$$\nabla f = \lambda \nabla g + \mu \nabla h$$

for some λ and μ. Since the points we seek also lie in both surfaces, their coordinates must satisfy the equations

$$g(x, y, z) = 0 \qquad \text{and} \qquad h(x, y, z) = 0,$$

which are the remaining requirements in (8).

EXAMPLE 5 The cone $z^2 = x^2 + y^2$ is cut by the plane $z = 1 + x + y$ in a conic section C. Find the points on C that are nearest to, and farthest from, the origin.

Solution We model this as a Lagrange multiplier problem in which we find the extreme values of

$$f(x, y, z) = x^2 + y^2 + z^2$$

(the square of the function that measures distance from the origin) subject to the two constraints

$$g(x, y, z) = x^2 + y^2 - z^2 = 0, \tag{9a}$$
$$h(x, y, z) = 1 + x + y - z = 0. \tag{9b}$$

The gradient equation (8a) gives

$$2x\mathbf{i} + 2y\mathbf{j} + 2z\mathbf{k} = \lambda(2x\mathbf{i} + 2y\mathbf{j} - 2z\mathbf{k}) + \mu(\mathbf{i} + \mathbf{j} - \mathbf{k}),$$

which leads to the scalar equations

$$\left.\begin{aligned} 2x &= 2x\lambda + \mu \\ 2y &= 2y\lambda + \mu \end{aligned}\right\} \Rightarrow x - y = (x - y)\lambda, \tag{10a}$$
$$\left.\begin{aligned} 2y &= 2y\lambda + \mu \\ 2z &= -2z\lambda - \mu \end{aligned}\right\} \Rightarrow y + z = (y - z)\lambda. \tag{10b}$$

The equation $x - y = (x - y)\lambda$ is satisfied if $x = y$ or if $x \neq y$ and $\lambda = 1$.

The case $\lambda = 1$ does not lead to a point on the cutting plane (that is, to a point whose coordinates satisfy Eq. 9b). For $\lambda = 1$ implies

$$y + z = y - z \qquad \text{(from Eq. 10b)}$$
$$2z = 0$$
$$z = 0$$

and Eq. (9a) then gives

$$x^2 + y^2 = 0, \qquad x = 0, \qquad y = 0.$$

The point (0, 0, 0) does not satisfy the constraint (9b).

Therefore $\lambda \neq 1$ and we have $x = y$. The plane $x = y$ meets the plane $z = 1 + x + y$ in a line that cuts the cone in just two points, as we can see by substituting $y = x$ and $z = 1 + 2x$ into the cone equations:

$$z^2 = x^2 + y^2$$
$$(1 + 2x)^2 = x^2 + x^2$$
$$2x^2 + 4x + 1 = 0$$
$$x = -1 \pm \frac{\sqrt{2}}{2}.$$

The points are

$$A = \left(-1 - \sqrt{\tfrac{1}{2}}, -1 - \sqrt{\tfrac{1}{2}}, -1 - \sqrt{2}\right), \tag{11a}$$
$$B = \left(-1 + \sqrt{\tfrac{1}{2}}, -1 + \sqrt{\tfrac{1}{2}}, -1 + \sqrt{2}\right). \tag{11b}$$

Now, we know that C is either an ellipse or a hyperbola. If it is an ellipse, we would conclude that B is the point on it nearest the origin, and A the point farthest from the origin. But if it is a hyperbola, then there is no point on it that is farthest away from the origin, and the points A and B are the points on the two branches that are nearest the origin. Problem 33 asks you to think about these possibilities and decide between them. It seems obvious that the critical points should satisfy the condition $x = y$ because all three of the functions f, g, and h treat x and y alike. □

PROBLEMS

1. Find the points on the ellipse $x^2 + 2y^2 = 1$ where $f(x, y) = xy$ has its extreme values.
2. Find the extreme values of $f(x, y) = xy$ subject to the constraint $g(x, y) = x^2 + y^2 - 10 = 0$.
3. Find the maximum value of $f(x, y) = 9 - x^2 - y^2$ on the line $x + 3y = 12$ (Fig. 15.42).
4. Find the minimum distance between the line $y = x + 1$ and the parabola $y^2 = x$.
5. Find the extreme values of $f(x, y) = x^2y$ on the line $x + y = 3$.
6. Find the points on the curve $x^2y = 2$ nearest the origin.
7. Use the method of Lagrange multipliers to find (a) the minimum value of $x + y$, subject to the constraint $xy = 16$; (b) the maximum value of xy, subject to the constraint $x + y = 16$. Comment on the geometry of each solution.
8. Find the points on the curve $x^2 + xy + y^2 = 1$ that are nearest to and farthest from the origin.
9. Find the dimensions of the closed circular can of smallest surface area whose volume is 16π cm^3.
10. Use the method of Lagrange multipliers to find the dimensions of the rectangle of greatest area that can be inscribed in the ellipse $x^2/16 + y^2/9 = 1$ with sides parallel to the coordinate axes.
11. The temperature at a point (x, y) on a metal plate is $T(x, y) = 4x^2 - 4xy + y^2$. An ant on the plate walks around the circle of radius 5 centered at the origin.

What are the highest and lowest temperatures encountered by the ant?

12. A horizontal water tank is to be constructed in the form of a cylinder with hemispherical ends. Find the diameter and the length of the cylindrical portion of the tank if the tank is to hold 8000 cubic feet of water and the least amount of material is to be used in constructing the tank.

13. Find the maximum and minimum values of $x^2 + y^2$ subject to the constraint $x^2 - 2x + y^2 - 4y = 0$.

14. A pentagon is made by mounting an isosceles triangle on top of a rectangle. What dimensions minimize the perimeter for a given area A?

15. Find the point on the plane $x + 2y + 3z = 13$ closest to the point $(1, 1, 1)$.

16. Find the maximum and minimum values of

$$f(x, y, z) = x - 2y + 5z$$

on the sphere

$$x^2 + y^2 + z^2 = 30.$$

17. Find the minimum distance from the surface $x^2 + y^2 - z^2 = 1$ to the origin.

18. Find the point on the surface $z = xy + 1$ nearest the origin.

19. Find the points on the surface $z^2 = xy + 4$ closest to the origin.

20. Find the points on the sphere $x^2 + y^2 + z^2 = 25$ where $f(x, y, z) = x + 2y + 3z$ has its maximum and minimum values.

21. The temperature T at any point (x, y, z) in space is $T = 400xyz^2$. Find the highest temperature on the unit sphere

$$x^2 + y^2 + z^2 = 1.$$

22. Find three real numbers whose sum is 9 and the sum of whose squares is as small as possible.

23. Find the largest product the numbers x, y, and z can have if $x + y + z^2 = 16$.

24. If a, b, and c are positive numbers, find the maximum value that $f(x, y, z) = ax + by + cz$ can take on the sphere $x^2 + y^2 + z^2 = 1$.

25. Find the maximum and minimum values of $f(x, y) = x^2 + 3y^2 + 2y$ on the unit disc $x^2 + y^2 \leq 1$.

26. Find the extreme values of $f(x, y, z) = x^2yz + 1$ on the intersection of the plane $z = 1$ with the sphere $x^2 + y^2 + z^2 = 2$.

27. A space probe in the shape of the ellipsoid

$$4x^2 + y^2 + 4z^2 = 16$$

enters the earth's atmosphere and its surface begins to heat. After one hour, the temperature at the point (x, y, z) on the probe's surface is

$$T(x, y, z) = 8x^2 + 4yz - 16z + 600.$$

Find the hottest point on the probe's surface.

28. You are in charge of erecting a radio telescope on a newly discovered planet. To minimize interference, you want to place it where the magnetic field of the planet is weakest. The planet is spherical with a radius of 6 units. The strength of the magnetic field is given by $M(x, y, z) = 6x - y^2 + xz + 60$ based on a coordinate system whose origin is at the center of the planet. Where should you locate the radio telescope?

29. Given n positive numbers $a_1, a_2, \ldots, a_n$, find the maximum value of the expression

$$w = a_1x_1 + a_2x_2 + \cdots + a_nx_n$$

if the variables $x_1, x_2, \ldots, x_n$ are restricted so that the sum of their squares is 1.

30. A plane of the form

$$z = Ax + By + C$$

is to be "fitted" to the following points (x_i, y_i, z_i):

$$(0, 0, 0), \quad (0, 1, 1), \quad (1, 1, 1), \quad (1, 0, -1).$$

Find the plane that minimizes the sum of squares of the deviations

$$\sum_{i=1}^{4} (Ax_i + By_i + C - z_i)^2.$$

31. Find the maximum value of $f(x, y, z) = x^2 + 2y - z^2$ subject to the constraints $2x - y = 0$ and $y + z = 0$.

32. a) Find the maximum value of $w = xyz$ among all points lying on the intersection of the two planes $x + y + z = 40$ and $z = x + y$.

b) Give a geometric argument supporting the fact that you have found a maximum (and not a minimum) value of xyz subject to the constraints.

33. In the solution of Example 5, two points A and B were located as candidates for maximum or minimum distances from the origin. Use cylindrical coordinates r, θ, z to express the equations of the cone, the plane $z = 1 + x + y$, and the cylinder that contains their curve of intersection and has elements parallel to the z-axis. Is this cylinder circular, elliptical, parabolic, or hyperbolic? (Consider its intersection with the xy-plane.) Express the distance from the origin to a point on the curve of intersection of the cone and the plane as a function of θ. Does it have a minimum? A maximum? What can you now say about the points A and B of Eqs. (11a, b) as solutions in Example 5?

34. In Example 5, the extrema for $f(x, y, z)$ on the cone $z^2 = x^2 + y^2$ and the plane $z = 1 + x + y$ were found to satisfy $y = x$ as well. Thus $z = 1 + 2x$, and the function to be made a maximum or minimum is $x^2 + y^2 + z^2 = x^2 + x^2 + (1 + 2x)^2$, which can also be written as $6(x + \frac{1}{3})^2 + \frac{1}{3}$. This is obviously a minimum when $x = -\frac{1}{3}$. But the point we get with

$$y = x, \qquad z = 1 + 2x, \qquad x = -\frac{1}{3},$$

is $(-\frac{1}{3}, -\frac{1}{3}, \frac{1}{3})$, which is not on the cone. What's wrong? (A sketch of the situation in the plane $y = x$ may throw some light on the question.)

35. In Example 5, we can determine that for all points on the cone $z^2 = x^2 + y^2$, the square of the distance from the origin to $P(x, y, z)$ is $w = 2(x^2 + y^2)$, which is a function of two independent variables, x and y. But if P is also to be on the plane

$$z = 1 + x + y$$

as well as the cone, then

$$(1 + x + y)^2 = x^2 + y^2.$$

Show that these points have coordinates that satisfy the equation

$$2xy + 2x + 2y + 1 = 0.$$

Interpret this equation in two ways:

a) as a curve in the xy-plane, and

b) as a set of points on a cylinder in 3-space.

Sketch the curve of (a) and find the point or points on it for which w is a minimum. Are there points on this curve for which w is a maximum? Use the information you now have to complete the discussion of Example 5.

36. *An argument for the two-variable version of the method of Lagrange multipliers.* Suppose that $f(x, y)$ and $g(x, y)$ have continuous first partial derivatives and that we wish to find the relative maximum and minimum values of f subject to the constraint that $g(x, y) = 0$. To find these values of f we find the values of x, y, and λ that satisfy the equations

$$\nabla f = \lambda \nabla g \qquad \text{and} \qquad g(x, y) = 0$$

simultaneously.

To prove this, let P_0 be a point on the curve $g(x, y) = 0$ at which f has a local maximum or minimum value relative to its values on the curve. Then carry out the following steps:

STEP 1. Let $x = x(t)$, $y = y(t)$ be a parameterization of the curve $g(x, y) = 0$ in an interval about P_0. Show that $df/dt = 0$ at P_0 and therefore that ∇f is perpendicular to the velocity vector of the curve at P_0. (Assume that neither vector vanishes at P_0.)

STEP 2. Combine the result in Step 1 above with the result of Problem 81 in Article 15.6 to show that ∇f and ∇g are parallel at P_0, and therefore that $\nabla f = \lambda \nabla g$ at P_0 for some number λ.

15.11 Exact Differentials

The Differential of a Function

If $f(x, y)$ and its first order partial derivatives are continuous, and $x = x(t)$, $y = y(t)$ are differentiable functions of t, then we know from the chain rule that f is a differentiable function of t and that

$$\frac{df}{dt} = \frac{\partial f}{\partial x}\frac{dx}{dt} + \frac{\partial f}{\partial y}\frac{dy}{dt}. \tag{1}$$

This equation is sometimes written in differential form as

$$df = \frac{\partial f}{\partial x}\,dx + \frac{\partial f}{\partial y}\,dy, \tag{2}$$

which is like the familiar

$$du = g'(x)\,dx \tag{3}$$

for the differential of a function $u = g(x)$ of a single variable.

In many problems we must recover u from $g'(x)$ by integrating both sides of Eq. (3) to get

$$u = \int g'(x)\,dx, \tag{4}$$

which determines $u = g(x)$ up to a constant. In like manner, we can determine $f(x, y)$ up to a constant from the equation

$$f = \int df = \int \frac{\partial f}{\partial x}\,dx + \int \frac{\partial f}{\partial y}\,dy, \tag{5}$$

in which the integration

$$\int \frac{\partial f}{\partial x}\,dx$$

reverses the partial differentiation with respect to x, and

$$\int \frac{\partial f}{\partial y}\,dy$$

reverses the partial differentiation with respect to y. The following example shows how this is done.

EXAMPLE 1 Find $f(x, y)$ if

$$df = (x^2 + y^2)\,dx + (2xy + 1)\,dy.$$

Discussion. We must try to recover f from the information that

$$\frac{\partial f}{\partial x} = x^2 + y^2 \qquad \text{and} \qquad \frac{\partial f}{\partial y} = 2xy + 1, \tag{6}$$

for this is all we know about f. The first two solutions we give reconstruct f by a method we shall use again in discussing Theorem 7 below. Once you have read them you will be ready for the third solution, which contains a shortcut that makes it the fastest method of all.

Solution 1 (Preview of the construction in Theorem 7.) The partial derivative

$$\frac{\partial f}{\partial x} = x^2 + y^2$$

was calculated by holding y constant and differentiating with respect to x. We may therefore find f by holding y at a constant value and integrating with respect to x. When we do so we get

$$f(x, y) = \int_{y\ \text{const.}} (x^2 + y^2)\,dx = \frac{x^3}{3} + y^2x + k(y), \tag{7}$$

where the constant of integration $k(y)$ is written as a function of y because its value may change with each new y.

To determine $k(y)$, we calculate $\partial f/\partial y$ from (7) and set the result equal to the given partial derivative, $2xy + 1$:

$$\begin{aligned} \frac{\partial}{\partial y}\left(\frac{x^3}{3} + y^2x + k(y)\right) &= 2xy + 1 \\ 2xy + k'(y) &= 2xy + 1 \\ k'(y) &= 1. \end{aligned} \tag{8}$$

This shows that

$$k(y) = y + C$$

and

$$f(x, y) = \frac{x^3}{3} + xy^2 + y + C.$$

We have recovered f up to a constant.

Solution 2 (Like Solution 1, but we integrate first with respect to y.) We begin our recovery of f by integrating $\partial f/\partial y = 2xy + 1$ with respect to y, holding x constant. This gives

$$f(x, y) = \int_{x\ \text{const.}} (2xy + 1)\, dy = xy^2 + y + h(x), \tag{9}$$

where the constant of integration $h(x)$ is now to be regarded as a function of x.

To determine $h(x)$, we set

$$\frac{\partial f}{\partial x} = x^2 + y^2$$

to obtain the equations

$$\begin{aligned} \frac{\partial}{\partial x}(xy^2 + y + h(x)) &= x^2 + y^2 \\ y^2 + h'(x) &= x^2 + y^2 \\ h'(x) &= x^2 \\ h(x) &= \frac{x^3}{3} + C. \end{aligned} \tag{10}$$

We conclude, as in Solution 1, that

$$f(x, y) = \frac{x^3}{3} + xy^2 + y + C.$$

Solution 3 *Fastest way:* Do both integrations first.

$(x^2 + y^2)dx$ + $(2xy + 1)dy$

⇓ Integrate with respect to x — ⇙ Integrate with respect to y

$\frac{x^3}{3} + xy^2 +$ y-terms + Const. — $xy^2 + y +$ x-terms + Const.

⇘ ⇙ Terms needed to account for both intermediate expressions.

$$\frac{x^3}{3} + xy^2 + y + C.$$

If you use this method, do not fall into the trap of doubling terms that appear in both intermediate expressions.

Right: $f(x, y) = \frac{x^3}{3} + xy^2 + y + C$

Wrong: $f(x, y) = \frac{x^3}{3} + 2xy^2 + y + C.$ □

Exactness of Differential Forms

An expression like

$$(x^2 + y^2)\,dx + (2xy + 1)\,dy$$

that has the form

$$M(x, y)\,dx + N(x, y)\,dy \tag{11}$$

is called a *differential form* in x and y. Such a form is said to be *exact* over a region R if throughout R it is the differential df of some function f. That is, (11) is exact on R if there exists a function $f(x, y)$ such that

$$\frac{\partial f}{\partial x} = M(x, y) \quad \text{and} \quad \frac{\partial f}{\partial y} = N(x, y)$$

for all (x, y) in R.

In terms of gradients, we may say that $M(x, y)\,dx + N(x, y)\,dy$ is exact over a region R if there exists a function $f(x, y)$ such that

$$\nabla f = M(x, y)\mathbf{i} + N(x, y)\mathbf{j}$$

for all (x, y) in R.

Differential Equations

Our interest in exact differential forms comes partly from a desire to solve differential equations like

$$(x^2 + y^2) + (2xy + 1)\frac{dy}{dx} = 0 \tag{12}$$

that have the form

$$M(x, y) + N(x, y)\frac{dy}{dx} = 0. \tag{13}$$

In the simplest cases we may be able to separate variables and rewrite (13) in the form

$$g(y)\frac{dy}{dx} = h(x),$$

which we can then solve if we are able to integrate both sides with respect to x. If we cannot separate the variables (for example, we cannot do so in Eq. 12), we may still be able to solve the equation by rewriting it in the form

$$M(x, y)\,dx + N(x, y)\,dy = 0. \tag{14}$$

For if the left-hand side turns out to be the differential of a function $f(x, y)$ when we do this, then

$$\frac{\partial f}{\partial x} = M(x, y), \qquad \frac{\partial f}{\partial y} = N(x, y),$$

and Eq. (14) becomes

$$df = 0,$$

and its general solution is

$$f(x, y) = C. \tag{15}$$

Equation (15) may be considered to be a solution of the original equation

$$M(x, y) + N(x, y)\frac{dy}{dx} = 0 \tag{16}$$

in the sense that the equation $f(x, y) = C$ defines y implicitly as one or more differentiable functions of x that solve Eq. (16). To see that this is so we differentiate both sides of $f(x, y) = C$ with respect to x, treating y as a differentiable function of x and applying the chain rule. This gives

$$\begin{aligned} f(x, y) &= C \\ \frac{d}{dx} f(x, y) &= \frac{d}{dx} C \\ \frac{\partial f}{\partial x}\frac{dx}{dx} + \frac{\partial f}{\partial y}\frac{dy}{dx} &= 0 \\ \frac{\partial f}{\partial x} + \frac{\partial f}{\partial y}\frac{dy}{dx} &= 0, \end{aligned} \tag{17}$$

or, returning to the original notation,

$$M(x, y) + N(x, y)\frac{dy}{dx} = 0,$$

which is Eq. (16). Thus, any differentiable function $y = y(x)$ defined implicitly by Eq. (15) is seen to satisfy Eq. (16).

EXAMPLE 2 Find the general solution $f(x, y) = C$ of the differential equation

$$(x^2 + y^2) + (2xy + 1)\frac{dy}{dx} = 0.$$

Then find the particular solution whose graph passes through the point (0, 1).

Solution It is not possible to separate the variables in the given equation, so we rewrite it in the form

$$(x^2 + y^2)\,dx + (2xy + 1)\,dy = 0.$$

We then seek a function $f(x, y)$ whose differential df gives the left-hand side of this equation:

$$df = (x^2 + y^2)\,dx + (2xy + 1)\,dy.$$

We know from Example 1 that

$$f(x, y) = \frac{x^3}{3} + xy^2 + y$$

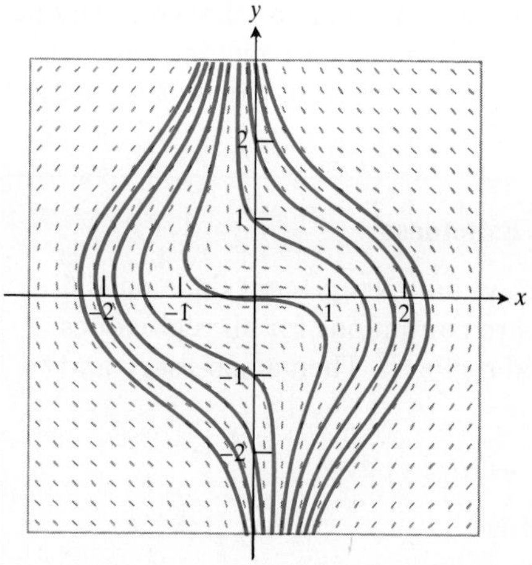

15.46 Solutions: $(x^3/3) + xy^2 + y = C$ for $C = 4, 3, 2, 1, 0, -1, -2, -3, -4$.

is such a function. Therefore, the general solution of the given equation is

$$f(x, y) = C, \qquad \text{or} \qquad \frac{x^3}{3} + xy^2 + y = C.$$

The particular solution whose graph passes through the point (0, 1) is found from substitution to be

$$\frac{x^3}{3} + xy^2 + y = 1.$$

See Fig. 15.46. □

The Test for Exactness

Had the differential form

$$(x^2 + y^2)\,dx + (2xy + 1)\,dy$$

not been exact, we could not have proceeded the way we did to solve the equation

$$(x^2 + y^2)\,dx + (2xy + 1)\,dy = 0$$

in Example 2. How can we tell when a form

$$M(x, y)\,dx + N(x, y)\,dy \tag{18}$$

is exact?

The answer lies in the observation that if

$$M(x, y)\,dx + N(x, y)\,dy = f_x\,dx + f_y\,dy = df$$

is the differential of a function $f(x, y)$ that has continuous first and second order partial derivatives, then

$$M_y = f_{xy} = f_{yx} = N_x.$$

These mixed partial derivatives have to be equal by Theorem 4 in Article 15.7. Thus,

$$M_y = N_x$$

is a *necessary* condition for (18) to be exact. We may therefore exclude from consideration any candidates that fail this test.

The amazing fact is that the condition $M_x = N_y$ is also a *sufficient* condition for exactness. That is, if $M_x = N_y$, then (18) is exact. Theorem 7 below gives the details.

Theorem 7 introduces a mild restriction on the geometric nature of the set R on which the functions $M(x, y)$ and $N(x, y)$ are defined. Specifically, R is required to be a simply connected region. The word *region* has the technical meaning here of "connected open set." A subset of the plane is *open* if each of its points is the center of a circle whose interior lies entirely in the set. An *open* subset of the plane is *connected* if every two of its points can be joined by a path that lies entirely in the set. A subset R of the plane is *simply connected* if no simple closed curve in R surrounds points not in R. When you read "simply connected," think "no holes." (See Fig. 15.47.) You need not be concerned with these technicalities now. The plane domains on which most of the functions in this book are defined really are regions or unions of regions, and most of them are simply

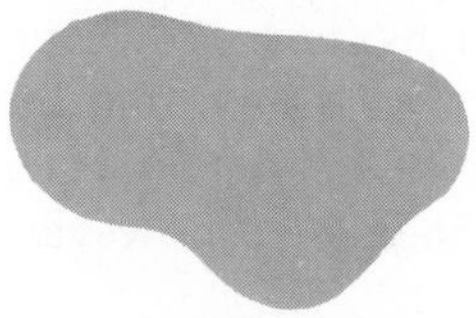

Connected and simply connected.

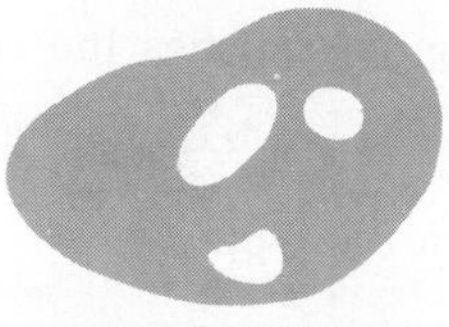

Connected but not simply connected.

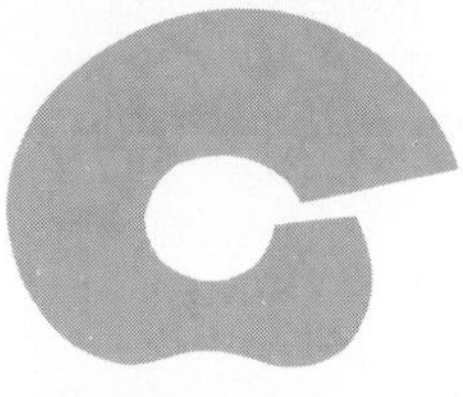

Connected and simply connected.

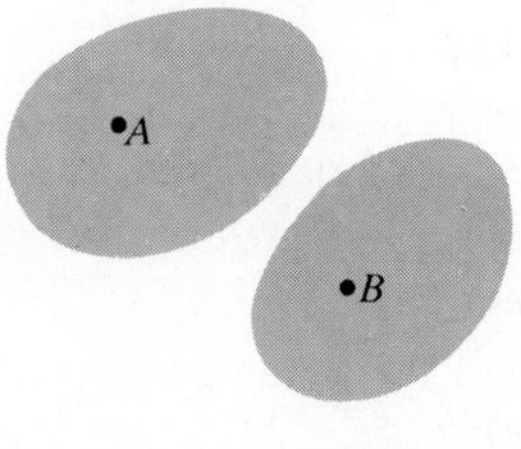

Simply connected but not connected.
No path from A to B lies entirely in the region.

Figure 15.47

connected. It will not be necessary to test every function that comes along. But without these conditions on R (which are usually met in practice), or others like them, some of the most useful theorems of calculus would not be true.

THEOREM 7

Test for Exactness

Suppose that the functions $M(x, y)$ and $N(x, y)$ and their partial derivatives M_x, M_y, N_x, and N_y are continuous for all real values of x and y in a simply connected region R. Then a necessary and sufficient condition for

$$M(x, y)\,dx + N(x, y)\,dy$$

to be an exact differential in R is

$$\frac{\partial M}{\partial y} = \frac{\partial N}{\partial x}.$$

Discussion. We saw before we stated Theorem 7 that $M_y = N_x$ is a necessary condition for exactness, and the argument did not require anything special of the region R. What we have yet to show, however, is that $M_y = N_x$ is sufficient to guarantee exactness. It is here that the simple connectivity comes in, but it is also here that the argument becomes too technical to present in full detail. What we can do here, however, is show how to use the equation

$$M_x = N_y$$

to construct a function $f(x, y)$ whose differential satisfies the equation

$$df = M\,dx + N\,dy. \tag{19}$$

The construction repeats the steps we took in Example 1.

For Eq. (19) to hold we must have

$$\frac{\partial f}{\partial x} = M(x, y), \qquad \frac{\partial f}{\partial y} = N(x, y).$$

Integrating the first of these with respect ot x, we find that for each value of y we have

$$f(x, y) = \int M(x, y)\,dx + k(y), \tag{20}$$

where $k(y)$ is a constant that may vary with the value of y. To see how f may satisfy the second condition, $\partial f/\partial y = N(x, y)$, we differentiate both sides of Eq. (20) with respect to y, obtaining

$$\frac{\partial f}{\partial y} = \frac{\partial}{\partial y}\int M(x, y)\,dx + k'(y). \tag{21}$$

We then set $\partial f/\partial y$ equal to $N(x, y)$ and have

$$\frac{\partial}{\partial y}\int M(x, y)\,dx + k'(y) = N(x, y)$$

$$k'(y) = N(x, y) - \frac{\partial}{\partial y}\int M(x, y)\,dx. \tag{22}$$

We use this differential equation to determine $k(y)$ and substitute the result back into Eq. (20).

The success of this construction of $f(x, y)$ depends on the fact that the right-hand side of Eq. (22) is a function of y alone, and it is exactly this independence of x that is guaranteed by the condition $M_y = N_x$. (If the right-hand side of Eq. (22) depended on x as well as y it could not be equal to $k'(y)$, which involves only y.)

We prove that the right-hand side of Eq. (22) is independent of x by showing that its derivative with respect to x is identical to zero. We calculate

$$\begin{aligned}\frac{\partial}{\partial x}\left(N(x, y) - \frac{\partial}{\partial y}\int M(x, y)\,dx\right) &= \frac{\partial N}{\partial x} - \frac{\partial^2}{\partial x\,\partial y}\int M(x, y)\,dx \\ &= \frac{\partial N}{\partial x} - \frac{\partial}{\partial y}\left(\frac{\partial}{\partial x}\int M(x, y)\,dx\right) \\ &= \frac{\partial N}{\partial x} - \frac{\partial}{\partial y}(M),\end{aligned}$$

and this last expression vanishes exactly when $M_y = N_x$ as we have assumed. ■

EXAMPLE 3 The following differential forms are defined over the region R that consists of the entire xy-plane. Test the forms for exactness on R.

a) $(x^2 + y^2)\,dx - 2xy\,dy$.

b) $(e^x \cos y)\,dx + (1 - e^x \sin y)\,dy$.

Solution We check to see whether $M_y = N_x$ at all points (x, y) in R.

a) Not exact on R because $M_y \neq N_x$ except when $y = 0$:

$$\frac{\partial}{\partial y}(x^2 + y^2) = 2y,$$

$$\frac{\partial}{\partial x}(-2xy) = -2y.$$

b) Exact on R because $M_y = N_x$ for all x and y:

$$\frac{\partial}{\partial y}(e^x \cos y) = -e^x \sin y,$$

$$\frac{\partial}{\partial x}(1 - e^x \sin y) = -e^x \sin y. \quad \square$$

PROBLEMS

In Problems 1–14, determine whether the given expression is or is not the differential of some function $f(x, y)$ over the region R that consists of the entire xy-plane. If it is, find f.

1. $2x(x^3 + y^3)\,dx + 3y^2(x^2 + y^2)\,dy$

2. $e^y\,dx + x(e^y + 1)\,dy$

3. $(2x + y)\,dx + (x + 2y)\,dy$

4. $(\cosh y + y \cosh x)\,dx + (\sinh x + x \sinh y)\,dy$

5. $(\sin y + y \sin x)\,dx + (\cos x + x \cos y)\,dy$

6. $(1 + e^x)\,dy + e^x(y - x)\,dx$

7. $e^{x+y}\,dx - e^{x-y}\,dy$

8. $ye^{xy}\,dx + xe^{xy}\,dy$

9. $(y \cos x + \sin y)\,dx + (\sin x + x \cos y - \sin y)\,dy$

10. $(y \cos x + y^2)\,dx + (\sin x)\,dy$

11. $(6xy^5 + 6y)\,dx + (15x^2y^4 + 6x - 10y)\,dy$

12. $(ye^x + y)\,dx + (e^x + 1)\,dy$

13. $(ye^{xy} + 10x)\,dx + xe^{xy}\,dy$

14. $(-2x + e^{2y} \cos 3x)\,dx + (e^y + e^{2y} \sin 3x)\,dy$

15. Find the general solution $f(x, y) = C$ of the differential equation

$$(12xy + 2y^2) + (6x^2 + 4xy)\frac{dy}{dx} = 0.$$

Then find the particular solution whose graph passes through the point $(1, -3)$.

16. Find the general solution $f(x, y) = C$ of the differential equation

$$(e^x \sin y + \cos x) + (e^x \cos y + 3e^{3y})\frac{dy}{dx} = 0.$$

Then find the particular solution whose graph passes through the point $(\pi/2, 0)$.

17. Find the general solution $f(x, y) = C$ of the differential equation

$$\left(\frac{y}{x} + e^y\right) + (\ln x + 2y + xe^y)\frac{dy}{dx} = 0.$$

Then find the particular solution whose graph passes through the point $(1, 0)$.

18. Find the general solution $f(x, y) = C$ of the differential equation

$$3x^2y^2 - (10y^4 - 2yx^3)\frac{dy}{dx} = 0.$$

Then find the particular solution that satisfies the condition $f(1, 1) = 0$.

19. Find the solution of the differential equation

$$(2x + e^{-y} \sin x) + (e^y + e^{-y} \cos x)\frac{dy}{dx} = 0$$

whose graph passes through the origin.

20. a) Find the value of the constant b that makes

$$(y \cos x + b \cos y)\,dx + (x \sin y + \sin x + y)\,dy$$

the differential of a function $f(x, y)$.

b) Find the function $f(x, y)$ that corresponds to the value of b you found in part (a) and satisfies the condition $f(0, 1) = 0$.

21. a) Find the value of α that makes

$$(y^5 + 3x^{1/2}y^3)\,dx + (5xy^4 + 3\alpha x^{3/2}y^2 + 2y)\,dy$$

the differential of a function $g(x, y)$.

b) Find the function $g(x, y)$ that corresponds to the value of α you found in part (a) and satisfies the condition $g(1, -1) = 0$.

22. Let $u = u(x, y)$, $v = v(x, y)$. Assuming that all the necessary partial derivatives exist, establish the following differential formulas:

a) $d(cu) = c\,du$ (any constant c),

b) $d(u^n) = n\,u^{n-1}$ (any number n),

c) $d(uv) = u\,dv + v\,du$,

d) $d\left(\frac{u}{v}\right) = \frac{v\,du - u\,dv}{v^2}$.

15.12
Least Squares

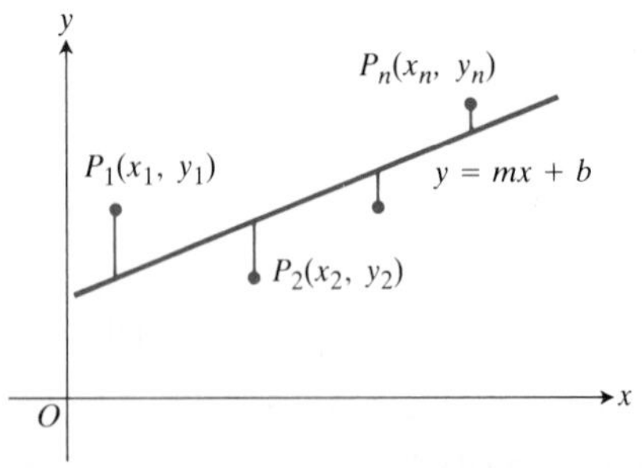

15.48 To fit a line to noncollinear points, we may choose a line that minimizes the sum of the squares of the deviations.

An important application of minimizing a function of two variables is the *method of least squares* for fitting a straight line

$$y = mx + b \tag{1}$$

to a set of experimentally observed points (x_1, y_1), (x_2, y_2), . . . , (x_n, y_n) (Fig. 15.48). Corresponding to each observed value of x there are two values of y, namely, the observed value y_{obs} and the value predicted by the straight line $y = mx_{obs} + b$. We call the difference

$$\text{Predicted value} - \text{Observed value} = (mx_{obs} + b) - y_{obs} \tag{2}$$

a *deviation*. Each deviation measures the amount by which the predicted value of y differs from the observed value. The set of deviations

$$d_1 = (mx_1 + b) - y_1, \ldots, d_n = (mx_n + b) - y_n \tag{3}$$

gives a picture of how closely the line of Eq. (1) fits the observed data. The line is a perfect fit if and only if all of the deviations are zero. But in general no straight line will give a perfect fit. Then we are confronted with the problem of finding a line that fits *best* in some sense or other. Here is where the method of least squares comes in.

For a straight line that comes *close* to fitting all of the observed points, some of the deviations will probably be positive and some will be negative. But their squares will all be positive, and the expression

$$f(m, b) = (mx_1 + b - y_1)^2 + (mx_2 + b - y_2)^2 + \cdots + (mx_n + b - y_n)^2 \quad (4)$$

counts a positive deviation d and a negative deviation $-d$ equally. This sum of squares of the deviations depends upon the choice of m and b. It is never negative and it can be zero only if m and b have values that produce a straight line that is a perfect fit.

Whether such a perfectly fitting line can be found or not, the method of least squares says, "take as the line $y = mx + b$ of best fit that one for which the sum of squares of the deviations

$$f(m, b) = d_1^2 + d_2^2 + \cdots + d_n^2$$

is a minimum." Thus we try to find the values of m and b where the surface

$$w = f(m, b)$$

in mbw-space has a low point (Fig. 15.49). To do this, we solve the equations

$$\frac{\partial f}{\partial m} = 0 \quad \text{and} \quad \frac{\partial f}{\partial b} = 0 \quad (5)$$

simultaneously.

These equations are equivalent (Problem 13) to

$$\begin{aligned} \left(\sum x_i^2\right)m + \left(\sum x_i\right)b &= \sum x_i y_i \quad \left(\frac{\partial f}{\partial m} = 0\right), \\ \left(\sum x_i\right)m + nb &= \sum y_i \quad \left(\frac{\partial f}{\partial b} = 0\right). \end{aligned} \quad (6)$$

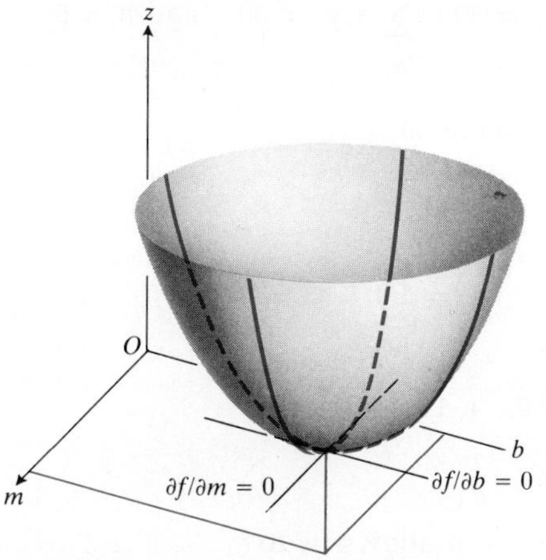

15.49 The sum $f(m, b)$ of the squares of the deviations has a minimum when $\partial f/\partial m$ and $\partial f/\partial b$ are both zero.

where the sums run from $i = 1$ to $i = n$. Note that the variables whose values we wish to determine from Eqs. (6) are m and b. The x's and y's are the coordinates of the data points, which are known, and n is the number of data points. Thus (6) is a system of two equations in the two unknowns m and b, which can be solved for m and b in the familiar way.

It can be proved that the function $f(m, b)$ has an absolute minimum value at the point (m, b) obtained by solving the equations in (6), but the algebra in showing that $f_{mm}f_{bb} - f_{mb}^2 > 0$ is complicated and we shall not go into it here.

Method of Least Squares

To find the *least squares line*

$$y = mx + b$$

for n data points

$$(x_1, y_1), (x_2, y_2), \ldots, (x_n, y_n),$$

solve the equations

$$\left(\sum x_i^2\right)m + \left(\sum x_i\right)b = \sum x_i y_i, \tag{7}$$

$$\left(\sum x_i\right)m + nb = \sum y_i \tag{8}$$

simultaneously for m and b.

EXAMPLE Find the least squares line for the points (0, 1), (1, 3), (2, 2), (3, 4), (4, 5).

Solution It is useful to organize the computations in a table, like this:

i	x_i	y_i	x_i^2	$x_i y_i$
1	0	1	0	0
2	1	3	1	3
3	2	2	4	4
4	3	4	9	12
5	4	5	16	20
Σ	10	15	30	39

We then solve Eqs. (7) and (8) with

$$\sum x_i = 10, \quad \sum y_i = 15, \quad \sum x_i^2 = 30, \quad \sum x_i y_i = 39, \quad \text{and } n = 5$$

to get

$$30m + 10b = 39$$
$$10m + 5b = 15$$

and the solution

$$m = 0.9, \qquad b = 1.2.$$

The least squares line is therefore

$$y = 0.9x + 1.2.$$

See Fig. 15.50. □

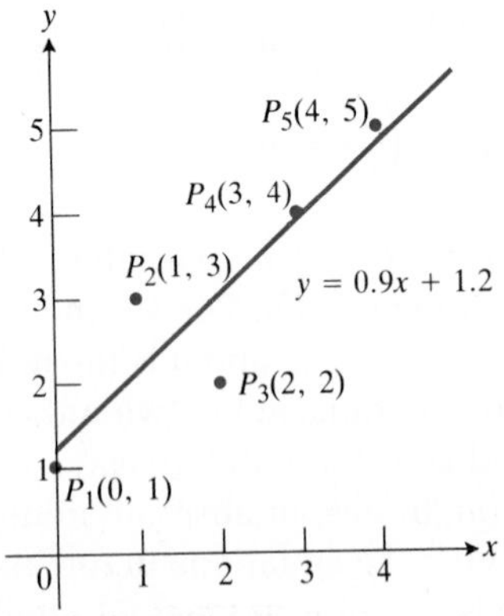

15.50 The least squares line for the data in the Example.

Finding a least squares line $y = mx + b$ allows us to

1. summarize data with a simple expression,
2. predict values of y for other, experimentally untried values of x,
3. handle data analytically.

The method of least squares is so convenient that when data are nonlinear it is common practice to transform one or both of the variables x and y to linearize the data, then fit a line and transform back. (See Problem 7.)

In many fields, the least squares line is called a *regression line* or *line of regression*.

One may also use least squares techniques to fit planes to three-dimensional data but we shall not treat this subject here. (However, see Problem 30 in Article 15.10.)

PROBLEMS

1. In practice, the values of m and b determined by Eqs. (7) and (8) are often calculated from the equations

$$m = \frac{\left(\sum x_i\right)\left(\sum y_i\right) - n\sum x_i y_i}{\left(\sum x_i\right)^2 - n\sum x_i^2}, \quad (9)$$

$$b = \frac{1}{n}\left(\sum y_i - m\sum x_i\right). \quad (10)$$

One finds m first, then uses the value of m in finding b. Show that these values of m and b solve Eqs. (7) and (8).

In Problems 2–6, find the least squares line for each set of data points. Use the linear equation you obtain to predict the value of y that would correspond to $x = 4$.

2. $(-1, 2)$, $(0, 1)$, $(3, -4)$

3. $(-2, 0)$, $(0, 2)$, $(2, 3)$

4. $(0, 0)$, $(1, 2)$, $(2, 3)$

5. $(0, 1)$, $(2, 2)$, $(3, 2)$

6. $(1, 1)$, $(2, 1)$, $(-3, 0)$

7. (*Calculator*) To determine the intermolecular potential between potassium ions and xenon gas, Budenholzer, Gislason, and Jorgensen (June 1, 1977, *Journal of Chemical Physics*, **66**, No. 11; p. 4832) accelerated a beam of potassium ions toward a cell containing xenon and measured the current I of ions leaving the cell as a percentage of the current I_0 entering the cell. This fraction, which is a function of xenon gas pressure, was recorded at five different pressures, with the results shown in Table 13.1.

Table 13.1 Scattering of potassium ions by xenon

x (pressure in millitorr)	$\frac{I}{I_0}$
0.165	0.940
0.399	0.862
0.573	0.810
0.930	0.712
1.281	0.622

As a step in their determination, the authors used the method of least squares to find the slope m and y-intercept b of a line $y = mx + b$, where $y = \ln(I/I_0)$. In particular, they hoped to find that, within the limits of experimental error, b was zero. (a) Write down the value of y for each x in the table and fit a least squares line to the (x, y) data points. Round m and b to three decimal places. (b) Express I/I_0 as a function of x.

8. (*Calculator*) Write a linear equation for the effect of irrigation on the yield of alfalfa by fitting a least squares line to the data in Table 13.2 (from the University of California Experimental Station, *Bulletin* No. 450, p. 8). Plot the data and draw the line.

Table 13.2 Growth of alfalfa

x (total seasonal depth of water applied (in.))	y (average alfalfa yield (tons/acre))
12	5.27
18	5.68
24	6.25
30	7.21
36	8.20
42	8.71

9. (*Calculator*) *Hubble's law* for the expansion of the universe is the linear equation

$$\text{Velocity} = \text{the Hubble constant} \cdot \text{distance}$$

or

$$v = Hx.$$

It says that the velocity with which a galaxy appears to move away from us is proportional to how far away the galaxy lies. The farther away it lies, the faster it recedes. If the velocity is measured in kilometers per second and the distance in millions of light-years, then Hubble's constant is given in kilometers per second per million light-years. H is the rate at which the universe appears to be expanding. For every extra million light-years of distance, the galaxies we can observe recede faster by H kilometers per second.

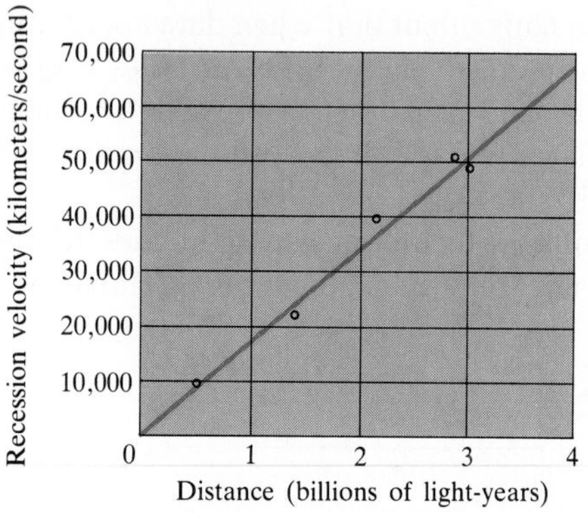

15.51 Velocity vs. distance observed for the galaxies in Table 13–3.

Table 13.3 lists the observed distances and velocities for five galaxies. Discover Hubble's constant H by fitting a least squares line to the data (Fig. 15.51). Round your answer to the nearest integer. You will find that the y-intercept of the line is 240 km/s when rounded to the nearest integer, and not zero. A discrepancy in the intercept is to be expected, given the uncertainties in measurement and, here, the small size of the sample. Note that the discrepancy is a small percentage of the observed recession velocities. (For more information, see Jastrow and Thompson's *Astronomy: Fundamentals and Frontiers*, Third Edition, John Wiley and Sons, Inc., 1977, Chapter 11.)

Table 13.3 Observed velocities and distances of five receding galaxies

Galaxy	Observed distance (10^6 l-yr)	Recession velocity (km/s)
A	500	9,000
B	1,400	22,000
C	2,100	39,000
D	2,900	51,000
E	3,000	49,000

10. (*Calculator*) *Craters of Mars.* One theory of crater formation suggests that the frequency of large craters should fall off as the square of the diameter (Marcus, *Science*, June 21, 1968, p. 1334). Pictures from Mariner IV show the frequencies listed in Table 13.4.

Table 13.4 Crater sizes on Mars

Diameter in km, D	$1/D^2$ (for left value of class interval)	Frequency, F
32–45	0.001	53
45–64	0.0005	22
64–90	0.00025	14
90–128	0.000125	3

Fit a line of the form $F = m(1/D^2) + b$ to the data. Plot the data and draw the line.

11. If $y = mx + b$ is the best-fitting straight line, in the sense of least squares, show that the sum of deviations

$$\sum_{i=1}^{n} (y_i - mx_i - b)$$

is zero. (This means that positive and negative deviations cancel.)

12. Show that the point

$$(\bar{x}, \bar{y}) = \left[\frac{1}{n}\left(\sum_{i=1}^{n} x_i\right), \frac{1}{n}\left(\sum_{i=1}^{n} y_i\right)\right]$$

lies on the straight line $y = mx + b$ that is determined by the method of least squares. (This means that the "best-fitting" line passes through the center of gravity of the n points.)

13. Expand the equations in (4) and (5) to show that they lead to the equations in (6).

REVIEW QUESTIONS AND EXERCISES

1. Let $z = f(x, y)$ be a function of two independent variables defined on a region R in the xy-plane. Describe two graphical ways to represent f.
2. Define level surface and contour line.
3. Give two equivalent definitions of

$$\lim_{(x, y)\to(x_0, y_0)} f(x, y) = L.$$

What is the basic theorem for calculating limits of sums, differences, products, and quotients of functions of two variables?

4. When is a function of two variables continuous at a point of its domain? Give an example of a function that is discontinuous at some point(s) of its domain.
5. When is a function of three variables continuous at a point of its domain? Give an example of a function of three independent variables that is continuous at some points of its domain and discontinuous at at least one place in its domain. Give an example that is discontinuous at all points of a surface $F(x, y, z) = 0$; at all points of a line.

6. Let $z = f(x, y)$. Define $\partial z/\partial x$ and $\partial z/\partial y$ at a point (x_0, y_0) in the domain of f.
7. What does the Increment Theorem for functions of two variables say?
8. State several chain rules for partial derivatives of functions of two and three variables, and give their tree diagrams.
9. Define the gradient of a function $f(x, y, z)$ and describe the role the gradient plays in defining directional derivatives, tangent planes, and normal lines. What is the relation between ∇f and the directions in which f changes most rapidly? What are the analogous results for functions of two variables?
10. Suppose $\mathbf{R}(t) = x(t)\mathbf{i} + y(t)\mathbf{j} + z(t)\mathbf{k}$ is a curve in the domain of a function $f(x, y, z)$. Describe the relation between df/dt, ∇f, and $\mathbf{v} = d\mathbf{R}/dt$. What can be said about ∇f and $\mathbf{v}$ at points of the curve where f has a local maximum or minimum relative to its values on the curve?
11. Why do we require functions defining surfaces to have continuous first partial derivatives before we define tangent planes?
12. What is the basic theorem about mixed second order partial derivatives of functions of two variables? Give an example. What about mixed partial derivatives of higher order?
13. Give three important partial differential equations from physics, along with sample solutions.
14. Give the basic formula for the linear approximation of a function $f(x, y)$ near a point (x_0, y_0), and give an upper bound for the error in the approximation. What formula may be used to estimate increments in f?
15. Describe a way to find the extreme values of a function $f(x, y)$ that is continuous and has continuous first and second order partial derivatives (a) at an interior point of its domain, and (b) at a boundary point. Give an example.
16. Describe the method of Lagrange multipliers and its geometric interpretation as it applies to the problem of maximizing or minimizing a function $f(x, y, z)$
    a) subject to one constraint,
    $$g(x, y, z) = 0;$$
    b) subject to two constraints,
    $$g(x, y, z) = 0 \quad \text{and} \quad h(x, y, z) = 0.$$
17. Sketch some of the level curves of $f(x, y) = 2x + 3y$ and find points on the curve $g(x, y) = x^2 + xy + y^2 - 3 = 0$ that (a) maximize f, (b) minimize f. Does it seem reasonable that the level curve of f at each of these points should be tangent to the curve $g(x, y) = 0$ at each of these points? Is it true?
18. Explain the test for the exactness of the differential form
    $$M(x, y)\,dx + N(x, y)\,dy.$$
    Give examples of forms that pass, and fail, the test. Find the solution of the differential equation
    $$(2x + y)\,dx + (2y + x)\,dy = 0$$
    whose graph passes through the point (1, 1).
19. Describe the method of least squares for fitting a line to a finite set of data points. To what use may we put the least squares line?

MISCELLANEOUS PROBLEMS

1. Let $f(x, y) = (x^2 - y^2)/(x^2 + y^2)$ for $x^2 + y^2 \neq 0$. Is it possible to define the value of f at $x = 0$, $y = 0$, in such a way that the function would be continuous at $x = 0$, $y = 0$? Why?
2. Let the function $f(x, y)$ be defined by the relations
   $$f(x, y) = \frac{\sin^2(x - y)}{|x| + |y|} \quad \text{for} \quad |x| + |y| \neq 0, \quad f(0, 0) = 0.$$
   Is f continuous at $x = 0$, $y = 0$?
3. Prove that if $f(x, y)$ is defined for all x, y, by
   $$f(x, y) = \begin{cases} \dfrac{2xy}{x^2 + y^2} & \text{if} \quad (x, y) \neq (0, 0), \\ 0 & \text{if} \quad x = y = 0, \end{cases}$$
   then
   a) for any fixed x, $f(x, y)$ is a continuous function of y;
   b) for any fixed y, $f(x, y)$ is a continuous function of x;
   c) $f(x, y)$ is not continuous at (0, 0);
   d) $\partial f/\partial x$ and $\partial f/\partial y$ exist at (0, 0) but are not continuous there.

   (This example shows that a function may possess partial derivatives at all points of a region, yet not be continuous in the region.) Contrast the case of a function of one variable, where the existence of a derivative implies continuity.
4. Let $f(x, y)$ be defined and continuous for all x, y (differentiability not assumed). Show that it is always

possible to find arbitrarily many points (x_1, y_1), $(x_2, y_2), \ldots, (x_n, y_n)$ such that the function has the same value at each of them.

5. Find the first partial derivatives of the following functions:
a) $(\sin xy)^2$, b) $\sin[(xy)^2]$.

6. Let α, β, γ be the angles that a line passing through the origin into the first octant makes with the positive coordinate axes. Consider γ as a function of α and β. Find the value of $\partial\gamma/\partial\alpha$ when $\alpha = \pi/4$, $\beta = \pi/3$, $\gamma = \pi/3$.

7. Let (r, θ) and (x, y) be polar coordinates and cartesian coordinates in the plane. Show geometrically why $\partial r/\partial x$ is not equal to $(\partial x/\partial r)^{-1}$, by appealing to the definitions of these derivatives.

8. Consider the surface whose equation is $x^3z + y^2x^2 + \sin(yz) + 54 = 0$. Give an equation of the tangent plane to the surface at the point $P(3, 0, -2)$, and give equations of the line through P normal to the surface.

9. a) Sketch and name the surface $x^2 - y^2 + z^2 = 4$.
b) Find a vector that is normal to this surface at $(2, -3, 3)$.
c) Find equations of the surface's tangent plane and normal line at $(2, -3, 3)$.

10. a) Find an equation of the plane tangent to the surface

$$x^3 + xy^2 + y^3 + z^3 + 1 = 0$$

at the point $(-2, 1, 2)$.
b) Find equations of the straight line perpendicular to the plane above at the point $(-2, 1, 2)$.

11. The directional derivative of a given function $f(x, y)$ at the point $P_0(1, 2)$ in the direction toward $P_1(2, 3)$ is $2\sqrt{2}$, and in the direction toward $P_2(1, 0)$ it is -3. Compute $\partial f/\partial x$ and $\partial f/\partial y$ at $P_0(1, 2)$, and compute the derivative of f at $P_0(1, 2)$ in the direction toward $P_3(4, 6)$.

12. Let $z = f(x, y)$ have continuous first partial derivatives. Let C be any curve lying on the surface and passing through (x_0, y_0, z_0). Prove that the tangent line to C at (x_0, y_0, z_0) must lie wholly in the plane determined by the tangent lines to the curves C_x and C_y, where C_x is the curve of intersection of $y = y_0$ and $z = f(x, y)$, and C_y is the curve of intersection of $x = x_0$ and the surface.

13. Let $u = xyz$. Show that if x and y are the independent variables (so u and z are functions of x and y), then

$$\partial u/\partial x = xy\,(\partial z/\partial x) + yz;$$

but that if x, y, and z are the independent variables,

$$\partial u/\partial x = yz.$$

14. Let

$$\mathbf{u} = u_1\mathbf{i} + u_2\mathbf{j} + u_3\mathbf{k} \quad \text{and} \quad \mathbf{v} = v_1\mathbf{i} + v_2\mathbf{j} + v_3\mathbf{k}$$

be given constant unit vectors and let $f(x, y, z)$ be a given scalar function. Compute
a) the directional derivative $D_{\mathbf{u}}f$, and
b) the directional derivative $D_{\mathbf{v}}(D_{\mathbf{u}}f)$,
in terms of derivatives of f and the components of $\mathbf{u}$ and $\mathbf{v}$.

15. Consider the function $w = xyz$.
a) Compute the directional derivative of w at the point $(1, 1, 1)$ in the direction of the vector $\mathbf{i} + \mathbf{j} + \mathbf{k}$.
b) Compute the largest value of the directional derivative of w at the point $(1, 1, 1)$.

16. The function $w = f(x, y)$ has, at the point $(1, 2)$, directional derivatives that are equal to $+2$ in the direction toward $(2, 2)$, and -2 in the direction toward $(1, 1)$. What is its directional derivative at $(1, 2)$ in the direction toward $(4, 6)$?

17. Given the function $f(x, y, z) = x^2 + y^2 - 3z$, what is the maximum value of the directional derivative of f at the point $(1, 3, 5)$?

18. Given the function

$$(x-1)^2 + 2(y+1)^2 + 3(z-2)^2 - 6.$$

Find the derivative of the function at the point $(2, 0, 1)$ in the direction of the vector $\mathbf{i} - \mathbf{j} + 2\mathbf{k}$.

19. Find the derivative of the function

$$f(x, y, z) = x^2 - 2y^2 + z^2$$

at the point $(3, 3, 1)$ in the direction of the vector $2\mathbf{i} + \mathbf{j} - \mathbf{k}$.

20. The two equations

$$e^u \cos v - x = 0 \quad \text{and} \quad e^u \sin v - y = 0$$

define u and v as functions of x and y, say $u = u(x, y)$ and $v = v(x, y)$. Show that the angle between the two vectors

$$\left(\frac{\partial u}{\partial x}\right)\mathbf{i} + \left(\frac{\partial u}{\partial y}\right)\mathbf{j} \quad \text{and} \quad \left(\frac{\partial v}{\partial x}\right)\mathbf{i} + \left(\frac{\partial v}{\partial y}\right)\mathbf{j}$$

is constant.

21. a) Find a vector $\mathbf{N}(x, y, z)$ normal to the surface $z = \sqrt{x^2 + y^2} + (x^2 + y^2)^{3/2}$ at the point (x, y, z) of the surface.
b) Find the cosine of the angle γ between $\mathbf{N}(x, y, z)$ and the z-axis. Find the limit of $\cos\gamma$ as $(x, y, z) \to (0, 0, 0)$.

22. Find all points (a, b, c) in space for which the spheres

$$(x-a)^2 + (y-b)^2 + (z-c)^2 = 1$$

and

$$x^2 + y^2 + z^2 = 1$$

will intersect orthogonally. (Their tangents are to be perpendicular at each point of intersection.)

23. a) Find the gradient of the function

$$f(x, y, z) = x^2 + 2xy - y^2 + z^2$$

at the point $P_0(1, -1, 3)$.

b) Find the plane that is tangent to the surface $x^2 + 2xy - y^2 + z^2 = 7$ at $P_0(1, -1, 3)$.

24. Find a unit vector normal to the surface $x^2 + y^2 = 3z$ at the point $(1, 3, \frac{10}{3})$.

25. In a flowing fluid, the density $\rho(x, y, z, t)$ depends on position and time. If

$$\mathbf{V} = \mathbf{V}(x, y, z, t)$$

is the velocity of the fluid particle at the point (x, y, z) at time t, then

$$\frac{d\rho}{dt} = \mathbf{V}\cdot\nabla\rho + \frac{\partial\rho}{\partial t} = V_1\frac{\partial\rho}{\partial x} + V_2\frac{\partial\rho}{\partial y} + V_3\frac{\partial\rho}{\partial z} + \frac{\partial\rho}{\partial t},$$

where $\mathbf{V} = V_1\mathbf{i} + V_2\mathbf{j} + V_3\mathbf{k}$. Explain the physical and geometrical meaning of this relation.

26. Find a constant a such that, at any point of intersection of the two spheres

$$(x - a)^2 + y^2 + z^2 = 3 \text{ and } x^2 + (y - 1)^2 + z^2 = 1,$$

their tangent planes will be perpendicular to each other.

27. If the gradient of a function $f(x, y, z)$ is always parallel to the vector $x\mathbf{i} + y\mathbf{j} + z\mathbf{k}$, show that the function must assume the same value at the points $(0, 0, a)$ and $(0, 0, -a)$.

28. Let $f(P)$ denote a function defined for points P in the plane; i.e., to each point P there is attached a real number $f(P)$. Explain how one could introduce the notions of continuity and differentiability of the function and define the vector ∇f *without* introducing a coordinate system. If one introduces a polar coordinate system r, θ, $\mathbf{U}_r$, $\mathbf{U}_\theta$, what form does the vector $\nabla f(r, \theta)$ take?

29. Show that the directional derivative of

$$r = \sqrt{x^2 + y^2 + z^2}$$

equals 1 in any direction at the origin, but that r does not have a gradient vector at the origin.

30. Let $\mathbf{R} = x\mathbf{i} + y\mathbf{j} + z\mathbf{k}$ and $r = |\mathbf{R}|$.

a) From its geometrical interpretation, show that $\nabla r = \mathbf{R}/r$.

b) Show that $\nabla(r^n) = nr^{n-2}\mathbf{R}$.

c) Find a function with gradient equal to $\mathbf{R}$.

d) Show that $\mathbf{R}\cdot d\mathbf{R} = r\,dr$.

e) If $\mathbf{A}$ is a constant vector, show that $\nabla(\mathbf{A}\cdot\mathbf{R}) = \mathbf{A}$.

31. If θ is the polar coordinate in the xy-plane, find the direction and magnitude of $\nabla\theta$.

32. If r_1, r_2 are the distances from the point $P(x, y)$ on an ellipse to its foci, show that the equation $r_1 + r_2 =$ const., satisfied by these distances, requires $\mathbf{U}\cdot\nabla(r_1 + r_2) = 0$, where $\mathbf{U}$ is a unit tangent to the curve. By geometrical interpretation, show that the tangent makes equal angles with the lines to the foci.

33. If A, B are fixed points and θ is the angle at $P(x, y, z)$ subtended by the line segment AB, show that $\nabla\theta$ is normal to the circle through A, B, P.

34. Find the general solution of the partial differential equations:

a) $af_x + bf_y = 0$, a, b constants, b) $yf_x - xf_y = 0$. (*Hint:* Consider the geometrical meaning of the equations.)

35. When y is eliminated from the two equations $z = f(x, y)$ and $g(x, y) = 0$, the result is expressible in the form $z = h(x)$. Express the derivative $h'(x)$ in terms of $\partial f/\partial x$, $\partial f/\partial y$, $\partial g/\partial x$, $\partial g/\partial y$. Check your formula by computing $h(x)$ and $h'(x)$ explicitly in the example where $f(x, y) = x^2 + y^2$ and $g(x, y) = x^3 + y^2 - x$.

36. Suppose the equation $F(x, y, z) = 0$ defines z as a function of x and y, say $z = f(x, y)$, with derivatives $\partial f/\partial x$ and $\partial f/\partial y$. Suppose also that the same equation $F(x, y, z) = 0$ defines x as a function of y and z, say $x = g(y, z)$, with derivatives $\partial g/\partial y$ and $\partial g/\partial z$. Prove that

$$\frac{\partial g}{\partial y} = -\frac{\partial f/\partial y}{\partial f/\partial x},$$

and also express $\partial g/\partial z$ in terms of $\partial f/\partial x$ and $\partial f/\partial y$.

37. Given $z = x \sin x - y^2$, $\cos y = y \sin z$, find dx/dz.

38. If

$$z = f\left(\frac{x - y}{y}\right),$$

show that $x(\partial z/\partial x) + y(\partial z/\partial y) = 0$.

39. If the substitution $u = (x - y)/2$, $v = (x + y)/2$, changes $f(u, v)$ into $F(x, y)$, express $\partial F/\partial x$ and $\partial F/\partial y$ in terms of the derivatives of $f(u, v)$ with respect to u and v.

40. Given $w = f(x, y)$ with $x = u + v$, $y = u - v$, show that

$$\frac{\partial^2 w}{\partial u\,\partial v} = \frac{\partial^2 w}{\partial x^2} - \frac{\partial^2 w}{\partial y^2}.$$

41. Suppose $f(x, y, z)$ is a function that has continuous partial derivatives and satisfies $f(tx, ty, tz) = t^n f(x, y, z)$ for every quadruple of numbers x, y, z, t (where n is a fixed integer). Show the identity

$$\frac{\partial f}{\partial x}x + \frac{\partial f}{\partial y}y + \frac{\partial f}{\partial z}z = nf.$$

(*Hint:* Differentiate with respect to t; then set $t = 1$.)

42. The substitution $u = x + y$, $v = xy^2$ changes the function $f(u, v)$ into $F(x, y)$. Express the partial derivative $\partial^2 F/\partial x\,\partial y$ in terms of x, y, and the partial derivatives of $f(u, v)$ with respect to u, v.

43. Given $z = u(x, y) \cdot e^{ax+by}$, where $u(x, y)$ is a function of x and y such that $\partial^2 u/\partial x\,\partial y = 0$, (a, b constants). Find values of a and b that will make the expression $\partial^2 z/\partial x\,\partial y - \partial z/\partial x - \partial z/\partial y + z$ identically zero.

44. Introducing polar coordinates, $x = r\cos\theta$, $y = r\sin\theta$, changes $f(x, y)$ into $g(r, \theta)$. Compute the value of the second derivative $\partial^2 g/\partial\theta^2$ at the point where $r = 2$ and $\theta = \pi/2$, given that

$$\frac{\partial f}{\partial x} = \frac{\partial f}{\partial y} = \frac{\partial^2 f}{\partial x^2} = \frac{\partial^2 f}{\partial y^2} = 1$$

at that point.

45. Let $w = f(u, v)$ be a function of u, v with continuous partial derivatives, where u, v in turn are functions of independent variables, x, y, z, with continuous partial derivatives. Show that if w is regarded as a function of x, y, z, its gradient at any point (x_0, y_0, z_0) lies in a common plane with the gradients of $u = u(x, y, z)$ and $v = v(x, y, z)$.

46. Show that if a function u has first derivatives that satisfy a relation of the form $F(u_x, u_y) = 0$, and if $\partial F/\partial u_x$ and $\partial F/\partial u_y$ are not both zero, then u also satisfies $u_{xx}u_{yy} - u_{xy}^2 = 0$. (*Hint:* Differentiate both sides of the equation $F = 0$ with respect to x and y.)

47. If $f(x, y) = 0$, find d^2y/dx^2.

48. If $f(x, y, z) = 0$ and $z = x + y$, find dz/dx.

49. The function $v(x, t)$ is defined for $0 \le x \le 1$, $0 \le t$ and satisfies the partial differential equation

$$v_t = v_x(v - x) + av_{xx}$$

(a = constant > 0) and the boundary conditions $v(0, t) = 0$, $v(1, t) = 1$. Suppose that for each fixed t, $v(x, t)$ is a strictly increasing function of x; that is, $v_x(x, t) > 0$. Show that v and t may be introduced as independent variables and x as dependent variable, and find the partial differential equation satisfied by the function $x(v, t)$. Also find the region of definition of $x(v, t)$ and boundary values that it satisfies. By considering level curves, show geometrically why the assumption $v_x(x, t) > 0$ is necessary for the success of this transformation.

50. Let $f(x, y, z)$ be a function depending only on $r = \sqrt{x^2 + y^2 + z^2}$; that is, $f(x, y, z) = g(r)$. Prove that if $f_{xx} + f_{yy} + f_{zz} = 0$, it follows that

$$f = \left(\frac{a}{r}\right) + b,$$

where a and b are constants.

51. A function $f(x, y)$, defined and differentiable for all x, y, is said to be homogeneous of degree n (a non-negative integer) if $f(tx, ty) = t^n f(x, y)$ for all t, x, and y. For such a function prove:

a) $x(\partial f/\partial x) + y(\partial f/\partial y) = nf(x, y)$ and express this in vector form;

b) $$x^2\left(\frac{\partial^2 f}{\partial x^2}\right) + 2xy\left(\frac{\partial^2 f}{\partial x\partial y}\right) + y^2\left(\frac{\partial^2 f}{\partial y^2}\right) = n(n-1)$$

if f has continuous second partial derivatives;

c) a homogeneous function of degree zero is a constant.

52. Prove the Mean Value Theorem for functions of two variables

$$f(x+h, y+k) - f(x, y) = f_x(x+\theta h, y+\theta k)h + f_y(x+\theta h, y+\theta k)k, \quad 0 < \theta < 1,$$

with suitable assumptions about f. What assumptions? (Apply the Mean Value Theorem for functions of one variable to $F(t) = f(x + ht, y + kt)$.)

53. Prove the theorem: If $f(x, y)$ is defined in a region R, and f_x, f_y exist and are bounded in R, then $f(x, y)$ is continuous in R. (The assumption of boundedness is essential.)

54. For each of the following three surfaces, find all the values of x and y for which z is a maximum or minimum (if there are any). Give complete reasonings.

a) $x^2 + y^2 + z^2 = 3$ b) $x^2 + y^2 = 2z$

c) $x^2 - y^2 = 2z$

55. Find the point(s) on the surface $xyz = 1$ whose distance from the origin is a minimum.

56. A closed rectangular box is to be made to hold a given volume, V in³. The cost of the material used in the box is a cents/in² for top and bottom, b cents/in² for front and back, c cents/in² for the remaining two sides. What dimensions make the total cost of materials a minimum?

57. Find the maximum value of the function $xye^{-(2x+3y)}$ in the first quadrant.

58. A surface is defined by $z = x^3 + y^3 - 9xy + 27$. Prove that the only possible maxima and minima of z occur at (0, 0) or (3, 3). Prove that (0, 0) is neither a maximum nor a minimum. Determine whether (3, 3) is a maximum or a minimum.

59. Find the minimum volume bounded by the planes $x = 0$, $y = 0$, $z = 0$, and a plane that is tangent to the ellipsoid

$$\frac{x^2}{a^2} + \frac{y^2}{b^2} + \frac{z^2}{c^2} = 1$$

at a point in the octant $x > 0$, $y > 0$, $z > 0$.

60. Let z be defined implicitly as a function of x and y by the equation $\sin(x + y) + \sin(y + z) = 1$. Compute $\partial^2 z/\partial x\,\partial y$ in terms of x, y, and z.

61. Given $z = xy^2 - y \sin x$, calculate the value of $y(\partial^2 z/\partial y\, \partial x) - \partial z/\partial x$.

62. Let $w = z \tan^{-1}(x/y)$. Compute

$$\frac{\partial^2 w}{\partial x^2} + \frac{\partial^2 w}{\partial y^2} + \frac{\partial^2 w}{\partial z^2}.$$

63. Show that the function satisfies the equation.

a) $\int_0^{x/2\sqrt{kt}} e^{-\sigma^2}\, d\sigma$, $\quad kf_{xx} - f_t = 0 \quad$ (k const.)

b) $\phi(x + at) + \psi(x - at)$, $\quad f_{tt} = a^2 f_{xx}$

64. Let

$$f(x, y) = \begin{cases} xy\dfrac{x^2 - y^2}{x^2 + y^2}, & (x, y) \neq (0, 0), \\ 0, & (x, y) = (0, 0). \end{cases}$$

Find $f_{yx}(0, 0)$ and $f_{xy}(0, 0)$. See Fig. 15.52.

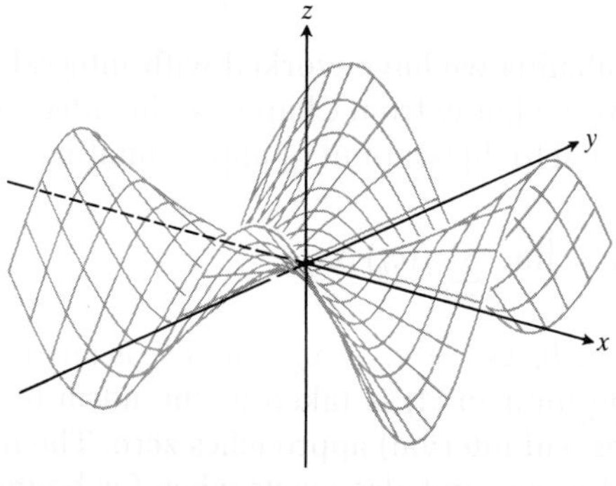

15.52 The surface of Problem 64.

65. Is $2x(x^3 + y^3)\, dx + 3y^2(x^2 + y^2)\, dy$ the differential df of a function $f(x, y)$? If so, find the function.

66. Find a function $w = f(x, y)$ such that $\partial w/\partial x = 1 + e^x \cos y$ and $\partial w/\partial y = 2y - e^x \sin y$, or else explain why no such function exists.

67. In thermodynamics the five quantities S, T, u, p, v are such that any two of them may be considered independent variables, the others then being determined. They are connected by the differential relation $T\, dS = du + p\, dv$. Show that

$$\left(\frac{\partial S}{\partial v}\right)_T = \left(\frac{\partial p}{\partial T}\right)_v \quad \text{and} \quad \left(\frac{\partial v}{\partial S}\right)_p = \left(\frac{\partial T}{\partial p}\right)_S.$$

68. Let

$$f(r, \theta) = \begin{cases} \dfrac{\sin 6r}{6r}, & r \neq 0 \\ 1, & r = 0. \end{cases}$$

(See Fig. 15.53.) Find (a) $\lim_{r \to 0} f(r, \theta)$, (b) $f_r(0, 0)$, (c) $f_\theta(r, \theta)$, $r \neq 0$.

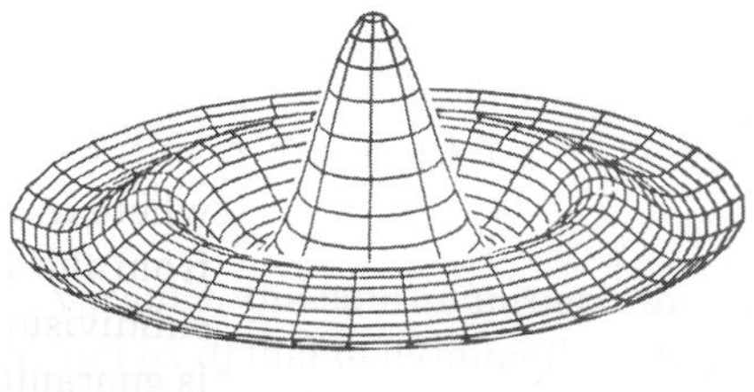

15.53 The surface of Problem 68.

I4. $\iint\limits_R f(x, y)\, dA \geq 0$ if $f(x, y) \geq 0$ on R

I5. $\iint\limits_R f(x, y)\, dA \geq \iint\limits_R g(x, y)\, dA$ if $f(x, y) \geq g(x, y)$ on R

These are like the properties I1–I5 in Article 4.5, and the proofs (which we omit) are similar.

There is also a "domain additivity" property:

I6. $\iint\limits_R f(x, y)\, dA = \iint\limits_{R_1} f(x, y)\, dA + \iint\limits_{R_2} f(x, y)\, dA,$

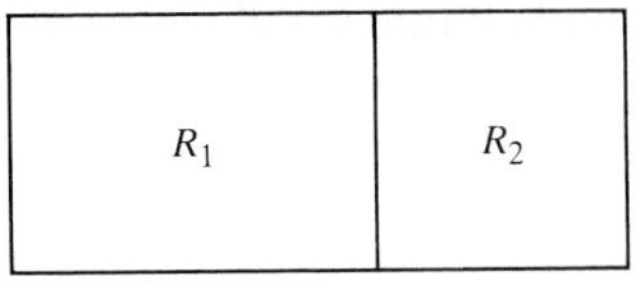

16.2 $\iint\limits_{R_1 \cup R_2} f(x, y)\, dA = \int\limits_{R_1} f(x, y)\, dA + \int\limits_{R_2} f(x, y)\, dA.$

which holds when R is the union of two nonoverlapping rectangles R_1 and R_2 as shown in Fig. 16.2. Again, we shall omit the proof.

Volume

When $f(x, y) > 0$, we may interpret $\iint_R f(x, y)\, dA$ as the volume of the solid enclosed by R, the planes $x = a$, $x = b$, $y = c$, $y = d$, and the surface $z = f(x, y)$, as shown in Fig. 16.3. Each term $f(x_k, y_k)\, \Delta A_k$ in the sum

$$S_n = \Sigma f(x_k, y_k)\, \Delta A_k$$

is the volume of a vertical rectangular prism that approximates the volume of the portion of the solid that stands directly above the base ΔA_k. The sum S_n thus approximates what we want to call the total volume of the solid, and we *define* this volume to be

$$\text{Volume} = \lim S_n = \iint\limits_R f(x, y)\, dA. \tag{4}$$

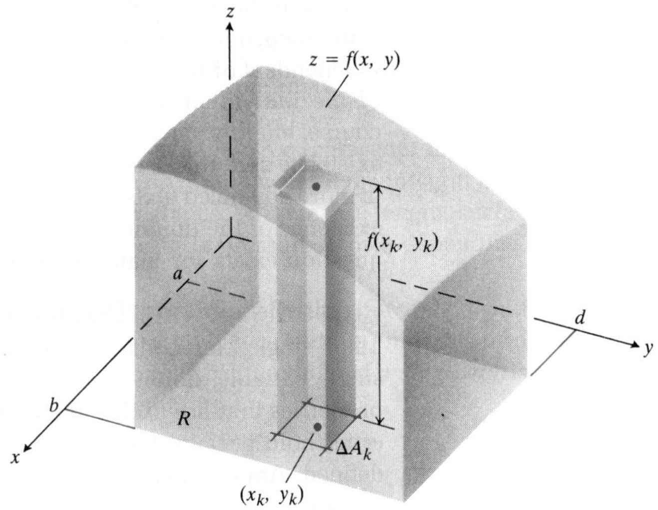

16.3 Approximating a solid with rectangular prisms leads to a definition of volume consistent with past definitions.

Fubini's Theorem for Calculating Double Integrals

We are now ready to calculate our first double integral.

Suppose we wish to calculate the volume under the plane $z = 4 - x - y$ over the region R: $0 \leq x \leq 2$, $0 \leq y \leq 1$ in the xy-plane. If

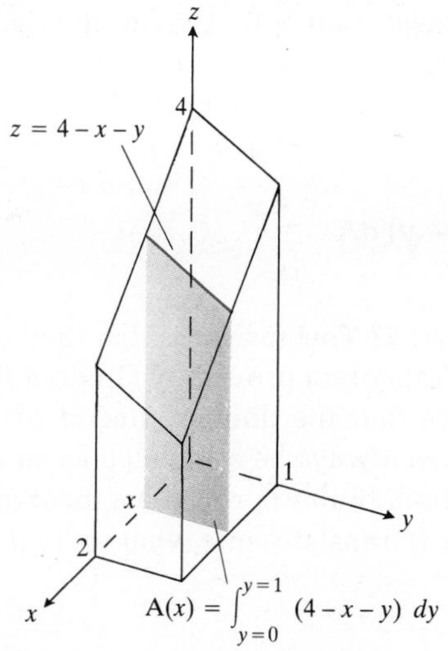

16.4 The cross-sectional area $A(x)$ is obtained by holding x fixed and integrating with respect to y.

we apply the method of slicing from Article 5.4, with slices perpendicular to the x-axis (Fig. 16.4), then the volume is

$$\int_{x=0}^{x=2} A(x)\, dx, \tag{5}$$

where $A(x)$ is the cross-sectional area at x. For each value of x we may calculate $A(x)$ as the integral

$$A(x) = \int_{y=0}^{y=1} (4 - x - y)\, dy, \tag{6}$$

which is the area under the curve $z = 4 - x - y$ in the plane of the cross section at x. In calculating $A(x)$, x is held fixed and the integration takes place with respect to y. Combining (5) and (6) we see that the volume of the entire solid is

$$\begin{aligned} \text{Volume} &= \int_{x=0}^{x=2} A(x)\, dx = \int_{x=0}^{x=2} \left(\int_{x=0}^{x=1} (4 - x - y)\, dy \right) dx \\ &= \int_{x=0}^{x=2} \left[4y - xy - \frac{y^2}{2} \right]_{y=0}^{y=1} dx \\ &= \int_{x=0}^{x=2} \left[\frac{7}{2} - x \right] dx = \frac{7}{2}x - \frac{x^2}{2} \bigg]_0^2 = 5. \end{aligned} \tag{7}$$

If we had just wanted to write instructions for calculating the volume, without carrying out any of the integrations, we could have written

$$\text{Volume} = \int_0^2 \int_0^1 (4 - x - y)\, dy\, dx.$$

The expression on the right, called an *iterated* or *repeated* integral, says that the volume is to be obtained by integrating $4 - x - y$ with respect to y from $y = 0$ to $y = 1$ holding x fixed, and then by integrating the resulting expression in x with respect to x from $x = 0$ to $x = 2$.

What would have happened if we had calculated the volume by slicing with planes perpendicular to the y-axis, as shown in Fig. 16.5? As a function of y the typical cross-sectional area is now

$$A(y) = \int_{x=0}^{x=2} (4 - x - y)\, dx = 4x - \frac{x^2}{2} - xy \bigg]_{x=0}^{x=2} = 6 - 2y. \tag{8}$$

The volume of the entire solid is therefore

$$\text{Volume} = \int_{y=0}^{y=1} A(y)\, dy = \int_{y=0}^{y=1} (6 - 2y)\, dy = 6y - y^2 \bigg]_0^1 = 5,$$

in agreement with our earlier calculation.

Again, we may give instructions for calculating the volume as an iterated integral by writing

$$\text{Volume} = \int_0^1 \int_0^2 (4 - x - y)\, dx\, dy.$$

The expression on the right says that the volume may be obtained by integrating $4 - x - y$ with respect to x from $x = 0$ to $x = 2$ (as we did in Eq. 8) and by integrating the result with respect to y from $y = 0$ to $y = 1$.

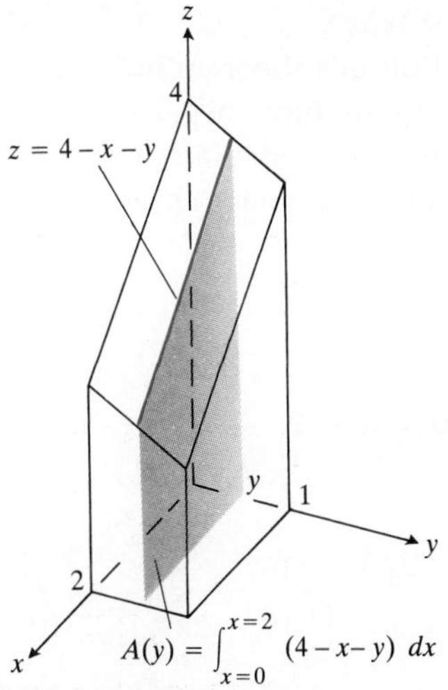

16.5 The cross-sectional area $A(y)$ is obtained by holding y fixed and integrating with respect to x.

In this iterated integral the order of integration is first x and then y, the reverse of the order we used in (7).

What do these two volume calculations with iterated integrals have to do with the double integral

$$\iint_R (4 - x - y)\,dA$$

over the rectangle R: $0 \le x \le 2$, $0 \le y \le 1$? The answer is that they both give the value of the double integral. A theorem proved by Guido Fubini (1879–1943) and published in 1907 says that the double integral of any continuous function over a rectangle can always be calculated as an iterated integral in either order of integration. (Fubini proved his theorem in much greater generality, but this is how it translates into what we're doing now.)

THEOREM 1

Fubini's Theorem (First Form)

If $f(x, y)$ is continuous on the rectangular region R: $a \le x \le b$, $c \le y \le d$, then

$$\iint_R f(x, y)\,dA = \int_c^d \int_a^b f(x, y)\,dx\,dy = \int_a^b \int_c^d f(x, y)\,dy\,dx.$$

Fubini's theorem says that double integrals over rectangles can always be calculated as iterated integrals. This means that we can evaluate a double integral by integrating one variable at a time, using the integration techniques we already know for functions of a single variable.

Fubini's theorem also says that we may calculate the double integral by integrating in *either* order (a genuine convenience, as we shall see in Example 3). In particular, when we calculate a volume by slicing, we may use either planes perpendicular to the x-axis or planes perpendicular to the y-axis. We get the same answer either way.

Even more important is the fact that Fubini's theorem holds for *any* continuous function $f(x, y)$. In particular, f may have negative values as well as positive values on R, and the integrals we calculate with Fubini's theorem may represent other things besides volumes (as we shall see later on).

EXAMPLE 1 Calculate $\iint_R f(x, y)\,dA$ for

$$f(x, y) = 1 - 6x^2y \quad \text{and} \quad R: 0 \le x \le 2,\ -1 \le y \le 1.$$

Solution By Fubini's theorem

$$\begin{aligned}\iint_R f(x, y)\,dA &= \int_{-1}^{1}\int_0^2 (1 - 6x^2y)\,dx\,dy = \int_{-1}^{1} [x - 2x^3y]_{x=0}^{x=2}\,dy \\ &= \int_{-1}^{1} [2 - 16y]\,dy = 2y - 8y^2]_{-1}^{1} \\ &= (2 - 8) - (-2 - 8) = 4.\end{aligned}$$

Reversing the order of integration gives the same answer:

$$\int_0^2 \int_{-1}^1 (1 - 6x^2y)\,dy\,dx = \int_0^2 [y - 3x^2y^2]_{y=-1}^{y=1}\,dx$$
$$= \int_0^2 [(1 - 3x^2 - (-1 - 3x^2)]\,dx$$
$$= \int_0^2 2\,dx = 4. \quad \square$$

Double Integrals for Bounded Nonrectangular Regions

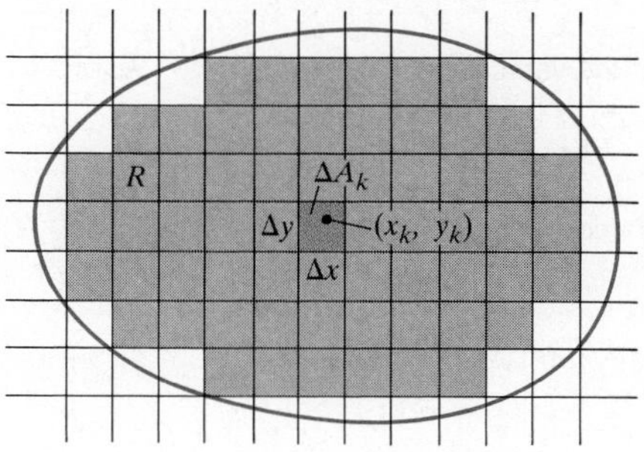

16.6 A rectangular grid subdividing a bounded nonrectangular region into cells.

To define the double integral of a function $f(x, y)$ over a bounded nonrectangular region like the one shown in Fig. 16.6, we again imagine R to be covered by a rectangular grid, but we include in the partial sum only the small pieces of area $\Delta A = \Delta x\,\Delta y$ that lie entirely within the region (shaded in the figure). We number the pieces in some order, choose an arbitrary point (x_k, y_k) in each ΔA_k, and form the sum

$$S_n = \sum_{k=1}^{n} f(x_k, y_k)\,\Delta A_k.$$

The only difference between this sum and the one in Eq. (2) is that now the areas ΔA_k may not cover the entire region R. But, as the mesh becomes increasingly fine and the number of terms in S_n increases, more and more of R is included. If f is continuous, and the boundary of R is made up of a finite number of line segments or smooth curves pieced together end to end, then the sums S_n will have a limit as Δx and Δy approach zero. We call the limit the double integral of f over R:

$$\iint_R f(x, y)\,dA = \lim_{\Delta A \to 0} \sum f(x_k, y_k)\,\Delta A_k.$$

This limit may also exist under less restrictive circumstances, but we shall not pursue this point here.

Double integrals of continuous functions over nonrectangular regions have the algebraic properties I1–I5 listed earlier for integrals over rectangular regions. The domain additivity property corresponding to I6 says that if R is decomposed into nonoverlapping regions R_1 and R_2 with boundaries that are again made of line segments or smooth curves (see Fig. 16.7 for an example), then

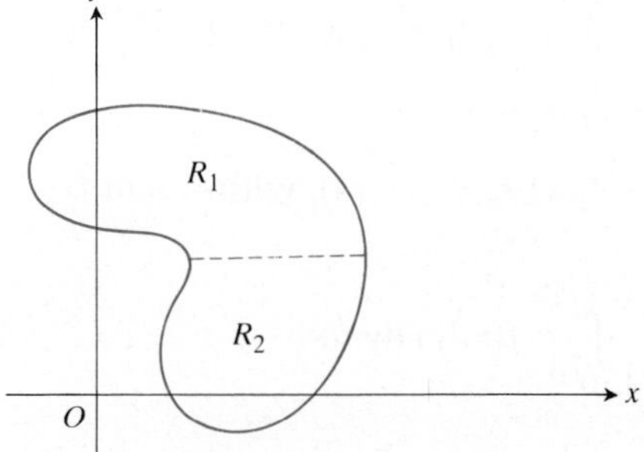

16.7 $\iint_R f(x, y)\,dA = \iint_{R_1} f(x, y)\,dA + \iint_{R_2} f(x, y)\,dA.$

I6′. $\iint_R f(x, y)\,dA = \iint_{R_1} f(x, y)\,dA + \iint_{R_2} f(x, y)\,dA.$

If $f(x, y)$ is positive and continuous over R (Fig. 16.8), we define the volume of the solid region between R and the surface $z = f(x, y)$ to be $\iint_R f(x, y)\,dA$, as before.

If R is a region like the one shown in the xy-plane in Fig. 16.9, bounded "above" and "below" by the curves $y = f_2(x)$ and $y = f_1(x)$, and on the sides by the lines $x = a$, $x = b$, we may again calculate the volume by the method of slicing. We first calculate the cross-sectional area

$$A(x) = \int_{y=f_1(x)}^{y=f_2(x)} f(x, y)\,dy$$

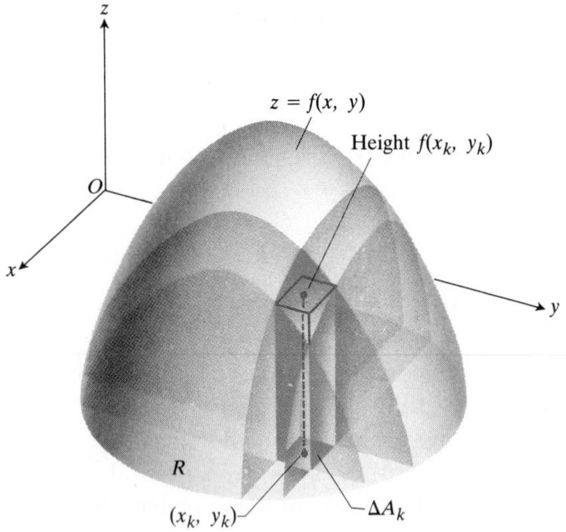

16.8 Volume $= \lim \Sigma f(x_k, y_k)\Delta A_k = \iint_R f(x, y)\, dA$.

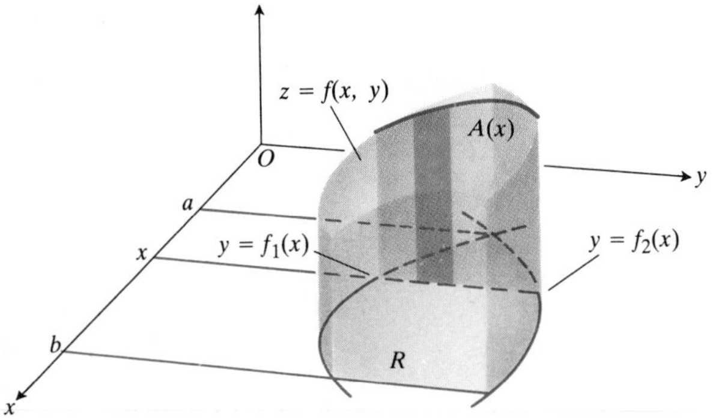

16.9 The area of the vertical slice shown here is

$$A(x) = \int_{f_1(x)}^{f_2(x)} f(x, y)\, dy.$$

This area is integrated from a to b to calculate the volume.

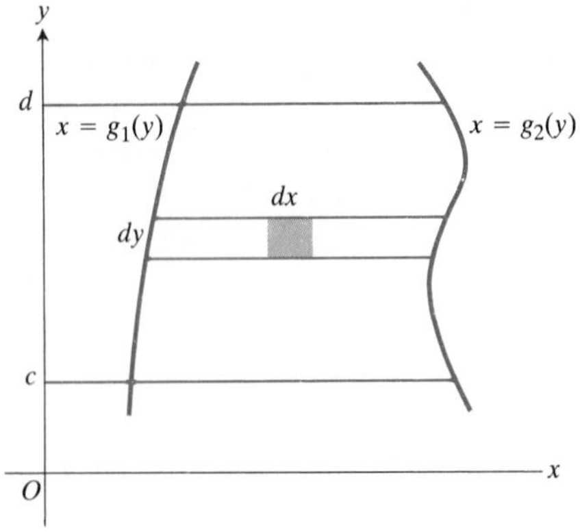

16.10 Area: $\int_c^d \int_{g_1(y)}^{g_2(y)} dx\, dy$.

and then integrate $A(x)$ from $x = a$ to $x = b$ to get the volume as an iterated integral:

$$V = \int_a^b A(x)\, dx = \int_a^b \int_{f_1(x)}^{f_2(x)} f(x, y)\, dy\, dx. \tag{9}$$

Similarly, if R is a region like the one shown in Fig. 16.10, bounded on the right by $x = g_2(y)$, on the left by $x = g_1(y)$, and below and above by the lines $y = c$ and $y = d$, then the volume calculated by slicing is given by the iterated integral

$$\text{Volume} = \int_c^d \int_{g_1(y)}^{g_2(y)} f(x, y)\, dx\, dy. \tag{10}$$

The fact that the iterated integrals in Eqs. (9) and (10) both give the volume that we defined to be the double integral of f over R is a consequence of the following stronger form of Fubini's theorem.

THEOREM 2

Fubini's Theorem (Stronger Form)

Let $f(x, y)$ be continuous on a region R.

1. If R is defined by $a \le x \le b$, $f_1(x) \le y \le f_2(x)$, with f_1 and f_2 continuous on $[a, b]$, then

$$\iint_R f(x, y)\, dA = \int_a^b \int_{f_1(x)}^{f_2(x)} f(x, y)\, dy\, dx.$$

2. If R is defined by $c \le y \le d$, $g_1(y) \le x \le g_2(y)$, with g_1 and g_2 continuous on $[c, d]$, then

$$\iint_R f(x, y)\, dA = \int_c^d \int_{g_1(y)}^{g_2(y)} f(x, y)\, dx\, dy.$$

EXAMPLE 2 Find the volume of the prism whose base is the triangle in the xy-plane bounded by the x-axis and the lines $y = x$ and $x = 1$, and whose top lies in the plane

$$z = f(x, y) = 3 - x - y.$$

Solution For any x between 0 and 1, y may vary from $y = 0$ to $y = x$ (Fig. 16.11). Hence,

$$V = \int_0^1 \int_0^x (3 - x - y)\, dy\, dx = \int_0^1 \left[3y - xy - \frac{y^2}{2}\right]_{y=0}^{y=x} dx$$

$$= \int_0^1 \left(3x - \frac{3x^2}{2}\right) dx = \frac{3x^2}{2} - \frac{x^3}{2}\Big]_{x=0}^{x=1} = 1.$$

When the order of integration is reversed, the integral for the volume is

$$V = \int_0^1 \int_y^1 (3 - x - y)\, dx\, dy = \int_0^1 \left[3x - \frac{x^2}{2} - xy\right]_{x=y}^{x=1} dy$$

$$= \int_0^1 \left(3 - \frac{1}{2} - y - 3y + \frac{y^2}{2} + y^2\right) dy$$

$$= \int_0^1 \left(\frac{5}{2} - 4y + \frac{3}{2}y^2\right) dy = \frac{5}{2}y - 2y^2 + \frac{y^3}{2}\Big]_{y=0}^{y=1} = 1.$$

The two integrals are equal, as they should be. □

While Fubini's theorem assures us that a double integral may be calculated as an iterated integral in either order of integration, the value of one integral may be easier to find than the value of the other. The next example shows how this can happen.

Figure 16.11

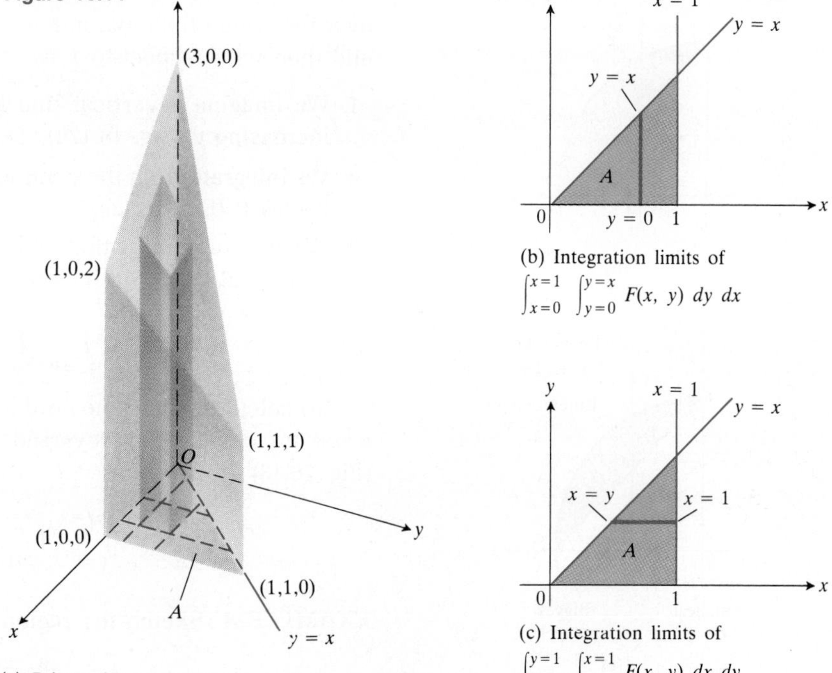

(a) Prism with a triangular base in the xy-plane.

(b) Integration limits of $\int_{x=0}^{x=1} \int_{y=0}^{y=x} F(x, y)\, dy\, dx$

(c) Integration limits of $\int_{y=0}^{y=1} \int_{x=y}^{x=1} F(x, y)\, dx\, dy$

EXAMPLE 3 Calculate

$$\iint_A \frac{\sin x}{x}\, dA,$$

where A is the triangle in the xy-plane bounded by the x-axis, the line $y = x$, and the line $x = 1$.

Solution The region of integration is the same as the one in Example 2. If we integrate first with respect to y and then with respect to x, we find

$$\int_0^1 \left(\int_0^x \frac{\sin x}{x}\, dy\right) dx = \int_0^1 \left(y \frac{\sin x}{x}\Big]_{y=0}^{y=x}\right) dy$$

$$= \int_0^1 \sin x\, dx = -\cos(1) + 1 \approx 0.46.$$

If we reverse the order of integration, and attempt to calculate

$$\int_0^1 \int_y^1 \frac{\sin x}{x}\, dx\, dy,$$

we are stopped by the fact that $\int (\sin x/x)\, dx$ cannot be expressed in terms of elementary functions. □

Determining the Limits of Integration

The hardest part of evaluating a double integral can be finding the limits of integration. Fortunately, there is a good procedure to follow.

If we want to evaluate

$$\iint_R f(x, y)\, dA$$

over the region R shown in Fig. 16.12(a), integrating first with respect to y and then with respect to x, we take the following steps:

1. We imagine a vertical line L cutting through R in the direction of increasing y (Fig. 16.12b).
2. We integrate from the y-value where L enters R to the y-value where L leaves R (Fig. 16.12c).
3. We choose x-limits that include all the vertical lines that pass through R (Fig. 16.12c). The integral is

$$\int_{x=0}^{x=1} \int_{y=1-x}^{y=\sqrt{1-x^2}} f(x, y)\, dy\, dx.$$

To calculate the same double integral as an iterated integral with the order of integration reversed, the procedure uses horizontal lines (Fig. 16.13) to give

$$\int_{y=0}^{y=1} \int_{x=1-y}^{x=\sqrt{1-y^2}} f(x, y)\, dx\, dy.$$

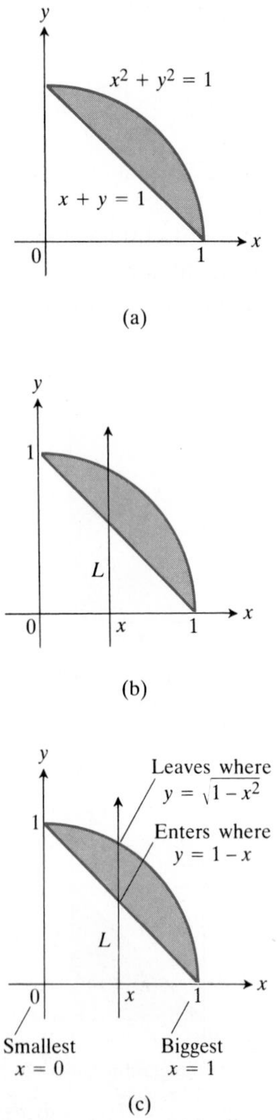

16.12 Finding limits of integration.

EXAMPLE 4 Sketch the region of integration of

$$\int_0^2 \int_{x^2}^{2x} f(x, y)\, dy\, dx$$

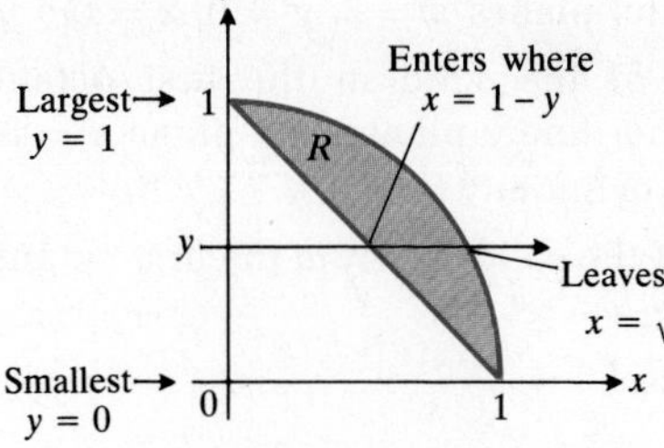

16.13 $\iint_R f(x, y)\, dA = \int_0^1 \int_{1-y}^{\sqrt{1-y^2}} f(x, y)\, dx\, dy.$

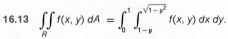

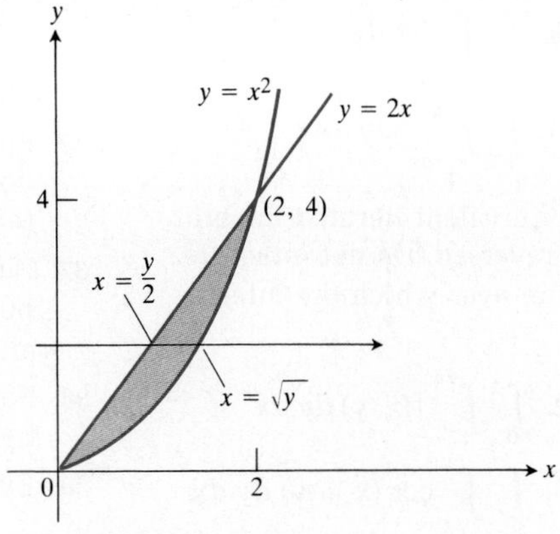

16.14 $\int_0^2 \int_{x^2}^{2x} f(x, y)\, dy\, dx = \int_0^4 \int_{y/2}^{\sqrt{y}} f(x, y)\, dx\, dy.$

and express the integral as an equivalent double integral with the order of integration reversed.

Solution The region of integration is given by the inequalities $x^2 \le y \le 2x$, $0 \le x \le 2$. It is therefore the region bounded by the curves $y = x^2$ and $y = 2x$ between the vertical lines $x = 0$ and $x = 2$, as shown in Fig. 16.14.

To find the limits for the integration in the reverse order we imagine a horizontal line passing from left to right through the region. It enters at $x = y/2$ and leaves at $x = \sqrt{y}$. To include all such lines we let y run from $y = 0$ to $y = 4$. The integral is

$$\int_0^4 \int_{y/2}^{\sqrt{y}} f(x, y)\, dx\, dy. \quad \square$$

PROBLEMS

Evaluate the following integrals and sketch the region over which each integration takes place.

1. $\int_0^3 \int_0^2 (4 - y^2)\, dy\, dx$

2. $\int_{-1}^0 \int_{-1}^1 (x + y + 1)\, dx\, dy$

3. $\int_0^3 \int_{-2}^0 (x^2y - 2xy)\, dy\, dx$

4. $\int_\pi^{2\pi} \int_0^\pi (\sin x + \cos y)\, dx\, dy$

5. $\int_0^\pi \int_0^x x \sin y\, dy\, dx$

6. $\int_1^{\ln 8} \int_0^{\ln y} e^{x+y}\, dx\, dy$

7. $\int_0^\pi \int_0^{\sin x} y\, dy\, dx$

8. $\int_1^2 \int_y^{y^2} dx\, dy$

Evaluate the following integrals.

9. $\int_{10}^1 \int_0^{1/y} ye^{xy}\, dx\, dy$

10. $\int_0^1 \int_0^{x^3} e^{y/x}\, dy\, dx$

In Problems 11–16, integrate the function $f(x, y)$ over the given region.

11. $f(x, y) = x/y$ over the region in the first quadrant bounded by the lines $y = x$, $y = 2x$, $x = 1$, $x = 2$.

12. $f(x, y) = x^2 + y^2$ over the triangular region whose vertices are (0, 0), (1, 0) and (0, 1).

13. $f(x, y) = y - \sqrt{x}$ over the triangular region cut from the first quadrant by the line $x + y = 1$.

14. $f(x, y) = x^2 + 3xy$ over the rectangle R: $0 \le x \le 1$, $0 \le y \le 1$.

15. $f(x, y) = 1/xy$ over the rectangle R: $1 \le x \le 2$, $1 \le y \le 2$.

16. $f(x, y) = y \cos xy$ over the rectangle R: $0 \le x \le \pi$, $0 \le y \le 1$.

In Problems 17–20, sketch the region over which the integration takes place and write an equivalent integral with the order of integration reversed. Evaluate both integrals.

17. $\int_0^2 \int_1^{e^x} dy\,dx$ **18.** $\int_0^1 \int_{\sqrt{y}}^1 dx\,dy$

19. $\int_0^{\sqrt{2}} \int_{-\sqrt{4-2y^2}}^{\sqrt{4-2y^2}} y\,dx\,dy$ **20.** $\int_{-2}^1 \int_{x^2+4x}^{3x+2} dy\,dx$

In Problems 21–26, write an equivalent iterated integral with the order of integration reversed. *Do not integrate.* It will help to sketch the region over which the integration takes place.

21. $\int_0^1 \int_{x^2}^x f(x, y)\,dy\,dx$ **22.** $\int_0^1 \int_x^{2x} f(x, y)\,dy\,dx$

23. $\int_0^1 \int_1^{e^x} dy\,dx$ **24.** $\int_0^1 \int_{\sqrt{x}}^1 \cos(x + y)\,dy\,dx$

25. $\int_0^2 \int_0^{x^3} f(x, y)\,dy\,dx$ **26.** $\int_0^1 \int_{-\sqrt{y}}^{\sqrt{y}} f(x, y)\,dx\,dy$

Evaluate the integrals in Problems 27–32 by integrating the equivalent integral obtained by reversing the order of integration.

27. $\int_0^\pi \int_x^\pi \frac{\sin y}{y}\,dy\,dx$ **28.** $\int_0^1 \int_{2y}^2 \cos(x^2)\,dx\,dy$

29. $\int_0^1 \int_y^1 x^2 e^{xy}\,dx\,dy$ **30.** $\int_0^2 \int_x^2 y^2 \sin xy\,dy\,dx$

31. $\int_0^8 \int_{\sqrt[3]{x}}^2 \frac{dy\,dx}{y^4 + 1}$ **32.** $\int_0^2 \int_0^{4-x^2} \frac{xe^{2y}}{4 - y}\,dy\,dx$

33. Find the volume of the solid whose base is the region in the xy-plane that is bounded by the parabola $y = 4 - x^2$ and the line $y = 3x$, while the top of the solid is bounded by the plane $z = x + 4$.

34. The base of a solid is the region in the xy-plane that is bounded by the circle $x^2 + y^2 = a^2$, while the top of the solid is bounded by the paraboloid $az = x^2 + y^2$. Find the volume.

35. Find the volume in the first octant bounded by the coordinate planes, the cylinder $x^2 + y^2 = 4$, and the plane $z + y = 3$.

36. Find the volume of the solid in the first octant bounded by the plane, the cylinder $y = x^2$, the surface $z = xy$, and the planes $x = 2$, $y = 0$, $z = 0$.

37. Find the volume of the solid in the first octant bounded by the coordinate planes, the plane $x = 3$, and the parabolic cylinder $z = 4 - y^2$.

38. Find the volume of the solid cut from the first octant by the surface $z = 4 - x^2 - y$.

39. Evaluate the integral

$$\int_0^2 \int_{y/2}^1 e^{x^2}\,dx\,dy.$$

40. The volume under the paraboloid $z = x^2 + y^2$ and above a certain region R in the xy-plane is

$$V = \int_0^1 \int_0^y (x^2 + y^2)\,dx\,dy + \int_1^2 \int_0^{2-y} (x^2 + y^2)\,dx\,dy.$$

Sketch the region and express the volume as an iterated integral with the order of integration reversed.

41. Evaluate the integral

$$\int_0^2 (\tan^{-1} \pi x - \tan^{-1} x)\,dx.$$

(*Hint:* Write the integrand as an integral.)

***Toolkit* programs**
Double Integral

16.3 Area

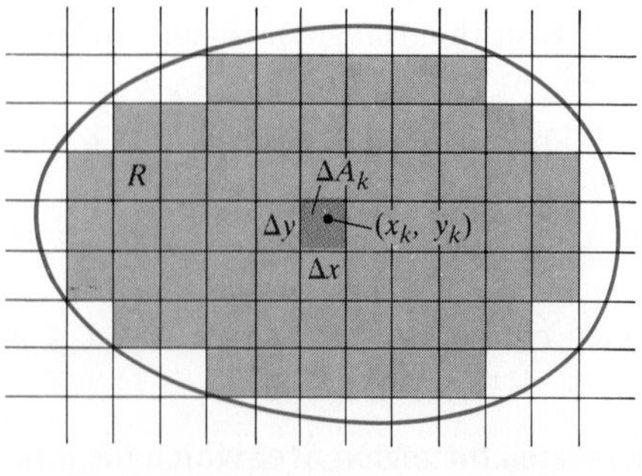

16.15 A rectangular grid subdividing a bounded nonrectangular region into cells.

If we take $f(x, y) = 1$ in the definition of the double integral over a region R in the preceding article, the partial sums reduce to

$$S_n = \sum_{k=1}^n f(x_k, y_k)\,\Delta A_k = \sum_{k=1}^n \Delta A_k \tag{1}$$

and give an approximation to what we would like to call the area of the region R. As Δx and Δy approach zero, the coverage of R by the ΔA_k's (Fig. 16.15) becomes increasingly complete, and we *define* the area of R to be the limit

$$\text{Area} = \lim \sum \Delta A_k = \iint_R dA. \tag{2}$$

To evaluate the area integral in (2), we integrate the constant function $f(x, y) \equiv 1$ over R.

EXAMPLE 1 Find the area of the region R bounded by $y = x$ and $y = x^2$ in the first quadrant.

Solution We sketch the region (Fig. 16.16) and calculate the area as

$$A = \int_0^1 \int_{x^2}^x dy\, dx = \int_0^1 (x - x^2)\, dx = \frac{x^2}{2} - \frac{x^3}{3}\Big]_0^1 = \frac{1}{6}. \quad \square$$

EXAMPLE 2 Find the area of the region R enclosed by the parabola $y = x^2$ and the line $y = x + 2$.

Solution If we divide R into the regions R_1 and R_2 shown in Fig. 16.17, we may calculate the area as

$$A = \iint_{R_1} dA + \iint_{R_2} = \int_0^1 \int_{-\sqrt{y}}^{\sqrt{y}} dx\, dy + \int_1^4 \int_{y-2}^{\sqrt{y}} dx\, dy.$$

On the other hand, reversing the order of integration gives

$$A = \int_{-1}^2 \int_{x^2}^{x+2} dy\, dx.$$

Clearly, this result is simpler and is the only one we would bother to write down in practice. Evaluation of this integral leads to the result

$$A = \int_{-1}^2 y\Big]_{x^2}^{x+2} dx = \int_{-1}^2 (x + 2 - x^2)\, dx = \frac{9}{2}. \quad \square$$

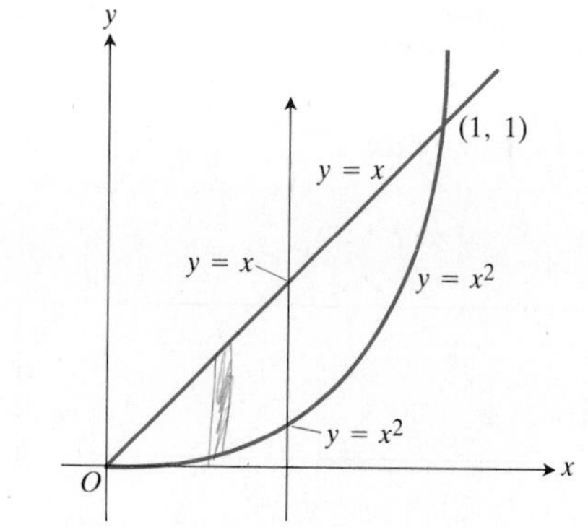

16.16 Area: $\int_0^1 \int_{x^2}^x dy\, dx$.

$\int_0^1 (x - x^2)\, dx$

$= \frac{x^2}{2} - \frac{x^3}{3}\Big)_0^1$

$= \frac{1}{2} - \frac{1}{3}$

$= \frac{3}{6} - \frac{2}{6} = \frac{1}{6}$

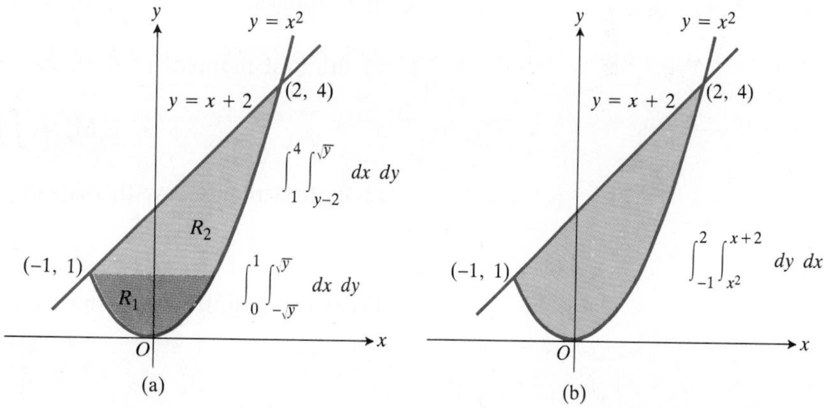

16.17 Calculating the area shown here takes (a) two integrals if the first integration is with respect to x, but (b) only one if the first integration is with respect to y.

PROBLEMS

In Problems 1–8, find the area of the region bounded by the given curves and lines by means of double integration.

1. The coordinate axes and the line $x + y = a$.

2. The x-axis, the curve $y = e^x$, and the lines $x = 0$, $x = 1$.

3. The y-axis, the line $y = 2x$, and the line $y = 4$.

4. The curve $y^2 + x = 0$, and the line $y = x + 2$.

5. The curves $x = y^2$, $x = 2y - y^2$.

6. The semicircle $y = \sqrt{a^2 - x^2}$, the lines $x = \pm a$, and the line $y = -a$.

7. The parabola $x = y - y^2$ and the line $x + y = 0$.

8. Above by $y = x^2$, below by $y = -1$, on the left by $x = -2$, and on the right by $y = 2x - 1$.

The integrals in Problems 9–14 give areas of regions in the xy-plane. Sketch each region. Label each bounding curve with its equation, and give the coordinates of the boundary points where the curves intersect.

9. $\int_0^1 \int_y^{\sqrt{y}} dx\,dy$

10. $\int_0^3 \int_{-x}^{x(2-x)} dy\,dx$

11. $\int_0^{\pi/4} \int_{\sin x}^{\cos x} dy\,dx$

12. $\int_{-1}^2 \int_{y^2}^{y+2} dx\,dy$

13. $\int_{-1}^0 \int_{-2x}^{1-x} dy\,dx + \int_0^2 \int_{-x/2}^{1-x} dy\,dx$

14. $\int_0^2 \int_{x^2-4}^0 dy\,dx + \int_0^4 \int_0^{\sqrt{x}} dy\,dx$

***Toolkit* programs**
Double Integral

16.4

Physical Applications

First and Second Moments

If the representative element of mass *dm* in a mass that is continuously distributed over some region *R* of the xy-plane is taken to be

$$dm = \delta(x, y)\,dy\,dx$$
$$= \delta(x, y)\,dA, \qquad (1)$$

where $\delta = \delta(x, y)$ is the density at the point (x, y) of *R* (Fig. 16.18), then double integration may be used to calculate

a) the mass, $$m = \iint \delta(x, y)\,dA, \qquad (2)$$

b) the first moment of the mass with respect to the x-axis,

$$M_x = \iint y\,\delta(x, y)\,dA, \qquad (3a)$$

c) its first moment with respect to the y-axis,

$$M_y = \iint x\,\delta(x, y)\,dA. \qquad (3b)$$

From (2) and (3) we get the coordinates of the center of mass,

$$\bar{x} = \frac{M_y}{m}, \qquad \bar{y} = \frac{M_x}{m}.$$

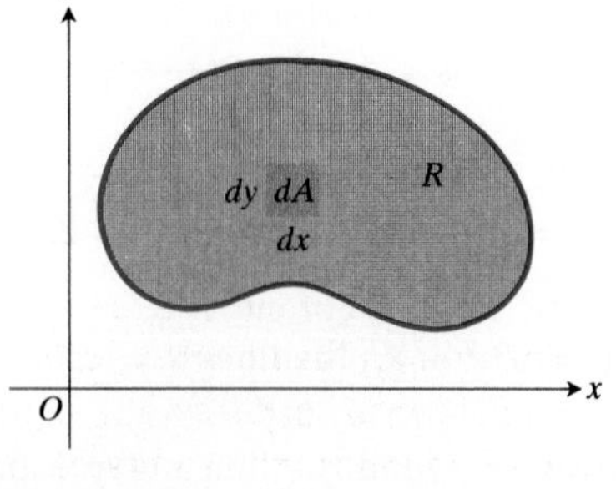

16.18 Area element $dA = dx\,dy$.

Other moments of importance in physical application are the *moments of inertia* of the mass. These are the *second* moments that we get by using the squares instead of the first powers of the "lever-arm" distances x and y. Thus the moment of inertia about the x-axis, denoted by I_x, is defined by

$$I_x = \iint y^2\,\delta(x, y)\,dA. \qquad (4)$$

The moment of inertia about the y-axis is

$$I_y = \iint x^2\,\delta(x, y)\,dA. \qquad (5)$$

Also of interest is the *polar moment of inertia* about the origin, I_0, given by

$$I_0 = \iint r^2\,\delta(x, y)\,dA = \iint (x^2 + y^2)\,\delta(x, y)\,dA = I_x + I_y. \tag{6}$$

Here $r^2 = x^2 + y^2$ is the square of the distance from the origin to the representative point (x, y) in the element of mass dm.

In all of these integrals, the same limits of integration are to be supplied as would be called for if one were calculating only the area of R.

EXAMPLE 1 A thin plate of uniform (constant) thickness and density δ covers the region in the xy-plane shown in Fig. 16.19. Find the center of mass and the inertial moments I_x, I_y, and I_0.

Solution Using the area computation in Example 1, Article 16.3, we find

$$m = \int_0^1 \int_{x^2}^{x} \delta\,dy\,dx = \delta \int_0^1 \int_{x^2}^{x} dy\,dx = \frac{\delta}{6}.$$

$$M_x = \int_0^1 \int_{x^2}^{x} y\,\delta\,dy\,dx = \delta \int_0^1 \left[\frac{y^2}{2}\right]_{y=x^2}^{y=x} dx$$

$$= \delta \int_0^1 \left(\frac{x^2}{2} - \frac{x^4}{2}\right) dx = \delta\left[\frac{x^3}{6} - \frac{x^5}{10}\right]_0^1 = \frac{\delta}{15}.$$

$$M_y = \int_0^1 \int_{x^2}^{x} x\,\delta\,dy\,dx = \delta \int_0^1 [xy]_{y=x^2}^{y=x}\,dx$$

$$= \delta \int_0^1 (x^2 - x^3)\,dx = \delta\left[\frac{x^3}{3} - \frac{x^4}{4}\right]_0^1 = \frac{\delta}{12}.$$

Center of mass $(\bar{x}, \bar{y})$: $\bar{x} = \dfrac{M_y}{m} = \dfrac{1}{2}$, $\bar{y} = \dfrac{M_x}{m} = \dfrac{2}{5}$

$$I_x = \int_0^1 \int_{x^2}^{x} y^2\,\delta\,dy\,dx = \delta \int_0^1 \left[\frac{y^3}{3}\right]_{y=x^2}^{y=x} dx$$

$$= \delta \int_0^1 \left(\frac{x^3}{3} - \frac{x^6}{3}\right) dx = \delta\left[\frac{x^4}{12} - \frac{x^7}{21}\right]_0^1 = \frac{\delta}{28}.$$

$$I_y = \int_0^1 \int_{x^2}^{x} x^2\,\delta\,dy\,dx = \delta \int_0^1 (x^3 - x^4)\,dx = \delta\left[\frac{x^4}{4} - \frac{x^5}{5}\right]_0^1 = \frac{\delta}{20}.$$

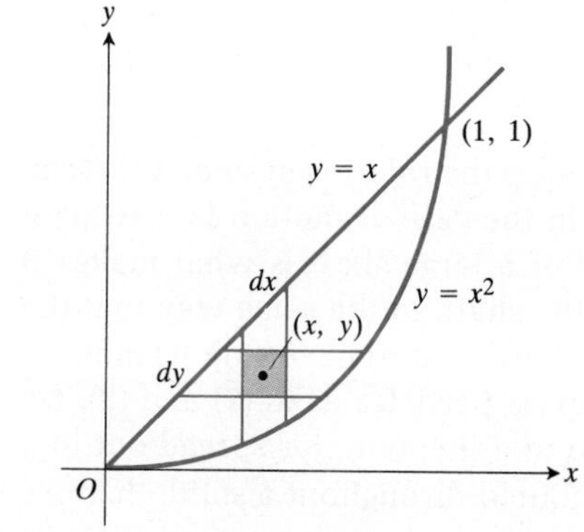

16.19 $I_x = \int_0^1 \int_{x^2}^{x} y^2\,dy\,dx.$

We therefore have

$$I_0 = I_x + I_y = \frac{\delta}{28} + \frac{\delta}{20} = \frac{3\delta}{35}.$$

Note that since the density δ in this problem is a constant we are able to move it outside the integral signs. If the density had been given instead as a variable function of x and y, we would have taken this into account by substituting this function for δ before integrating. □

Geometric Figures

Unless otherwise specified, geometric figures in the plane will be treated as objects with constant density $\delta = 1$. The moments of such a figure are then called *moments of area*, and the center of mass is called the figure's *centroid*. Thus, physical bodies have centers of mass (or centers of gravity), while geometric figures have centroids.

Relation of Moments to Kinetic Energy

When a particle of mass m is rotating about an axis in a circle of radius r with angular velocity ω and linear velocity $v = \omega r$, its kinetic energy is

$$\tfrac{1}{2}mv^2 = \tfrac{1}{2}mr^2\omega^2.$$

If a system of particles of masses $m_1, m_2, \ldots, m_n$, all rotate about the same axis with the same angular velocity ω but their respective distances from the axis of rotation are $r_1, r_2, \ldots, r_n$, then the kinetic energy of the system of particles is

$$\text{K.E.} = \tfrac{1}{2}(m_1v_1^2 + \cdots + m_nv_n^2) = \tfrac{1}{2}\omega^2 \sum_{k=1}^{n} m_kr_k^2 = \tfrac{1}{2}\omega^2 I, \tag{7}$$

where

$$I = \sum_{k=1}^{n} m_kr_k^2 \tag{8}$$

is the *moment of inertia* of the system about the axis in question. I depends only upon the magnitudes m_k of the masses and their distances r_k from the axis. When a mass m is moving in a straight line with velocity v, its kinetic energy is $\frac{1}{2}mv^2$, and an amount of work equal to this must be expended to stop the object and bring it to rest. Similarly, when a system of mass is moving in a *rotational* motion (like a turning shaft), the kinetic energy it possesses is

$$\text{K.E.} = \tfrac{1}{2}I\omega^2, \tag{9}$$

and this amount of work is required to stop the rotating system. It is seen that I here plays the role that m plays in the case of motion in a straight line. In a sense, the *moment of inertia* of a large shaft is what makes it hard to start or to stop the rotation of the shaft, in the same way that the *mass* of an automobile is what makes it hard to start or to stop its motion.

If, instead of a system of discrete mass particles as in (7) and (8), we have a continuous distribution of mass in a fine wire, or spread out in a thin film or plate over an area, or distributed throughout a solid, then we may divide the total mass into small elements of mass Δm such that if r represents the distance of some *one* point of the element Δm from an axis, then *all* points of that element will be within a distance $r \pm \varepsilon$ of the axis, where $\varepsilon \to 0$ as the largest dimension of the elements $\Delta m \to 0$. Then we define the moment of inertia of the total mass about the axis in question to be

$$I = \lim_{\Delta m \to 0} \sum r^2 \,\Delta m = \int r^2 \,dm. \tag{10}$$

Thus, for example, the polar moment of inertia, given by Eq. (6), is the moment of inertia with respect to a z-axis through O perpendicular to the xy-plane.

Deflection of Beams

In addition to its importance in connection with the kinetic energy of rotating bodies, the moment of inertia plays an important part in the theory of the deflection of beams under transverse loading, where the "stiffness factor" is given by EI, where E is Young's modulus, and I is the moment of inertia of a cross section of the beam with respect to a horizontal axis through its center of gravity. The greater the value of I, the stiffer the beam and the less it will deflect. This fact is exploited in so-called I-beams, where the flanges at the top and bottom of the beam are at relatively large distances from the center and hence correspond to large values of r^2 in Eq. (10), thereby contributing a larger amount to the moment of inertia than would be the case if the same mass were all distributed uniformly, say in a beam with a square cross section.

Radius of Gyration

The equation

$$I_y = mR_y^2$$

defines the number

$$R_y = \sqrt{I_y/m},$$

called the *radius of gyration* of the mass m with respect to the y-axis. It tells how far from the y-axis the entire mass might be concentrated and still give the same I_y. It gives a convenient way to express the moment of inertia of the mass of a body in terms of the mass and a length.

The radii of gyration of a mass m with respect to the x-axis and the origin are

$$R_x = \sqrt{I_x/m} \quad \text{and} \quad R_0 = \sqrt{I_0/m}.$$

EXAMPLE 2 The radii of gyration of the mass in Example 1 are

$$R_x = \sqrt{I_x/m} = \sqrt{3/14}, \qquad R_y = \sqrt{I_y/m} = \sqrt{3/10},$$
$$R_0 = \sqrt{I_0/m} = \sqrt{18/35}. \square$$

Mass: $m = \iint \delta(x, y)\, dA$

First moments: $M_x = \iint y\delta(x, y)\, dA, \qquad M_y = \iint x\delta(x, y)\, dA$

Center of mass: $\bar{x} = \dfrac{M_y}{m}, \qquad \bar{y} = \dfrac{M_x}{m}$

Moments of interia (second moments):

$$I_x = \iint y^2\delta(x, y)\, dA, \qquad I_y = \iint x^2\delta(x, y)\, dA$$

$$I_0 = \iint (x^2 + y^2)\, \delta\, (x, y)\, dA = I_x + I_y$$

Radii of gyration:

$$R_x = \sqrt{I_x/m}, \qquad R_y = \sqrt{I_y/m}, \qquad R_0 = \sqrt{I_0/m}$$

PROBLEMS

1. Find the first moment about the y-axis of a thin plate of uniform density δ bounded by $x = 0$, $y = x$, and $y = 2 - x^2$.
2. Find the centroid of the region in the first quadrant bounded by the x-axis, the parabola $y^2 = 2x$, and the line $x + y = 4$.
3. Find the centroid of the triangular region cut from the first quadrant by the line $x + y = a$, $a > 0$.
4. Find the centroid of the region bounded by the curve $y^2 + x = 0$ and the line $y = x + 2$.
5. Find the center of mass of a thin plate bounded by

the semicircle $y = \sqrt{a^2 - x^2}$, the lines $x = \pm a$, and the line $y = -a$ if $\delta(x, y) = y + a$.

6. The area of the region in the first quadrant bounded by $y = 6x - x^2$ and $y = x$ is 125/6 square units. Find $\bar{y}$.

7. The area of the region in the first quadrant bounded by $y = 4x - x^2$ and $y = x$ is 9/2. Find $\bar{x}$.

8. Find the center of mass of a thin plate bounded by $y = x$, $y = 2 - x$, and the x-axis if $\delta(x, y) = 1 + 2x + y$.

9. Find the centroid of the region cut from the first quadrant by the circle $x^2 + y^2 = a^2$. (*Hint:* $\bar{x} = \bar{y}$ and the area is $\pi a^2/4$.)

10. A thin plate bounded by $x^2 + 4y^2 = 12$ and $x = 4y^2$ has a variable density given by $\delta(x, y) = kx$ (k a constant). Find the plate's mass.

11. Find the moment of inertia about the x-axis of the region bounded by the x-axis, the curve $y = e^x$, and the lines $x = 0$, $x = 1$. (For a region, we take $\delta = 1$.)

12. Find the moment of inertia about the x-axis of a thin plate bounded by the curves $x = y^2$, $x = 2y - y^2$ if its density at the point (x, y) is $\delta(x, y) = y + 1$.

13. Find the moment of inertia about the x-axis of a thin plate bounded by the parabola $x = y - y^2$ and the line $x + y = 0$ if $\delta(x, y) = x + y$.

14. Find the moment of inertia with respect to the y-axis of the area bounded by the curve $y = (\sin^2 x)/x^2$ and the interval $\pi \le x \le 2\pi$ on the x-axis.

15. Find the polar moment of inertia about the origin of the triangular region bounded by the y-axis, the line $y = 2x$, and the line $y = 4$.

16. Find the radius of gyration of a uniform slender rod with constant density δ, and of length L, with respect to an axis

a) perpendicular to the axis of the rod through the rod's center of mass,

b) perpendicular to the axis of the rod at one end,

c) parallel to the rod at a distance d from the axis of the rod. Assume that d is very large compared to the radius of the rod.

17. Find the moment of inertia and radius of gyration about the x-axis of each of the following figures.

a) The rectangular region $0 \le x \le b$, $0 \le y \le h$.

b) Any triangular region with base the interval $0 \le x \le b$ on the x-axis and opposite vertex somewhere on the line $y = x$ above the x-axis. (They all have the same moment and radius.)

c) The disc enclosed by the circle $x^2 + y^2 = a^2$.

d) The region cut from the first quadrant by the circle $x^2 + y^2 = a^2$.

e) The region bounded by the ellipse $\dfrac{x^2}{a^2} + \dfrac{y^2}{b^2} = 1$.

18. Find the centroid of the region in the second quadrant bounded by the two axes and the curve $y = e^x$.

19. Find the moment with respect to the y-axis of the region in the first quadrant under the curve $y = e^{-x^2/2}$.

20. Find the centroid of the region in the xy-plane bounded by the curves $y = 1/\sqrt{1 - x^2}$, $y = -1/\sqrt{1 - x^2}$, and the lines $x = 0$, $x = 1$. (*Hint:* Note that $\bar{y} = 0$, by symmetry.)

21. A horizontal cylindrical tank 10 ft in diameter is half full of oil weighing 50 lb/ft³. Find the pressure exerted by the oil on one end of the tank.

22. The average value of a function $f(x, y)$ over a region R is defined to be

$$\frac{1}{\text{area } R} \iint_R f(x, y)\, dA.$$

a) Calculate the average value of the derivative of the function $w = \frac{1}{2}(x^2 + y^2)$ in the direction of the unit vector $\mathbf{u} = u_1\mathbf{i} + u_2\mathbf{j}$ over the region enclosed by the triangle whose vertices are (0, 0), (0, 1) and (1, 0).

b) Show in general that if $w = \frac{1}{2}(x^2 + y^2)$, then the average value of $D_{\mathbf{u}}w$ over a region R is the value of $D_{\mathbf{u}}w$ at the centroid of R.

***Toolkit* programs**

Double Integral

16.5 Changing to Polar Coordinates

When we defined the integral of a function $f(x, y)$ over a region R we divided R with rectangles. Rectangles are easy to describe in rectangular coordinates, and their areas easy to compute. When we work in polar coordinates, however, it is more natural to subdivide R into "polar rectangles," in the way we now describe.

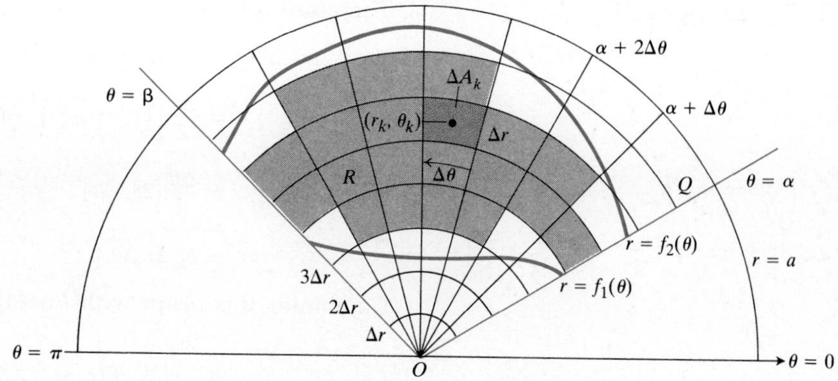

16.20 The region R: $f_1(\theta) \leq r \leq f_2(\theta)$, $\alpha \leq \theta \leq \beta$ is contained in the fan-shaped region Q: $0 \leq r \leq a$, $\alpha \leq \theta \leq \beta$. In the subdivision of Q in polar coordinates, $\Delta A_k = r_k \, \Delta\theta \, \Delta r$.

Suppose that the function $F(r, \theta)$ is defined over a region R bounded by the rays $\theta = \alpha$, $\theta = \beta$, and the continuous curve $r = f_1(\theta)$, $r = f_2(\theta)$, as shown in Fig. 16.20. Suppose that $0 \leq f_1(\theta) \leq f_2(\theta) \leq a$ for all θ between α and β. Then R is contained in the fan-shaped region Q: $0 \leq r \leq a$, $\alpha \leq \theta \leq \beta$. We may cover Q by a grid of circular arcs with centers at 0 and radii

$$\Delta r, 2\,\Delta r, \ldots, m\,\Delta r,$$

where $\Delta r = a/m$, and rays through 0 along

$$\theta = \alpha, \alpha + \Delta\theta, \alpha + 2\,\Delta\theta, \ldots, \alpha + n\,\Delta\theta = \beta,$$

where $\Delta\theta = (\beta - \alpha)/n$. These arcs and rays divide the fan-shaped region Q into small patches called polar rectangles. We number the rectangles that lie inside R, calling their areas

$$\Delta A_1, \Delta A_2, \ldots, \Delta A_n.$$

We let (r_k, θ_k) be the center of the rectangle with area ΔA_k (see Fig. 16.20). By "center" we mean the point that lies halfway between the circular arcs on the ray that bisects them. We then form the sum

$$S_n = \sum_{k=1}^{n} F(r_k, \theta_k)\,\Delta A_k. \tag{1}$$

We may express ΔA_k in terms of Δr and $\Delta\theta$ in the following way. The radius of the inner arc bounding ΔA_k is $r_k - \frac{1}{2}\Delta r$. The area of the circular sector subtended by this arc at the origin is therefore

$$\frac{\Delta\theta}{2\pi} \cdot \pi\left(r_k - \frac{1}{2}\Delta r\right)^2 = \frac{1}{2}\left(r_k - \frac{1}{2}\Delta r\right)^2 \Delta\theta.$$

Similarly, the radius of the outer boundary of ΔA_k is $r_k + \frac{1}{2}\Delta r$, and the area of the sector it subtends is

$$\frac{\Delta\theta}{2\pi} \cdot \pi\left(r_k + \frac{1}{2}\Delta r\right)^2 = \frac{1}{2}\left(r_k + \frac{1}{2}\Delta r\right)^2 \Delta\theta. \tag{2}$$

Therefore,

$$\begin{aligned}\Delta A_k &= \text{area of larger sector} - \text{area of smaller sector}\\ &= \frac{\Delta\theta}{2}\left[\left(r_k + \frac{\Delta r}{2}\right)^2 - \left(r_k - \frac{\Delta r}{2}\right)^2\right]\\ &= \frac{\Delta\theta}{2}[2r_k\,\Delta r]\\ &= r_k\,\Delta r\,\Delta\theta.\end{aligned}$$

Combining this result with Eq. (1) gives

$$S_n = \sum_{k=1}^{n} F(r_k, \theta_k)\,\Delta A_k = \sum_{k=1}^{n} F(r_k, \theta_k) r_k\,\Delta r\,\Delta\theta. \tag{3}$$

If F is continuous on R, then these sums approach a limit

$$\lim S_n = \int_R F(r, \theta)\,dA$$

as Δr and $\Delta\theta$ approach zero, and a version of Fubini's theorem says that this limit may be evaluated by repeated single integrations with respect to r and θ as

$$\iint F(r, \theta)\,dA = \int_{\theta=\alpha}^{\theta=\beta}\int_{r=f_1(\theta)}^{r=f_2(\theta)} F(r, \theta)\,r\,dr\,d\theta. \tag{4}$$

If $F(r, \theta) \equiv 1$ is the constant function whose value is one, then the value of the integral of F over R is the area of R (in agreement with our earlier definition, although we shall not prove this fact). Thus,

$$\text{Area of } R = \iint_R r\,dr\,d\theta. \tag{5}$$

Finding the Limits of Integration

The procedure we used for finding limits of integration for integrals in rectangular coordinates also works for polar coordinates.

EXAMPLE 1 Find the limits of integration for integrating a function $F(r, \theta)$ over the region R that lies inside the cardioid $r = a(1 + \cos\theta)$ and outside the circle $r = a$.

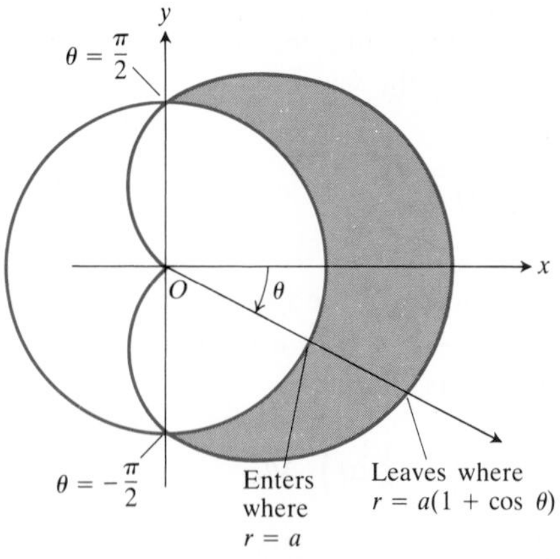

16.21 The shaded area between cardioid and circle is

$$\int_{-\pi/2}^{\pi/2}\int_a^{a(1+\cos\theta)} r\,dr\,d\theta.$$

Solution We graph the cardioid and circle (Fig. 16.21) and carry out the following steps:

1. Hold θ fixed, and let r increase to trace a ray out from the origin.
2. Integrate from the r-value where the ray enters R to the r-value where it leaves R.
3. Choose θ-limits to include all the rays from the origin that intersect R.

The result is the integral

$$\int_{-\pi/2}^{\pi/2}\int_{r=a}^{r=a(1+\cos\theta)} F(r, \theta)\,r\,dr\,d\theta. \quad \square$$

EXAMPLE 2 Find the area enclosed by the lemniscate $r^2 = 2a^2\cos 2\theta$.

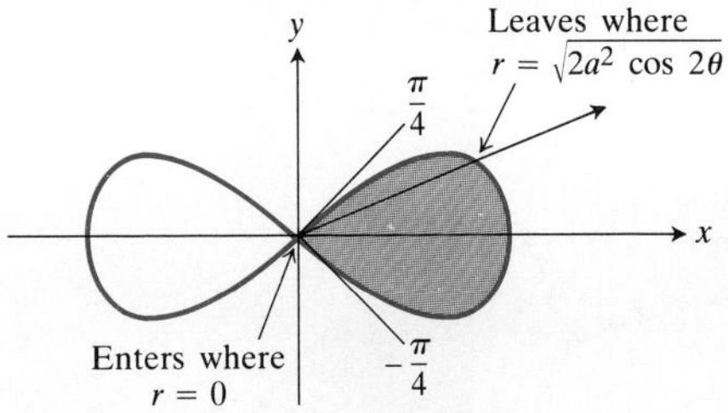

16.22 Limits for integrating over the shaded region bounded by the lemniscate $r^2 = 2a^2 \cos 2\theta$.

Solution We graph the lemniscate (Fig. 16.22) and calculate the area of the right-hand half to be

$$\int_{-\pi/4}^{\pi/4} \int_0^{\sqrt{2a^2 \cos 2\theta}} r\, dr\, d\theta = \int_{-\pi/4}^{\pi/4} \left[\frac{r^2}{2}\right]_{r=0}^{r=\sqrt{2a^2 \cos 2\theta}} d\theta$$

$$= \int_{-\pi/4}^{\pi/4} a^2 \cos 2\theta d\theta = \frac{a^2}{2} \sin 2\theta \bigg]_{-\pi/4}^{\pi/4}$$

$$= \frac{a^2}{2}[1 - (-1)] = a^2.$$

The total area is therefore $2a^2$. □

Changing Coordinates

If a region G in the uv-plane is transformed into the region R in the xy-plane by differentiable equations of the form

$$x = f(u, v), \qquad y = g(u, v)$$

(see Fig. 16.23), then a function $\phi(x, y)$ defined on R can be thought of as a function

$$\phi(f(u, v), g(u, v))$$

defined on G. It is a theorem from advanced calculus that if all the functions involved are continuous and have continuous first derivatives, then the integral of $\phi(x, y)$ over R and the integral of $\phi(f(u, v), g(u, v))$ over G are related by the equation

$$\iint_R \phi(x, y)\, dx\, dy = \iint_G \phi[f(u, v), g(u, v)] \frac{\partial(x, y)}{\partial(u, v)}\, du\, dv, \tag{6}$$

where $\partial(x, y)/\partial(u, v)$ denotes the determinant of partial derivatives

$$\frac{\partial(x, y)}{\partial(u, v)} = \begin{vmatrix} \dfrac{\partial x}{\partial u} & \dfrac{\partial x}{\partial v} \\ \dfrac{\partial y}{\partial u} & \dfrac{\partial y}{\partial v} \end{vmatrix}. \tag{7}$$

This determinant is called the *Jacobian* of the coordinate transformation $x = f(u, v)$, $y = g(u, v)$.

In the case of polar coordinates, we have r and θ in place of u and v,

$$x = r \cos \theta, \qquad y = r \sin \theta,$$

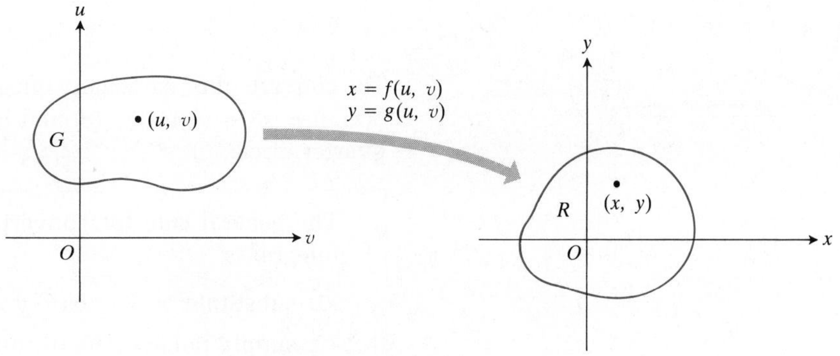

16.23 The equations $x = f(u, v)$, $y = g(u, v)$ allow us to rewrite an integral over R as an equivalent integral over G.

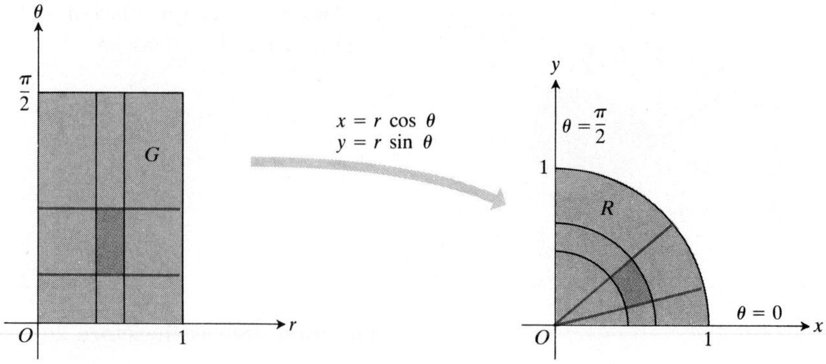

16.24 The equations $x = r \cos \theta$, $y = r \sin \theta$ transform G into R.

and

$$\frac{\partial(x, y)}{\partial(r, \theta)} = \begin{vmatrix} \cos \theta & -r \sin \theta \\ \sin \theta & r \cos \theta \end{vmatrix} = r(\cos^2 \theta + \sin^2 \theta) = r.$$

Hence, Eq. (6) becomes

$$\iint_R \phi(x, y)\, dx\, dy = \iint_G \phi(r \cos \theta, r \sin \theta) r\, dr\, d\theta, \tag{8}$$

which corresponds to Eq. (4).

Figure 16.24 shows how the equations $x = r \cos \theta$, $y = r \sin \theta$ transform the rectangle G: $0 \le r \le 1$, $0 \le 0 \le \pi/2$ into the quarter circle R bounded by $x^2 + y^2 = 1$ in the first quadrant of the xy-plane.

The object of using the equality of the integrals in Eq. (6) is to simplify integration.

EXAMPLE 3 Find the polar moment of inertia about the origin of a thin plate of density $\delta = 1$ bounded by the quarter circle $x^2 + y^2 = 1$ in the first quadrant.

Solution With reference to Fig. 16.24, Eq. (8) gives

$$\iint_{\substack{\text{quarter} \\ \text{circle } R}} (x^2 + y^2)\, dx\, dy = \iint_{\substack{\text{rectangle} \\ G}} (r^2)\, r\, dr\, d\theta,$$

or

$$\int_0^1 \int_0^{\sqrt{1-x^2}} (x^2 + y^2)\, dy\, dx = \int_0^{\pi/2} \int_0^1 (r^2)\, r\, dr\, d\theta$$
$$= \int_0^{\pi/2} \left[\frac{r^4}{4}\right]_{r=0}^{r=1} d\theta = \int_0^{\pi/2} \frac{1}{4}\, d\theta = \frac{\pi}{8}.$$

To convert the cartesian integral to the polar integral, we took $\phi(x, y) = x^2 + y^2$ in Eq. (8) and used the polar limits that described the quarter circle. □

The general rule for converting a cartesian integral to a polar integral is

1. substitute: $x = r \cos \theta$, $y = r \sin \theta$, $dy\, dx = r\, dr\, d\theta$,
2. supply polar limits of integration.

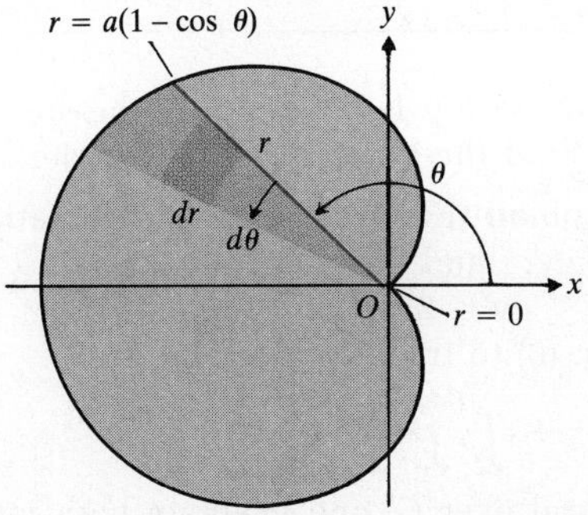

16.25 The cardioid $r = a(1 - \cos\theta)$.

EXAMPLE 4 Find the moment of inertia about the y-axis of the region enclosed by the cardioid

$$r = a(1 - \cos\theta).$$

Solution We take $\delta(x, y) = 1$ when working with a geometric figure. Thus

$$I_y = \iint x^2\delta(x, y)\, dA = \iint x^2\, dA.$$

With

$$x = r\cos\theta, \qquad dA = r\, dr\, d\theta,$$

and limits of integration determined from Fig. 16.25, we have

$$I_y = \int_0^{2\pi}\int_0^{a(1-\cos\theta)} r^3\cos^2\theta\, dr\, d\theta$$
$$= \int_0^{2\pi} \frac{a^4}{4}\cos^2\theta(1 - \cos\theta)^4\, d\theta.$$

The evaluation of the integrals

$$\int_0^{2\pi} \cos^n\theta\, d\theta \qquad (n = 2, 3, 4, 5, 6)$$

that now arise is made easier by use of the reduction formula

$$\int_0^{2\pi} \cos^n\theta\, d\theta = \frac{\cos^{n-1}\theta\sin\theta}{n}\Big]_0^{2\pi} + \frac{n-1}{n}\int_0^{2\pi}\cos^{n-2}\theta\, d\theta$$

or, since $\sin\theta$ vanishes at both limits,

$$\int_0^{2\pi}\cos^n\theta\, d\theta = \frac{n-1}{n}\int_0^{2\pi}\cos^{n-2}\theta\, d\theta.$$

Thus

$$\int_0^{2\pi}\cos^2\theta\, d\theta = \frac{1}{2}\int_0^{2\pi} d\theta = \pi,$$

$$\int_0^{2\pi}\cos^3\theta\, d\theta = \frac{2}{3}\int_0^{2\pi}\cos\theta\, d\theta = \frac{2}{3}\sin\theta\Big]_0^{2\pi} = 0,$$

$$\int_0^{2\pi}\cos^4\theta\, d\theta = \frac{3}{4}\int_0^{2\pi}\cos^2\theta\, d\theta = \frac{3\pi}{4},$$

$$\int_0^{2\pi}\cos^5\theta\, d\theta = \frac{4}{5}\int_0^{2\pi}\cos^3\theta\, d\theta = 0,$$

$$\int_0^{2\pi}\cos^6\theta\, d\theta = \frac{5}{6}\int_0^{2\pi}\cos^4\theta\, d\theta = \frac{5\pi}{8},$$

Therefore

$$I_y = \frac{a^4}{4}\int_0^{2\pi}(\cos^2\theta - 4\cos^3\theta + 6\cos^4\theta - 4\cos^5\theta + \cos^6\theta)\, d\theta$$
$$= \frac{a^4}{4}\left[1 + \frac{18}{4} + \frac{5}{8}\right]\pi = \frac{49\pi a^4}{32}. \ \square$$

PROBLEMS

In Problems 1–8, change the cartesian integral into an equivalent polar integral and evaluate the polar integral.

1. $\int_{-a}^{a}\int_{-\sqrt{a^2-x^2}}^{\sqrt{a^2-x^2}} dy\,dx$

2. $\int_{0}^{a}\int_{0}^{\sqrt{a^2-y^2}} (x^2+y^2)\,dx\,dy$

3. $\int_{0}^{a/\sqrt{2}}\int_{y}^{\sqrt{a^2-y^2}} x\,dx\,dy$

4. $\int_{0}^{\infty}\int_{0}^{\infty} e^{-(x^2+y^2)}\,dx\,dy$

5. $\int_{0}^{2}\int_{0}^{x} y\,dy\,dx$

6. $\int_{0}^{2a}\int_{0}^{\sqrt{2ax-x^2}} x^2\,dy\,dx$

7. $\int_{0}^{3}\int_{0}^{\sqrt{3}x} \frac{dy\,dx}{\sqrt{x^2+y^2}}$

8. $\int_{0}^{2}\int_{0}^{\sqrt{4-x^2}} \frac{xy}{\sqrt{x^2+y^2}}\,dy\,dx$

9. Find the area of the region cut from the first quadrant by the cardioid $r = 1 + \sin\theta$.

10. Find the area of the region common to the cardioids $r = 1 + \cos\theta$ and $r = 1 - \cos\theta$.

11. Find the area cut from the first quadrant by the curve $r = (2 - \sin 2\theta)^{1/2}$.

12. Integrate the function $f(x, y) = 1/(1 - x^2 - y^2)$ over the disc $x^2 + y^2 \le 3/4$.

13. Find $M_x = \iint y\,dA$ for the region bounded below by the x-axis and above by the cardioid $r = 1 - \cos\theta$.

14. A thin plate in the first quadrant is bounded by the coordinate axes and the circle $x^2 + y^2 = 1$. Find the moment of inertia of the plate with respect to the x-axis if the density varies as the square of the distance from the origin.

15. Find the centroid of the region that lies inside the cardioid $r = a(1 + \cos\theta)$ and outside the circle $r = a$.

16. Find the polar moment of inertia with respect to the origin for the region in Problem 15.

17. The region in Problem 15 is the base of a solid right cylinder whose top lies in the plane $z = x$. Find the volume of the cylinder.

18. Find the area enclosed by one leaf of the rose $r = \cos 3\theta$.

19. The lemniscate $r^2 = 2a^2 \cos 2\theta$ is the base of a solid right cylinder whose top is bounded by the sphere $z = \sqrt{2a^2 - r^2}$. Find the volume of the cylinder.

20. a) Find the Jacobian (Eq. 7) of the transformation $x = u$, $y = uv$, and sketch the region G: $1 \le u \le 2$, $1 \le uv \le 2$ in the uv-plane.
    b) Then use Eq. (6) to transform the integral
$$\int_{1}^{2}\int_{1}^{2} \frac{y}{x}\,dy\,dx$$
into an integral over G, and evaluate both integrals.

21. Let R be the region in the first quadrant of the xy-plane bounded by the hyperbolas $xy = 1$, $xy = 9$ and the lines $y = x$, $y = 4x$. Use the transformation $x = u/v$, $y = uv$ with $u > 0$ and $v > 0$ to rewrite $\iint_R dx\,dy$ as an integral over an appropriate region G in the uv-plane. Then evaluate the uv-integral over G.

22. The area πab of the ellipse $x^2/a^2 + y^2/b^2 = 1$ can be found by integrating the function $f(x, y) \equiv 1$ over the region bounded by the ellipse in the xy-plane. Evaluating the integral directly requires a trigonometric substitution. An easier way to evaluate the integral is to use the transformation $x = au$, $y = bv$ and evaluate the transformed integral over the disc G: $u^2 + v^2 \le 1$ in the uv-plane. Find the area this way.

23. A thin plate of uniform thickness and density covers the region bounded by the ellipse $x^2/a^2 + y^2/b^2 = 1$ in the xy-plane. Find I_0, the polar moment of the plane about the origin. (*Hint:* Use the transformation $x = ar\cos\theta$, $y = br\sin\theta$.)

24. Use the transformation $x = u + (1/2)v$, $y = v$ to evaluate the integral
$$\int_{0}^{2}\int_{y/2}^{(y+4)/2} y^3(2x - y)e^{(2x-y)^2}\,dx\,dy$$
by first writing it as an equivalent iterated integral over a region G in the uv-plane.

25. In Article 10.4 we derived the formula $A = \int_\alpha^\beta \frac{1}{2}r^2 d\theta$ for the area of the region swept out by the radius vector $\overrightarrow{OP}$ as P moves along a curve $r = f(\theta)$ from $\theta = \alpha$ to $\theta = \beta$. Show that this formula is also a consequence of Eq. (5) in the present article.

16.6
Triple Integrals in Rectangular Coordinates

If $F(x, y, z)$ is a function defined on a bounded region D in space (a solid ball or truncated cone, for example, or something resembling a swiss cheese, or a finite union of such objects), then the integral of F over D may

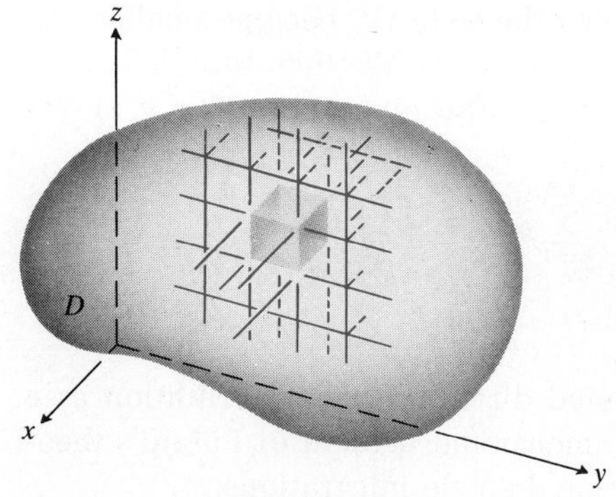

16.26 Partitioning a solid with rectangular cells of volume ΔV.

be defined in the following way. We partition a rectangular region about D into rectangular cells by planes parallel to the coordinate planes, as shown in Fig. 16.26. The cells have dimensions Δx by Δy by Δz. We number the cells that lie inside D in some order

$$\Delta V_1, \Delta V_2, \ldots, \Delta V_n,$$

choose a point (x_k, y_k, z_k) in each ΔV_k, and form the sum

$$S_n = \sum_{k=1}^{n} F(x_k, y_k, z_k)\,\Delta V_k. \tag{1}$$

If F is continuous and the bounding surface of D is made of smooth surfaces joined along continuous curves, then as Δx, Δy, and Δz all approach zero the sums S_n will approach a limit

$$\lim S_n = \iiint_D F(x, y, z)\,dV. \tag{2}$$

We call this limit the *triple integral of F over D*. The limit also exists for some discontinuous functions.

Triple integrals share many algebraic properties with double and single integrals. Writing F for $F(x, y, z)$ and G for $G(x, y, z)$ we have the following.

I1. $\displaystyle\iiint_D kF\,dV = k\iiint_D F\,dV$ (any number k)

I2. $\displaystyle\iiint_D [F + G]\,dV = \iiint_D F\,dV + \iiint_D G\,dV$

I3. $\displaystyle\iiint_D [F - G]\,dV = \iiint_D F\,dV - \iiint_D G\,dV$

I4. $\displaystyle\iiint_D F\,dV \geq 0$ if $F \geq 0$ on D

I5. $\displaystyle\iiint_D F\,dV \geq \iiint_D G\,dV$ if $F \geq G$ on D

The integrals have these properties because the sums that approximate them have these properties.

Triple integrals also have a domain additivity property that proves useful in physics and engineering as well as in mathematics. If the domain D of a continuous function F is partitioned by smooth surfaces into a finite number of cells $D_1, D_2, \ldots, D_n$, then

I6. $\displaystyle\iiint_D F\,dV = \iiint_{D_1} F\,dV + \iiint_{D_2} F\,dV + \cdots + \iiint_{D_n} F\,dV.$

Volume

If $F(x, y, z) \equiv 1$ is the constant function whose value is one, then the sums in Eq. (1) reduce to

$$S_n = \sum_{k=1}^{n} 1 \cdot \Delta V_k = \sum_{k=1}^{n} \Delta V_k. \tag{3}$$

As Δx, Δy, and Δz all approach zero, the cells ΔV_k become smaller and more numerous and fill up more and more of D. We therefore define the volume of D to be the triple integral of the constant function $F(x, y, z) \equiv 1$ over D:

$$\text{Volume of } D = \lim \sum_{k=1}^{n} \Delta V_k = \iiint_D dV. \tag{4}$$

Evaluation

The triple integral is seldom evaluated directly from its definition as a limit. Instead, one applies a three-dimensional version of Fubini's theorem to evaluate the integral by repeated single integrations.

For example, suppose we want to integrate a continuous function $F(x, y, z)$ over a region D that is bounded below by a surface $z = f_1(x, y)$, above by the surface $z = f_2(x, y)$, and on the side by a cylinder C parallel to the z-axis (Fig. 16.27). Let R denote the vertical projection of D onto the xy-plane, which is the region in the xy-plane enclosed by C. The integral of F over D is then evaluated as

$$\iiint_D F(x, y, z)\, dV = \iint_R \left(\int_{f_1(x, y)}^{f_2(x, y)} F(x, y, z)\, dz \right) dy\, dx,$$

or

$$\iiint_D F(x, y, z)\, dV = \iint_R \int_{f_1(x, y)}^{f_2(x, y)} F(x, y, z)\, dz\, dx\, dy, \tag{5}$$

if we omit the parentheses. The z-limits of integration indicate that for every (x, y) in the region R, z may extend from the lower surface $z = f_1(x, y)$ to the upper surface $z = f_2(x, y)$. The y- and x-limits of integration have not been given explicitly in Eq. (5) but are to be determined in the usual way from the boundaries of R.

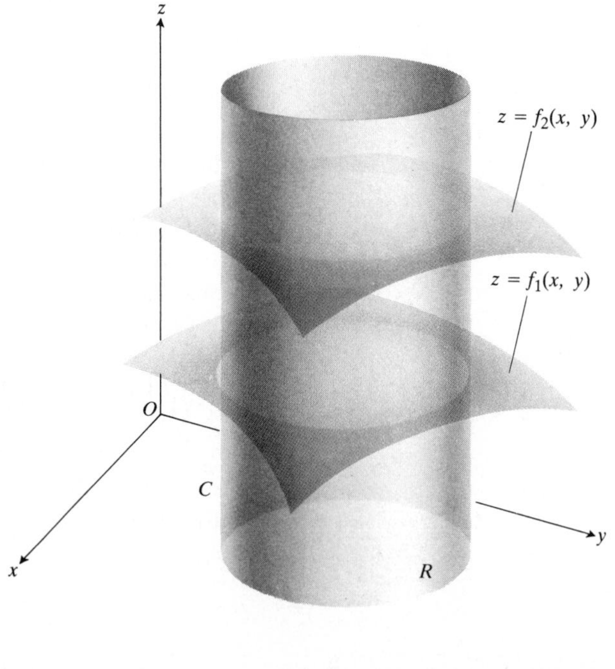

16.27 The enclosed volume can be found by evaluating

$$V = \iint_R \int_{f_1(x, y)}^{f_2(x, y)} dz\, dy\, dx.$$

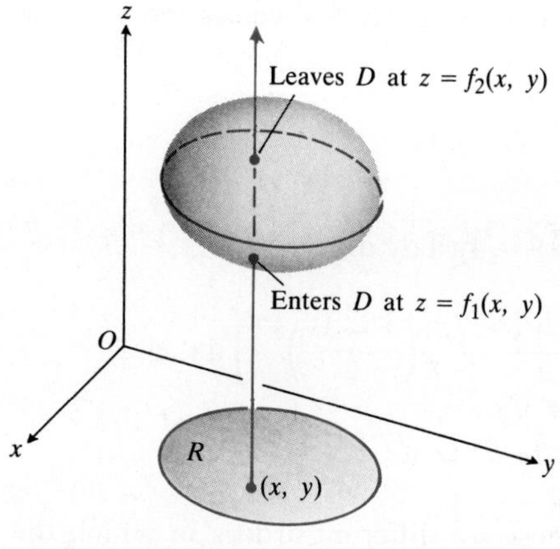

16.28 Finding limits of integration. The boundary of R is defined by the equation $f_1(x, y) = f_2(x, y)$.

In case the lateral surface of the cylinder reduces to zero, as in Fig. 16.28 and the example that follows, we may find the equation of the boundary of R by eliminating z between the two equations $z = f_1(x, y)$ and $z = f_2(x, y)$. This gives

$$f_1(x, y) = f_2(x, y),$$

an equation that contains no z and that defines the boundary of R in the xy-plane.

To supply the z-limits of integration in any particular instance we may use a procedure like the one for double integrals. We imagine a line L through a point (x, y) in R and parallel to the z-axis. As z increases, the line enters D at $z = f_1(x, y)$ and leaves D at $z = f_2(x, y)$. These give the lower and upper limits of the integration with respect to z. The result of this integration is now a function of x and y alone, which we integrate over R, supplying limits in the familiar way.

EXAMPLE 1 Find the volume enclosed between the two surfaces

$$z = x^2 + 3y^2 \qquad \text{and} \qquad z = 8 - x^2 - y^2.$$

Solution The two surfaces (Fig. 16.29) intersect on the elliptic cylinder

$$x^2 + 3y^2 = 8 - x^2 - y^2,$$

or

$$x^2 + 2y^2 = 4.$$

The volume projects into the region R (in the xy-plane) that is enclosed by the ellipse having this same equation. In the double integral with respect to y and x over R, if we integrate first with respect to y, holding x fixed, y

16.29 The volume between two paraboloids.

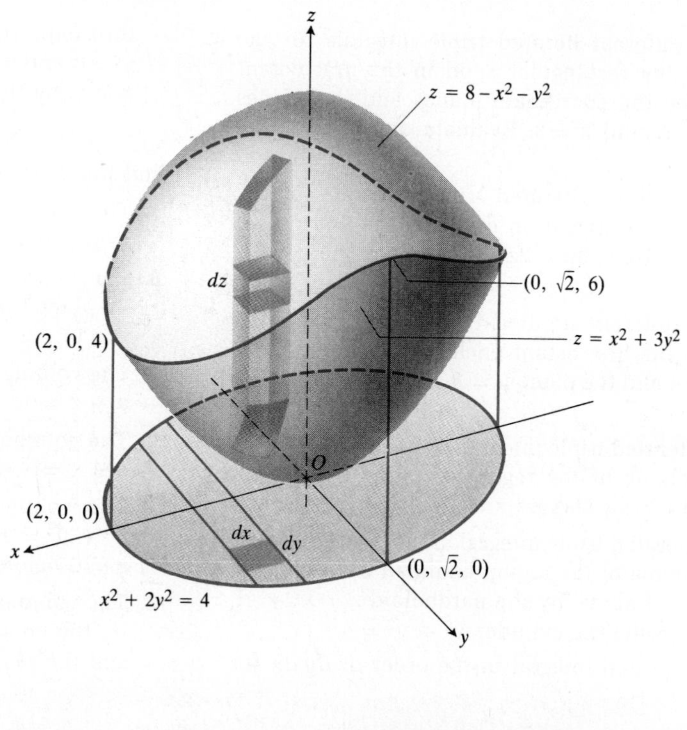

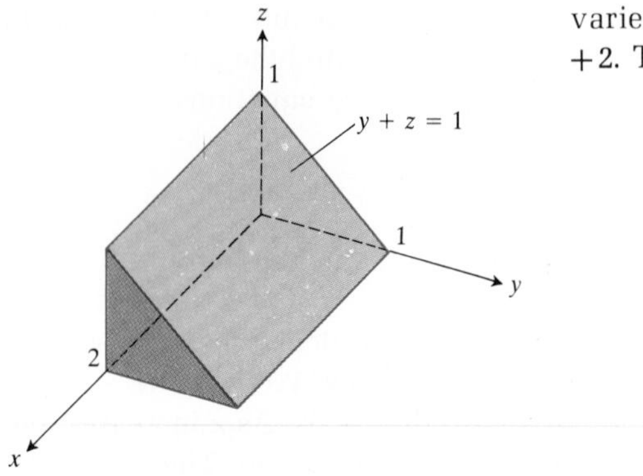

16.30 Example 2 shows how to calculate the volume of this prism with six different iterated triple integrals.

varies from $-\sqrt{(4-x^2)/2}$ to $+\sqrt{(4-x^2)/2}$. Then x varies from -2 to $+2$. Thus we have

$$\begin{aligned} V &= \int_{-2}^{2}\int_{-\sqrt{(4-x^2)/2}}^{\sqrt{(4-x^2)/2}}\int_{x^2+3y^2}^{8-x^2-y^2} dz\,dy\,dx \\ &= \int_{-2}^{2}\int_{-\sqrt{(4-x^2)/2}}^{\sqrt{(4-x^2)/2}} (8-2x^2-4y^2)\,dy\,dx \\ &= \int_{-2}^{2}\left[2(8-2x^2)\sqrt{\frac{4-x^2}{2}} - \frac{8}{3}\left(\frac{4-x^2}{2}\right)^{3/2}\right]dx \\ &= \frac{4\sqrt{2}}{3}\int_{-2}^{2}(4-x^2)^{3/2}\,dx = 8\pi\sqrt{2}. \ \square \end{aligned}$$

As we know, there are sometimes two different orders in which the single integrations that evaluate a double integral may be worked (but not always). For triple integrals there are sometimes (but not always) as many as *six* workable orders of integration. The next example shows an extreme case in which all six are possible.

EXAMPLE 2 Each of the following integrals gives the volume of the solid shown in Fig. 16.30.

a) $\int_0^1\int_0^{1-z}\int_0^2 dx\,dy\,dz$ b) $\int_0^2\int_0^{1-y}\int_0^2 dx\,dz\,dy$

c) $\int_0^1\int_0^2\int_0^{1-z} dy\,dx\,dz$ d) $\int_0^2\int_0^1\int_0^{1-z} dy\,dz\,dx$

e) $\int_0^1\int_0^2\int_0^{1-y} dz\,dx\,dy$ f) $\int_0^2\int_0^1\int_0^{1-y} dz\,dy\,dx$ □

PROBLEMS

1. Write six different iterated triple integrals for the volume of the rectangular solid in the first octant bounded by the coordinate planes and the planes $x = 1$, $y = 2$, and $z = 3$. Evaluate one of the integrals.
2. Write six different iterated triple integrals for the volume of the tetrahedron cut from the first octant by the plane $6x + 3y + 2z = 6$. Evaluate one of the integrals.
3. Write six different iterated triple integrals for the volume in the first octant enclosed by the cylinder $x^2 + z^2 = 4$ and the plane $y = 3$. Evaluate one of the integrals.
4. Write an iterated triple integral in the order *dz dx dy* for the volume of the region bounded by the xy-plane and the paraboloid $z = 4 - x^2 - y^2$.
5. Write an iterated triple integral in the order *dz dy dx* for the volume of the region bounded below by the xy-plane and above by the paraboloid $z = x^2 + y^2$, and lying inside the cylinder $x^2 + y^2 = 4$.
6. Write an iterated integral in the order *dz dy dx* for the volume of the region that lies between the planes $x = 0$ and $x = 1$ in the first octant and is bounded above by the plane $x + y + z = 2$.

Find the volumes in Problems 7–20.

7. The volume of the tetrahedron bounded by the plane $x/a + y/b + z/c = 1$ and the coordinate planes.
8. The volume between the cylinder $z = y^2$ and the xy-plane that is bounded by the four vertical planes $x = 0$, $x = 1$, $y = -1$, $y = 1$.
9. The volume in the first octant bounded by the planes $x + z = 1$, $y + 2z = 2$.
10. The volume in the first octant bounded by the cylinder $x = 4 - y^2$ and the planes $z = y$, $x = 0$, $z = 0$.
11. The volume of the wedge cut from the cylinder $x^2 + y^2 = 1$ by the plane $z = y$ above and the plane $z = 0$ below.
12. The volume of the region in the first octant bounded by the coordinate planes, the cylinder $x^2 + y^2 = 4$, and the plane $x + z = 3$.

13. The volume of the region in the first octant bounded by the coordinate planes, above by the cylinder $x^2 + z = 1$, and on the right by the paraboloid $y = x^2 + z^2$. (*Hint:* Integrate first with respect to y.)
14. The volume enclosed by the cylinders $z = 5 - x^2$, $z = 4x^2$, and the planes $y = 0$, $x + y = 1$.
15. The volume enclosed by the cylinder $y^2 + 4z^2 = 16$ and the planes $x = 0$, $x + y = 4$.
16. The volume bounded below by the plane $z = 0$, laterally by the elliptic cylinder $x^2 + 4y^2 = 4$, and above by the plane $z = x + 2$.
17. The volume common to the two cylinders $x^2 + y^2 = a^2$ and $x^2 + z^2 = a^2$. (See Fig. 16.31.)
18. The volume bounded by the elliptic paraboloids $z = x^2 + 9y^2$ and $z = 18 - x^2 - 9y^2$.
19. The volume of an ellipsoid of semiaxes a, b, c.
20. The volume of the region in the first octant bounded by the coordinate planes, by the plane $z = 3$ below, and by the surface $z = 4 - x^2 - y$ above.
21. Sketch the domain of integration of the integral

$$\int_{-1}^{1}\int_{x^2}^{1}\int_{0}^{1-y} dz\, dy\, dx.$$

Then rewrite the integral as an equivalent iterated integral in the order
a) dy dz dx
b) dy dx dz
c) dx dy dz
d) dx dz dy
e) dz dx dy

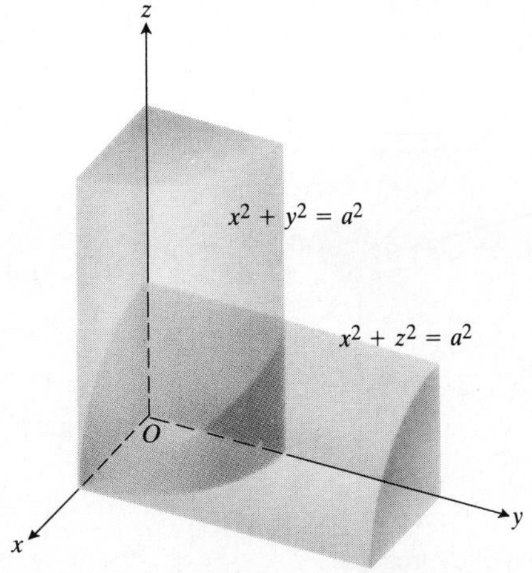

16.31 One eighth of the volume common to the cylinders $x^2 + y^2 = a^2$, $x^2 + z^2 = a^2$.

22. Repeat Problem 21 for the integral

$$\int_{0}^{1}\int_{0}^{1}\int_{0}^{y^2} dz\, dy\, dx.$$

***Toolkit* programs**
Double Integral

16.7 Physical Applications in Three Dimensions

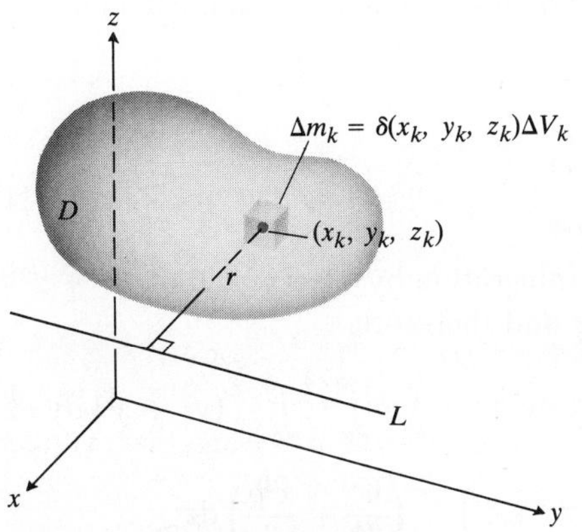

16.32 To define I_L we first imagine D to be subdivided into a finite number of mass elements Δm_k.

If $F(x, y, z) = \delta(x, y, z)$ is the density of an object occupying a region D in space, and we imagine D to be subdivided as in Fig. 16.26 in the preceding article, then the integral of the density,

$$m = \lim \sum_k \delta(x_k, y_k, z_k)\Delta V_k = \iiint_D \delta(x, y, z)\, dV$$

gives the mass of the object.

If $r(x, y, z)$ is the distance from the point (x, y, z) in D to a line L, then the moment of inertia about L of the mass element

$$\Delta m_k = \delta(x_k, y_k, z_k)\Delta V_k$$

shown in Fig. 16.32 is approximately

$$\Delta I_k = r^2(x_k, y_k, z_k)\Delta m_k$$

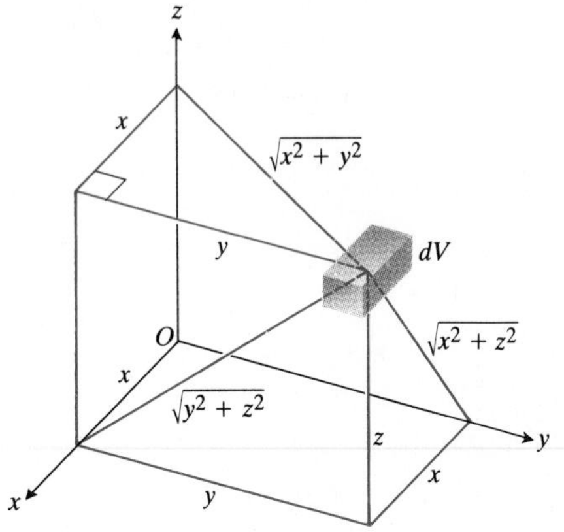

16.33 Distances of dV from the coordinate planes and axes.

and the moment of inertia I_L of the object about L is

$$I_L = \lim \sum_k \Delta I_k$$

$$= \lim \sum_k r^2(x_k, y_k, z_k)\delta(x_k, y_k, z_k)\Delta V_k = \iiint_D r^2\delta \, dV.$$

If L is the x-axis, then $r^2 = y^2 + z^2$ (Fig. 16.33) and

$$I_x = \iiint_D (y^2 + z^2)\delta \, dV.$$

Similarly,

$$I_y = \iiint_D (x^2 + z^2)\delta \, dV \quad \text{and} \quad I_z = \iiint_D (x^2 + y^2)\delta \, dV.$$

These and other useful formulas are summarized in the following list.

Mass: $m = \iiint_D \delta \, dV \quad (\delta = \text{density})$

First moments about the coordinate planes:

$$M_{yz} = \iiint_D x\,\delta \, dV, \qquad M_{xz} = \iiint_D y\,\delta \, dV, \qquad M_{xy} = \iiint_D z\,\delta \, dV$$

Center of mass:

$$\bar{x} = \frac{\iiint x\,\delta \, dV}{m}, \qquad \bar{y} = \frac{\iiint y\,\delta \, dV}{m}, \qquad \bar{z} = \frac{\iiint z\,\delta \, dV}{m}$$

Moments of inertia (second moments):

$$I_x = \iiint (y^2 + z^2)\delta \, dV, \qquad I_y = \iiint (x^2 + z^2)\delta \, dV,$$
$$I_z = \iiint (x^2 + y^2)\delta \, dV \qquad I_L = \iiint r^2\delta \, dV,$$

$r(x, y, z) =$ distance from point (x, y, z) to line L

Radius of gyration about a line L: $R = \sqrt{I_L/m}$

EXAMPLE 1 Find I_x, I_y, I_z for the rectangular solid of uniform density δ shown in Fig. 16.34.

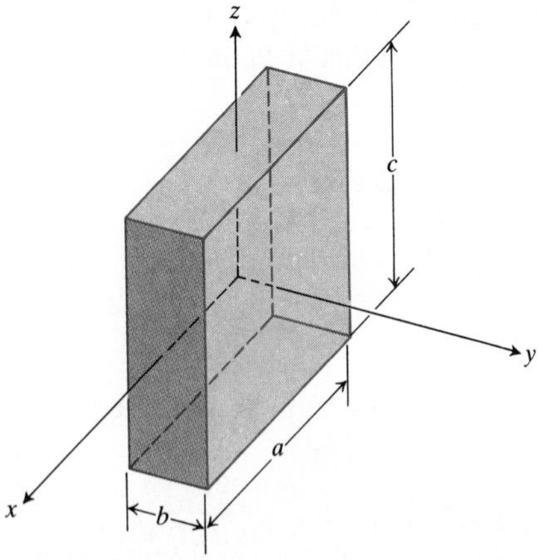

16.34 Example 1 calculates I_x, I_y, and I_z for the block shown here.

Solution

$$I_x = \int_{-c/2}^{c/2}\int_{-b/2}^{b/2}\int_{-a/2}^{a/2} (y^2 + z^2)\delta \, dx\, dy\, dz. \tag{1}$$

We can avoid some of the work of integration by observing that $(y^2 + z^2)\delta$ is an even function of x, y, and z and therefore

$$I_x = 8\int_0^{c/2}\int_0^{b/2}\int_0^{a/2} (y^2 + z^2)\delta \, dx\, dy\, dz = 4a\delta \int_0^{c/2}\int_0^{b/2} (y^2 + z^2)\, dy\, dz$$

$$= 4a\delta \int_0^{c/2} \left[\frac{y^3}{3} + z^2 y\right]_{y=0}^{y=b/2} dz = 4a\delta \int_0^{c/2} \left(\frac{b^3}{24} + \frac{z^2 b}{2}\right) dz$$

$$= 4a\delta\left(\frac{b^3 c}{48} + \frac{c^3 b}{48}\right) = \frac{abc\delta}{12}(b^2 + c^2) = \frac{m}{12}(b^2 + c^2).$$

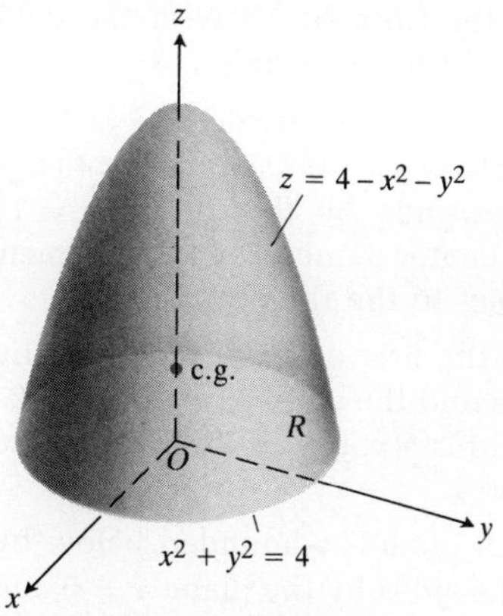

16.35 Example 2 calculates the coordinates of the center of gravity of this solid.

Similarly,

$$I_y = \frac{m}{12}(a^2 + c^2) \qquad \text{and} \qquad I_z = \frac{m}{12}(a^2 + b^2). \ \square$$

EXAMPLE 2 Find the center of gravity of a solid of uniform density δ bounded below by the disc R: $x^2 + y^2 \le 4$ in the plane $z = 0$ and above by the paraboloid $z = 4 - x^2 - y^2$ (Fig. 16.35).

Solution By symmetry, $\bar{x} = \bar{y} = 0$. To find $\bar{z}$ we first calculate

$$M_{xy} = \iint_R \int_{z=0}^{z=4-x^2-y^2} z\,\delta\,dz\,dy\,dx$$

$$= \iint_R \left[\frac{z^2}{2}\right]_{z=0}^{z=4-x^2-y^2} \delta\,dy\,dx$$

$$= \frac{\delta}{2}\iint_R (4 - x^2 - y^2)^2\,dy\,dx$$

$$= \frac{\delta}{2}\int_0^{2\pi}\int_0^2 (4 - r^2)^2\,r\,dr\,d\theta \qquad \text{(polar coordinates)}$$

$$= \frac{\delta}{2}\int_0^{2\pi}\left[-\frac{1}{6}(4 - r^2)^3\right]_{r=0}^{r=2} d\theta$$

$$= \frac{16\delta}{3}\int_0^{2\pi} d\theta$$

$$= \frac{32\pi\delta}{3}.$$

A similar calculation gives

$$m = \iint_R \int_0^{4-x^2-y^2} \delta\,dz\,dy\,dx = 8\pi\delta\,.$$

Therefore,

$$\bar{z} = \frac{M_{xy}}{m} = \frac{4}{3},$$

and the center of gravity is $(\bar{x}, \bar{y}, \bar{z}) = (0, 0, \frac{4}{3})$. $\square$

Geometric Figures

In moment calculations, geometric figures in space are treated as objects with constant density $\delta = 1$, and the resulting centers of mass are called *centroids*.

PROBLEMS

1. Evaluate the integral for I_x in Eq. (1) directly to show that the shortcut in Example 1 gives the same answer. Use the results in Example 1 to find the radius of gyration of the rectangular solid about each coordinate axis.

2. The axes shown in Fig. 16.36 run through the centroid of the solid wedge parallel to its edges. Using the dimensions shown, find I_x, I_y, and I_z.

3. Find the moments of inertia of the rectangular solid

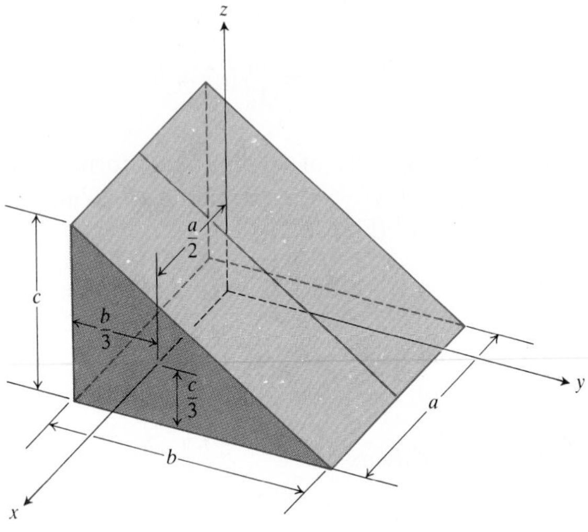

16.36 Figure for Problem 2.

shown in Fig. 16.37 with respect to its edges by calculating I_x, I_y, and I_z.

4. a) Find the centroid and the moments of inertia I_x, I_y, and I_z of the tetrahedron whose vertices are the points (0, 0, 0), (1, 0, 0), (0, 1, 0) and (0, 0, 1).
   b) Find the radius of gyration of the tetrahedron

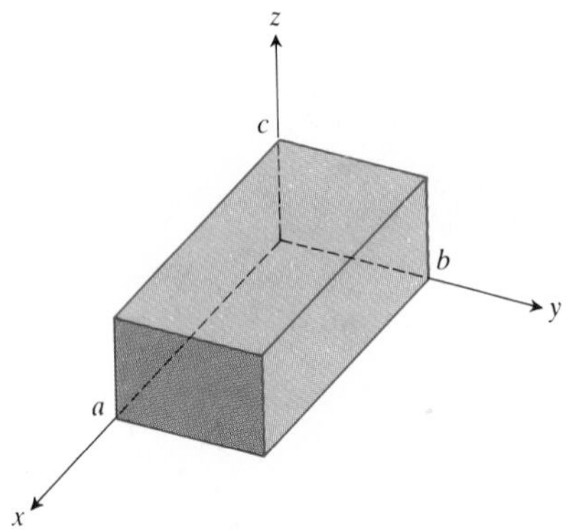

16.37 Figure for Problem 3.

about the x-axis. Compare it with the distance from the centroid to the x-axis.

5. A solid "trough" of uniform density δ is bounded below by the surface $z = 4y^2$, above by the plane $z = 4$, and on the ends by the planes $x = 1$ and $x = -1$. Find the center of mass, and the moments of inertia with respect to the three axes.

6. A solid region in the first octant is bounded by the coordinate planes and the plane $x + y + z = 2$. The density of the solid is $\delta(x, y, z) = 2x$. Find the center of mass.

7. A solid in the first octant is bounded below by the plane $z = 0$, on the sides by the plane $y = 0$ and the surface $x = y^2$, and above by the surface $z = 4 - x^2$. The density is $\delta(x, y, z) = kxy$, where k is a constant. Find the mass.

8. Write a triple integral for the mass of a solid hemisphere that is bounded below by the plane $z = 0$, and above by the sphere $x^2 + y^2 + z^2 = 4$, if the density at any point is proportional to the distance of the point from the z-axis. (Do not evaluate the integral.)

9. The density of a solid enclosed by the ellipsoid $9x^2 + 4y^2 + 36z^2 = 36$ is $\delta(x, y, z) = kx$, where k is a constant. Write, but do not evaluate, integral expressions for the solid's mass m and for the moments M_{yz} and I_y.

10. Find the x-coordinate of the centroid of the region bounded below by the plane $z = 0$, laterally by the elliptic cylinder $x^2 + 4y^2 = 4$, and above by the plane $z = x + 2$.

11. A torus of mass m is generated by rotating a circle of radius a about an axis in its plane at distance b from the center (b greater than a). Find its moment of inertia about the axis of revolution.

Toolkit programs

Double Integral

16.8 Integrals in Cylindrical and Spherical Coordinates

Cylindrical Coordinates

Cylindrical coordinates are useful in applications that involve cylinders along the z-axis and planes that contain or are perpendicular to the z-axis, because these surfaces have simple equations of constant coordinate

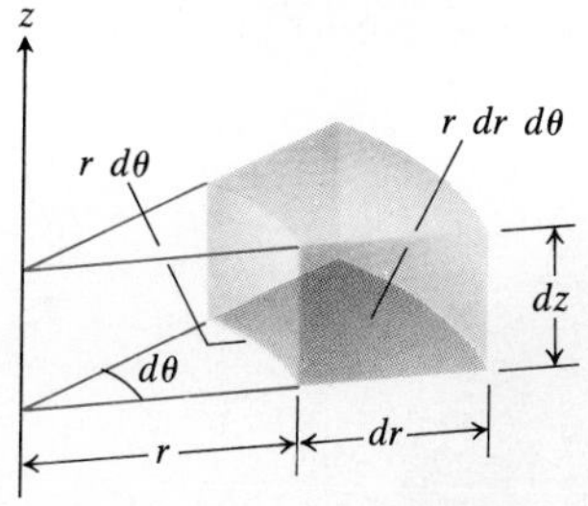

16.38 The volume element $dV = dz\, r\, dr\, d\theta$.

value like

$$r = 4, \qquad \theta = \frac{\pi}{3}, \qquad z = 2.$$

If we are working with a solid that has an axis of symmetry we can often simplify our calculations by taking the axis of symmetry to be the z-axis.

The volume element for subdividing a region in space with cylindrical coordinates is taken to be

$$dV = dz\, r\, dr\, d\theta, \tag{1}$$

as shown in Fig. 16.38. Triple integrals in cylindrical coordinates are then evaluated as iterated integrals, as in the examples that follow.

EXAMPLE 1 Find the centroid of a solid hemisphere of radius a.

Solution We may choose the origin at the center of the sphere and consider the hemisphere that lies above the xy-plane. (See Fig. 16.39.) The equation of the hemispherical surface is

$$z = \sqrt{a^2 - x^2 - y^2}$$

or, in terms of cylindrical coordinates,

$$z = \sqrt{a^2 - r^2}.$$

By symmetry we have

$$\bar{x} = \bar{y} = 0.$$

We calculate $\bar{z}$:

$$\bar{z} = \frac{\iiint z\, dV}{\iiint dV} = \frac{\int_0^{2\pi}\int_0^{a}\int_0^{\sqrt{a^2-r^2}} z\, dz\, r\, dr\, d\theta}{\frac{2}{3}\pi a^3} = \frac{3a}{8}. \ \square$$

EXAMPLE 2 Find the moment of inertia I_z of the solid that is bounded below by the xy-plane, above by the sphere $x^2 + y^2 + z^2 = 4a^2$, and laterally by the cylinder $r = 2a \cos \theta$.

16.39 The volume element in cylindrical coordinates is $dV = dz\, r\, dr\, d\theta$.

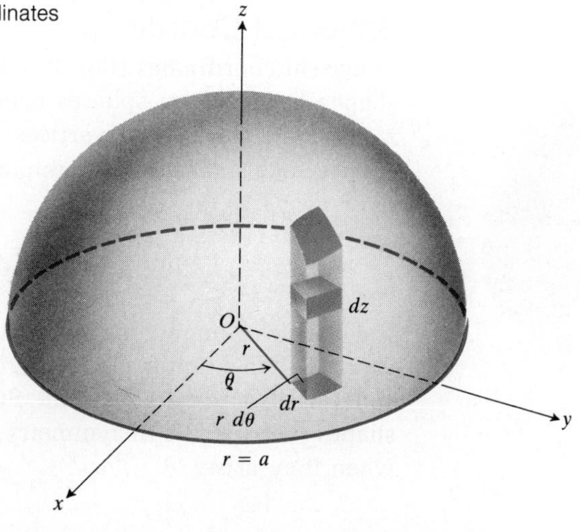

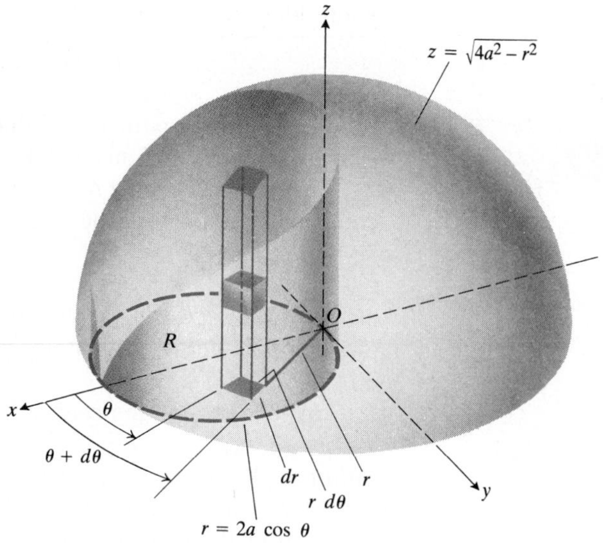

16.40 The solid bounded by the xy-plane, the sphere $x^2 + y^2 + z^2 = 4a^2$, and the cylinder $r = 2a \cos\theta$.

Solution The solid lies in front of the yz-plane, as shown in Fig. 16.40. It projects vertically onto the region R bounded by the circle $r = 2a\cos\theta$ in the xy-plane. Since

$$I_z = \iiint (x^2 + y^2)\, dV$$

(from the list of formulas in Article 16.7), we have

$$I_z = \iiint r^2\, dz\, r\, dr\, d\theta.$$

Integrating first from $z = 0$ to $z = \sqrt{4a^2 - x^2 - y^2} = \sqrt{4a^2 - r^2}$, we find

$$I_z = \iint_R \int_{z=0}^{z=\sqrt{4a^2-r^2}} dz\, r^3\, dr\, d\theta = \int_{-\pi/2}^{\pi/2} \int_0^{2a\cos\theta} \int_0^{\sqrt{4a^2-r^2}} dz\, r^3\, dr\, d\theta$$

$$= \int_{-\pi/2}^{\pi/2} \int_0^{2a\cos\theta} r^3\sqrt{4a^2 - r^2}\, dr\, d\theta = \frac{64a^5}{15}\left(\pi - \frac{26}{15}\right). \quad \square$$

Spherical Coordinates

Spherical coordinates (Fig. 16.41) are useful in applications that involve shapes bounded by spheres centered at the origin, planes through the z-axis, and cones with vertices at the origin whose axes lie along the z-axis. Such surfaces have simple equations of constant coordinate value like

$$\rho = r, \qquad \phi = \frac{\pi}{3}, \qquad \theta = \frac{\pi}{3}.$$

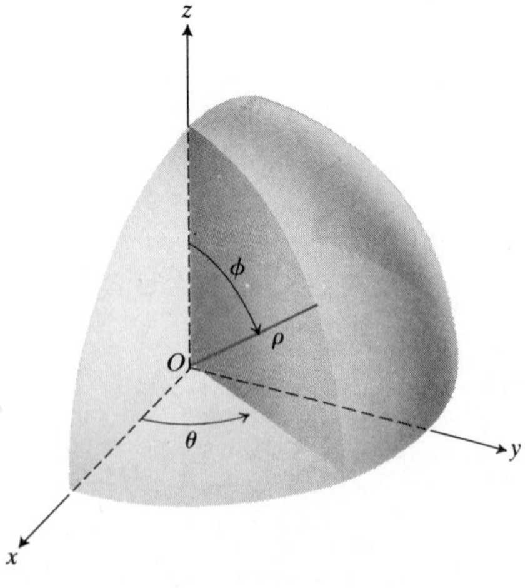

16.41 Spherical coordinates.

If we are working with a shape that is symmetric with respect to a point, we can often simplify our work by choosing that point as the origin of a spherical coordinate system. Such shapes arise less frequently than shapes with an axis of symmetry, but it is good to be able to handle them when they arise.

The volume element in spherical coordinates is

$$dV = \rho^2 \sin\phi \, d\rho \, d\phi \, d\theta \tag{2}$$

(as shown in Fig. 16.42) and triple integrals take the form

$$\iiint F(\rho, \phi, \theta) dV = \iiint F(\rho, \phi, \theta) \rho^2 \sin\phi \, d\rho \, d\phi \, d\theta. \tag{3}$$

To evaluate these integrals we first integrate with respect to ρ. The procedure for finding the limits of integration for a region D in space is therefore the following:

1. Hold ϕ and θ fixed and let ρ increase. This gives a ray out from the origin.
2. Integrate from the ρ-value where the ray first enters D to the ρ-value where the ray leaves D. This gives the limits for ρ.
3. Hold θ fixed and let ϕ increase. (This gives a family of rays that make a "fan.") Integrate over the ϕ-values for which the rays pass through D.
4. Choose θ-limits that include all the fans that intersect D.

EXAMPLE 3 Find the volume cut from the sphere $\rho = a$ by the cone $\phi = \alpha$. (See Fig. 16.43.)

Solution The volume is given by

$$V = \int_0^{2\pi} \int_0^{\alpha} \int_0^{a} \rho^2 \sin\phi \, d\rho \, d\phi \, d\theta = \frac{2\pi a^3}{3}(1 - \cos\alpha).$$

As a check, we note that the special cases $\alpha = \pi/2$ and $\alpha = \pi$ correspond to the cases of a hemisphere and a sphere, of volumes $2\pi a^3/3$ and $4\pi a^3/3$, respectively. □

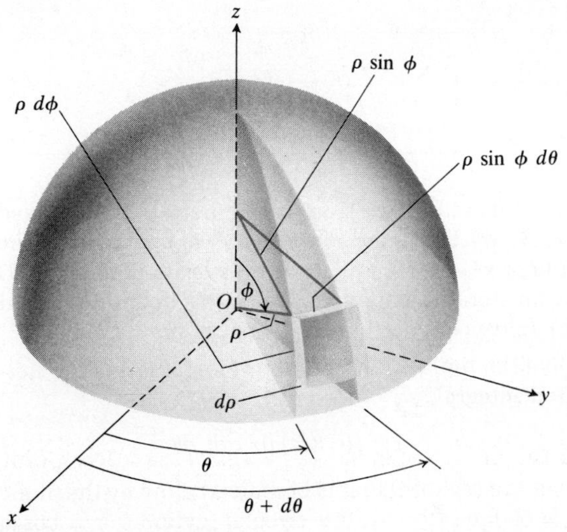

16.42 The volume element in spherical coordinates is $dV = d\rho \cdot \rho \, d\phi \cdot \rho \sin\phi \, d\theta$.

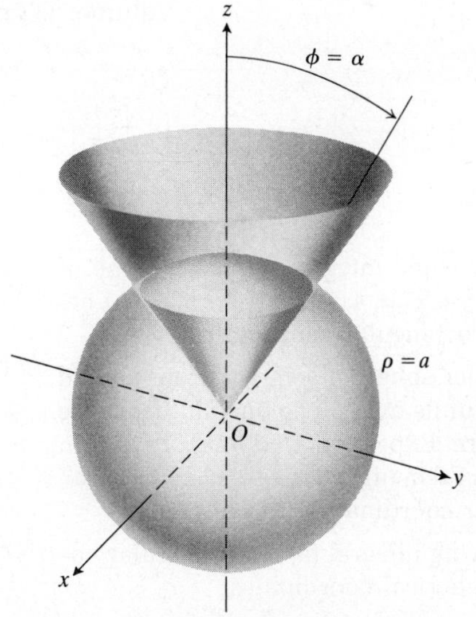

16.43 The volume cut from the sphere $\rho = a$ by the cone $\phi = \alpha$.

EXAMPLE 4 Find I_z for the region in Example 3 if the equation for the cone is $\phi = \pi/3$.

Solution With

$$I_z = \iiint (x^2 + y^2)\, dV$$

and

$$x^2 + y^2 = (\rho \sin\phi \cos\theta)^2 + (\rho \sin\phi \sin\theta)^2 = \rho^2 \sin^2\phi,$$

we have

$$\begin{aligned} I_z &= \int_0^{2\pi}\int_0^{\pi/3}\int_0^a (\rho^2 \sin^2\phi)\rho^2 \sin\phi \, d\rho\, d\phi\, d\theta \\ &= \int_0^{2\pi}\int_0^{\pi/3}\int_0^a \rho^4 \sin^3\phi\, d\rho\, d\phi\, d\theta \\ &= \frac{a^5}{5}\int_0^{2\pi}\int_0^{\pi/3} \sin^3\phi\, d\phi\, d\theta = \frac{a^5}{5}\int_0^{2\pi}\int_0^{\pi/3}(1-\cos^2\phi)\sin\phi\, d\phi\, d\theta \\ &= \frac{a^5}{5}\int_0^{2\pi}\left[-\cos\phi + \frac{\cos^3\phi}{3}\right]_{\phi=0}^{\phi=\pi/3} d\theta = \frac{a^5}{5}\int_0^{2\pi}\left[\frac{5}{24}\right] d\theta = \frac{\pi a^5}{12}. \quad \square \end{aligned}$$

Coordinate Conversion Formulas

Polar to rectangular	Spherical to cylindrical	Spherical to rectangular
$x = r\cos\theta$	$r = \rho\sin\phi$	$x = \rho\sin\phi\cos\theta$
$y = r\sin\theta$	$z = \rho\cos\phi$	$y = \rho\sin\phi\sin\theta$
$z = z$	$\theta = \theta$	$z = \rho\cos\phi$

(Details in Article 13.4.)

Volume: $\iiint dx\, dy\, dz = \iiint dz\, r\, dr\, d\theta = \iiint \rho^2 \sin\phi\, d\rho\, d\phi\, d\theta$

PROBLEMS

1. Set up an iterated triple integral for the volume of the sphere $x^2 + y^2 + z^2 = 4$ in (a) spherical, (b) cylindrical, and (c) rectangular coordinates.

2. Let D be the smaller spherical cap cut from a solid ball of radius two units by a plane one unit from the center of the sphere. Express the volume of D as an iterated triple integral in (a) spherical, (b) cylindrical, and (c) rectangular coordinates.

3. Convert the following integral (a) to rectangular coordinates, (b) to spherical coordinates:

$$\int_0^{2\pi}\int_0^1\int_0^{\sqrt{4-r^2}} r\, dz\, r\, dr\, d\theta.$$

4. Let D denote the region in the first octant enclosed below by the cone $z^2 = x^2 + y^2$ and above by the sphere $x^2 + y^2 + z^2 = 8$. Express the volume V of D as an iterated triple integral in (a) cylindrical, and (b) spherical coordinates. Then (c) find V.

5. Give the limits of integration for evaluating the iterated integral

$$\iiint f(r, \theta, z)\, dz\, r\, dr\, d\theta$$

over the region D that is bounded below by the plane $z = 0$, laterally by the circular cylinder $r = \cos\theta$, and above by the paraboloid $z = 3r^2$.

6. Express $I_z = \iiint (x^2 + y^2)\delta\, dV$ as an iterated triple integral over an arbitrary region D in (a) cylindrical and (b) spherical coordinates.

7. Express the moment of inertia I_z of the solid bounded below by the plane $z = 0$ and above by the sphere $x^2 + y^2 + z^2 = 1$ as an iterated triple integral in (a) cylindrical and (b) spherical coordinates.

8. Express $M_{xy} = \iiint z\delta dV$ as an iterated triple integral over an arbitrary region D in (a) cylindrical and (b) spherical coordinates.

Cylindrical coordinate problems

9. Find the volume bounded below by the plane $z = 0$, laterally by the cylinder $x^2 + y^2 = 1$, and above by the paraboloid $z = x^2 + y^2$.

10. Find the volume bounded below by the paraboloid $z = x^2 + y^2$, laterally by the cylinder $x^2 + y^2 = 1$, and above by the paraboloid $z = x^2 + y^2 + 1$.

11. Find the volume bounded below by the plane $z = 0$ and above by the paraboloid $z = 4 - x^2 - y^2$.

12. Find the volume enclosed by the cylinder $x^2 + y^2 = 4$ and the planes $z = 0$ and $y + z = 4$. (*Hint:* In cylindrical coordinates, $z = 4 - y$ becomes $z = 4 - r \sin \theta$.)

13. Find the volume bounded above by the paraboloid $z = 5 - x^2 - y^2$ and below by the paraboloid $z = 4x^2 + 4y^2$.

14. Find the volume that is bounded above by the paraboloid $z = 9 - x^2 - y^2$, below by the xy-plane, and that lies *outside* the cylinder $x^2 + y^2 = 1$.

15. Find the volume cut from the sphere $x^2 + y^2 + z^2 = 4a^2$ by the cylinder $x^2 + y^2 = a^2$.

16. Find the volume bounded below by the paraboloid $z = x^2 + y^2$ and above by the plane $z = 2y$.

17. Find the volume bounded above by the sphere $x^2 + y^2 + z^2 = 2a^2$ and below by the paraboloid $az = x^2 + y^2$.

18. Find the volume in the first octant bounded by the cylinder $x^2 + y^2 = a^2$ and the planes $x = a$, $y = a$, $z = 0$, $z = x + y$.

19. A hemispherical bowl of radius 5 cm is filled with water to within 3 cm of the top. Find the volume of water in the bowl.

20. A solid of uniform density δ in the first octant is bounded above by the cone $z^2 = x^2 + y^2$, below by the plane $z = 0$, and on the sides by the cylinder $x^2 + y^2 = 4$ and the planes $x = 0$ and $y = 0$. Find the center of gravity of the solid. (*Hint:* $\bar{x} = \bar{y}$.)

21. Find I_z for the region cut from the sphere $x^2 + y^2 + z^2 = 4a^2$ by the cylinder $x^2 + y^2 = a^2$.

22. Find I_z for the region bounded below by the paraboloid $z = x^2 + y^2$ and above by the plane $z = 2y$.

23. Find the centroid of the region bounded above by the sphere $x^2 + y^2 + z^2 = 2a^2$ and below by the paraboloid $az = x^2 + y^2$.

24. Find the moment of inertia of a solid circular cylinder of radius r and height h (a) about the axis of the cylinder, (b) about a line through the centroid perpendicular to the axis of the cylinder.

25. Use cylindrical coordinates to find the moment of inertia of a sphere of radius a and mass m about a diameter.

26. Find the volume generated by rotating the cardioid $r = a(1 - \cos \theta)$ about the x-axis. [*Hint:* Use double integration. Rotate an area element dA around the x-axis to generate a volume element dV.]

27. Find the moment of inertia, about the x-axis, of the volume of Problem 26.

28. Find the moment of inertia of a right circular cone of base radius a, altitude h, and mass m about an axis through the vertex and parallel to the base.

29. Find the moment of inertia of a sphere of radius a and mass m with respect to a tangent line.

30. Find the centroid of that portion of the volume of the sphere $r^2 + z^2 = a^2$ that lies between the planes $\theta = -(\pi/4)$ and $\theta = \pi/4$.

31. Find the volume in the first octant that lies between the cylinders $r = 1$ and $r = 2$ and that is bounded below by the xy-plane and above by the surface $z = xy$.

32. Convert the integral

$$\int_{-1}^{1} \int_{0}^{\sqrt{1-y^2}} \int_{0}^{x} (x^2 + y^2)\, dz\, dx\, dy$$

to an equivalent integral in cylindrical coordinates and evaluate the result.

Spherical coordinates

33. Find the volume of the solid bounded above by the sphere $\rho = a$ and below by the cone $\phi = \pi/3$. (The solid resembles a filled ice cream cone.)

34. Find the volume inside the sphere $\rho = a$ that lies between the cones $\phi = \pi/3$ and $\phi = 2\pi/3$.

35. Find the volume cut from the sphere $\rho = a$ by the planes $\theta = 0$ and $\theta = \pi/6$ in the first octant.

36. Find the centroid of the solid in Problem 33.

37. Find the smaller volume cut from the sphere $\rho = 2$ by the plane $z = \sqrt{2}$.

38. Find the volume enclosed by the surface $\rho = a(1 - \cos \phi)$. Compare with Problem 26.

39. Find the volume of the region bounded inside by the surface $\rho = 1 + \cos \phi$ and outside by the sphere $\rho = 2$.

40. The region bounded by the circle $\rho = 2 \sin \phi, \theta = \pi/2$ (circle of radius 1 in the yz-plane tangent to the z-axis at the origin and lying to the right of the z-axis) is revolved about the z-axis to sweep out a solid. Find the volume of the solid. (The first theorem of Pappus in Article 5.11 gives a quick way to check your result.)

41. Find the center of gravity of a homogeneous solid hemisphere of radius a and constant density δ.

42. a) Find the moment of inertia of a solid sphere of radius a and uniform density δ about a diameter of the sphere.

b) Find the radius of gyration of the sphere about the diameter.

43. Find the radius of gyration, with respect to the diameter, of a spherical shell of mass m bounded by the spheres $\rho = a$ and $\rho = 2a$ if the density is $\delta = \rho^2$.

44. Show that the centroid of a solid circular cone lies on the axis of the cone and one fourth of the way from the base to the vertex.

45. Find the moment of inertia of a solid circular cone of base radius r and height h about the axis of the cone.

16.9 Surface Area

Figure 16.44 shows a piece S of the surface $F(x, y, z) = c$ lying above its "shadow" R on a ground plane directly beneath it. In this article we shall show how to define and calculate the area of such a surface from the formula for F as an integral over R when F and its first partial derivatives are continuous. We shall also investigate the special case in which the surface is given by an equation of the form $z = f(x, y)$.

The first step in defining the area of S is to divide the region R into small rectangles ΔA_k of the kind we would use if we were defining an integral over R. Directly above each ΔA_k there lies a patch of surface $\Delta\sigma_k$ that we may approximate with a portion of ΔP_k of the tangent plane. To be specific, we suppose that ΔP_k is a portion of the plane that is tangent to the surface at the point (x_k, y_k, z_k) directly above the back corner C_k of ΔA_k. If the tangent plane is parallel to R, then ΔP_k will be congruent to ΔA_k. Otherwise it will be a parallelogram whose area is somewhat larger than that of ΔA_k.

Figure 16.45 gives a magnified view of $\Delta\sigma_k$ and ΔP_k, showing the vector $\nabla F(x_k, y_k, z_k)$ and a vector $\mathbf{n}$ that is a unit vector normal to the ground plane. The vector $\mathbf{n}$ is included in the figure because the angle γ it makes with ∇F will prove to be important in later calculations. The other vectors in this picture, $\mathbf{u}$ and $\mathbf{v}$, are the vectors that lie along the edges of the patch ΔP_k in the tangent plane. Both $\mathbf{u} \times \mathbf{v}$ and $\nabla F(x_k, y_k, z_k)$ are normal to ΔP_k.

At this point we need a fact that we haven't used since Article 13.9, namely that $|(\mathbf{u} \times \mathbf{v}) \cdot \mathbf{n}|$ is the area of the orthogonal projection of the parallelogram determined by $\mathbf{u}$ and $\mathbf{v}$ onto a plane whose normal is $\mathbf{n}$. In our case, this translates into the statement

$$|(\mathbf{u} \times \mathbf{v}) \cdot \mathbf{n}| = \Delta A_k. \tag{1}$$

Now, $|\mathbf{u} \times \mathbf{v}|$ itself is the area ΔP_k (standard fact about cross products) so that Eq. (1) becomes

$$\underbrace{|\mathbf{u} \times \mathbf{v}|}_{\Delta P_k}\ \underbrace{|\mathbf{n}|}_{1}\ \underbrace{|\cos\,(\text{angle between } \mathbf{u} \times \mathbf{v} \text{ and } \mathbf{n})|}_{\substack{\text{Same as } |\cos\gamma| \text{ because} \\ \nabla F \text{ and } \mathbf{u}\times\mathbf{v} \text{ are both} \\ \text{normal to the tangent plane}}} = \Delta A_k \tag{2}$$

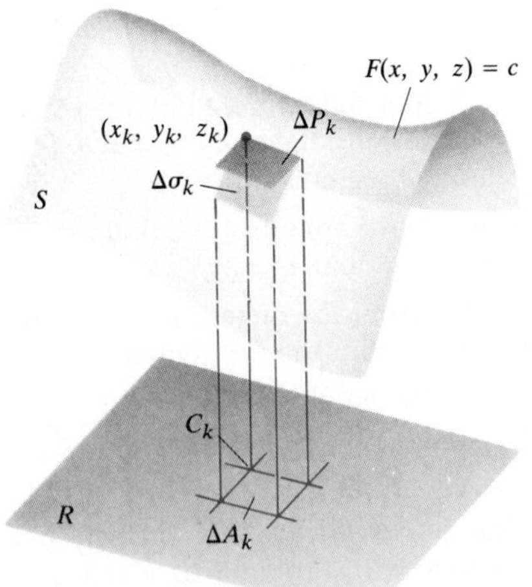

16.44 The tangent plate ΔP_k approximates the surface patch $\Delta\sigma_k$ on the level surface $F(x, y, z) = c$, lying above ΔA_k.

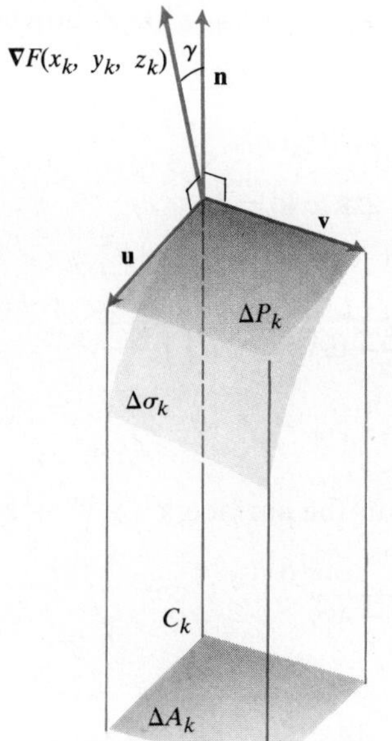

16.45 Magnified view from Fig. 16.44. The vector $\mathbf{u} \times \mathbf{v}$ (not shown) is parallel to ∇F because both vectors are normal to ΔP_k.

or

$$\Delta P_k |\cos \gamma| = \Delta A_k$$

or

$$\Delta P_k = \frac{\Delta A_k}{|\cos \gamma|},$$

provided $\cos \gamma \neq 0$. We will have $\cos \gamma \neq 0$ as long as ∇F is not parallel to the ground plane, or as long as $\nabla F \cdot \mathbf{n} \neq 0$.

Since the patches ΔP_k approximate the surface patches $\Delta\sigma_k$, which fit together to make S, the sum

$$\sum \Delta P_k = \sum \frac{\Delta A_k}{|\cos \gamma|} \tag{3}$$

looks like an approximation of what we might like to call the surface area of S. It also looks as if the approximation would improve if we refined the subdivision of R. In fact, the sums on the right-hand side of Eq. (3) are approximating sums for the double integral

$$\iint_R \frac{1}{|\cos \gamma|}\, dA. \tag{4}$$

We therefore define the area of S to be the value of this integral whenever the integral exists.

For any particular surface $F(x, y, z) = c$ we have

$$|\nabla F \cdot \mathbf{n}| = |\nabla F|\,|\mathbf{n}|\,|\cos \gamma|, \qquad \text{or} \qquad \frac{1}{|\cos \gamma|} = \frac{|\nabla F|\,|\mathbf{n}|}{|\nabla F \cdot \mathbf{n}|} = \frac{|\nabla F|}{|\nabla F \cdot \mathbf{n}|},$$

which combines with Eq. (4) to give the following formula.

$$\text{Surface area} = \iint_R \frac{|\nabla F|}{|\nabla F \cdot \mathbf{n}|}\, dA \tag{5}$$

We reached Eq. (5) under the assumptions that $\nabla F \cdot \mathbf{n} \neq 0$ and that F and its first partial derivatives were continuous over R (so that ∇F would be defined and continuous over R). However, whenever the quotient $|\nabla F|/|\nabla F \cdot \mathbf{n}|$ is integrable over R we may define the value of the integral in Eq. (5) to be the surface area of the portion of the surface $F(x, y, z) = c$ that lies over R.

The surface area defined in Eq. (5) agrees with our earlier definitions of surface area. We shall not prove this, but see Problems 12, 13, 15, and 21.

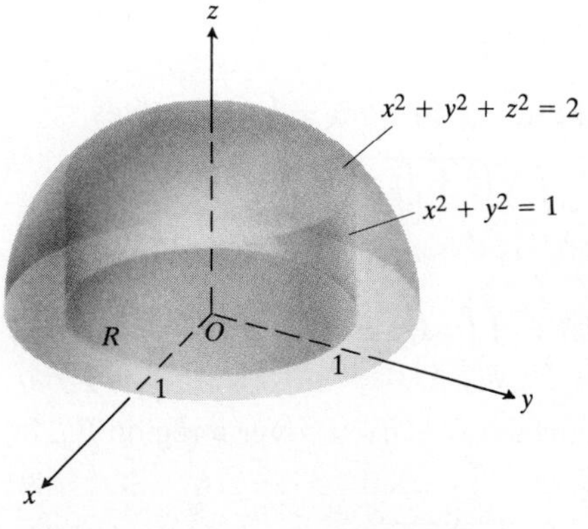

16.46 The cap cut from the hemisphere by the cylinder projects one to one onto the disc R: $x^2 + y^2 \leq 1$.

EXAMPLE 1 Find the area of the upper cap cut from the sphere $x^2 + y^2 + z^2 = 2$ by the cylinder $x^2 + y^2 = 1$ (Fig. 16.46).

Solution The cap, part of the surface $F(x, y, z) = x^2 + y^2 + z^2 = 2$, projects onto the disc R: $x^2 + y^2 \leq 1$ in the xy-plane. At any point (x, y, z) in space we have

$$F = x^2 + y^2 + z^2,$$
$$\nabla F = 2x\mathbf{i} + 2y\mathbf{j} + 2z\mathbf{k},$$
$$|\nabla F| = \sqrt{(2x)^2 + (2y)^2 + (2z)^2} = 2\sqrt{x^2 + y^2 + z^2}.$$

At points on the level surface $F(x, y, z) = x^2 + y^2 + z^2 = 2$, however,

$$|\nabla F| = 2\sqrt{x^2 + y^2 + z^2} = 2\sqrt{2}.$$

Taking $\mathbf{n} = \mathbf{k}$ as the unit vector normal to R gives

$$\nabla F \cdot \mathbf{n} = (2x\mathbf{i} + 2y\mathbf{j} + 2z\mathbf{k}) \cdot \mathbf{k} = 2z.$$

The surface area S is therefore

$$S = \iint_R \frac{|\nabla F|}{|\nabla F \cdot \mathbf{n}|}\, dA = \iint_R \frac{2\sqrt{2}}{2z}\, dA = \sqrt{2}\iint_R \frac{dA}{z}.$$

What do we do about the z?

Since z is the coordinate of a point on the surface $x^2 + y^2 + z^2 = 2$, we may express z in terms of x and y as

$$z = \sqrt{2 - x^2 - y^2}.$$

Thus,

$$\begin{aligned}
\text{Surface area} &= \sqrt{2} \iint_{x^2 + y^2 \le 1} \frac{dA}{\sqrt{2 - x^2 - y^2}} \\
&= \sqrt{2} \int_0^{2\pi} \int_0^1 \frac{r\, dr\, d\theta}{\sqrt{2 - r^2}} \qquad \text{(polar coordinates)} \\
&= \sqrt{2} \int_0^{2\pi} [-(2 - r^2)^{1/2}]_{r=0}^{r=1}\, d\theta \\
&= \sqrt{2} \int_0^{2\pi} (\sqrt{2} - 1)\, d\theta \\
&= 2\pi(2 - \sqrt{2}). \quad \square
\end{aligned}$$

Special Cases

In the case of a smooth surface S given by an equation $z = f(x, y)$ defined over a region R_{xy} in the xy-plane, we may write

$$F(x, y, z) = f(x, y) - z$$

and regard S as the level surface

$$F(x, y, z) = 0$$

of the function F. Taking the normal to R_{xy} to be $\mathbf{n} = \mathbf{k}$ then gives

$$|\nabla F| = |f_x\mathbf{i} + f_y\mathbf{j} - \mathbf{k}| = \sqrt{f_x^2 + f_y^2 + 1},$$

$$|\nabla F \cdot \mathbf{n}| = |-\mathbf{k} \cdot \mathbf{k}| = 1,$$

$$\text{Area of } S = \iint_{R_{xy}} \frac{|\nabla F|}{|\nabla F \cdot \mathbf{n}|}\, dA = \iint_{R_{xy}} \sqrt{f_x^2 + f_y^2 + 1}\, dy\, dx. \tag{6a}$$

Similarly, for the area of a smooth surface $x = f(y, z)$ over a region R_{yz} in the yz-plane we get

$$\text{Area of } S = \iint_{R_{yz}} \sqrt{f_y^2 + f_z^2 + 1}\, dy\, dz \tag{6b}$$

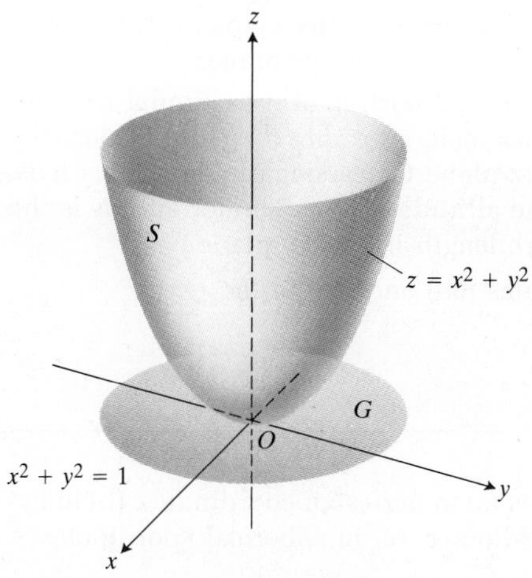

16.47 The area of the parabolic surface above is calculated in Example 2.

and for the area of a smooth surface $y = f(x, z)$ over a region R_{xz} in the xz-plane we get

$$\text{Area of } S = \iint_{R_{xz}} \sqrt{f_x^2 + f_z^2 + 1}\, dx\, dz. \tag{6c}$$

EXAMPLE 2 Find the area of the surface cut from the bottom of the paraboloid $z = x^2 + y^2$ by the plane $z = 1$.

Solution The surface projects onto the disk R: $x^2 + y^2 \le 1$ in the xy-plane (Fig. 16.47). We apply Eq. (6a) with

$$z = f(x, y) = x^2 + y^2$$

to get

$$\begin{aligned}\text{Area} &= \iint_{R_{xy}} \sqrt{f_x^2 + f_y^2 + 1}\, dx\, dy = \iint_{x^2+y^2\le 1} \sqrt{4x^2 + 4y^2 + 1}\, dx\, dy \\ &= \int_0^{2\pi}\int_0^1 \sqrt{4r^2 + 1}\, r\, dr\, d\theta \qquad \text{(polar coordinates)} \\ &= \frac{\pi}{6}(5\sqrt{5} - 1). \; \square\end{aligned}$$

PROBLEMS

1. Find the area of the surface cut from the paraboloid $z = x^2 + y^2$ by the plane $z = 4$.
2. Find the area of the surface cut from the paraboloid $z = 4 - x^2 - y^2$ by the xy-plane.
3. Find the area cut from the paraboloid $z = 1 - x^2 - y^2$ by the plane $z = -3$. (*Hint:* Integrate over the disc $x^2 + y^2 \le 4$ in the xy-plane.)
4. Find the area of the portion of the surface $z = 9 - x^2 - y^2$ above the plane $z = 5$.
5. Find the area cut from the plane $x + y + z = 5$ by the cylinder whose walls are $x = y^2$ and $x = 2 - y^2$.
6. Find the area cut from the paraboloid $y = 1 - x^2 - z^2$ (a) by the plane $y = 0$, (b) by the plane $y = 3$. (*Hint:* Project onto the xz-plane.)
7. Find the area of the portion of the surface $2z = x^2$ that lies above the triangle bounded by the lines $x = 1$, $y = 0$, and $y = x$ in the xy-plane.
8. Find the area of the surface $y + z = x^2/2$ that lies above the triangle bounded by the lines $x = 2$, $y = 0$, and $y = x$ in the xy-plane.
9. Find the area of the surface $z = 9 - x^2 - y^2$ that lies above the ring R: $1 \le x^2 + y^2 \le 9$ in the xy-plane.
10. Find the area of the surface $z = x^2 + \sqrt{15}\, y - 2 \ln x$ above the square R: $0 \le y \le 1$, $1 \le x \le 2$ in the xy-plane.
11. Write a double iterated integral for the area cut from the upper half of the sphere $x^2 + y^2 + z^2 = 9$ by the elliptic cylinder $x^2 + 4y^2 = 9$.
12. Whenever we replace an old definition with a new one it is a good idea to try out the new one on familiar objects to see that it gives the answers we expect. For instance, is the surface area of a circular cylinder of base radius r and height h still $2\pi rh$? Find out by using Eq. (5) with $\mathbf{n} = \mathbf{k}$ to calculate the area of the surface cut from the cylinder $y^2 + z^2 = r^2$ by the planes $x = 0$ and $x = h > 0$. (Find the area above the xy-plane, and double.)
13. Show that the surface area of the sphere $x^2 + y^2 + z^2 = a^2$ is $4\pi a^2$.
14. Find the area cut from the paraboloid $z = 9 - x^2 - y^2$ by the planes $z = 0$ and $z = 8$.
15. Find the area of the triangle cut from the plane $x/a + y/b + z/c = 1$ (a, b, and $c > 0$) by the coordinate planes. Check your answer by vector methods.
16. Find the area of that portion of the sphere $x^2 + y^2 + z^2 = 2a^2$ that is cut out by the upper nappe of the cone $x^2 + y^2 = z^2$.
17. Find the area cut from the plane $z = cx$ by the cylinder $x^2 + y^2 = a^2$.
18. Find the area of that portion of the cylinder $x^2 + z^2 = a^2$ that lies between the planes $y = \pm a/2$, $x = \pm a/2$.

19. Find the area cut from the surface $az = y^2 - x^2$ by the cylinder $x^2 + y^2 = a^2$.

20. Find the area of that portion of the sphere $x^2 + y^2 + z^2 = 4a^2$ that lies inside the cylinder $x^2 + y^2 = 2ax$. (Figure 16.40 shows the top half.)

21. Find the area of the portion of the sphere $x^2 + y^2 + z^2 = a^2$ that lies in the first octant (a) by integrating over a region in the xy-plane, (b) by integrating over a region in the yz-plane.

22. Find the area of that portion of the cylinder in Problem 20 that lies inside the sphere. (*Hint:* Project the area into the xz-plane. Or use single integration, $\int h\,ds$, where h is the altitude of the cylinder and ds is the element of arc length in the xy-plane.)

23. Derive formulas (6b) and (6c) in the text.

REVIEW QUESTIONS AND EXERCISES

1. Define the double integral of a function of two variables. What geometric interpretation may be given to the integral?

2. List four applications of multiple integration.

3. Define *moment of inertia* and *radius of gyration*.

4. How does a double integral in polar coordinates differ from a double integral in cartesian coordinates? In what way are they alike?

5. What are the fundamental volume elements for triple integrals (a) in cartesian coordinates, (b) in cylindrical coordinates, (c) in spherical coordinates?

6. Describe the general procedures for determining limits of integration in iterated double and triple integrals.

7. How is surface area defined? Which formula in Article 16.9 is the most general one for computing surface area, in the sense that it includes the others as special cases?

MISCELLANEOUS PROBLEMS

1. Reverse the order of integration and evaluate

$$\int_0^4 \int_{-\sqrt{4-y}}^{(y-4)/2} dx\,dy.$$

2. Sketch the region over which the integral

$$\int_0^1 \int_{\sqrt{y}}^{2-\sqrt{y}} xy\,dx\,dy$$

is to be evaluated and find its value.

3. The integral

$$\int_{-1}^1 \int_{x^2}^1 dy\,dx$$

represents the area of a region of the xy-plane. Sketch the region and express the same area as a double integral with the order of integration reversed.

4. The base of a pile of sand covers the region in the xy-plane that is bounded by the parabola $x^2 + y = 6$ and the line $y = x$. The depth of the sand above the point (x, y) is x^2. Sketch the base of the sand pile and a representative element of volume dV, and find the volume of sand in the pile by double integration.

5. In setting up a double integral for the volume V under the paraboloid $z = x^2 + y^2$ and above a certain region R of the xy-plane, the following sum of iterated integrals was obtained:

$$V = \int_0^1 \left(\int_0^y (x^2 + y^2)\,dx\right) dy + \int_1^2 \left(\int_1^{2-y} (x^2 + y^2)\,dx\right) dy.$$

Sketch the region R in the xy-plane and express V as an iterated integral in which the order of integration is reversed.

6. By change of order of integration, show that the following double integral can be reduced to a single integral

$$\int_0^x \int_0^u e^{m(x-t)} f(t)\,dt\,du = \int_0^x (x - t)e^{m(x-t)} f(t)\,dt.$$

Similarly, it can be shown that

$$\int_0^x \int_0^v \int_0^u e^{m(x-t)} f(t)\,dt\,du\,dv = \int_0^x \frac{(x-t)^2}{2!} e^{m(x-t)} f(t)\,dt.$$

Evaluate integrals for the case $f(t) = \cos at$. (This example illustrates that such reductions usually make calculation easier.)

7. Sometimes a multiple integral with variable limits may be changed into one with constant limits. By changing the order of integration, show that

$$\int_0^1 f(x)\left(\int_0^x \log(x-y) f(y)\, dy\right) dx$$
$$= \int_0^1 f(y)\left(\int_y^1 \log(x-y)\, f(x)\, dx\right) dy$$
$$= \frac{1}{2}\int_0^1 \int_0^1 \log|x-y|\, f(x)\, f(y)\, dx\, dy.$$

8. Evaluate the integral

$$\int_0^\infty \frac{e^{-ax} - e^{-bx}}{x}\, dx.$$

(*Hint:* Use the relation

$$\frac{e^{-ax} - e^{-bx}}{x} = \int_a^b e^{-xy}\, dy$$

to form a double integral, and evaluate it by changing the order of integration.)

9. By double integration, find the centroid of that part of the area of the circle $x^2 + y^2 = a^2$ contained in the first quadrant.

10. Determine the centroid of the plane region that is given in polar coordinates by $0 \leq r \leq a$, $-\alpha \leq \theta \leq \alpha$.

11. Find the centroid of the region bounded by the lines $\theta = 0°$ and $\theta = 45°$, and the circles $r = 1$ and $r = 2$.

12. By double integration, find the centroid of the area between the parabola $x + y^2 - 2y = 0$ and the line $x + 2y = 0$.

13. For a solid body of constant density, having its center of gravity at the origin, show that the moment of inertia about an axis through (x_0, y_0) parallel to the z-axis is equal to the moment of inertia about the z-axis plus $m(x_0^2 + y_0^2)$, where m is the mass of the body.

14. Find the moment of inertia of the angle section shown in Fig. 16.48 (a) with respect to the horizontal base, (b) with respect to a horizontal line through its centroid.

15. Show that, for a uniform elliptic lamina of semiaxes a, b, the moment of inertia about an axis in its plane through the center of the ellipse making an angle α with the axis of length $2a$ is $\frac{1}{4}m(a^2 \sin^2 \alpha + b^2 \cos^2 \alpha)$, where m is the mass of the lamina.

16. A counterweight of a flywheel has the form of the smaller segment cut from a circle of radius a by a chord at a distance b from the center ($b < a$). Find the area of this counterweight and its polar moment of inertia about the center of the circle.

17. Consider an ellipse $(x^2/a^2) + (y^2/b^2) = 1$ revolving about the x-axis to generate an ellipsoid. Find the radius of gyration of the ellipsoid with respect to the x-axis.

18. Find the radii of gyration about $\theta = 0$ and $\theta = \pi/2$ for the area of a loop of the curve $r^2 = a^2 \cos 2\theta$, $(a > 0)$.

19. The hydrostatic pressure at a depth y in a fluid is wy. Taking the x-axis in the surface of the fluid and the y-axis vertically downward, consider a semicircular lamina, radius a, completely immersed with its bounding diameter horizontal, uppermost, and at a depth c. Show that the depth of the center of pressure is

$$\frac{3\pi a^2 + 32ac + 12\pi c^2}{4(4a + 3\pi c)}.$$

The center of pressure is defined as the point where the entire hydrostatic force could be concentrated so as to produce the same first moment of force.

20. Show that

$$\iint \frac{\partial^2 F(x, y)}{\partial x\, \partial y}\, dx\, dy$$

over the rectangle $x_0 \leq x \leq x_1$, $y_0 \leq y \leq y_1$, is

$$F(x_1, y_1) - F(x_0, y_1) - F(x_1, y_0) + F(x_0, y_0).$$

21. Change the following double integral to an equivalent double integral in polar coordinates, and sketch the region of integration.

$$\int_{-a}^{a} \int_0^{\sqrt{a^2 - y^2}} x\, dx\, dy.$$

22. A customary method of evaluating the improper integral $I = \int_0^\infty e^{-x^2}\, dx$ is to calculate its square,

$$I^2 = \left(\int_0^\infty e^{-x^2}\, dx\right)\left(\int_0^\infty e^{-y^2}\, dy\right)$$
$$= \int_0^\infty \int_0^\infty e^{-(x^2+y^2)}\, dx\, dy.$$

Introduce polar coordinates in the last expression and show that

$$I = \int_0^\infty e^{-x^2}\, dx = \frac{\sqrt{\pi}}{2}.$$

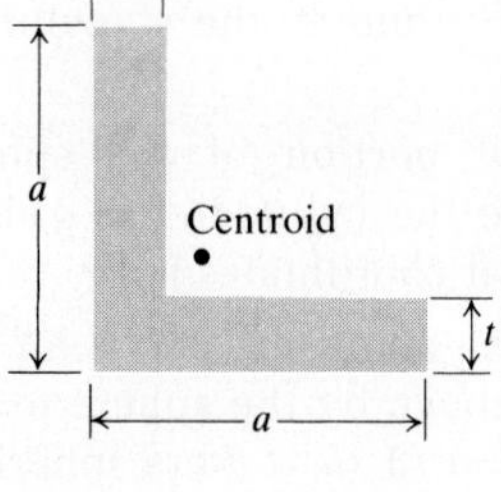

Figure 16.48

23. By transformation of variables $u = x - y$, $v = y$, show that

$$\int_0^\infty e^{-sx}\,dx \int_0^x f(x - y, y)\,dy = \int_0^\infty \int_0^\infty e^{-s(u+v)} f(u, v)\,du\,dv.$$

24. How must a, b, c be chosen in order that $\int_{-\infty}^{\infty}\int_{-\infty}^{\infty} e^{-(ax^2+2bxy+cy^2)}\,dx\,dy = 1$? (*Hint:* Introduce the transformation

$$s = \alpha x + \beta y, \qquad t = \gamma x + \delta y$$

where $(\alpha\delta - \beta\gamma)^2 = ac - b^2$; then

$$ax^2 + 2bxy + cy^2 = s^2 + t^2.)$$

25. Find the area enclosed by the lemniscate $r^2 = 2a^2 \cos 2\theta$. Also find the moment of inertia of this area about the y-axis.

26. Evaluate the integral

$$\iint \frac{dx\,dy}{(1 + x^2 + y^2)^2}$$

taken (a) over one loop of the lemniscate $(x^2 + y^2)^2 - (x^2 - y^2) = 0$, (b) over the triangle with vertices $(0, 0)$, $(2, 0)$, $(1, \sqrt{3})$. (*Hint:* Transform to polar coordinates.)

27. Show, by transforming to polar coordinates, that

$$K(a) = \int_0^{a \sin\beta} \int_{y \cot\beta}^{\sqrt{a^2-y^2}} \ln(x^2 + y^2)\,dx\,dy = a^2\beta\left(\ln a - \frac{1}{2}\right),$$

where $0 < \beta < \pi/2$. Changing the order of integration, what expression do you obtain?

28. Find the volume bounded by the cylinder $y = \cos x$ and the planes

$$z = y, \qquad x = 0, \qquad x = \pi/2, \qquad \text{and} \qquad z = 0.$$

29. Find the center of mass of the homogeneous pyramid whose base is the square enclosed by the lines $x = 1$, $x = -1$, $y = 1$, $y = -1$, in the plane $z = 0$, and whose vertex is at the point $(0, 0, 1)$.

30. Find the volume bounded above by the sphere $x^2 + y^2 + z^2 = 2a^2$ and below by the paraboloid $az = x^2 + y^2$.

31. Find the volume bounded by the surfaces

$$z = x^2 + y^2 \qquad \text{and} \qquad z = \frac{1}{2}(x^2 + y^2 + 1).$$

32. Determine by triple integration the volume enclosed by the two surfaces $x = y^2 + z^2$ and $x = 1 - y^2$.

33. Find the moment of inertia, with respect to the z-axis, of a solid that is bounded below by the paraboloid $3az = x^2 + y^2$ and above by the sphere $x^2 + y^2 + z^2 = 4a^2$, if its density δ is constant.

34. Find by integration the volume of the ellipsoid

$$\frac{x^2}{a^2} + \frac{y^2}{b^2} + \frac{z^2}{c^2} = 1.$$

35. Evaluate the integral $\iiint |xyz|\,dx\,dy\,dz$ taken throughout the ellipsoid $x^2/a^2 + y^2/b^2 + z^2/c^2 \le 1$. (*Hint:* Introduce new coordinates:

$$x = au, \qquad y = bv, \qquad z = cw.)$$

36. The volume of a certain solid is given by the triple integral

$$\int_0^2 \left[\int_0^{\sqrt{2x-x^2}} \left(\int_{-\sqrt{4-x^2-y^2}}^{\sqrt{4-x^2-y^2}} dz\right) dy\right] dx.$$

a) Describe the solid by giving the equations of all the surfaces that form its boundary.
b) Express the volume as a triple integral in cylindrical coordinates. Give the limits of integration explicitly, but do not evaluate the integral.

37. WARNING: Hard problem. Setting up the integral is straightforward, but integrating the result takes hours. (It took MACSYMA 20 minutes.)

A square hole of side $2b$ is cut symmetrically through a sphere of radius a ($a > b\sqrt{2}$). Find the volume removed.

38. A hole is bored through a sphere, the axis of the hole being a diameter of the sphere. The volume of the solid remaining is given by the integral

$$V = 2\int_0^{2\pi} \int_0^{\sqrt{3}} \int_1^{\sqrt{4-z^2}} r\,dr\,dz\,d\theta.$$

a) By inspecting the given integral, determine the radius of the hole and the radius of the sphere.
b) Calculate the numerical value of the integral.

39. Set up an equivalent triple integral in rectangular coordinates. (Arrange the order so that the first integration is with respect to z, the second with respect to y, and the last with respect to x.)

$$\int_0^{\pi/2} \int_1^{\sqrt{3}} \int_1^{\sqrt{4-r^2}} r^3 \sin\theta \cos\theta\, z^2\,dz\,dr\,d\theta.$$

40. Find the volume bounded by the plane $z = 0$, the cylinder $x^2 + y^2 = a^2$, and the cylinder $az = a^2 - x^2$.

41. Find the volume of that portion of the sphere $r^2 + z^2 = a^2$ that is inside the cylinder $r = a \sin\theta$. (Here r, θ, z are cylindrical coordinates.)

42. Find the moment of inertia, about the z-axis, of the volume that is bounded above by the sphere $\rho = a$ and below by the cone $\phi = \pi/3$. (ρ, ϕ, θ are spherical coordinates.)

43. Find the volume enclosed by the surface $\rho = a \sin \phi$, in spherical coordinates.

44. Find the moment of inertia of the solid of constant density δ bounded by two concentric spheres of radii a and b $(a < b)$, about a diameter.

45. Let S be a solid homogeneous sphere of radius a, constant density δ, mass $M = \frac{4}{3}\pi a^3 \delta$. Let P be a particle of mass m situated at distance b $(b > a)$ from the center of S. According to Newton, the force of gravitational attraction of the sphere for P is given by the equation

$$\mathbf{F} = \gamma m \iiint \frac{\mathbf{u}\,\delta\,dV}{r^2},$$

where γ is the gravitational constant, $\mathbf{u}$ is a unit vector in the direction from P toward the volume element dV in S, r^2 is the square of the distance from P to dV, and the integration is extended throughout S. Take the origin at the center of the sphere and P at $(0, 0, b)$ on the z-axis, and show that $\mathbf{F} = -(\gamma M m/b^2)\mathbf{k}$. (*Remark.* This result shows that the force is the same as it would be if all the mass of the sphere were concentrated at its center.)

46. The density at P, a point of a solid sphere of radius a and center O, is given to be

$$\rho_0\{1 + \varepsilon \cos \theta + \frac{1}{2}\varepsilon^2(3 \cos \theta - 1)\},$$

where θ is the angle OP makes with a fixed radius OQ, and ρ_0 and ε are constants. Find the average density of the sphere.

47. Find the area of the surface $y^2 + z^2 = 2x$ cut off by the plane $x = 1$.

48. Find the area cut from the plane $x + y + z = 1$ by the cylinder $x^2 + y^2 = 1$.

49. Find the area above the xy-plane cut from the cone $x^2 + y^2 = z^2$ by the cylinder $x^2 + y^2 = 2ax$.

50. Find the surface area of that portion of the sphere $r^2 + z^2 = a^2$ that is inside the cylinder $r = a \sin \theta$. ((r, θ, z) are cylindrical coordinates.)

51. Obtain a double integral expressing the surface area cut from the cylinder $z = a^2 - y^2$ by the cylinder $x^2 + y^2 = a^2$, and reduce this double integral to a definite single integral with respect to the variable y.

52. A square hole of side $2\sqrt{2}$ is cut symmetrically through a sphere of radius 2. Show that the area of the surface removed is $16\pi(\sqrt{2} - 1)$.

53. A torus surface is generated by moving a sphere of unit radius whose center travels on a closed plane circle of radius 2. Calculate the area of this surface.

54. Calculate the area of the surface $(x^2 + y^2 + z^2)^2 = x^2 - y^2$. (*Hint:* Use polar coordinates.)

55. Calculate the area of the spherical part of the boundary of the region

$$x^2 + y^2 + z^2 = r^2,$$
$$x^2 + y^2 - rx \geq 0,$$
$$x^2 + y^2 + rx \geq 0.$$

(*Hint:* Integrate first with respect to x and y.)

56. Prove that the potential of a circular disc of mass m per unit area, and of radius a, at a point distant h from the center and on the normal to the disc through the center, is $2\pi m(\sqrt{(a^2 + h^2)} - h)$. (The potential at a point P due to a mass Δm at Q is $\Delta m/r$, where r is the distance from P to Q.)

57. Find the attraction, at the vertex, of a solid right circular cone of mass M, height h, and radius of base a. (The attraction at P due to a mass Δm at Q is $(\Delta m/r^2)\mathbf{u}$, where r is the distance from P to Q, and $\mathbf{u}$ is a unit vector in the direction of $\overrightarrow{PQ}$.)

17 Vector Analysis

17.1 Vector Fields

Suppose that a certain region G of 3-space is occupied by a moving fluid: air, for example, or water. We may imagine that the fluid is made up of an infinite number of particles, and that at time t, the particle that is in position P at that instant has a velocity $\mathbf{v}$. If we stay at P and observe new particles that pass through it, we shall probably see that they have different velocities. This would surely be true, for example, in turbulent motions caused by high winds or stormy seas. Again, if we could take a picture of the velocities of particles at different places at the same instant, we would expect to find that these velocities vary from place to place. Thus the velocity at position P at time t is, in general, a function of both position and time:

$$\mathbf{v} = \mathbf{F}(x, y, z, t). \tag{1}$$

Equation (1) indicates that the velocity $\mathbf{v}$ is a vector function $\mathbf{F}$ of the four independent variables x, y, z, and t. Such functions have many applications, particularly in treatments of flows of material. In hydrodynamics, for example, if $\delta = \delta(x, y, z, t)$ is the *density* of the fluid at (x, y, z) at time t, and we take $\mathbf{F} = \mathbf{i}u + \mathbf{j}v + \mathbf{k}w$ to be the velocity expressed in terms of components, then we are able to derive the Euler partial differential equation of continuity of motion:

$$\frac{\partial \delta}{\partial t} + \frac{\partial(\delta u)}{\partial x} + \frac{\partial(\delta v)}{\partial y} + \frac{\partial(\delta w)}{\partial z} = 0$$

(Article 17.6). Such functions are also applied in physics and electrical engineering; for example, in the study of propagation of electromagnetic waves. Also, much current research activity in applied mathematics has to do with such functions.

Steady-State Flows

In this chapter, we shall deal mainly with flows for which the velocity function, Eq. (1), does not depend on the time t. Such flows are called steady-state flows. They exemplify *vector fields*.

DEFINITION

> If, to each point P in some region G, a vector $\mathbf{F}(P)$ is assigned, the collection of all such vectors is called a *vector field* on G.

In addition to the vector fields that are associated with fluid flows, there are vector *force* fields that are associated with gravitational attraction, magnetic force fields, electric fields, and purely mathematical fields.

EXAMPLE 1 Imagine an ideal fluid flowing with steady-state flow in a long cylindrical pipe of radius a, so that particles at distance r from the central axis are moving parallel to the axis with speed $|\mathbf{v}| = a^2 - r^2$ (Fig. 17.1). Describe this field by a formula for $\mathbf{v}$.

Solution Let the z-axis lie along the axis of the pipe, with positive direction in the direction of the flow. Then, in the usual way, introduce a right-handed cartesian coordinate system with unit vectors along the axes. Since all particles move parallel to the z-axis, the $\mathbf{k}$-component of the flow is the only one different from zero. Therefore

$$\mathbf{v} = (a^2 - r^2)\mathbf{k} = (a^2 - x^2 - y^2)\mathbf{k}$$

for points inside the pipe. This vector field is not defined outside the cylinder $x^2 + y^2 = a^2$. If we were to draw the velocity vectors at all points in the disc

$$x^2 + y^2 \leq a^2, \qquad z = 0,$$

their tips would describe the surface

$$z = a^2 - r^2$$

(cylindrical coordinates) for $z \geq 0$. Since this field does not depend on z, a similar figure would illustrate the flow field across any cross section of the pipe made by a plane perpendicular to its axis. □

EXAMPLE 2 A fluid is rotating about the z-axis with constant angular velocity ω. Every particle at a distance r from the z-axis and in a plane

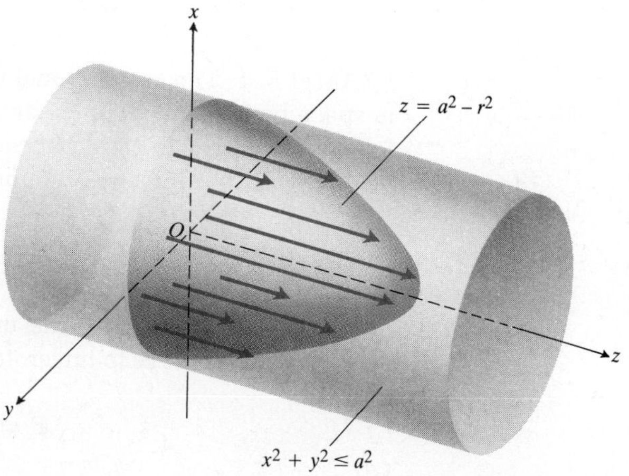

17.1 The flow of fluid in a long cylindrical pipe. The vectors $\mathbf{v} = (a^2 - r^2)\mathbf{k}$ inside the cylinder that have their bases in the xy-plane have their tips on the paraboloid $z = a^2 - r^2$.

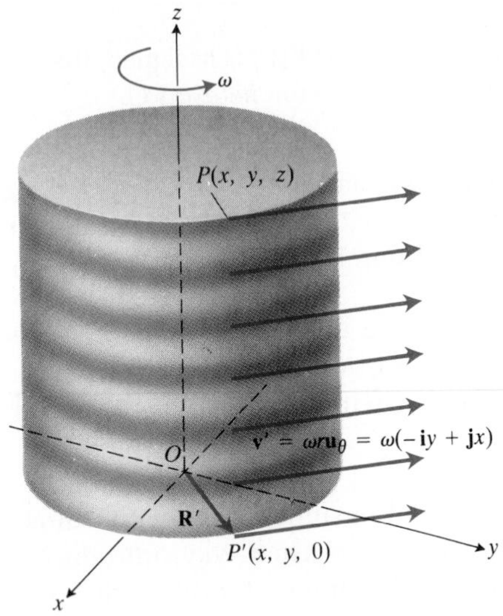

17.2 A steady rotational flow parallel to the xy-plane, with constant angular velocity ω in the positive (counterclockwise) direction.

perpendicular to the z-axis traces a circle of radius r, and each such particle has constant speed $|\mathbf{v}| = \omega r$. Describe this field by writing an equation for the velocity at $P(x, y, z)$.

Solution (See Fig. 17.2.) Each particle travels in a circle parallel to the xy-plane. Therefore it is convenient to begin by looking at the projection of such a circle onto this plane. The point $P(x, y, z)$ in space projects onto the image point $P'(x, y, 0)$, and the velocity vector $\mathbf{v}$ of a particle at P projects onto the velocity vector $\mathbf{v}'$ of a particle at P'. We assume that the motion is in the positive, or counterclockwise, direction, as indicated in the figure. The position vector of P' is $\mathbf{R}' = \mathbf{i}x + \mathbf{j}y$, and the vectors $-\mathbf{i}y + \mathbf{j}x$ and $\mathbf{i}y - \mathbf{j}x$ are both perpendicular to $\mathbf{R}'$. All three of these vectors have magnitude

$$\sqrt{x^2 + y^2} = r.$$

The velocity vector we want has magnitude ωr, is perpendicular to $\mathbf{R}'$, and points in the direction of motion. When x and y are both positive (that is, when the particle is in the first quadrant), the velocity has a negative $\mathbf{i}$-component and a positive $\mathbf{j}$-component. The vector that has these properties is

$$\mathbf{v}' = \omega r \mathbf{u}_\theta = \omega(-\mathbf{i}y + \mathbf{j}x). \tag{2a}$$

This formula can be verified for P' in the other three quadrants as well. For example, in the third quadrant both x and y are negative, so Eq. (2a) gives a vector with a positive $\mathbf{i}$-component and a negative $\mathbf{j}$-component, which is correct. Also, because the motion of P is in a circle parallel to the circle traced by P', and has the same velocity, we have

$$\mathbf{v} = \omega(-\mathbf{i}y + \mathbf{j}x) \tag{2b}$$

for any point in the fluid. □

EXAMPLE 3 At every point in space, a fluid has a velocity vector that is the sum of a constant velocity vector parallel to the z-axis and a rotational velocity vector given by Eq. (2b). Describe the field.

Solution Let the constant component parallel to the z-axis be $c\mathbf{k}$. Then the resultant field is

$$\mathbf{v} = \omega(-\mathbf{i}y + \mathbf{j}x) + c\mathbf{k}. \ \square \tag{3}$$

EXAMPLE 4 The gravitational force field induced at the point $P(x, y, z)$ in space by a mass M that is taken to lie at an origin is defined to be the force with which M would attract a particle of *unit* mass at P. Describe this field mathematically, assuming the inverse-square law

$$\mathbf{F} = -\frac{GMm}{r^2}\mathbf{r}.$$

Solution Because M and the unit mass at P are assumed to be *point* masses, we don't have to integrate anything; we just write down the force:

$$\mathbf{F} = -\frac{GM(1)}{|\overrightarrow{OP}|^2}\mathbf{r} \tag{4a}$$

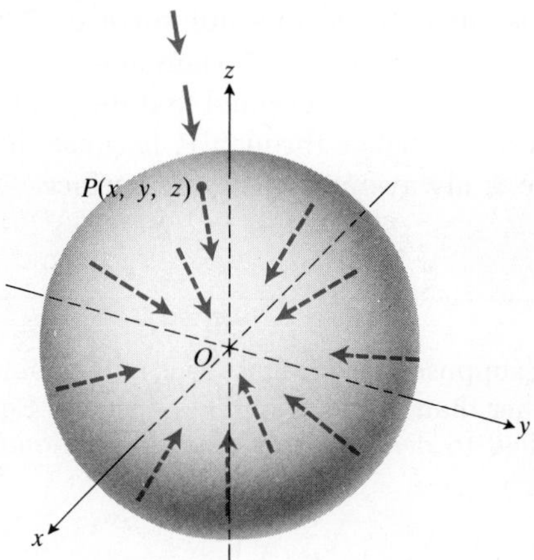

17.3 Some of the vectors of the gravitational field of Example 4.

where G is the gravitational constant, and

$$\mathbf{r} = \frac{\overrightarrow{OP}}{|\overrightarrow{OP}|}.$$

The position vector of P is $\overrightarrow{OP} = \mathbf{i}x + \mathbf{j}y + \mathbf{k}z$, so we find that

$$\mathbf{F} = \frac{-GM(\mathbf{i}x + \mathbf{j}y + \mathbf{k}z)}{(x^2 + y^2 + z^2)^{3/2}} \tag{4b}$$

gives the gravitational force field in question. Its graph would consist of infinitely many vectors, one starting from each point P (except the origin), and pointing straight toward the origin. If P is near the origin, the associated vector is longer than for points farther away from O. For points P on a ray through the origin, the $\mathbf{F}$-vectors would lie along that same ray, and decrease in length in proportion to the square of the distance from O. Figure 17.3 is a partial representation of this field. As you look at the figure, however, you should also imagine that an $\mathbf{F}$-vector is attached to *every* point $P \neq O$, and not just to those shown. At points on the surface of the sphere $|\overrightarrow{OP}| = a$, the vectors all have the same length, and all point toward the center of the sphere. □

Vector fields arise naturally in mathematics when we apply the gradient operator ∇ to a scalar function.

EXAMPLE 5 Suppose that the temperature T at each point $P(x, y, z)$ in some region of space is

$$T = 100 - x^2 - y^2 - z^2, \tag{5}$$

and that

$$\mathbf{F} = \nabla T.$$

Describe this vector field and discuss some of its properties.

Solution We have

$$\begin{aligned}\mathbf{F} = \nabla T = \text{grad } T &= \mathbf{i}\frac{\partial T}{\partial x} + \mathbf{j}\frac{\partial T}{\partial y} + \mathbf{k}\frac{\partial T}{\partial z}\\ &= -2x\mathbf{i} - 2y\mathbf{j} - 2z\mathbf{k}\\ &= -2\mathbf{R},\end{aligned}$$

where

$$\mathbf{R} = \overrightarrow{OP} = \mathbf{i}x + \mathbf{j}y + \mathbf{k}z$$

is the position vector of $P(x, y, z)$. This field is like a central force field, all vectors $\mathbf{F}$ being directed toward the origin. At points on a sphere with $|\overrightarrow{OP}|$ equal to a constant, the magnitude of the field vectors is a constant equal to twice the radius of the sphere. So, to represent the field, we could construct any sphere with center at O and draw a vector from any point P on the surface straight through O to the other side of the sphere. The collection of all such vectors, for points in the domain of the function T of Eq. (5), constitutes the *gradient field* of this particular scalar function.

Any surface on which T is constant is an *isothermal* surface. The isothermal surfaces here are spheres with center at the origin and radius $\sqrt{100 - T}$. Our calculation of $\mathbf{F} = \nabla T = -2\mathbf{R}$ has verified that the gradient of T at P is normal to the isothermal surface through P, because the diameter of such a spherical surface is always normal to the surface. □

PROBLEMS

1. In Example 1, where is a particle's speed (a) the greatest? (b) the least?
2. A vector field $\mathbf{F}$ in space is the field of a force directed toward the origin whose magnitude at each point $P(x, y, z)$ is a constant k times the inverse of the fourth power of the distance from P to the origin. Express $\mathbf{F}$ in terms of $\mathbf{i}$, $\mathbf{j}$, and $\mathbf{k}$ and the coordinates of P.
3. In Example 2, the position vector of P is

$$\mathbf{R} = \mathbf{R}' + z\mathbf{k}.$$

Show that for the motion described, it is correct to say that

$$\frac{d\mathbf{R}}{dt} = \mathbf{v} = \frac{d\mathbf{R}'}{dt} = \mathbf{v}'.$$

4. Describe, in words, the motion of the fluid discussed in Example 3. What path in space is described by a particle of the fluid that goes through the point $A(a, 0, 0)$ at time $t = 0$? Prove your result by taking $x = \cos \omega t$, $y = \sin \omega t$ and integrating the vector equation

$$\frac{d\mathbf{R}}{dt} = \omega(-\mathbf{i}y + \mathbf{j}x) + c\mathbf{k}.$$

with respect to t from $t = 0$ to $t = \tau$.

5. In Example 4, suppose that the mass M is at the point (x_0, y_0, z_0) rather than at the origin. How should Eq. (4b) be modified to describe this new gravitational force field?

In Problems 6–10, find the gradient fields $\mathbf{F}(x, y, z) = \nabla f$ for the given functions f.

6. $f(x, y, z) = x^2 \exp(2y + 3z)$
7. $f(x, y, z) = \ln(x^2 + y^2 + z^2)$
8. $f(x, y, z) = \tan^{-1}(xy/z)$
9. $f(x, y, z) = 2x - 3y + 5z$
10. $f(x, y, z) = (x^2 + y^2 + z^2)^{n/2}$
11. Suppose the density of the fluid in Example 1 is δ = constant at $P(x, y, z)$. Explain why the double integral

$$\int_0^a \int_0^{2\pi} \delta(a^2 - r^2)r \, d\theta \, dr$$

represents the *mass transport* (amount of mass per unit of time) flowing across the disc

$$x^2 + y^2 \le a^2, \qquad z = 0.$$

Evaluate the integral.

17.2 Surface Integrals

In this article we show how to integrate a continuous function over a piecewise smooth surface in space. This generalizes the notion of an integral over a region in a plane, which is a flat surface. As we shall see, these "surface integrals" are closely related to the concept of surface area. Surface integrals have important applications in engineering and physics as well as in mathematics.

Suppose, for example, that we have an electrical charge distributed over a surface $F(x, y, z) = c$ like the one shown in Fig. 17.4, and that the function $g(x, y, z)$ gives the charge density (charge per unit area) at each point on S. Then we may calculate the total charge on S as an integral in the following way.

We subdivide the shadow region R on the ground plane beneath the surface into small rectangles ΔA_k of the kind we would use if we were

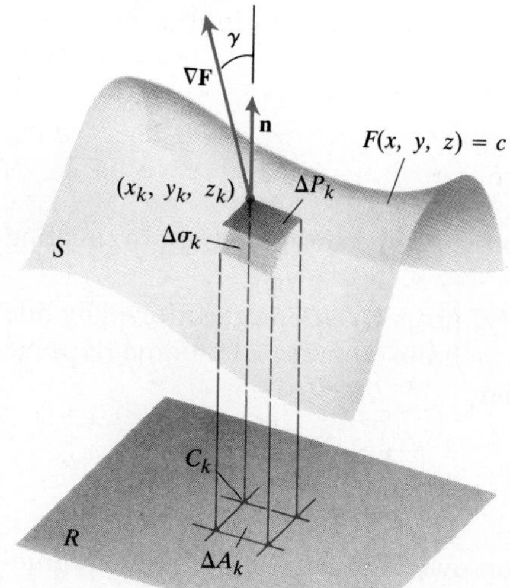

17.4 The integral of a function $g(x, y, z)$ over S is the limit of the sums

$$\sum g(x_k, y_k, z_k)\,\Delta P_k = \sum g(x_k, y_k, z_k)\frac{\Delta A_k}{|\cos\gamma|}.$$

defining the surface area of S. Then directly above each ΔA_k there lies a patch of surface $\Delta\sigma_k$ that we may approximate with a parallelogram-shaped portion of tangent plane, ΔP_k. To be specific we may suppose that ΔP_k is tangent to the surface at the point (x_k, y_k, z_k) directly above the back corner C_k of ΔA_k.

Up to this point the construction proceeds as in the definition of surface area in Article 16.9, but now we take one additional step: We evaluate g at (x_k, y_k, z_k) and approximate the total charge on the surface patch $\Delta\sigma_k$ by the product

$$g(x_k, y_k, z_k)\cdot\Delta P_k.$$

The rationale is that when the subdivision of R is sufficiently fine the value of g throughout $\Delta\sigma_k$ is nearly constant and ΔP_k is nearly the same as $\Delta\sigma_k$. The total charge over S is then approximated by the sum

$$\begin{aligned}\text{Total charge} &\approx \sum g(x_k, y_k, z_k)\,\Delta P_k\\ &\approx \sum g(x_k, y_k, z_k)\frac{\Delta A_k}{|\cos\gamma|},\end{aligned} \tag{1}$$

where γ is the angle between the surface normal ∇F and a unit normal $\mathbf{n}$ to R, as in Article 16.9.

If F, the function defining the surface S, and its first partial derivatives are continuous, and if g is continuous over S, then the sums on the right-hand side of Eq. (1) will approach the limit

$$\iint_R g(x, y, z)\frac{dA}{|\cos\gamma|} = \iint_R g(x, y, z)\frac{|\nabla F|}{|\nabla F\cdot\mathbf{n}|}\,dA \tag{2}$$

as the rectangular subdivision of R is refined in the usual way. This limit is called the *integral of g over the surface S* and is often written as

$$\iint_R g(x, y, z)\frac{|\nabla F|}{|\nabla F\cdot\mathbf{n}|}\,dA = \iint_S g\,d\sigma, \tag{3}$$

where the *surface area differential* $d\sigma$ is short for

$$d\sigma = \frac{|\nabla F|}{|\nabla F\cdot\mathbf{n}|}\,dA.$$

As we might expect, the formula in Eq. (3) is used to define the integral of *any* function g over the surface S (as long as the integral exists) and the integral's value takes on different meanings in different applications. For instance, if $g(x, y, z) \equiv 1$ is the constant function whose value is one, then the integral of g over S in Eq. (3) is the surface area of S.

Surface integrals have the same algebraic properties as other kinds of integrals. Writing g for $g(x, y, z)$ and h for $h(x, y, z)$ we have

I1. $\displaystyle\iint_S kg\,d\sigma = k\iint_S g\,d\sigma$ (any number k)

I2. $\displaystyle\iint_S (g + h)\,d\sigma = \iint_S g\,d\sigma + \iint_S h\,d\sigma$

I3. $\displaystyle\iint_S (g - h)\,d\sigma = \iint_S g\,d\sigma - \iint_S h\,d\sigma$

I4. $\iint_S h \, d\sigma \geq 0$ if $h \geq 0$ on S

I5. $\iint_S h \, d\sigma \geq \iint_S g \, d\sigma$ if $h \geq g$ on S.

As usual, the integrals have these properties because their approximating sums have these properties.

There is also a "surface additivity" property. If S is partitioned by line segments and smooth curves into a finite number of nonoverlapping smooth patches $S_1, S_2, \ldots, S_n$, then

I6. $\iint_S h \, d\sigma = \iint_{S_1} h \, d\sigma + \iint_{S_2} h \, d\sigma + \cdots + \iint_{S_n} h \, d\sigma.$

Thus we can integrate a function over the surface of a cube by integrating it over each face and adding the results. We can integrate over any "turtle shell" of welded plates one plate at a time and add the results when we are through.

Special Cases

If a smooth surface S is given by an equation in the form $z = f(x, y)$ defined over a region R_{xy} in the xy-plane, then the formula in Eq. (3) specializes to

$$\iint_S g \, d\sigma = \iint_{R_{xy}} g(x, y, z) \sqrt{f_x^2 + f_y^2 + 1} \, dx \, dy, \tag{4}$$

as we may deduce from the derivation of Eq. (6a) in Article 16.9. Similarly, the surface integral of g over a smooth surface $x = f(y, z)$ that is defined over a region R_{yz} in the yz-plane is

$$\iint_S g \, d\sigma = \iint_{R_{yz}} g(x, y, z) \sqrt{f_y^2 + f_z^2 + 1} \, dy \, dz, \tag{5}$$

and the surface integral of g over a smooth surface $y = f(x, z)$ that is defined over a region R_{xz} in the xz-plane is

$$\iint_S g \, d\sigma = \iint_{R_{xz}} g(x, y, z) \sqrt{f_x^2 + f_z^2 + 1} \, dx \, dz. \tag{6}$$

EXAMPLE 1 Evaluate $\iint z \, d\sigma$ over the hemisphere

$$F(x, y, z) = x^2 + y^2 + z^2 = a^2, \qquad z \geq 0.$$

(This integral gives the first moment of the hemisphere with respect to the xy-plane.)

Solution With $\mathbf{n} = \mathbf{k}$ and base region the disc R: $x^2 + y^2 \leq a^2$, we have

$$|\nabla F| = |2x\mathbf{i} + 2y\mathbf{j} + 2z\mathbf{k}| = 2\sqrt{x^2 + y^2 + z^2} = 2a,$$

$$|\nabla F \cdot \mathbf{n}| = |2z| = 2z \qquad \text{(because } z \geq 0\text{)},$$

$$d\sigma = \frac{|\nabla F|}{|\nabla F \cdot \mathbf{n}|} \, dA = \frac{a}{z} \, dA,$$

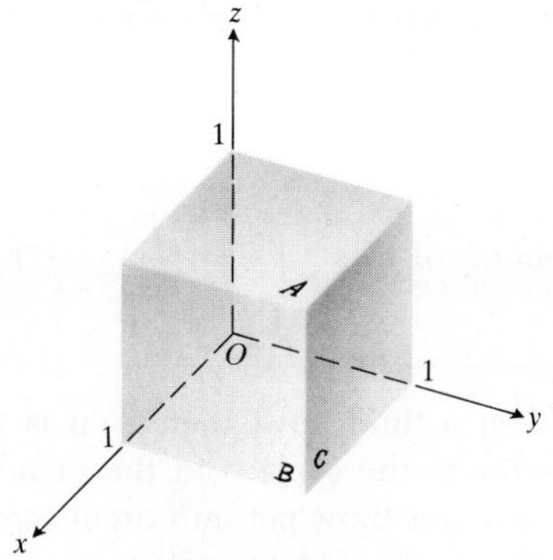

17.5 To integrate a function over the surface of a cube we integrate over each face and add the results.

and

$$\iint_S g\,d\sigma = \iint_{x^2+y^2 \le a^2} z\frac{a}{z}\,dA = a \cdot \pi a^2 = \pi a^3. \quad \square$$

EXAMPLE 2 Integrate $g(x, y, z) = xyz$ over the cube in the first octant bounded by the coordinate planes and the planes $x = 1$, $y = 1$, $z = 1$.

Solution The cube is shown in Fig. 17.5. We integrate $g = xyz$ over each face and add the results. Since $g = 0$ on the three faces that lie in the coordinate planes, the integral of g over each of these faces is zero and the integral of g over the cube reduces to

$$\iint_{\text{cube}} xyz\,d\sigma = \iint_{\text{face }A} xyz\,d\sigma + \iint_{\text{face }B} xyz\,d\sigma = \iint_{\text{face }C} xyz\,d\sigma.$$

As a surface, face A is given by the equation

$$z = f(x, y) = 1$$

over the square R_{xy}: $0 \le x \le 1$, $0 \le y \le 1$. Therefore

$$d\sigma = \sqrt{f_x^2 + f_y^2 + 1}\,dx\,dy = \sqrt{0 + 0 + 1}\,dx\,dy = dx\,dy.$$

Also,

$$g(x, y, z) = xyz = xy \cdot 1 = xy$$

on face A, so that

$$\iint_{\text{face }A} g\,d\sigma = \iint_{R_{xy}} xy\,dx\,dy = \int_0^1\int_0^1 xy\,dx\,dy = \int_0^1 \frac{y}{2}\,dy = \frac{1}{4}.$$

Similar calculations show

$$\iint_{\text{face }B} g\,d\sigma = \int_0^1\int_0^1 yz\,dy\,dz = \frac{1}{4}, \qquad \iint_{\text{face }C} g\,d\sigma = \int_0^1\int_0^1 xz\,dx\,dz = \frac{1}{4}.$$

Hence,

$$\iint_{\text{cube}} g\,d\sigma = \frac{1}{4} + \frac{1}{4} + \frac{1}{4} = \frac{3}{4}. \quad \square$$

The Surface Integral for Flux

If we can choose a unit normal vector **n** on a surface S in such a way that **n** varies continuously as its initial point moves about the surface, we call the surface *orientable*. Spheres and other closed surfaces in space (surfaces that enclose a solid) are orientable, and by convention we choose **n** on a closed surface to point outward. In any case, once **n** is chosen we call the surface an *oriented surface*. The direction of **n** at any point is called the *positive* direction. Any patch or subportion of an orientable surface is also orientable. The Möbius band shown in Fig. 17.27 in Article 17.7 is a nonorientable surface.

If **F** is a continuous vector field and S is an oriented surface we call the integral over S of $\mathbf{F}\cdot\mathbf{n}$, the component of **F** in the direction of **n**, the flux of **F** across S in the positive direction:

DEFINITION

The *flux* of **F** across S in the direction of **n** is $\iint_S \mathbf{F}\cdot\mathbf{n}\,d\sigma$. (7)

For instance, if **F** is the velocity field in a fluid flow, then $\mathbf{F}\cdot\mathbf{n}$ is the component of the velocity perpendicular to the surface in the positive direction and the flux of **F** is the total flow rate (flow per unit time) across S in the positive direction. We shall discuss this interpretation in detail later in the chapter.

If S is part of a level surface $G(x, y, z) = c$, then **n** may be taken to be one of the two vectors

$$\mathbf{n} = +\frac{\nabla G}{|\nabla G|} \quad \text{or} \quad \mathbf{n} = -\frac{\nabla G}{|\nabla G|},$$

depending on which vector gives the preferred direction.

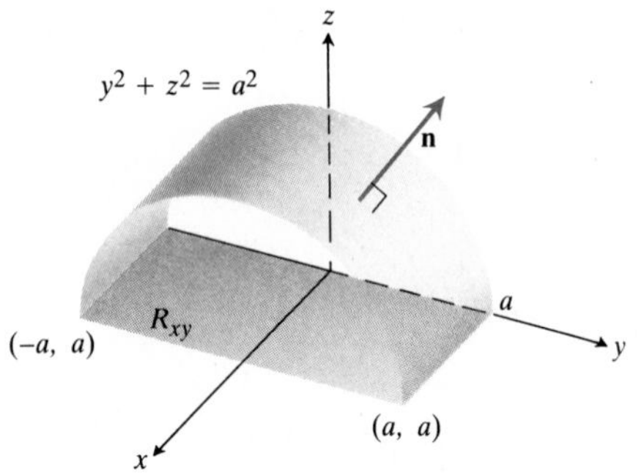

17.6 Example 3 calculates the flux of a vector field outward through this surface. The area of the shadow region R_{xy} is $2a^2$.

EXAMPLE 3 Find the flux of $\mathbf{F} = yz\mathbf{j} + z^2\mathbf{k}$ outward through the surface S cut from the cylinder $y^2 + z^2 = a^2$, $z \geq 0$ by the planes $x = 0$ and $x = a$ (Fig. 17.6).

Solution The outward unit normal to S may be calculated from the gradient of $G(x, y, z) = y^2 + z^2$ to be

$$\mathbf{n} = +\frac{\nabla G}{|\nabla G|} = \frac{2y\mathbf{j} + 2z\mathbf{k}}{\sqrt{4y^2 + 4z^2}} = \frac{2y\mathbf{j} + 2z\mathbf{k}}{2a} = \frac{y}{a}\mathbf{j} + \frac{z}{a}\mathbf{k}.$$

We also have

$$d\sigma = \frac{|\nabla G|}{|\nabla G\cdot\mathbf{k}|}\,dA = \frac{2a}{|2z|}\,dA = \frac{a}{z}\,dA,$$

where we drop the absolute value signs because $z \geq 0$ on S.

On the surface S, the value $\mathbf{F}\cdot\mathbf{n}$ is given by the formula

$$\begin{aligned}\mathbf{F}\cdot\mathbf{n} &= (yz\mathbf{j} + z^2\mathbf{k})\cdot\left(\frac{y}{a}\mathbf{j} + \frac{z}{a}\mathbf{k}\right)\\ &= \frac{y^2z + z^3}{a} = \frac{z}{a}(y^2 + z^2)\\ &= \frac{z}{a}(a^2) \qquad (y^2 + z^2 = a^2 \text{ on } S)\\ &= az.\end{aligned}$$

Therefore, the flux of **F** outward through S is

$$\iint_S \mathbf{F}\cdot\mathbf{n}\,d\sigma = \iint_S (az)\frac{a}{z}\,dA = \iint_{R_{xy}} a^2\,dx\,dy = a^2\cdot\text{area}\,(R_{xy}) = 2a^4. \;\square$$

Other Applications

The mass of a thin shell S of material whose density is given by $\delta(x, y, z)$ is

$$m = \text{mass}(S) = \iint_S \delta(x, y, z)\, d\sigma.$$

The first moments of S about the coordinate planes are

$$M_{yz} = \iint_S x\delta\, d\sigma, \qquad M_{xz} = \iint_S y\delta\, d\sigma, \qquad M_{xy} = \iint_S z\delta\, d\sigma.$$

The coordinates of the shell's center of mass are

$$\bar{x} = \frac{\iint_S x\delta\, d\sigma}{m}, \qquad \bar{y} = \frac{\iint_S y\delta\, d\sigma}{m}, \qquad \bar{z} = \frac{\iint_S z\delta\, d\sigma}{m}.$$

The moments of inertia of the shell about the coordinate axes are

$$I_x = \frac{\iint_S (y^2 + z^2)\delta\, d\sigma}{m},$$

$$I_y = \frac{\iint_S (x^2 + z^2)\delta\, d\sigma}{m},$$

$$I_z = \frac{\iint_S (x^2 + y^2)\delta\, d\sigma}{m}.$$

Radii of gyration are defined in the usual way.

The *average value* of a function $g(x, y, z)$ over a surface S is

$$\text{Average value} = \frac{1}{\text{Area}(S)} \iint_S g\, d\sigma = \frac{\iint_S g\, d\sigma}{\iint_S d\sigma}.$$

PROBLEMS

1. Integrate $g(x, y, z) = x + y + z$ over the surface of the unit cube shown in Fig. 17.5.
2. Integrate $h(x, y, z) = y + z$ over the surface of the wedge shown in Fig. 16.30.
3. Integrate $g(x, y, z) = xyz$ over the surface of the rectangular solid shown in Fig. 16.37.
4. Integrate $g(x, y, z) = xyz$ over the surface of the rectangular solid shown in Fig. 16.34.
5. Evaluate $\iint z\, d\sigma$ over the hemisphere
$$x^2 + y^2 + z^2 = a^2, \quad z \leq 0.$$
6. Evaluate $\iint z\, d\sigma$ over the sphere $x^2 + y^2 + z^2 = a^2$.

In Problems 7 and 8, let $h(x, y, z) = x + y + z$ and let S be the portion of the plane $z = 2x + 3y$ for which $x \geq 0$, $y \geq 0$, $x + y \leq 2$.

7. Evaluate $\iint_S h\, d\sigma$ by projecting S into the xy-plane. Sketch the projection.
8. Evaluate $\iint_S h\, d\sigma$ by projecting S into the yz-plane. Sketch the projection.
9. Find the center of mass of a thin hemispherical shell of radius a and uniform density δ. (Place the shell with its base on the xy-plane.)
10. Find the average height of the hemisphere $x^2 + y^2 + z^2 = a^2$, $z \geq 0$ above the xy-plane.
11. Find the moment of inertia about the z-axis of the surface cut from the cone $z^2 = x^2 + y^2$ by the cylinder $x^2 + y^2 = 2x$.
12. Find the moment of inertia about the z-axis of the hemisphere $x^2 + y^2 + z^2 = a^2$, $z \geq 0$.

In Problems 13–18, find the flux of the given vector field **F** through the portion of the sphere $x^2 + y^2 + z^2 = a^2$ that lies in the first octant in the outward direction away from the origin.

13. $\mathbf{F} = \mathbf{n}$
14. $\mathbf{F} = -\mathbf{i}y + \mathbf{j}x$
15. $\mathbf{F} = z\mathbf{k}$
16. $\mathbf{F} = \mathbf{i}x + \mathbf{j}y$
17. $\mathbf{F} = y\mathbf{i} - x\mathbf{j} + \mathbf{k}$
18. $\mathbf{F} = zx\mathbf{i} + zy\mathbf{j} + z^2\mathbf{k}$.
19. Let S be the portion of the cylinder $y = e^x$ in the first octant whose projection parallel to the x-axis onto

the yz-plane is the rectangle R_{yz}: $1 \le y \le e$, $0 \le z \le 1$. Let **n** be the unit vector normal to S that points away from the yz-plane. Find the flux of $\mathbf{F} = -2\mathbf{i} + 2y\mathbf{j} + z\mathbf{k}$ through S in the direction of **n**.

20. Let S be the portion of the cylinder $y = \ln x$ in the first octant whose projection parallel to the y-axis onto the xz-plane is the rectangle R_{xz}: $1 \le x \le e$, $0 \le z \le 1$. Let **n** be the unit vector normal to S that points away from the xz-plane. Find the flux of $\mathbf{F} = 2y\mathbf{j} + z\mathbf{k}$ through S in the direction of **n**.

21. Find the moment of inertia about the z-axis of the surface of the cube shown in Fig. 17.5. ($\delta = 1$ for a geometric figure.)

22. The sphere $x^2 + y^2 + z^2 = 25$ is cut by the plane $z = 3$, the smaller portion cut off forming a solid that is bounded by a closed surface S made up of a spherical cap S_1 and a flat disc S_2. Find

$$\iint_S \mathbf{F}\cdot\mathbf{n}\,d\sigma = \iint_{S_1} \mathbf{F}\cdot\mathbf{n}\,d\sigma + \iint_{S_2} \mathbf{F}\cdot\mathbf{n}\,d\sigma$$

if $\mathbf{F} = xz\mathbf{i} + yz\mathbf{j} + \mathbf{k}$ and in each integral on the right **n** is taken to be the outward-pointing normal to the surface.

23. (*Spherical coordinates.*) Suppose that the surface of the hemisphere in Example 1 is subdivided by arcs of great circles on which the spherical coordinate θ remains constant (meridians of longitude), and by circles parallel to the xy-plane on which ϕ remains constant (parallels of latitude). Let the angular spacings be $\Delta\theta$ and $\Delta\phi$, respectively. Express the integral $\iint_S z\,d\sigma$ of Example 1 in the form

$$\lim_{\substack{\Delta\theta\to 0\\ \Delta\phi\to 0}} \sum F(\theta, \phi)\,\Delta\theta\,\Delta\phi = \iint F(\theta, \phi)\,d\theta\,d\phi,$$

with appropriate limits of integration, and evaluate. (*Hint:* You should get

$$d\sigma = (r\,d\theta)\cdot(\rho\,d\phi) = \rho^2 \sin\phi\,d\theta\,d\phi = a^2 \sin\phi\,d\theta\,d\phi,$$

where ρ, ϕ, θ are spherical coordinates.)

24. Use the result $d\sigma = a^2 \sin\phi\,d\theta\,d\phi$ from Problem 23 to find the average great-circle distance of a point on the hemisphere $x^2 + y^2 + z^2 = a^2$, $z \ge 0$ from the north pole $N(0, 0, a)$.

***Toolkit* Programs**

Double Integral

17.3

Line Integrals and Work

In mechanics, the work done by a constant force **F** when the point of application undergoes a displacement $\Delta\mathbf{R}$ is defined to be $\mathbf{F}\cdot\Delta\mathbf{R}$. When the force **F** varies with position, however, and the point of application moves along a curve

$$\mathbf{R}(t) = x(t)\mathbf{i} + y(t)\mathbf{j} + z(t)\mathbf{k}$$

in space from point A to point B, the work done along the curve is defined as the integral

$$\int_A^B \mathbf{F}\cdot d\mathbf{R} = \int_A^B \mathbf{F}\cdot\frac{d\mathbf{R}}{ds}\,ds. \tag{1}$$

This integral is one of the so-called line integrals that are the subject of this article. After defining these integrals and calculating some of them, we shall return to the notion of work and show that in some fields the work done by the force along a curve depends only on the points at which the curve begins and ends, and not on the particular route taken between the points.

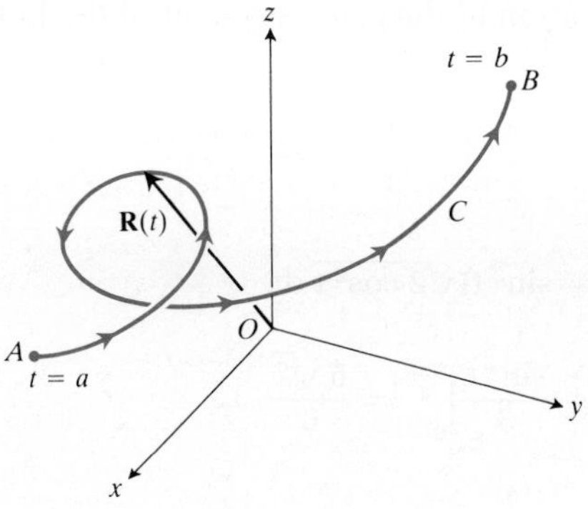

17.7 The directed curve C from A to B.

Line Integrals

Suppose that $w(x, y, z)$ is a function whose domain contains a piecewise smooth space curve C whose position vector

$$\mathbf{R}(t) = x(t)\mathbf{i} + y(t)\mathbf{j} + z(t)\mathbf{k}, \quad a \le t \le b$$

moves from point A to point B as t increases from a to b (Fig. 17.7). We can then define the integral of w over C in much the same way as we would define the integral of w over a piecewise smooth surface in its domain. We subdivide the curve into small pieces of length

$$\Delta s_1, \Delta s_2, \ldots, \Delta s_n,$$

evaluate w at a point (x_k, y_k, z_k) in each piece, form the sums

$$\sum_{k=1}^{n} w(x_k, y_k, z_k)\,\Delta s_k,$$

and define the integral of w over C from A to B to be the limit of these sums as the subdivision of C is refined so that the largest Δs_k approaches zero:

$$\int_C w\,ds = \lim_{n\to\infty} \sum_{k=1}^{n} w(x_k, y_k, z_k)\,\Delta s_k. \tag{2}$$

This limit will exist if w is continuous and the derivatives $x'(t)$, $y'(t)$, and $z'(t)$ are bounded and have only a finite number of discontinuities on the interval $a \le t \le b$. The limit may exist in other cases as well, and whenever it exists we call it the *line integral of w along C from A to B*. Note that if $w(x, y, z) \equiv 1$ is the constant function whose value is one, then the line integral gives the length of C from A to B.

It is a theorem from advanced calculus that the value of the line integral in (2) may be calculated from any parameterization of the curve for which $ds/dt = \sqrt{x'(t)^2 + y'(t)^2 + z'(t)^2}$ is positive from the formula

$$\int_C w\,ds = \int_{t=a}^{t=b} w(x(t), y(t), z(t))\sqrt{x'(t)^2 + y'(t)^2 + z'(t)^2}\,dt, \tag{3}$$

provided the derivatives of $x(t)$, $y(t)$, and $z(t)$ are bounded and piecewise continuous over the interval $a \le t \le b$. Thus the value of the line integral depends only on the nature of w and the geometry of the curve and may be calculated from any convenient parameterization.

EXAMPLE 1 Let C be the line segment from $A(0, 0)$ to $B(1, 1)$, and let $w = x + y^2$. Evaluate $\int_C w\,ds$ for two different parameterizations of C.

Solution 1 If we let

$$x = t \quad \text{and} \quad y = t, \quad 0 \le t \le 1,$$

then we get

$$\int_C w\,ds = \int_0^1 (t + t^2)\sqrt{1 + 1}\,dt = \sqrt{2}\left[\frac{t^2}{2} + \frac{t^3}{3}\right]_0^1 = \frac{5\sqrt{2}}{6}.$$

Solution 2 As a second parameterization of the given segment of the line $y = x$, we let

$$x = \sin t \quad \text{and} \quad y = \sin t, \qquad 0 \le t \le \pi/2,$$

and get

$$\int_C w\, ds = \int_0^{\pi/2} (\sin t + \sin^2 t)\sqrt{2\cos^2 t}\, dt$$

$$= \sqrt{2}\left[\frac{\sin^2 t}{2} + \frac{\sin^3 t}{3}\right]_0^{\pi/2} = \frac{5\sqrt{2}}{6}. \ \square$$

Algebraic Properties

Line integrals inherit the usual algebraic properties from their approximating sums (properties analogous to I1–I6 in the preceding article). The additivity property, as we shall see, is particularly useful. If the curve C is made by joining a finite number of piecewise smooth curves $C_1, C_2, \ldots, C_n$ together end to end, then

I6. $\int_C w\, ds = \int_{C_1} w\, ds + \int_{C_2} w\, ds + \cdots + \int_{C_n} w\, ds.$

Thus the integral of w along a polygon is the sum of the integrals over the polygon's sides, and so on.

Line integrals also have the property that changing the direction of integration reverses the sign of the integral:

I7. $\int_B^A w\, ds = -\int_A^B w\, ds.$

Thus if we integrate along a curve from A to B and then integrate from B back to A again, the net result is zero:

$$\int_{ABA} w\, ds = \int_A^B w\, ds + \int_B^A w\, ds = \int_A^B w\, ds - \int_A^B w\, ds = 0.$$

Work

To calculate the work done by a force

$$\mathbf{F} = M(x, y, z)\mathbf{i} + N(x, y, z)\mathbf{j} + P(x, y, z)\mathbf{k}$$

whose point of application moves along a curve

$$\mathbf{R}(t) = x(t)\mathbf{i} + y(t)\mathbf{j} + z(t)\mathbf{k}, \qquad a \le t \le b$$

from a point A to a point B in space, we integrate the scalar product

$$w = \mathbf{F} \cdot \frac{d\mathbf{R}}{ds}$$

along the curve from $t = a$ to $t = b$. Thus,

$$\text{Work} = \int \left(\mathbf{F} \cdot \frac{d\mathbf{R}}{ds}\right) ds$$

$$= \int_{t=a}^{t=b} (\mathbf{F} \cdot \mathbf{T})\sqrt{x'(t)^2 + y'(t)^2 + z'(t)^2}\, dt, \qquad (4)$$

where the second equality comes from the fact that $d\mathbf{R}/ds$ is the unit tangent vector $\mathbf{T}$.

Equation (4) emphasizes the fact that the work is the value of the line integral along the curve of the tangential component of the force field **F**. If we write

$$d\mathbf{R} = \mathbf{i}\,dx + \mathbf{j}\,dy + \mathbf{k}\,dz,$$

we obtain Eq. (4) in the alternative forms

$$\text{Work} = \int_C \mathbf{F} \cdot \frac{d\mathbf{R}}{ds}\,ds = \int_C \mathbf{F} \cdot d\mathbf{R} = \int_C M\,dx + N\,dy + P\,dz, \tag{5}$$

which are all in common use.

In general the amount of work done by a force **F** along a curve from point A to point B depends on the path as well as on the endpoints A and B.

EXAMPLE 2 The point of application of the force

$$\mathbf{F} = \mathbf{i}(x^2 - y) + \mathbf{j}(y^2 - z) + \mathbf{k}(z^2 - x)$$

moves from the origin O to the point $A(1, 1, 1)$,

a) along the straight line OA, and
b) along the curve

$$x = t, \qquad y = t^2, \qquad z = t^3, \qquad 0 \le t \le 1.$$

Find the work done in each case.

Solution a) Equations for the line OA are

$$x = y = z.$$

The integral to be evaluated is

$$W = \int_C (x^2 - y)\,dx + (y^2 - z)\,dy + (z^2 - x)\,dz$$

which, for the path OA, becomes

$$W = \int_0^1 3(x^2 - x)\,dx = -\frac{1}{2}.$$

b) Along the curve, we get

$$W = \int_0^1 2(t^4 - t^3)t\,dt + 3(t^6 - t)t^2\,dt = -\frac{29}{60}. \quad \square$$

Path Independence

Under certain conditions, the line integral between two points A and B is independent of the path C joining them. That is, the integral in Eq. (5) has the same value for any two paths C_1 and C_2 joining A and B. This happens when the force field **F** is a *gradient field,* that is, when

$$\mathbf{F}(x, y, z) = \nabla f = \mathbf{i}\frac{\partial f}{\partial x} + \mathbf{j}\frac{\partial f}{\partial y} + \mathbf{k}\frac{\partial f}{\partial z},$$

for some differentiable function f. We state this as a formal theorem and prove the sufficiency and necessity of the conditions, with some interpolated remarks.

THEOREM 1

Let **F** be a vector field with components M, N, P, that are continuous throughout some connected region D. Then a necessary and sufficient condition for the integral

$$\int_A^B \mathbf{F} \cdot d\mathbf{R}$$

to be independent of the path joining the points A and B in D is that there exist a differentiable function f such that

$$\mathbf{F} = \nabla f = \mathbf{i}\frac{\partial f}{\partial x} + \mathbf{j}\frac{\partial f}{\partial y} + \mathbf{k}\frac{\partial f}{\partial z} \tag{6}$$

throughout D. Furthermore, if the integral is independent of the path from A to B, then its value is

$$\int_A^B \mathbf{F} \cdot d\mathbf{R} = f(B) - f(A). \tag{7}$$

Proof of sufficiency. First, we suppose that Eq. (6) is satisfied, and then consider A and B to be two points in D (see Fig. 17.8). Suppose that C is any piecewise smooth curve joining A and B:

$$x = x(t), \qquad y = y(t), \qquad z = z(t), \qquad t_1 \leq t \leq t_2.$$

Along C, $f = f[x(t), y(t), z(t)]$ is a function of t to which we may apply the chain rule to differentiate with respect to t:

$$\begin{aligned}\frac{df}{dt} &= \frac{\partial f}{\partial x}\frac{dx}{dt} + \frac{\partial f}{\partial y}\frac{dy}{dt} + \frac{\partial f}{\partial z}\frac{dz}{dt}\\ &= \nabla f \cdot \left(\mathbf{i}\frac{dx}{dt} + \mathbf{j}\frac{dy}{dt} + \mathbf{k}\frac{dz}{dt}\right)\\ &= \nabla f \cdot \frac{d\mathbf{R}}{dt}.\end{aligned} \tag{8}$$

Because Eq. (6) holds, we also have

$$\mathbf{F} \cdot d\mathbf{R} = \nabla f \cdot d\mathbf{R} = \nabla f \cdot \frac{d\mathbf{R}}{dt}\,dt = \frac{df}{dt}\,dt. \tag{9}$$

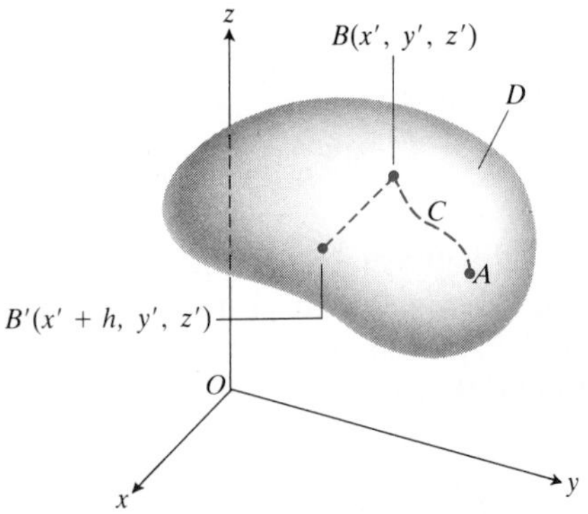

17.8 A piecewise smooth curve C joining points A and B in the region D of Theorem 1.

We now use this result to integrate $\mathbf{F} \cdot d\mathbf{R}$ along C from A to B:

$$\begin{aligned}\int_C \mathbf{F} \cdot d\mathbf{R} &= \int_{t_1}^{t_2} \frac{df}{dt}\,dt\\ &= \int_{t_1}^{t_2} \frac{d}{dt} f(x(t), y(t), z(t))\,dt\\ &= f(x(t), y(t), z(t))\Big]_{t_1}^{t_2}\\ &= f(x(t_2), y(t_2), z(t_2)) - f(x(t_1), y(t_1), z(t_1))\\ &= f(B) - f(A).\end{aligned}$$

Therefore, if $\mathbf{F} = \nabla f$ we have

$$\int_A^B \mathbf{F} \cdot d\mathbf{R} = \int_A^B \nabla f \cdot d\mathbf{R} = f(B) - f(A). \tag{10}$$

The value of the integral $f(B) - f(A)$ does not depend on the path C at all. Equation (10) is the space analog of the First Fundamental Theorem of Integral Calculus (see Article 4.7):

$$\int_a^b f'(x)\,dx = f(b) - f(a).$$

The only difference is that we have $\nabla f \cdot d\mathbf{R}$ in place of $f'(x)\,dx$. This analogy suggests that perhaps there is also a space analog of the fact that any continuous function of a single real variable is the derivative with respect to x of its integral from a to x (again, see Article 4.7). In other words, if we define a function f by the rule

$$f(x', y', z') = \int_A^{(x', y', z')} \mathbf{F} \cdot d\mathbf{R}, \tag{11}$$

perhaps it will be true that

$$\nabla f = \mathbf{F}. \tag{12}$$

Equation (12) is indeed true when the right-hand side of Eq. (11) is path-independent, and the proof of this fact will complete our theorem.

Proof of necessity. We now assume that the line integral in (11) is path-independent, and prove that $\mathbf{F} = \nabla f$ for the function f defined by Eq. (11). We first write $\mathbf{F}$ in terms of its **i**-, **j**-, and **k**-components:

$$\mathbf{F}(x, y, z) = \mathbf{i}M(x, y, z) + \mathbf{j}N(x, y, z) + \mathbf{k}P(x, y, z), \tag{13}$$

and fix the points A and $B(x', y', z')$ in D. To establish Eq. (12), we need to show that the equalities

$$\frac{\partial f}{\partial x} = M, \qquad \frac{\partial f}{\partial y} = N, \qquad \frac{\partial f}{\partial z} = P \tag{14}$$

hold at each point of D. In what follows, we either assume that D is an open set, so that all of its points are interior points, or we restrict our attention to interior points.

The point $B(x', y', z')$ is the center of some small sphere whose interior lies entirely inside D. We let $h \neq 0$ be small enough so that all points on the ray from B to $B'(x' + h, y', z')$ lie in D (Fig. 17.8), and consider the difference quotient

$$\frac{f(x' + h, y', z') - f(x', y', z')}{h} = \frac{1}{h}\int_B^{B'} \mathbf{F} \cdot d\mathbf{R}. \tag{15}$$

Since the integral does not depend on a particular path, we choose one convenient for our purpose:

$$x = x' + th, \qquad y = y', \qquad z = z', \qquad 0 \leq t \leq 1,$$

along which neither y nor z varies, and along which $d\mathbf{R} = \mathbf{i}\,dx = \mathbf{i}h\,dt$. When this is substituted into Eq. (15), along with $\mathbf{F}$ from Eq. (13), we get

$$\begin{aligned}\frac{f(x' + h, y', z') - f(x', y', z')}{h} &= \frac{1}{h}\int_0^1 M(x' + ht, y', z')h\,dt \\ &= \int_0^1 M(x' + ht, y', z')\,dt. \end{aligned} \tag{16}$$

By hypothesis, $\mathbf{F}$ is continuous, so each component is a continuous function. Thus, given any $\varepsilon > 0$, there is a $\delta > 0$ such that

$$|M(x' + ht, y', z') - M(x', y', z')| < \varepsilon \qquad \text{when} \qquad |ht| < \delta.$$

This implies that when $|h| < \delta$, the integral in Eq. (16) also differs from

$$\int_0^1 M(x', y', z')\,dt = M(x', y', z') \tag{17}$$

by less than ε. The equality in (17) follows from the fact that the integrand is a constant, and

$$\int_0^1 dt = 1.$$

Therefore, as $h \to 0$ in Eq. (16), the right-hand side has as limit $M(x', y', z')$, so the left-hand side must have the same limit. That, however, is just the partial derivative of f with respect to x at $B(x', y', z')$. Therefore

$$\frac{\partial f}{\partial x} = M \tag{18}$$

holds at each interior point of D.

Equation (18) is the first of the three equalities, Eqs. (14), that are needed to establish Eq. (12). Proofs of the remaining two equalities in (14) are very similar, and you are asked to prove one of them in Problem 42. One sets up a difference quotient like Eq. (15), but takes

$$B' = (x', y' + h, z') \qquad \text{or} \qquad B' = (x', y', z' + h).$$

In other words, from B, one integrates along a path parallel to the y-axis or parallel to the z-axis. On the former path, $d\mathbf{R} = \mathbf{j}h\,dt$, and on the latter, $d\mathbf{R} = \mathbf{k}h\,dt$. This concludes the proof. ■

EXAMPLE 3 Find a function f such that if $\mathbf{F} = 2x\mathbf{i} + 2y\mathbf{j} + 2z\mathbf{k}$, then $\mathbf{F} = \nabla f$.

Solution We might be lucky and guess

$$f(x, y, z) = x^2 + y^2 + z^2, \tag{19}$$

because $2x$, $2y$, $2z$ are its partial derivatives with respect to x, y, and z. But if we weren't so inspired, then we might try something like Eq. (11). In the first place, the functions $2x$, $2y$, $2z$ are everywhere continuous, so the region D can be all of space. Of course we don't know, until *after* we find that $\mathbf{F}$ is a gradient, that the integral in (11) is path-independent, but we proceed on faith (or at least with hope). The choice of A is up to us, so we make life easy for ourselves by taking $A = (0, 0, 0)$. For the path of integration from A to $B(x', y', z')$, we take the line segment

$$x = x't, \qquad y = y't, \qquad z = z't, \qquad 0 \le t \le 1,$$

along which

$$d\mathbf{R} = (x'\mathbf{i} + y'\mathbf{j} + z'\mathbf{k})\,dt$$

and

$$\begin{aligned} \mathbf{F} \cdot d\mathbf{R} &= (2xx' + 2yy' + 2zz')\,dt \\ &= (2x'^2 + 2y'^2 + 2z'^2)t\,dt. \end{aligned}$$

Therefore, when we substitute into Eq. (11), we get

$$\begin{aligned} f(x', y', z') &= \int_{(0,0,0)}^{(x',y',z')} \mathbf{F} \cdot d\mathbf{R} \\ &= [x'^2 + y'^2 + z'^2] \int_0^1 2t\,dt \\ &= x'^2 + y'^2 + z'^2. \end{aligned}$$

If we delete the primes, this equation is identical with Eq. (19). □

A comment on notation. The upper limit of integration in Eq. (11) is an arbitrary point in the domain of **F**, but we use (x', y', z') to designate it, rather than (x, y, z), because the latter is used for the running point that covers the arc C from A to B during the integration. After we have completed the computation of $f(x', y', z')$, we then delete the primes to express the result as $f(x, y, z)$. The analog in one dimension would be

$$\ln x' = \int_1^{x'} \frac{1}{x}\,dx.$$

We must be careful not to confuse the variable of integration x with the limit of integration x'. We distinguished between these two things in a slightly different manner in Article 6.4, Eq. (4), where we wrote

$$\ln x = \int_1^{x} \frac{1}{t}\,dt.$$

Our purpose there was the same as it is here, however: to maintain a notational difference between the variable *upper limit* and the *variable of integration*.

So far we have only the criterion

$$\mathbf{F} = \nabla f$$

for deciding whether

$$\int_A^B \mathbf{F} \cdot d\mathbf{R}$$

is path-independent. We shall discover another criterion in Eqs. (24) below. If we follow the method indicated by Eq. (11) and illustrated in Example 3 for a field **F** that is *not* path-independent, then we should discover, on trying to verify that $\mathbf{F} = \nabla f$, that it isn't so. The next example illustrates exactly this situation.

EXAMPLE 4 Show that there is no function f such that

$$\mathbf{F} = \nabla f \quad \text{if} \quad \mathbf{F} = y\mathbf{i} - x\mathbf{j}.$$

Solution Here's one way to show it: If there were such a function f, then

$$\frac{\partial f}{\partial x} = y \quad \text{and} \quad \frac{\partial f}{\partial y} = -x,$$

from which we would get

$$\frac{\partial^2 f}{\partial y\,\partial x} = \frac{\partial(y)}{\partial y} = 1 \neq \frac{\partial^2 f}{\partial x\,\partial y} = \frac{\partial(-x)}{\partial x} = -1.$$

or

$$\frac{\partial g(y, z)}{\partial y} = 0. \tag{27}$$

Integrating Eq. (27) with respect to y, holding z constant, and adding an arbitrary function $h(z)$ as constant of integration, we obtain

$$g(y, z) = h(z). \tag{28}$$

We substitute this into Eq. (26) and then calculate $\partial f/\partial z$, which we compare with the third of Eqs. (25). We find that

$$xy + z = xy + \frac{dh(z)}{dz} \quad \text{or} \quad \frac{dh(z)}{dz} = z,$$

so that

$$h(z) = \frac{z^2}{2} + C.$$

Hence we may write Eq. (26) as

$$f(x, y, z) = e^x \cos y + xyz + (z^2/2) + C.$$

Then, for this function, it is easy to see that

$$\mathbf{F} = \nabla f. \ \square$$

A function $f(x, y, z)$ that has the property that its gradient gives the force vector **F** is called a "potential" function. (Sometimes a minus sign is introduced. For example, the electric intensity of a field is the negative of the potential gradient in the field.

PROBLEMS

1. In Example 1, let C be given by

$$x = t^2, \quad y = t^2, \quad 0 \le t \le 1,$$

and evaluate $\int_C w\, ds$ for $w = x + y^2$.

2. In Example 1, let C be given by $x = f(t) = y$, where $f(0) = 0$ and $f(1) = 1$. Show that if $f'(t)$ is continuous on [0, 1], then $\int_C w\, ds = (5\sqrt{2})/6$, no matter what the particular function f may be.

3. Evaluate $\int \mathbf{F} \cdot d\mathbf{R}$ around the circle

$$x = \cos t, \quad y = \sin t, \quad z = 0, \quad 0 \le t \le 2\pi$$

for the force given in Example 2.

4. In Example 3, evaluate $\int \mathbf{F} \cdot d\mathbf{R}$ along a curve C lying on the sphere $x^2 + y^2 + z^2 = a^2$. Do you need to know anything more about C? Why?

5. Assume $\mathbf{F} = y\mathbf{i} - x\mathbf{j}$, as in Example 4, and take $A = (0, 0, 0)$, $B = (1, 1, 0)$. Evaluate $\int \mathbf{F} \cdot d\mathbf{R}$ for:
a) $x = y = t$, $0 \le t \le 1$;
b) $x = t$, $y = t^2$, $0 \le t \le 1$.
Comment on the meaning of your answers.

6. Evaluate $\int_C \mathbf{F} \cdot d\mathbf{R}$, where the point of application of the force $\mathbf{F} = xy\mathbf{i} - x^2\mathbf{j}$ follows the path C in the xy-plane that consists of
a) the line segment on the x-axis from (1, 0) to (−1, 0),
b) the upper half of the circle $x^2 + y^2 = 1$ from (1, 0) to (−1, 0),
c) the line segment from (1, 0) to (0, 1) followed by the line segment from (0, 1) to (−1, 0).

In Problems 7–16, find the work done by the given force **F** as the point of application moves from (0, 0, 0) to (1, 1, 1)
a) along the straight line $x = y = z$,
b) along the curve $x = t$, $y = t^2$, $z = t^4$, and
c) along the x-axis to (1, 0, 0), then in a straight line to (1, 1, 0), and from there in a straight line to (1, 1, 1).

7. $\mathbf{F} = 2x\mathbf{i} + 3y\mathbf{j} + 4z\mathbf{k}$

8. $\mathbf{F} = 3y\mathbf{i} + 2x\mathbf{j} + 4z\mathbf{k}$

9. $\mathbf{F} = \dfrac{1}{x^2 + 1}\mathbf{j}$

10. $\mathbf{F} = \sqrt{z}\,\mathbf{i} - 2x\mathbf{j} + \sqrt{y}\,\mathbf{k}$

11. $\mathbf{F} = xy\mathbf{i} + yz\mathbf{j} + xz\mathbf{k}$

12. $\mathbf{F} = 3x(x - 1)\mathbf{i} + 3z\mathbf{j} + \mathbf{k}$

13. $\mathbf{F} = \mathbf{i}x \sin y + \mathbf{j} \cos y + \mathbf{k}(x + y)$

14. $\mathbf{F} = \mathbf{i}(y + z) + \mathbf{j}(z + x) + \mathbf{k}(x + y)$

15. $\mathbf{F} = e^{y+2z}(\mathbf{i} + \mathbf{j}x + 2\mathbf{k}x)$

16. $\mathbf{F} = \mathbf{i}y \sin z + \mathbf{j}x \sin z + \mathbf{k}xy \cos z$

Evaluate $\int_C \mathbf{F} \cdot d\mathbf{R}$ for the fields and curves in Problems 17–20.

17. $\mathbf{F} = xy\mathbf{i} + y\mathbf{j} - yz\mathbf{k}$
$\mathbf{R} = t\mathbf{i} + t^2\mathbf{j} + t\mathbf{k}, \quad 0 \le t \le 1$

18. $\mathbf{F} = 2y\mathbf{i} + 3x\mathbf{j} + (x + y)\mathbf{k}$
$\mathbf{R} = \mathbf{i} \cos t + \mathbf{j} \sin t + \frac{t}{6}\mathbf{k}, \quad 0 \le t \le 2\pi$

19. $\mathbf{F} = z\mathbf{i} + x\mathbf{j} + y\mathbf{k}$
$\mathbf{R} = \mathbf{i} \sin t + \mathbf{j} \cos t + t\mathbf{k}, \quad 0 \le t \le 2\pi$

20. $\mathbf{F} = 6z\mathbf{i} + y^2\mathbf{j} + 12x\mathbf{k}$
$\mathbf{R} = \mathbf{i} \sin t + \mathbf{j} \cos t + \frac{t}{6}\mathbf{k}, \quad 0 \le t \le 2\pi$

21. Find the work done by the force $\mathbf{F} = -4xy\mathbf{i} + 8y\mathbf{j} + 2\mathbf{k}$ as the point of application moves along the parabola $y = x^2$, $z = 1$ from $A(0, 0, 1)$ to $B(2, 4, 1)$.

22. Evaluate $\int_C y\, ds$ along the curve $y = 2\sqrt{x}$ from (1, 2) to (4, 4).

In Problems 23–26, find a function $f(x, y, z)$ such that $\mathbf{F} = \nabla f$.

23. $\mathbf{F} = 2x\mathbf{i} + 3y\mathbf{j} + 4z\mathbf{k}$

24. $\mathbf{F} = \mathbf{i}(y + z) + \mathbf{j}(z + x) + \mathbf{k}(x + y)$

25. $\mathbf{F} = e^{y+2z}(\mathbf{i} + x\mathbf{j} + 2x\mathbf{k})$

26. $\mathbf{F} = \mathbf{i}y \sin z + \mathbf{j}x \sin z + \mathbf{k}xy \cos z$

27. If A and B are given, prove that the line integral

$$\int_A^B (z^2\, dx + 2y\, dy + 2xz\, dz)$$

is independent of the path of integration.

28. If $\mathbf{F} = y\mathbf{i} + x\mathbf{j}$, evaluate the line integral $\int_A^B \mathbf{F} \cdot d\mathbf{R}$ along the straight line from $A(1, 1, 1)$ to $B(3, 3, 3)$.

29. If $\mathbf{F} = \mathbf{i}x^2 + \mathbf{j}yz + \mathbf{k}y^2$, compute $\int_A^B \mathbf{F} \cdot d\mathbf{R}$, where $A = (0, 0, 0)$, $B = (0, 3, 4)$, along the straight line connecting these points.

30. Let C denote the plane curve whose vector equation is

$$\mathbf{R}(t) = (e^t \cos t)\mathbf{i} + (e^t \sin t)\mathbf{j}.$$

Evaluate the line integral

$$\int \frac{x\, dx + y\, dy}{(x^2 + y^2)^{3/2}}$$

along that arc of C from the point (1, 0) to the point $(e^{2\pi}, 0)$.

Evaluate the line integrals in Problems 31–38.

31. $\int_C 2x \cos y\, dx - x^2 \sin y\, dy$, where C is the path from (1, 0) to (0, 1) on the curve $\mathbf{R} = \mathbf{i} \cos^3 t + \mathbf{j} \sin^3 t$.

32. $\int_C 2x \sin y\, dx + x^2 \cos y\, dy$, where C is the path from (1, 0) to (0, 1) on the curve $\mathbf{R} = t\mathbf{i} + (t - 1)^2\mathbf{j}$.

33. $\int_C (x^2 + y)\, dx + (y^2 + x)\, dy$, where C is the line segment from (1, 1) to (2, 3).

34. $\int_C (y^2x + y)\, dx + (x^2y + x)\, dy$, where C is the line segment from (1, 1) to (2, 3).

35. $\int_C yz\, dx + xz\, dy + xy\, dz$, where C is the line segment from (1, 1, 2) to (3, 5, 0).

36. $$\int_{(0,0,0)}^{(1,2,3)} 2xy\, dx + (x^2 + z^2)\, dy + 2yz\, dz.$$

37. $$\int_{(0,1,1)}^{(2,2,1)} 3x^2\, dx + \frac{z^2}{y}\, dy + 2z \ln y\, dz.$$

38. $$\int_{(1,0,1)}^{(0,1,1)} \sin y \cos x\, dx + \cos y \sin x\, dy + dz$$

39. If the density $\rho(x, y, z)$ of a fluid is a function of the pressure $p(x, y, z)$, and

$$\phi(x, y, z) = \int_{p_0}^{p} (dp/\rho),$$

where p_0 is constant, show that $\nabla\phi = \nabla p/\rho$.

40. If $\mathbf{F} = y\mathbf{i}$, show that the line integral $\int_A^B \mathbf{F} \cdot d\mathbf{R}$ along an arc AB in the xy-plane is equal to an area bounded by the x-axis, the arc, and the ordinates at A and B.

REMARK Despite similarity of appearance and identity of value, the integral of this problem and the integral of earlier calculus are conceptually distinct. The latter is a line integral for which the path lies along the x-axis.

41. In Example 4, when we considered $\mathbf{F} = y\mathbf{i} - x\mathbf{j}$, we found a function $f(x', y', z') = 0$ which expresses the value of the integral

$$\int_{(0,0,0)}^{(x',y',z')} \mathbf{F} \cdot d\mathbf{R}$$

along the line segment from (0, 0, 0) to an arbitrary point (x', y', z'). Using this result, and the first half of the proof of Theorem 1, prove that if $\mathbf{F}$ were a gradient field, say $\mathbf{F} = \nabla g$, then $g - f =$ constant. From this, show that no such g exists for the given $\mathbf{F}$.

42. Using the notations of Eqs. (11) and (13), show that $\partial f/\partial y = N$ holds at each point of D if $\mathbf{F}$ is continuous and if the integral in (11) is path-independent in D.

43. Let $\rho = (x^2 + y^2 + z^2)^{1/2}$. Show that

$$\nabla(\rho^n) = n\rho^{n-2}\mathbf{R},$$

where $\mathbf{R} = \mathbf{i}x + \mathbf{j}y + \mathbf{k}z$. Is there a value of n for which $\mathbf{F} = \nabla(\rho^n)$ represents the "inverse-square law" field? If so, what is this value of n?

44. The "curl" of a vector field

$$\mathbf{F} = \mathbf{i}f(x, y, z) + \mathbf{j}g(x, y, z) + \mathbf{k}h(x, y, z)$$

is defined to be del cross $\mathbf{F}$; that is,

$$\operatorname{curl} \mathbf{F} \equiv \nabla \times \mathbf{F} \equiv \begin{vmatrix} \mathbf{i} & \mathbf{j} & \mathbf{k} \\ \dfrac{\partial}{\partial x} & \dfrac{\partial}{\partial y} & \dfrac{\partial}{\partial z} \\ f & g & h \end{vmatrix},$$

or

$$\operatorname{curl} \mathbf{F} \equiv \mathbf{i}\left(\frac{\partial h}{\partial y} - \frac{\partial g}{\partial z}\right) + \mathbf{j}\left(\frac{\partial f}{\partial z} - \frac{\partial h}{\partial x}\right) + \mathbf{k}\left(\frac{\partial g}{\partial x} - \frac{\partial f}{\partial y}\right),$$

and the "divergence" of a vector field

$$\mathbf{V} = \mathbf{i}u(x, y, z) + \mathbf{j}v(x, y, z) + \mathbf{k}w(x, y, z)$$

is defined to be del dot $\mathbf{V}$; that is,

$$\operatorname{div} \mathbf{V} \equiv \nabla \cdot \mathbf{V} \equiv \frac{\partial u}{\partial x} + \frac{\partial v}{\partial y} + \frac{\partial w}{\partial z}.$$

If the components f, g, h of $\mathbf{F}$ are functions that possess continuous mixed partial derivatives

$$\frac{\partial^2 h}{\partial x\, \partial y}, \quad \ldots,$$

show that

$$\operatorname{div}(\operatorname{curl} \mathbf{F}) = 0.$$

***Toolkit* programs**	
Scalar Fields	Vector Fields

17.4

Two-dimensional Fields. Flux Across a Plane Curve

In this article, we turn our attention to two-dimensional vector fields of the form

$$\mathbf{F} = \mathbf{i}M(x, y) + \mathbf{j}N(x, y). \tag{1}$$

Figure 17.9 shows how such a two-dimensional vector field might look in space. In the figure, for example, $\mathbf{F}$ might represent a fluid flow in which each particle travels in a circle parallel to the xy-plane in such a way that all particles on a given line perpendicular to the xy-plane travel with the same velocity. Example 1 provides another instance, that of an electric field with field strength

$$\mathbf{E} = \frac{\mathbf{i}x + \mathbf{j}y}{x^2 + y^2}. \tag{2}$$

Note that this formulation is like the right-hand side of Eq. (1), with

$$M(x, y) = \frac{x}{x^2 + y^2}, \qquad N(x, y) = \frac{y}{x^2 + y^2}.$$

The essential features of a two-dimensional field are (1) the vectors in $\mathbf{F}$ are all parallel to one plane, which we have taken to be the xy-plane, and (2) in every plane parallel to the xy-plane, the field is the same as it is in that plane. In Eq. (1), the field has a zero $\mathbf{k}$-component everywhere, which makes the vectors parallel to the xy-plane. The $\mathbf{i}$- and $\mathbf{j}$-components do not depend on z, so they are the same in all planes parallel to the xy-plane.

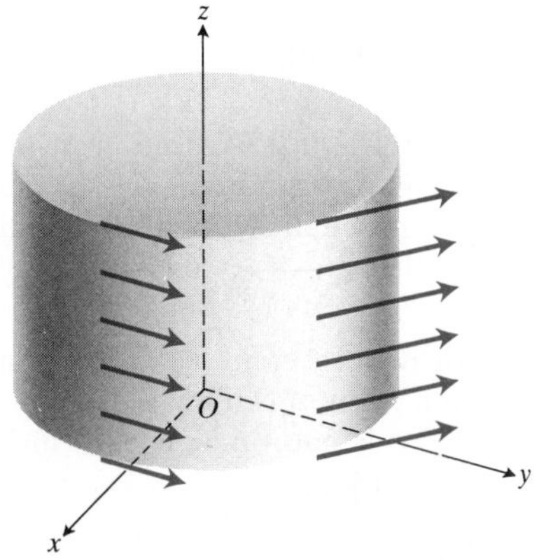

17.9 A two-dimensional vector field in space.

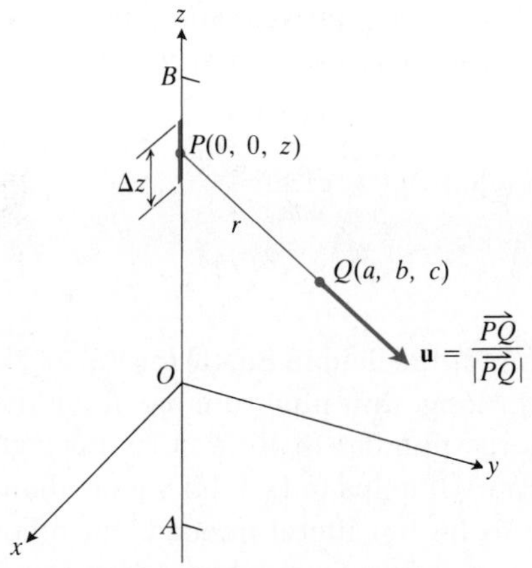

17.10 According to Coulomb's law, a charge $\delta_0\,\Delta z$ at $P(0, 0, z)$ produces a field $\Delta\mathbf{F}$ on a test charge q_0 at $Q(a, b, c)$.

EXAMPLE 1 An infinitely long, thin, straight wire has a uniform electric charge density δ_0. Using Coulomb's law, find the electric field intensity around the wire due to this charge.

Solution From a physics textbook, we find that *Coulomb's law* is an inverse-square law. It says that the force acting on a positive test charge q_0 placed at a distance r from a positive *point charge* q is directed away from q and has magnitude $(4\pi\varepsilon_0)^{-1}(qq_0)/r^2$, where ε_0 is a constant called the *permittivity*. For the charged wire, we have a distributed charge instead of a point charge, but we handle it in the familiar way: We replace this distributed charge by a large number of tiny elements, add, and take limits.

To be specific, suppose that the wire runs along the z-axis from $-\infty$ to $+\infty$. Take a long but finite piece of the wire and divide it into a lot of small segments. One of these is indicated in Fig. 17.10, with its center at $P(0, 0, z)$, and its length equal to Δz. We assume that Δz is so small that we can treat the charge $\delta_0\,\Delta z$ on this segment of the wire as a point charge at P. Now let $Q(a, b, c)$ be any point not on the z-axis, and let Δz denote the force on the test charge at Q due to the point charge at P:

$$\Delta\mathbf{F} = \frac{\delta_0 q_0}{4\pi\varepsilon_0}\frac{\Delta z\,\mathbf{u}}{a^2 + b^2 + (c - z)^2}, \tag{3a}$$

where $\mathbf{u}$ is a unit vector in the direction of $\overrightarrow{PQ}$:

$$\mathbf{u} = \frac{\mathbf{i}a + \mathbf{j}b + \mathbf{k}(c - z)}{[a^2 + b^2 + (c - z)^2]^{1/2}}. \tag{3b}$$

When we add the vector forces $\Delta\mathbf{F}$ for pieces of wire between $z = A$ and $z = B$ and take the limit as $\Delta z \to 0$, we get an integral of the form

$$\frac{\delta_0 q_0}{4\pi\varepsilon_0}\int_A^B \frac{\mathbf{i}a + \mathbf{j}b + \mathbf{k}(c - z)}{[a^2 + b^2 + (c - z)^2]^{3/2}}\,dz. \tag{3c}$$

The denominator of the integrand behaves about like $|z|^3$ as $|z| \to \infty$, so the three component integrals in (3c) converge as $A \to -\infty$ and $B \to \infty$. As our final integral representation of the field, therefore, we get

$$\mathbf{F} = \frac{\delta_0 q_0}{4\pi\varepsilon_0}\int_{-\infty}^{+\infty} \frac{\mathbf{i}a + \mathbf{j}b + \mathbf{k}(c - z)}{[a^2 + b^2 + (c - z)^2]^{3/2}}\,dz. \tag{4}$$

There are three separate integrals, but two are essentially the same except for the constant coefficients a and b, and the third is zero. It is easy to do the integration by making the substitutions

$$\sqrt{a^2 + b^2} = m,$$

$$z - c = m\tan\theta, \qquad \theta = \tan^{-1}\left(\frac{z - c}{m}\right),$$

$$dz = m\sec^2\theta\,d\theta,$$

and observing that the limits for θ are from $-\pi/2$ to $\pi/2$. In Problem 1 you are asked to finish these calculations and to show that the result is

$$\mathbf{F} = \frac{\delta_0 q_0}{2\pi\varepsilon_0}\left(\frac{\mathbf{i}a + \mathbf{j}b}{a^2 + b^2}\right). \tag{5}$$

Note that $\mathbf{F}$ is a two-dimensional field that does not depend on $\mathbf{k}$ or on c. If we write the resultant field strength $\mathbf{E} = \mathbf{F}/q_0$ as a function of position (x, y, z) instead of (a, b, c), we have

$$\mathbf{E} = k_0\left(\frac{\mathbf{i}x + \mathbf{j}y}{x^2 + y^2}\right), \qquad \text{where } k_0 = \frac{\delta_0}{2\pi\varepsilon_0}. \ \square \tag{6}$$

Mass Transport

There is also a fluid-flow interpretation for the field in Eq. (6) (or Eq. 2). To arrive at that interpretation, imagine a long, thin pipe running along the z-axis, and perforated with a very large number of little holes through which it supplies fluid at a constant rate. (It helps to be a bit vague about the actual physics of this: Don't try to be too literal-minded.) In other words, the z-axis is to be thought of as a *source* from which water flows radially outward to produce a velocity field

$$\mathbf{v} = f(r)\mathbf{u}_r, \tag{7a}$$

where $\mathbf{u}_r$ is the usual unit vector associated with the cylindrical coordinates r, θ, z, and $f(r)$ is a function of r alone. Thus the velocity is all perpendicular to the z-axis, and it is independent of both θ and z. (These are our present interpretations of the phrases "at a constant rate" and "radially outward.")

Now consider the amount of fluid that flows out through the cylinder $r = a$ between the planes $z = 0$ and $z = 1$ in a short interval of time, from t to $t + \Delta t$. According to the law expressed by Eq. (7a), every particle of water that is on the surface $r = a$ moves radially outward a distance Δr, which is approximately $f(a)\,\Delta t$. Thus the volume of fluid that crosses the boundary $r = a$ between $z = 0$ and $z = 1$ in this time interval is approximately the volume between the cylinders $r = a$ and $r = a + f(a)\,\Delta t$, or $2\pi a f(a)\,\Delta t$. If we multiply this by the density δ, we get the *mass* transported through the wall of a unit length of the cylinder $r = a$ in the interval from t to $t + \Delta t$:

$$\Delta m \approx \delta 2\pi a f(a)\,\Delta t. \tag{7b}$$

If we divide both sides of Eq. (7b) by Δt and take the limit as $\Delta t \to 0$, we get the *rate* at which fluid is flowing across the unit length of the cylinder $r = a$:

$$\frac{dm}{dt} = \delta 2\pi a f(a). \tag{7c}$$

For an incompressible fluid such as water, all fluid that flows across the cylinder $r = a$ flows across the cylinder $r = b$ as well (there are no other sources or sinks between the two cylinders). Therefore, for the model under discussion, the rate of mass transport given by Eq. (7c) is independent of a, and its value for any radius $r = a$ is the same as for $r = 1$:

$$\delta 2\pi f(1) = \delta 2\pi a f(a). \tag{7d}$$

From Eq. (7d), we get

$$f(a) = f(1)/a \qquad \text{for any } a > 0.$$

Writing r in place of a and substituting C for $f(1)$, we can rewrite the velocity field (7a) in the form

$$\mathbf{v} = \frac{C}{r}\mathbf{u}_r. \tag{8}$$

If we recall that the position vector in cylindrical coordinates is

$$\mathbf{R} = \overrightarrow{OP} = r\mathbf{u}_r + \mathbf{k}z$$

and that this must also be equal to $\mathbf{R} = \mathbf{i}x + \mathbf{j}y + \mathbf{k}z$, then we conclude that

$$r\mathbf{u}_r = \mathbf{i}x + \mathbf{j}y, \quad \text{or} \quad \mathbf{u}_r = \frac{\mathbf{i}x + \mathbf{j}y}{r},$$

Thus,

$$\mathbf{v} = \frac{C}{r}\mathbf{u}_r = \frac{C}{r}\frac{\mathbf{i}x + \mathbf{j}y}{r} = C\frac{\mathbf{i}x + \mathbf{j}y}{x^2 + y^2}, \tag{9}$$

and Eqs. (6) and (8) describe the same field if $C = k_0$.

Plane versus Space

Instead of interpreting the two-dimensional vector fields as we have done in 3-space, we can interpret them simply as fields in the xy-plane itself. Then r in Eqs. (8) and (9) is just the distance from the origin to the point $P(x, y)$ in the plane, and the unit vector is

$$\mathbf{u}_r = \frac{(\mathbf{i}x + \mathbf{j}y)}{r} = \mathbf{i}\cos\theta + \mathbf{j}\sin\theta,$$

where $r > 0$ and θ measures the angle from the positive x-axis to the position vector $\overrightarrow{OP}$. Equation (8) then describes a vector field in the plane that is directed radially outward and whose strength decreases like $1/r$ as r increases. We still use the language of flow across boundaries, but in this interpretation the boundary would be a *curve* in the plane, rather than a unit length of a cylinder. Equation (7d) would be interpreted as saying that the amount of fluid flowing across the unit circle $r = 1$ per unit time is equal to the amount of fluid flowing across the circle $r = a$ in the same time. (This describes conditions after the flow has reached steady state, not during the transient phase.)

We can easily go back and forth between the two interpretations of two-dimensional fields, but henceforth we shall usually treat them as existing just in the xy-plane and ignore the fact that we can project the field onto any plane parallel to the xy-plane and thereby go to the 3-space view.

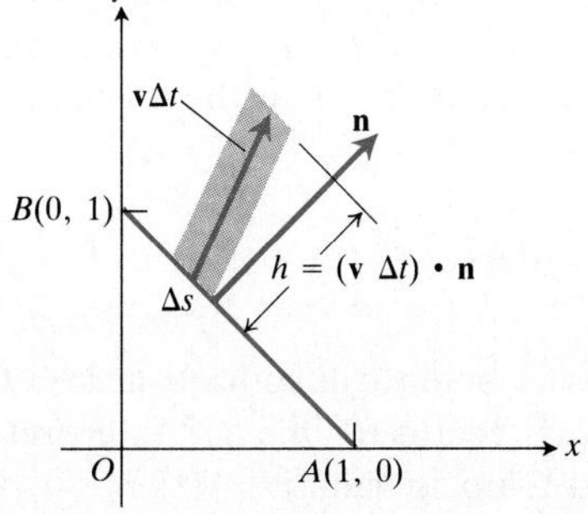

17.11 The fluid that flows across Δs in short time Δt fills a "parallelogram" whose altitude is $h = (\mathbf{v}\,\Delta t)\cdot\mathbf{n}$. The area of the parallelogram is therefore $\mathbf{v}\cdot\mathbf{n}\,\Delta t\,\Delta s$ and the mass of the fluid covering it is $\delta\mathbf{v}\cdot\mathbf{n}\,\Delta t\,\Delta s$.

EXAMPLE 2 Given the velocity field $\mathbf{v} = (\mathbf{i}x + \mathbf{j}y)/r^2$, calculate the mass-transport rate across the line segment AB joining the points $A(1, 0)$ and $B(0, 1)$.

Solution Let δ denote mass per unit area, the density factor by which we multiply the area of a region to get the mass of fluid in that region. (We assume the density to be constant.) Consider a segment of AB having length Δs, with its center at $P(x, y)$ on AB. From Fig. 17.11 we see that the mass of fluid Δm that flows across the segment in time Δt is given, approximately, by

$$\Delta m \approx \delta(\mathbf{v}\cdot\mathbf{n})(\Delta t\,\Delta s), \tag{10}$$

where **n** is a unit vector normal to the line AB at P, and pointing away from the origin:

$$\mathbf{n} = \frac{\mathbf{i}+\mathbf{j}}{\sqrt{2}}.$$

If we divide both sides of Eq. (10) by Δt, then sum for all the pieces Δs of the segment AB, and take the limit as $\Delta s \to 0$, we get the *average rate* of mass transport across AB:

$$\frac{\Delta M}{\Delta t} \approx \int_{AB} \delta(\mathbf{v}\cdot\mathbf{n})\,ds.$$

Finally, letting $\Delta t \to 0$ and substituting for **v**, **n**, and ds, with

$$x = t, \qquad y = 1 - t, \qquad 0 \le t \le 1$$

as parameterization of the segment AB, we get as the *instantaneous* mass-transport rate

$$\frac{dM}{dt} = \delta \int_0^1 \frac{x+y}{r^2\sqrt{2}}(\sqrt{2}\,dt) = \delta \int_0^1 \frac{dt}{t^2+(1-t)^2} = \delta(\pi/2). \tag{11}$$

(In Problem 2 you are asked to verify this integration.) □

Flux

As Fig. 17.12 shows, the right-hand side of Eq. (10) also describes a flow of mass due to a velocity field **v** across any smooth curve in the plane in a unit of time Δt. The quantity $\mathbf{v}\,\Delta t$ is, very nearly, the vector displacement of all particles of fluid that were on the segment Δs at time t; hence those particles have swept over a "parallelogram" with dimensions Δs and $|\mathbf{v}\,\Delta t|$. The **n**-component of $\mathbf{v}\,\Delta t$ is the altitude h of this parallelogram. Its area is therefore approximately

$$(\mathbf{v}\,\Delta t)\cdot\mathbf{n} \text{ times } \Delta s,$$

and the mass of fluid that fills this parallelogram is what flows across the tiny segment Δs between t and $t + \Delta t$.

This same line of reasoning would apply to flows in the xy-plane in general. It leads to the result

$$\frac{dM}{dt} = \int_C \delta(\mathbf{v}\cdot\mathbf{n})\,ds, \tag{12}$$

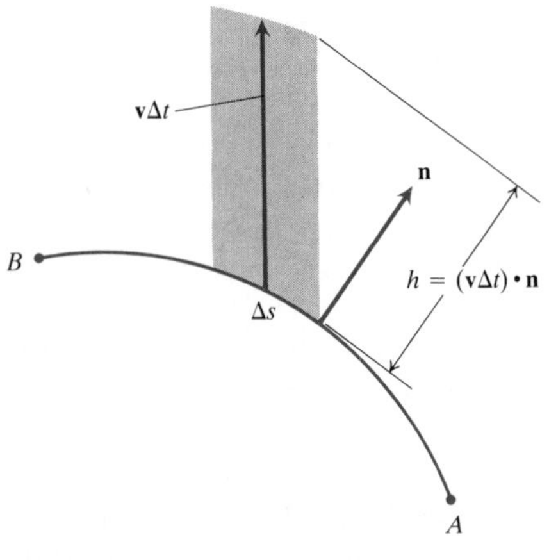

17.12 The fluid that flows across Δs in a short time Δt fills a "parallelogram" whose area is approximately base × height = $\mathbf{v}\cdot\mathbf{n}\,\Delta t\,\Delta s$.

where dM/dt is the *rate* at which mass is being transported across the curve C, in the direction of the unit normal vector **n.** One can interpret M as the amount of mass that has crossed C up to time t.

If the oppositely directed normal $\mathbf{n}' = -\mathbf{n}$ is substituted in place of **n** in the integral in Eq. (12), the sign of the answer changes. This just means that if flow in one direction across C is considered to be in the positive sense, then flow in the opposite direction is then considered to be negative. If C is a simple closed curve, we usually choose **n** to point outward. In Eq. (11), we chose the normal to point away from the origin. We got a positive answer because the flow in the first quadrant is generally upward and to the right for the given **v**.

To simplify further discussion, we take

$$\delta \mathbf{v} = \mathbf{F}(x, y)$$

in Eq. (12), and call the resulting integral the *flux* of **F** across C:

$$\text{Flux of } \mathbf{F} \text{ across } C = \int_C \mathbf{F} \cdot \mathbf{n}\, ds. \tag{13}$$

We shall use this terminology even when the field **F** has nothing to do with a fluid flow, but you may wish to keep the fluid-flow interpretation in mind too.

The curve C in Eq. (13) is to have a direction along it called the *positive* direction, and arc length s, measured from some arbitrary starting point, is to increase in this direction. We also assume that C is smooth enough to have a tangent vector except at a finite number of points where there may be corners or cusps. Thus, $d\mathbf{R}/ds = \mathbf{T}$ exists almost everywhere and points in the positive direction along C.

Because we often want the flux integral to represent flow outward from a region R bounded by a simple closed curve C, we choose the counterclockwise direction on C as positive, and choose the *outward-pointing* unit normal vector as **n**. From Fig. 17.13, we can see that, because

$$\mathbf{T} = \frac{dx}{ds}\mathbf{i} + \frac{dy}{ds}\mathbf{j}, \tag{14a}$$

for the indicated choice of **n** we should choose $\mathbf{n} = \mathbf{T} \times \mathbf{k}$, so that

$$\mathbf{n} = \frac{dy}{ds}\mathbf{i} - \frac{dx}{ds}\mathbf{j}. \tag{14b}$$

As a check, it is easy to see that the dot product of the vectors in Eqs. (14a) and (14b) is zero, that both have unit length, and that when **T** points upward and to the right, **n** has a positive **i**-component and a negative **j**-component. If we proceed around the curve C in the direction in which **T** points, with the interior toward our left, then **n**, as given by Eq. (14b), points to our right, as it should.

We now use Eq. (14b) to write the flux of

$$\mathbf{F}(x, y) = \mathbf{i}M(x, y) + \mathbf{j}N(x, y)$$

across C:

$$\begin{aligned}\text{Flux across } C &= \int_C \mathbf{F} \cdot \mathbf{n}\, ds \\ &= \int_C \left(M\frac{dy}{ds} - N\frac{dx}{ds}\right) ds \\ &= \int_C (M\, dy - N\, dx).\end{aligned} \tag{15}$$

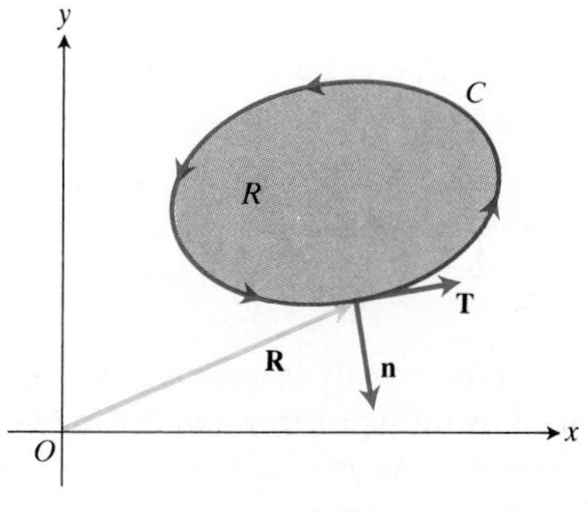

17.13 The position vector $\mathbf{R} = \mathbf{i}x + \mathbf{j}y$, the unit tangent vector $\mathbf{T} = d\mathbf{R}/ds$, and the unit normal vector $\mathbf{n} = \mathbf{T} \times \mathbf{k}$.

The virtue of the final integral in Eq. (15) is this: It can be evaluated using *any* reasonable parameterization of C; we aren't restricted to using the arc length s, provided we integrate in the positive direction along C.

EXAMPLE 3 Find the flux of the field

$$\mathbf{F} = 2x\mathbf{i} - 3y\mathbf{j}$$

outward across the ellipse

$$x = \cos t, \qquad y = 4 \sin t, \qquad 0 \le t \le 2\pi.$$

Solution By Eq. (15),

$$\begin{aligned}\int_C (M\,dy - N\,dx) &= \int_C (2x\,dy + 3y\,dx)\\ &= \int_0^{2\pi} (8\cos^2 t - 12\sin^2 t)\,dt\\ &= \int_0^{2\pi} [4(1 + \cos 2t) - 6(1 - \cos 2t)]\,dt\\ &= -4\pi.\end{aligned}$$

The negative answer means that the net flux is inward. □

Flux and Work

The integral that we have just evaluated was set up as a flux integral, but it can also be interpreted as a work integral of the form

$$\int_C \mathbf{G}\cdot d\mathbf{R} = \int_C (3y\mathbf{i} + 2x\mathbf{j})\cdot(\mathbf{i}\,dx + \mathbf{j}\,dy).$$

Conversely, any work integral in the plane can be reinterpreted as a flux integral of a related field. Problem 7 asks you to supply the details.

PROBLEMS

1. In Example 1, complete the calculations that lead from Eq. (4) to Eq. (5).

2. Evaluate

$$\int_0^1 \frac{dt}{2t^2 - 2t + 1}$$

by changing it to

$$\int_0^1 \frac{2\,dt}{(2t-1)^2 + 1}$$

and making a change of variables. This will verify Eq. (11).

3. Find the rate of mass transport outward across the circle $r = a$ for the (velocity) flow field of Example 2. (The calculations are trivial if you interpret Eq. (12) correctly.)

4. In a three-dimensional velocity field, let S be a closed surface bounding a region D in its interior. Explain why

$$\iint_S \delta(\mathbf{v}\cdot\mathbf{n})\,d\sigma$$

can be interpreted as the rate of mass transport outward through S if $\mathbf{n}$ is the outward-pointing unit vector normal to S.

In Problems 5 and 6, use the result of Problem 4 to find the rate of mass transport outward through the surface given, if the flow vector $\mathbf{F} = \delta\mathbf{v}$ is

a) $\mathbf{F} = -\mathbf{i}y + \mathbf{j}x$,
b) $\mathbf{F} = (x^2 + y^2 + z^2)^{-3/2}(\mathbf{i}x + \mathbf{j}y + \mathbf{k}z)$.

5. S is the sphere $x^2 + y^2 + z^2 = a^2$.

6. S is the closed cylinder $x^2 + y^2 = a^2$, $-h \le z \le h$, plus the end discs $z = \pm h$, $x^2 + y^2 \le a^2$.

7. Suppose that C is a directed curve with unit tangent and normal vectors **T** and **n** related as in the text, and suppose that **F** and **G** are two-dimensional fields:

$$\mathbf{F}(x, y) = \mathbf{i}M(x, y) + \mathbf{j}N(x, y),$$
$$\mathbf{G}(x, y) = -\mathbf{i}N(x, y) + \mathbf{j}M(x, y).$$

Show that

$$\int_C \mathbf{F} \cdot \mathbf{n}\, ds = \int_C \mathbf{G} \cdot \mathbf{T}\, ds.$$

Which integral represents work? Which represents flux?

***Toolkit* programs**

Scalar Fields

17.5 Green's Theorem

Green's theorem asserts that under suitable conditions the line integral

$$\oint (M\, dx + N\, dy) \tag{1}$$

around a simple closed curve C in the counterclockwise direction in the xy-plane is equal to the double integral

$$\iint_R \left(\frac{\partial N}{\partial x} - \frac{\partial M}{\partial y}\right) dx\, dy \tag{2}$$

over the region R that lies inside C.

NOTATION. The symbol $\oint$ is used only when the curve C is *closed*.

EXAMPLE 1 Let C be the circle $x = a \cos\theta$, $y = a \sin\theta$, $0 \le \theta \le 2\pi$, and let $M = -y$, $N = x$. Then the integral in (1) is

$$\oint_C (-y\, dx + x\, dy) = \int_0^{2\pi} a^2(\sin^2\theta + \cos^2\theta)\, d\theta = 2\pi a^2,$$

and the integral in (2) is

$$\iint_{x^2+y^2 \le a^2} 2\, dx\, dy = 2\int_0^{2\pi}\int_0^a r\, dr\, d\theta = 2\pi a^2.$$

Both integrals equal twice the area inside the circle. □

THEOREM

Green's Theorem

Let C be a simple closed curve in the xy-plane such that a line parallel to either axis cuts C in at most two points. Let M, N, $\partial N/\partial x$, and $\partial M/\partial y$ be continuous functions of (x, y) inside and on C. Let R be the region inside C. Then

$$\oint M\, dx + N\, dy = \iint_R \left(\frac{\partial N}{\partial x} - \frac{\partial M}{\partial y}\right) dx\, dy. \tag{3}$$

The line integral on the left side of Eq. (3) is to be taken counterclockwise, as indicated by the arrow on the small circle. We did this automati-

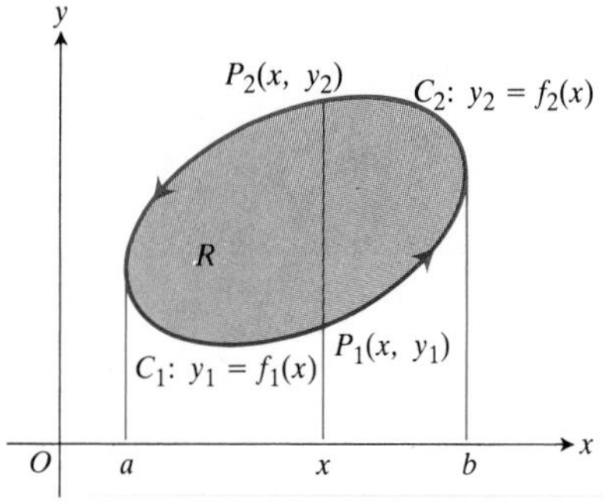

17.14 The boundary curve C, made up of $C_1: y = f_1(x)$ and $C_2: y = f_2(x)$.

cally in Example 1 by letting θ vary from 0 to 2π in the parametric representation of the circle. Figure 17.14 shows a curve C made up of two parts,

$$C_1:\quad a \le x \le b, \qquad y_1 = f_1(x),$$
$$C_2:\quad b \ge x \ge a, \qquad y_2 = f_2(x).$$

We use this notation in the proof.

Proof. Consider the double integral of $\partial M/\partial y$ over R in Fig. 17.14. For any x between a and b, we first integrate with respect to y from $y_1 = f_1(x)$ to $y_2 = f_2(x)$ and obtain

$$\int_{y_1}^{y_2} \frac{\partial M}{\partial y}\, dy = M(x, y)\Big]_{y=f_1(x)}^{y=f_2(x)} = M(x, f_2(x)) - M(x, f_1(x)). \tag{4}$$

Next we integrate this with respect to x from a to b:

$$\begin{aligned}\int_a^b \int_{f_1(x)}^{f_2(x)} \frac{\partial M}{\partial y}\, dy\, dx &= \int_a^b \{M(x, f_2(x)) - M(x, f_1(x))\}\, dx \\ &= -\int_b^a M(x, f_2(x))\, dx - \int_a^b M(x, f_1(x))\, dx \\ &= -\int_{C_2} M\, dx - \int_{C_1} M\, dx \\ &= -\oint_C M\, dx.\end{aligned}$$

Therefore

$$\oint M\, dx = \iint_R \left(-\frac{\partial M}{\partial y}\right) dx\, dy. \tag{5}$$

Equation (5) is half the result we need for Eq. (3). Problem 1 asks you to derive the other half, by integrating $\partial N/\partial x$ first with respect to x and then with respect to y, as suggested by Fig. 17.15. This shows the curve C of Fig. 17.14, decomposed into the two directed parts,

$$C_1':\quad c \le y \le d, \qquad x = g_1(y),$$
$$C_2':\quad d \ge y \ge c, \qquad x = g_2(y).$$

The result of this double integration is expressed by

$$\oint_C N\, dy = \iint_R \frac{\partial N}{\partial x}\, dx\, dy. \tag{6}$$

Combining Eqs. (5) and (6), we get Eq. (3). This concludes the proof. ■

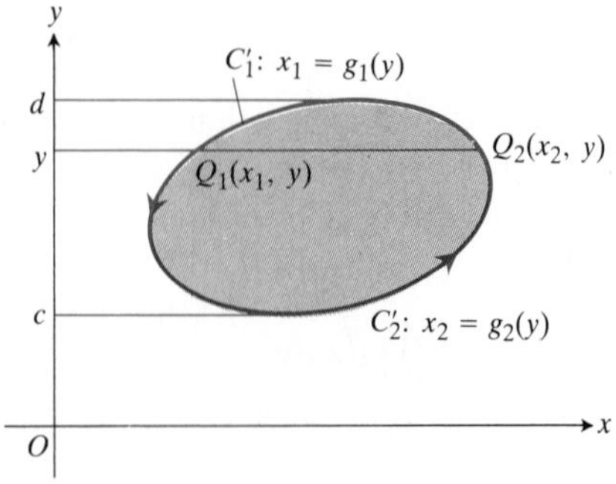

17.15 The boundary curve C', made up of $C'_1: x_1 = g_1(y)$ and $C'_2: x_2 = g_2(y)$.

EXAMPLE 2 Use Green's theorem to find the area enclosed by the ellipse

$$x = a\cos\theta, \qquad y = b\sin\theta, \qquad 0 \le \theta \le 2\pi.$$

Solution If we take $M = -y$, $N = x$, as in Example 1, and apply Green's theorem, we obtain

$$\begin{aligned}\oint M\, dx + N\, dy &= \oint -y\, dx + x\, dy \\ &= \int_0^{2\pi} ab(\sin^2\theta + \cos^2\theta)\, d\theta \\ &= \int_0^{2\pi} ab\, d\theta = 2\pi ab,\end{aligned}$$

and

$$\iint_R \left(\frac{\partial N}{\partial x} - \frac{\partial M}{\partial y}\right) dx\, dy = \iint_R 2\, dx\, dy$$

$$= 2 \times (\text{area inside ellipse}).$$

Therefore

$$\text{Area inside ellipse} = \frac{1}{2}\oint (-\,y\, dx + x\, dy)$$

$$= \pi ab. \;\square$$

COROLLARY

Area Corollary to Green's Theorem

If C is a simple closed curve such that a line parallel to either axis cuts it in at most two points, then the area enclosed by C is equal to

$$\frac{1}{2}\oint_C (x\, dy - y\, dx).$$

Proof. If we take $M = -y/2$, $N = x/2$ in Eq. (3), we obtain

$$\oint_C \left(\frac{1}{2}x\, dy - \frac{1}{2}y\, dx\right) = \iint_R \left(\frac{1}{2} + \frac{1}{2}\right) dx\, dy \tag{7}$$

$$= \text{Area of } R. \;\blacksquare$$

Green's Theorem for Other Curves and Regions

Green's theorem may apply to curves and regions that don't meet all of the requirements stated in it. For example, C could be a rectangle, as shown in Fig. 17.16. Here C is considered as composed of four directed parts:

$$C_1:\; y = c, \qquad a \le x \le b,$$
$$C_2:\; x = b, \qquad c \le y \le d,$$
$$C_3:\; y = d, \qquad b \ge x \ge a,$$
$$C_4:\; x = a, \qquad d \ge y \ge c.$$

The lines $x = a$ and $x = b$ intersect C in more than two points, and so do the boundaries $y = c$ and $y = d$.

Proceeding as in the proof of Eq. (6), we have

$$\int_{y=c}^{d}\int_{x=a}^{b} \frac{\partial N}{\partial x}\, dx\, dy = \int_c^d [N(b, y) - N(a, y)]\, dy$$

$$= \int_c^d N(b, y)\, dy + \int_d^c N(a, y)\, dy$$

$$= \int_{C_2} N\, dy + \int_{C_4} N\, dy. \tag{8}$$

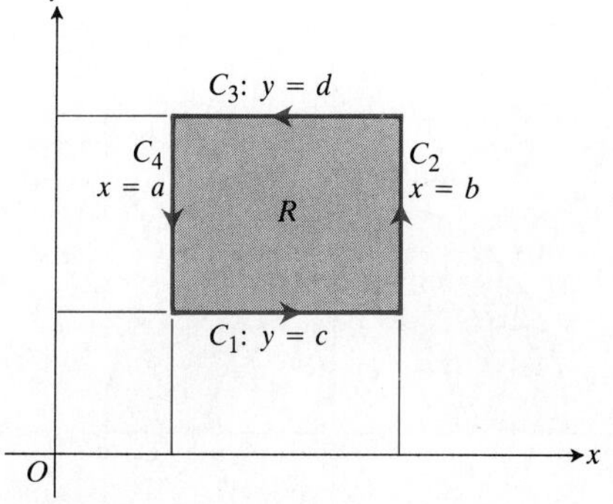

17.16 The rectangle made up of the four segments C_1, C_2, C_3, C_4.

Because y is constant along C_1 and C_3,

$$\int_{C_1} N\, dy = \int_{C_3} N\, dy = 0,$$

so we can add

$$\int_{C_1} N\, dy + \int_{C_3} N\, dy$$

to the right-hand side of Eq. (8) without changing the equality. Doing so, we have

$$\int_c^d \int_a^b \frac{\partial N}{\partial x}\, dx\, dy = \oint_C N\, dy. \tag{9}$$

Similarly we could show that

$$\int_a^b \int_c^d \frac{\partial M}{\partial y}\, dy\, dx = -\oint_C M\, dx. \tag{10}$$

Subtracting (10) from (9), we again arrive at

$$\oint_C M\, dx + N\, dy = \iint_R \left(\frac{\partial N}{\partial x} - \frac{\partial M}{\partial y}\right) dx\, dy.$$

Regions such as those shown in Fig. 17.17 can be handled with no greater difficulty. Equation (3) still applies. It also applies to the horseshoe-shaped region R shown in Fig. 17.18, as we see by putting together the regions R_1 and R_2 and their boundaries. Green's theorem applies to C_1, R_1, and to C_2, R_2, yielding

$$\int_{C_1} M\, dx + N\, dy = \iint_{R_1} \left(\frac{\partial N}{\partial x} - \frac{\partial M}{\partial y}\right) dx\, dy,$$

$$\int_{C_2} M\, dx + N\, dy = \iint_{R_2} \left(\frac{\partial N}{\partial x} - \frac{\partial M}{\partial y}\right) dx\, dy.$$

When we add, the line integral along the y-axis from b to a for C_1 cancels the integral over the same segment but in the opposite direction for C_2. Hence

$$\oint_C (M\, dx + N\, dy) = \iint_R \left(\frac{\partial N}{\partial x} - \frac{\partial M}{\partial y}\right) dx\, dy,$$

where C consists of the two segments of the x-axis from $-b$ to $-a$ and from a to b, and of the two semicircles, and where R is the region inside C.

The device of adding line integrals over separate boundaries to build up an integral over a single boundary can be extended to any finite number of subregions. In Fig. 17.19(a), let C_1 be the boundary of the region R_1

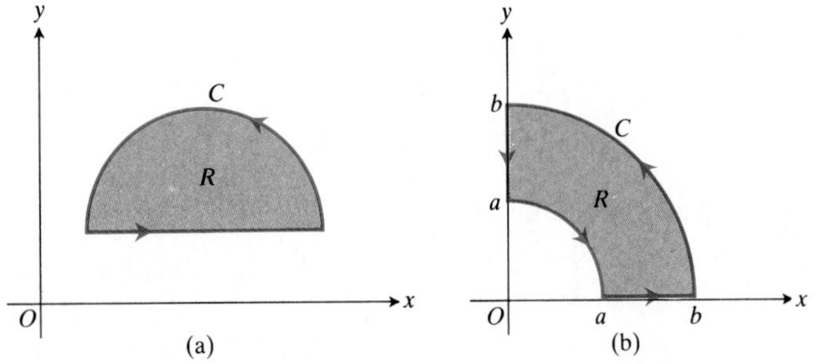

17.17 Other regions to which Green's theorem applies.

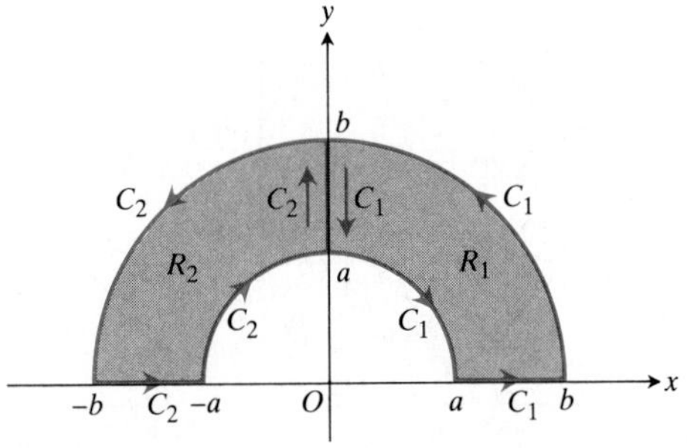

17.18 A region R that combines regions R_1 and R_2.

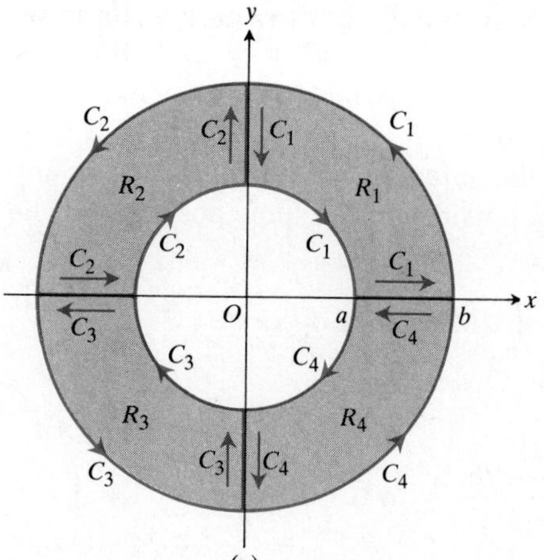

(a)

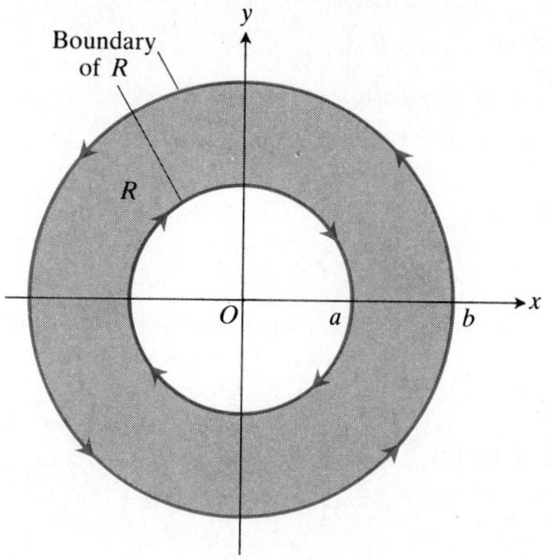

(b)

17.19 The annular region R combines four smaller regions.

in the first quadrant. Similarly for the other three quadrants: C_i is the boundary of the region R_i, $i = 1, 2, 3, 4$. By Green's theorem,

$$\oint_{C_i} M\,dx + N\,dy = \iint_{R_i} \left(\frac{\partial N}{\partial x} - \frac{\partial M}{\partial y}\right) dx\,dy. \tag{11}$$

We add Eqs. (11) for $i = 1, 2, 3, 4$, and get

$$\oint_{r=b} (M\,dx + N\,dy) + \oint_{r=a} (M\,dx + N\,dy) = \iint_{a \le r \le b} \left(\frac{\partial N}{\partial x} - \frac{\partial M}{\partial y}\right) dx\,dy. \tag{12}$$

Equation (12) says that the double integral of

$$\left(\frac{\partial N}{\partial x} - \frac{\partial M}{\partial y}\right) dx\,dy$$

over the annular ring R is equal to the line integral of $M\,dx + N\,dy$ over the *entire* boundary of R, in the *direction* along the boundary that keeps the region R on our left as we progress. Figure 17.19(b) shows R and its boundary (two concentric circles) and the positive direction on the boundary.

EXAMPLE 3 Verify Eq. (3) for

$$M = \frac{-y}{x^2 + y^2}, \qquad N = \frac{x}{x^2 + y^2},$$

$$R = \{(x, y): h^2 \le x^2 + y^2 \le 1\},$$

where $0 < h < 1$.

Solution (See Fig. 17.20.) The boundary of R consists of the circle C_1:

$$x = \cos\theta, \qquad y = \sin\theta, \qquad 0 \le \theta \le 2\pi,$$

around which we shall integrate counterclockwise, and the circle C_h:

$$x = h\cos\phi, \qquad y = h\sin\phi, \qquad 2\pi \ge \phi \ge 0,$$

around which we shall integrate in the clockwise direction. Note that the origin is not included in R, because h is positive. For all $(x, y) \ne (0, 0)$, the functions M and N and their partial derivatives are continuous. Moreover,

$$\frac{\partial M}{\partial y} = \frac{(x^2 + y^2)(-1) + y(2y)}{(x^2 + y^2)^2} = \frac{y^2 - x^2}{(x^2 + y^2)^2} = \frac{\partial N}{\partial x},$$

so

$$\iint_R \left(\frac{\partial N}{\partial x} - \frac{\partial M}{\partial y}\right) dx\,dy = \iint_R 0\,dx\,dy = 0.$$

The line integral is

$$\int_C M\,dx + N\,dy = \oint_{C_1} \frac{x\,dy - y\,dx}{x^2 + y^2} + \oint_{C_h} \frac{x\,dy - y\,dx}{x^2 + y^2}$$

$$= \int_0^{2\pi} (\cos^2\theta + \sin^2\theta)\,d\theta + \int_{2\pi}^0 \frac{h^2(\cos^2\phi + \sin^2\phi)\,d\phi}{h^2}$$

$$= 2\pi - 2\pi = 0. \ \square$$

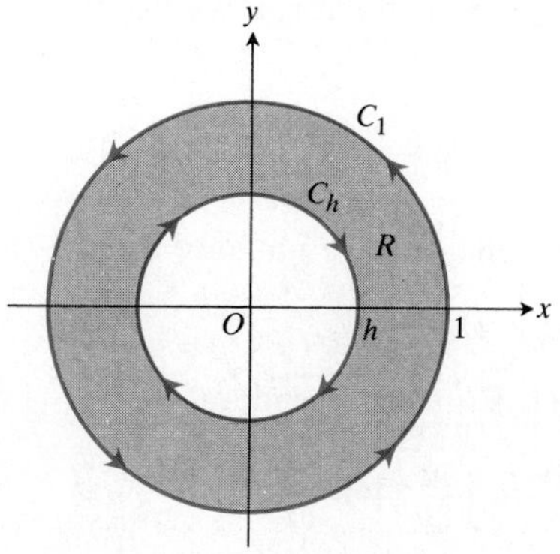

17.20 Green's theorem may be applied to the annulus R by integrating along the boundaries as shown.

In Example 3, the functions M and N are discontinuous at (0, 0), so we cannot immediately apply Green's theorem to C_1: $x^2 + y^2 = 1$, and all of the region inside it. We must delete the origin, which we did by excluding points inside C_h.

In Example 3, we could replace the outer circle C_1 by an ellipse or any other simple closed curve Γ that lies outside C_h (for some positive h). The result would be

$$\oint_{\Gamma} (M\,dx + N\,dy) + \oint_{C_h} (M\,dx + N\,dy) = 0,$$

which leads to the conclusion

$$\oint_{\Gamma} \frac{x\,dy - y\,dx}{x^2 + y^2} = 2\pi.$$

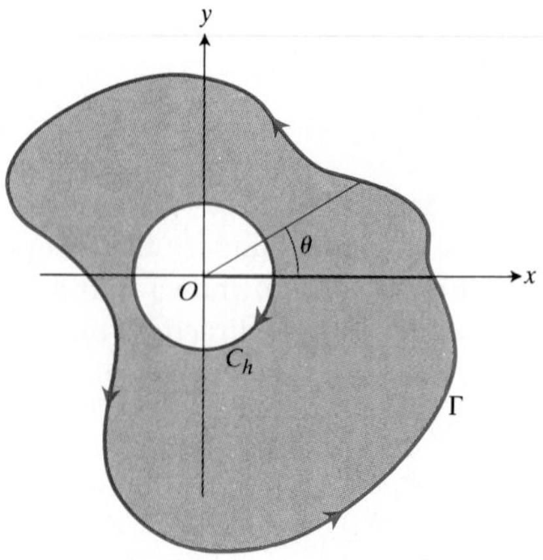

17.21 The region bounded by the circle C_h and the curve Γ.

This result is easily accounted for if we change to polar coordinates for Γ:

$$x = r\cos\theta, \qquad y = r\sin\theta,$$
$$dx = -r\sin\theta\,d\theta + \cos\theta\,dr,$$
$$dy = r\cos\theta\,d\theta + \sin\theta\,dr.$$

For then

$$\frac{x\,dy - y\,dx}{x^2 + y^2} = \frac{r^2(\cos^2\theta + \sin^2\theta)\,d\theta}{r^2} = d\theta;$$

and θ increases by 2π as we progress once around Γ counterclockwise (see Fig. 17.21).

Green's Theorem in Vector Form

Let

$$\mathbf{F} = M\mathbf{i} + N\mathbf{j} + P\mathbf{k} \qquad \text{and} \qquad \mathbf{R} = x\mathbf{i} + y\mathbf{j}.$$

Then the left-hand side of Eq. (3) is given by

$$\oint_C \mathbf{F}\cdot d\mathbf{R} = \oint_C (M\,dx + N\,dy).$$

To express the right-hand side of Eq. (3) in vector form, we use the symbolic vector operator

$$\nabla = \mathbf{i}\frac{\partial}{\partial x} + \mathbf{j}\frac{\partial}{\partial y} + \mathbf{k}\frac{\partial}{\partial z}.$$

We met the del operator in Article 15.6, where we saw that if

$$w = f(x, y, z)$$

is a differentiable scalar function, then ∇w is the gradient of w:

$$\text{grad } w = \nabla w = \mathbf{i}\frac{\partial w}{\partial x} + \mathbf{j}\frac{\partial w}{\partial y} + \mathbf{k}\frac{\partial w}{\partial z}.$$

Other uses of the del operator are given in Problem 44, Article 17.3, where the curl of a vector $\mathbf{F}$ is defined as del cross $\mathbf{F}$. Thus, if $\mathbf{F} = M\mathbf{i} + N\mathbf{j} + P\mathbf{k}$, then

$$\text{curl } \mathbf{F} = \nabla \times \mathbf{F} = \begin{vmatrix} \mathbf{i} & \mathbf{j} & \mathbf{k} \\ \dfrac{\partial}{\partial x} & \dfrac{\partial}{\partial y} & \dfrac{\partial}{\partial z} \\ M & N & P \end{vmatrix}$$

$$= \mathbf{i}\left(\frac{\partial P}{\partial y} - \frac{\partial N}{\partial z}\right) + \mathbf{j}\left(\frac{\partial M}{\partial z} - \frac{\partial P}{\partial x}\right) + \mathbf{k}\left(\frac{\partial N}{\partial x} - \frac{\partial M}{\partial y}\right).$$

The component of curl $\mathbf{F}$ that is normal to a region R in the xy-plane is

$$(\nabla \times \mathbf{F}) \cdot \mathbf{k} = \frac{\partial N}{\partial x} - \frac{\partial M}{\partial y}.$$

Hence Green's theorem can be written in vector form as

$$\oint_C \mathbf{F} \cdot d\mathbf{R} = \iint_R \text{curl } \mathbf{F} \cdot \mathbf{k} \, dx \, dy = \iint_R (\nabla \times \mathbf{F}) \cdot d\mathbf{A}, \qquad (13)$$

where $d\mathbf{A} = \mathbf{k}\, dx\, dy$ is a vector normal to the region R and of magnitude $|d\mathbf{A}| = dx\, dy$. In words, Green's theorem states that the integral around C of the tangential component of $\mathbf{F}$ is equal to the integral, over the region R bounded by C, of the component of curl $\mathbf{F}$ that is normal to R. This integral is the flux of curl $\mathbf{F}$ through R. We shall later extend this result to more general curves and surfaces in a formulation known as Stokes's theorem.

There is a second, *normal,* vector form for Green's theorem. It involves the gradient operator ∇ in another form, one that produces the *divergence.* Now the integrand of the line integral of Eq. (13) is the *tangential* component of the field $\mathbf{F}$ because

$$\mathbf{F} \cdot d\mathbf{R} = \left(\mathbf{F} \cdot \frac{d\mathbf{R}}{ds}\right) ds = (\mathbf{F} \cdot \mathbf{T})\, ds.$$

As in Article 17.4, if we let

$$\mathbf{F} = \mathbf{i}M(x, y) + \mathbf{j}N(x, y),$$

and let $\mathbf{G}$ be the orthogonal field given by

$$\mathbf{G} = \mathbf{i}N(x, y) - \mathbf{j}M(x, y),$$

then it follows that

$$\mathbf{F} \cdot \mathbf{T} = \mathbf{G} \cdot \mathbf{n} = M\frac{dx}{ds} + N\frac{dy}{ds}$$

because

$$\mathbf{T} = \mathbf{i}\frac{dx}{ds} + \mathbf{j}\frac{dy}{ds}, \qquad \mathbf{n} = \mathbf{i}\frac{dy}{ds} - \mathbf{j}\frac{dx}{ds}.$$

Therefore Green's theorem, which says that

$$\oint_C (M\, dx + N\, dy) = \iint_R \left[\frac{\partial N}{\partial x} - \frac{\partial M}{\partial y}\right] dx\, dy,$$

also says that

$$\oint_C \mathbf{G}\cdot\mathbf{n}\,ds = \iint_R \nabla\cdot\mathbf{G}\,dx\,dy, \qquad (14)$$

where

$$\nabla\cdot\mathbf{G} = \text{div}\,\mathbf{G} = \frac{\partial N}{\partial x} + \frac{\partial(-M)}{\partial y}.$$

In words, Eq. (14) says that the line integral of the *normal* component of any vector field **G** around the boundary of a region R in which **G** is continuous and has continuous partial derivatives is equal to the double integral of the *divergence* of **G** over R. In the next article, we shall extend this result to three-dimensional vector fields and discuss the physical interpretation of the divergence. For a three-dimensional vector field

$$\mathbf{F}(x, y, z) = \mathbf{i}M(x, y, z) + \mathbf{j}N(x, y, z) + \mathbf{k}P(x, y, z),$$

the *divergence* is defined to be

$$\text{div}\,\mathbf{F} = \nabla\cdot\mathbf{F} = \frac{\partial M}{\partial x} + \frac{\partial N}{\partial y} + \frac{\partial P}{\partial z}. \qquad (15)$$

PROBLEMS

Verify Green's formula, Eq. (3), with C the circle $x^2 + y^2 = a^2$ and R the disc $x^2 + y^2 \leq a^2$, for the functions M and N given in Problems 1–6.

1. $M = x, \quad N = y$ **2.** $M = N = xy$

3. $M = -x^2y, \quad N = xy^2$ **4.** $M = N = e^{x+y}$

5. $M = y, \quad N = 0$ **6.** $M = -y, \quad N = x$

7. Verify Green's theorem for the vector field $\mathbf{F} = xy\mathbf{i} + y^2\mathbf{j}$ and the region **R** enclosed between the parabola $y = x^2$ and the line $y = x$ in the first quadrant.

8. Use Green's theorem to find the area between the ellipse $x = 3\cos t$, $y = 2\sin t$ and the circle $x = \cos t$, $y = \sin t$.

9. Find the area enclosed by the hypocycloid $x = a\cos^3 t$, $y = a\sin^3 t$. (Figure 13.21 shows a portion of this curve.)

10. Let C be the boundary of a region on which Green's theorem holds. Use Green's theorem to calculate

a) $\oint_C f(x)\,dx + g(y)\,dy,$

b) $\oint_C k\,y\,dx + h\,x\,dy$ (k and h constants).

Apply Green's theorem to evaluate the line integrals in Problems 11–16.

11. C is the triangle bounded by $x = 0$, $x + y = 1$, $y = 0$:

$$\oint_C (y^2\,dx + x^2\,dy).$$

12. C is the boundary of $0 \leq x \leq \pi$, $0 \leq y \leq \sin x$:

$$\oint_C (3y\,dx + 2x\,dy).$$

13. C is the circle $(x - 2)^2 + (y - 3)^2 = 4$:

$$\oint_C (6y + x)\,dx + (y + 2x)\,dy.$$

14. C is any simple closed curve in the plane for which Green's theorem holds:

$$\oint_C (2x + y^2)\,dx + (2xy + 3y)\,dy.$$

15. C is the boundary of the "triangular" region in the first quadrant enclosed by the x-axis, the line $x = 1$, and the curve $y = x^3$:

$$\oint_C 2xy^3\,dx + 4x^2y^2\,dy.$$

16. C is the circle $(x - 2)^2 + (y - 2)^2 = 4$:

$$\oint_C (4x - 2y)\,dx + (2x - 4y)\,dy.$$

17. Show that $\oint_C 4x^3y\,dx + x^4\,dy = 0$ for any closed curve C to which Green's theorem applies.

18. Show that $\oint_C -y^3\,dx + x^3\,dy$ is positive for any closed curve C to which Green's theorem applies.

19. Show that the value of $\oint_C xy^2\,dx + (x^2y + 2x)\,dy$ around any square depends only on the size of the square and not on its location in the plane.

20. Assuming that all the necessary derivatives exist and are continuous, show that if $f(x, y)$ satisfies the Laplace equation

$$\frac{\partial^2 f}{\partial x^2} + \frac{\partial^2 f}{\partial y^2} = 0,$$

then

$$\oint_C \frac{\partial f}{\partial y}\,dx - \frac{\partial f}{dx}\,dy = 0$$

for all closed curves C to which Green's theorem applies. (The converse is also true: if the line integral is always zero, then f satisfies the Laplace equation.)

21. Let

$$\mathbf{F} = \left(\frac{1}{4}x^2y + \frac{1}{3}y^3\right)\mathbf{i} + x\mathbf{j}.$$

Among all smooth simple closed curves in the plane, oriented counterclockwise, find the curve around which the work done by **F** is the greatest. (*Hint:* Where is (curl **F**) · **k** positive?)

22. Supply the details necessary to establish Eq. (6).

23. Supply the steps necessary to establish Eq. (10).

24. Suppose that

$$R = \{(x, y)\colon 0 \le y \le \sqrt{a^2 - x^2},\quad -a \le x \le a\},$$

and that C is the boundary of R.

a) Sketch R and C.

b) Write out the proof of Green's theorem for this region.

Definition. A region R is said to be simply connected if every simple closed curve lying in R can be continuously contracted to a point without its touching any part of the boundary of R. Examples are the interiors of circles, ellipses, cardioids, and rectangles; and, in three dimensions, the region between two concentric spheres. (The annular ring in Fig. 17.20 is not simply connected. Also, see Fig. 15.47.)

25. Show, by a geometric argument, that Green's formula, Eq. (3), holds for any simply connected region R whose boundary is a simple closed curve C, provided R can be decomposed into a finite number of nonoverlapping regions $R_1, R_2, \ldots, R_n$ with boundaries $C_1, C_2, \ldots, C_n$ of a type for which the formula (3) is true for each R_i and C_i, $i = 1, \ldots, n$.

26. Suppose R is a region in the xy-plane, C is its boundary, and the area of R is given by

$$A(R) = \oint_C \frac{1}{2}(x\,dy - y\,dx).$$

Suppose the equations $x = f(u, v)$, $y = g(u, v)$ map R and C in a continuous and one-to-one manner onto a region R' and curve C', respectively, in the uv-plane. Use Green's formula to show that

$$\iint_R dx\,dy = \iint_{R'} \begin{vmatrix} f_u & f_v \\ g_u & g_v \end{vmatrix} du\,dv$$

$$= \iint_{R'} \left(\frac{\partial f}{\partial u}\frac{\partial g}{\partial v} - \frac{\partial f}{\partial v}\frac{\partial g}{\partial u}\right) du\,dv.$$

(*Hint:* Note that

$$\iint_R dx\,dy = \frac{1}{2}\int_C (x\,dy - y\,dx)$$

$$= \frac{1}{2}\int_{C'} \left[f(u, v)\left(\frac{\partial g}{\partial u}\,du + \frac{\partial g}{\partial v}\,dv\right)\right.$$

$$\left. - g(u, v)\left(\frac{\partial f}{\partial u}\,du + \frac{\partial f}{\partial v}\,dv\right)\right],$$

and apply Green's formula to C' and R'.)

27. Rewrite Eq. (14) in nonvector notation for a vector field $\mathbf{F} = \mathbf{i}M(x, y) + \mathbf{j}N(x, y)$ in place of **G**. (In other words, first write it in vector form with **F** in place of **G**, and then translate the result into nonvector notation.)

***Toolkit* programs**

Scalar Fields   Vector Fields

17.6 The Divergence Theorem

The divergence theorem states that under appropriate conditions the triple integral

$$\iiint_D \operatorname{div} \mathbf{F}\,dV \qquad (1)$$

is equal to the double integral

$$\iint_S \mathbf{F}\cdot\mathbf{n}\, d\sigma. \tag{2}$$

Here $\mathbf{F} = \mathbf{i}M + \mathbf{j}N + \mathbf{k}P$, with M, N, and P continuous functions of (x, y, z) that have continuous first-order partial derivatives;

$$\operatorname{div}\mathbf{F} = \frac{\partial M}{\partial x} + \frac{\partial N}{\partial y} + \frac{\partial P}{\partial z};$$

$\mathbf{n}\, d\sigma$ is a vector element of surface area directed along the unit outer normal vector $\mathbf{n}$; and S is the surface enclosing the region D. We shall first show that (1) and (2) are equal if D is some convex region with no holes, such as the interior of a sphere, or a cube, or an ellipsoid, and if S is a piecewise smooth surface. In addition, we assume that the projection of D into the xy-plane is a simply connected region R_{xy} and that any line perpendicular to the xy-plane at an interior point of R_{xy} intersects the surface S in at most two points, producing surfaces S_1 and S_2:

$$S_1:\quad z_1 = f_1(x, y),\quad (x, y) \text{ in } R_{xy},$$
$$S_2:\quad z_2 = f_2(x, y),\quad (x, y) \text{ in } R_{xy},$$

with $z_1 \le z_2$. Similarly for the projection of D onto the other coordinate planes.

If we write the unit normal vector $\mathbf{n}$ in terms of its direction cosines, as

$$\mathbf{n} = \mathbf{i}\cos\alpha + \mathbf{j}\cos\beta + \mathbf{k}\cos\gamma,$$

then

$$\mathbf{F}\cdot(\mathbf{n}\, d\sigma) = (\mathbf{F}\cdot\mathbf{n})\, d\sigma = (M\cos\alpha + N\cos\beta + P\cos\gamma)\, d\sigma; \tag{3}$$

and the divergence theorem states that

$$\iiint_D \left(\frac{\partial M}{\partial x} + \frac{\partial N}{\partial y} + \frac{\partial P}{\partial z}\right) dx\, dy\, dz = \iint_S (M\cos\alpha + N\cos\beta + P\cos\gamma)\, d\sigma. \tag{4}$$

We see that both sides of Eq. (4) are additive with respect to M, N, and P, and that our task is accomplished if we prove

$$\iiint \frac{\partial M}{\partial x}\, dx\, dy\, dz = \iint M\cos\alpha\, d\sigma, \tag{5a}$$

$$\iiint \frac{\partial N}{\partial y}\, dx\, dy\, dz = \iint N\cos\beta\, d\sigma, \tag{5b}$$

$$\iiint \frac{\partial P}{\partial z}\, dx\, dy\, dz = \iint P\cos\gamma\, d\sigma. \tag{5c}$$

We shall establish (5c) in detail.

Figure 17.22 illustrates the projection of D into the xy-plane. The surface S consists of the *upper part*

$$S_2:\quad z = f_2(x, y),\quad (x, y) \text{ in } R_{xy},$$

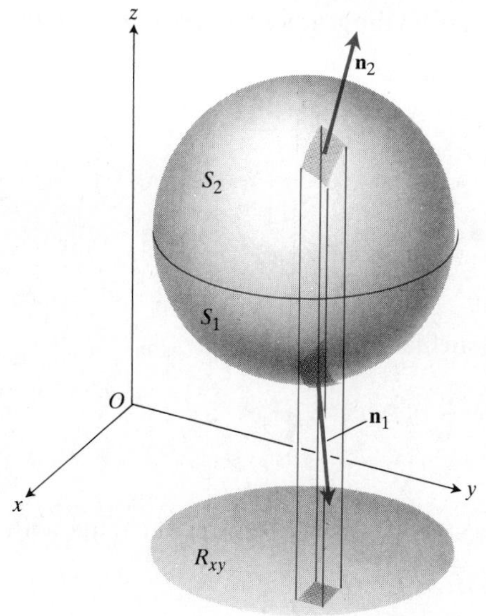

17.22 Regions S_1 and S_2 project onto R_{xy}.

and the *lower part:*

$$S_1: \quad z = f_1(x, y), \quad (x, y) \text{ in } R_{xy}.$$

On the surface S_2, the outer normal has a positive **k**-component, and

$$\cos \gamma_2 \, d\sigma_2 = dx\, dy \tag{6a}$$

is the projection of $d\sigma$ into R_{xy}. On the surface S_1, the outer normal has a negative **k**-component, and

$$\cos \gamma_1 \, d\sigma_1 = -dx\, dy. \tag{6b}$$

Therefore we can evaluate the surface integral on the right-hand side of Eq. (5c) in the following way:

$$\begin{aligned}\iint_S P \cos \gamma \, d\sigma &= \iint_{S_2} P_2 \cos \gamma_2 \, d\sigma_2 + \iint_{S_1} P_1 \cos \gamma_1 \, d\sigma_1 \\ &= \iint_{R_{xy}} P(x, y, z_2)\, dx\, dy - \iint_{R_{xy}} P(x, y, z_1)\, dx\, dy \\ &= \iint_{R_{xy}} [P(x, y, z_2) - P(x, y, z_1)]\, dx\, dy \\ &= \iint_{R_{xy}} \left[\int_{z_1}^{z_2} \frac{\partial P}{\partial z} dz \right] dx\, dy \\ &= \iiint_D \frac{\partial P}{\partial z} dz\, dx\, dy. \end{aligned} \tag{7}$$

Thus we have established Eq. (5c). Proofs for (5a) and (5b) follow the same pattern; or just permute x, y, z; M, N, P; α, β, γ, in order, and get those results from (5c) by renaming the axes. Finally, by addition of (5a, b, c), we get Eq. (4):

THEOREM

Divergence Theorem

$$\iiint_D \operatorname{div} \mathbf{F}\, dV = \iint_S \mathbf{F} \cdot \mathbf{n}\, d\sigma. \tag{8}$$

EXAMPLE 1 Verify Eq. (8) for the sphere

$$x^2 + y^2 + z^2 = a^2$$

if

$$\mathbf{F} = \mathbf{i}x + \mathbf{j}y + \mathbf{k}z.$$

Solution

$$\operatorname{div} \mathbf{F} = \frac{\partial x}{\partial x} + \frac{\partial y}{\partial y} + \frac{\partial z}{\partial z} = 3,$$

so

$$\iiint_D \operatorname{div} \mathbf{F}\, dV = \iiint_D 3\, dV = 3\left(\frac{4}{3}\pi a^3\right) = 4\pi a^3.$$

The outer unit normal to S, calculated from the gradient of $f(x, y, z) = x^2 + y^2 + z^2 - a^2$, is

$$\mathbf{n} = \frac{2(x\mathbf{i} + y\mathbf{j} + z\mathbf{k})}{\sqrt{4(x^2 + y^2 + z^2)}} = \frac{x\mathbf{i} + y\mathbf{j} + z\mathbf{k}}{a},$$

and

$$\mathbf{F} \cdot \mathbf{n}\, d\sigma = \frac{x^2 + y^2 + z^2}{a}\, d\sigma = \frac{a^2}{a}\, d\sigma = a\, d\sigma,$$

because $x^2 + y^2 + z^2 = a^2$ on the surface. Therefore

$$\iint_S \mathbf{F} \cdot d\sigma = \iint_S a\, d\sigma = a(4\pi a^2) = 4\pi a^3. \quad \square$$

EXAMPLE 2 Show that the divergence theorem holds for the cube with faces in the planes

$$\begin{aligned} x &= x_0, & x &= x_0 + h, \\ y &= y_0, & y &= y_0 + h, \\ z &= z_0, & z &= z_0 + h, \end{aligned}$$

where h is a positive constant.

Solution We compute $\iint \mathbf{F} \cdot \mathbf{n}\, d\sigma$ as the sum of the integrals over the six faces separately. We begin with the two faces perpendicular to the x-axis. For the face $x = x_0$ and the face $x = x_0 + h$, respectively, we have the first and second lines of the following table.

Range of integration	**Outward unit normal**	$\iint(\mathbf{F} \cdot \mathbf{n})\, d\sigma$
$y_0 \le y \le y_0 + h,\ z_0 \le z \le z_0 + h$	$-\mathbf{i}$	$-\iint M(x_0, y, z)\, dy\, dz$
$y_0 \le y \le y_0 + h,\ z_0 \le z \le z_0 + h$	$\mathbf{i}$	$+\iint M(x_0 + h, y, z)\, dy\, dz$

The sum of the surface integrals over these two faces is

$$\begin{aligned} \iint (\mathbf{F} \cdot \mathbf{n})\, d\sigma &= \iint [M(x_0 + h, y, z) - M(x_0, y, z)]\, dy\, dz \\ &= \iint \left(\int_{x_0}^{x_0+h} \frac{\partial M}{\partial x}\, dx \right) dy\, dz \\ &= \iiint_D \frac{\partial M}{\partial x}\, dV. \end{aligned}$$

Similarly the sum of the surface integrals over the two faces perpendicular to the y-axis is equal to

$$\iiint_D (\partial N/\partial y)\, dV,$$

and the sum of the surface integrals over the other two faces is equal to

$$\iiint_D \left(\frac{\partial P}{\partial z}\right) dV.$$

Hence the surface integral over the six faces is equal to the sum of three volume integrals, and Eq. (8) holds for the cube:

$$\iint_S \mathbf{F}\cdot\mathbf{n}\,d\sigma = \iiint_D \left(\frac{\partial M}{\partial x} + \frac{\partial N}{\partial y} + \frac{\partial P}{\partial z}\right) dV$$

$$= \iiint_D \operatorname{div}\mathbf{F}\,dV. \square$$

The Divergence Theorem for Other Regions

The divergence theorem can be extended to more complex regions that can be split up into a finite number of simple regions of the type discussed, and to regions that can be defined as limits of simpler regions in certain ways. For example, suppose D is the region between two concentric spheres, and $\mathbf{F}$ has continuously differentiable components throughout D and on the bounding surfaces. Split D by an equatorial plane and apply the divergence theorem to each half separately. The top half, D_1, is shown in Fig. 17.23. The surface that bounds D_1 consists of an outer hemisphere, a plane washer-shaped base, and an inner hemisphere. The divergence theorem says that

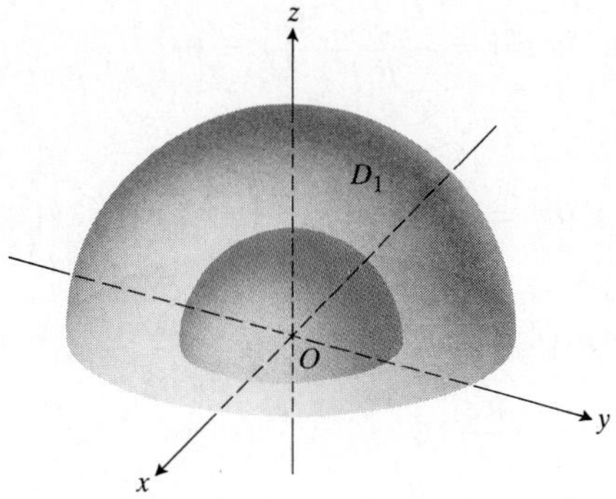

17.23 Upper half of the region between two spheres.

$$\iiint_{D_1} \operatorname{div}\mathbf{F}\,dV_1 = \iint_{S_1} \mathbf{F}\cdot\mathbf{n}_1\,d\sigma_1. \tag{9a}$$

The unit normal $\mathbf{n}_1$ that points outward from D_1 points away from the origin along the outer surface, points down along the flat base, and points toward the origin along the inner surface. Next apply the divergence theorem to D_2, as shown in Fig. 17.24:

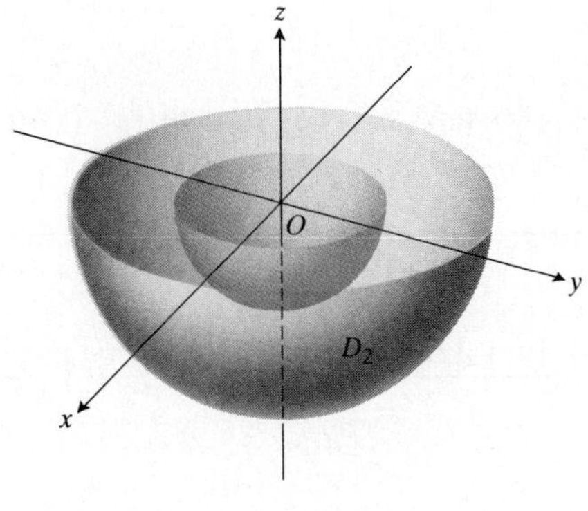

17.24 Lower half of the region between two spheres.

$$\iiint_{D_2} \operatorname{div}\mathbf{F}\,dV_2 = \iint_{S_2} \mathbf{F}\cdot\mathbf{n}_2\,d\sigma_2. \tag{9b}$$

As we follow $\mathbf{n}_2$ over S_2, pointing outward from D_2, we see that $\mathbf{n}_2$ points upward along the flat surface in the xy-plane, points away from the origin on the outer sphere, and points toward the origin on the inner sphere. When we add (9a) and (9b), the surface integrals over the flat base cancel because of the opposite signs of $\mathbf{n}_1$ and $\mathbf{n}_2$. We thus arrive at the result

$$\iiint_D \operatorname{div}\mathbf{F}\,dV = \iint_S \mathbf{F}\cdot\mathbf{n}\,d\sigma, \tag{10}$$

with D the region between the spheres, S the boundary of D consisting of two spheres, and $\mathbf{n}$ the unit normal to S directed outward from D.

EXAMPLE 3 Verify Eq. (10) for the region

$$1 \le x^2 + y^2 + z^2 \le 4$$

if

$$\mathbf{F} = -\frac{\mathbf{i}x + \mathbf{j}y + \mathbf{k}z}{\rho^3}, \qquad \rho = \sqrt{x^2 + y^2 + z^2}.$$

Solution Observe that

$$\frac{\partial \rho}{\partial x} = \frac{x}{\rho}$$

and

$$\frac{\partial}{\partial x}(x\rho^{-3}) = \rho^{-3} - 3x\rho^{-4}\frac{\partial \rho}{\partial x} = \frac{1}{\rho^3} - \frac{3x^2}{\rho^5}.$$

Thus, throughout the region $1 \leq \rho \leq 2$, all functions considered are continuous, and

$$\text{div } \mathbf{F} = \frac{-3}{\rho^3} + \frac{3}{\rho^5}(x^2 + y^2 + z^2) = -\frac{3}{\rho^3} + \frac{3\rho^2}{\rho^5} = 0.$$

Therefore

$$\iiint_D \text{div } \mathbf{F} \, dV = 0. \tag{11}$$

On the outer sphere ($\rho = 2$), the positive unit normal is

$$\mathbf{n} = \frac{\mathbf{i}x + \mathbf{j}y + \mathbf{k}z}{\rho},$$

and

$$\mathbf{F} \cdot \mathbf{n} \, d\sigma = -\frac{x^2 + y^2 + z^2}{\rho^4} \, d\sigma = -\frac{1}{\rho^2} \, d\sigma.$$

Hence

$$\iint_{\rho=2} \mathbf{F} \cdot \mathbf{n} \, d\sigma = -\frac{1}{4}\iint_{\rho=2} d\sigma = -\frac{1}{4} \cdot 4\pi\rho^2 = -\pi\rho^2 = -4\pi. \tag{12a}$$

On the inner sphere ($\rho = 1$), the positive unit normal points toward the origin; its equation is

$$\mathbf{n} = \frac{-(\mathbf{i}x + \mathbf{j}y + \mathbf{k}z)}{\rho}.$$

Hence

$$\mathbf{F} \cdot \mathbf{n} \, d\sigma = +\frac{x^2 + y^2 + z^2}{\rho^4} = \frac{1}{\rho^2} \, d\sigma.$$

Thus

$$\iint_{\rho=1} \mathbf{F} \cdot \mathbf{n} \, d\sigma = \iint_{\rho=1} \frac{1}{\rho^2} \, d\sigma = \frac{1}{\rho^2} \cdot 4\pi\rho^2 = 4\pi. \tag{12b}$$

The sum of (12a) and (12b) is the surface integral over the complete boundary of D:

$$-4\pi + 4\pi = 0,$$

which agrees with (11), as it should. □

The Continuity Equation of Hydrodynamics

If $\mathbf{v}(x, y, z)$ is the velocity vector of a differentiable fluid flow through a region D in space, $\delta = \delta(x, y, z, t)$ is the density of the fluid at each point (x, y, z) at time t, and $\mathbf{F} = \delta\mathbf{v}$, then the *continuity equation* is the statement that

$$\text{div } \mathbf{F} + \frac{\partial \delta}{\partial t} = 0. \tag{13}$$

The continuity equation evolves naturally from the divergence theorem

$$\iiint_D \text{div } \mathbf{F} \, dV = \iint_S \mathbf{F} \cdot \mathbf{n} \, d\sigma$$

if the functions involved are continuous, as we shall now show.

First of all, the integral

$$\iint_S \mathbf{F} \cdot \mathbf{n} \, d\sigma$$

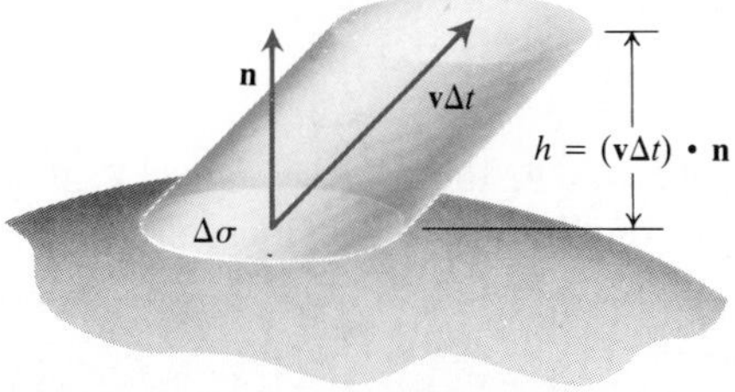

17.25 The fluid that flows upward through the patch $\Delta\sigma$ in a short time Δt fills a "cylinder" whose volume is approximately base × height = $\mathbf{v} \cdot \mathbf{n} \, \Delta t \, \Delta\sigma$.

is the rate at which mass leaves D across S (leaves, because $\mathbf{n}$ is the outer normal). To see why, consider a patch of area $\Delta\sigma$ on the surface (Fig. 17.25). In a short time interval Δt, the volume ΔV of fluid that flows across the patch is approximately equal to the volume of a cylinder with base area $\Delta\sigma$ and height $(\mathbf{v} \, \Delta t) \cdot \mathbf{n}$, where $\mathbf{v}$ is a velocity vector rooted at a point of the patch:

$$\Delta V \approx \mathbf{v} \cdot \mathbf{n} \, \Delta\sigma \, \Delta t.$$

The mass of this volume of fluid is about

$$\Delta m \approx \delta\mathbf{v} \cdot \mathbf{n} \, \Delta\sigma \, \Delta t,$$

so that the rate at which mass is flowing out of D across the patch is about

$$\frac{\Delta m}{\Delta t} \approx \delta\mathbf{v} \cdot \mathbf{n} \, \Delta\sigma.$$

This leads to the approximation

$$\frac{\Sigma \, \Delta m}{\Delta t} \approx \sum \delta\mathbf{v} \cdot \mathbf{n} \, \Delta\sigma \tag{14}$$

as an estimate of the average rate at which mass flows across S. Finally, letting $\Delta\sigma \to 0$ and $\Delta t \to 0$ gives the instantaneous rate at which mass leaves D across S as

$$\frac{dm}{dt} = \iint_S \delta\mathbf{v} \cdot \mathbf{n} \, d\sigma, \tag{15}$$

which, for our particular flow, is

$$\frac{dm}{dt} = \iint_S \mathbf{F} \cdot \mathbf{n} \, d\sigma. \tag{16}$$

Now let B be a ball centered at a point Q in the flow. The average value of div **F** over the ball is

$$\frac{1}{\text{volume}} \iiint_B \operatorname{div} \mathbf{F}\, dV. \tag{17}$$

It is a consequence of the continuity of div **F** that div **F** actually takes on this value at some point P in B. Thus,

$$\begin{aligned}(\operatorname{div} \mathbf{F})_P &= \frac{1}{\text{volume}} \iiint_B \operatorname{div} \mathbf{F}\, dV \\ &= \frac{\iint_S \mathbf{F} \cdot \mathbf{n}\, d\sigma}{\text{volume}} \\ &= \frac{\text{rate at which mass leaves } B \text{ across its surface } S}{\text{volume of } B}. \end{aligned} \tag{18}$$

The fraction on the right describes decrease in mass per unit volume.

Now let the radius of the ball B approach 0 while the center Q stays fixed. The left-hand side of Eq. (18) converges to $(\operatorname{div} \mathbf{F})_Q$, the right side to $(-\partial\delta/\partial t)_Q$. The equality of these two limits is the continuity equation

$$\operatorname{div} \mathbf{F} = -\frac{\partial \delta}{\partial t}. \tag{19}$$

The continuity equation "explains" div **F**: the divergence of **F** at a point is the rate at which the density of the fluid is decreasing there. If the fluid is incompressible, its density is constant and div $\mathbf{F} = 0$.

The divergence theorem

$$\iiint_D \operatorname{div} \mathbf{F}\, dV = \iint_S \mathbf{F} \cdot \mathbf{n}\, d\sigma$$

now says that the net decrease in density of the region D is accounted for by the mass transported across the surface S. In a way, this is a statement about conservation of mass.

PROBLEMS

In Problems 1–5, verify the divergence theorem for the cube with center at the origin and faces in the planes $x = \pm 1$, $y = \pm 1$, $z = \pm 1$, and **F** as given.

1. $\mathbf{F} = 2\mathbf{i} + 3\mathbf{j} + 4\mathbf{k}$

2. $\mathbf{F} = \mathbf{i}x + \mathbf{j}y + \mathbf{k}z$

3. $\mathbf{F} = \mathbf{i}yz + \mathbf{j}xz + \mathbf{k}xy$

4. $\mathbf{F} = \mathbf{i}(x - y) + \mathbf{j}(y - z) + \mathbf{k}(x - y)$

5. $\mathbf{F} = \mathbf{i}x^2 + \mathbf{j}y^2 + \mathbf{k}z^2$

In Problems 6–10, compute both

$$\iiint_D \operatorname{div} \mathbf{F}\, dV \quad \text{and} \quad \iint_S \mathbf{F} \cdot \mathbf{n}\, d\sigma$$

directly. Compare the results with the divergence theorem expressed by Eq. (8), given that

$$\mathbf{F} = \mathbf{i}(x + y) + \mathbf{j}(y + z) + \mathbf{k}(z + x),$$

and given that S bounds the region D given in the problem.

6. $0 \le z \le 4 - x^2 - y^2,\quad 0 \le x^2 + y^2 \le 4$

7. $-4 + x^2 + y^2 \le z \le 4 - x^2 - y^2,\quad 0 \le x^2 + y^2 \le 4$

8. $0 \le x^2 + y^2 \le 9,\quad 0 \le z \le 5$

9. $0 \le x^2 + y^2 + z^2 \le a^2$

10. $|x| \le 1,\quad |y| \le 1,\quad |z| \le 1$

Use the divergence theorem to evaluate the surface integral

$$\iint_S \mathbf{F}\cdot\mathbf{n}\,d\sigma$$

for the surfaces and fields given in Problems 11–14. Take $\mathbf{n}$ to be the outer unit normal in each case.

11. S is the sphere $x^2+y^2+z^2=4$, and $\mathbf{F}=2x\mathbf{i}+xz\mathbf{j}+z\mathbf{k}$.

12. S is the surface of the solid in the first octant bounded by the coordinate planes and the sphere $x^2+y^2+z^2=4$, and $\mathbf{F}=x^2\mathbf{i}-2xy\mathbf{j}+3xz\mathbf{k}$.

13. S is the surface of the solid in the first octant bounded by the coordinate planes, the cylinder $x^2+y^2=4$, and the plane $z=4$, and $\mathbf{F}=(6x^2+2xy)\mathbf{i}+(2y+x^2z)\mathbf{j}+4x^2y^3\mathbf{k}$.

14. S is the surface of the wedge in the first octant bounded by the coordinate planes $x=0$ and $z=0$, the plane $z=y$, and the elliptical cylinder $x^2+4y^2=16$, and $\mathbf{F}=2xz\mathbf{i}+y\mathbf{j}-z^2\mathbf{k}$.

15. Find the rate of mass transport dm/dt outward through the surface of the closed cylinder $x^2+y^2\le 4$, $-1\le z\le 1$, if the flow vector $\mathbf{F}=\delta\mathbf{v}$ is given by $\mathbf{F}=-y\mathbf{i}+x\mathbf{j}+z\mathbf{k}$.

16. Let S be the spherical cap $x^2+y^2+z^2=2a^2$, $z\ge a$, together with its base $x^2+y^2\le a^2$, $z=a$. Find the flux of $\mathbf{F}=xz\mathbf{i}-yz\mathbf{j}+y^2\mathbf{k}$ outward through S (a) by evaluating $\iint_S \mathbf{F}\cdot\mathbf{n}\,d\sigma$ directly, (b) by applying the divergence theorem.

17. Let S be the closed cube-like surface shown in Fig. 17.26, with its base the unit square in the xy-plane, its four sides lying in the planes $x=0$, $x=1$, $y=0$, $y=1$, and its top an arbitrary smooth surface whose identity is unknown. Let $\mathbf{F}=x\mathbf{i}-2y\mathbf{j}+(z+3)\mathbf{k}$. Suppose that the outward flux of $\mathbf{F}$ through side A is 1 and through side B is -3. Calculate the outward flux of $\mathbf{F}$ through the top.

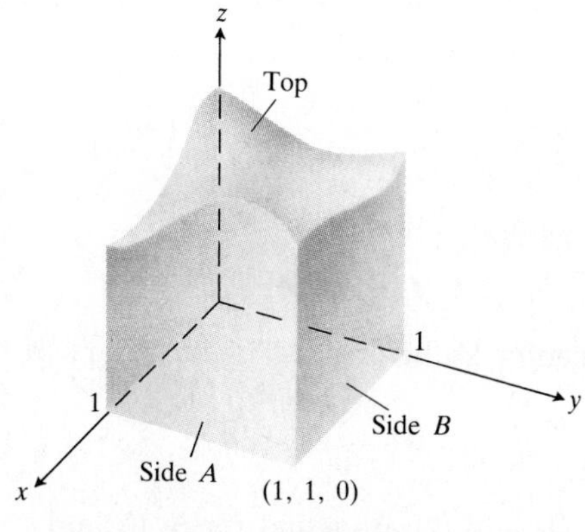

17.26 Problem 17 calculates the flux of a vector field through the top of this solid.

18. Let $\mathbf{F}(x, y, z)$ be a vector field with components that are continuous and differentiable on a portion of space containing a (finite) region D and its bounding surface S. Suppose that the length of the vector $\mathbf{F}(x, y, z)$ never exceeds 1 when (x, y, z) is a point of S. What bound can be placed on the numerical size of the integral

$$\iiint_D \operatorname{div}\mathbf{F}\,dV?$$

Explain.

19. a) Show that the flux of the position vector $\mathbf{F}=x\mathbf{i}+y\mathbf{j}+z\mathbf{k}$ outward through a piecewise smooth closed surface is three times the volume contained by the surface.
b) Verify part (a) for the cube bounded by the planes $x=\pm1$, $y=\pm1$, $z=\pm1$.
c) Verify part (a) for the sphere $x^2+y^2+z^2=a^2$.
d) Let $\mathbf{n}$ be the outward unit normal to a piecewise smooth closed surface S. Show that it is impossible for the position vector $\mathbf{F}=x\mathbf{i}+y\mathbf{j}+z\mathbf{k}$ to be orthogonal to $\mathbf{n}$ at every point of S.

20. Let $\mathbf{F}=M(x, y, z)\mathbf{i}+N(x, y, z)\mathbf{j}+P(x, y, z)\mathbf{k}$ be a vector field whose components M, N, and P are continuous and have continuous second partial derivatives of all kinds.
a) Show by direct computation that $\operatorname{div}(\operatorname{curl}\mathbf{F})=0$.
b) Use the result of part (a) to show that

$$\iint_S (\operatorname{curl}\mathbf{F})\cdot\mathbf{n}\,d\sigma=0$$

for any surface to which the divergence theorem applies.

21. Verify the divergence theorem, Eq. (8), for $\mathbf{F}=x\mathbf{i}+y\mathbf{j}+z\mathbf{k}$ over the region D: $4\le x^2+y^2+z^2\le 9$.

22. *Identities involving div, grad, and curl.*
a) Prove that if ϕ is a scalar function of x, y, z, then

$$\operatorname{curl}(\operatorname{grad}\phi)=\nabla\times(\nabla\phi)=\mathbf{0}.$$

b) State in terms of the vector field $\nabla\times\mathbf{F}$ how would you express the condition that $\mathbf{F}\cdot d\mathbf{R}$ be an exact differential.

Prove the following results: If

$$\mathbf{r}=x\mathbf{i}+y\mathbf{j}+z\mathbf{k},$$

c) $\operatorname{div}(\phi\mathbf{F})\equiv\nabla\cdot(\phi\mathbf{F})=\phi\nabla\cdot\mathbf{F}+\mathbf{F}\cdot\nabla\phi$;
d) $\nabla\times(\phi\mathbf{F})=\phi\nabla\times\mathbf{F}+(\nabla\phi)\times\mathbf{F}$;
e) $\nabla\cdot(\mathbf{F}_1\times\mathbf{F}_2)=\mathbf{F}_2\cdot\nabla\times\mathbf{F}_1-\mathbf{F}_1\cdot\nabla\times\mathbf{F}_2$;
f) $\nabla\cdot\mathbf{R}=3$ and $\nabla\times\mathbf{R}=\mathbf{0}$.

23. A function f is said to be *harmonic* in a region D if throughout D it satisfies the Laplace equation

$$\frac{\partial^2 f}{\partial x^2}+\frac{\partial^2 f}{\partial y^2}+\frac{d^2 f}{\partial z^2}=0.$$

Suppose f is harmonic throughout D, S is the boundary of D, $\mathbf{n}$ is the positive unit normal on S, and $\partial f/\partial n$ is the directional derivative of f in the direction of $\mathbf{n}$. This derivative is called the normal derivative of f. Prove that

$$\iint_S \frac{\partial f}{\partial n}\, d\sigma = 0.$$

(*Hint:* Let $\mathbf{F} = \text{grad } f$.)

24. Prove that if f is harmonic in D (see Problem 23), then

$$\iint_S f\frac{\partial f}{\partial n}\, d\sigma = \iiint_D |\text{grad } f|^2\, dV.$$

(*Hint:* Let $\mathbf{F} = f\,\text{grad } f$.)

25. Let S be the eighth of the sphere $x^2 + y^2 + z^2 = a^2$ lying in the first octant, and let $f(x, y, z) = \ln\sqrt{x^2 + y^2 + z^2}$. Calculate

$$\iint_S \frac{\partial f}{\partial n}\, d\sigma.$$

(See Problem 23.)

***Toolkit* programs**

Vector Fields

17.7 Stokes's Theorem

Stokes's theorem is an extension of Green's theorem in vector form to surfaces and curves in three dimensions. It says that the line integral

$$\oint \mathbf{F}\cdot d\mathbf{R} \tag{1}$$

is equal to the surface integral

$$\iint_S (\text{curl } \mathbf{F})\cdot \mathbf{n}\, d\sigma, \tag{2}$$

under suitable restrictions (i) on the vector

$$\mathbf{F} = \mathbf{i}M + \mathbf{j}N + \mathbf{k}P,$$

(ii) on the simple closed curve C:

$$x = f(t), \quad y = g(t), \quad z = h(t), \quad 0 \le t \le 1,$$

(iii) on the surface

$$S:\ \phi(x, y, z) = 0$$

bounded by C.

EXAMPLE 1 Let S be the hemisphere

$$z = \sqrt{4 - x^2 - y^2}, \qquad 0 \le x^2 + y^2 \le 4,$$

lying above the xy-plane, with center at the origin. The boundary of this hemisphere is the circle C:

$$z = 0, \qquad x^2 + y^2 = 4.$$

Show that the integrals in Eqs. (1) and (2) are equal for S, C, and

$$\mathbf{F} = \mathbf{i}y - \mathbf{j}x.$$

Solution The integrand in the line integral (1) is

$$\begin{aligned}\mathbf{F}\cdot d\mathbf{R} &= \mathbf{F}\cdot(\mathbf{i}\,dx + \mathbf{j}\,dy + \mathbf{k}\,dz)\\ &= y\,dx - x\,dy.\end{aligned}$$

By Green's theorem for *plane* curves and surfaces, we have

$$\oint \mathbf{F}\cdot d\mathbf{R} = \oint_C (y\,dx - x\,dy) = \iint\limits_{x^2+y^2\le 4} -2\,dx\,dy = -8\pi. \tag{3}$$

To evaluate the surface integral (2), we compute

$$\operatorname{curl}\mathbf{F} = \mathbf{i}\left(\frac{\partial P}{\partial y} - \frac{\partial N}{\partial z}\right) + \mathbf{j}\left(\frac{\partial M}{\partial z} - \frac{\partial P}{\partial x}\right) + \mathbf{k}\left(\frac{\partial N}{\partial x} + \frac{\partial M}{\partial y}\right),$$

with

$$M = y, \qquad N = -x, \qquad P = 0,$$

to get

$$\operatorname{curl}\mathbf{F} = -2\mathbf{k}.$$

The unit outer normal to the hemisphere is

$$\mathbf{n} = \frac{\mathbf{i}x + \mathbf{j}y + \mathbf{k}z}{\sqrt{x^2+y^2+z^2}} = \frac{\mathbf{i}x + \mathbf{j}y + \mathbf{k}z}{2}.$$

Therefore,

$$\operatorname{curl}\mathbf{F}\cdot\mathbf{n}\,d\sigma = -z\,d\sigma. \tag{4}$$

For element of surface area $d\sigma$ (Article 17.2), we use

$$d\sigma = \sqrt{1 + \left(\frac{\partial z}{\partial x}\right)^2 + \left(\frac{\partial z}{\partial y}\right)^2}\,dx\,dy, \tag{5}$$

with

$$\frac{\partial z}{\partial x} = \frac{-x}{\sqrt{4-x^2-y^2}} \quad\text{and}\quad \frac{\partial z}{\partial y} = \frac{-y}{\sqrt{4-x^2-y^2}},$$

or

$$\frac{\partial z}{\partial x} = \frac{-x}{z} \quad\text{and}\quad \frac{\partial z}{\partial y} = \frac{-y}{z}. \tag{6}$$

From (4), (5), and (6), we get

$$\begin{aligned}\operatorname{curl}\mathbf{F}\cdot\mathbf{n}\,d\sigma &= -z\,d\sigma\\ &= -z\sqrt{1 + \frac{x^2}{z^2} + \frac{y^2}{z^2}}\,dx\,dy\\ &= -\sqrt{x^2+y^2+z^2}\,dx\,dy\\ &= -\sqrt{4}\,dx\,dy = -2\,dx\,dy.\end{aligned} \tag{7}$$

Therefore

$$\iint\limits_S \operatorname{curl}\mathbf{F}\cdot\mathbf{n}\,d\sigma = \iint\limits_{x^2+y^2\le 4} -2\,dx\,dy = -8\pi, \tag{8}$$

which agrees with the value of the line integral in Eq. (3). □

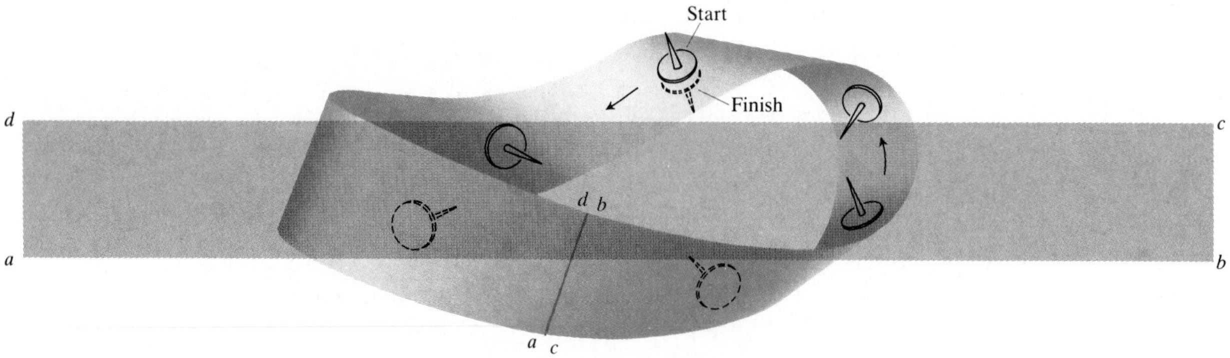

17.27 The Möbius strip (colored band) can be constructed by taking a rectangular strip of paper *abcd*, giving the end *bc* a single twist to interchange the positions of the vertices *b* and *c*, and then pasting the ends of the strip together so as to bring vertices *a* and *c* together, and also *b* and *d*. The Möbius strip is a nonorientable surface.

In Example 1, the surface integral (8) taken over the *hemisphere* turns out to have the same value as a surface integral taken over the *plane base* of that hemisphere. The underlying reason for this equality is that both surface integrals are equal to the line integral around the circle that is their common boundary. (See Problem 14.)

In Stokes's theorem, we require that the surface be *orientable*. By "orientable," we mean that it is possible to consistently assign a unique direction, called *positive*, at each point of S, and that there exists a unit normal $\mathbf{n}$ pointing in this direction. As we move about over the surface S without touching its boundary, the direction cosines of the unit vector $\mathbf{n}$ should vary continuously. Also, when we return to the starting position, $\mathbf{n}$ should return to its initial direction. This rules out a surface like the Möbius strip shown in Fig. 17.27. This surface is nonorientable because a unit normal vector (think of the shaft of a thumbtack) can be continuously moved around the surface without its touching the boundary of the surface, and in such a way that when it is returned to its initial position it will point in a direction *exactly opposite* to its initial direction.

We also want C to have a positive direction that is related to the positive direction on S. We imagine a simple closed curve Γ on S, near the boundary C (see Fig. 17.28), and let $\mathbf{n}$ be normal to S at some point inside Γ. We then assign to Γ a positive direction, the counterclockwise direction as viewed by an observer who is at the end of $\mathbf{n}$ and looking down. (Note that choosing this direction keeps the interior of Γ on our left as we progress around Γ. We could equally well have specified $\mathbf{n}$'s direction by this condition.) Now we move Γ about on S until it touches and is tangent to C. The direction of the positive tangent to Γ at this point of common tangency we shall take to be the positive direction along C. It is a consequence of the orientability of S that a consistent assignment of positive direction along C is induced by this process. The same positive direction is assigned all the way around C, no matter where on S the process is begun. This would not be true of the (nonorientable) Möbius strip.

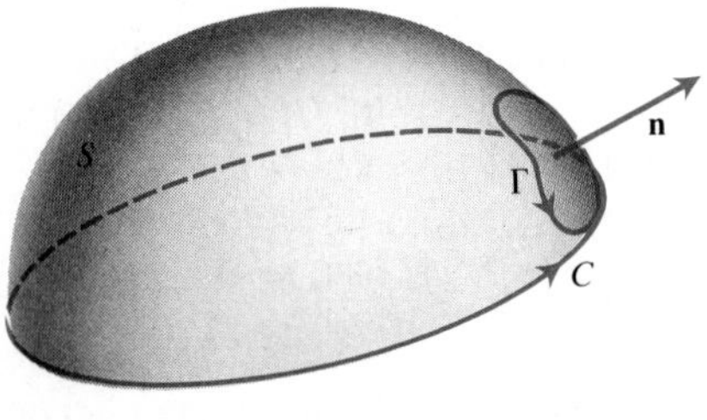

17.28 Orientation of the boundary of an oriented surface.

THEOREM

Stokes's Theorem

Let S be a smooth orientable surface bounded by a closed curve C. Let

$$\mathbf{F} = \mathbf{i}M + \mathbf{j}N + \mathbf{k}P,$$

where M, N, and P and their first order partial derivatives are continuous throughout a region D containing S and C in its interior. Let $\mathbf{n}$ be a positive unit vector normal to S, and let the positive direction around C be the one induced by the positive orientation of S. Then

$$\oint_C \mathbf{F} \cdot d\mathbf{R} = \iint_S \text{curl } \mathbf{F} \cdot \mathbf{n} \, d\sigma, \tag{9}$$

where

$$d\mathbf{R} = \mathbf{i}\, dx + \mathbf{j}\, dy + \mathbf{k}\, dz = \mathbf{T}\, ds$$

and

$$\mathbf{n}\, d\sigma = (\mathbf{i} \cos \alpha + \mathbf{j} \cos \beta + \mathbf{k} \cos \gamma)\, d\sigma.$$

Proof for a polyhedral surface S. Let the surface S be a polyhedral surface consisting of a finite number of plane regions. (Think of one of Buckminster Fuller's geodesic domes.) We apply Green's theorem to each separate panel of S. There are two types of panels:

1. those that are surrounded on all sides by other panels, and
2. those that have one or more edges that are not adjacent to other panels.

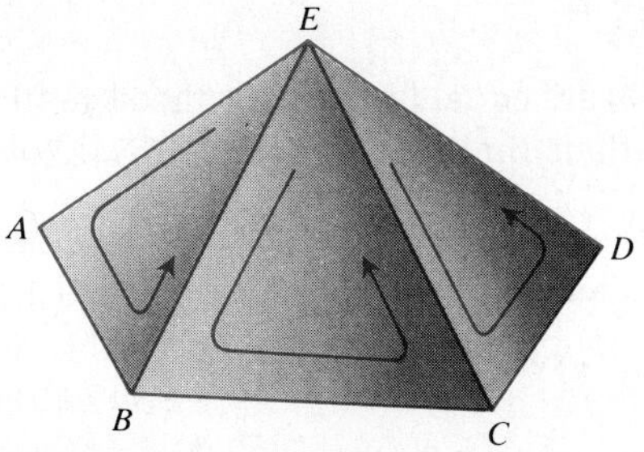

17.29 Part of a polyhedral surface.

Let Δ be part of the boundary of S that consists of those edges of the type 2 panels that are not adjacent to other panels. In Fig. 17.29, the triangles ABE, BCE, and CDE represent a part of S, with $ABCD$ part of the boundary Δ. Applying Green's theorem to the three triangles in turn and adding the results, we get

$$\left(\oint_{ABE} + \oint_{BCE} + \oint_{CDE}\right) \mathbf{F} \cdot d\mathbf{R} = \left(\iint_{ABE} + \iint_{BCE} + \iint_{CDE}\right) \text{curl } \mathbf{F} \cdot \mathbf{n}\, d\sigma. \tag{10}$$

The three line integrals on the left-hand side of Eq. (10) combine into a single line integral taken around the periphery $ABCDE$, because the integrals along interior segments cancel in pairs. For example, the integral along the segment BE in triangle ABE is opposite in sign to the integral along the same segment in triangle EBC. Similarly for the segment CE. Hence (10) reduces to

$$\oint_{ABCDE} \mathbf{F} \cdot d\mathbf{R} = \iint_{ABCDE} \text{curl } \mathbf{F} \cdot \mathbf{n}\, d\sigma.$$

In general, when we apply Green's theorem to all the panels and add the results, we get

$$\oint_{\Delta} \mathbf{F} \cdot d\mathbf{R} = \iint_{S} \operatorname{curl} \mathbf{F} \cdot \mathbf{n}\, d\sigma. \tag{11}$$

This is Stokes's theorem for a polyhedral surface S. ■

A rigorous proof of Stokes's theorem for more general surfaces is beyond the level of a beginning calculus course.† However, the following intuitive argument shows why one would expect Eq. (9) to be true. Imagine a sequence of polyhedral surfaces

$$S_1, \quad S_2, \quad \ldots,$$

and their corresponding boundaries $\Delta_1, \Delta_2, \ldots$. The surface S_n is constructed in such a way that its boundary Δ_n is inscribed in or tangent to C, the boundary of S, and so that the length of Δ_n approaches the length of C as $n \to \infty$. (C needs to be smooth enough to have a length.) The faces of S_n might be polygonal regions, approximating pieces of S, and such that the area of S_n approaches the area of S as $n \to \infty$. S also needs to have finite area. Assuming that M, N, P, and their partial derivatives are continuous in a region D containing S and C, it is plausible to expect that

$$\oint_{\Delta_n} \mathbf{F} \cdot d\mathbf{R} \quad \text{approaches} \quad \oint_{C} \mathbf{F} \cdot d\mathbf{R}$$

and that

$$\iint_{S_n} \operatorname{curl} \mathbf{F} \cdot \mathbf{n}\, d\sigma_n \quad \text{approaches} \quad \iint_{S} \operatorname{curl} \mathbf{F} \cdot \mathbf{n}\, d\sigma$$

as $n \to \infty$. But if

$$\oint_{\Delta_n} \mathbf{F} \cdot d\mathbf{R} \to \oint_{C} \mathbf{F} \cdot d\mathbf{R} \tag{12}$$

and

$$\iint_{S_n} \operatorname{curl} \mathbf{F} \cdot \mathbf{n}\, d\sigma \to \iint_{S} \operatorname{curl} \mathbf{F} \cdot \mathbf{n}\, d\sigma, \tag{13}$$

and if the left-hand sides of (12) and (13) are equal by Stokes's theorem for polyhedra, we then have equality of their limits.

EXAMPLE 2 Let S be the portion of the paraboloid $z = 4 - x^2 - y^2$ that lies above the plane $z = 0$. Let C be their curve of intersection, and let

$$\mathbf{F} = \mathbf{i}(z - y) + \mathbf{j}(z + x) - \mathbf{k}(x + y).$$

Show that

$$\oint_{C} \mathbf{F} \cdot d\mathbf{R} = \iint_{S} \operatorname{curl} \mathbf{F} \cdot \mathbf{n}\, d\sigma.$$

Solution The curve C is the circle $x^2 + y^2 = 4$ in the xy-plane. Along C, where $z = 0$ and $d\mathbf{R} = \mathbf{i}\,dx + \mathbf{j}\,dy$, we have

$$\begin{aligned}\mathbf{F} \cdot d\mathbf{R} &= (z - y)\,dx + (z + x)\,dy - (x + y) \cdot 0\\ &= -y\,dx + x\,dy,\end{aligned}$$

† See, for example, Buck, *Advanced Calculus*, 3d ed. (New York: McGraw-Hill, 1978), or Apostol, *Mathematical Analysis*, 2d ed. (Reading, Mass.: Addison-Wesley, 1974).

whose integral around C is twice the area of the circle:

$$\oint_C \mathbf{F} \cdot d\mathbf{R} = 8\pi$$

(see Article 17.5).

For the surface integral, we compute

$$\text{curl } \mathbf{F} = \begin{vmatrix} \mathbf{i} & \mathbf{j} & \mathbf{k} \\ \frac{\partial}{\partial x} & \frac{\partial}{\partial y} & \frac{\partial}{\partial z} \\ z - y & z + x & -x - y \end{vmatrix} = -2\mathbf{i} + 2\mathbf{j} + 2\mathbf{k}.$$

For a positive unit normal (positive for the chosen orientation of C) on the surface

$$S: \quad f(x, y, z) = z - 4 + x^2 + y^2 = 0,$$

we take

$$\mathbf{n} = \frac{\text{grad } f}{|\text{grad } f|} = \frac{2x\mathbf{i} + 2y\mathbf{j} + \mathbf{k}}{\sqrt{4x^2 + 4y^2 + 1}}.$$

The projection of S onto the xy-plane is the region $x^2 + y^2 \leq 4$, and for the element of surface area $d\sigma$, we take

$$d\sigma = \sqrt{\left(\frac{\partial z}{\partial x}\right)^2 + \left(\frac{\partial z}{\partial y}\right)^2 + 1}\, dx\, dy = \sqrt{4x^2 + 4y^2 + 1}\, dx\, dy.$$

Thus

$$\iint_S \text{curl } \mathbf{F} \cdot \mathbf{n}\, d\sigma = \iint_{x^2 + y^2 \leq 4} (-4x + 4y + 2)\, dx\, dy \tag{a}$$

$$= \iint_{x^2 + y^2 \leq 4} 2\, dx\, dy = 8\pi, \tag{b}$$

where (b) follows from (a) because odd powers of x or y integrate to zero over the interior of the circle. □

Stokes's Theorem for Surfaces with Holes

Stokes's theorem can also be extended to a surface S that has one or more holes in it (like a curved slice of Swiss cheese), in a way exactly analogous to Green's theorem: The surface integral over S of the *normal component* of curl **F** is equal to the sum of the line integrals around all the boundaries of S (including boundaries of the holes) of the *tangential component* of **F**, where the boundary curves are to be traced in the positive direction induced by the positive orientation of S.

Circulation

Stokes's theorem provides the following vector interpretation for curl **F**. As in the discussion of divergence, let **v** be the velocity field of a moving fluid, δ the density, and $\mathbf{F} = \delta\mathbf{v}$. Then

$$\oint_C \mathbf{F} \cdot \mathbf{T}\, ds$$

is a measure of the *circulation* of fluid around the closed curve C. By Stokes's theorem, this circulation is also equal to the flux of curl **F** through a surface S spanning C:

$$\oint_C \mathbf{F} \cdot d\mathbf{R} = \iint_S \text{curl } \mathbf{F} \cdot \mathbf{n} \, d\sigma.$$

Suppose we fix a point Q and a direction **u** at Q. Let C be a circle of radius ρ, with center at Q, whose plane is normal to **u.** If curl **F** is continuous at Q, then the average value of the **u**-component of curl **F** over the circular disc bounded by C approaches the **u**-component of curl **F** at Q as $\rho \to 0$:

$$(\text{curl } \mathbf{F} \cdot \mathbf{u})_Q = \lim_{\rho \to 0} \frac{1}{\pi\rho^2} \iint_S \text{curl } \mathbf{F} \cdot \mathbf{u} \, d\sigma. \tag{14}$$

If we replace the double integral on the right-hand side of Eq. (14) by the circulation, we get

$$(\text{curl } \mathbf{F} \cdot \mathbf{u})_Q = \lim_{\rho \to 0} \frac{1}{\pi\rho^2} \oint_C \mathbf{F} \cdot d\mathbf{R}. \tag{15}$$

The left-hand side of Eq. (15) is a maximum at Q when **u** has the same direction as curl **F.** When ρ is small, the right-hand side of Eq. (15) is approximately equal to

$$\frac{1}{\pi\rho^2} \oint_C \mathbf{F} \cdot d\mathbf{R},$$

which is the circulation around C divided by the area of the disc (*circulation density*). Suppose that a small paddle wheel, of radius ρ, is introduced into the fluid at Q, with its axle directed along **u.** The circulation of the fluid around C will affect the rate of spin of the paddle wheel. The wheel will spin fastest when the circulation integral is maximized; therefore it will spin fastest when the axle of the paddle wheel points in the direction of curl **F.** (See Fig. 17.30.)

EXAMPLE 3 A fluid of constant density δ rotates around the z-axis with velocity $\mathbf{v} = \omega(\mathbf{j}x - \mathbf{i}y)$, where ω is a positive constant. If $\mathbf{F} = \delta\mathbf{v}$, find curl **F**, and show its relation to the circulation density.

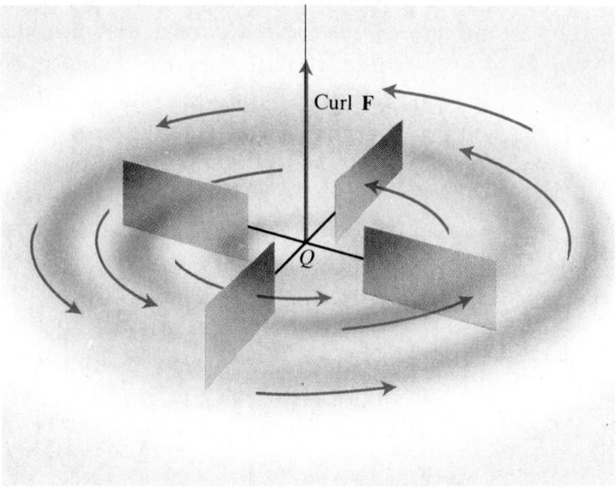

17.30 The paddle-wheel interpretation of curl **F**.

Solution

$$\mathbf{F} = \delta\omega(\mathbf{j}x - \mathbf{i}y),$$

and

$$\text{curl } \mathbf{F} = \begin{vmatrix} \mathbf{i} & \mathbf{j} & \mathbf{k} \\ \dfrac{\partial}{\partial x} & \dfrac{\partial}{\partial y} & \dfrac{\partial}{\partial z} \\ -\delta\omega y & \delta\omega x & 0 \end{vmatrix} = 2\,\delta\omega\mathbf{k}.$$

The work done by a force equal to $\mathbf{F}$, as the point of application moves around a circle C of radius ρ, is

$$\oint_C \mathbf{F} \cdot d\mathbf{R}.$$

If C lies in a plane parallel to the xy-plane, Stokes's theorem gives

$$\oint_C \mathbf{F} \cdot d\mathbf{R} = \iint_S \text{curl } \mathbf{F} \cdot \mathbf{n}\, d\sigma = \iint 2\,\delta\omega\mathbf{k} \cdot \mathbf{k}\, dx\, dy = (2\,\delta\omega)(\pi\rho^2).$$

Thus,

$$(\text{curl } \mathbf{F}) \cdot \mathbf{k} = 2\,\delta\omega = \frac{1}{\pi\rho^2} \oint_C \mathbf{F} \cdot d\mathbf{R},$$

in agreement with Eq. (15) with $\mathbf{u} = \mathbf{k}$. □

PROBLEMS

1. Verify Stokes's theorem, Eq. (9), when S is the hemisphere $x^2 + y^2 + z^2 = 1$, $z \geq 0$, and C is its boundary, for

a) $\mathbf{F} = x\mathbf{i} + y\mathbf{j} + z\mathbf{k}$,

b) $\mathbf{F} = y\mathbf{i} + z\mathbf{j} + x\mathbf{k}$.

2. Let S be the cylinder $x^2 + y^2 = a^2$, $0 \leq z \leq h$, together with its top, $x^2 + y^2 \leq a^2$, $z = h$. Let $\mathbf{F} = -y\mathbf{i} + x\mathbf{j} + x^2\mathbf{k}$. Use Stokes's theorem to calculate the flux of curl $\mathbf{F}$ outward through S.

3. Verify Stokes's theorem, Eq. (9), for $\mathbf{F} = y\mathbf{i} + xz\mathbf{j} + x^2\mathbf{k}$ and S the triangular plate whose corners are $(1, 0, 0)$, $(0, 1, 0)$, $(0, 0, 1)$. That is, choose one of the two possible unit normals $\mathbf{n}$ and a corresponding orientation for the boundary of S. One equation for the plane containing the plate is $x + y + z = 1$.

4. Verify Stokes's theorem, Eq. (9), for $\mathbf{F} = 2y\mathbf{i} + 3x\mathbf{j} - z^2\mathbf{k}$ and S the hemisphere $x^2 + y^2 + z^2 = 9$, $z \geq 0$. Use $\mathbf{n} = (x/3)\mathbf{i} + (y/3)\mathbf{j} + (z/3)\mathbf{k}$.

In Problems 5–8, verify the result of Stokes's theorem for the vector

$$\mathbf{F} = \mathbf{i}(y^2 + z^2) + \mathbf{j}(x^2 + z^2) + \mathbf{k}(x^2 + y^2)$$

for the given surface S and boundary C.

5. S: $z = \sqrt{1 - x^2}$, $-1 \leq x \leq 1$, $-2 \leq y \leq 2$,
$y = 2$, $0 \leq z \leq \sqrt{1 - x^2}$, $-1 \leq x \leq 1$,
$y = -2$, $0 \leq z \leq \sqrt{1 - x^2}$, $-1 \leq x \leq 1$;
C: $z = 0$,
$x = \pm 1$, $-2 \leq y \leq 2$,
$y = \pm 2$, $-1 \leq x \leq 1$

6. The surface S is the surface of the upper half of the cube with one vertex at $(1, 1, 1)$, center at the origin, and edges parallel to the axes; the curve C is the intersection of S with the xy-plane.

7. The surface S is as in Problem 6, with a hole cut out of the top face by the circular disc whose cylindrical coordinates satisfy

$$z = 1, \qquad 0 \leq r \leq \cos\theta, \qquad -\frac{1}{2}\pi \leq \theta \leq \frac{1}{2}\pi.$$

(The circle $z = 1$, $r = \cos\theta$ becomes part of the boundary of S.)

8. The surface S is the surface (excluding the face in the yz-plane) of a pyramid with vertices at the origin O and at $A(1, 0, 0)$, $B(0, 1, 0)$, and $D(0, 0, 1)$; the boundary curve C is the triangle OBD in the yz-plane.

9. Let S be the region bounded by the ellipse C: $4x^2 + y^2 = 4$ in the plane $z = 1$, let $\mathbf{n} = \mathbf{k}$, and let $\mathbf{F} = x^2\mathbf{i} + 2x\mathbf{j} + z^2\mathbf{k}$. Find the value of

$$\oint_C \mathbf{F} \cdot d\mathbf{R}.$$

10. Let $\mathbf{n}$ be the outer normal of the elliptical shell

$$S:\ 4x^2 + 9y^2 + 36z^2 = 36, \qquad z \geq 0,$$

and let

$$\mathbf{F} = y\mathbf{i} + x^2\mathbf{j} + (x^2 + y^4)^{3/2} \sin e^{\sqrt{xyz}}\mathbf{k}.$$

Use Stokes's theorem to find the value of

$$\iint_S \text{curl } \mathbf{F} \cdot \mathbf{n}\, d\sigma.$$

(*Hint:* One parameterization of the ellipse at the base of the shell is $x = 3 \cos t$, $y = 2 \sin t$, $0 \leq t \leq 2\pi$.)

11. Suppose $\mathbf{F} = \text{grad } \phi$ is the gradient of a scalar function ϕ having continuous second order partial derivatives

$$\frac{\partial^2 \phi}{\partial x^2}, \quad \frac{\partial^2 \phi}{\partial x\, \partial y}, \quad \ldots$$

throughout a simply connected region D that contains the surface S and its boundary C in the interior of D. What constant value does

$$\oint_C \mathbf{F} \cdot d\mathbf{R}$$

have in such circumstances? Explain.

12. Let $\phi = (x^2 + y^2 + z^2)^{-1/2}$, let $\mathbf{F} = \text{grad } \phi$, and let C be the circle $x^2 + y^2 = a^2$, $z = 0$ in the xy-plane. Show that

$$\oint_C \mathbf{F} \cdot d\mathbf{R} = 0$$

a) by direct evaluation of the integral, and

b) by applying Stokes's theorem with S the hemisphere

$$z = \sqrt{a^2 - x^2 - y^2}, \qquad x^2 + y^2 \leq a^2.$$

13. Suppose that the components of $\mathbf{F}$ are continuous and have continuous second order partial derivatives of all types. Use the divergence theorem and the fact that div (curl $\mathbf{F}$) = 0 (Problem 20, Article 17.6) to show that

$$\iint_S \text{curl } \mathbf{F} \cdot \mathbf{n}\, d\sigma = 0$$

if S is a closed surface like a sphere, an ellipsoid, or a cube.

14. By Stokes's theorem, if S_1 and S_2 are two oriented surfaces having the same positively oriented curve C as boundary, then

$$\iint_{S_1} \text{curl } \mathbf{F} \cdot \mathbf{n}_1\, d\sigma_1 = \oint_C \mathbf{F} \cdot d\mathbf{R} = \iint_{S_2} \text{curl } \mathbf{F} \cdot \mathbf{n}_2\, d\sigma_2.$$

Deduce that

$$\iint_S \text{curl } \mathbf{F} \cdot \mathbf{n}\, d\sigma$$

has the same value for all oriented surfaces S that span C and that induce the same positive direction on C.

15. Use Stokes's theorem to deduce that if curl $\mathbf{F} = \mathbf{0}$ throughout a simply connected region D, then

$$\int_{P_1}^{P_2} \mathbf{F} \cdot d\mathbf{R}$$

has the same value for all simple paths lying in D and joining P_1 and P_2. In other words, $\mathbf{F}$ is conservative.

Toolkit programs

Vector Fields

REVIEW QUESTIONS AND EXERCISES

1. What is a vector field? Give an example of a two-dimensional vector field; of a three-dimensional field.

2. What is the velocity vector field for a fluid rotating about the x-axis if the angular velocity is a constant ω and the flow is counterclockwise as viewed by an observer at (1, 0, 0) looking toward (0, 0, 0)?

3. Give examples of gradient fields
a) in the plane, b) in space.
State a property that gradient fields have and other fields do not.

4. If $\mathbf{F} = \nabla f$ is a gradient field, S is a level surface for f, and C is a curve on S, why is it true (or not true) that $\int_C \mathbf{F} \cdot d\mathbf{R} = 0$?

5. Suppose that S is a portion of a level surface of a function $f(x, y, z)$. How could you select an orientation (if S is orientable) on S such that

$$\iint_S \nabla f \cdot \mathbf{n}\, d\sigma = \iint_S |\nabla f|\, d\sigma?$$

Why is the hypothesis that S is a level surface of f important?

6. If $f(x, y, z) = 2x - 3y + e^z$, and C is any smooth curve from $A(0, 0, 0)$ to $(1, 2, \ln 3)$, what is the value of $\int_C \nabla f \cdot d\mathbf{R}$? Why doesn't the answer depend on C?
7. Write a formula for a vector field $\mathbf{F}(x, y, z)$ such that $\mathbf{F} = f(\rho)\mathbf{R}$, where $\rho = |\mathbf{R}|$ and $\mathbf{R} = \mathbf{i}x + \mathbf{j}y + \mathbf{k}z$, if it is true that
   a) $\mathbf{F}$ is directed radially outward from the origin and $|\mathbf{F}| = \rho^{-n}$,
   b) $\mathbf{F}$ is directed toward the origin and $|\mathbf{F}| = 2$,
   c) $\mathbf{F}$ is a gravitational attraction field for a mass M at the origin in which an inverse-*cube* law applies. (Don't worry about the possible nonexistence of such a field.)
8. Give one physical and one geometrical interpretation for a surface integral

$$\iint_S h\, d\sigma.$$

9. State both the normal form and the tangential form of Green's theorem in the plane.
10. State the divergence theorem and show that it applies to the region D described by $1 \le |\mathbf{R}| \le 2$, assuming that S is the total boundary of this region, $\mathbf{n}$ is directed away from D at each point, and $\mathbf{F} = \nabla(1/|\mathbf{R}|)$, $\mathbf{R} = \mathbf{i}x + \mathbf{j}y + \mathbf{k}z$.

MISCELLANEOUS PROBLEMS

In Problems 1–4, describe the vector fields in words and with graphs.

1. $\mathbf{F} = x\mathbf{i} + y\mathbf{j} + z\mathbf{k}$
2. $\mathbf{F} = -x\mathbf{i} - y\mathbf{j} - z\mathbf{k}$
3. $\mathbf{F} = (x - y)\mathbf{i} + (x + y)\mathbf{j}$
4. $\mathbf{F} = (x\mathbf{i} + y\mathbf{j})/(x^2 + y^2)$

In Problems 5–8, evaluate the surface integrals

$$\iint_S h\, d\sigma$$

for the given functions and surfaces.

5. The surface S is the hemisphere $x^2 + y^2 + z^2 = a^2$, $z \ge 0$, and $h(x, y, z) = x + y$.
6. The surface S is the portion of the plane $z = x + y$ for which $x \ge 0$, $y \ge 0$, $z \le \pi$, and

$$h(x, y, z) = \sin z.$$

7. The surface S is $z = 4 - x^2 - y^2$, $z \ge 0$, and $h = z$.
8. The surface S is the sphere $\rho = a$, and $h = z^2$. (You might do the upper and lower hemispheres separately and add; or use spherical coordinates and not project the surface.)

For the functions and surfaces given in Problems 9–12, evaluate

$$\iint_S \frac{\partial f}{\partial n}\, d\sigma,$$

where $\partial f/\partial n$ is the directional derivative of f in the direction of the normal $\mathbf{n}$ in the sense specified.

9. The surface S is the sphere $x^2 + y^2 + z^2 = a^2$, $\mathbf{n}$ is directed outward on S, and $f = x^2 + y^2 + z^2$.
10. The surface and normal are as in Problem 9, and $f = (x^2 + y^2 + z^2)^{-1}$.
11. The surface S is the portion of the plane

$$z = 2x + 3y$$

for which $x \ge 0$, $y \ge 0$, $z \le 5$, $\mathbf{n}$ has a positive $\mathbf{k}$-component, and $f = x + y + z$.
12. The surface S is the one-eighth of the sphere $x^2 + y^2 + z^2 = a^2$ that lies in the first octant, $\mathbf{n}$ is directed inward with respect to the sphere, and

$$f = \ln (x^2 + y^2 + z^2)^{1/2}.$$

In Problems 13–20, evaluate the line integrals

$$\int_C \mathbf{F} \cdot d\mathbf{R}$$

for the given fields $\mathbf{F}$ and paths C.

13. The field $\mathbf{F} = x\mathbf{i} + y\mathbf{j}$ and the circle C:

$$x = \cos t, \qquad y = \sin t, \qquad 0 \le t \le 2\pi.$$

14. The field $\mathbf{F} = -y\mathbf{i} + x\mathbf{j}$ and C as in Problem 13.
15. The field $\mathbf{F} = (x - y)\mathbf{i} + (x + y)\mathbf{j}$ and the circle C in Problem 13.
16. The field $\mathbf{F} = x\mathbf{i} + y\mathbf{j} + z\mathbf{k}$ and the ellipse C, in which the plane $z = 2x + 3y$ cuts the cylinder $x^2 + y^2 = 12$, counterclockwise as viewed from the positive end of the z-axis looking toward the origin.
17. The field $\mathbf{F} = \nabla(xy^2z^3)$ and C as in Problem 16.
18. The field $\mathbf{F} = \nabla \times (x\mathbf{i} + y\mathbf{j} + z\mathbf{k})$ and C as in Problem 13.
19. The field $\mathbf{F}$ as in Problem 18 and C the line segment from the origin to the point $(1, 2, 3)$.
20. The field $\mathbf{F}$ as in Problem 17 and C the line segment from $(1, 1, 1)$ to $(2, 1, -1)$.
21. Heat flows from a hotter body to a cooler body. In three-dimensional heat flow, the fundamental equa-

tion for the rate at which heat flows out of D is

$$\iint_S K \frac{\partial u}{\partial n}\, d\sigma = \iiint_D c\delta \frac{\partial u}{\partial t}\, dV. \qquad (1)$$

The symbolism in this equation is as follows:

$u = u(x, y, z, t)$	the temperature at the point (x, y, z) at time t,
K	the thermal conductivity coefficient,
δ	the mass density,
c	the specific heat coefficient. This is the amount of heat required to raise one unit of mass of the material of the body one degree,
S	the boundary surface of the region D,
$\frac{\partial u}{\partial n}$	the directional derivative in the direction of the outward normal to S.

How is $\partial u/\partial n$ related to the *gradient* of the temperature? In which direction (described in words) does ∇u point? Why does the left-hand side of Eq. (1) appear to make sense as a measure of the rate of flow? Now look at the right-hand side of Eq. (1): If ΔV is a small volume element in D, what does $\delta\, \Delta V$ represent? If the temperature of this element changes by an amount Δu in time Δt, what is

a) the amount, b) the average rate

of change of heat in the element? In words, what does the right-hand side of Eq. (1) represent physically? Is it reasonable to interpret Eq. (1) as saying that the rate at which heat flows out through the boundary of D is equal to the rate at which heat is being supplied from D?

22. Assuming Eq. (1), Problem 21, and assuming that there is no heat source or sink in D, derive the equation

$$\nabla \cdot (K\, \nabla u) = c\delta \frac{\partial u}{\partial t} \qquad (2)$$

as the equation that must be satisfied at each point in D.

Suggestion. Apply the divergence theorem to the left-hand side of Eq. (1), and make D be a sphere of radius ε; then let $\varepsilon \to 0$.

23. Assuming the result of Problem 22, and assuming that K, c, and δ are constants, deduce that the condition for steady-state temperature in D is Laplace's equation

$$\nabla^2 u = 0, \quad \text{or} \quad \text{div (grad } u) = 0.$$

In higher mathematics, the symbol Δ is used for the *Laplace operator:*

$$\Delta u = \frac{\partial^2 u}{\partial x^2} + \frac{\partial^2 u}{\partial y^2} + \frac{\partial^2 u}{\partial z^2}.$$

Thus, in this notation,

$$\Delta u = \nabla^2 u = \nabla \cdot \nabla u = \text{div (grad } u).$$

Using the divergence theorem, and assuming that the functions u and v and their first and second order partial derivatives are continuous in the regions considered, verify the formulas in Problems 24–27. Assume that S is the boundary surface of the simply connected region D.

24. $\displaystyle\iint_S u\, \nabla v \cdot d\sigma = \iiint_D [u\, \Delta v + (\nabla u) \cdot (\nabla v)]\, dV$

25. $\displaystyle\iint_S \left(u\frac{\partial v}{\partial n} - v\frac{\partial u}{\partial n}\right) d\sigma = \iiint_D (u\, \Delta v - v\, \Delta u)\, dV$

Suggestion. Use the result of Problem 24 as given and in the form you get by interchanging u and v.

26. $\displaystyle\iint_S u\frac{\partial u}{\partial n}\, d\sigma = \iiint_D [u\, \Delta u + |\nabla u|^2]\, dV$

Suggestion. Use the result of Problem 24 with $v = u$.

27. $\displaystyle\iint_S \frac{\partial u}{\partial n}\, d\sigma = \iiint_D \Delta u\, dV$

Suggestion. Use $v = -1$ in the result of Problem 25.

28. A function u is *harmonic* in a region D if and only if it satisfies *Laplace's equation* $\Delta u = 0$ throughout D. Use the identity in Problem 26 to deduce that if u is harmonic in D and either $u = 0$ or $\partial u/\partial n = 0$ at all points on the surface S that is the boundary of D, then $\nabla u = \mathbf{0}$ throughout D, and therefore u is constant throughout D.

29. The result of Problem 28 can be used to establish the uniqueness of solutions of Laplace's equation in D, provided that either (i) the value of u is prescribed at each point on S, or (ii) the value of $\partial u/\partial n$ is prescribed at each point of S. This is done by supposing that u_1 and u_2 are harmonic in D and that both satisfy the same boundary conditions, and then letting $u = u_1 - u_2$. Complete this uniqueness proof.

30. Problems 21 through 23 deal with heat flow. Assume that K, c, and δ are constant and that the temperature $u = u(x, y, z)$ does not vary with time. Use the results of Problems 23 and 27 to conclude that the net rate of outflow of heat through the surface S is zero.

Note. This result might apply, for example, to the region D between two concentric spheres if the inner one were maintained at 100° and the outer one at 0°, so that heat would flow into D through the inner surface and out through the outer surface at the same rate.

31. Let $\rho = (x^2 + y^2 + z^2)^{1/2}$. Show that

$$u = C_1 + \frac{C_2}{\rho}$$

is harmonic where $\rho > 0$, if C_1 and C_2 are constants. Find values of C_1 and C_2 so that the following boundary conditions are satisfied:

a) $u = 100$ when $\rho = 1$, and $u = 0$ when $\rho = 2$;

b) $u = 100$ when $\rho = 1$, and $\partial u/\partial n = 0$ when $\rho = 2$.

Note. Part (a) refers to a steady-state heat flow problem that is like the one discussed at the end of Problem 30; part (b) refers to an insulated boundary on the sphere $\rho = 2$.

18 Differential Equations

18.1 Introduction

A differential equation is an equation that involves one or more derivatives, or differentials. Differential equations are classified by

a) type (namely, *ordinary* or *partial*),

b) order (that of the highest order derivative that occurs in the equation), and

c) degree (the exponent of the highest power of the highest order derivative, after the equation has been cleared of fractions and radicals in the dependent variable and its derivatives).

For example,

$$\left(\frac{d^3y}{dx^3}\right)^2 + \left(\frac{d^2y}{dx^2}\right)^5 + \frac{y}{x^2+1} = e^x \tag{1}$$

is an ordinary differential equation, of order three and degree two.

Only "ordinary" derivatives occur when the dependent variable y is a function of a single independent variable x. On the other hand, if the dependent variable y is a function of two or more independent variables, say

$$y = f(x, t),$$

where x and t are independent variables, then partial derivatives of y may occur. For example, the wave equation

$$\frac{\partial^2 y}{\partial t^2} = a^2\frac{\partial^2 y}{\partial x^2} \tag{2}$$

(Article 15.7, Problems 32–38) is a partial differential equation, of order two and degree one. (A systematic treatment of partial differential equations lies beyond the scope of this book. For a discussion of partial differential equations, including the wave equation, and solutions of associated physical problems, see Kaplan, *Advanced Calculus,* Chapter 10.)

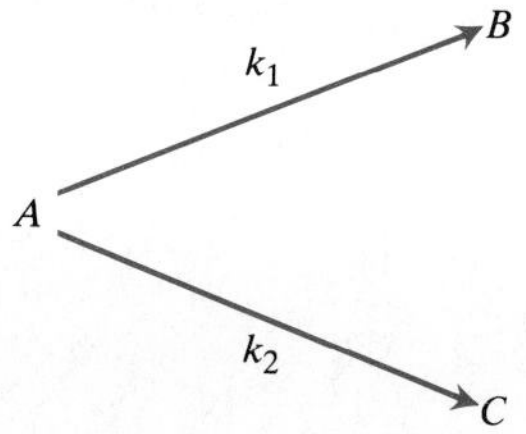

18.1 The reactant *A* yields products *B* and *C* at rates proportional to the amount of *A* present.

Many physical problems, when formulated in mathematical terms, lead to differential equations. In Article 13.2, for example, we discussed and solved the system of differential equations

$$m\frac{d^2x}{dt^2} = 0, \qquad m\frac{d^2y}{dt^2} = -mg \tag{3}$$

which described the motion of a projectile (neglecting air resistance). Indeed, one of the chief sources of differential equations in mechanics is Newton's second law:

$$\mathbf{F} = \frac{d}{dt}(m\mathbf{v}), \tag{4}$$

where $\mathbf{F}$ is the resultant of the forces acting on a body of mass m and $\mathbf{v}$ is its velocity.

An example from the field of chemical kinetics may be represented by a reactant A that undergoes parallel transformations into products B and C at rates that are proportional to the amount of A that is present at time t. (See Fig. 18.1.) If x, y, and z denote the amounts of A, B, and C present at time t, then the differential equations that describe the process are

$$\frac{dx}{dt} = -(k_1 + k_2)x, \qquad \frac{dy}{dt} = k_1x, \qquad \frac{dz}{dt} = k_2x. \tag{5}$$

If, at time $t = 0$, only A is present, the initial conditions are

$$x(0) = x_0, \qquad y(0) = 0, \qquad z(0) = 0. \tag{6}$$

Equations (5) and (6) together provide an example of an *initial value problem*. The first of Eqs. (5) can easily be solved for x to yield

$$x = x_0e^{-(k_1+k_2)t}. \tag{7a}$$

When this is substituted into the remaining Eqs. (5), they can be integrated also, to give

$$y = \frac{k_1}{k_1 + k_2}(1 - e^{-(k_1+k_2)t})\, x_0, \tag{7b}$$

$$z = \frac{k_2}{k_1 + k_2}(1 - e^{-(k_1+k_2)t})\, x_0. \tag{7c}$$

Together, Eqs. (7*a*, *b*, *c*) give a solution of the initial value problem of Eqs. (5) and (6), as can be verified by direct substitition.

Differential equations enter naturally as models for many other phenomena in physics, chemistry, biology, economics, and engineering. Many of these phenomena are complex and require more detailed knowledge of the specific subjects than we can assume here. The models that we shall study will be simple, including examples of exponential growth and decay, simple electrical circuits, vibrations, and motion.

18.2 Solutions

A function

$$y = f(x)$$

is said to be a *solution* of a differential equation if the latter is satisfied when y and its derivatives are replaced throughout by $f(x)$ and its corresponding derivatives. For example, if c_1 and c_2 are any constants, then

$$y = c_1 \cos x + c_2 \sin x \tag{1a}$$

is a solution of the differential equation

$$\frac{d^2y}{dx^2} + y = 0. \tag{1b}$$

A physical problem that translates into a differential equation usually involves additional conditions not expressed by the differential equation itself. In mechanics, for example, the initial position and velocity of the moving body are usually prescribed, as well as the forces. The differential equation, or equations, of motion will usually have solutions in which certain arbitrary constants occur, as in (1a) above. Specific values are then assigned to these arbitrary constants to meet the prescribed initial conditions. (See the examples worked out in Articles 6.9 and 13.2.)

A differential equation of order n will generally have a solution involving n arbitrary constants. (There is a more precise mathematical theorem, which we shall neither state nor prove.) This solution is called the *general* solution. Once this general solution is known, it is only a matter of algebra to determine specific values of the constants if initial conditions are also prescribed. Hence we shall devote our attention to the problem of finding the general solutions of certain types of differential equations.

The subject of differential equations is complex, and there are many textbooks on differential equations and advanced calculus where you can find a more extensive treatment than we shall present here. The field is also the subject of a great deal of current research; and, with the widespread availability of computers, numerical methods for solving differential equations play an important role.

In the remainder of this chapter, the following topics will be treated. Throughout, only *ordinary* differential equations will be considered.

1. First order.
   a) Variables separable.
   b) Homogeneous.
   c) Linear.
   d) Exact differentials.
2. Special types of second order.
3. Linear equations with constant coefficients.
   a) Homogeneous.
   b) Nonhomogeneous.

4. Approximation techniques.
   a) Power series.
   b) Numerical methods.

Several new terms occur in this list, and they will be defined in the appropriate places.

PROBLEMS

Show that each function is a solution of the accompanying differential equation.

1. $xy'' - y' = 0$,
   a) $y = x^2 + 3$ b) $y = C_1x^2 + C_2$
2. $x^3y''' + 4x^2y'' + xy' + y = x$, $y = \frac{1}{2}x$
3. $yy'' = 2(y')^2 - 2y'$,
   a) $y = C$ b) $C_1y = \tan(C_1x + C_2)$
4. $y' + \dfrac{1}{x}y = 1$, $y = \dfrac{C}{x} + \dfrac{x}{2}$
5. $2y' + 3y = e^{-x}$, $y = e^{-x} + Ce^{-(3/2)x}$
6. $(x \sin x)y' + (\sin x + x \cos x)y = e^x/x$,
$$y = \frac{1}{x \sin x}\int_{-1/2}^{x} \frac{e^t}{t}\,dt$$

18.3 First Order: Variables Separable

A first order differential equation can be solved by integration if it is possible to collect all y terms with dy and all x terms with dx. That is, if it is possible to write the equation in the form

$$f(y)\,dy + g(x)\,dx = 0,$$

then the general solution is

$$\int f(y)\,dy + \int g(x)\,dx = C,$$

where C is an arbitrary constant.

EXAMPLE Solve the equation

$$(x + 1)\frac{dy}{dx} = x(y^2 + 1).$$

Solution We change to differential form, separate the variables, and integrate:

$$(x + 1)\,dy = x(y^2 + 1)\,dx,$$

$$\frac{dy}{y^2 + 1} = \frac{x\,dx}{x + 1},$$

$$\tan^{-1} y = x - \ln|x + 1| + C. \quad \square$$

Some models of population growth assume that the rate of change of the population at time t is proportional to the number y of individuals present at that time. This leads to the equation

$$\frac{dy}{dt} = ky, \tag{1}$$

where k is a constant that is positive if the population is increasing and negative if it is decreasing. To solve Eq. (1), we separate variables and integrate, to obtain

$$\int \frac{dy}{y} = \int k\,dt,$$

or

$$\ln y = kt + C_1$$

(remember that y is positive). It follows that

$$y = e^{kt+C_1} = Ce^{kt},$$

where $C = e^{C_1}$. If y_0 denotes the population when $t = 0$, then $C = y_0$ and

$$y = y_0 e^{kt}. \tag{2}$$

This equation is called the *law of exponential growth.*

PROBLEMS

Separate the variables and solve the following differential equations.

1. $x(2y - 3)\,dx + (x^2 + 1)\,dy = 0$
2. $x^2(y^2 + 1)\,dx + y\sqrt{x^3 + 1}\,dy = 0$
3. $\dfrac{dy}{dx} = e^{x-y}$
4. $\sqrt{2xy}\dfrac{dy}{dx} = 1$
5. $\sin x \dfrac{dx}{dy} + \cosh 2y = 0$
6. $\ln x \dfrac{dx}{dy} = \dfrac{x}{y}$
7. $xe^y\,dy + \dfrac{x^2+1}{y}\,dx = 0$
8. $y\sqrt{2x^2 + 3}\,dy + x\sqrt{4 - y^2}\,dx = 0$
9. $\sqrt{1 + x^2}\,dy + \sqrt{y^2 - 1}\,dx = 0$
10. $x^2y\dfrac{dy}{dx} = (1 + x)\csc y$
11. *Continuous compounding.* An investor has \$1000 with which to open an account and plans to add \$1000 per year. All funds in the account will earn 10% interest per year, continuously compounded. Assuming that the added deposits are also made continuously, show that $dx/dt = 1000 + 0.10x$, and $x(0) = 1000$, where x is the number of dollars in the account at time t. How many years would it take for the account to reach \$100,000? For more on compound interest, see Article 6.10.
12. *The snow plow problem.* One morning in January it began to snow and kept on snowing at a constant rate. A snow plow began to plow at noon. It plowed clean and at a constant rate (volume per unit time). From one o'clock until two o'clock it went only half as far as it had gone between noon and one o'clock. When did the snow begin falling?
13. *Newton's law of cooling* assumes that the temperature T of a small hot object placed in a medium of temperature T_a decreases at a rate proportional to $T - T_a$. If an object cools from 100 °C to 80 °C in twenty minutes when the surrounding temperature is 20 °C, how long does it take to cool from 100 °C to 60 °C?

(For additional problems, see Article 6.9.)

18.4
First Order: Homogeneous

Occasionally a differential equation whose variables cannot be separated can be transformed by a change of variable into an equation whose variables can be separated. This is the case with any equation that can be put into the form

$$\frac{dy}{dx} = F\left(\frac{y}{x}\right). \tag{1}$$

Such an equation is called *homogeneous*.

To transform Eq. (1) into an equation whose variables may be separated, we introduce the new variable

$$v = \frac{y}{x}. \tag{2}$$

Then

$$y = vx, \qquad \frac{dy}{dx} = v + x\frac{dv}{dx},$$

and (1) becomes

$$v + x\frac{dv}{dx} = F(v). \tag{3}$$

Equation (3) can be solved by separation of variables:

$$\frac{dx}{x} + \frac{dv}{v - F(v)} = 0. \tag{4}$$

After (4) is solved, the solution of the original equation is obtained when we replace v by y/x.

EXAMPLE Show that the equation

$$(x^2 + y^2)\,dx + 2xy\,dy = 0$$

is homogeneous, and solve it.

Solution From the given equation, we have

$$\frac{dy}{dx} = -\frac{x^2 + y^2}{2xy} = -\frac{1 + (y/x)^2}{2(y/x)}.$$

This has the form of Eq. (1), with

$$F(v) = -\frac{1 + v^2}{2v}, \qquad \text{where} \quad v = \frac{y}{x}.$$

Then Eq. (4) becomes

$$\frac{dx}{x} + \frac{dv}{v + \dfrac{1 + v^2}{2v}} = 0, \qquad \text{or} \qquad \frac{dx}{x} + \frac{2v\,dv}{1 + 3v^2} = 0.$$

The solution of this is

$$\ln|x| + \frac{1}{3}\ln(1 + 3v^2) = \frac{1}{3}\ln C,$$

so that

$$x^3(1 + 3v^2) = \pm C.$$

In terms of y and x, the solution is

$$x(x^2 + 3y^2) = C. \ \square$$

PROBLEMS

Show that the following equations are homogeneous, and solve.

1. $(x^2 + y^2)\,dx + xy\,dy = 0$
2. $x^2\,dy + (y^2 - xy)\,dx = 0$
3. $(xe^{y/x} + y)dx - x\,dy = 0$
4. $(x + y)\,dy + (x - y)\,dx = 0$
5. $y' = \dfrac{y}{x} + \cos\dfrac{y - x}{x}$
6. $\left(x \sin\dfrac{y}{x} - y\cos\dfrac{y}{x}\right) dx + x\cos\dfrac{y}{x}\,dy = 0$
7. Solve the equation

$$(x + y + 1)\,dx + (y - x - 3)\,dy = 0$$

by making a change of variable of the form

$$x = r + a, \qquad y = s + b,$$

and choosing the constants a and b so that the resulting equation is

$$(r + s)\,dr + (r - s)\,ds = 0.$$

Then solve this equation and express its solution in terms of x and y.

8. Use the substitution $u = x + y$ to solve the equation $y' = (x + y)^2$.

Orthogonal trajectories. If every member of a family of curves is a solution of the differential equation

$$M(x, y)\,dx + N(x, y)\,dy = 0,$$

while every member of a second family of curves is a solution of the related equation

$$N(x, y)\,dx - M(x, y)\,dy = 0,$$

then each curve of the one family is orthogonal to every curve of the other family. Each family is said to be a family of *orthogonal trajectories* of the other. In Problems 9 and 10, find the family of solutions of the given differential equation and the family of orthogonal trajectories. Sketch both families.

9. $x\,dy - 2y\,dx = 0$
10. $2xy\,dy + (x^2 - y^2)\,dx = 0$
11. Find the orthogonal trajectories of the family of curves $xy = c$.

18.5
First Order: Linear

The complexity of a differential equation depends primarily upon the way in which the *dependent* variable and its derivatives occur. Of particular importance are those equations that are linear. In a linear differential equation, each term of the equation is of degree one or zero, where, in computing the degree of a term, we add the exponents of the dependent variable and of any of its derivatives that occur. Thus, for example, (d^2y/dx^2) is of the first degree, while $y(dy/dx)$ is of the second degree because we must add 1 for the exponent of y, and 1 for the exponent of dy/dx.

A differential equation of first order, which is also linear, can always be put into the standard form

$$\frac{dy}{dx} + Py = Q, \tag{1}$$

where P and Q are functions of x.

One method for solving Eq. (1) is to find a function $\rho = \rho(x)$ such that if the equation is multiplied by ρ, the left side becomes the derivative of the product ρy. That is, we multiply (1) by ρ,

$$\rho\frac{dy}{dx} + \rho Py = \rho Q, \tag{1'}$$

and then try to impose upon ρ the condition that

$$\rho\frac{dy}{dx} + \rho P y = \frac{d}{dx}(\rho y). \tag{2}$$

When we expand the right-hand side of (2) and cancel terms, we obtain, as the condition to be satisfied by ρ,

$$\frac{d\rho}{dx} = \rho P. \tag{3}$$

In Eq. (3), $P = P(x)$ is a known function, so we can separate the variables and solve for ρ:

$$\frac{d\rho}{\rho} = P\,dx, \qquad \ln|\rho| = \int P\,dx + \ln C,$$

$$\rho = \pm C e^{\int P\,dx}. \tag{4}$$

Since we do not require the most general function ρ, we may take $\pm C = 1$ in (4) and use

$$\rho = e^{\int P\,dx}. \tag{5}$$

This function is called an *integrating factor* for Eq. (1). With its help, (1′) becomes

$$\frac{d}{dx}(\rho y) = \rho Q,$$

whose solution is

$$\rho y = \int \rho Q\,dx + C. \tag{6}$$

Since ρ is given by (5), while P and Q are known from the given differential equation (1), we have, in Eqs. (5) and (6), a summary of all that is required to solve (1).

EXAMPLE 1 Solve the equation

$$\frac{dy}{dx} + y = e^x.$$

Solution

$$P = 1, \qquad Q = e^x,$$

$$\rho = e^{\int dx} = e^x,$$

$$e^x y = \int e^{2x}\,dx + C = \frac{1}{2}e^{2x} + C,$$

$$y = \frac{1}{2}e^x + Ce^{-x}. \ \square$$

EXAMPLE 2 Solve the equation

$$x\frac{dy}{dx} - 3y = x^2.$$

Solution We put the equation in standard form,

$$\frac{dy}{dx} - \frac{3}{x}y = x,$$

and then read off

$$P = -\frac{3}{x}, \qquad Q = x.$$

Hence

$$\rho = e^{\int -(3/x)\,dx} = e^{-3\ln x} = \frac{1}{e^{3\ln x}} = \frac{1}{x^3},$$

and

$$\frac{1}{x^3}\,y = \int \frac{x}{x^3}\,dx + C = -\frac{1}{x} + C,$$

$$y = -x^2 + Cx^3. \ \square$$

Note that whenever $\int P\,dx$ involves logarithms, as in Example 2, it is profitable to simplify the expression for $e^{\int P\,dx}$ before substituting into Eq. (6). To simplify the expression, we use the properties of the logarithmic and exponential functions:

$$e^{\ln A} = A, \qquad e^{m\ln A} = A^m, \qquad e^{n+m\ln A} = A^m e^n.$$

A differential equation that is linear in y and dy/dx may also be separable, or homogeneous. In such cases, we have a choice of methods of solution. Observe also that an equation that is linear in x and dx/dy can be solved by the technique of this article; one need only interchange the roles of x and y in Eqs. (1), (5), and (6).

PROBLEMS

Solve Problems 1–10.

1. $\dfrac{dy}{dx} + 2y = e^{-x}$
2. $2\dfrac{dy}{dx} - y = e^{x/2}$
3. $x\dfrac{dy}{dx} + 3y = \dfrac{\sin x}{x^2}$
4. $x\,dy + y\,dx = \sin x\,dx$
5. $x\,dy + y\,dx = y\,dy$
6. $(x-1)^3\dfrac{dy}{dx} + 4(x-1)^2 y = x + 1$
7. $\cosh x\,dy + (y\sinh x + e^x)\,dx = 0$
8. $e^{2y}\,dx + 2(xe^{2y} - y)\,dy = 0$
9. $(x - 2y)\,dy + y\,dx = 0$
10. $(y^2 + 1)\,dx + (2xy + 1)\,dy = 0$
11. *Blood sugar.* If glucose is fed intravenously at a constant rate, the change in the overall concentration $c(t)$ of glucose in the blood with respect to time may be described by the differential equation

$$\frac{dc}{dt} = \frac{G}{100V} - kc.$$

In this equation, G, V, and k are positive constants, G being the rate at which glucose is admitted, in milligrams per minute, and V the volume of blood in the body, in liters (around 5 liters for an adult). The concentration $c(t)$ is measured in milligrams per centiliter. The term $-kc$ is included because the glucose is assumed to be changing continually into other molecules at a rate proportional to its concentration.

a) Solve the equation for $c(t)$, using c_0 to denote $c(0)$.
b) Find the steady state concentration, $\lim_{t\to\infty} c(t)$.

12. *An investment program.* An investor has a salary of \$20,000 per year with expected increases of \$1000 per year. If an initial deposit of \$1000 is invested in a program that pays 8% per annum, compounded continuously, and additional deposits are continuously added at the rate of 5% of salary, a model for the investment x at the end of t years is of the form $dx/dt = a + bx + ct$, where a, b, and c are positive constants, and $x(0) = 1000$:

$$\frac{dx}{dt} = 0.08x + 0.05(20{,}000 + 1000t)$$
$$= 1000 + 0.08x + 50t.$$

What is x at time t? (If you have a programmable calculator you can easily compute x(5), x(10), x(20), and x(40) to see how the investment grows.)

18.6

First Order: Exact

An equation that can be written in the form

$$M(x, y)\,dx + N(x, y)\,dy = 0, \tag{1}$$

and having the property that

$$\frac{\partial M}{\partial y} = \frac{\partial N}{\partial x}, \tag{2}$$

is said to be *exact*, because its left-hand side is an exact differential. The technique of solving an exact equation consists in finding a function $f(x, y)$ such that

$$df = M\,dx + N\,dy. \tag{3}$$

Then (1) becomes

$$df = 0$$

and the solution is

$$f(x, y) = C,$$

where C is an arbitrary constant. The method of finding $f(x, y)$ to satisfy (3) is discussed and illustrated in Article 15.11.

It can be proved that every first order differential equation

$$P(x, y)\,dx + Q(x, y)\,dy = 0$$

can be made exact by multiplication by a suitable *integrating factor* $\rho(x, y)$. Such an integrating factor has the property that

$$\frac{\partial}{\partial y}[\rho(x, y)P(x, y)] = \frac{\partial}{\partial x}[\rho(x, y)Q(x, y)].$$

Unfortunately, it is not easy to determine ρ from this equation. In fact, there is no general technique by which even a single integrating factor can be produced for an arbitrary differential equation, and the search for one can be a frustrating experience. However, one can often recognize certain combinations of differentials that can be made exact by "ingenious devices."

EXAMPLE Solve the equation

$$x\,dy - y\,dx = xy^2\,dx.$$

Solution The combination $x\,dy - y\,dx$ may ring a bell in our memories and cause us to recall the formula

$$d\left(\frac{u}{v}\right) = \frac{v\,du - u\,dv}{v^2} = -\left[\frac{u\,dv - v\,du}{v^2}\right].$$

Therefore we might divide the given equation by x^2, or change signs and divide by y^2. Clearly, the latter approach will be more profitable, so we proceed as follows:

$$x\,dy - y\,dx = xy^2\,dx, \qquad \frac{y\,dx - x\,dy}{y^2} = -x\,dx,$$

$$d\left(\frac{x}{y}\right) + x\,dx = 0, \qquad \frac{x}{y} + \frac{x^2}{2} = C.$$

The same result would be obtained if we wrote our equation in the form

$$(xy^2 + y)\,dx - x\,dy = 0$$

and multiplied by the integrating factor $1/y^2$. This would give

$$\left(x + \frac{1}{y}\right) dx - \left(\frac{x}{y^2}\right) dy = 0,$$

which is exact, since

$$\frac{\partial}{\partial y}\left(x + \frac{1}{y}\right) = \frac{\partial}{\partial x}\left(-\frac{x}{y^2}\right). \ \square$$

PROBLEMS

In Problems 1–3, use the given integrating factors to make differential equations exact. Then solve the equations.

1. $(x + 2y)\,dx - x\,dy = 0, \quad 1/x^3$
2. $y\,dx + x\,dy = 0,$

   a) $\dfrac{1}{xy}$ b) $\dfrac{1}{(xy)^2}$
3. $y\,dx - x\,dy = 0,$

   a) $\dfrac{1}{y^2}$ b) $\dfrac{1}{x^2}$

   c) $\dfrac{1}{xy}$ d) $\dfrac{1}{x^2 + y^2}$

Solve Problems 4–13.

4. $(x + y)\,dx + (x + y^2)\,dy = 0$
5. $(2xe^y + e^x)\,dx + (x^2 + 1)e^y\,dy = 0$
6. $(2xy + y^2)\,dx + (x^2 + 2xy - y)\,dy = 0$
7. $(x + \sqrt{y^2 + 1})\,dx - \left(y - \dfrac{xy}{\sqrt{y^2 + 1}}\right) dy = 0$
8. $x\,dy - y\,dx + x^3\,dx = 0$
9. $x\,dy - y\,dx = (x^2 + y^2)\,dx$
10. $(x^2 + x - y)\,dx + x\,dy = 0$
11. $\left(e^x + \ln y + \dfrac{y}{x}\right) dx + \left(\dfrac{x}{y} + \ln x + \sin y\right) dy = 0$
12. $\left(\dfrac{y^2}{1 + x^2} - 2y\right) dx$

    $+ (2y \tan^{-1} x - 2x + \sinh y)\,dy = 0$
13. $dy + \dfrac{y - \sin x}{x}\,dx = 0$
14. If a, b, and c are constants and

    $$(ax^2 + by^2)\,dx + cxy\,dy = 0$$

    is an exact equation, what relation(s) among the constants must hold? Solve the equation subject to such conditions.
15. If a, b, and c are constants and

    $$(ax^2 + by^2)\,dx + cxy\,dy = 0$$

    is not exact, but has the integrating factor $1/x^2$, what relation(s) among the constants must hold? Solve the equation subject to such conditions.

18.7 Special Types of Second Order Equations

Certain types of second order differential equations, of which the general form is

$$F\left(x, y, \frac{dy}{dx}, \frac{d^2y}{dx^2}\right) = 0, \tag{1}$$

can be reduced to first order equations by a suitable change of variables.

Type 1. *Equations with dependent variable missing.* When Eq. (1) has the special form

$$F\left(x, \frac{dy}{dx}, \frac{d^2y}{dx^2}\right) = 0, \tag{2}$$

we can reduce it to a first order equation by substituting

$$p = \frac{dy}{dx}, \qquad \frac{d^2y}{dx^2} = \frac{dp}{dx}.$$

Then Eq. (2) takes the form

$$F\left(x, p, \frac{dp}{dx}\right) = 0,$$

which is of the first order in p. If this can be solved for p as a function of x, say

$$p = \phi(x, C_1),$$

then we can find y by one additional integration:

$$y = \int (dy/dx)\, dx = \int p\, dx = \int \phi(x, C_1)\, dx + C_2.$$

The differential equation

$$\frac{d^2y}{dx^2} = \frac{w}{H}\sqrt{1 + \left(\frac{dy}{dx}\right)^2}$$

was solved by this technique in Article 9.5.

Type 2. *Equations with independent variable missing.* When Eq. (1) does not contain x explicitly but has the form

$$F\left(y, \frac{dy}{dx}, \frac{d^2y}{dx^2}\right) = 0, \tag{3}$$

the substitutions to use are

$$p = \frac{dy}{dx} \quad \text{and} \quad \frac{d^2y}{dx^2} = p\frac{dp}{dy}.$$

Then Eq. (3) takes the form

$$F\left(y, p, p\frac{dp}{dy}\right) = 0,$$

which is of the first order in p. Its solution gives p in terms of y, and then a further integration gives the solution of Eq. (3).

EXAMPLE Solve the equation

$$\frac{d^2y}{dx^2} + y = 0.$$

Solution Let

$$\frac{dy}{dx} = p, \qquad \frac{d^2y}{dx^2} = \frac{dp}{dx} = \frac{dp}{dy}\frac{dy}{dx} = \frac{dp}{dy}p.$$

Then we proceed as follows:

$$p\frac{dp}{dy} + y = 0, \qquad p\,dp + y\,dy = 0,$$

$$\frac{p^2}{2} + \frac{y^2}{2} = \frac{C_1^2}{2}, \qquad p = \frac{dy}{dx} = \pm\sqrt{C_1^2 - y^2},$$

$$\frac{dy}{\sqrt{C_1^2 - y^2}} = \pm dx, \qquad \sin^{-1}\frac{y}{C_1} = \pm(x + C_2),$$

$$y = C_1 \sin[\pm(x + C_2)] = \pm C_1 \sin(x + C_2).$$

Since C_1 is arbitrary, there is no need for the $\pm$ sign, and we have

$$y = C_1 \sin(x + C_2)$$

as the general solution. □

PROBLEMS

Solve Problems 1–8.

1. $\dfrac{d^2y}{dx^2} + \dfrac{dy}{dx} = 0$

2. $\dfrac{d^2y}{dx^2} + y\dfrac{dy}{dx} = 0$

3. $\dfrac{d^2y}{dx^2} + x\dfrac{dy}{dx} = 0$

4. $x\dfrac{d^2y}{dx^2} + \dfrac{dy}{dx} = 0$

5. $\dfrac{d^2y}{dx^2} - y = 0$

6. $\dfrac{d^2y}{dx^2} + \omega^2 y = 0 \qquad (\omega = \text{constant} \neq 0)$

7. $xy''' - 2y'' = 0$. (*Hint:* Substitute $y'' = q$.)

8. $2y'' - (y')^2 + 1 = 0$

9. *Hooke's law.* A mass m is suspended from one end of a vertical spring whose other end is attached to a rigid support. The body is allowed to come to rest and is then pulled down an additional slight amount A and released. Find its motion. (*Hint:* Assume Newton's second law of motion and Hooke's law, which says that the tension in the spring is proportional to the amount it is stretched. Let x denote the displacement of the body at time t, measured from the equilibrium position. Then $m(d^2x/dt^2) = -kx$, where k, the "spring constant," is the proportionality factor in Hooke's law.)

10. A jeep suspended from a parachute falls through space under the pull of gravity. If air resistance produces a retarding force proportional to the jeep's velocity and the jeep starts from rest at time $t = 0$, find the distance it falls in time t.

18.8

Linear Equations with Constant Coefficients

An equation of the form

$$\frac{d^ny}{dx^n} + a_1\frac{d^{n-1}y}{dx^{n-1}} + a_2\frac{d^{n-2}y}{dx^{n-2}} + \cdots + a_{n-1}\frac{dy}{dx} + a_n y = F(x), \tag{1}$$

which is linear in y and its derivatives, is called a *linear* equation of order n. If $F(x)$ is identically zero, the equation is said to be *homogeneous;* otherwise it is called *nonhomogeneous.* The equation is called linear even when the coefficients $a_1, a_2, \ldots, a_n$ are functions of x. However, we shall consider only the case where these coefficients are *constants.*

It is convenient to introduce the symbol D to represent the operation of differentiation with respect to x. That is, we write $Df(x)$ to mean $(d/dx)f(x)$. Furthermore, we define powers of D to mean taking successive derivatives:

$$D^2f(x) = D\{Df(x)\} = \frac{d^2f(x)}{dx^2},$$

$$D^3f(x) = D\{D^2f(x)\} = \frac{d^3f(x)}{dx^3},$$

and so on. A polynomial in D is to be interpreted as an operator that, when applied to $f(x)$, produces a linear combination of f and its successive derivatives. For example,

$$\begin{aligned}(D^2 + D - 2)f(x) &= D^2f(x) + Df(x) - 2f(x)\\ &= \frac{d^2f(x)}{dx^2} + \frac{df(x)}{dx} - 2f(x).\end{aligned}$$

Such a polynomial in D is called a *linear differential operator* and may be denoted by the single letter L. If L_1 and L_2 are two such linear operators, their sum and product are defined by the equations

$$\begin{aligned}(L_1 + L_2)f(x) &= L_1f(x) + L_2f(x),\\ L_1L_2f(x) &= L_1(L_2f(x)).\end{aligned}$$

Linear differential operators that are polynomials in D with constant coefficients satisfy basic algebraic laws that make it possible to treat them like ordinary polynomials so far as addition, multiplication, and factoring are concerned. Thus,

$$\begin{aligned}(D^2 + D - 2)f(x) &= (D + 2)(D - 1)f(x)\\ &= (D - 1)(D + 2)f(x).\end{aligned} \tag{2}$$

Since Eq. (2) holds for any twice-differentiable function f, we also write the equality between operators:

$$D^2 + D - 2 = (D + 2)(D - 1) = (D - 1)(D + 2). \tag{3}$$

18.9

Linear Second Order Homogeneous Equations with Constant Coefficients

Suppose, now, we wish to solve a differential equation of order two, say

$$\frac{d^2y}{dx^2} + 2a\frac{dy}{dx} + by = 0, \tag{1}$$

where a and b are constants. In operator notation, this becomes

$$(D^2 + 2aD + b)y = 0. \tag{1'}$$

Associated with this differential equation is the algebraic equation

$$r^2 + 2ar + b = 0, \tag{1''}$$

which we get by replacing D by r and suppressing y. This is called the *characteristic equation* of the differential equation. Suppose the roots of (1″) are r_1 and r_2. Then

$$r^2 + 2ar + b = (r - r_1)(r - r_2)$$

and

$$D^2 + 2aD + b = (D - r_1)(D - r_2).$$

Hence Eq. (1′) is equivalent to

$$(D - r_1)(D - r_2)y = 0. \tag{2}$$

If we now let

$$(D - r_2)y = u \tag{3a}$$

and

$$(D - r_1)u = 0, \tag{3b}$$

we can solve Eq. (1′) in two steps. From Eq. (3b), which is separable, we find

$$u = C_1 e^{r_1 x}.$$

We substitute this into (3a), which becomes

$$(D - r_2)y = C_1 e^{r_1 x}$$

or

$$\frac{dy}{dx} - r_2 y = C_1 e^{r_1 x}.$$

This equation is linear. Its integrating factor is

$$\rho = e^{-r_2 x},$$

(see Article 18.5), and its solution is

$$e^{-r_2 x} y = C_1 \int e^{(r_1 - r_2)x}\, dx + C_2. \tag{4}$$

How we proceed at this point depends on whether r_1 and r_2 are equal.

CASE 1. If $r_1 \neq r_2$, the evaluation of the integral in Eq. (4) leads to

$$e^{-r_2 x} y = \frac{C_1}{r_1 - r_2} e^{(r_1 - r_2)x} + C_2$$

or

$$y = \frac{C_1}{r_1 - r_2} e^{r_1 x} + C_2 e^{r_2 x}.$$

Since C_1 is an arbitrary constant, so is $C_1/(r_1 - r_2)$, and the solution of Eq. (2) can be written simply as

$$y = C_1 e^{r_1 x} + C_2 e^{r_2 x}, \qquad \text{if} \qquad r_1 \neq r_2. \tag{5}$$

CASE 2. If $r_1 = r_2$, then $e^{(r_1 - r_2)x} = e^0 = 1$, and Eq. (4) reduces to

$$e^{-r_2 x}y = C_1 x + C_2$$

or

$$y = (C_1 x + C_2)e^{r_2 x}, \qquad \text{if} \qquad r_1 = r_2. \tag{6}$$

EXAMPLE 1 Solve the equation

$$\frac{d^2y}{dx^2} + \frac{dy}{dx} - 2y = 0.$$

Solution $r^2 + r - 2 = 0$ has roots $r_1 = 1$, $r_2 = -2$. Hence, by Eq. (5), the solution of the differential equation is

$$y = C_1 e^x + C_2 e^{-2x}. \ \square$$

EXAMPLE 2 Solve the equation

$$\frac{d^2y}{dx^2} + 4\frac{dy}{dx} + 4y = 0.$$

Solution

$$r^2 + 4r + 4 = (r + 2)^2,$$
$$r_1 = r_2 = -2,$$
$$y = (C_1 x + C_2)e^{-2x}. \ \square$$

Imaginary Roots

If the coefficients a and b in Eq. (1) are real, the roots of the characteristic Eq. (1″) either will be real, or will be a pair of complex conjugate numbers:

$$r_1 = \alpha + i\beta, \qquad r_2 = \alpha - i\beta. \tag{7}$$

If $\beta \neq 0$, Eq. (5) applies, with the result

$$\begin{aligned} y &= c_1 e^{(\alpha + i\beta)x} + c_2 e^{(\alpha - i\beta)x} \\ &= e^{\alpha x}[c_1 e^{i\beta x} + c_2 e^{-i\beta x}]. \end{aligned} \tag{8}$$

By Euler's formula,

$$e^{i\beta x} = \cos \beta x + i \sin \beta x,$$

$$e^{-i\beta x} = \cos \beta x - i \sin \beta x.$$

Hence, Eq. (8) may be replaced by

$$y = e^{\alpha x}[(c_1 + c_2) \cos \beta x + i(c_1 - c_2) \sin \beta x]. \tag{9}$$

Finally, if we introduce new arbitrary constants

$$C_1 = c_1 + c_2,$$
$$C_2 = i(c_1 - c_2),$$

Eq. (9) takes the form

$$y = e^{\alpha x}[C_1 \cos \beta x + C_2 \sin \beta x],$$

if

$$r_1 = \alpha + i\beta, \qquad r_2 = \alpha - i\beta. \tag{9'}$$

The arbitrary constants C_1 and C_2 in (9′) will be real provided the constants c_1 and c_2 in (9) are complex conjugates:

$$c_1 = \frac{1}{2}(C_1 - iC_2), \qquad c_2 = \frac{1}{2}(C_1 + iC_2).$$

To solve a problem where the roots of the characteristic equation are complex conjugates, we simply write down the appropriate version of Eq. (9′).

EXAMPLE 3 Solve the equation

$$\frac{d^2y}{dx^2} + 2\frac{dy}{dx} + 2y = 0.$$

Solution The characteristic equation $r^2 + 2r + 2 = 0$ has roots $r_1 = -1 + i$, $r_2 = -1 - i$. Hence, in Eq. (9′), we take

$$\alpha = -1, \qquad \beta = 1,$$

and obtain the solution

$$y = e^{-x}[C_1 \cos x + C_2 \sin x]. \ \square$$

EXAMPLE 4 Solve the equation

$$\frac{d^2y}{dx^2} + \omega^2 y = 0, \qquad \omega \neq 0.$$

Solution The characteristic equation $r^2 + \omega^2 = 0$ has roots $r_1 = i\omega$, $r_2 = -i\omega$. Hence we take $\alpha = 0$, $\beta = \omega$ in Eq. (9′), and obtain the solution

$$y = C_1 \cos \omega x + C_2 \sin \omega x. \ \square$$

PROBLEMS

Solve the following equations.

1. $\dfrac{d^2y}{dx^2} + 2\dfrac{dy}{dx} = 0$
2. $\dfrac{d^2y}{dx^2} + 5\dfrac{dy}{dx} + 6y = 0$
3. $\dfrac{d^2y}{dx^2} + 6\dfrac{dy}{dx} + 5y = 0$
4. $\dfrac{d^2y}{dx^2} - 2\dfrac{dy}{dx} - 3y = 0$
5. $\dfrac{d^2y}{dx^2} + \dfrac{dy}{dx} + y = 0$
6. $\dfrac{d^2y}{dx^2} - 4\dfrac{dy}{dx} + 4y = 0$
7. $\dfrac{d^2y}{dx^2} + 6\dfrac{dy}{dx} + 9y = 0$
8. $\dfrac{d^2y}{dx^2} - 6\dfrac{dy}{dx} + 10y = 0$
9. $\dfrac{d^2y}{dx^2} - 2\dfrac{dy}{dx} + 4y = 0$
10. $\dfrac{d^2y}{dx^2} - 10\dfrac{dy}{dx} + 16y = 0$

18.10 Linear Second Order Nonhomogeneous Equations with Constant Coefficients

In Article 18.9, we learned how to solve the homogeneous equation

$$\frac{d^2y}{dx^2} + 2a\frac{dy}{dx} + by = 0. \tag{1}$$

We can now describe a method for solving the nonhomogeneous equation

$$\frac{d^2y}{dx^2} + 2a\frac{dy}{dx} + by = F(x). \tag{2}$$

To solve Eq. (2), we first obtain the general solution of the related homogeneous Eq. (1) obtained by replacing $F(x)$ by zero. Denote this solution by

$$y_h = C_1u_1(x) + C_2u_2(x), \tag{3}$$

where C_1 and C_2 are arbitrary constants and $u_1(x)$, $u_2(x)$ are functions of one or more of the following forms:

$$e^{rx}, \qquad xe^{rx}, \qquad e^{\alpha x}\cos \beta x, \qquad e^{\alpha x}\sin \beta x.$$

Now we might, by inspection or by inspired guesswork, be able to discover *one* particular function $y = y_p(x)$ that satisfies Eq. (2). In this case, we would be able to solve Eq. (2) completely, as

$$y = y_h(x) + y_p(x).$$

REMARK. If $y_1(x)$ and $y_2(x)$ are any solutions of the same nonhomogeneous Eq. (2), then their difference $y(x) = y_1(x) - y_2(x)$ satisfies the homogeneous Eq. (1). It is for this reason that every solution of Eq. (2) is included if we add the general solution of the homogeneous equation to any particular solution of the nonhomogeneous equation. In this regard, the general homogeneous solution is analogous to the constant of integration that we add to a particular solution of $y' = f(x)$ to get the general solution.

EXAMPLE 1 Solve the equation

$$\frac{d^2y}{dx^2} + 2\frac{dy}{dx} - 3y = 6.$$

Solution y_h satisfies

$$\frac{d^2y_h}{dx^2} + 2\frac{dy_h}{dx} - 3y_h = 0.$$

The characteristic equation is

$$r^2 + 2r - 3 = 0,$$

and its roots are

$$r_1 = -3, \qquad r_2 = 1.$$

Hence

$$y_h = C_1 e^{-3x} + C_2 e^{x}.$$

Now, to find a particular solution of the original equation, observe that $y =$ constant would do, provided $-3y = 6$. Hence,

$$y_p = -2$$

is one particular solution. The complete solution is

$$y = y_p + y_h = -2 + C_1 e^{-3x} + C_2 e^{x}. \ \square$$

Variation of Parameters

Fortunately, there is a general method for finding the solution of the nonhomogeneous Eq. (2) once the general solution of the corresponding homogeneous equation is known. The method is known as the method of *variation of parameters*. It consists in replacing the constants C_1 and C_2 in Eq. (3) by functions $v_1 = v_1(x)$ and $v_2 = v_2(x)$, and requiring (in a way to be explained) that the resulting expression satisfy Eq. (2). There are two functions to be determined, and requiring that Eq. (2) be satisfied is only one condition. As a second condition, we also require that

$$v_1' u_1 + v_2' u_2 = 0. \tag{4}$$

Then we have

$$y = v_1 u_1 + v_2 u_2,$$

$$\frac{dy}{dx} = v_1 u_1' + v_2 u_2',$$

$$\frac{d^2y}{dx^2} = v_1 u_1'' + v_2 u_2'' + v_1' u_1' + v_2' u_2'.$$

If we substitute these expressions into the left-hand side of Eq. (2), we obtain

$$v_1\left[\frac{d^2u_1}{dx^2} + 2a\frac{du_1}{dx} + bu_1\right] + v_2\left[\frac{d^2u_2}{dx^2} + 2a\frac{du_2}{dx} + bu_2\right] + v_1' u_1' + v_2' u_2' = F(x).$$

The two bracketed terms are zero, since u_1 and u_2 are solutions of the homogeneous Eq. (1). Hence Eq. (2) is satisfied if, in addition to Eq. (4), we require that

$$v_1' u_1' + v_2' u_2' = F(x). \tag{5}$$

Equations (4) and (5) may be solved together as a pair,

$$v_1' u_1 + v_2' u_2 = 0,$$
$$v_1' u_1' + v_2' u_2' = F(x),$$

for the unknown functions v_1' and v_2'. Cramer's rule gives

$$v_1' = \frac{\begin{vmatrix} 0 & u_2 \\ F(x) & u_2' \end{vmatrix}}{\begin{vmatrix} u_1 & u_2 \\ u_1' & u_2' \end{vmatrix}} = \frac{-u_2F(x)}{D},$$

$$v_2' = \frac{\begin{vmatrix} u_1 & 0 \\ u_1' & F(x) \end{vmatrix}}{\begin{vmatrix} u_1 & u_2 \\ u_1' & u_2' \end{vmatrix}} = \frac{u_1F(x)}{D}, \tag{6}$$

where

$$D = \begin{vmatrix} u_1 & u_2 \\ u_1' & u_2' \end{vmatrix}.$$

Now v_1 and v_2 can be found by integration.

In applying the method of variation of parameters to solve the equation

$$\frac{d^2y}{dx^2} + 2a\frac{dy}{dx} + by = F(x), \tag{2}$$

we can work directly with the equations in (6). It is not necessary to rederive them. The steps are as follows:

1. Solve the associated homogeneous equation,

$$\frac{d^2y}{dx^2} + 2a\frac{dy}{dx} + by = 0,$$

   to find the functions u_1 and u_2.
2. Calculate D, v_1', and v_2' from (6).
3. Integrate v_1' and v_2' to find v_1 and v_2.
4. Write down the general solution of (2) as

$$y = v_1u_1 + v_2u_2.$$

EXAMPLE 2 Solve the equation

$$\frac{d^2y}{dx^2} + 2\frac{dy}{dx} - 3y = 6$$

of Example 1 by variation of parameters.

Solution We first solve the associated homogeneous equation

$$\frac{d^2y}{dx^2} + 2\frac{dy}{dx} - 3y = 0$$

as in Example 1 to find

$$u_1(x) = e^{-3x}, \qquad u_2(x) = e^{x}.$$

Then, from the equations in (6), we have

$$D = \begin{vmatrix} e^{-3x} & e^{x} \\ -3e^{-3x} & e^{x} \end{vmatrix} = e^{-2x} + 3e^{-2x} = 4e^{-2x},$$

$$v_1' = \frac{-6e^{x}}{4e^{-2x}} = -\frac{3}{2}e^{3x}, \qquad v_2' = \frac{6e^{-3x}}{4e^{-2x}} = \frac{3}{2}e^{-x}. \tag{7}$$

Hence

$$v_1 = \int -\frac{3}{2}e^{3x}\,dx = -\frac{1}{2}e^{3x} + C_1,$$

$$v_2 = \int \frac{3}{2}e^{-x}\,dx = -\frac{3}{2}e^{-x} + C_2,$$

and

$$\begin{aligned} y &= v_1u_1 + v_2u_2 \\ &= (-\tfrac{1}{2}e^{3x} + C_1)e^{-3x} + (-\tfrac{3}{2}e^{-x} + C_2)e^{x} \\ &= -2 + C_1e^{-3x} + C_2e^{x}. \ \square \end{aligned}$$

Undetermined Coefficients

The method of variation of parameters is completely general, for it will produce a particular solution of the nonhomogeneous equation (2) for any continuous function $F(x)$. But the calculations involved can be complicated, and in special cases there may be easier methods to use. For instance, we do not need to use variation of parameters to find a particular solution of

$$\frac{d^2y}{dx^2} - \frac{dy}{dx} + 5y = 3, \tag{8}$$

if we can find the particular solution $y_p = 3/5$ by inspection first. And even for an equation like

$$\frac{d^2y}{dx^2} + 3y = e^{x}, \tag{9}$$

we might guess that there is a solution of the form

$$y_p = Ae^{x}$$

and substitute $y_p = Ae^{x}$ and its second derivative into Eq. (9) to find A. If we do so, we find that the function $y_p = \frac{1}{4}e^{x}$ is a solution of Eq. (9).

Again, we might guess that the equation

$$\frac{d^2y}{dx^2} + y = 3x^2 + 4 \tag{10}$$

has a particular solution of the form

$$y_p = Cx^2 + Dx + E.$$

When we substitute this polynomial and its second derivative into Eq. (10) to see whether appropriate values for the constants C, D, and E can be found, Eq. (10) becomes

$$\begin{aligned} 2C + (Cx^2 + Dx + E) &= 3x^2 + 4, \\ Cx^2 + Dx + 2C + E &= 3x^2 + 4. \end{aligned} \tag{11}$$

This latter equation will hold for all values of x if its two sides are identical as polynomials in x; that is, if

$$C = 3, \qquad D = 0, \qquad \text{and} \qquad 2C + E = 4, \tag{12}$$

or,

$$C = 3, \qquad D = 0, \qquad E = -2.$$

We conclude that

$$y_p = 3x^2 + 0x - 2 = 3x^2 - 2 \tag{13}$$

is a solution of Eq. (10).

In each of the foregoing examples, the particular solution we found resembled the function $F(x)$ on the right side of the given differential equation. This was no accident, for we guessed the form of the particular solution by looking at $F(x)$ first. The method of first guessing the form of the solution up to certain undetermined constants, and then determining the values of these constants by using the differential equation, is known as the *method of undetermined coefficients.* It depends for its success upon our ability to recognize the form of a particular solution, and for this reason, among others, it lacks the generality of the method of variation of parameters. Nevertheless, its simplicity makes it the method of choice in a number of special cases.

We shall limit our discussion of the method of undetermined coefficients to selected equations in which the function $F(x)$ in Eq. (2) is the sum of one or more terms like

$$e^{rx}, \qquad \cos kx, \qquad \sin kx, \qquad ax^2 + bx + c.$$

Even so, we will find that the particular solutions of Eq. (2) do not always resemble $F(x)$ as closely as the ones we have seen.

EXAMPLE 3 Find a particular solution of

$$\frac{d^2y}{dx^2} - \frac{dy}{dx} = 2 \sin x. \tag{14}$$

Solution If we try to find a particular solution of the form

$$y_p = A \sin x,$$

and substitute the derivatives of y_p in the given equation, we find that A must satisfy the equation

$$-A \sin x + A \cos x = 2 \sin x \tag{15}$$

for all values of x. Since this requires A to be equal to -2 and 0 at the same time, we conclude that Eq. (14) has no solution of the form $A \sin x$.

It turns out that the required form is the sum

$$y_p = A \sin x + B \cos x. \tag{16}$$

The result of substituting the derivatives of this new candidate into Eq. (14) is

$$-A\sin x - B\cos x - (A\cos x - B\sin x) = 2\sin x.$$
$$(B - A)\sin x - (A + B)\cos x = 2\sin x. \tag{17}$$

Equation (17) will be an identity if

$$B - A = 2 \quad \text{and} \quad A + B = 0,$$

or,

$$A = -1, \quad B = 1.$$

Our particular solution is

$$y_p = \cos x - \sin x. \ \square$$

EXAMPLE 4 Find a particular solution of

$$\frac{d^2y}{dx^2} - 3\frac{dy}{dx} + 2y = 5e^x. \tag{18}$$

Solution If we substitute

$$y_p = Ae^x$$

and its derivatives in Eq. (18), we find that

$$Ae^x - 3Ae^x + 2Ae^x = 5e^x,$$
$$0 = 5e^x.$$

The trouble can be traced to the fact that $y = e^x$ is already a solution of the related homogeneous equation,

$$\frac{d^2y}{dx^2} - 3\frac{dy}{dx} + 2y = 0. \tag{19}$$

The characteristic equation of Eq. (19) is

$$r^2 - 3r + 2 = (r - 1)(r - 2) = 0,$$

which has $r = 1$ as a simple root. We may therefore expect Ae^x to "vanish" when substituted into the left-hand side of (18).

The appropriate way to modify the trial solution in this case is to replace Ae^x by Axe^x. When we substitute

$$y_p = Axe^x$$

and its derivatives into (18), we obtain

$$(Axe^x + 2Ae^x) - 3(Axe^x + Ae^x) + 2Axe^x = 5e^x$$
$$-Ae^x = 5e^x$$
$$A = -5.$$

The function

$$y_p = -5xe^x$$

is a particular solution of Eq. (19). $\square$

EXAMPLE 5 Find a particular solution of

$$\frac{d^2y}{dx^2} - 6\frac{dy}{dx} + 9y = e^{3x}. \tag{20}$$

Solution The characteristic equation,

$$r^2 - 6r + 9 = (r - 3)^2 = 0,$$

has $r = 3$ as a *double* root. The appropriate choice for y_p in this case is neither Ae^{3x} nor Axe^{3x}, but Ax^2e^{3x}. When we substitute

$$y_p = Ax^2e^{3x}$$

and its derivatives in the given differential equation, we get

$$\begin{aligned}(9Ax^2e^{3x} + 12Axe^{3x} + 2Ae^{3x}) - 6(3Ax^2e^{3x} + 2Axe^{3x}) + 9Ax^2e^{3x} &= e^{3x}.\\ 2Ae^{3x} &= e^{3x}\\ A &= \tfrac{1}{2}.\end{aligned}$$

The function

$$y_p = \tfrac{1}{2}x^2e^{3x}$$

is a solution of Eq. (20). □

When we wish to find a particular solution of Eq. (2), and the function $F(x)$ is the sum of two or more terms, we choose a trial function for each term in $F(x)$, and add them.

EXAMPLE 6 Solve the equation

$$\frac{d^2y}{dx^2} - \frac{dy}{dx} = 5e^x - \sin 2x. \tag{21}$$

Solution We first check the characteristic equation, which is

$$r^2 - r = 0.$$

Its roots are

$$r_1 = 1, \qquad r_2 = 0,$$

so the general solution of the related homogeneous equation is

$$y_h = C_1e^x + C_2.$$

We now seek a particular solution y_p. That is, we seek a function that will produce $5e^x - \sin 2x$ when substituted into the left side of Eq. (21). One part of y_p is to produce $5e^x$, the other $-\sin 2x$.

Since any function of the form $C_1 e^x$ is a solution of the related homogeneous equation, we choose our trial y_p to be the sum

$$y_p = Axe^x + B\cos 2x + C\sin 2x,$$

including xe^x where we might otherwise have included e^x. When the derivatives of y_p are substituted in Eq. (21), the resulting equations are

$$\begin{aligned}&(Axe^x + 2Ae^x - 4B\cos 2x - 4C\sin 2x)\\ &\qquad - (Axe^x + Ae^x - 2B\sin 2x + 2C\cos 2x) = 5e^x - \sin 2x,\\ &Ae^x - (4B + 2C)\cos 2x + (2B - 4C)\sin 2x = 5e^x - \sin 2x.\end{aligned}$$

These equations will hold if

$$A = 5, \qquad (4B + 2C) = 0, \qquad (2B - 4C) = -1,$$

or,

$$A = 5, \qquad B = -\tfrac{1}{10}, \qquad C = \tfrac{1}{5}.$$

Our particular solution is

$$y_p = 5xe^x - \tfrac{1}{10}\cos 2x + \tfrac{1}{5}\sin 2x.$$

The complete solution of Eq. (21) is

$$y = y_h + y_p = C_1e^x + C_2 + 5xe^x - \tfrac{1}{10}\cos 2x + \tfrac{1}{5}\sin 2x. \quad \square$$

You may find the following table helpful in solving the problems at the end of this article.

Table 18.1 The method of undetermined coefficients for selected equations of the form

$$\frac{d^2y}{dx^2} + 2a\frac{dy}{dx} + by = F(x)$$

If F(x) has a term that is a constant multiple of . . .	And if	Then include this expression in the trial function for y_p.
e^{rx}	r is not a root of the characteristic equation	Ae^{rx}
	r is a single root of the characteristic equation	Axe^{rx}
	r is a double root of the characteristic equation	Ax^2e^{rx}
$\sin kx$, $\cos kx$	ki is not a root of the characteristic equation	$B\cos kx + C\sin kx$
$ax^2 + bx + c$	0 is not a root of the characteristic equation	$Dx^2 + Ex + F$ (chosen to match the degree of $ax^2 + bx + c$)
	0 is a single root of the characteristic equation	$Dx^3 + Ex^2 + Fx$ (degree one higher than the degree of $ax^2 + bx + c$)
	0 is a double root of the characteristic equation	$Dx^4 + Ex^3 + Fx^2$ (degree two higher than the degree of $ax^2 + bx + c$)

PROBLEMS

Solve the equations in Problems 1–12 by variation of parameters.

1. $\dfrac{d^2y}{dx^2} + \dfrac{dy}{dx} = x$

2. $\dfrac{d^2y}{dx^2} + y = \tan x, \quad -\dfrac{\pi}{2} < x < \dfrac{\pi}{2}$

3. $\dfrac{d^2y}{dx^2} + y = \sin x$

4. $\dfrac{d^2y}{dx^2} + 2\dfrac{dy}{dx} + y = e^x$

5. $\dfrac{d^2y}{dx^2} + 2\dfrac{dy}{dx} + y = e^{-x}$

6. $\dfrac{d^2y}{dx^2} - y = x$

7. $\dfrac{d^2y}{dx^2} - y = e^x$

8. $\dfrac{d^2y}{dx^2} - y = \sin x$

9. $\dfrac{d^2y}{dx^2} + 4\dfrac{dy}{dx} + 5y = 10$

10. $\dfrac{d^2y}{dx^2} - \dfrac{dy}{dx} = 2^x$

11. $\dfrac{d^2y}{dx^2} + y = \sec x, \quad -\dfrac{\pi}{2} < x < \dfrac{\pi}{2}$

12. $\dfrac{d^2y}{dx^2} - \dfrac{dy}{dx} = e^x \cos x, \quad x > 0$

Solve the equations in Problems 13–28 by the method of undetermined coefficients.

13. $\dfrac{d^2y}{dx^2} - 3\dfrac{dy}{dx} - 10y = -3$

14. $\dfrac{d^2y}{dx^2} - 3\dfrac{dy}{dx} - 10y = 2x - 3$

15. $\dfrac{d^2y}{dx^2} - \dfrac{dy}{dx} = \sin x$

16. $\dfrac{d^2y}{dx^2} + 2\dfrac{dy}{dx} + y = x^2$

17. $\dfrac{d^2y}{dx^2} + y = \cos 3x$

18. $\dfrac{d^2y}{dx^2} + y = e^{2x}$

19. $\dfrac{d^2y}{dx^2} - \dfrac{dy}{dx} - 2y = 20 \cos x$

20. $\dfrac{d^2y}{dx^2} + y = 2x + 3e^x$

21. $\dfrac{d^2y}{dx^2} - y = e^x + x^2$

22. $\dfrac{d^2y}{dx^2} + 2\dfrac{dy}{dx} + y = 6 \sin 2x$

23. $\dfrac{d^2y}{dx^2} - \dfrac{dy}{dx} - 6y = e^{-x} - 7 \cos x$

24. $\dfrac{d^2y}{dx^2} + 3\dfrac{dy}{dx} + 2y = e^{-x} + e^{-2x} - x$

25. $\dfrac{d^2y}{dx^2} + 5\dfrac{dy}{dx} = 15x^2$

26. $\dfrac{d^2y}{dx^2} - \dfrac{dy}{dx} = -8x + 3$

27. $\dfrac{d^2y}{dx^2} - 3\dfrac{dy}{dx} = e^{3x} - 12x$

28. $\dfrac{d^2y}{dx^2} + 7\dfrac{dy}{dx} = 42x^2 + 5x + 1$

In each of Problems 29–31, the given differential equation has a particular solution y_p of the form given. Determine the coefficients in y_p. Then solve the differential equation.

29. $\dfrac{d^2y}{dx^2} - 5\dfrac{dy}{dx} = xe^{5x}, \quad y_p = Ax^2e^{5x} + Bxe^{5x}$

30. $\dfrac{d^2y}{dx^2} - \dfrac{dy}{dx} = \cos x + \sin x, \quad y_p = A \cos x + B \sin x$

31. $\dfrac{d^2y}{dx^2} + y = 2 \cos x + \sin x, \quad y_p = Ax \cos x + Bx \sin x$

In Problems 32–35, solve the given differential equations (a) by variation of parameters, and (b) by the method of undetermined coefficients.

32. $\dfrac{d^2y}{dx^2} - 4\dfrac{dy}{dx} + 4y = 2e^{2x}$

33. $\dfrac{d^2y}{dx^2} - \dfrac{dy}{dx} = e^x + e^{-x}$

34. $\dfrac{d^2y}{dx^2} - 9\dfrac{dy}{dx} = 9e^{9x}$

35. $\dfrac{d^2y}{dx^2} - 4\dfrac{dy}{dx} - 5y = e^x + 4$

Solve the differential equations in Problems 36–45. Some of the equations can be solved by the method of undetermined coefficients, but others cannot.

36. $\dfrac{d^2y}{dx^2} + y = \csc x, \quad 0 < x < \pi$

37. $\dfrac{d^2y}{dx^2} + y = \cot x, \quad 0 < x < \pi$

38. $\dfrac{d^2y}{dx^2} + 4y = \sin x$

39. $\dfrac{d^2y}{dx^2} - 8\dfrac{dy}{dx} = e^{8x}$

40. $\dfrac{d^2y}{dx^2} + 4\dfrac{dy}{dx} + 5y = x + 2$

41. $\dfrac{d^2y}{dx^2} - \dfrac{dy}{dx} = x^3$

42. $\dfrac{d^2y}{dx^2} + 9y = 9x - \cos x$

43. $\dfrac{d^2y}{dx^2} + 2\dfrac{dy}{dx} = x^2 - e^x$

44. $\dfrac{d^2y}{dx^2} - 3\dfrac{dy}{dx} + 2y = e^x - e^{2x}$

45. $\dfrac{d^2y}{dx^2} + y = \sec x \tan x, \quad -\dfrac{\pi}{2} < x < \dfrac{\pi}{2}$

The method of undetermined coefficients can sometimes be used to solve first order ordinary differential equations. Use the method to solve the equations in Problems 46–49.

46. $\dfrac{dy}{dx} + 4y = x$

47. $\dfrac{dy}{dx} - 3y = e^x$

48. $\dfrac{dy}{dx} + y = \sin x$

49. $\dfrac{dy}{dx} - 3y = 5e^{3x}$

Solve the differential equations in Problems 50 and 51 subject to the given initial conditions.

50. $\dfrac{d^2y}{dx^2} + y = e^{2x}; \quad y(0) = 0, \quad y'(0) = \dfrac{2}{5}$

51. $\dfrac{d^2y}{dx^2} + y = \sec^2 x, \quad -\dfrac{\pi}{2} < x < \dfrac{\pi}{2}; \quad y(0) = y'(0) = 1$

52. *Bernoulli's equation of order 2.* Solve the equation

$$\frac{dy}{dx} + y = (xy)^2$$

by carrying out the following steps: (1) divide both sides of the equation by y^2; (2) make the change of variable $u = y^{-1}$; (3) solve the resulting equation for u in terms of x; (4) let $y = u^{-1}$.

53. Solve the integral equation

$$y(x) + \int_0^x y(t)\, dt = x.$$

(*Hint:* Differentiate.)

18.11 Higher Order Linear Equations with Constant Coefficients

The methods of Articles 18.9 and 18.10 can be extended to equations of higher order. The characteristic algebraic equation associated with the differential equation

$$(D^n + a_1D^{n-1} + \cdots + a_{n-1}D + a_n)y = F(x) \tag{1}$$

is

$$r^n + a_1r^{n-1} + \cdots + a_{n-1}r + a_n = 0. \tag{2}$$

If its roots $r_1, r_2, \ldots, r_n$ are all distinct, the solution of the homogeneous equation obtained by replacing $F(x)$ by 0 in Eq. (1) is

$$y_h = c_1e^{r_1x} + c_2e^{r_2x} + \cdots + c_ne^{r_nx}.$$

Pairs of complex conjugate roots $\alpha \pm i\beta$ can be grouped together, and the corresponding part of y_h can be written in terms of the functions

$$e^{\alpha x} \cos \beta x \qquad \text{and} \qquad e^{\alpha x} \sin \beta x.$$

In case the roots of Eq. (2) are not all distinct, the portion of y_h which corresponds to a root r of multiplicity m is to be replaced by

$$(C_1x^{m-1} + C_2x^{m-2} + \cdots + C_m)e^{rx}.$$

Note that the polynomial in parentheses contains m arbitrary constants.

EXAMPLE Solve the equation

$$\frac{d^4y}{dx^4} - 3\frac{d^3y}{dx^3} + 3\frac{d^2y}{dx^2} - \frac{dy}{dx} = 0.$$

Solution $r^4 - 3r^3 + 3r^2 - r = r(r-1)^3$. The roots of the characteristic equation are

$$r_1 = 0, \qquad r_2 = r_3 = r_4 = 1.$$

The solution is

$$y = C_1 + (C_2x^2 + C_3x + C_4)e^x. \ \square$$

Variation of Parameters If the general solution of the homogeneous equation is

$$y_h = C_1u_1 + C_2u_2 + \cdots + C_nu_n,$$

then

$$y = v_1u_1 + v_2u_2 + \cdots + v_nu_n$$

will be a solution of the nonhomogeneous Eq. (1), provided

$$\begin{aligned}
v_1'u_1 + v_2'u_2 + \cdots + v_n'u_n &= 0,\\
v_1'u_1' + v_2'u_2' + \cdots + v_n'u_n' &= 0,\\
&\vdots\\
v_1'u_1^{(n-2)} + v_2'u_2^{(n-2)} + \cdots + v_n'u_n^{(n-2)} &= 0,\\
v_1'u_1^{(n-1)} + v_2'u_2^{(n-1)} + \cdots + v_n'u_n^{(n-1)} &= F(x).
\end{aligned}$$

These equations may be solved for $v_1', v_2', \ldots, v_n'$ by Cramer's rule, and the results integrated to give $v_1, v_2, \ldots, v_n$.

PROBLEMS

Solve the following equations.

1. $\dfrac{d^3y}{dx^3} - 3\dfrac{d^2y}{dx^2} + 2\dfrac{dy}{dx} = 0$
2. $\dfrac{d^3y}{dx^3} - y = 0$
3. $\dfrac{d^4y}{dx^4} - 4\dfrac{d^2y}{dx^2} + 4y = 0$
4. $\dfrac{d^4y}{dx^4} - 16y = 0$
5. $\dfrac{d^4y}{dx^4} + 16y = 0$
6. $\dfrac{d^3y}{dx^3} - 3\dfrac{dy}{dx} + 2y = e^x$
7. $\dfrac{d^4y}{dx^4} - 4\dfrac{d^3y}{dx^3} + 6\dfrac{d^2y}{dx^2} - 4\dfrac{dy}{dx} + y = 7$
8. $\dfrac{d^4y}{dx^4} + y = x + 1$

18.12 Vibrations

A spring of natural length L has its upper end fastened to a rigid support at A (Fig. 18.2). A weight W, of mass m, is suspended from the spring. The weight stretches the spring to a length $L + s$ when the system is allowed to come to rest in a new equilibrium position. By Hooke's law, the tension in the spring is ks, where k is the so-called spring constant. The force of gravity pulling down on the weight is $W = mg$. Equilibrium requires

$$ks = mg. \tag{1}$$

Suppose now that the weight is pulled down an additional amount a beyond the equilibrium position, and released. We shall discuss its motion.

Let x, positive direction downward, denote the displacement of the weight away from equilibrium at any time t after the motion has started. Then the forces acting upon the weight are

$+mg$, due to gravity,

$-k(s + x)$, due to the spring tension.

The resultant of these forces is, by Newton's second law, also equal to

$$m\frac{d^2x}{dt^2}.$$

Therefore

$$m\frac{d^2x}{dt^2} = mg - ks - kx. \tag{2}$$

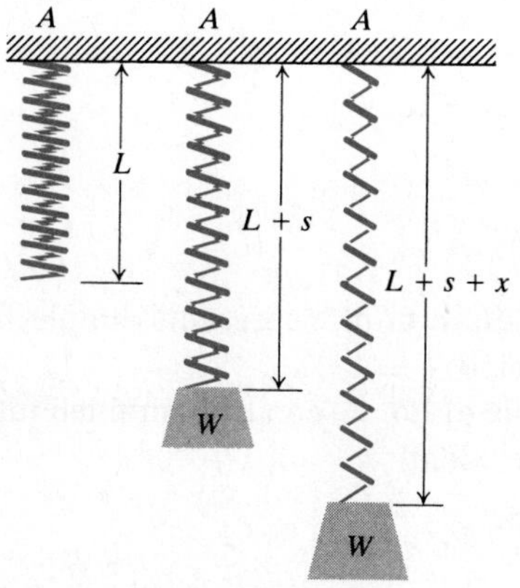

18.2 A spring stretched beyond its natural length by a weight.

By Eq. (1), $mg - ks = 0$, so that (2) becomes

$$m\frac{d^2x}{dt^2} + kx = 0. \tag{3}$$

In addition to the differential equation (3), the motion satisfies the initial conditions:

$$\text{At } t = 0: \qquad x = a \quad \text{and} \quad \frac{dx}{dt} = 0. \tag{4}$$

Let $\omega = \sqrt{k/m}$. Then Eq. (3) becomes

$$\frac{d^2x}{dt^2} + \omega^2 x = 0$$

or

$$(D^2 + \omega^2)x = 0,$$

where

$$D = \frac{d}{dt}.$$

The roots of the characteristic equation

$$r^2 + \omega^2 = 0$$

are complex conjugates

$$r = \pm\omega i.$$

Hence

$$x = c_1 \cos \omega t + c_2 \sin \omega t \tag{5}$$

is the general solution of the differential equation. To fit the initial conditions, we also compute

$$\frac{dx}{dt} = -c_1\omega \sin \omega t + c_2\omega \cos \omega t,$$

and then substitute from (4). This yields

$$a = c_1, \qquad 0 = c_2\omega.$$

Therefore

$$c_1 = a, \qquad c_2 = 0,$$

and

$$x = a \cos \omega t \tag{6}$$

describes the motion of the weight. Equation (6) represents simple harmonic motion of amplitude a and period $T = 2\pi/\omega$.

The two terms on the right-hand side of Eq. (5) can be combined into a single term by using the trigonometric identity

$$\sin(\omega t + \phi) = \cos \omega t \sin \phi + \sin \omega t \cos \phi.$$

To apply the identity, we take

$$c_1 = C \sin \phi, \qquad c_2 = C \cos \phi, \tag{7a}$$

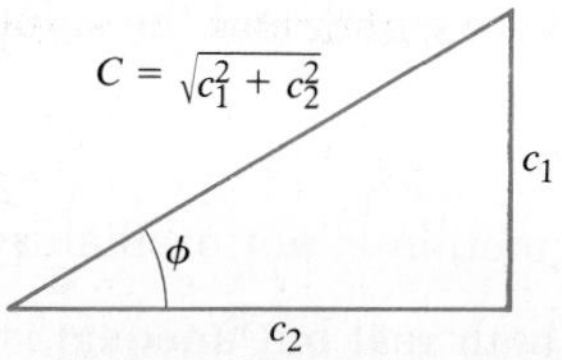

18.3 $c_1 = C \sin\phi$ and $c_2 = C \cos\phi$.

where

$$C = \sqrt{c_1^2 + c_2^2}, \qquad \phi = \tan^{-1}\frac{c_1}{c_2}, \tag{7b}$$

as in Fig. 18.3. Then Eq. (5) can be written in the alternative form

$$x = C \sin(\omega t + \phi). \tag{8}$$

Here C and ϕ may be taken as two new arbitrary constants, replacing the two constants c_1 and c_2 of Eq. (5). Equation (8) represents simple harmonic motion of amplitude C and period $T = 2\pi/\omega$. The angle $\omega t + \phi$ is called the *phase angle*, and ϕ may be interpreted as the initial value of the phase angle. A graph of Eq. (8) is given in Fig. 18.4.

Equation (3) assumes that there is no friction in the system. Next, consider the case where the motion of the weight is retarded by a friction force $c(dx/dt)$ proportional to the velocity, where c is a positive constant. Then the differential equation is

$$m\frac{d^2x}{dt^2} = -kx - c\frac{dx}{dt},$$

or

$$\frac{d^2x}{dt^2} + 2b\frac{dx}{dt} + \omega^2 x = 0, \tag{9}$$

where

$$2b = \frac{c}{m} \qquad \text{and} \qquad \omega = \sqrt{\frac{k}{m}}.$$

If we introduce the operator $D = d/dt$, Eq. (9) becomes

$$(D^2 + 2bD + \omega^2)x = 0.$$

The characteristic equation is

$$r^2 + 2br + \omega^2 = 0$$

with roots

$$r = -b \pm \sqrt{b^2 - \omega^2}. \tag{10}$$

Three cases now present themselves, depending upon the relative sizes of b and ω.

18.4 Undamped vibration.

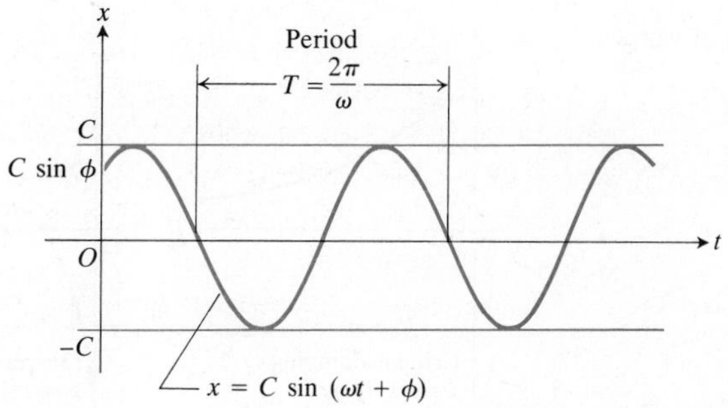

CASE 1. If $b = \omega$, the two roots in Eq. (10) are equal, and the solution of (9) is

$$x = (c_1 + c_2 t)e^{-\omega t}. \tag{11}$$

As time goes on, x approaches zero. The motion is not oscillatory.

CASE 2. If $b > \omega$, then the roots (10) are both real but unequal, and

$$x = c_1 e^{r_1 t} + c_2 e^{r_2 t}, \tag{12}$$

where

$$r_1 = -b + \sqrt{b^2 - \omega^2} \quad \text{and} \quad r_2 = -b - \sqrt{b^2 - \omega^2}.$$

Here again the motion is not oscillatory. Both r_1 and r_2 are negative, and x approaches zero as time goes on.

CASE 3. If $b < \omega$, let

$$\omega^2 - b^2 = \alpha^2.$$

Then

$$r_1 = -b + \alpha i, \qquad r_2 = -b - \alpha i$$

and

$$x = e^{-bt}[c_1 \cos \alpha t + c_2 \sin \alpha t]. \tag{13a}$$

If we introduce the substitutions (7), we may also write Eq. (13a) in the equivalent form

$$x = Ce^{-bt} \sin (\alpha t + \phi). \tag{13b}$$

This equation represents damped vibratory motion. It is analogous to simple harmonic motion, of period $T = 2\pi/\alpha$, except that the amplitude is not constant but is given by Ce^{-bt}. Since this tends to zero as t increases, the vibrations tend to die out as time goes on. Observe, however, that Eq. (13b) reduces to Eq. (8) in the absence of friction. The effect of friction is twofold:

1. $b = c/(2m)$ appears as a coefficient in the exponential *damping factor* e^{-bt}. The larger b is, the more quickly do the vibrations tend to become unnoticeable.
2. The period $T = 2\pi/\alpha = 2\pi/\sqrt{\omega^2 - b^2}$ is greater than the period $T_0 = 2\pi/\omega$ in the friction-free system.

Curves representing solutions of Eq. (9) in typical cases are shown in Figs. 18.4 and 18.5. The size of b, relative to ω, determines the kind of

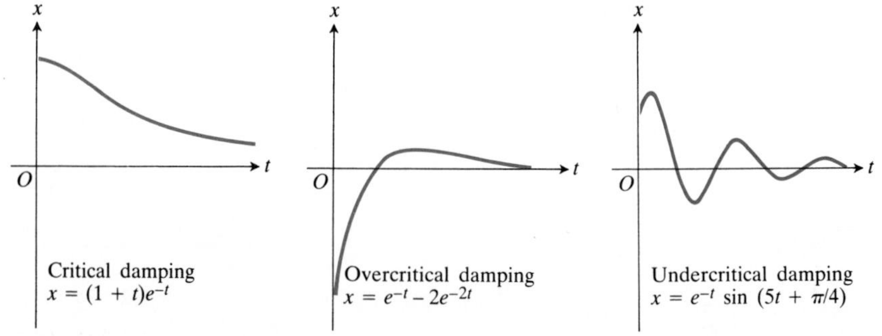

18.5 Three kinds of damping of vibration.

solution, and b also determines the rate of damping. It is therefore customary to say that there is

a) critical damping if $b = \omega$;

b) overcritical damping if $b > \omega$;

c) undercritical damping if $0 < b < \omega$;

d) no damping if $b = 0$.

PROBLEMS

1. Suppose the motion of the weight in Fig. 18.2 is described by the differential equation (3). Find the motion if $x = x_0$ and $dx/dt = v_0$ at $t = 0$. Express the answer in two equivalent forms (Eqs. 5 and 6).

2. A 5-lb weight is suspended from the lower end of a spring whose upper end is attached to a rigid support. The weight extends the spring by 2 in. If, after the weight has come to rest in its new equilibrium position, it is struck a sharp blow that starts it downward with a velocity of 4 ft/s, find its subsequent motion, assuming there is no friction.

3. A simple electrical circuit shown in Fig. 18.6 contains a capacitor of capacitance C farads, a coil of inductance L henrys, a resistance of R ohms, and a generator that produces an electromotive force E volts, in series. If the current intensity at time t at some point of the circuit is I amperes, the differential equation describing the current I is

$$L\frac{d^2I}{dt^2} + R\frac{dI}{dt} + \frac{1}{C}I = \frac{dE}{dt}.$$

Find I as a function of t if

a) $R = 0,\quad 1/(LC) = \omega^2,\quad E = \text{constant}$,

b) $R = 0,\quad 1/(LC) = \omega^2,\quad E = A \sin \alpha t$; $\alpha = \text{constant} \neq \omega$,

c) $R = 0,\quad 1/(LC) = \omega^2,\quad E = A \sin \omega t$,

d) $R = 50,\quad L = 5,\quad C = 9 \times 10^{-6},\quad E = \text{constant}$.

4. A simple pendulum of length l makes an angle θ with the vertical. As it swings back and forth, its motion, neglecting friction, is described by the differential equation

$$\frac{d^2\theta}{dt^2} = -\frac{g}{l}\sin\theta,$$

where g (the acceleration due to gravity, $g \approx 32$ ft/s²) is a constant. Solve the differential equation of motion, under the assumption that θ is so small that $\sin\theta$ may be replaced by θ without appreciable error. Assume that $\theta = \theta_0$ and $d\theta/dt = 0$ when $t = 0$.

5. A circular disc of mass m and radius r is suspended by a thin wire attached to the center of one of its flat faces. If the disc is twisted through an angle θ, torsion in the wire tends to turn the disc back in the opposite direction. The differential equation for the motion is

$$\frac{1}{2}mr^2\frac{d^2\theta}{dt^2} = -k\theta,$$

where k is the coefficient of torsion of the wire. Find the motion if $\theta = \theta_0$ and $d\theta/dt = v_0$ at $t = 0$.

6. A cylindrical spar buoy, diameter 1 foot, weight 100 lb, floats partially submerged, in an upright position. When it is depressed slightly from its equilibrium position and released, it bobs up and down according to the differential equation

$$\frac{100}{g}\frac{d^2x}{dt^2} = -16\pi x - c\frac{dx}{dt}.$$

Here $c(dx/dt)$ is the frictional resistance of the water. Find c if the period of oscillation is observed to be 1.6 s. (Take $g = 32$ ft/s².)

7. Suppose the upper end of the spring in Fig. 18.2 is attached, not to a rigid support at A, but to a member that itself undergoes up and down motion given by a function of the time t, say $y = f(t)$. If the positive direction of y is downward, the differential equation of motion is

$$m\frac{d^2x}{dt^2} + kx = kf(t).$$

Let $x = x_0$ and $dx/dt = 0$ when $t = 0$, and solve for x

a) if $f(t) = A \sin \alpha t$ and $\alpha \neq \sqrt{k/m}$,

b) if $f(t) = A \sin \alpha t$ and $\alpha = \sqrt{k/m}$.

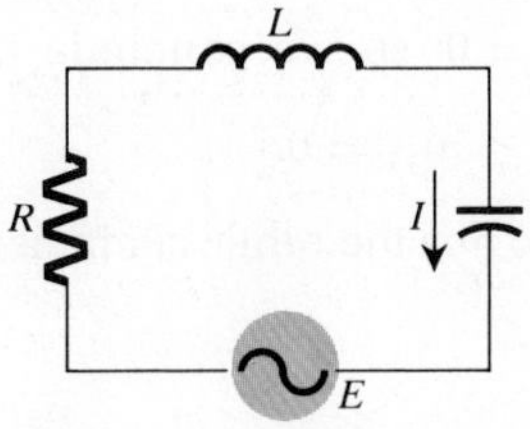

18.6 An R–L–C series circuit.

Toolkit **programs**	
Damped Oscillator	Forced Damped Oscillator

18.13

Approximation Methods: Power Series

What can we do if a given differential equation fails to fit any of the categories we know how to handle? We may go to more advanced textbooks, or treatises on the subject, but even this won't always help because some simple-looking equations have no solution in terms of finite combinations of the elementary functions that we know. The next example illustrates how we might be able to solve such an equation by a power series. (3)

EXAMPLE 1 Find a power series solution for

$$u'' + x^2u = 0. \tag{1}$$

Solution Assume that there is a solution of the form

$$u = a_0 + a_1x + a_2x^2 + \cdots + a_nx^n + \cdots \tag{2}$$

where the coefficients a_k are to be determined to satisfy Eq. (1). If we differentiate the series twice, we get

$$u'' = 2a_2 + 3 \cdot 2a_3x + \cdots + n(n-1)a_nx^{n-2} + \cdots. \tag{3}$$

The series for x^2u is just x^2 times the right-hand side of Eq. (2):

$$x^2u = a_0x^2 + a_1x^3 + a_2x^4 + \cdots + a_nx^{n+2} + \cdots. \tag{4}$$

The series for $u'' + x^2u$ is the sum of the series in Eqs. (3) and (4):

$$\begin{aligned} u'' + x^2u = 2a_2 + 6a_3x &+ (12a_4 + a_0)x^2 + (20a_5 + a_1)x^3 + \cdots \\ &+ [n(n-1)a_n + a_{n-4}]x^{n-2} + \cdots. \end{aligned} \tag{5}$$

Notice that the coefficient of x^{n-2} in Eq. (4) is a_{n-4}. To satisfy Eq. (1), the coefficients of the individual powers of x in Eq. (5) must all be zero:

$$2a_2 = 0, \qquad 6a_3 = 0, \qquad 12a_4 + a_0 = 0, \qquad 20a_5 + a_1 = 0,$$

and for all $n \geq 4$,

$$n(n-1)a_n + a_{n-4} = 0. \tag{7}$$

We can see immediately from Eq. (2) that

$$a_0 = u(0), \qquad a_1 = u'(0). \tag{8}$$

In other words, the first two coefficients are the values of u and of u' at $x = 0$. The recursion formulas (6) and (7) enable us to evaluate all the other coefficients in terms of these two.

The first two of Eqs. (6) give

$$a_2 = 0, \qquad a_3 = 0.$$

Equation (7) shows that if $a_{n-4} = 0$, then $a_n = 0$; so we conclude that

$$a_6 = 0, \qquad a_7 = 0, \qquad a_{10} = 0, \qquad a_{11} = 0,$$

and whenever $n = 4k + 2$ or $4k + 3$, a_n is zero. For the other coefficients we have

$$a_n = \frac{-a_{n-4}}{n(n-1)}$$

so that

$$a_4 = \frac{-a_0}{4 \cdot 3}, \qquad a_8 = \frac{-a_4}{8 \cdot 7} = \frac{a_0}{3 \cdot 4 \cdot 7 \cdot 8},$$

$$a_{12} = \frac{-a_8}{11 \cdot 12} = \frac{-a_0}{3 \cdot 4 \cdot 7 \cdot 8 \cdot 11 \cdot 12};$$

and

$$a_5 = \frac{-a_1}{5 \cdot 4}, \qquad a_9 = \frac{-a_5}{9 \cdot 8} = \frac{a_1}{4 \cdot 5 \cdot 8 \cdot 9},$$

$$a_{13} = \frac{-a_9}{12 \cdot 13} = \frac{-a_1}{4 \cdot 5 \cdot 8 \cdot 9 \cdot 12 \cdot 13}.$$

The answer is best expressed as the sum of two separate series—one multiplied by a_0, the other by a_1:

$$\begin{aligned} u = a_0\left(1 - \frac{x^4}{3 \cdot 4} + \frac{x^8}{3 \cdot 4 \cdot 7 \cdot 8} - \frac{x^{12}}{3 \cdot 4 \cdot 7 \cdot 8 \cdot 11 \cdot 12} + \cdots\right) \\ + a_1\left(x - \frac{x^5}{4 \cdot 5} + \frac{x^9}{4 \cdot 5 \cdot 8 \cdot 9} - \frac{x^{13}}{4 \cdot 5 \cdot 8 \cdot 9 \cdot 12 \cdot 13} + \cdots\right). \end{aligned} \tag{9}$$

Both series converge absolutely for all values of x as is readily seen by the ratio test. □

REMARK. If the differential equation $u'' + x^2u = 0$ were of sufficient importance to justify the effort, one could tabulate values of the two series for selected values of x, or program the computations for a calculator or computer.

PROBLEMS

Find power series solutions for the following differential equations. (Some of the equations can be solved directly without series, but they are included for practice.) Find at least four nonzero terms in each series.

1. $y' = y$

2. $y' + y = 0$

3. $y' = 2y$

4. $y' + 2y = 0$

5. $y'' = y$

6. $y'' + y = 0$

7. $y'' + y = x$

8. $y'' - y = x$

9. $y'' + x = y$

10. $y'' + y = 2x$

11. $y'' + y = \sin x$

12. $y'' - x^2y = 0$

13. $y'' + x^2y = x$

14. $y'' - x^2y = e^x$

18.14 Direction Fields and Picard's Theorem

Direction Fields and Isoclines

Associated with the differential equation $y' = f(x, y)$, for a given function f, is its *direction field*. Figure 18.7 shows a part of such a field for the equation

$$y' = x + y. \tag{1}$$

This figure was done by a computer, but it can be visualized as a pattern of iron filings that have been sprinkled onto a piece of paper and have

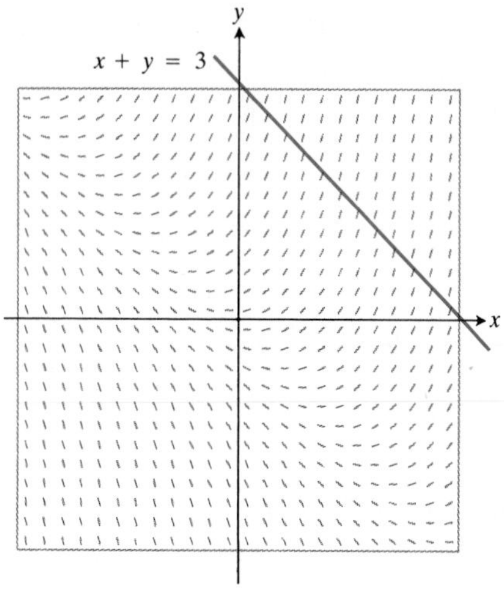

18.7 Direction field for $y' = x + y$, $-3 \leq x \leq 3$, $-3 \leq y \leq 3$.

arranged themselves so that the bit at the point $P(x, y)$ has slope $x + y$. Since iron filings won't actually do that for us, we resort to other approaches.

One simple approach is to first construct a set of *isoclines*. As the name implies, an isocline is a curve (in this case a straight line) at each point of which $f(x, y)$ is a fixed number.

For Eq. (1), each isocline has the form

$$x + y = C,$$

where C is a constant. Through some point on the line $x + y = 3$, for example, we draw a short line segment of slope 3. Having drawn one such segment, we then draw others parallel to it. Fig. 18.7 shows twelve such segments on that line. We repeat this process on each of the isoclines, starting with a short segment of slope C at a point on the line $x + y = C$. The resulting pattern is a portion of the direction field for the differential equation $y' = x + y$.

Notice one special line—the line $x + y = -1$. This line itself has slope -1 and satisfies the equation $x + y = y'$. At each point of this line, the direction field follows the line itself. Above this line, any solution curve is concave upward because $y'' = 1 + y' = 1 + x + y$ is positive if $x + y > -1$. Below the line, any solution curve must be concave downward.

Fig. 18.8 is a composite that shows two solution curves as well as a set of isoclines and part of the direction field for $y' = x + y$, in the region $-3 \leq x \leq 3$, $-3 \leq y \leq 3$. The upper solution curve is for the initial value problem

$$y' = x + y, \qquad y(0) = 1; \qquad y = -1 - x + 2e^x$$

and the lower curve is for

$$y' = x + y, \qquad y(-2) = 0; \qquad y = -1 - x - e^{x+2}.$$

In this example it is easy to find the general solution of

$$y' = x + y, \qquad y(x_0) = y_0.$$

Problem 15 asks you to show that the general solution is

$$y = -1 - x + (1 + x_0 + y_0)e^{x-x_0}.$$

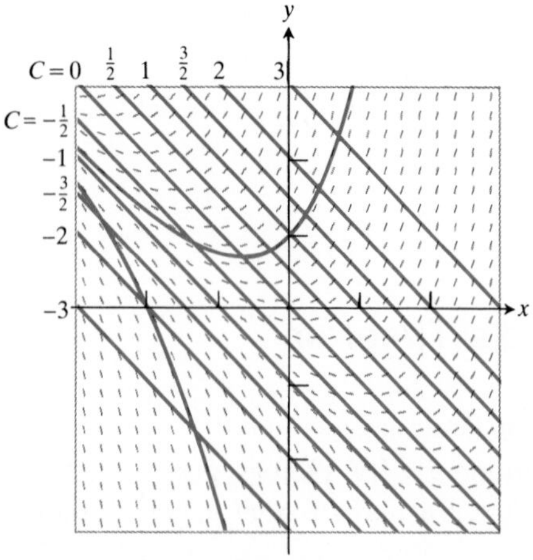

18.8 Direction field, isoclines $x + y = C$, and specific solutions for the equation $y' = x + y$, $-3 \leq x \leq 3$, $-3 \leq y \leq 3$.

Picard's Theorem

Suppose we are given an initial value problem

$$y' = f(x, y), \qquad y(x_0) = y_0 \tag{2}$$

where f is defined and continuous inside a rectangle R in the xy-plane. Does the initial value problem always have at least one solution if (x_0, y_0) is inside R? (See Fig. 18.9.) Might it have more than one solution? The next example shows that the answer to the second question is yes, unless something more is required of f.

EXAMPLE 1 The initial value problem

$$y' = y^{4/5}, \qquad y(0) = 0, \tag{3}$$

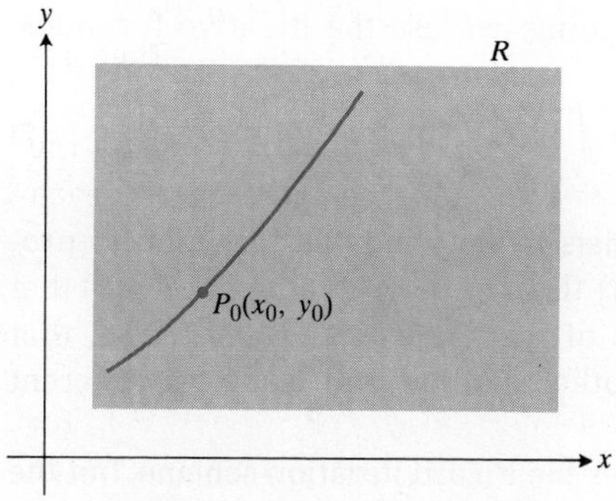

18.9 When is there exactly one solution of $y' = f(x, y)$, with $y(x_0) = y_0$?
Ans. When f and f_y are continuous throughout a rectangular region R that contains $P_0(x_0, y_0)$ in its interior.

has the obvious solution $y \equiv 0$. Another solution is found by separating the variables and integrating:

$$y = \left(\frac{x}{5}\right)^5.$$

There are two more solutions (see Fig. 18.10):

$$y = \begin{cases} 0 & \text{for } x \leq 0, \\ \left(\frac{x}{5}\right)^5 & \text{for } x > 0, \end{cases}$$

and

$$y = \begin{cases} \left(\frac{x}{5}\right)^5 & \text{for } x \leq 0, \\ 0 & \text{for } x > 0. \end{cases}$$

In this example, $f(x, y) = y^{4/5}$ is continuous in the entire xy-plane. But, its partial derivative $f_y = \frac{4}{5}y^{-1/5}$ is not continuous where $y = 0$. □

There is a useful theorem, due to Picard, that gives a sufficient condition for the existence and uniqueness of a solution of the initial value problem of Eqs. (2).

THEOREM

Picard's Theorem

Let $f(x, y)$ and its first order partial derivative f_y be continuous throughout the interior and on the boundary of a rectangle R. If $P(x_0, y_0)$ is any point in the interior of R, then there exists a positive number r such that the initial value problem

$$\begin{aligned} y' &= f(x, y), \\ y(x_0) &= y_0 \end{aligned} \tag{4}$$

has a unique solution, $y = y(x)$, for $x_0 - r \leq x \leq x_0 + r$.

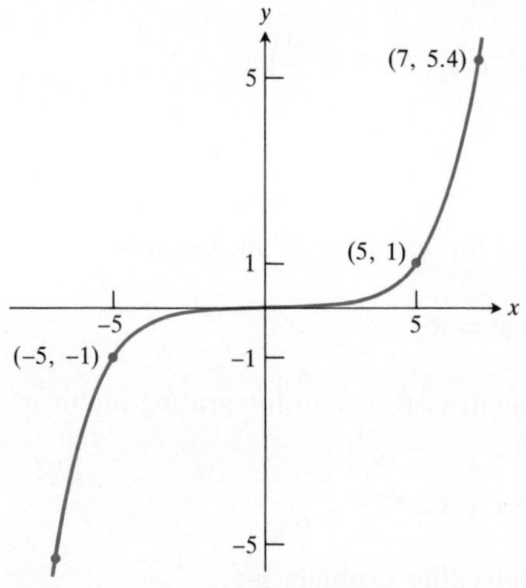

18.10 Part of graph of $y = (x/5)^5$, one of the solutions of $y' = y^{4/5}$, $y(0) = 0$. Another solution is $y = 0$.

Although we shall not prove this theorem, we relate it to Picard's iteration scheme that was mentioned in Article 2.10. In the first place, any function $y(x)$ that satisfies Eqs. (4) must also satisfy the integral equation

$$y(x) = y_0 + \int_{x_0}^{x} f(t, y(t))\, dt \tag{5}$$

because

$$\int_{x_0}^{x} \frac{dy}{dt}\, dt = y(x) - y(x_0).$$

The converse is also true: if $y(x)$ satisfies Eq. (5), then $y' = f(x, y(x))$ and $y(x_0) = y_0$. So Eqs. (4) may be replaced by Eq. (5). This sets the stage for Picard's iteration method: In the integrand in Eq. (5), replace $y(t)$ by the constant y_0, then integrate and call the resulting right-hand side of Eq. (5) $y_1(x)$:

$$y_1(x) = y_0 + \int_{x_0}^{x} f(t, y_0)\, dt. \tag{6}$$

This starts the process. To keep it going, we use the iterative formulas:

$$y_{n+1}(x) = y_0 + \int_{x_0}^{x} f(t, y_n(t))\, dt. \tag{7}$$

The proof of Picard's theorem consists in showing that this process produces a sequence of functions $\{y_n(x)\}$ that converge to a function $y(x)$ that satisfies Eqs. (4) and (5), for values of x sufficiently near x_0. (And, that the solution is unique: that is, no other method will lead to a different solution.)

The following example illustrates the Picard iteration scheme, but the computations soon become too burdensome to continue.

EXAMPLE 2 Illustrate the Picard iteration scheme for the initial value problem

$$y' = x - y, \qquad y(0) = 1. \tag{8}$$

Solution For the problem at hand, Eq. (6) becomes

$$\begin{aligned} y_1(x) &= 1 + \int_0^x (t - 1)\, dt \\ &= 1 + \frac{x^2}{2} - x. \end{aligned} \tag{9a}$$

If we now use Eq. (7) with $n = 1$, we get

$$\begin{aligned} y_2(x) &= 1 + \int_0^x \left(t - 1 - \frac{t^2}{2} + t\right) dt \\ &= 1 - x + x^2 - \frac{x^3}{6}. \end{aligned} \tag{9b}$$

The next iteration, with $n = 2$, gives

$$\begin{aligned} y_3(x) &= 1 + \int_0^x \left(t - 1 + t - t^2 + \frac{t^3}{6}\right) dt \\ &= 1 - x + x^2 - \frac{x^3}{3} + \frac{x^4}{4!}. \end{aligned} \tag{9c}$$

In this example, it is possible to find the exact solution, because

$$\frac{dy}{dx} + y = x$$

is a first order equation that is linear in y. It has an integrating factor e^x and the general solution is

$$y = x - 1 + Ce^{-x}.$$

The solution of our particular initial value problem is

$$y = x - 1 + 2e^{-x}. \tag{10}$$

If we substitute the Maclaurin series for e^{-x} in Eq. (10), we get

$$y = x - 1 + 2\left(1 - x + \frac{x^2}{2!} - \frac{x^3}{3!} + \frac{x^4}{4!} - \cdots\right)$$

$$= 1 - x + x^2 - \frac{x^3}{3} + 2\left(\frac{x^4}{4!} - \frac{x^5}{5!} + \cdots\right),$$

and we see that the Picard method has given us the first four terms of this expansion in Eq. (9c) in $y_3(x)$. □

In the next example, we cannot find a solution in terms of elementary functions. The Picard scheme is one way we could get an idea of how the solution behaves near the initial point.

EXAMPLE 3 Find $y_n(x)$ for $n = 0, 1, 2,$ and 3 for the initial value problem

$$y' = x^2 + y^2, \qquad y(0) = 0.$$

Solution By definition, $y_0(x) = y(0) = 0$. The other functions $y_n(x)$ are generated by the integral representation

$$y_{n+1}(x) = 0 + \int_0^x [t^2 + (y_n(t))^2]\,dt$$

$$= \frac{x^3}{3} + \int_0^x (y_n(t))^2\,dt.$$

We successively calculate

$$y_1(x) = \frac{x^3}{3}, \qquad y_2(x) = \frac{x^3}{3} + \frac{x^7}{63},$$

$$y_3(x) = \frac{x^3}{3} + \frac{x^7}{63} + \frac{2x^{11}}{2079} + \frac{x^{15}}{59535}. \ \square$$

REMARK. In the next article we shall introduce numerical methods for solving initial value problems like Examples 2 and 3 above. If we program such a numerical solution for a calculator or computer, it is helpful to have an independent check on the program. For x near zero, we would expect $y_2(x)$, or $y_3(x)$, to provide such a check.

PROBLEMS

In Problems 1–6, sketch some of the isoclines and part of the direction field. Using the direction field, sketch a portion of the graph of the solution that includes the point $P(1, -1)$.

1. $y' = x$ **2.** $y' = y$ **3.** $y' = 1/x$

4. $y' = 1/y$ **5.** $y' = xy$ **6.** $y' = x^2 + y^2$

7. Sketch some of the direction field for the equation $y' = (x + y)^2$. Show isoclines where the slope is 0, $\frac{1}{4}$, 1, 4. Sketch (roughly) the solutions of the equation that include (a) $P(0, 0)$, (b) $Q(-1, 1)$, and (c) $R(1, 0)$. Now make the substitution $u = x + y$ and find the general solution of $y' = (x + y)^2$. What is the equation of the solution that passes through the origin? Sketch that solution more accurately.

8. Show that every solution of the equation $y' = (x + y)^2$ has a graph that has a point of inflection, but no maximum or minimum. (The graph also has vertical asymptotes as you may discover by letting $x + y = u$.)

9. Use isoclines to sketch part of the direction field for $y' = x - y$. Include some part of all four quadrants.
a) Which isocline is also a solution of the differential equation?
b) Use the differential equation to determine where the solution curves are concave upward; concave downward.

Use Picard's iteration scheme to find $y_n(x)$ for $n = 0, 1, 2, 3$ in Problems 10–15.

10. $y' = x, \quad y(1) = 2$

11. $y' = y, \quad y(0) = 1$

12. $y' = xy, \quad y(1) = 1$

13. $y' = x + y, \quad y(0) = 0$

14. $y' = x + y, \quad y(0) = 1$

15. $y' = 2x - y, \quad y(-1) = 1$

16. Show that the general solution of

$$y' = x + y, \quad y(x_0) = y_0$$

is

$$y = -1 - x + (1 + x_0 + y_0)e^{x-x_0}.$$

17. In Example 3, verify the correctness of the equation for $y_3(x)$. (Find the mistake if there is one.)

***Toolkit* programs**

Antiderivatives and Direction Fields

18.15 Numerical Methods

The Euler Method

The initial value problem

$$y' = f(x, y), \qquad y(x_0) = y_0, \tag{1}$$

provides us with a starting point, $P(x_0, y_0)$, and a slope $f(x_0, y_0)$. We know that the graph of the solution must be a curve through P with that slope. If we use the tangent through P to approximate the actual solution curve, the approximation may be fairly good from $x_0 - h$ to $x_0 + h$, for small values of h. Thus, we might choose $h = 0.1$, say, and move along the tangent line from P to $P'(x_1, y_1)$ where $x_1 = x_0 + h$ and $y_1 = y_0 + hf(x_0, y_0)$. If we think of P' as a new starting point, we can move from P' to $P''(x_2, y_2)$ where $x_2 = x_1 + h$ and $y_2 = y_1 + hf(x_1, y_1)$. (This method is due to Euler.) If we replace h by $-h$, we move to the left from P instead of to the right. The process can be continued, but the errors are likely to accumulate as we take more steps. In order to see how the process works and to gain some idea of the errors, we illustrate for the initial value problem $y' = 1 + y$, $y(0) = 1$.

EXAMPLE 1 Take $h = 0.1$ and investigate the accuracy of the Euler approximation method for the initial value problem

$$y' = 1 + y, \qquad y(0) = 1, \tag{2}$$

by letting

$$x_{n+1} = x_n + h, \quad y_{n+1} = y_n + hf(x_n, y_n). \tag{3}$$

Continue to $x = 1$.

Solution The exact solution of Eqs. (2) is

$$y = 2e^x - 1. \tag{4}$$

The following table shows the results using Eqs. (3) and the exact results rounded to four decimals for comparison.

x	y(approx)	y(exact)	Error = y(exact) − y(approx)
0	1	1	0
0.1	1.2	1.2103	0.0103
0.2	1.42	1.4428	0.0228
0.3	1.662	1.6997	0.0377
0.4	1.9282	1.9836	0.0554
0.5	2.22102	2.2974	0.0764
0.6	2.543122	2.6442	0.1011
0.7	2.8974	3.0275	0.1301
0.8	3.2872	3.4511	0.1639
0.9	3.7159	3.9192	0.2033
1.0	4.1875	4.4366	0.2491

By the time we get to $x = 1$, the error is about 5.6%. □

The Improved Euler Method

This method first gets an estimate of y_{n+1}, as in the original Euler method, but calls the result z_{n+1} and then takes the average of $f(x_n, y_n)$ and $f(x_n, z_{n+1})$ in place of $f(x_n, y_n)$. Thus

$$z_{n+1} = y_n + hf(x_n, y_n), \tag{5a}$$

$$y_{n+1} = y_n + \frac{h}{2}[f(x_n, y_n) + f(x_n, z_{n+1})]. \tag{5b}$$

If we apply this improved method to Example 1, again with $h = 0.1$, we get the following results at $x = 1$:

$$\begin{aligned} y(\text{approx}) &= 4.428161693, \\ y(\text{exact}) &= 4.436563656, \\ \text{error} = y(\text{exact}) - y(\text{approx}) &= 0.008401963, \end{aligned}$$

and the error is less than 2/10 of 1%.

The Runge-Kutta Method

The improved Euler method just described corresponds to approximating an integral by the trapezoidal formula. The method now to be described is analogous to using Simpson's approximation to an integral. It requires four intermediate calculations, as given in the following equations:

$$\begin{aligned} k_1 &= hf(x_n, y_n) \\ k_2 &= hf\left(x_n + \frac{h}{2}, y_n + \frac{k_1}{2}\right) \\ k_3 &= hf\left(x_n + \frac{h}{2}, y_n + \frac{k_2}{2}\right) \\ k_4 &= hf(x_n + h, y_n + k_3). \end{aligned} \tag{6}$$

We then calculate y_{n+1} from the formula

$$y_{n+1} = y_n + \frac{1}{6}(k_1 + 2k_2 + 2k_3 + k_4). \tag{7}$$

When we apply this method to the problem of estimating $y(1)$ for the problem $y' = 1 + y$, $y(0) = 1$, still using $h = 0.1$, we get

$$y(\text{approx}) = 4.436559490$$

with an error 0.000004166, which is less than 1/10,000 of 1%. This is clearly the most accurate of the three methods and is not difficult to program for a calculator or computer.

The next example shows that the error in the Runge-Kutta approximation did not continue to increase as the process was continued. In fact, with $h = 0.1$, the difference between the exact solutions and the approximations remained less than 10^{-6} for the two initial value problems:

a) $y' = x - y, \quad y(0) = 1,$

b) $y' = x - y, \quad y(0) = -2.$

REMARK. The fact that the differential equation is linear in y is significant. Such accuracy is not attained for the initial value problem

$$y' = x^2 + y^2, \quad y(0) = 0,$$

which becomes infinite at $x = 2^+$. (See Problem 11.) The Runge-Kutta approximation to $y(2.1)$, using $h = 0.1$, is 1.47×10^{11}.

EXAMPLE 2 The following table shows the comparison of $y(x)$ as estimated by the Runge-Kutta method with $h = 0.1$ and the true value, for solutions of $y' = x - y$ (a) with $y(0) = 1$ and (b) with $y(0) = -2$.

x	y(Runge-Kutta)	y(true value)	Difference
0	1	1	0
0.5	0.713061869	0.713061319	5.50×10^{-7}
1.0	0.735759549	0.735758882	6.67×10^{-7}
1.5	0.946260927	0.946260320	6.07×10^{-7}
2.0	1.270671057	1.270670566	4.91×10^{-7}
2.5	1.664170370	1.664169997	3.73×10^{-7}
3.0	2.099574407	2.099574137	2.70×10^{-7}
0	−2	−2	0
0.5	−1.106530935	−1.106530660	-2.75×10^{-7}
1.0	−0.367879775	−0.367879441	-3.34×10^{-7}
1.5	+0.276869537	+0.276869840	-3.03×10^{-7}
2.0	0.864664472	0.864664717	-2.46×10^{-7}
2.5	1.417914816	1.417915001	-1.85×10^{-7}
3.0	1.950212796	1.950212932	-1.36×10^{-7}

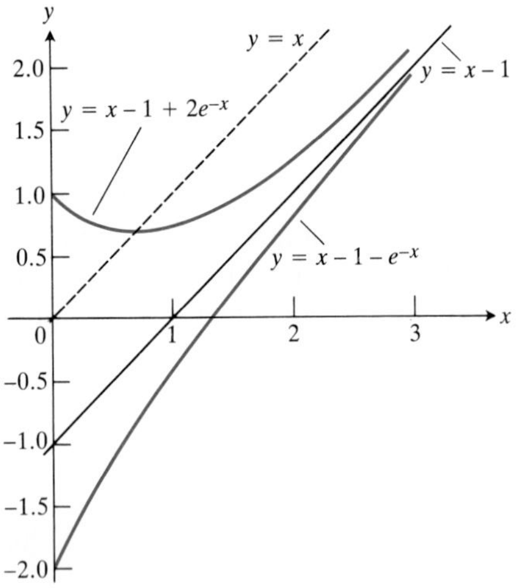

18.11 Two solutions of $y' = x - y$.
a) $y(0) = 1$, $y = x - 1 + 2e^{-x}$
b) $y(0) = -2$, $y = x - 1 - e^{-x}$

More points were actually computed and plotted to give the graphs shown in Fig. 18.11. Notice that the upper curve, $y = x - 1 + 2e^{-x}$ is concave upward and has a minimum when $x = y = \ln 2$. The lower curve is concave downward, is always rising as x increases, and crosses the x-axis at a value of x near 1.3. Both curves approach the line $y = x - 1$ as an asymptote as $x \to \infty$. □

For the initial value problem

$$y' = x^2 + y^2, \quad y(0) = 0, \tag{8}$$

some of the Runge-Kutta approximations we got with $h = 0.1$ are shown below. Part of the graph is shown in Fig. 18.12.

x	y(Runge-Kutta)	y(actual)
0	0	0
0.5	0.041791	0.041791146
1.0	0.350234	0.350231844
1.5	1.517473	1.517447544
2.0	71.578996	317.224400
2.1	1.470011×10^{11}	
2.2	9.999999×10^{99}	(meaning, "you broke the bank!")

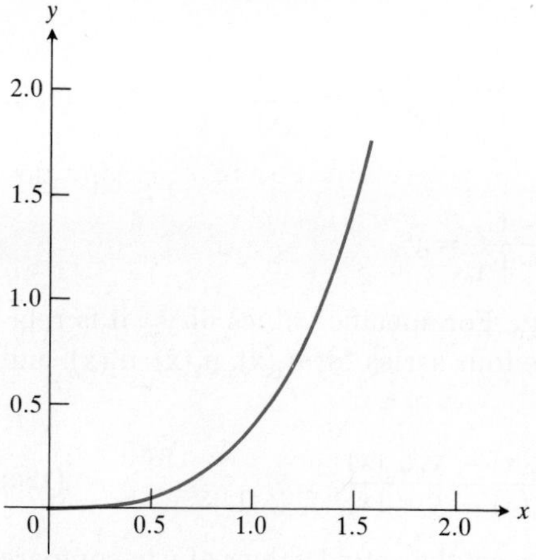

18.12 Approximate solution of $y' = x^2+y^2$, $y(0) = 0$.

Although there is no elementary solution of the equation $y' = x^2 + y^2$, it is possible to get an answer in terms of series. In the first place, the Maclaurin series that corresponds to the initial value $y(0) = 0$ begins this way:

$$y = \frac{x^3}{3} + \frac{x^7}{63} + \frac{2x^{11}}{6237} + \cdots. \tag{9}$$

In theory, one can get many more terms of the series, but the coefficients do not follow a simple pattern.

A more productive method involves an ingenious substitution

$$y = -\frac{u'}{u}, \tag{10a}$$

which transforms the equation $y' = x^2 + y^2$ into the second order equation

$$u'' + x^2u = 0. \tag{10b}$$

In Article 18.13, Eq. (9), we found the solution of Eq. (10b) in the form of two power series:

$$u_1(x) = 1 - \frac{x^4}{3 \cdot 4} + \frac{x^8}{3 \cdot 4 \cdot 7 \cdot 8} - \frac{x^{12}}{3 \cdot 4 \cdot 7 \cdot 8 \cdot 11 \cdot 12} + \cdots \tag{11a}$$

and

$$u_2(x) = x - \frac{x^5}{4 \cdot 5} + \frac{x^9}{4 \cdot 5 \cdot 8 \cdot 9} - \frac{x^{13}}{4 \cdot 5 \cdot 8 \cdot 9 \cdot 12 \cdot 13} + \cdots. \tag{11b}$$

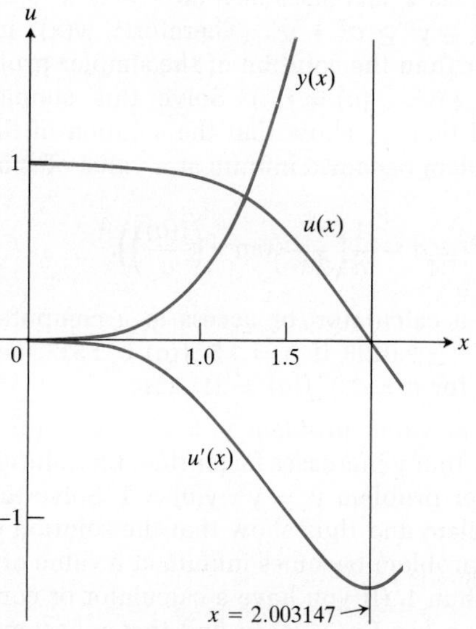

18.13 Graphs of $y(x)$, $u(x)$, and $u'(x)$ for the initial value problems

$$y' = x^2 + y^2,\ y(0) = 0$$
$$u'' + x^2u = 0,\ u(0) = 1,\ u'(0) = 0$$
$$y(x) = -u'/u.$$

The y-curve has an asymptote at $x = 2.003147$ where $u = 0$.

The general solution of Eq. (10b) is

$$u = C_1u_1 + C_2u_2.$$

Notice that

$$u_1(0) = 1, \quad u_2(0) = 0, \quad u_1'(0) = 0, \quad u_2'(0) = 1.$$

Therefore, if we wish to solve the general initial value problem

$$y' = x^2 + y^2, \qquad y(0) = y_0, \tag{12a}$$

using

$$y(x) = -\frac{u'(x)}{u(x)} = -\frac{C_1u_1'(x) + C_2u_2'(x)}{C_1u_1(x) + C_2u_2(x)}, \tag{12b}$$

we need to have

$$y(0) = -\frac{C_2}{C_1} = y_0.$$

We therefore take $C_1 = 1$, $C_2 = -y_0$. For specific values of y_0, it is relatively easy to compute values of the four series for $u_1(x)$, $u_2(x)$, $u_1'(x)$, and $u_2'(x)$ and combine them to obtain

$$y(x) = -\frac{u_1'(x) - y_0u_2'(x)}{u_1(x) - y_0u_2(x)}. \tag{12c}$$

We used this approach with $y_0 = 0$ to get the actual values of y to compare with the values obtained by using the Runge-Kutta method in the table following Eq. (8). (See Fig. 18.13.)

Problems 11 and 12 invite you to consider how you could predict that the solution of Eqs. (12a) becomes infinite near $x = 2$ if $y_0 = 0$, and before $x = 1$ if $y_0 = 1$.

PROBLEMS

1. Use the Euler method with $h = 1/5$ to estimate $y(1)$ if $y' = y$ and $y(0) = 1$. What is the exact value of $y(1)$?
2. Show that the Euler method leads to the estimate $(1 + (1/n))^n$ for $y(1)$ if $h = 1/n$, $y' = y$, and $y(0) = 1$. What is the limit as $n \to \infty$?
3. Use the improved Euler method with $h = 1/5$ to estimate $y(1)$ if $y' = y$ and $y(0) = 1$.
4. Use the Runge-Kutta method with $h = 1/5$ to estimate $y(1)$ if $y' = y$ and $y(0) = 1$.

In Problems 5–8, write an equivalent first order differential equation and initial condition for y.

5. $y = 1 + \int_0^x y(t)\,dt$
6. $y = -1 + \int_1^x t - y(t)\,dt$
7. $y = \int_1^x \frac{1}{t}\,dt$
8. $y = 2 - \int_0^x (1 + y(t))\sin t\,dt$
9. What integral equation is equivalent to the initial value problem $y' = f(x)$, $y(x_0) = y_0$?
10. If $f(x, y)$ does not depend upon y so that $f(x, y) = F(x)$, show that the Runge-Kutta method for approximating y_{n+1} is equivalent to using Simpson's approximation for

$$\int_{x_n}^{x_{n+1}} F(x)\,dx.$$

11. For the initial value problem $y' = x^2 + y^2$, $y(0) = 0$, let $y = f(a)$ be the value obtained at $x = a$, where a is positive. (For example, we could take $a = 1.5$, $f(a) = 1.517$.) As x increases beyond $x = a$, $x^2 \geq a^2$ and $y' = x^2 + y^2 \geq a^2 + y^2$. Therefore, $y(x)$ increases faster than the solution of the simpler problem $y' = a^2 + y^2$, $y(a) = f(a)$. Solve this simpler problem and thereby show that the solution of the original problem becomes infinite at a value of x no greater than

$$x^* = a + \frac{1}{a}\left(\frac{\pi}{2} - \tan^{-1}\left(\frac{f(a)}{a}\right)\right).$$

If you have a calculator, or access to a computer, show that $x^* = 2.0198$ if $a = 1.5$, $f(a) = 1.517$, and calculate x^* for $a = 2.0$, $f(a) = 317.2244$.

12. For the initial value problem $y' = x^2 + y^2$, $y(0) = y_0 = 1$, show that y increases faster than the solution of the simpler problem $y' = y^2$, $y(0) = 1$. Solve this simpler problem and thus show that the solution of the original problem becomes infinite at a value of x not greater than 1. (If you have a calculator or computer, you can use Eq. (12c) to find that y becomes infinite as x approaches 0.969810654, where $u_1 = u_2 = 0.927439876$.)
13. Find the first five nonzero terms in the Maclaurin series for the solution of the initial value problem $y' = x^2 + y^2$, $y(0) = 1$. Recall from Chapter 12 that if

$$y = a_0 + a_1x + a_2x^2 + \cdots + a_nx^n + \cdots$$

and

$$y^2 = c_0 + c_1x + c_2x^2 + \cdots + c_nx^n + \cdots,$$

then

$$c_n = \sum_{k=0}^{n} a_k a_{n-k},$$

and

$$y' = a_1 + 2a_2x + 3a_3x^2 + \cdots + na_nx^{n-1} + \cdots.$$

To satisfy the differential equation, we must have $a_1 = c_0$, $2a_2 = c_1$, $3a_3 = 1 + c_2$, and $na_n = c_{n-1}$ for $n \geq 4$. You can now determine a_0 (from the initial value), c_0, a_1, c_1, a_2, c_2, a_3, and so on.

14. Solve the problem $y' = 1 + y^2$, $y(0) = 0$, (a) by separating the variables, and (b) by using the substitution $y = -u'/u$ and solving the equivalent problem for u. (Note the similarity with the problem $y' = x^2 + y^2$, $y(0) = 0$. In the present problem we can find $u(x)$ without using series.)

Toolkit **programs**	
First Order Initial Value Problem	Second Order Initial Value Problem

REVIEW QUESTIONS AND EXERCISES

1. List some differential equations (having physical interpretations) that you have come across in your courses in chemistry, physics, engineering, or life and social sciences; or look for some in the articles on differential equations, dynamics, electromagnetic waves, hydromechanics, quantum mechanics, or thermodynamics in the *Encyclopaedia Britannica*.

2. How are differential equations classified?

3. What is meant by a "solution" of a differential equation?

4. Review methods for solving ordinary, first order, and first degree differential equations
a) when the variables are separable,
b) when the equation is homogeneous,
c) when the equation is linear in one variable,
d) when the equation is exact.
Illustrate each with an example.

5. Review methods of solving second order equations
a) with dependent variable missing,
b) with independent variable missing.
Illustrate each with an example.

6. Review methods for solving linear differential equations with constant coefficients
a) in the homogeneous case,
b) in the nonhomogeneous case.
Illustrate each with an example.

7. If an external force F acts upon a system whose mass varies with time, Newton's law of motion is

$$\frac{d(mv)}{dt} = F + (v + u)\frac{dm}{dt}.$$

In this equation, m is the mass of the system at time t, v its velocity, and $v + u$ is the velocity of the mass that is entering (or leaving) the system at the rate dm/dt. Suppose that a rocket of initial mass m_0 starts from rest, but is driven upward by firing some of its mass directly backward at the constant rate of $dm/dt = -b$ units per second and at constant speed relative to the rocket $u = -c$. The only external force acting on the rocket is $F = -mg$ due to gravity. Under these assumptions, show that the height of the rocket above the ground at the end of t seconds (t small compared to m_0/b) is

$$y = c\left[t + \frac{m_0 - bt}{b}\ln\frac{m_0 - bt}{m_0}\right] - \frac{1}{2}gt^2.$$

8. Which of the following differential equations *cannot* be solved as a first order, exact differential equation?
a) $y^2\,dx + 2xy\,dy = 0$
b) $(2x \sin y + y^3e^x)\,dx + (x^2 \cos y + 3y^2e^x)\,dy = 0$
c) $(2x \cos y + 3x^2y)\,dx + (x^3 - x^2 \sin y - y)\,dy = 0$
d) $(y \ln y - e^{-xy})\,dx + \left(\frac{1}{y} + x \ln y\right)dy = 0$

9. The differential equation $(x^2 - y^3)y' = 2xy$ has an integrating factor of the form y^n. Find n and solve the equation.

10. If y is a solution of $(D^2 + 4)y = e^x$, show that it satisfies $(D - 1)(D^2 + 4)y = 0$, because $(D - 1)e^x = 0$, where $D = d/dx$. Find the general solution of $(D - 1)(D^2 + 4)y = 0$ and use the result to solve $(D^2 + 4)y = e^x$.

11. If you were to solve $y' = y^2$, $y(0) = 1$, by finding the Maclaurin series for $y(x)$, for what values of x would you expect the series to converge? Why?

Suggestion. Solve the initial value problem without the use of series, then expand your answer in a series.

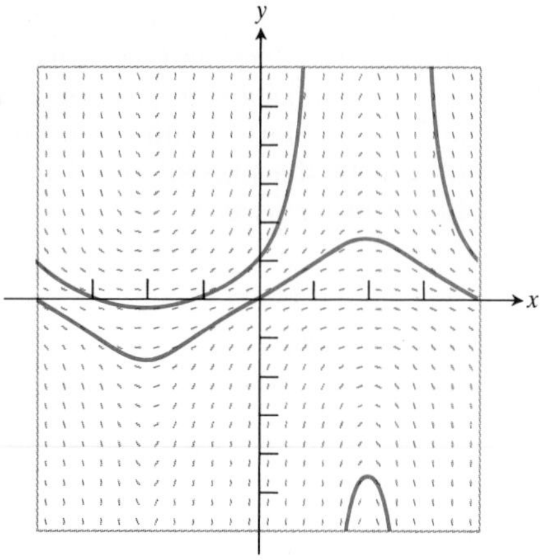

18.14 Solutions: $y = \tan(\sin x + D)$ for $D = 0, \pi/4$

12. A student of mathematics and computer science used a computer to graph solutions of

$$y' = (1 + y^2)\cos x,$$

a) for $y(0) = 0$, and
b) for $y(0) = 1$.

One curve was continuous and very well behaved. The other blew up! By solving the equation for an arbitrary initial value $y(0) = y_0$, find the set of values of y_0 for which the solution remains bounded. If a solution does not remain bounded, where are the asymptotes of its graph? See Fig. 18.14.

MISCELLANEOUS PROBLEMS

Solve the following differential equations.

1. $y \ln y\, dx + (1 + x^2)\, dy = 0$
2. $\dfrac{dy}{dx} = \dfrac{y^2 - y - 2}{x^2 + x}$
3. $e^{x+2y}\, dy - e^{y-2}\, dx = 0$
4. $\sqrt{1 + \left(\dfrac{dy}{dx}\right)^2} = ky$
5. $y\, dy = \sqrt{1 + y^4}\, dx$
6. $(2x + y)\, dx + (x - 2y)\, dy = 0$
7. $\dfrac{dy}{dx} = \dfrac{x^2 + y^2}{2xy}$
8. $x \dfrac{dy}{dx} = y + \sqrt{x^2 + y^2}$
9. $x\, dy = \left(y + x \cos^2 \dfrac{y}{x}\right) dx$
10. $x(\ln y - \ln x)\, dy = y(1 + \ln y - \ln x)\, dx$
11. $x\, dy + (2y - x^2 - 1)\, dx = 0$
12. $\cos y\, dx + (x \sin y - \cos^2 y)\, dy = 0$
13. $\cosh x\, dy - (y + \cosh x) \sinh x\, dx = 0$
14. $(x + 1)\, dy + (2y - x)\, dx = 0$
15. $(1 + y^2)\, dx + (2xy + y^2 + 1)\, dy = 0$
16. $(x^2 + y)\, dx + (e^y + x)\, dy = 0$
17. $(x^2 + y^2)\, dx + (2xy + \cosh y)\, dy = 0$
18. $(e^x + \ln y)\, dx + \dfrac{x + y}{y}\, dy = 0$
19. $x(1 + e^y)\, dx + \frac{1}{2}(x^2 + y^2)e^y\, dy = 0$
20. $\left(\sin x + \tan^{-1} \dfrac{y}{x}\right) dx - (y - \ln \sqrt{x^2 + y^2})\, dy = 0$
21. $\dfrac{d^2y}{dx^2} - 2y \dfrac{dy}{dx} = 0$
22. $\dfrac{d^2x}{dy^2} + 4x = 0$
23. $\dfrac{d^2y}{dx^2} = 1 + \left(\dfrac{dy}{dx}\right)^2$
24. $\dfrac{d^2x}{dy^2} = 1 - \left(\dfrac{dx}{dy}\right)^2$
25. $x^2 \dfrac{d^2y}{dx^2} + x \dfrac{dy}{dx} = 1$
26. $\dfrac{d^2y}{dx^2} - 4 \dfrac{dy}{dx} + 3y = 0$
27. $\dfrac{d^3y}{dx^3} - 2 \dfrac{d^2y}{dx^2} + \dfrac{dy}{dx} = 0$
28. $\dfrac{d^2y}{dx^2} + 4y = \sec 2x$
29. $\dfrac{d^2y}{dx^2} - \dfrac{dy}{dx} - 2y = e^{2x}$
30. $\dfrac{d^2y}{dx^2} - 2 \dfrac{dy}{dx} + 5y = e^{-x}$
31. Find the *general solution* of the differential equation $4x^2y'' + 4xy' - y = 0$, given that there is a particular solution of the form $y = x^c$ for some constant c.
32. Show that the only plane curves that have constant curvature are circles and straight lines. (Assume the appropriate derivatives exist and are continuous.)
33. Find the orthogonal trajectories of the family of curves $x^2 = Cy^3$. (*Caution:* The differential equation should not contain the arbitrary constant C.)
34. Find the orthogonal trajectories of the family of circles $(x - C)^2 + y^2 = C^2$.
35. Find the orthogonal trajectories of the family of parabolas $y^2 = 4C(C - x)$.
36. The equation $d^2y/dt^2 + 100y = 0$ represents a simple harmonic motion. Find the general solution of the equation and determine the constants of integration if $y = 10$, $dy/dt = 50$, when $t = 0$. Find the period and the amplitude of the motion.

Problems 37–39 refer to the differential equation $dy/dx = x + \sin y$. (See Fig. 18.15.)

37. For solutions near $y = 0$, we might use the approximation $\sin y \approx y$ and replace the original problem by $dy/dx = x + y$. Solve this equation with $y(0) = 0$. What is the Maclaurin series for your answer?

38. (*Series*) Given the initial value problem $y' = x + \sin y$, $y(0) = 0$. By implicit differentiation we can calculate successive derivatives. For example, $y'' = 1 + (\cos y)y'$, $y''' = -(\sin y)(y')^2 + (\cos y)y''$. These can, in turn, be evaluated at $x = 0$ by using the given initial value $y(0) = 0$. Using this procedure, find the terms of the Maclaurin series for $y(x)$ through x^4. (Compare the answer to Problem 37.)

39. Using the method of the preceding problem, find the terms through x^4 in the Maclaurin series for $y(x)$ if y satisfies $y' = x + \sin y$, and (a) $y(0) = \pi/2$. (b) Repeat for $y(0) = -\pi/2$.

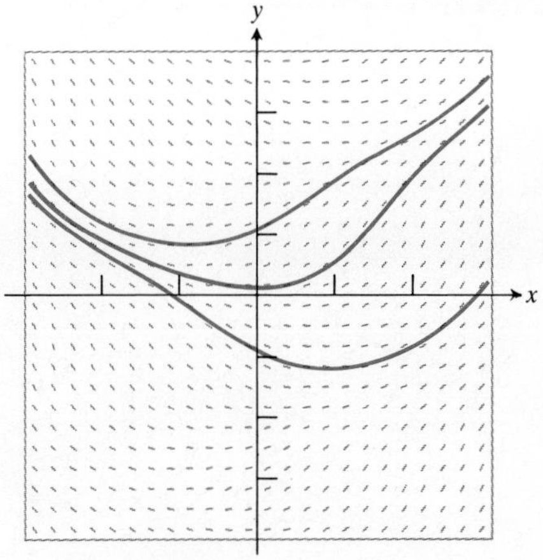

18.15 Solutions to: $dy/dx = x + \sin y$ passing through: $(0, 0)$, $(0, \pi/2)$, $(0, -\pi/2)$

APPENDIXES

A.1 Determinants and Cramer's Rule

A rectangular array of numbers like

$$A = \begin{bmatrix} 2 & 1 & 3 \\ 1 & 0 & -2 \end{bmatrix}$$

is called a *matrix*. We call A a "2 by 3" matrix because it has two rows and three columns. An "m by n" matrix has m rows and n columns, and the *entry* or *element* (number) in the ith row and jth column is often denoted by a_{ij}:

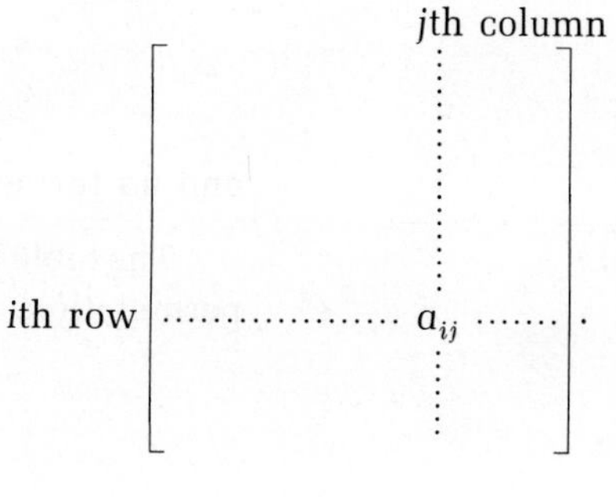

In the matrix A above,

$$a_{11} = 2, \quad a_{12} = 1, \quad a_{13} = 3,$$
$$a_{21} = 1, \quad a_{22} = 0, \quad a_{23} = -2.$$

A matrix with the same number of rows as columns is a *square* matrix. It is a *matrix of order n* if the number of rows and columns is n.

With each square matrix A we associate a number $\det(A)$ or $|a_{ij}|$ called the *determinant* of A, calculated from the entries of A in the following way. (The vertical bars in the notation $|a_{ij}|$ do not mean absolute value.) For $n = 1$ and $n = 2$ we define

$$\det [a] = a, \tag{1}$$

$$\det \begin{bmatrix} a_{11} & a_{12} \\ a_{21} & a_{22} \end{bmatrix} = a_{11}a_{22} - a_{21}a_{12}. \tag{2}$$

For a matrix of order 3, we define

$$\det(A) = \det\begin{bmatrix} a_{11} & a_{12} & a_{13} \\ a_{21} & a_{22} & a_{23} \\ a_{31} & a_{32} & a_{33} \end{bmatrix} = \begin{array}{l}\text{Sum of all signed products} \\ \text{of the form } \pm a_{1i}a_{2j}a_{3k},\end{array} \tag{3}$$

where i, j, k is a permutation of 1, 2, 3 in some order. There are $3! = 6$ such permutations, so there are six terms in the sum. Half of these have plus signs and the other half have minus signs, according to the index of the permutation, where the index is a number we next define.

DEFINITION

Index of a Permutation

Given any permutation of the numbers 1, 2, 3, . . . , n, denote the permutation by $i_1, i_2, i_3, \ldots, i_n$. In this arrangement, some of the numbers following i_1 may be less than i_1, and however many of these there are is called the *number of inversions* in the arrangement pertaining to i_1. Likewise, there is a number of inversions pertaining to each of the other i's; it is the number of indices that come after that particular one in the arrangement and are less than it. The *index* of the permutation is the sum of all of the numbers of inversions pertaining to the separate indices.

EXAMPLE 1 For $n = 5$, the permutation

5 3 1 2 4

has

4 inversions pertaining to the first element, 5,

2 inversions pertaining to the second element, 3,

and no further inversions, so the index is $4 + 2 = 6$. □

The table below shows the permutations of 1, 2, 3 and the index of each permutation. The signed product in the determinant of Eq. (3) is also shown.

Permutation	*Index*	*Signed product*
1 2 3	0	$+a_{11}a_{22}a_{33}$
1 3 2	1	$-a_{11}a_{23}a_{32}$
2 1 3	1	$-a_{12}a_{21}a_{33}$
2 3 1	2	$+a_{12}a_{23}a_{31}$
3 1 2	2	$+a_{13}a_{21}a_{32}$
3 2 1	3	$-a_{13}a_{22}a_{31}$

(4)

The sum of the six signed products is

$$a_{11}(a_{22}a_{33} - a_{23}a_{32}) - a_{12}(a_{21}a_{33} - a_{23}a_{31}) + a_{13}(a_{21}a_{32} - a_{22}a_{31})$$

$$= a_{11}\begin{vmatrix} a_{22} & a_{23} \\ a_{32} & a_{33} \end{vmatrix} - a_{12}\begin{vmatrix} a_{21} & a_{23} \\ a_{31} & a_{33} \end{vmatrix} + a_{13}\begin{vmatrix} a_{21} & a_{22} \\ a_{31} & a_{32} \end{vmatrix} \tag{5}$$

$$= \begin{vmatrix} a_{11} & a_{12} & a_{13} \\ a_{21} & a_{22} & a_{23} \\ a_{31} & a_{32} & a_{33} \end{vmatrix},$$

and the formula

$$\begin{vmatrix} a_{11} & a_{12} & a_{13} \\ a_{21} & a_{22} & a_{23} \\ a_{31} & a_{32} & a_{33} \end{vmatrix} = a_{11}\begin{vmatrix} a_{22} & a_{23} \\ a_{32} & a_{33} \end{vmatrix} - a_{12}\begin{vmatrix} a_{21} & a_{23} \\ a_{31} & a_{33} \end{vmatrix} + a_{13}\begin{vmatrix} a_{21} & a_{22} \\ a_{31} & a_{32} \end{vmatrix} \quad (6)$$

is often used to calculate 3 by 3 determinants.

Equation (6) reduces the calculation of a 3 by 3 determinant to the calculation of three 2 by 2 determinants.

Many people prefer to remember the following scheme for calculating the six signed products in the determinant of a 3 by 3 matrix:

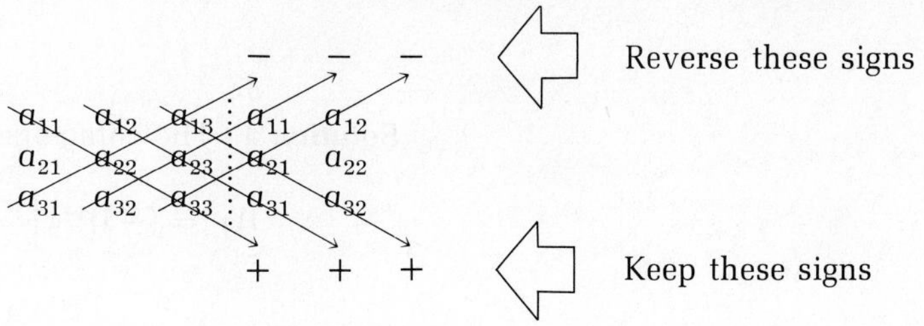

(7)

Minors and Cofactors

The second order determinants on the right-hand side of Eq. (6) are called the *minors* (short for minor determinant) of the entries they multiply. Thus,

$$\begin{vmatrix} a_{22} & a_{23} \\ a_{32} & a_{33} \end{vmatrix} \text{ is the minor of } a_{11},$$

$$\begin{vmatrix} a_{21} & a_{23} \\ a_{31} & a_{33} \end{vmatrix} \text{ is the minor of } a_{12},$$

and so on. The minor of the element a_{ij} in a matrix A is the determinant of the matrix that remains when the row and column containing a_{ij} are deleted:

$$\begin{vmatrix} a_{11} & a_{12} & a_{13} \\ a_{21} & a_{22} & a_{23} \\ a_{31} & a_{32} & a_{33} \end{vmatrix}. \quad \text{The minor of } a_{22} \text{ is } \begin{vmatrix} a_{11} & a_{13} \\ a_{31} & a_{33} \end{vmatrix}.$$

$$\begin{vmatrix} a_{11} & a_{12} & a_{13} \\ a_{21} & a_{22} & a_{23} \\ a_{31} & a_{32} & a_{33} \end{vmatrix}. \quad \text{The minor of } a_{23} \text{ is } \begin{vmatrix} a_{11} & a_{12} \\ a_{31} & a_{32} \end{vmatrix}.$$

The *cofactor* of a_{ij} is the determinant A_{ij} that is $(-1)^{i+j}$ times the minor of a_{ij}. Thus,

$$A_{22} = (-1)^{2+2}\begin{vmatrix} a_{11} & a_{13} \\ a_{31} & a_{33} \end{vmatrix} = \begin{vmatrix} a_{11} & a_{13} \\ a_{31} & a_{33} \end{vmatrix},$$

$$A_{23} = (-1)^{2+3}\begin{vmatrix} a_{11} & a_{12} \\ a_{31} & a_{32} \end{vmatrix} = -\begin{vmatrix} a_{11} & a_{12} \\ a_{31} & a_{32} \end{vmatrix}.$$

The effect of the factor $(-1)^{i+j}$ is to change the sign of the minor when the sum $i + j$ is odd. There is a checkerboard pattern for remembering these sign changes:

$$\begin{matrix} + & - & + \\ - & + & - \\ + & - & + \end{matrix}$$

In the upper left corner, $i = 1$, $j = 1$ and $(-1)^{1+1} = +1$. In going from any cell to an adjacent cell in the same row or column, we change i by 1 or j by 1, but not both, so we change the exponent from even to odd or from odd to even, which changes the sign from + to − or from − to +.

When we rewrite Eq. (6) in terms of cofactors we get

$$\det(A) = a_{11}A_{11} + a_{12}A_{12} + a_{13}A_{13}. \tag{8}$$

EXAMPLE 2 Find the determinant of the matrix

$$A = \begin{bmatrix} 2 & 1 & 3 \\ 3 & -1 & -2 \\ 2 & 3 & 1 \end{bmatrix}.$$

Solution 1 The cofactors are

$$A_{11} = (-1)^{1+1}\begin{vmatrix} -1 & -2 \\ 3 & 1 \end{vmatrix}, \qquad A_{12} = (-1)^{1+2}\begin{vmatrix} 3 & -2 \\ 2 & 1 \end{vmatrix},$$

$$A_{13} = (-1)^{1+3}\begin{vmatrix} 3 & -1 \\ 2 & 3 \end{vmatrix}.$$

To find $\det(A)$, we multiply each element of the first row of A by its cofactor and add:

$$\det(A) = 2\begin{vmatrix} -1 & -2 \\ 3 & 1 \end{vmatrix} + (-1)\begin{vmatrix} 3 & -2 \\ 2 & 1 \end{vmatrix} + 3\begin{vmatrix} 3 & -1 \\ 2 & 3 \end{vmatrix}$$

$$= 2(-1 + 6) - 1(3 + 4) + 3(9 + 2) = 10 - 7 + 33 = 36.$$

Solution 2 From Eq. (7) we find

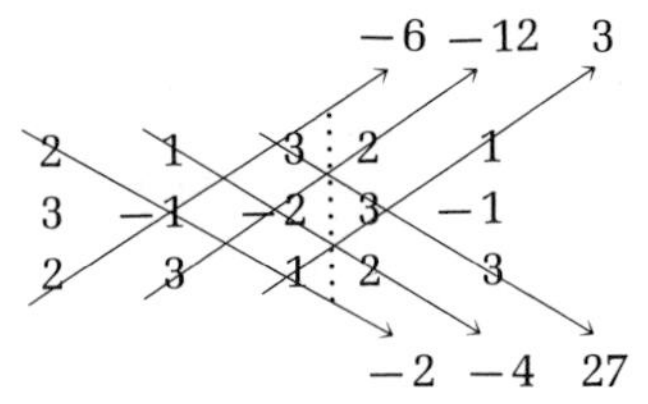

$$\det(A) = -(-6) - (-12) - 3 + (-2) + (-4) + 27 = 36. \ \square$$

Expanding by Columns or by Other Rows

The determinant of a square matrix can be calculated from the cofactors of any row or any column.

If we were to expand the determinant in Example 2 by cofactors according to elements of its third column, say, we would get

$$+3\begin{vmatrix} 3 & -1 \\ 2 & 3 \end{vmatrix} - (-2)\begin{vmatrix} 2 & 1 \\ 2 & 3 \end{vmatrix} + 1\begin{vmatrix} 2 & 1 \\ 3 & -1 \end{vmatrix}$$

$$= 3(9 + 2) + 2(6 - 2) + 1(-2 - 3) = 33 + 8 - 5 = 36.$$

Helpful Properties of Determinants

We now state some facts about determinants. You should know and be able to use these facts, but we omit the proofs.

FACT 1. If two rows of a matrix are identical, the determinant is zero.

FACT 2. If two rows of a matrix are interchanged, the determinant changes sign.

FACT 3. The determinant of a matrix is the sum of the products of the elements of the ith row (or column) by their cofactors, for any i.

FACT 4. The determinant of the transpose of a matrix is equal to the original determinant. ("Transpose" means write the rows as columns.)

FACT 5. If each element of some row (or column) of a matrix is multiplied by a constant c, the determinant is multiplied by c.

FACT 6. If all elements of a matrix above the main diagonal (or all below it) are zero, the determinant of the matrix is the product of the elements on the main diagonal.

EXAMPLE 3

$$\begin{vmatrix} 3 & 4 & 7 \\ 0 & -2 & 5 \\ 0 & 0 & 5 \end{vmatrix} = (3)(-2)(5) = -30. \square$$

FACT 7. If the elements of any row of a matrix are multiplied by the cofactors of the corresponding elements of a different row and these products are summed, the sum is zero.

EXAMPLE 4 If A_{11}, A_{12}, A_{13} are the cofactors of the elements of the first row of $A = (a_{ij})$, then the sums

$$a_{21}A_{11} + a_{22}A_{12} + a_{23}A_{13}$$

(elements of second row times cofactors of elements of first row) and

$$a_{31}A_{11} + a_{32}A_{12} + a_{33}A_{13}$$

are both zero. A similar result holds for columns. □

FACT 8. If the elements of any column of a matrix are multiplied by the cofactors of the corresponding elements of a different column and these products are summed, the sum is zero.

FACT 9. If each element of a row of a matrix is multiplied by a constant c and the results added to a different row, the determinant is not changed.

EXAMPLE 5 Evaluate the fourth order determinant

$$\begin{vmatrix} 1 & -2 & 3 & 1 \\ 2 & 1 & 0 & 2 \\ -1 & 2 & 1 & -2 \\ 0 & 1 & 2 & 1 \end{vmatrix}.$$

Solution By adding appropriate multiples of row 1 to rows 2 and 3, we can get the equal determinant

$$\begin{vmatrix} 1 & -2 & 3 & 1 \\ 0 & 5 & -6 & 0 \\ 0 & 0 & 4 & -1 \\ 0 & 1 & 2 & 1 \end{vmatrix}.$$

We could further reduce this to a triangular matrix (one with zeros below the main diagonal), or expand it by cofactors of its first column. By multiplying the elements of the first column by their respective cofactors, we get the third order determinant

$$\begin{vmatrix} 5 & -6 & 0 \\ 0 & 4 & -1 \\ 1 & 2 & 1 \end{vmatrix} = 5(4 + 2) - (-6)(0 + 1) + 0 = 36. \ \square$$

Cramer's Rule

Cramer's rule is a rule for solving a system of linear equations like

$$\begin{aligned} a_{11}x + a_{12}y &= b_1, \\ a_{21}x + a_{22}y &= b_2, \end{aligned} \tag{9}$$

when the determinant of the coefficient matrix

$$D = \det(A) = \begin{vmatrix} a_{11} & a_{12} \\ a_{21} & a_{22} \end{vmatrix}$$

is different from zero. The solution of (9) is unique when $D \neq 0$ and may be calculated from the formulas

$$x = \frac{\begin{vmatrix} b_1 & a_{12} \\ b_2 & a_{22} \end{vmatrix}}{D}, \qquad y = \frac{\begin{vmatrix} a_{11} & b_1 \\ a_{21} & b_2 \end{vmatrix}}{D}. \tag{10}$$

The numerator in the formula for x comes from replacing the first column in A (the x-column) by the column of constants b_1 and b_2 (the b-column). Replacing the y-column by the b-column gives the numerator of the y-solution.

For a system of three equations in three unknowns,

$$\begin{aligned} a_{11}x + a_{12}y + a_{13}z &= b_1, \\ a_{21}x + a_{22}y + a_{23}z &= b_2, \\ a_{31}x + a_{32}y + a_{33}z &= b_3, \end{aligned}$$

Cramer's rule gives

$$x = \frac{1}{D}\begin{vmatrix} b_1 & a_{12} & a_{13} \\ b_2 & a_{22} & a_{23} \\ b_3 & a_{32} & a_{33} \end{vmatrix},$$

$$y = \frac{1}{D}\begin{vmatrix} a_{11} & b_1 & a_{13} \\ a_{21} & b_2 & a_{23} \\ a_{31} & b_3 & a_{33} \end{vmatrix}, \tag{11}$$

$$z = \frac{1}{D}\begin{vmatrix} a_{11} & a_{12} & b_1 \\ a_{21} & a_{22} & b_2 \\ a_{31} & a_{32} & b_3 \end{vmatrix},$$

where

$$D = \det(A) = \begin{vmatrix} a_{11} & a_{12} & a_{13} \\ a_{21} & a_{22} & a_{23} \\ a_{31} & a_{32} & a_{33} \end{vmatrix}.$$

The pattern continues in all higher dimensions: If $AX = B$, and $\det(A) \neq 0$, then

$$x_i = \frac{|U_i|}{|A|}, \qquad i = 1, 2, \ldots, n \tag{12}$$

where U_i is the matrix obtained from A by replacing the ith column in A by the b-column.

Other Formulas that Use Determinants

1. The equation of a line through two points (x_1, y_1) and (x_2, y_2) in the plane is

$$\begin{vmatrix} x & y & 1 \\ x_1 & y_1 & 1 \\ x_2 & y_2 & 1 \end{vmatrix} = 0.$$

2. The area of a triangle with vertices (x_1, y_1), (x_2, y_2), (x_3, y_3) is

$$\pm\frac{1}{2}\begin{vmatrix} x_1 & y_1 & 1 \\ x_2 & y_2 & 1 \\ x_3 & y_3 & 1 \end{vmatrix}.$$

3. The cross product of $\mathbf{A} = a_1\mathbf{i} + a_2\mathbf{j} + a_3\mathbf{k}$ and $\mathbf{B} = b_1\mathbf{i} + b_2\mathbf{j} + b_3\mathbf{k}$ is

$$\mathbf{A} \times \mathbf{B} = \begin{vmatrix} \mathbf{i} & \mathbf{j} & \mathbf{k} \\ a_1 & a_2 & a_3 \\ b_1 & b_2 & b_3 \end{vmatrix}.$$

4. The volume of the parallelepiped spanned by the vectors **A, B,** and **C** is

$$\pm\mathbf{A} \cdot (\mathbf{B} \times \mathbf{C}) = \pm\begin{vmatrix} a_1 & a_2 & a_3 \\ b_1 & b_2 & b_3 \\ c_1 & c_2 & c_3 \end{vmatrix}.$$

5. If $F = M(x, y, z)\mathbf{i} + N(x, y, z)\mathbf{j} + P(x, y, z)\mathbf{k}$, then

$$\text{curl}\,\mathbf{F} = \nabla \times \mathbf{F} = \begin{vmatrix} \mathbf{i} & \mathbf{j} & \mathbf{k} \\ \dfrac{\partial}{\partial x} & \dfrac{\partial}{\partial y} & \dfrac{\partial}{\partial z} \\ M & N & P \end{vmatrix}.$$

A Reduction Formula for Evaluating Determinants

The following reduction formula is derived in E. Miller's article "Evaluating an nth order determinant in n easy steps," *MATYC Journal* 12 (1978), 123–128. Evaluating determinants with this formula is relatively fast and readily programmable. For a short FORTRAN IV program, see Alban J. Rogues's article, "Determinants: A Short Program," *Two-Year College Mathematics Journal* Vol. 10, No. 5 (November 1979), pp. 340–342.

The formula for the determinant of an n by n matrix $A = (a_{ij})$ is

$$\det(A) = \left(\frac{1}{a_{11}}\right)^{n-2} \begin{vmatrix} \begin{vmatrix} a_{11} & a_{12} \\ a_{21} & a_{22} \end{vmatrix} & \begin{vmatrix} a_{11} & a_{13} \\ a_{21} & a_{23} \end{vmatrix} & \begin{vmatrix} a_{11} & a_{14} \\ a_{21} & a_{24} \end{vmatrix} & \cdots & \begin{vmatrix} a_{11} & a_{1n} \\ a_{21} & a_{2n} \end{vmatrix} \\ \begin{vmatrix} a_{11} & a_{12} \\ a_{31} & a_{32} \end{vmatrix} & \begin{vmatrix} a_{11} & a_{13} \\ a_{31} & a_{33} \end{vmatrix} & \begin{vmatrix} a_{11} & a_{14} \\ a_{31} & a_{34} \end{vmatrix} & \cdots & \begin{vmatrix} a_{11} & a_{1n} \\ a_{31} & a_{3n} \end{vmatrix} \\ \begin{vmatrix} a_{11} & a_{12} \\ a_{41} & a_{42} \end{vmatrix} & \begin{vmatrix} a_{11} & a_{13} \\ a_{41} & a_{43} \end{vmatrix} & \begin{vmatrix} a_{11} & a_{14} \\ a_{41} & a_{44} \end{vmatrix} & \cdots & \begin{vmatrix} a_{11} & a_{1n} \\ a_{41} & a_{4n} \end{vmatrix} \\ \vdots & \vdots & \vdots & & \vdots \\ \begin{vmatrix} a_{11} & a_{12} \\ a_{n1} & a_{n2} \end{vmatrix} & \begin{vmatrix} a_{11} & a_{13} \\ a_{n1} & a_{n3} \end{vmatrix} & \begin{vmatrix} a_{11} & a_{14} \\ a_{n1} & a_{n4} \end{vmatrix} & \cdots & \begin{vmatrix} a_{11} & a_{1n} \\ a_{n1} & a_{nn} \end{vmatrix} \end{vmatrix}$$

EXAMPLE 6

$$\begin{vmatrix} 1 & 0 & 2 & -1 \\ 3 & -2 & 6 & 4 \\ 5 & 4 & 3 & 0 \\ 2 & 2 & -5 & 6 \end{vmatrix} = 1^2 \begin{vmatrix} \begin{vmatrix} 1 & 0 \\ 3 & -2 \end{vmatrix} & \begin{vmatrix} 1 & 2 \\ 3 & 6 \end{vmatrix} & \begin{vmatrix} 1 & -1 \\ 3 & 4 \end{vmatrix} \\ \begin{vmatrix} 1 & 0 \\ 5 & 4 \end{vmatrix} & \begin{vmatrix} 1 & 2 \\ 5 & 3 \end{vmatrix} & \begin{vmatrix} 1 & -1 \\ 5 & 0 \end{vmatrix} \\ \begin{vmatrix} 1 & 0 \\ 2 & 2 \end{vmatrix} & \begin{vmatrix} 1 & 2 \\ 2 & -5 \end{vmatrix} & \begin{vmatrix} 1 & -1 \\ 2 & 6 \end{vmatrix} \end{vmatrix}$$

$$= \begin{vmatrix} -2 & 0 & 7 \\ 4 & -7 & 5 \\ 2 & -9 & 8 \end{vmatrix} = \left(-\frac{1}{2}\right) \begin{vmatrix} 14 & -38 \\ 18 & -30 \end{vmatrix} = -132. \ \square$$

PROBLEMS

Evaluate the following determinants.

1. $\begin{vmatrix} 2 & 3 & 1 \\ 4 & 5 & 2 \\ 1 & 2 & 3 \end{vmatrix}$

2. $\begin{vmatrix} 2 & -1 & -2 \\ -1 & 2 & 1 \\ 3 & 0 & -3 \end{vmatrix}$

3. $\begin{vmatrix} 1 & 2 & 3 & 4 \\ 0 & 1 & 2 & 3 \\ 0 & 0 & 2 & 1 \\ 0 & 0 & 3 & 2 \end{vmatrix}$

4. $\begin{vmatrix} 1 & -1 & 2 & 3 \\ 2 & 1 & 2 & 6 \\ 1 & 0 & 2 & 3 \\ -2 & 2 & 0 & -5 \end{vmatrix}$

Evaluate the following determinants by expanding according to the cofactors of (a) the third row, and (b) the the second column.

5. $\begin{vmatrix} 2 & -1 & 2 \\ 1 & 0 & 3 \\ 0 & 2 & 1 \end{vmatrix}$

6. $\begin{vmatrix} 1 & 0 & -1 \\ 0 & 2 & -2 \\ 2 & 0 & 1 \end{vmatrix}$

7. $\begin{vmatrix} 1 & 1 & 0 & 0 \\ 0 & 0 & -2 & 1 \\ 0 & -1 & 0 & 7 \\ 3 & 0 & 2 & 1 \end{vmatrix}$

8. $\begin{vmatrix} 0 & 1 & 0 & 0 \\ 0 & 1 & 1 & 0 \\ 1 & 1 & 1 & 1 \\ 1 & 1 & 0 & 0 \end{vmatrix}$

Solve the following systems of equations by Cramer's rule.

9. $\begin{aligned} x + 8y &= 4 \\ 3x - y &= -13 \end{aligned}$

10. $\begin{aligned} 2x + 3y &= 5 \\ 3x - y &= 2 \end{aligned}$

11. $\begin{aligned} 4x - 3y &= 6 \\ 3x - 2y &= 5 \end{aligned}$

12. $\begin{aligned} x + y + z &= 2 \\ 2x - y + z &= 0 \\ x + 2y - z &= 4 \end{aligned}$

13. $\begin{aligned} 2x + y - z &= 2 \\ x - y + z &= 7 \\ 2x + 2y + z &= 4 \end{aligned}$

14. $\begin{aligned} 2x - 4y &= 6 \\ x + y + z &= 1 \\ 5y + 7z &= 10 \end{aligned}$

15. $\begin{aligned} x - z &= 3 \\ 2y - 2z &= 2 \\ 2x + z &= 3 \end{aligned}$

16. $\begin{aligned} x_1 + x_2 - x_3 + x_4 &= 2 \\ x_1 - x_2 + x_3 + x_4 &= -1 \\ x_1 + x_2 + x_3 - x_4 &= 2 \\ x_1 + x_3 + x_4 &= -1 \end{aligned}$

A.2

Matrices and Linear Equations

A rectangular array of numbers like

$$A = \begin{bmatrix} 2 & 1 & 3 \\ 1 & 0 & -2 \end{bmatrix}$$

is called a *matrix*. We call A a "2 by 3" matrix because it has two rows and three columns. An "m by n" matrix has m rows and n columns, and the *entry* or *element* (number) in the ith row and jth column is often denoted by a_{ij}:

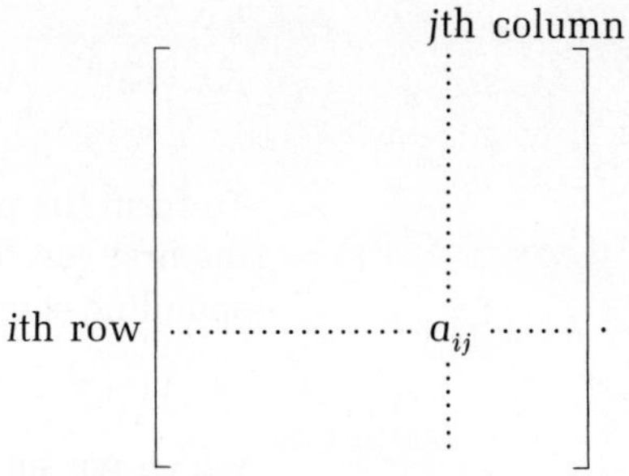

In the matrix A above,

$$\begin{aligned} a_{11} &= 2, & a_{12} &= 1, & a_{13} &= 3, \\ a_{21} &= 1, & a_{22} &= 0, & a_{23} &= -2. \end{aligned}$$

Matrix Addition and Multiplication

Two matrices are equal if (and only if) they have the same numbers in the same positions. For example,

$$B = \begin{bmatrix} b_{11} & b_{12} & b_{13} \\ b_{21} & b_{22} & b_{23} \end{bmatrix} = \begin{bmatrix} 2 & 1 & 3 \\ 1 & 0 & -2 \end{bmatrix}$$

if and only if

$$\begin{aligned} b_{11} &= 2, & b_{12} &= 1, & b_{13} &= 3, \\ b_{21} &= 1, & b_{22} &= 0, & b_{23} &= -2. \end{aligned}$$

For matrices A and B to be equal they must have the same *shape* (same number of rows and same number of columns) and must have

$$a_{ij} = b_{ij} \qquad \text{for all } i \text{ and } j.$$

Two matrices with the same shape can be added by adding corresponding elements. For example,

$$\begin{bmatrix} 2 & 1 & 3 \\ 1 & 0 & -2 \end{bmatrix} + \begin{bmatrix} 1 & -2 & 2 \\ 2 & 3 & -1 \end{bmatrix} = \begin{bmatrix} 3 & -1 & 5 \\ 3 & 3 & -3 \end{bmatrix}.$$

To multiply a matrix by a number c, we multiply each element by c. For example,

$$7\begin{bmatrix} 2 & 1 & 3 \\ 1 & 0 & -2 \end{bmatrix} = \begin{bmatrix} 14 & 7 & 21 \\ 7 & 0 & -14 \end{bmatrix}.$$

A system of simultaneous linear equations

$$\begin{aligned} a_{11}x + a_{12}y + a_{13}z &= b_1, \\ a_{21}x + a_{22}y + a_{23}z &= b_2. \end{aligned} \tag{1}$$

can be written in matrix form as

$$\begin{bmatrix} a_{11} & a_{12} & a_{13} \\ a_{21} & a_{22} & a_{23} \end{bmatrix} \begin{bmatrix} x \\ y \\ z \end{bmatrix} = \begin{bmatrix} b_1 \\ b_2 \end{bmatrix}, \tag{2}$$

or, more compactly, as

$$AX = B, \tag{3}$$

where

$$A = \begin{bmatrix} a_{11} & a_{12} & a_{13} \\ a_{21} & a_{22} & a_{23} \end{bmatrix}, \quad X = \begin{bmatrix} x \\ y \\ z \end{bmatrix}, \quad B = \begin{bmatrix} b_1 \\ b_2 \end{bmatrix}. \tag{4}$$

To form the product indicated by AX in Eq. (3), we take the elements of the first row of A in order from left to right and multiply by the corresponding elements of X from the top down, and add these products to get

$$a_{11}x + a_{12}y + a_{13}z,$$

which we set equal to b_1. The result is the first equation in (1). We then repeat the process with the second row in (2) to obtain the second equation in (1).

An m by n matrix A can multiply an n by p matrix B from the left to give an m by p matrix $C = AB$:

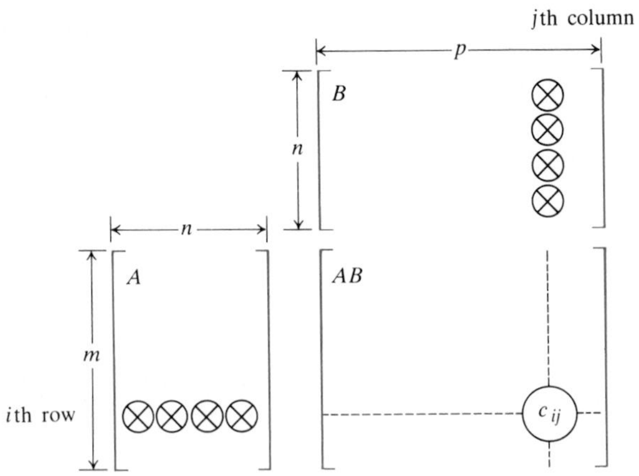

The element in the ith row and jth column of AB is the sum

$$\begin{gathered} c_{ij} = a_{i1}b_{1j} + a_{i2}b_{2j} + \cdots + a_{in}b_{nj} = \sum_{k=1}^{n} a_{ik}b_{kj}, \\ i = 1, 2, \ldots, m \quad \text{and} \quad j = 1, 2, \ldots, p. \end{gathered} \tag{6}$$

In words, Eq. (6) says "to get the element in the ith row and jth column of AB, multiply the individual entries in the ith row of A, one after the other from left to right, by the corresponding entries in the jth column of B from top to bottom, and add these products: their sum is a single number c_{ij}."

For example,

$$\begin{array}{c} \begin{bmatrix} a & b & c \\ d & e & f \\ u & v & w \end{bmatrix} \\ \begin{bmatrix} 2 & -1 & 3 \\ 3 & 2 & 2 \end{bmatrix} \begin{bmatrix} 2a - d + 3u & 2b - e + 3v & 2c - f + 3w \\ 3a + 2d + 2u & 3b + 2e + 2v & 3c + 2f + 2w \end{bmatrix}. \end{array} \tag{7}$$

In most cases it saves space to write the matrices in a product side by side. Thus we usually write the multiplication in (7) as

$$\begin{bmatrix} 2 & -1 & 3 \\ 3 & 2 & 2 \end{bmatrix} \begin{bmatrix} a & b & c \\ d & e & f \\ u & v & w \end{bmatrix} = \begin{bmatrix} 2a - d + 3u & 2b - e + 3v & 2c - f + 3w \\ 3a + 2d + 2u & 3b + 2e + 2v & 3c + 2f + 2w \end{bmatrix}.$$

Elementary Row Operations and Row Reduction

Two systems of linear equations are called *equivalent* if they have the same set of solutions. To solve a system of linear equations it is often possible to transform it step by step into an equivalent system of equations that is so simple it can be solved by inspection. We shall illustrate such a sequence of steps by transforming the system

$$\begin{aligned} 2x + 3y - 4z &= -3, \\ x + 2y + 3z &= 3, \\ 3x - y - z &= 6, \end{aligned} \tag{8}$$

into the equivalent system

$$\begin{aligned} x \qquad &= 2, \\ y \quad &= -1, \\ z &= 1. \end{aligned} \tag{9}$$

EXAMPLE 1 Solve the system of equations (8).

Solution The system (8) is the same as

$$AX = B, \qquad A = \begin{bmatrix} 2 & 3 & -4 \\ 1 & 2 & 3 \\ 3 & -1 & -1 \end{bmatrix}, \quad B = \begin{bmatrix} -3 \\ 3 \\ 6 \end{bmatrix}. \tag{10}$$

We start with the 3 by 4 matrix $[A \mid B]$ whose first three columns are the columns of A, and whose fourth column is B. That is,

$$[A \mid B] = \left[\begin{array}{ccc|c} 2 & 3 & -4 & -3 \\ 1 & 2 & 3 & 3 \\ 3 & -1 & -1 & 6 \end{array}\right]. \tag{11}$$

We are going to transform this augmented matrix with a sequence of so-called *elementary row operations*. These operations, which are to be performed on the rows of the matrix, are of three kinds:

1. Multiply any row by a constant different from 0.
2. Add a constant multiple of any row to another row.
3. Interchange two rows.

Our goal is to replace the matrix $[A \mid B]$ by the matrix $[I \mid S]$, where

$$I = \begin{bmatrix} 1 & 0 & 0 \\ 0 & 1 & 0 \\ 0 & 0 & 1 \end{bmatrix} \quad \text{and} \quad S = \begin{bmatrix} s_1 \\ s_2 \\ s_3 \end{bmatrix}. \tag{12}$$

If we succeed, the matrix $[I \mid S]$ will be the matrix of the system

$$\begin{aligned} x \qquad\qquad &= s_1, \\ y \qquad &= s_2, \\ z &= s_3. \end{aligned} \tag{13}$$

The virtue of this system is that its solution, $x = s_1$, $y = s_2$, $z = s_3$, is the same as the solution of (8).

Our systematic approach will be to get a 1 in the upper left corner and use Type 2 operations to get zeros elsewhere in the first column. That will make the first column the same as the first column of I. Then we shall use Type 1 or Type 3 operations to get a 1 in the second position in the second row, and follow that by Type 2 operations to get the second column to be what we want: namely, like the second column of I. Then we will work on the third column.

STEP 1. Interchange rows 1 and 2 and get

$$\left[\begin{array}{ccc|c} 1 & 2 & 3 & 3 \\ 2 & 3 & -4 & -3 \\ 3 & -1 & -1 & 6 \end{array}\right]. \tag{14}$$

STEP 2. Add -2 times row 1 to row 2.

STEP 3. Add -3 times row 1 to row 3.

The result of steps 2 and 3 is

$$\left[\begin{array}{ccc|c} 1 & 2 & 3 & 3 \\ 0 & -1 & -10 & -9 \\ 0 & -7 & -10 & -3 \end{array}\right]. \tag{15}$$

STEP 4. Multiply row 2 by -1; then

STEP 5. Add -2 times row 2 to row 1, and

STEP 6. Add 7 times row 2 to row 3.

The combined result of these steps is

$$\left[\begin{array}{ccc|c} 1 & 0 & -17 & -15 \\ 0 & 1 & 10 & 9 \\ 0 & 0 & 60 & 60 \end{array}\right]. \tag{16}$$

STEP 7. Multiply row 3 by $\frac{1}{60}$.

STEP 8. Add 17 times row 3 to row 1.

STEP 9. Add -10 times row 3 to row 2.

The final result is

$$[I \mid S] = \left[\begin{array}{ccc|c} 1 & 0 & 0 & 2 \\ 0 & 1 & 0 & -1 \\ 0 & 0 & 1 & 1 \end{array}\right]. \tag{17}$$

This represents the system (9). The solution of this system, and therefore of the system (8), is $x = 2$, $y = -1$, $x = 1$. To check the solution, we substitute these values in (8) and find that

$$\begin{aligned} 2(2) + 3(-1) - 4(1) &= -3, \\ (2) + 2(-1) + 3(1) &= 3, \\ 3(2) - (-1) - (1) &= 6. \; \square \end{aligned} \tag{18}$$

The method of using elementary row operations to reduce the augmented matrix of a system of linear equations to a simpler form is sometimes called the *method of row reduction.* It works because at each step the system of equations represented by the transformed matrix is equivalent to the original system. Thus, in Example 1, when we finally arrived at the matrix (17), which represented the system (9) whose solution could be found by inspection, we knew that this solution was also the solution of (8). Note that we checked the solution anyhow. It is always a good idea to do that.

Inverses

The matrix I in Eq. (12) is the *multiplicative identity* for all 3 by 3 matrices. That is, if M is any 3 by 3 matrix, then

$$IM = MI = M. \tag{19}$$

If P is a matrix with the property that

$$PM = I,$$

then we call P the *inverse* of M, and use the alternative notation

$$P = M^{-1},$$

pronounced "M inverse."

The sequence of row operations that we used to find the solution of the system $AX = B$ in Example 1 can be used to find the inverse of the matrix A. We start with the 3 by 6 matrix $[A \,|\, I]$ whose first three columns are the columns of A and whose last three columns are the columns of I, namely

$$[A \,|\, I] = \left[\begin{array}{rrr|rrr} 2 & 3 & -4 & 1 & 0 & 0 \\ 1 & 2 & 3 & 0 & 1 & 0 \\ 3 & -1 & -1 & 0 & 0 & 1 \end{array}\right]. \tag{20}$$

We then carry out Steps 1 through 9 of Example 1 on the augmented matrix $[A \,|\, I]$. The final result is

$$[I \,|\, A^{-1}] = \left[\begin{array}{rrr|rrr} 1 & 0 & 0 & \frac{1}{60} & \frac{7}{60} & \frac{17}{60} \\ 0 & 1 & 0 & \frac{10}{60} & \frac{10}{60} & -\frac{10}{60} \\ 0 & 0 & 1 & -\frac{7}{60} & \frac{11}{60} & \frac{1}{60} \end{array}\right]. \tag{21}$$

The 3 by 3 matrix in the last three columns is

$$A^{-1} = \frac{1}{60}\begin{bmatrix} 1 & 7 & 17 \\ 10 & 10 & -10 \\ -7 & 11 & 1 \end{bmatrix}. \tag{22}$$

By direct matrix multiplication, we verify our answer:

$$A = \frac{1}{60}\begin{bmatrix} 1 & 7 & 17 \\ 10 & 10 & -10 \\ -7 & 11 & 1 \end{bmatrix}\begin{bmatrix} 2 & 3 & -4 \\ 1 & 2 & 3 \\ 3 & -1 & -1 \end{bmatrix}$$

$$= \frac{1}{60}\begin{bmatrix} 2+7+51 & 3+14-17 & -4+21-17 \\ 20+10-30 & 30+20+10 & -40+30+10 \\ -14+11+3 & -21+22-1 & 28+33-1 \end{bmatrix} \quad (23)$$

$$= \frac{1}{60}\begin{bmatrix} 60 & 0 & 0 \\ 0 & 60 & 0 \\ 0 & 0 & 60 \end{bmatrix} = \begin{bmatrix} 1 & 0 & 0 \\ 0 & 1 & 0 \\ 0 & 0 & 1 \end{bmatrix}.$$

Knowing A^{-1} provides a second way to solve the system of equations in (8). We write the system in the form given in (10), and then multiply B on the left by A^{-1} to find the solution matrix X. Thus,

$$X = IX = (A^{-1}A)X = A^{-1}(AX) = A^{-1}B = \frac{1}{60}\begin{bmatrix} 1 & 7 & 17 \\ 10 & 10 & -10 \\ -7 & 11 & 1 \end{bmatrix}\begin{bmatrix} -3 \\ 3 \\ 6 \end{bmatrix}$$

$$= \frac{1}{60}\begin{bmatrix} 120 \\ -60 \\ 60 \end{bmatrix} = \begin{bmatrix} 2 \\ -1 \\ 1 \end{bmatrix}.$$

> If the coefficient matrix A of a system
>
> $$AX = B$$
>
> of n equations and n unknowns has an inverse, then the solution of the system is
>
> $$X = A^{-1}B.$$

Only square matrices can have inverses. If an n by n matrix A has an inverse, the method shown above for $n = 3$ will give it: Put the n by n identity matrix alongside A and use the row operations to get an n by n identity matrix in place of A. The n by n matrix that is then beside that is A^{-1}.

Not every n by n matrix has an inverse. For example, the 2 by 2 matrix

$$\begin{bmatrix} 1 & 1 \\ a & a \end{bmatrix}$$

has no inverse. The method we outlined above would reduce this to

$$\begin{bmatrix} 1 & 1 \\ 0 & 0 \end{bmatrix},$$

which cannot be further changed by elementary row operations into the 2 by 2 identity matrix. We shall have more to say about inverses in a moment.

A system of m equations in n unknowns may have no solutions, only one solution, or infinitely many solutions. If there are any solutions, they can be found by the method of row reduction described above.

Determinants and the Inverse of a Matrix

Another way to find the inverse of a square matrix A (assuming A has one) depends on the fact that A has an inverse if and only if $\det A \neq 0$. We describe the method, give an example, and then indicate why the method works. We illustrate with 3 by 3 matrices, but the method works for any square matrix whose determinant is not zero.

To find the inverse of a matrix whose determinant is not zero:

1. Construct the matrix of cofactors of A:

$$\operatorname{cof} A = \begin{bmatrix} A_{11} & A_{12} & A_{13} \\ A_{21} & A_{22} & A_{23} \\ A_{31} & A_{32} & A_{33} \end{bmatrix}.$$

2. Construct the transposed matrix of cofactors (called the *adjoint* of A):

$$\operatorname{adj} A = (\operatorname{cof} A)^T = \begin{bmatrix} A_{11} & A_{21} & A_{31} \\ A_{12} & A_{22} & A_{32} \\ A_{13} & A_{23} & A_{33} \end{bmatrix}.$$

("Transpose" means write the rows as columns.)

3. Then,

$$A^{-1} = \frac{1}{\det A} \operatorname{adj} A.$$

EXAMPLE 2 Let us take the same matrix A that we used in illustrating the method of elementary row operations:

$$A = \begin{bmatrix} 2 & 3 & -4 \\ 1 & 2 & 3 \\ 3 & -1 & -1 \end{bmatrix}.$$

You can verify that the matrix of minors is

$$\begin{bmatrix} 1 & -10 & -7 \\ -7 & 10 & -11 \\ 17 & 10 & 1 \end{bmatrix}.$$

We next apply the sign corrections according to the checkerboard pattern $(-1)^{i+j}$ to get the matrix of cofactors

$$\operatorname{cof} A = \begin{bmatrix} 1 & 10 & -7 \\ 7 & 10 & 11 \\ 17 & -10 & 1 \end{bmatrix}.$$

The adjoint of A is the transposed cofactor matrix

$$\operatorname{adj} A = \begin{bmatrix} 1 & 7 & 17 \\ 10 & 10 & -10 \\ -7 & 11 & 1 \end{bmatrix}.$$

We get the determinant of A by multiplying the first row of A and the first column of adj A (which is the first row of the matrix of cofactors, so we are multiplying the elements of the first row of A by their own cofactors):

$$\det A = 2(1) + 3(10) + (-4)(-7) = 2 + 30 + 28 = 60.$$

Therefore, when we divide adj A by det A, we get

$$A^{-1} = \frac{1}{60}\begin{bmatrix} 1 & 7 & 17 \\ 10 & 10 & -10 \\ -7 & 11 & 1 \end{bmatrix},$$

which agrees with our previous work. □

Why does the method work? Let us take a closer look at the products A adj A and (adj A) A for $n = 3$. Because adj A is the transposed cofactor matrix, we get

$$A\,(\text{adj }A) = \begin{bmatrix} a_{11} & a_{12} & a_{13} \\ a_{21} & a_{22} & a_{23} \\ a_{31} & a_{32} & a_{33} \end{bmatrix}\begin{bmatrix} A_{11} & A_{21} & A_{31} \\ A_{12} & A_{22} & A_{32} \\ A_{13} & A_{23} & A_{33} \end{bmatrix} = \begin{bmatrix} \det A & 0 & 0 \\ 0 & \det A & 0 \\ 0 & 0 & \det A \end{bmatrix}.$$

(Here A_{ij} = cofactor of a_{ij}.)

An element on the main diagonal in the final product is the product of a row of A and the corresponding column of adj A. This is the same as the sum of the products of the elements of a row of A and the cofactors of the same row, which is just det A. For those elements not on the main diagonal in the product, we are adding products of elements of some row of A by the cofactors of the corresponding elements of a different row of A, and that sum is zero.

If we were to multiply in the other order, (adj A) A, we would again get

$$(\text{adj }A)\,A = \begin{bmatrix} \det A & 0 & 0 \\ 0 & \det A & 0 \\ 0 & 0 & \det A \end{bmatrix},$$

because we are multiplying the elements of the jth column of A, say, by the cofactors of the corresponding elements of the ith column, in order to get the entry in the ith row and jth column in the product. The result is det A when $i = j$ and is 0 when $i \neq j$.

PROBLEMS

1. a) Write the following system of linear equations in matrix form $AX = B$.

$$\begin{aligned} 2x - 3y + 4z &= -19 \\ 6x + 4y - 2z &= 8 \\ x + 5y + 4z &= 23 \end{aligned}$$

b) Show that

$$X = \begin{bmatrix} -2 \\ 5 \\ 0 \end{bmatrix}$$

is a solution of the system in part (a).

2. Let A be an arbitrary matrix with 3 rows and 3 columns and let I be the 3 by 3 matrix that has 1's on the main diagonal and zeros elsewhere:

$$I = \begin{bmatrix} 1 & 0 & 0 \\ 0 & 1 & 0 \\ 0 & 0 & 1 \end{bmatrix}.$$

Show that $IA = A$ and also that $AI = A$. This will show that I is the multiplicative identity matrix for all 3 by 3 matrices.

3. Let A be an arbitrary 3 by 3 matrix and let R_{12} be the matrix that is obtained from the 3 by 3 identity matrix by interchanging rows 1 and 2:

$$R_{12} = \begin{bmatrix} 0 & 1 & 0 \\ 1 & 0 & 0 \\ 0 & 0 & 1 \end{bmatrix}.$$

Compute $R_{12}A$ and show that you would get the same result by interchanging rows 1 and 2 of A.

4. Let A and R_{12} be as in Problem 3 above. Compute AR_{12} and show that the result is what you would get by interchanging columns 1 and 2 of A. (Note that R_{12} is also the result of interchanging columns 1 and 2 of the 3 by 3 identity matrix I.)

Solve the following systems of equations by row reduction.

5. $\begin{aligned} x + 8y &= 4 \\ 3x - y &= -13 \end{aligned}$

6. $\begin{aligned} 2x + 3y &= 5 \\ 3x - y &= 2 \end{aligned}$

7. $\begin{aligned} 4x - 3y &= 6 \\ 3x - 2y &= 5 \end{aligned}$

8. $\begin{aligned} x + y + z &= 2 \\ 2x - y + z &= 0 \\ x + 2y - z &= 4 \end{aligned}$

9. $\begin{aligned} 2x + y - z &= 2 \\ x - y + z &= 7 \\ 2x + 2y + z &= 4 \end{aligned}$

10. $\begin{aligned} 2x - 4y &= 6 \\ x + y + z &= 1 \\ 5y + 7z &= 10 \end{aligned}$

11. $\begin{aligned} x - z &= 3 \\ 2y - 2z &= 2 \\ 2x + z &= 3 \end{aligned}$

12. $\begin{aligned} x_1 + x_2 - x_3 + x_4 &= 2 \\ x_1 - x_2 + x_3 + x_4 &= -1 \\ x_1 + x_2 + x_3 - x_4 &= 2 \\ x_1 + x_3 + x_4 &= -1 \end{aligned}$

13. Verify that the inverse of

$$A = \begin{bmatrix} a & b \\ c & d \end{bmatrix}$$

is

$$A^{-1} = \frac{1}{ad - bc}\begin{bmatrix} d & -b \\ -c & a \end{bmatrix}.$$

That is, show that

$$AA^{-1} = A^{-1}A = \begin{bmatrix} 1 & 0 \\ 0 & 1 \end{bmatrix}.$$

14. a) Use the result in Problem 13 to write down the inverses of

$$A = \begin{bmatrix} 2 & -1 \\ 3 & 1 \end{bmatrix} \quad \text{and} \quad B = \begin{bmatrix} 2 & 3 \\ -1 & 1 \end{bmatrix}.$$

b) In part (a), B is the transpose of A. Is B^{-1} the transpose of A^{-1}?

15. Use the result in Problem 13 to solve the system of equations in Problem 5.

16. Given

$$A = \begin{bmatrix} 1 & 0 & -1 \\ 0 & 2 & -2 \\ 2 & 0 & 1 \end{bmatrix} \quad \text{and} \quad A^{-1} = \begin{bmatrix} \frac{1}{3} & 0 & \frac{1}{3} \\ -\frac{2}{3} & \frac{1}{2} & \frac{1}{3} \\ -\frac{2}{3} & 0 & \frac{1}{3} \end{bmatrix},$$

solve the system

$$A\begin{bmatrix} x \\ y \\ z \end{bmatrix} = \begin{bmatrix} 3 \\ 2 \\ 3 \end{bmatrix}.$$

17. a) Find the inverse of the matrix

$$\begin{bmatrix} 1 & 8 & 9 \\ 0 & 4 & 6 \\ 0 & 0 & 3 \end{bmatrix}.$$

b) Solve the following system of equations.

$$\begin{aligned} x + 8y + 9z &= 10 \\ 4y + 6z &= 10 \\ 3z &= -10 \end{aligned}$$

18. Solve the system

$$\begin{aligned} 2x - y + 2z &= 5 \\ 3x + y - 3z &= 7 \end{aligned} \quad \text{or} \quad \begin{bmatrix} 2 & -1 \\ 3 & 1 \end{bmatrix}\begin{bmatrix} x \\ y \end{bmatrix} = \begin{bmatrix} 5 - 2z \\ 7 + 3z \end{bmatrix}$$

for x and y in terms of z. Each equation on the left represents a plane. In how many points do the planes intersect?

19. Expanding the quotient

$$\frac{ax + b}{(x - r_1)(x - r_2)}$$

by partial fractions calls for finding the values of C and D that make the equation

$$\frac{ax + b}{(x - r_1)(x - r_2)} = \frac{C}{x - r_1} + \frac{D}{x - r_2}$$

hold for all x.

a) Find a system of linear equations that determines C and D.

b) Under what circumstances does the system of equations in part (a) have a unique solution? That is, when is the determinant of the coefficient matrix of the system different from zero?

A.3

Proofs of the Limit Theorems of Article 1.9

We now prove Theorems 1 and 4 of Article 1.9. Proofs of Theorems 2 and 3 of Article 1.9 can be found in Problems 1–5 at the end of this appendix.

THEOREM 1

If $\lim_{t\to c} F_1(t) = L_1$ and $\lim_{t\to c} F_2(t) = L_2$, then

i) $\lim\,[F_1(t) + F_2(t)] = L_1 + L_2$

ii) $\lim\,[F_1(t) - F_2(t)] = L_1 - L_2$

iii) $\lim F_1(t)F_2(t) = L_1L_2$

iv) $\lim kF_2(t) = kL_2$, (any number k)

v) $\lim \dfrac{F_1(t)}{F_2(t)} = \dfrac{L_1}{L_2}$, provided $L_2 \neq 0$.

The limits are all to be taken as $t \to c$.

i) $\lim\,[F_1(t) + F_2(t)] = L_1 + L_2$

To prove that the sum $F_1(t) + F_2(t)$ has the limit $L_1 + L_2$ as $t \to c$ we must show that for any $\varepsilon > 0$ there exists a $\delta > 0$ such that for all t

$$0 < |t - c| < \delta \Rightarrow |F_1(t) + F_2(t) - (L_1 + L_2)| < \varepsilon. \qquad (1a)$$

If ε is positive, then so is $\varepsilon/2$, and because $\lim_{t\to c} F_1(t) = L_1$ we know that there is a $\delta_1 > 0$ such that for all t

$$0 < |t - c| < \delta_1 \Rightarrow |F_1(t) - L_1| < \varepsilon/2. \qquad (1b)$$

Likewise, there is a $\delta_2 > 0$ such that for all t

$$0 < |t - c| < \delta_2 \Rightarrow |F_2(t) - L_2| < \varepsilon/2. \qquad (1c)$$

Now let δ be the minimum of δ_1 and δ_2. Then δ is a positive number, and the ε inequalities in (1b) and (1c) both hold when $0 < |t - c| < \delta$. Thus, for all t, the inequality $0 < |t - c| < \delta$ implies

$$\begin{aligned}|F_1(t) + F_2(t) - (L_1 + L_2)| &= |F_1(t) - L_1 + F_2(t) - L_2| \\ &\le |F_1(t) - L_1| + |F_2(t) - L_2| \\ &< \frac{\varepsilon}{2} + \frac{\varepsilon}{2} = \varepsilon.\end{aligned}$$

This establishes (1a) and proves part (i) of the theorem.

iii) $\lim\,[F_1(t)\cdot F_2(t)] = L_1 \cdot L_2$

Let ε be an arbitrary positive number, and write

$$F_1(t) = L_1 + (F_1(t) - L_1), \qquad F_2(t) = L_2 + (F_2(t) - L_2).$$

When we multiply these expressions and subtract L_1L_2, we get

$$\begin{aligned}F_1(t)\cdot F_2(t) - L_1L_2 &= L_1(F_2(t) - L_2) + L_2(F_1(t) - L_1) \\ &\quad + (F_1(t) - L_1)\cdot(F_2(t) - L_2). \qquad (3a)\end{aligned}$$

The numbers $\sqrt{\varepsilon/3}$, $\varepsilon/[3(1 + |L_1|)]$, and $\varepsilon/[3(1 + |L_2|)]$ are all positive, and, because $F_1(t)$ has limit L_1 and $F_2(t)$ has limit L_2, there are positive numbers δ_1, δ_2, δ_3, and δ_4 such that

$$\begin{aligned}
&|F_1(t) - L_1| < \sqrt{\varepsilon/3} && \text{when } 0 < |t - c| < \delta_1,\\
&|F_2(t) - L_2| < \sqrt{\varepsilon/3} && \text{when } 0 < |t - c| < \delta_2,\\
&|F_1(t) - L_1| < \varepsilon/[3(1 + |L_2|)] && \text{when } 0 < |t - c| < \delta_3,\\
&|F_2(t) - L_2| < \varepsilon/[3(1 + |L_1|)] && \text{when } 0 < |t - c| < \delta_4,
\end{aligned}$$

We now let δ be the minimum of the four positive numbers δ_1, δ_2, δ_3, δ_4. Then δ is a positive number and all four of the inequalities above are satisfied when $0 < |t - c| < \delta$. By taking absolute values in Eq. (3a) and applying the triangle inequality, we get

$$\begin{aligned}
|F_1(t)F_2(t) - L_1L_2| &\le |L_1| \cdot |F_2(t) - L_2| + |L_2| \cdot |F_1(t) - L_1|\\
&\quad + |F_1(t) - L_1| \cdot |F_2(t) - L_2|\\
&< \frac{\varepsilon}{3} + \frac{\varepsilon}{3} + \frac{\varepsilon}{3} = \varepsilon \qquad \text{when} \quad 0 < |t - c| < \delta. \qquad (3b)
\end{aligned}$$

This completes the proof of part (iii).

iv) $\lim kF_2(t) = kL_2$

This is a special case of (iii) with $F_1(t) = k$, the function whose output value is the constant k for all values of t.

ii) $\lim [F_1(t) - F_2(t)] = L_1 - L_2$

This can be deduced from (i) and (iv) in the following way:

$$\begin{aligned}
\lim [F_1(t) - F_2(t)] &= \lim [F_1(t) + (-1)F_2(t)]\\
&= \lim F_1(t) + \lim (-1)F_2(t)\\
&= \lim F_1(t) + (-1)\lim F_2(t)\\
&= \lim F_1(t) - \lim F_2(t)\\
&= L_1 - L_2.
\end{aligned}$$

v) $\lim \dfrac{F_1(t)}{F_2(t)} = \dfrac{L_1}{L_2}$ if $L_2 \neq 0$

Since L_2 is not zero, $|L_2|$ is a positive number, and because $F_2(t)$ has L_2 as limit when t approaches c, we know that there is a positive number δ_1 such that

$$|F_2(t) - L_2| < \frac{|L_2|}{2} \qquad \text{when} \quad 0 < |t - c| < \delta_1. \qquad (5a)$$

Now, for any numbers A and B, it can be shown that

$$|A| - |B| \le |A - B| \qquad \text{and} \qquad |B| - |A| \le |A - B|,$$

from which it follows that

$$||A| - |B|| \le |A - B|. \qquad (5b)$$

Taking $A = F_2(t)$ and $B = L_2$ in (5b), we can deduce from (5a) that

$$-\frac{1}{2}|L_2| < |F_2(t)| - |L_2| < \frac{1}{2}|L_2| \qquad \text{when} \quad 0 < |t - c| < \delta_1.$$

By adding $|L_2|$ to the three terms of the foregoing inequality we get

$$\frac{1}{2}|L_2| < |F_2(t)| < \frac{3}{2}|L_2|,$$

from which it follows that

$$\left|\frac{1}{F_2(t)} - \frac{1}{L_2}\right| = \left|\frac{L_2 - F_2(t)}{L_2F_2(t)}\right| \leq \frac{2}{|L_2|^2}|L_2 - F_2(t)| \tag{5c}$$

when $0 < |t - c| < \delta_1$.

All that we have done so far is to show that the difference between the reciprocals of $F_2(t)$ and L_2, at the left-hand side of (5c) is no greater in absolute value than a constant times $|L_2 - F_2(t)|$, when t is close enough to c. The fact that L_2 is the limit of $F_2(t)$ has not yet been used with full force.

But now let ε be an arbitrary positive number. Then $\frac{1}{2}|L_2|^2\varepsilon$ is also a positive number and there is a positive number δ_2 such that

$$|L_2 - F_2(t)| < \frac{\varepsilon}{2}|L_2|^2 \quad \text{when} \quad 0 < |t - c| < \delta_2. \tag{5d}$$

We now let $\delta = \min\{\delta_1, \delta_2\}$ and have a positive number δ such that the inequalities in (5c) and (5d) combine to produce the result

$$\left|\frac{1}{F_2(t)} - \frac{1}{L_2}\right| < \varepsilon \quad \text{when} \quad 0 < |t - c| < \delta.$$

What we have just shown is that

If $\lim F_2(t) = L_2$ as t approaches c, and $L_2 \neq 0$, then

$$\lim \frac{1}{F_2(t)} = \frac{1}{L_2}.$$

Having already proved the product law, we get the final quotient law by applying the product law to $F_1(t)$ and $1/F_2(t)$ as follows:

$$\lim \frac{F_1(t)}{F_2(t)} = \lim \left[F_1(t) \cdot \frac{1}{F_2(t)}\right] = [\lim F_1(t)] \cdot \left[\lim \frac{1}{F_2(t)}\right] = L_1 \cdot \frac{1}{L_2}. \blacksquare$$

THEOREM 4

The Sandwich Theorem

Suppose that

$$f(t) \leq g(t) \leq h(t)$$

for all values of $t \neq c$ in some interval about c, and that $f(t)$ and $h(t)$ approach the same limit L as t approaches c. Then,

$$\lim_{t\to c} g(t) = L.$$

Proof for right limits. If

$$\lim_{t\to c^+} f(t) = \lim_{t\to c^+} h(t) = L,$$

then for any $\varepsilon > 0$ there exists a $\delta > 0$ such that for all t

$$c < t < c + \delta \Rightarrow L - \varepsilon < f(t) < L + \varepsilon \quad \text{and} \quad L - \varepsilon < h(t) < L + \varepsilon.$$

We combine the ε-inequalities on the right with the inequality $f(t) \leq g(t) \leq h(t)$ to obtain

$$L - \varepsilon < f(t) \leq g(t) \leq h(t) < L + \varepsilon$$

and thereby

$$L - \varepsilon < g(t) < L + \varepsilon.$$

Therefore, for all t,

$$c < t < c + \delta \Rightarrow |g(t) - L| < \varepsilon.$$

This shows that

$$\lim_{t \to c^+} g(t) = L. \blacksquare$$

Proof for left limits. Given $\varepsilon > 0$, there exists a $\delta > 0$ such that for all t

$$c - \delta < t < c \Rightarrow L - \varepsilon < f(t) < L + \varepsilon \quad \text{and} \quad L - \varepsilon < h(t) < L + \varepsilon.$$

As before, we conclude that for all t

$$c - \delta < t < c \Rightarrow L - \varepsilon < g(t) < L + \varepsilon$$

and therefore that

$$\lim_{t \to c^-} g(t) = L. \blacksquare$$

Proof for two-sided limits. If $\lim_{t \to c} f(t) = L$ and $\lim_{t \to c} h(t) = L$, then f and h approach L as $t \to c^-$ and as $t \to c^+$. Therefore, as we have shown above, $g(t) \to L$ as $t \to c^-$ and $t \to c^+$. Since the right and left limits of g as t approaches c exist and are both equal to L,

$$\lim_{t \to c} g(t) = L. \blacksquare$$

PROBLEMS

1. If $F_1(t)$, $F_2(t)$, and $F_3(t)$ have limits L_1, L_2, and L_3, respectively, as t approaches c, prove that their sum has limit $L_1 + L_2 + L_3$. Generalize the result to the sum of any finite number of functions.

2. If n is any positive integer greater than 1, and $F_1(t)$, $F_2(t), \ldots, F_n(t)$ have the finite limits $L_1, L_2, \ldots, L_n$, respectively, as t approaches c, prove that the product of the n functions has limit $L_1 \cdot L_2 \cdot \cdots \cdot L_n$. (Use part (iii) of Theorem 1, and induction on n.)

3. Use the results of Article 1.9, Example 2, and Problem 2 above to deduce that $\lim_{t \to c} t^n = c^n$ for any positive integer n.

4. Use the result of Example 3 in Article 1.9 and the results of Problems 1 and 3 above to prove that $\lim_{t \to c} f(t) = f(c)$ for any polynomial function

$$f(t) = a_0 t^n + a_1 t^{n-1} + \cdots + a_n.$$

5. Use Theorem 1 and the result of Problem 4 to prove that if $f(t)$ and $g(t)$ are polynomials, and if $g(c) \neq 0$, then

$$\lim_{t \to c} \frac{f(t)}{g(t)} = \frac{f(c)}{g(c)}.$$

6. Figure A.1 (p. A-22) gives a diagram for a proof that the composite of two continuous functions is continuous. Reconstruct the proof from the diagram. The statement to be proved is this:

If $f(t)$ is continuous at $t = c$ and $g(x)$ is continuous at $x = f(c)$, then the composite $g(f(t))$ is continuous at $t = c$.

Assume also c is an interior point of the domain of f, and that $f(c)$ is an interior point of the domain of g. This will make all the limits involved two-sided. (The proofs for the cases in which one or both of c and $f(c)$ are endpoints of the domains of f and g are similar to the argument that assumes both to be interior points.)

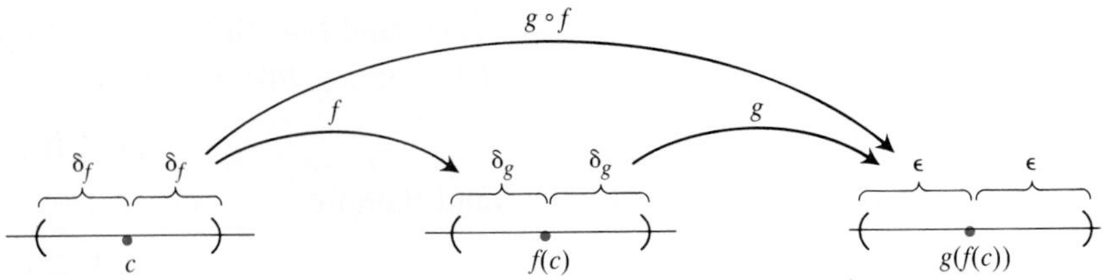

A.1 The diagram for a proof that the composite of two continuous functions is continuous. The continuity of composites holds for any finite number of functions. The only requirement is that each function be continuous where it is applied. In the diagram above, f is to be continuous at c, and g is to be continuous at $f(c)$.

A.4

The Increment and Mixed-Derivative Theorems

This appendix gives proofs for the *Increment Theorem* (Article 15.3, Theorem 2) and the *Mixed-derivative Theorem* (Article 15.7, Theorem 4) for functions of two variables.

THEOREM 2

Increment Theorem for Functions of Two Variables

Suppose that $z = f(x, y)$ is continuous and has partial derivatives throughout a region

$$R: \quad |x - x_0| < h, \qquad |y - y_0| < k$$

in the xy-plane. Suppose also that Δx *and* Δy are small enough for the point $(x_0 + \Delta x, y_0 + \Delta y)$ to lie in R. If f_x and f_y are continuous at (x_0, y_0), then the increment

$$\Delta z = f(x_0 + \Delta x, y_0 + \Delta y) - f(x_0, y_0) \tag{1}$$

can be written as

$$\Delta z = f_x(x_0, y_0)\,\Delta x + f_y(x_0, y_0)\,\Delta y + \varepsilon_1\,\Delta x + \varepsilon_2\,\Delta y, \tag{2}$$

where

$$\varepsilon_1, \varepsilon_2 \to 0 \quad \text{as} \quad \Delta x, \Delta y \to 0.$$

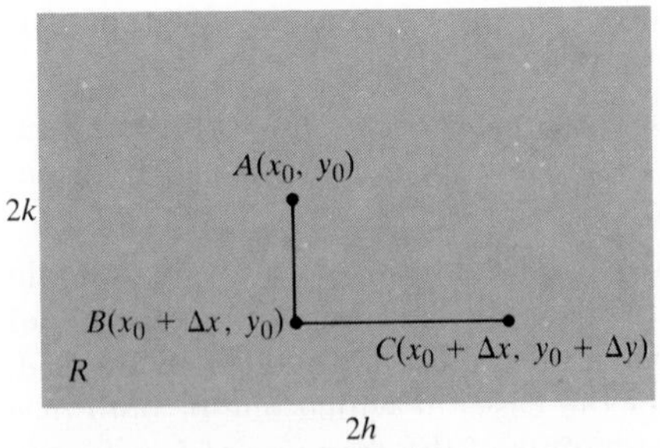

A.2 The rectangular region R in the proof of Theorem 2.

Proof. The region R (Fig. A.2) is a rectangle centered at $A(x_0, y_0)$ with dimensions $2h$ by $2k$. Since $C(x_0 + \Delta x, y_0 + \Delta y)$ lies in R, the point $B(x_0 + \Delta x, y_0)$ and the line segments AB and BC also lie in R. Thus f is continuous and has partial derivatives f_x and f_y at each point of these segments.

We may think of Δz as the sum

$$\Delta z = \Delta z_1 + \Delta z_2 \tag{3}$$

of two increments, where

$$\Delta z_1 = f(x_0 + \Delta x, y_0) - f(x_0, y_0) \tag{4}$$

is the change from A to B and

$$\Delta z_2 = f(x_0 + \Delta x, y_0 + \Delta y) - f(x_0 + \Delta x, y_0) \tag{5}$$

is the change from B to C (Fig. A.3). Note that the sum $\Delta z_1 + \Delta z_2$ equals Δz as it should:

$$\begin{aligned} \Delta z_1 + \Delta z_2 &= [f(x_0 + \Delta x, y_0) - f(x_0, y_0)] \\ &\quad + [f(x_0 + \Delta x, y_0 + \Delta y) - f(x_0 + \Delta x, y_0)] \\ &= f(x_0 + \Delta x, y_0 + \Delta y) - f(x_0, y_0) \\ &= \Delta z. \end{aligned} \tag{6}$$

The function

$$F(x) = f(x, y_0)$$

is a continuous and differentiable function of x on the closed interval joining x_0 and $x_0 + \Delta x$, with derivative

$$F'(x) = f_x(x, y_0).$$

Hence, by the Mean Value Theorem of Article 3.8, there is a point c between x and x + Δx such that

$$F(x_0 + \Delta x) - F(x_0) = F'(c)\,\Delta x$$

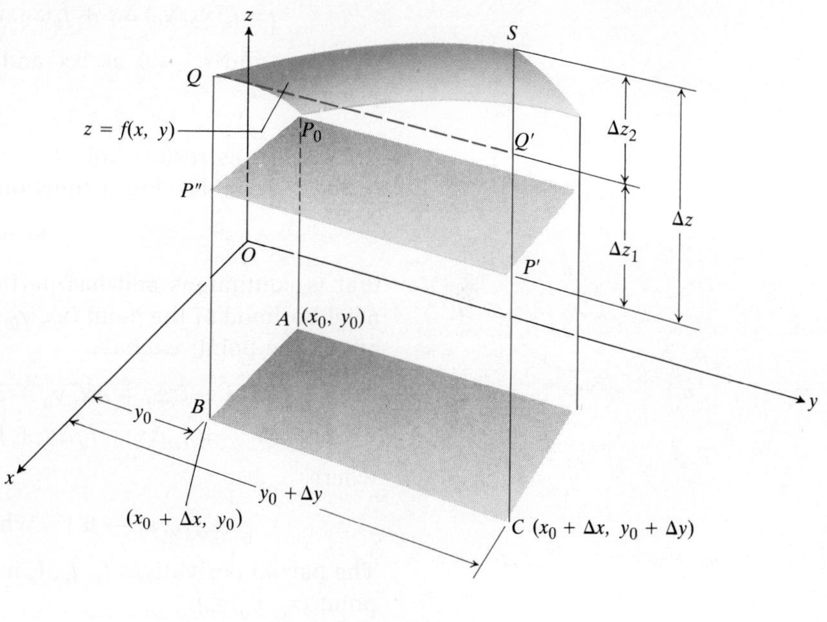

A.3 Part of the surface $z = f(x, y)$ near $P_0(x_0, y_0, f(x_0, y_0))$. The points P_0, P', and P'' have the same height $z_0 = f(x_0, y_0)$ above the xy-plane. The change in z is $\Delta z = P'S$. The change

$$\Delta z_2 = f(x_0 + \Delta x, y_0) - f(x_0, y_0),$$

shown as $P''Q = P'Q'$, is caused by changing x from x_0 to $x_0 + \Delta x$ while holding y equal to y_0. Then, with x held equal to $x_0 + \Delta x$,

$$\Delta z_2 = f(x_0 + \Delta x, y_0 + \Delta y) - f(x_0 + \Delta x, y_0)$$

is the change in z caused by changing y from y_0 to $y_0 + \Delta y$. This is represented by $Q'S$. The total change in z is the sum of Δz_1 and Δz_2.

or

$$f(x_0 + \Delta x, y_0) - f(x_0, y_0) = f_x(c, y_0)\,\Delta x$$

or

$$\Delta z_1 = f_x(c, y_0)\,\Delta x. \tag{7}$$

Similarly, the function

$$G(y) = f(x_0 + \Delta x, y)$$

is a continuous and differentiable function of y on the closed interval joining y_0 and $y_0 + \Delta y$, with derivative

$$G'(y) = f_y(x_0 + \Delta x, y).$$

Hence there is a point d between y_0 and $y_0 + \Delta y$ such that

$$G(y_0 + \Delta y) - G(y_0) = G'(d)\,\Delta y$$

or

$$f(x_0 + \Delta x, y_0 + \Delta y) - f(x_0 + \Delta x, y_0) = f_y(x_0 + \Delta x, d)\,\Delta y$$

or

$$\Delta z_2 = f_y(x_0 + \Delta x, d)\,\Delta y. \tag{8}$$

Now, as Δx and $\Delta y \to 0$, we know $c \to x_0$ and $d \to y_0$. Therefore, if f_x and f_y are continuous at (x_0, y_0), the quantities

$$\begin{aligned} \varepsilon_1 &= f_x(c, y_0) - f_x(x_0, y_0), \\ \varepsilon_2 &= f_y(x_0 + \Delta x, d) - f_y(x_0, y_0) \end{aligned} \tag{9}$$

both approach zero as Δx and $\Delta y \to 0$.

Finally,

$$\begin{aligned} \Delta z &= \Delta z_1 + \Delta z_2 \\ &= f_x(c, y_0)\,\Delta x + f_y(x_0 + \Delta x, d)\,\Delta y \qquad \text{(from (7) and (8))} \\ &= [f_x(x_0, y_0) + \varepsilon_1]\,\Delta x + [f_y(x_0, y_0) + \varepsilon_2]\,\Delta y \qquad \text{(from (9))} \\ &= f_x(x_0, y_0)\,\Delta x + f_y(x_0, y_0)\,\Delta y + \varepsilon_1\,\Delta x + \varepsilon_2\,\Delta y, \end{aligned}$$

where ε_1 and $\varepsilon_2 \to 0$ as Δx and $\Delta y \to 0$. This is what we set out to prove. ■

Analogous results hold for functions of any finite number of independent variables. For a function of three variables

$$w = f(x, y, z),$$

that is continuous and has partial derivatives f_x, f_y, f_z at and in some neighborhood of the point (x_0, y_0, z_0), and whose derivatives are continuous at the point, we have

$$\begin{aligned} \Delta w &= f(x_0 + \Delta x, y_0 + \Delta y, z_0 + \Delta z) - f(x_0, y_0, z_0) \\ &= f_x\,\Delta x + f_y\,\Delta y + f_z\,\Delta z + \varepsilon_1\,\Delta x + \varepsilon_2\,\Delta y + \varepsilon_3\,\Delta z, \end{aligned} \tag{10}$$

where

$$\varepsilon_1, \varepsilon_2, \varepsilon_3 \to 0 \qquad \text{when} \quad \Delta x, \Delta y, \text{ and } \Delta z \to 0.$$

The partial derivatives f_x, f_y, f_z in this formula are to be evaluated at the point (x_0, y_0, z_0).

Note. The result (10) may be proved by treating Δw as the sum of three increments,

$$\Delta w_1 = f(x_0 + \Delta x, y_0, z_0) - f(x_0, y_0, z_0), \tag{11a}$$
$$\Delta w_2 = f(x_0 + \Delta x, y_0 + \Delta y, z_0) - f(x_0 + \Delta x, y_0, z_0), \tag{11b}$$
$$\Delta w_3 = f(x_0 + \Delta x, y_0 + \Delta y, z_0 + \Delta z) - f(x_0 + \Delta x, y_0 + \Delta y, z_0), \tag{11c}$$

and applying the Mean Value Theorem to each of these separately. Note that two coordinates remain constant and only one varies in each of these partial increments Δw_1, Δw_2, Δw_3. For example, in (11b), only y varies, since x is held equal to $x_0 + \Delta x$ and z is held equal to z_0. Since the function $f(x_0 + \Delta x, y, z_0)$ is a continuous function of y with a derivative f_y, it is subject to the Mean Value Theorem, and we have

$$\Delta w_2 = f_y(x_0 + \Delta x, y_1, z_0)\,\Delta y$$

for some y_1 between y_0 and $y_0 + \Delta y$.

THEOREM 4

The Mixed-Derivative Theorem

If $f(x, y)$ and its partial derivatives f_x, f_y, f_{xy}, and f_{yx} are defined in a region containing a point (a, b) and are all continuous at (a, b), then

$$f_{xy}(a, b) = f_{yx}(a, b). \tag{12}$$

Proof. The equality of $f_{xy}(a, b)$ and $f_{yx}(a, b)$ can be established by four applications of the Mean Value Theorem. By hypothesis, the point (a, b) lies in the interior of a rectangle R in the xy-plane on which f, f_x, f_y, f_{xy}, and f_{yx} are all continuous. We let h and k be numbers such that the point $(a + h, b + k)$ also lies in the rectangle R, and we consider the difference

$$\Delta = F(a + h) - F(a), \tag{13}$$

where we define $F(x)$ in terms of $f(x, y)$ by the equation

$$F(x) = f(x, b + k) - f(x, b). \tag{14}$$

We apply the Mean Value Theorem to the function $F(x)$ (which is continuous because it is differentiable), and Eq. (13) becomes

$$\Delta = hF'(c_1), \tag{15}$$

where c_1 lies between a and $a + h$. From Eq. (14),

$$F'(x) = f_x(x, b + k) - f_x(x, b),$$

so Eq. (15) becomes

$$\Delta = h[f_x(c_1, b + k) - f_x(c_1, b)]. \tag{16}$$

Now we apply the Mean Value Theorem to the function $g(y) = f_x(c_1, y)$ and have

$$g(b + k) - g(b) = kg'(d_1) \tag{17a}$$

or

$$f_x(c_1, b + k) - f_x(c_1, b) = kf_{xy}(c_1, d_1), \tag{17b}$$

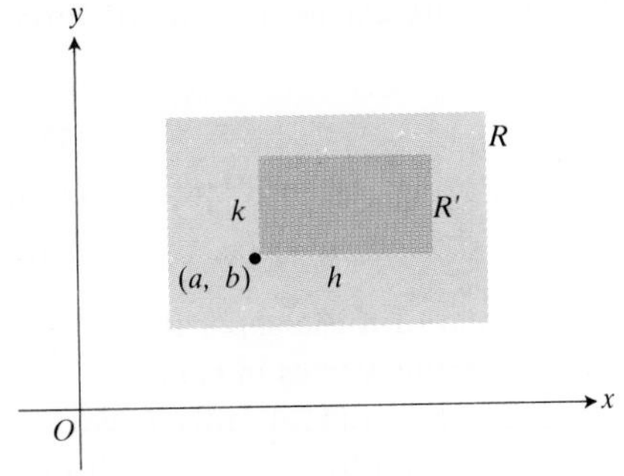

A.4 The key to proving $f_{xy}(a, b) = f_{yx}(a, b)$ is the fact that no matter how small R' is, f_{xy} and f_{yx} take on equal values somewhere inside R' (although not necessarily at the same point of R').

for some d_1 between b and $b + k$. By substituting this into Eq. (16), we get

$$\Delta = hkf_{xy}(c_1, d_1), \tag{18}$$

for some point (c_1, d_1) in the rectangle R' whose vertices are the four points (a, b), $(a + h, b)$, $(a + h, b + k)$, and $(a, b + k)$. (See Fig. A.4.)

On the other hand, by substituting from Eq. (14) into Eq. (13), we may also write

$$\begin{aligned}\Delta &= f(a + h, b + k) - f(a + h, b) - f(a, b + k) + f(a, b)\\ &= [f(a + h, b + k) - f(a, b + k)] - [f(a + h, b) - f(a, b)]\\ &= \phi(b + k) - \phi(b),\end{aligned} \tag{19}$$

where

$$\phi(y) = f(a + h, y) - f(a, y). \tag{20}$$

The Mean Value Theorem applied to Eq. (19) now gives

$$\Delta = k\phi'(d_2), \tag{21}$$

for some d_2 between b and $b + k$. By Eq. (20),

$$\phi'(y) = f_y(a + h, y) - f_y(a, y). \tag{22}$$

Substituting from Eq. (22) into Eq. (21), we have

$$\Delta = k[f_y(a + h, d_2) - f_y(a, d_2)]. \tag{23}$$

Finally, we apply the Mean Value Theorem to the expression in brackets and get

$$\Delta = khf_{yx}(c_2, d_2), \tag{24}$$

for some c_2 between a and $a + h$.

A comparison of Eqs. (18) and (24) shows that

$$f_{xy}(c_1, d_1) = f_{yx}(c_2, d_2), \tag{25}$$

where (c_1, d_1) and (c_2, d_2) both lie in the rectangle R' (Fig. A.4). Equation (25) is not quite the result we want, since it says only that the mixed derivative f_{xy} has the same value at (c_1, d_1) that the derivative f_{yx} has at (c_2, d_2). But the numbers h and k in our discussion may be made as small as we wish. The hypothesis that f_{xy} and f_{yz} are both continuous at (a, b) then means that $f_{xy}(c_1, d_1) = f_{xy}(a, b) + \varepsilon_1$ and $f_{yx}(c_2, d_2) = f_{yx}(a, b) + \varepsilon_2$, where $\varepsilon_1, \varepsilon_2 \to 0$ as $h, k \to 0$. Hence, if we let h and $k \to 0$, we have $f_{xy}(a, b) = f_{yx}(a, b)$. ■

REMARK. The equality of $f_{xy}(a, b)$ and $f_{yx}(a, b)$ can be proved with weaker hypotheses than the ones we assumed in Theorem 4. For example, it is enough for f, f_x, and f_y to exist in R and for f_{xy} to be continuous at (a, b). Then f_{yx} will exist at (a, b) and be equal to f_{xy} at that point.

A.5 Mathematical Induction

Many formulas, like

$$1 + 2 + \cdots + n = \frac{n(n+1)}{2},$$

that hold for all positive integers n can be proved by applying an axiom called *the mathematical induction principle.* A proof that uses this axiom is called a proof by mathematical induction, or a proof by induction.

The steps in proving a formula by induction are

1. check that it holds for $n = 1$, and
2. prove that if it holds for any positive integer $n = k$, then it also holds for $n = k + 1$.

Once these steps are completed (the axiom says), we know the formula holds for all positive integers n. By step 1 it holds for $n = 1$. By step 2 it holds for $n = 2$, and therefore by step 2 also for $n = 3$, and by step 2 again for $n = 4$, and so on. If the first domino falls, and the kth domino always knocks over the $(k + 1)$-st when it falls, all the dominoes fall.

From another point of view, suppose we have a sequence of statements

$$S_1, S_2, \ldots, S_n, \ldots,$$

one for each positive integer. Suppose we can show that assuming any one of the statements to be true implies that the next statement in line is true. Suppose that we can also show that S_1 is true. Then we may conclude that all the statements are true from S_1 on.

EXAMPLE 1 Show that

$$1 + 2 + \cdots + n = \frac{n(n+1)}{2}$$

for all positive integers n.

Solution We carry out the two steps of mathematical induction.

1. The formula holds for $n = 1$ because

$$1 = \frac{1(1+1)}{2}.$$

2. If

$$1 + 2 + \cdots + k = \frac{k(k+1)}{2},$$

then

$$\begin{aligned} 1 + 2 + \cdots + k + (k+1) &= \frac{k(k+1)}{2} + (k+1) \\ &= \frac{k^2 + k + 2k + 2}{2} \\ &= \frac{(k+1)(k+2)}{2} \\ &= \frac{(k+1)((k+1)+1)}{2}. \end{aligned}$$

The last expression in this string of equalities is the expression $n(n + 1)/2$ for $n = (k + 1)$.

The mathematical induction principle now guarantees the formula for all positive integers n.

Note that all we have to do here is carry out steps 1 and 2. The mathematical induction principle does the rest. □

EXAMPLE 2 Show that

$$\frac{1}{2^1} + \frac{1}{2^2} + \cdots + \frac{1}{2^n} = 1 - \frac{1}{2^n}$$

for all positive integers n.

Solution We carry out the two steps of mathematical induction.

1. The formula holds for $n = 1$ because

$$\frac{1}{2^1} = 1 - \frac{1}{2^1}.$$

2. If

$$\frac{1}{2^1} + \frac{1}{2^2} + \cdots + \frac{1}{2^k} = 1 - \frac{1}{2^k},$$

then

$$\begin{aligned}\frac{1}{2^1} + \frac{1}{2^2} + \cdots + \frac{1}{2^k} + \frac{1}{2^{k+1}} &= 1 - \frac{1}{2^k} + \frac{1}{2^{k+1}} \\ &= 1 - \frac{1 \cdot 2}{2^k \cdot 2} + \frac{1}{2^{k+1}} \\ &= 1 - \frac{2}{2^{k+1}} + \frac{1}{2^{k+1}} \\ &= 1 - \frac{1}{2^{k+1}}.\end{aligned}$$

The mathematical induction principle now guarantees the formula for all positive integers n. □

Instead of starting at $n = 1$, some induction arguments start at another integer. The steps for such an argument are

1. check that the formula holds for $n = n_1$ (whatever the appropriate first integer is), and
2. prove that if it holds for any integer $n = k \geq n_1$, then it also holds for $n = k + 1$.

Once these steps are completed, the mathematical induction principle will guarantee the formula for all $n \geq n_1$.

EXAMPLE 3 Show that $n! > 3^n$ for n sufficiently large.

Solution How large? We experiment:

n	1	2	3	4	5	6	7
$n!$	1	2	6	24	120	720	5040
3^n	3	9	27	81	243	729	2187

It looks as if $n! > 3^n$ for $n \geq 7$. To be sure, we apply mathematical induction. We take $n_1 = 7$ in step 1, and try for step 2.

Suppose $k! > 3^k$ for some $k \geq 7$. Then

$$(k + 1)! = (k + 1)(k!) > (k + 1)3^k > 8 \cdot 3^k > 3^{k+1}.$$

Thus, for $k \geq 7$,

$$k! > 3^k \Rightarrow (k + 1)! > 3^{k+1}.$$

The mathematical induction principle now guarantees $n! \geq 3^n$ for all $n \geq 7$. □

PROBLEMS

1. Assuming the triangle inequality $|a + b| \leq |a| + |b|$, show that

$$|x_1 + x_2 + \cdots + x_n| \leq |x_1| + |x_2| + \cdots + |x_n|$$

for any n numbers.

2. Show that if $r \neq 1$, then

$$1 + r + r^2 + \cdots + r^n = \frac{1 - r^{n+1}}{1 - r}$$

for all positive integers n.

3. Use the product rule

$$\frac{d}{dx}(uv) = u\frac{dv}{dx} + v\frac{du}{dx}$$

and the fact that

$$\frac{d}{dx}(x) = 1$$

to show that

$$\frac{d}{dx}(x^n) = nx^{n-1}$$

for all positive integers n.

4. Suppose that a function $f(x)$ has the property that $f(x_1x_2) = f(x_1) + f(x_2)$ for any two positive numbers x_1 and x_2. Show that

$$f(x_1x_2 \cdots x_n) = f(x_1) + f(x_2) + \cdots + f(x_n)$$

for the product of any n positive numbers x_1, x_2, ..., x_n.

5. Show that

$$\frac{2}{3^1} + \frac{2}{3^2} + \cdots + \frac{2}{3^n} = 1 - \frac{1}{3^n}$$

for all positive integers n.

6. Show that $n! > n^3$ for n sufficiently large.
7. Show that $2^n > n^2$ for n sufficiently large.
8. Show that $2^n \geq \frac{1}{8}$ for $n \geq -3$.

A.6 The Law of Cosines and the Addition Formulas from Trigonometry

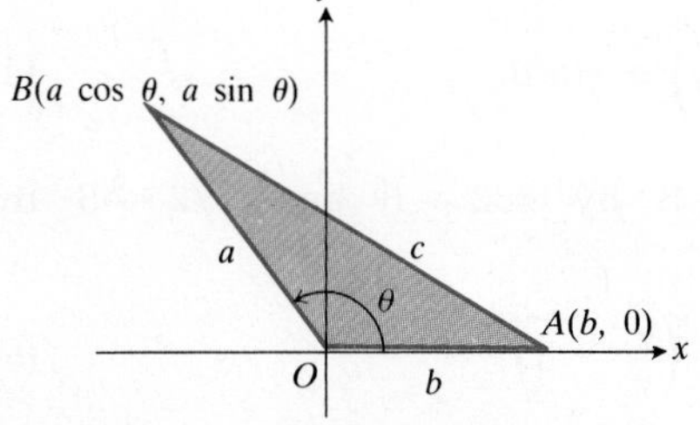

A.5 To derive the law of cosines, compute the distance between A and B, and square.

In Fig. A.5 triangle OAB has been placed with O at the origin and A on the x-axis at $A(b, 0)$. The third vertex B has coordinates

$$x = a\cos\theta, \qquad y = a\sin\theta. \tag{1}$$

The angle AOB has measure θ. By the formula for the distance between two points, the square of the distance c between A and B is

$$c^2 = (a\cos\theta - b)^2 + (a\sin\theta)^2 = a^2(\cos^2\theta + \sin^2\theta) + b^2 - 2ab\cos\theta.$$

or

$$c^2 = a^2 + b^2 - 2ab\cos\theta. \tag{2}$$

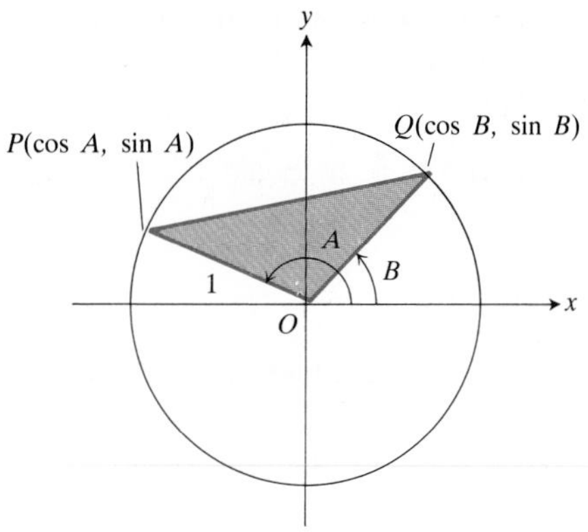

A.6 Diagram for cos $(A - B)$.

Equation (2) is called the *law of cosines*. In words, it says: "The square of any side of a triangle is equal to the sum of the squares of the other two sides minus twice the product of those two sides and the cosine of the angle between them." When the angle θ is a right angle, its cosine is zero, and Eq. (2) reduces to the theorem of Pythagoras. Equation (2) holds for a general angle θ, since it is based solely on the distance formula and on Eqs. (1). The same equation works with the exterior angle $(2\pi - \theta)$, or the opposite of $(2\pi - \theta)$, in place of θ, because

$$\cos(2\pi - \theta) = \cos(\theta - 2\pi) = \cos\theta.$$

It is still a valid formula when B is on the x-axis and $\theta = \pi$ or $\theta = 0$, as we can easily verify if we remember that $\cos 0 = 1$ and $\cos\pi = -1$. In these special cases, the right-hand side of Eq. (2) becomes $(a - b)^2$ or $(a + b)^2$.

Proofs of the Addition Formulas

Equation (11e) of Article 2.7 follows from the law of cosines applied to the triangle OPQ in Fig. A.6. We take $OP = OQ = r = 1$. Then the coordinates of P are

$$x_P = \cos A, \qquad y_P = \sin A$$

and of Q,

$$x_Q = \cos B, \qquad y_Q = \sin B.$$

Hence the square of the distance between P and Q is

$$\begin{aligned}(PQ)^2 &= (x_Q - x_P)^2 + (y_Q - y_P)^2\\ &= (x_Q^2 + y_Q^2) + (x_P^2 + y_P^2) - 2(x_Q x_P + y_Q y_P)\\ &= 2 - 2(\cos A\cos B + \sin A\sin B).\end{aligned}$$

But angle $QOP = A - B$, and the law of cosines gives

$$(PQ)^2 = (OP)^2 + (OQ)^2 - 2(OP)(OQ)\cos(A - B) = 2 - 2\cos(A - B).$$

When we equate these two expressions for $(PQ)^2$ we obtain

$$\cos(A - B) = \cos A\cos B + \sin A\sin B, \tag{11e}$$

which is Eq. (11e) of Article 2.7.

We now deduce Eqs. (a, b, d) of Article 2.7 from Eq. (11e). We shall also need the results

$$\begin{aligned}&\sin 0 = 0, \quad &&\sin(\pi/2) = 1, \quad &&\sin(-\pi/2) = -1,\\ &\cos 0 = 1, \quad &&\cos(\pi/2) = 0, \quad &&\cos(-\pi/2) = 0.\end{aligned} \tag{3}$$

1. In Eq. (11e), we put $A = \pi/2$ and use Eqs. (3) to get

$$\cos\left(\frac{\pi}{2} - B\right) = \sin B. \tag{4}$$

In this equation, if we replace B by $\pi/2 - B$ and $\pi/2 - B$ by $\pi/2 - (\pi/2 - B)$, we get

$$\cos B = \sin\left(\frac{\pi}{2} - B\right). \tag{5}$$

Equations (4) and (5) express the familiar results that the sine and cosine of an angle are the cosine and sine, respectively, of the complementary angle.

2. We next put $B = -\pi/2$ in Eq. (11e) and use Eqs. (3) to get

$$\cos\left(A + \frac{\pi}{2}\right) = -\sin A. \tag{6}$$

3. We can get the formula for $\cos(A + B)$ from Eq. (11e) by substituting $-B$ for B everywhere:

$$\begin{aligned}\cos(A + B) &= \cos[A - (-B)]\\ &= \cos A \cos(-B) + \sin A \sin(-B)\\ &= \cos A \cos B - \sin A \sin B. \end{aligned}\tag{11b}$$

where the final equality also uses Eqs. (11c) in Article 2.7.

4. To derive formulas for $\sin(A \pm B)$, we use Eq. (4) with B replaced by $A + B$, and then use Eq. (11e). Thus we have

$$\begin{aligned}\sin(A + B) &= \cos[\pi/2 - (A + B)]\\ &= \cos[(\pi/2 - A) - B]\\ &= \cos(\pi/2 - A)\cos B + \sin(\pi/2 - A)\sin B\\ &= \sin A \cos B + \cos A \sin B.\end{aligned}\tag{11a}$$

Equation (11d) of Article 2.7 follows from this if we replace B by $-B$:

$$\sin(A - B) = \sin A \cos B - \cos A \sin B. \tag{11d}$$

A.7 Formulas from Elementary Mathematics

ALGEBRA

1. Laws of Exponents

$$a^m a^n = a^{m+n}, \quad (ab)^m = a^m b^m, \quad (a^m)^n = a^{mn}, \quad a^{m/n} = \sqrt[n]{a^m}.$$

If $a \neq 0$,

$$\frac{a^m}{a^n} = a^{m-n}, \quad a^0 = 1, \quad a^{-m} = \frac{1}{a^m}.$$

2. Zero

$$a \cdot 0 = 0 \cdot a = 0 \text{ for any finite number } a.$$

If $a \neq 0$,

$$\frac{0}{a} = 0, \quad 0^a = 0, \quad a^0 = 1.$$

Division by zero is not defined.

3. Fractions

$$\frac{a}{b} + \frac{c}{d} = \frac{ad + bc}{bd}, \quad \frac{a}{b}\cdot\frac{c}{d} = \frac{ac}{bd}, \quad \frac{a/b}{c/d} = \frac{a}{b}\cdot\frac{d}{c}, \quad \frac{-a}{b} = -\frac{a}{b} = \frac{a}{-b}.$$

$$\frac{(a/b) + (c/d)}{(e/f) + (g/h)} = \frac{(a/b) + (c/d)}{(e/f) + (g/h)} \cdot \frac{bdfh}{bdfh} = \frac{(ad + bc)fh}{(eh + fg)bd}.$$

4. Binomial Theorem, for n = Positive Integer

$$(a + b)^n = a^n + na^{n-1}b + \frac{n(n-1)}{1 \cdot 2}a^{n-2}b^2 + \frac{n(n-1)(n-2)}{1 \cdot 2 \cdot 3}a^{n-3}b^3 + \cdots + nab^{n-1} + b^n.$$

For instance:

$$\begin{aligned}(a + b)^1 &= a + b,\\ (a + b)^2 &= a^2 + 2ab + b^2,\\ (a + b)^3 &= a^3 + 3a^2b + 3ab^2 + b^3,\\ (a + b)^4 &= a^4 + 4a^3b + 6a^2b^2 + 4ab^3 + b^4.\end{aligned}$$

5. Difference of Like Integer Powers, n > 1

$$a^n - b^n = (a - b)(a^{n-1} + a^{n-2}b + a^{n-3}b^2 + \cdots + ab^{n-2} + b^{n-1}),$$

For instance:

$$\begin{aligned}a^2 - b^2 &= (a - b)(a + b),\\ a^3 - b^3 &= (a - b)(a^2 + ab + b^2)\\ a^4 - b^4 &= (a - b)(a^3 + a^2b + ab^2 + b^3).\end{aligned}$$

6. Proportionality

The statements "y varies directly as x" and "y is directly proportional to x" mean the same thing, namely that

$$y = kx$$

for some constant k. This constant is called the *proportionality factor* of the equation.

Similarly, "y varies inversely as x" and "y is inversely proportional to x" both mean that

$$y = k\frac{1}{x}$$

for some constant k. Again, k is the porportionality factor of the equation.

7. Remainder Theorem and Factor Theorem

If the polynomial $f(x)$ is divided by $x - r$ until a remainder R independent of x is obtained, then $R = f(r)$. In particular, $x - r$ is a *factor* of $f(x)$ if and only if r is a *root* of the equation $f(x) = 0$.

8. Completing the Square

The equation

$$\begin{aligned}ax^2 + bx &= a\left(x^2 + \frac{b}{a}x\right)\\ &= a\left(x^2 + \frac{b}{a}x + \left(\frac{b}{2a}\right)^2 - \left(\frac{b}{2a}\right)^2\right)\\ &= a\left(\left(x + \frac{b}{2a}\right)^2 - \left(\frac{b}{2a}\right)^2\right)\end{aligned}$$

shows how to write $ax^2 + bx$ as a constant times the difference of two squares when $a \neq 0$.

9. Quadratic Formula

By completing the square on the first two terms of the equation

$$ax^2 + bx + c = 0$$

and solving the resulting equation for x (details omitted), one obtains the formula

$$x = \frac{-b \pm \sqrt{b^2 - 4ac}}{2a}.$$

10. Horner's Method

The fastest way (usually) to calculate the value of a polynomial

$$p(x) = a_0 + a_1x + a_2x^2 + \cdots a_nx^n$$

at $x = c$ is to calculate

$$p(c) = a_0 + c(a_1 + c(a_2 + c(a_3 + \cdots + c(a_{n-1} + ca_n))) \cdots),$$

working from the inside out. Arranged this way, the calculation does not require parenthesis keys. All we do is multiply and add n times, for a total of $2n$ operations (fewer if any of the coefficients are zero).

For example, the value of

$$p(x) = 3x^3 - 6x^2 + 4x + 5$$

at $x = 2$ is

$$p(2) = 5 + 2(4 + 2(-6 + 2 \cdot 3)) = 13.$$

Start here, work back

GEOMETRY (A = area, B = area of base, C = circumference, S = lateral area or surface area, V = volume.)

1. Triangle

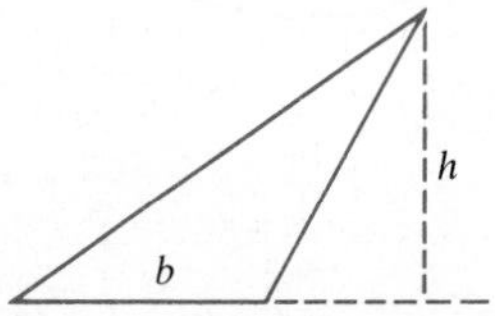

$$A = \tfrac{1}{2}bh$$

2. Similar Triangles

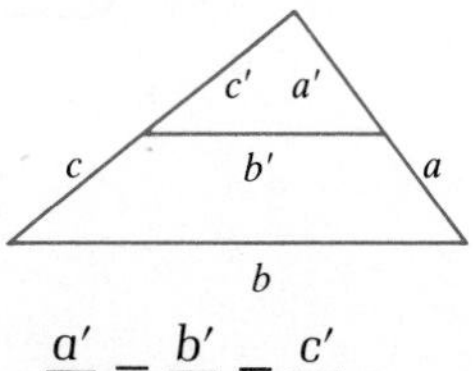

$$\frac{a'}{a} = \frac{b'}{b} = \frac{c'}{c}$$

3. Theorem of Pythagoras

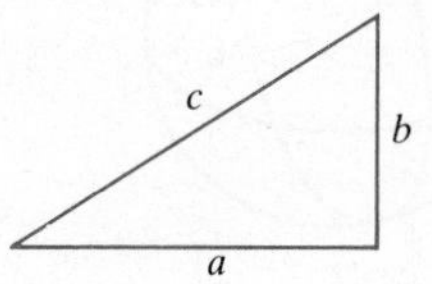

$$c^2 = a^2 + b^2$$

4. Parallelogram

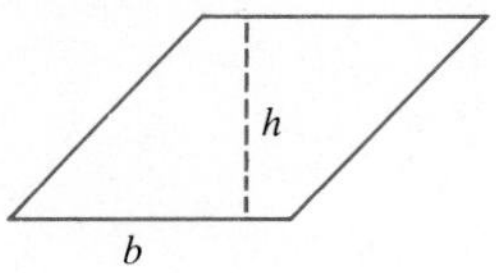

$$A = bh$$

5. Trapezoid

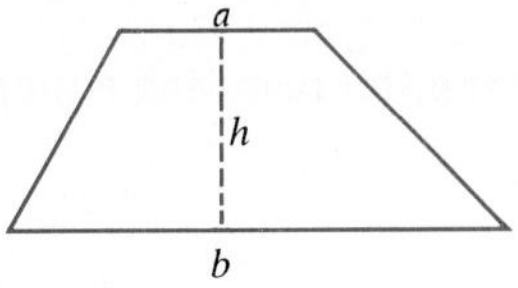

$A = \frac{1}{2}(a + b)h$

6. Circle

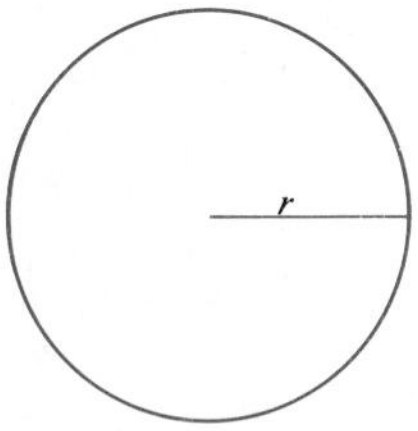

$A = \pi r^2, \qquad C = 2\pi r$

7. Any Cylinder or Prism with Parallel Bases

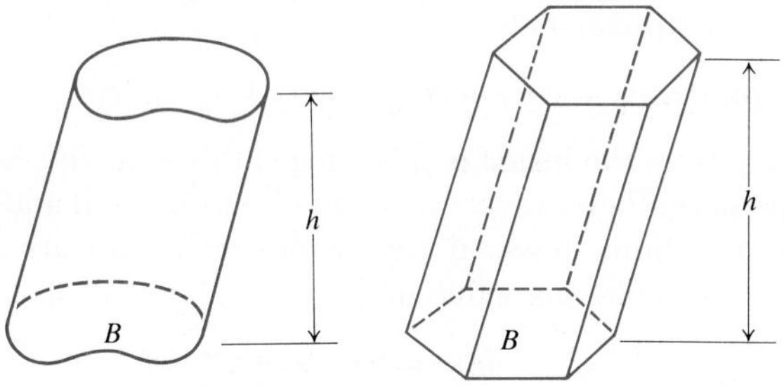

$V = Bh$

8. Right Circular Cylinder

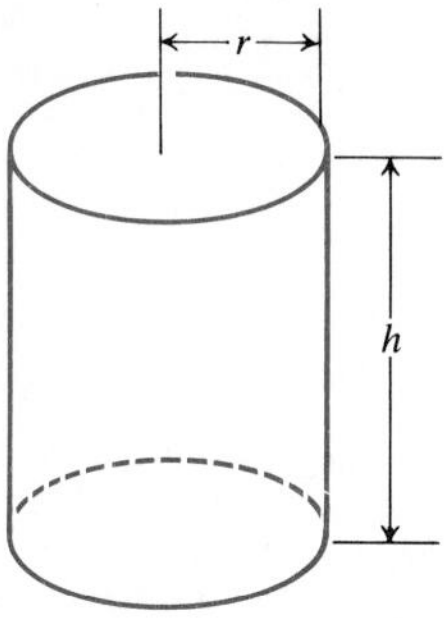

$V = \pi r^2 h, \qquad S = 2\pi rh$

9. Any Cone or Pyramid

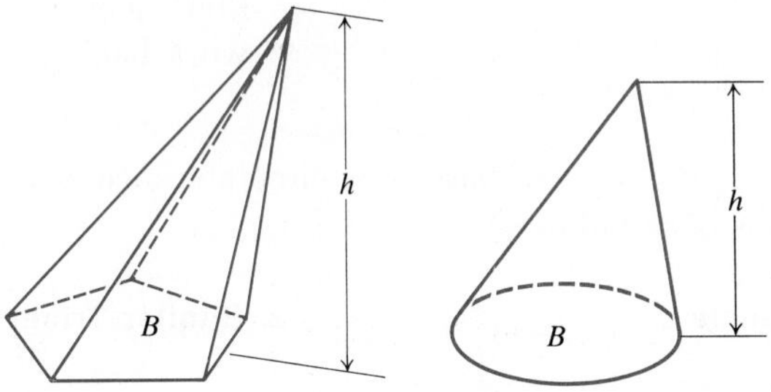

$V = \frac{1}{3}Bh$

10. Right Circular Cone

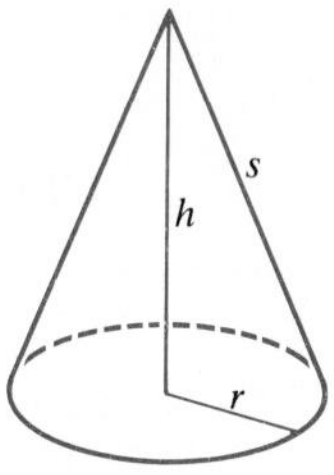

$V = \frac{1}{3}\pi r^2 h, \qquad S = \pi rs$

11. Sphere

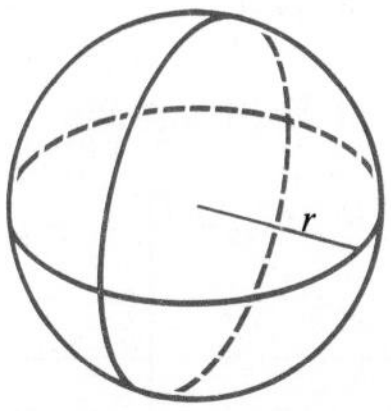

$V = \frac{4}{3}\pi r^3, \qquad S = 4\pi r^2$

TRIGONOMETRY

1. Definitions and Fundamental Identities

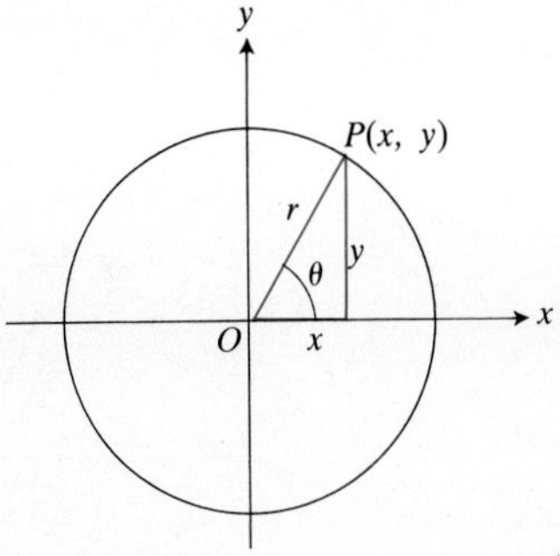

Sine: $\sin\theta = \dfrac{y}{r} = \dfrac{1}{\csc\theta}$

Cosine: $\cos\theta = \dfrac{x}{r} = \dfrac{1}{\sec\theta}$

Tangent: $\tan\theta = \dfrac{y}{x} = \dfrac{1}{\cot\theta}$

$$\sin(-\theta) = -\sin\theta,$$
$$\cos(-\theta) = \cos\theta$$

$$\sin^2\theta + \cos^2\theta = 1$$
$$\sec^2\theta = 1 + \tan^2\theta$$
$$\csc^2\theta = 1 + \cot^2\theta$$

$$\sin 2\theta = 2\sin\theta\cos\theta,$$
$$\cos 2\theta = \cos^2\theta - \sin^2\theta$$

$$\cos^2\theta = \frac{1 + \cos 2\theta}{2},$$
$$\sin^2\theta = \frac{1 - \cos 2\theta}{2}$$

$$\sin(A + B) = \sin A\cos B + \cos A\sin B$$
$$\sin(A - B) = \sin A\cos B - \cos A\sin B$$
$$\cos(A + B) = \cos A\cos B - \sin A\sin B$$
$$\cos(A - B) = \cos A\cos B + \sin A\sin B$$

$$\tan(A + B) = \frac{\tan A + \tan B}{1 - \tan A\tan B}$$
$$\tan(A - B) = \frac{\tan A - \tan B}{1 + \tan A\tan B}$$

$$\sin\left(A - \frac{\pi}{2}\right) = -\cos A, \qquad \cos\left(A - \frac{\pi}{2}\right) = \sin A$$
$$\sin\left(A + \frac{\pi}{2}\right) = \cos A, \qquad \cos\left(A + \frac{\pi}{2}\right) = -\sin A$$

$$\sin A\sin B = \tfrac{1}{2}\cos(A - B) - \tfrac{1}{2}\cos(A + B)$$
$$\cos A\cos B = \tfrac{1}{2}\cos(A - B) + \tfrac{1}{2}\cos(A + B)$$
$$\sin A\cos B = \tfrac{1}{2}\sin(A - B) + \tfrac{1}{2}\sin(A + B)$$

$$\sin A + \sin B = 2\sin\tfrac{1}{2}(A + B)\cos\tfrac{1}{2}(A - B)$$
$$\sin A - \sin B = 2\cos\tfrac{1}{2}(A + B)\sin\tfrac{1}{2}(A - B)$$

$$\cos A + \cos B = 2\cos\tfrac{1}{2}(A + B)\cos\tfrac{1}{2}(A - B)$$
$$\cos A - \cos B = -2\sin\tfrac{1}{2}(A + B)\sin\tfrac{1}{2}(A - B)$$

2. Common Reference Triangles

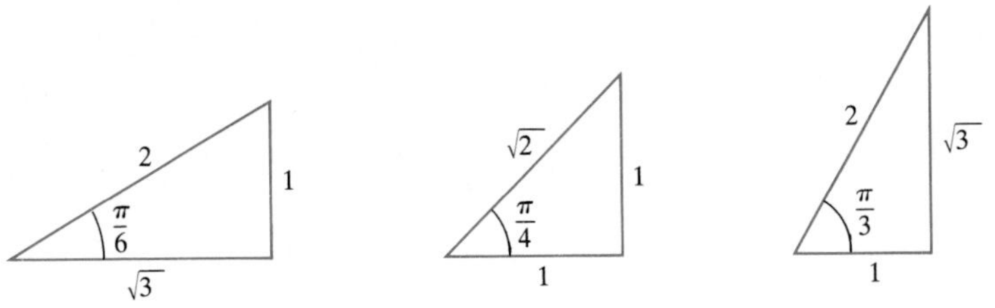

3. Angles and Sides of a Triangle

Law of cosines: $c^2 = a^2 + b^2 - 2ab \cos C$

Law of sines: $\dfrac{\sin A}{a} = \dfrac{\sin B}{b} = \dfrac{\sin C}{c}$

Area $= \frac{1}{2}bc \sin A = \frac{1}{2}ac \sin B = \frac{1}{2}ab \sin C$

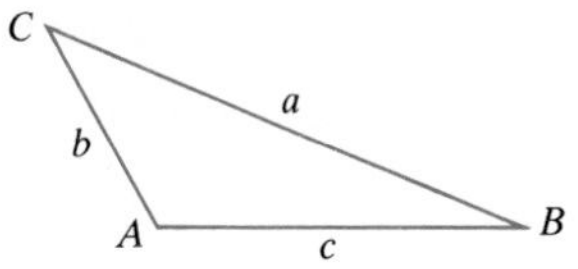

***Toolkit* programs**

Function Evaluator   Super * Grapher

A.8
Invented Number Systems. Complex Numbers

In this appendix we shall discuss complex numbers. These are expressions of the form $a + ib$, where a and b are "real" numbers and i is a symbol for $\sqrt{-1}$. Unfortunately, the words "real" and "imaginary" have connotations that somehow place $\sqrt{-1}$ in a less favorable position than $\sqrt{2}$ in our minds. As a matter of fact, a good deal of imagination, in the sense of *inventiveness,* has been required to construct the *real* number system, which forms the basis of the calculus. In this article we shall review the various stages of this invention. The further invention of a complex number system will then not seem so strange. It is fitting for us to study such a system, since modern engineering has found therein a convenient language for describing vibratory motion, harmonic oscillation, damped vibrations, alternating currents, and other wave phenomena.

The earliest stage of number development was the recognition of the *counting numbers* 1, 2, 3, . . . , which we now call the *natural numbers,* or the *positive integers.* Certain simple arithmetical operations can be performed with these numbers without getting outside the system. That is, the system of positive integers is *closed* with respect to the operations of *addition* and *multiplication*. By this we mean that if m and n are any positive integers, then

$$m + n = p \quad \text{and} \quad mn = q \tag{1}$$

are also positive integers. Given the two positive integers on the *left-hand side* of either equation in (1), we can find the corresponding positive integer on the right. More than this, we may sometimes specify the positive integers m and p and find a positive integer n such that $m + n = p$. For instance, $3 + n = 7$ can be *solved* when the only numbers we know are the positive integers. But the equation $7 + n = 3$ cannot be solved unless the number system is enlarged. The number concepts that we denote by zero and the *negative* integers were invented to solve equations like that. In a civilization that recognizes all the integers

$$\ldots, -3, -2, -1, 0, 1, 2, 3, \ldots, \tag{2}$$

an educated person may always find the missing integer that solves the equation $m + n = p$ when given the other two integers in the equation.

Suppose our educated people also know how to multiply any two integers of the set in (2). If, in Eqs. (1), they are given m and q, they discover that sometimes they can find n and sometimes they can't. If their *imagination* is still in good working order, they may be inspired to invent still more numbers and introduce fractions, which are just ordered pairs m/n of integers m and n. The number zero has special properties that may bother them for a while, but they ultimately discover that it is handy to have all ratios of integers m/n, excluding only those having zero in the denominator. This system, called the set of *rational numbers*, is now rich enough for them to perform the so-called *rational operations* of arithmetic:

1. a) addition
   b) subtraction
2. a) multiplication
   b) division

on any two numbers in the system, *except that they cannot divide by zero*.

The geometry of the unit square (Fig. A.7) and the Pythagorean Theorem showed that they could construct a geometric line segment which, in terms of some basic unit of length, has length equal to $\sqrt{2}$. Thus they could solve the equation

$$x^2 = 2$$

by a geometric construction. But then they discovered that the line segment representing $\sqrt{2}$ and the line segment representing the unit of length 1 were incommensurable quantities. This means that the ratio $\sqrt{2}/1$ cannot be expressed as the ratio of two *integral* multiples of some other, presumably more fundamental, unit of length. That is, our educated people could not find a rational number solution of the equation $x^2 = 2$.

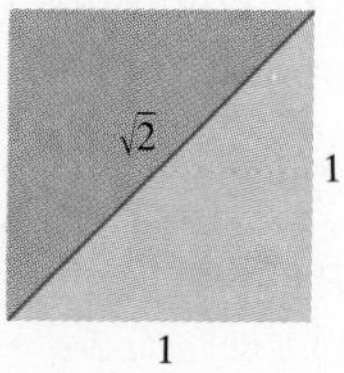

A.7 Segments of irrational length can be constructed with straightedge and compass.

There is a nice algebraic argument that there is no rational number whose square is 2. Suppose that there were such a rational number. Then we could find integers p and q with no common factor other than 1, and such that

$$p^2 = 2q^2. \tag{3}$$

Since p and q are integers, p must then be even, say $p = 2p_1$, where p_1 is an integer. This leads to $2p_1^2 = q^2$, which says that q must also be even, say $q = 2q_1$, where q_1 is also an integer. But this is contrary to our choice of p and q as integers having no common factor other than 1. Hence there is no rational number whose square is 2.

Our educated people *could*, however, get a *sequence* of rational numbers

$$\frac{1}{1}, \frac{7}{5}, \frac{41}{29}, \frac{239}{169}, \ldots, \tag{4}$$

whose squares form a sequence

$$\frac{1}{1}, \frac{49}{25}, \frac{1681}{841}, \frac{57{,}121}{28{,}561}, \ldots, \tag{5}$$

that converges to 2 as its *limit*. This time their *imagination* suggested that they needed the concept of a *limit of a sequence* of rational numbers. If we accept the fact that an increasing sequence that is bounded from above always approaches a limit, and observe that the sequence in (4) has these properties, then we want it to have a limit L. This would also mean, from (5), that $L^2 = 2$, and hence L is *not* one of our rational numbers. If to the *rational* numbers we further add the *limits* of all bounded increasing sequences of rational numbers, we arrive at the system of all "real" numbers. The word *real* is placed in quotes because there is nothing that is either "more real" or "less real" about this system than there is about any other well-defined mathematical system.

Imagination was called upon at many stages during the development of the real number system from the system of positive integers. In fact, the art of invention was needed at least three times in constructing the systems we have discussed so far:

1. The *first invented* system; the set of *all integers* as constructed from the counting numbers.
2. The *second invented* system; the set of *rational* numbers m/n as constructed from the integers.
3. The *third invented* system; the set of all "*real*" numbers x as constructed from the rational numbers.

These invented systems form a hierarchy in which each system contains the previous system. Each system is also richer than its predecessor in that it permits additional operations to be performed without going outside the system. Expressed in algebraic terms, we may say that

1. In the system of all integers, we can solve all equations of the form

$$x + a = 0, \tag{6}$$

   where a may be any integer.

2. In the system of all rational numbers, we can solve all equations of the form

$$ax + b = 0 \tag{7}$$

   provided a and b are rational numbers and $a \neq 0$.

3. In the system of all real numbers, we can solve all of the Eqs. (6) and (7) and, in addition, all quadratic equations

$$ax^2 + bx + c = 0 \quad \text{having} \quad a \neq 0 \quad \text{and} \quad b^2 - 4ac \geq 0. \tag{8}$$

Every student of algebra is familiar with the formula that gives the solutions of (8), namely,

$$x = \frac{-b \pm \sqrt{b^2 - 4ac}}{2a}, \tag{9}$$

and familiar with the further fact that when the discriminant, $d = b^2 - 4ac$, is *negative*, the solutions in (9) do *not* belong to any of the systems discussed above. In fact, the very simple quadratic equation

$$x^2 + 1 = 0 \tag{10}$$

is impossible to solve if the only number systems that can be used are the three invented systems mentioned so far.

Thus we come to the *fourth invented* system, the set of all complex numbers $a + ib$. We could, in fact, dispense entirely with the symbol i and use a notation like (a, b). We would then speak simply of a pair of real numbers a and b. Since, under algebraic operations, the numbers a and b are treated somewhat differently, it is essential to keep the *order* straight. We therefore might say that *the complex number system consists of the set of all ordered pairs of real numbers* (a, b), together with the rules by which they are to be equated, added, multiplied, and so on, listed below. We shall use both the (a, b) notation and the notation $a + ib$. We call a the "real part" and b the "imaginary part" of (a, b). We make the following definitions.

Equality	
$a + ib = c + id$ if and only if $a = c$ and $b = d$.	Two complex numbers (a, b) and (c, d) are *equal* if and only if $a = c$ and $b = d$.
Addition	
$(a + ib) + (c + id)$ $= (a + c) + i(b + d)$	The sum of the two complex numbers (a, b) and (c, d) is the complex number $(a + c, b + d)$.
Multiplication	
$(a + ib)(c + id)$ $= (ac - bd) + i(ad + bc)$	The product of two complex numbers (a, b) and (c, d) is the complex number $(ac - bd, ad + bc)$.
$c(a + ib) = ac + i(bc)$	The product of a real number c and the complex number (a, b) is the complex number (ac, bc).

The set of all complex numbers (a, b) in which the second number is zero has all the properties of the set of ordinary "real" numbers a. For example, addition and multiplication of $(a, 0)$ and $(c, 0)$ give

$$(a, 0) + (c, 0) = (a + c, 0),$$
$$(a, 0) \cdot (c, 0) = (ac, 0),$$

which are numbers of the same type with "imaginary part" equal to zero. Also, if we multiply a "real number" $(a, 0)$ and the "complex number" (c, d), we get

$$(a, 0) \cdot (c, d) = (ac, cd) = a(c, d).$$

In particular, the complex number $(0, 0)$ plays the role of zero in the complex number system and the complex number $(1, 0)$ plays the role of unity.

The number pair $(0, 1)$, which has "real part" equal to zero and "imaginary part" equal to one has the property that its square,

$$(0, 1)(0, 1) = (-1, 0),$$

has "real part" equal to minus one and "imaginary part" equal to zero. Therefore, in the system of complex numbers (a, b), there is a number $x = (0, 1)$ whose square can be added to unity $= (1, 0)$ to produce zero $= (0, 0)$; that is,

$$(0, 1)^2 + (1, 0) = (0, 0).$$

The equation

$$x^2 + 1 = 0$$

therefore has a solution $x = (0, 1)$ in this new number system.

You are probably more familiar with the $a + ib$ notation than you are with the notation (a, b). And since the laws of algebra for the ordered pairs enable us to write

$$(a, b) = (a, 0) + (0, b) = a(1, 0) + b(0, 1),$$

while $(1, 0)$ behaves like unity and $(0, 1)$ behaves like a square root of minus one, we need not hesitate to write $a + ib$ in place of (a, b). The i associated with b is like a tracer element that tags the "imaginary part" of $a + ib$. We can pass at will from the realm of ordered pairs (a, b) to the realm of expressions $a + ib$, and conversely. But there is nothing less "real" about the symbol $(0, 1) = i$ than there is about the symbol $(1, 0) = 1$, once we have learned the laws of algebra in the complex number system (a, b).

To reduce any rational combination of complex numbers to a single complex number, we need only apply the laws of elementary algebra, replacing i^2 wherever it appears by -1. Of course, we cannot divide by the complex number $(0, 0) = 0 + i0$. But if $a + ib \neq 0$, then we may carry out a division as follows:

$$\frac{c + id}{a + ib} = \frac{(c + id)(a - ib)}{(a + ib)(a - ib)} = \frac{(ac + bd) + i(ad - bc)}{a^2 + b^2}.$$

The result is a complex number $x + iy$ with

$$x = \frac{ac + bd}{a^2 + b^2}, \qquad y = \frac{ad - bc}{a^2 + b^2},$$

and $a^2 + b^2 \neq 0$, since $a + ib = (a, b) \neq (0, 0)$.

The number $a - ib$ that is used as multiplier to clear the i out of the denominator is called the *complex conjugate* of $a + ib$. It is customary to use $\bar{z}$ (read "z bar") to denote the complex conjugate of z; thus

$$z = a + ib, \qquad \bar{z} = a - ib.$$

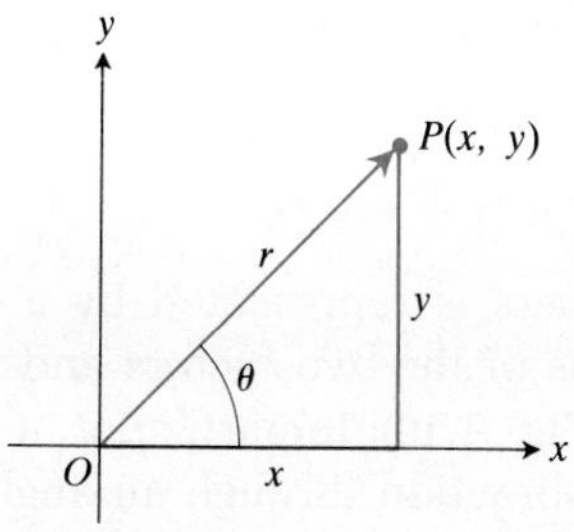

A.8 An Argand diagram representing $z = x + iy$ both as a point $P(x, y)$ and as a vector $\overrightarrow{OP}$.

Thus, we multiplied the numerator and denominator of the complex fraction $(c + id)/(a + ib)$ by the complex conjugate of the denominator. This will always replace the denominator by a real number.

Argand Diagrams There are two geometric representations of the complex number $z = x + iy$:

a) as the point $P(x, y)$ in the xy-plane, and

b) as the vector $\overrightarrow{OP}$ from the origin to P.

In each representation, the x-axis is called the "axis of reals" and the y-axis is the "imaginary axis." Both representations are called *Argand diagrams* (Fig. A.8).

In terms of the polar coordinates of x and y, we have

$$x = r\cos\theta, \qquad y = r\sin\theta,$$

and

$$z = x + iy = r(\cos\theta + i\sin\theta). \tag{11}$$

We define the *absolute value* of a complex number $x + iy$ to be the length r of a vector $\overrightarrow{OP}$ from the origin to $P(x, y)$. We denote the absolute value by vertical bars, thus:

$$|x + iy| = \sqrt{x^2 + y^2}. \tag{12a}$$

If we always choose the polar coordinates r and θ so that r is nonnegative, then we have

$$r = |x + iy|. \tag{12b}$$

The polar angle θ is called the *argument* of z and written $\theta = \arg z$. Of course, any integral multiple of 2π may be added to θ to produce another appropriate angle. The *principal value* of the argument will, in this book, be taken to be that value of θ for which $-\pi < \theta \leq +\pi$.

The following equation gives a useful formula connecting a complex number z, its conjugate $\bar{z}$, and its absolute value $|z|$, namely,

$$z \cdot \bar{z} = |z|^2. \tag{13}$$

The identity

$$e^{i\theta} = \cos\theta + i\sin\theta \tag{14}$$

introduced in Article 12.2 leads to the following rules for calculating products, quotients, powers, and roots of complex numbers. It also leads to Argand diagrams for $e^{i\theta}$. Since $\cos\theta + i\sin\theta$ is what we get from Eq. (11) by taking $r = 1$, we can say that $e^{i\theta}$ is represented by a unit vector that makes an angle θ with the positive x-axis, as shown in Fig. A.9.

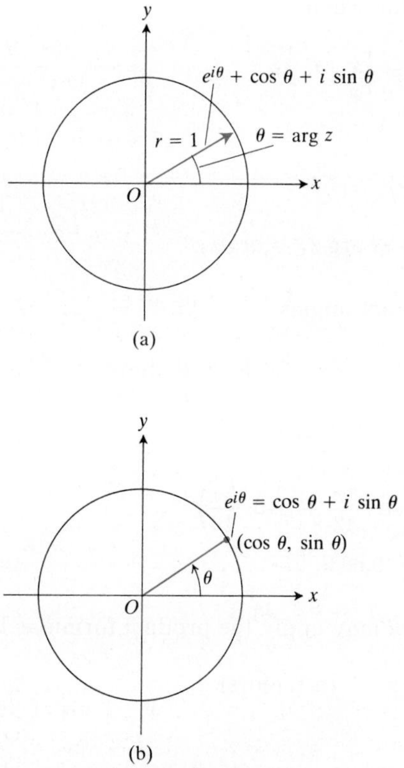

A.9 Argand diagrams for $e^{i\theta} = \cos\theta + i\sin\theta$ (a) as a vector, (b) as a point.

Products Let

$$z_1 = r_1 e^{i\theta_1} \qquad z_2 = r_2 e^{i\theta_2} \tag{15}$$

so that

$$|z_1| = r_1, \quad \arg z_1 = \theta_1; \qquad |z_2| = r_2, \quad \arg z_2 = \theta_2. \tag{16}$$

Then

$$z_1 z_2 = r_1 e^{i\theta_1} \cdot r_2 e^{i\theta_2} = r_1 r_2 e^{i(\theta_1 + \theta_2)}$$

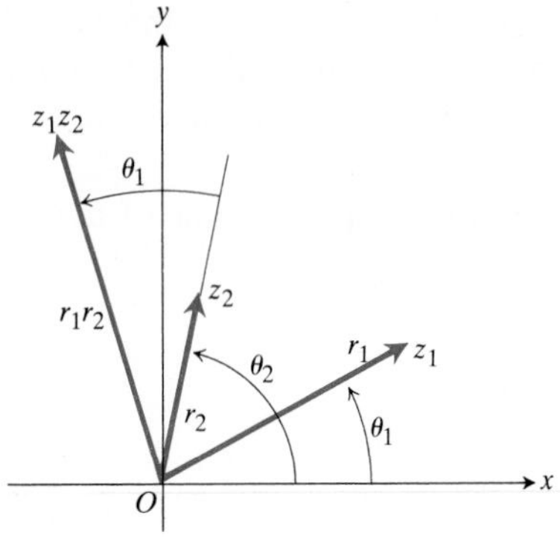

A.10 When z_1 and z_2 are multiplied, $|z_1z_2| = r_1 \cdot r_2$, and $\arg(z_1z_2) = \theta_1 + \theta_2$.

and hence

$$|z_1z_2| = r_1r_2 = |z_1| \cdot |z_2|,$$
$$\arg(z_1z_2) = \theta_1 + \theta_2 = \arg z_1 + \arg z_2. \tag{17}$$

Thus the product of two complex numbers is represented by a vector whose length is the product of the lengths of the two factors and whose argument is the sum of their arguments (Fig. A.10). In particular, a vector may be rotated in the counterclockwise direction through an angle θ by simply multiplying it by $e^{i\theta}$. Multiplication by i rotates 90°, by -1 rotates 180°, by $-i$ rotates 270°, etc.

EXAMPLE 1 Let

$$z_1 = 1 + i, \qquad z_2 = \sqrt{3} - i.$$

We plot these complex numbers in an Argand diagram (Fig. A.11) from which we read off the polar representations

$$z_1 = \sqrt{2}e^{i\pi/4}, \qquad z_2 = 2e^{-i\pi/6}.$$

Then

$$z_1z_2 = 2\sqrt{2}\exp\left(\frac{i\pi}{4} - \frac{i\pi}{6}\right) = 2\sqrt{2}\exp\left(\frac{i\pi}{12}\right)$$
$$= 2\sqrt{2}\left(\cos\frac{\pi}{12} + i\sin\frac{\pi}{12}\right) \approx 2.73 + 0.73i. \ \square$$

Quotients Suppose $r_2 \neq 0$ in Eq. (15). Then

$$\frac{z_1}{z_2} = \frac{r_1e^{i\theta_1}}{r_2e^{i\theta_2}} = \frac{r_1}{r_2}e^{i(\theta_1-\theta_2)}.$$

Hence

$$\left|\frac{z_1}{z_2}\right| = \frac{r_1}{r_2} = \frac{|z_1|}{|z_2|},$$
$$\arg\left(\frac{z_1}{z_2}\right) = \theta_1 - \theta_2 = \arg z_1 - \arg z_2.$$

That is, we divide lengths and subtract angles.

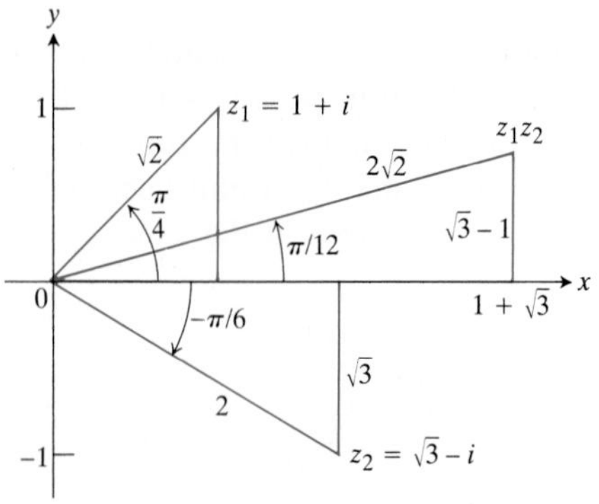

A.11 To multiply two complex numbers, one multiplies their absolute values, and adds their arguments.

EXAMPLE 2 Let $z_1 = 1 + i$ and $z_2 = \sqrt{3} - i$, as in Example 1. Then,

$$\frac{1+i}{\sqrt{3}-i} = \frac{\sqrt{2}e^{i\pi/4}}{2e^{-i\pi/6}} = \frac{\sqrt{2}}{2}e^{5\pi i/12}$$
$$\approx 0.707\left(\cos\frac{5\pi}{12} + i\sin\frac{5\pi}{12}\right)$$
$$\approx 0.183 + 0.683i. \ \square$$

Powers If n is a positive integer, we may apply the product formulas in (17) to find

$$z^n = z \cdot z \cdot \cdots \cdot z \qquad (n \text{ factors}).$$

With $z = re^{i\theta}$, we obtain

$$z^n = (re^{i\theta})^n = r^ne^{i(\theta+\theta+\cdots+\theta)} \qquad (n \text{ summands})$$
$$= r^ne^{in\theta}. \tag{18}$$

The length $r = |z|$ is raised to the nth power and the angle $\theta = \arg z$ is multiplied by n.

In particular, if we place $r = 1$ in Eq. (18), we obtain De Moivre's theorem.

THEOREM

De Moivre's Theorem

$$(\cos\theta + i\sin\theta)^n = \cos n\theta + i\sin n\theta. \tag{19}$$

If we expand the left-hand side of this equation by the binomial theorem and reduce it to the form $a + ib$, we obtain formulas for $\cos n\theta$ and $\sin n\theta$ as polynomials of degree n in $\cos\theta$ and $\sin\theta$.

EXAMPLE 4 If $n = 3$ in Eq. (19), we have

$$(\cos\theta + i\sin\theta)^3 = \cos 3\theta + i\sin 3\theta.$$

The left-hand side of this equation is

$$\cos^3\theta + 3i\cos^2\theta\sin\theta - 3\cos\theta\sin^2\theta - i\sin^3\theta.$$

The real part of this must equal $\cos 3\theta$ and the imaginary part must equal $\sin 3\theta$. Therefore,

$$\cos 3\theta = \cos^3\theta - 3\cos\theta\sin^2\theta,$$
$$\sin 3\theta = 3\cos^2\theta\sin\theta - \sin^3\theta. \;\square$$

Roots If $z = re^{i\theta}$ is a complex number different from zero and n is a positive integer, then there are precisely n different complex numbers w_0, $w_1, \ldots, w_{n-1}$, that are nth roots of z. To see why, let $w = \rho e^{i\alpha}$ be an nth root of $z = re^{i\theta}$, so that

$$w^n = z$$

or

$$\rho^n e^{in\alpha} = re^{i\theta}. \tag{20}$$

Then

$$\rho = \sqrt[n]{r} \tag{21}$$

is the real, positive, nth root of r. As regards the angle, although we cannot say that $n\alpha$ and θ must be equal, we can say that they may differ only by an integral multiple of 2π. That is,

$$n\alpha = \theta + 2k\pi, \qquad k = 0, \pm 1, \pm 2, \ldots \tag{22}$$

Therefore

$$\alpha = \frac{\theta}{n} + k\frac{2\pi}{n}.$$

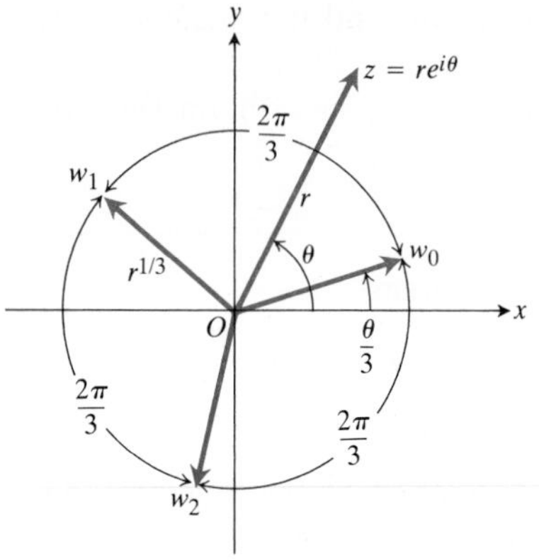

A.12 The three cube roots of $z = re^{i\theta}$.

Hence all the nth roots of $z = re^{i\theta}$ are given by

$$\sqrt[n]{re^{i\theta}} = \sqrt[n]{r}\,\exp\left(\frac{\theta}{n} + k\frac{2\pi}{n}\right), \qquad k = 0, \pm 1, \pm 2, \ldots . \quad (23)$$

There might appear to be infinitely many different answers corresponding to the infinitely many possible values of k. But one readily sees that $k = n + m$ gives the same answer as $k = m$ in Eq. (23). Thus we need only take n consecutive values for k to obtain all the different nth roots of z. For convenience, we may take

$$k = 0, 1, 2, \ldots, n - 1.$$

All the nth roots of $re^{i\theta}$ lie on a circle centered at the origin O and having radius equal to the real, positive nth root of r. One of them has argument $\alpha = \theta/n$. The others are uniformly spaced around the circle, each being separated from its neighbors by an angle equal to $2\pi/n$. Figure A.12 illustrates the placement of the three cube roots, w_0, w_1, w_2, of the complex number $z = re^{i\theta}$,

EXAMPLE 5 Find the four fourth roots of -16.

Solution As our first step, we plot the given number in an Argand diagram (Fig. A.13) and determine its polar representation $re^{i\theta}$. Here,

$$z = -16, \qquad r = +16, \qquad \theta = \pi.$$

One of the fourth roots of $16\,e^{i\pi}$ is $2\,e^{i\pi/4}$. We obtain others by successive additions of $2\pi/4 = \pi/2$ to the argument of this first one. Hence

$$\sqrt[4]{16 \exp i\pi} = 2 \exp i\left(\frac{\pi}{4}, \frac{3\pi}{4}, \frac{5\pi}{4}, \frac{7\pi}{4}\right),$$

and the four roots are

$$w_0 = 2\left[\cos\frac{\pi}{4} + i\sin\frac{\pi}{4}\right] = \sqrt{2}(1 + i),$$

$$w_1 = 2\left[\cos\frac{3\pi}{4} + i\sin\frac{3\pi}{4}\right] = \sqrt{2}(-1 + i),$$

$$w_2 = 2\left[\cos\frac{5\pi}{4} + i\sin\frac{5\pi}{4}\right] = \sqrt{2}(-1 - i),$$

$$w_3 = 2\left[\cos\frac{7\pi}{4} + i\sin\frac{7\pi}{4}\right] = \sqrt{2}(1 - i). \ \square$$

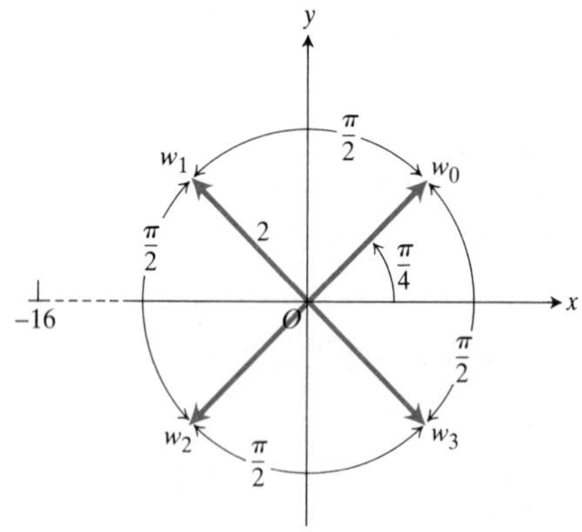

A.13 The four fourth roots of -16.

The Fundamental Theorem of Algebra One may well say that the invention of $\sqrt{-1}$ is all well and good and leads to a number system that is richer than the real number system alone; but where will this process end? Are we also going to invent still more systems so as to obtain $\sqrt[4]{-1}$, $\sqrt[6]{-1}$, and so on? By now it should be clear that this is not necessary. These numbers are already expressible in terms of the complex number system $a + ib$. In fact, the Fundamental Theorem of Algebra says that with the introduction of the complex numbers we now have enough numbers to factor every polynomial into a product of linear factors and hence enough numbers to solve every possible polynomial equation.

THEOREM

The Fundamental Theorem of Algebra

Every polynomial equation of the form

$$a_0z^n + a_1z^{n-1} + a_2z^{n-2} + \cdots + a_{n-1}z + a_n = 0,$$

in which the coefficients $a_0, a_1, \ldots, a_n$ are any complex numbers, whose degree n is greater than or equal to one, and whose leading coefficient a_0 is not zero, has exactly n roots in the complex number system, provided each multiple root of multiplicity m is counted as m roots.

This theorem is rather difficult to prove, and we can only state it here.

PROBLEMS

1. Find $(a, b) \cdot (c, d) = (ac - bd, ad + bc)$.
a) $(2, 3) \cdot (4, -2)$ b) $(2, -1) \cdot (-2, 3)$
c) $(-1, -2) \cdot (2, 1)$
(*Note:* This is how complex numbers are multiplied by computers.)

2. Solve the following equations for the real numbers x and y.
a) $(3 + 4i)^2 - 2(x - iy) = x + iy$
b) $\left(\frac{1+i}{1-i}\right)^2 + \frac{1}{x + iy} = 1 + i$
c) $(3 - 2i)(x + iy) = 2(x - 2iy) + 2i - 1$

3. Show with an Argand diagram that the law for adding complex numbers is the same as the parallelogram law for adding vectors.

4. How may the following complex numbers be obtained from $z = x + iy$ geometrically? Sketch.
a) $\bar{z}$ b) $\overline{(-z)}$
c) $-z$ d) $1/z$

5. Show that the conjugate of the sum (product, or quotient) of two complex numbers z_1 and z_2 is the same as the sum (product, or quotient) of their conjugates.

6. a) Extend the results of Problem 5 to show that

$f(\bar{z}) = \overline{f(z)}$ if

$$f(z) = a_0z^n + a_1z^{n-1} + \cdots + a_{n-1}z + a_n$$

is a polynomial with real coefficients $a_0, \ldots, a_n$.
b) If z is a root of the equation $f(z) = 0$, where $f(z)$ is a polynomial with real coefficients as in part (a) above, show that the conjugate $\bar{z}$ is also a root of the equation. (*Hint:* Let $f(z) = u + iv = 0$; then both u and v are zero. Now use the fact that $f(\bar{z}) = \overline{f(z)} = u - iv$.)

7. Show that $|\bar{z}| = |z|$.

8. If z and $\bar{z}$ are equal, what can you say about the location of the point z in the complex plane?

9. Let $R(z)$, $I(z)$ denote respectively the real and imaginary parts of z. Show that
a) $z + \bar{z} = 2R(z)$, b) $z - \bar{z} = 2iI(z)$,
c) $|R(z)| \leq |z|$,
d) $|z_1 + z_2|^2 = |z_1|^2 + |z_2|^2 + 2R(z_1\bar{z}_2)$,
e) $|z_1 + z_2| \leq |z_1| + |z_2|$.

10. Show that the distance between the two points z_1 and z_2 in an Argand diagram is equal to $|z_1 - z_2|$.

In Problems 11–15, graph the points $z = x + iy$ that satisfy the given conditions.

11. a) $|z| = 2$ b) $|z| < 2$ c) $|z| > 2$

12. $|z - 1| = 2$

13. $|z + 1| = 1$

14. $|z + 1| = |z - 1|$

15. $|z + i| = |z - 1|$

Express the answer to each of the Problems 16–19 in the form $re^{i\theta}$, with $r \geq 0$ and $-\pi < \theta \leq \pi$. Sketch.

16. $(1 + \sqrt{-3})^2$

17. $\frac{1+i}{1-i}$

18. $\frac{1 + i\sqrt{3}}{1 - i\sqrt{3}}$

19. $(2 + 3i)(1 - 2i)$

20. Use De Moivre's theorem (Eq. 19) to express $\cos 4\theta$ and $\sin 4\theta$ as polynomials in $\cos\theta$ and $\sin\theta$.

21. Find the three cube roots of unity.

22. Find the two square roots of i.

23. Find the three cube roots of $-8i$.

24. Find the six sixth roots of 64.

25. Find the four roots of the equation $z^4 - 2z^2 + 4 = 0$.

26. Find the six roots of the equation $z^6 + 2z^3 + 2 = 0$.

27. Find all roots of the equation $x^4 + 4x^2 + 16 = 0$.

28. Solve: $x^4 + 1 = 0$.

***Toolkit* programs**

Complex Number Calculator

A.9
Tables

Table 1 Natural trigonometric functions

Angle					Angle				
Degree	**Radian**	**Sine**	**Cosine**	**Tangent**	**Degree**	**Radian**	**Sine**	**Cosine**	**Tangent**
0°	0.000	0.000	1.000	0.000					
1°	0.017	0.017	1.000	0.017	46°	0.803	0.719	0.695	1.036
2°	0.035	0.035	0.999	0.035	47°	0.820	0.731	0.682	1.072
3°	0.052	0.052	0.999	0.052	48°	0.838	0.743	0.669	1.111
4°	0.070	0.070	0.998	0.070	49°	0.855	0.755	0.656	1.150
5°	0.087	0.087	0.996	0.087	50°	0.873	0.766	0.643	1.192
6°	0.105	0.105	0.995	0.105	51°	0.890	0.777	0.629	1.235
7°	0.122	0.122	0.993	0.123	52°	0.908	0.788	0.616	1.280
8°	0.140	0.139	0.990	0.141	53°	0.925	0.799	0.602	1.327
9°	0.157	0.156	0.988	0.158	54°	0.942	0.809	0.588	1.376
10°	0.175	0.174	0.985	0.176	55°	0.960	0.819	0.574	1.428
11°	0.192	0.191	0.982	0.194	56°	0.977	0.829	0.559	1.483
12°	0.209	0.208	0.978	0.213	57°	0.995	0.839	0.545	1.540
13°	0.227	0.225	0.974	0.231	58°	1.012	0.848	0.530	1.600
14°	0.244	0.242	0.970	0.249	59°	1.030	0.857	0.515	1.664
15°	0.262	0.259	0.966	0.268	60°	1.047	0.866	0.500	1.732
16°	0.279	0.276	0.961	0.287	61°	1.065	0.875	0.485	1.804
17°	0.297	0.292	0.956	0.306	62°	1.082	0.883	0.469	1.881
18°	0.314	0.309	0.951	0.325	63°	1.100	0.891	0.454	1.963
19°	0.332	0.326	0.946	0.344	64°	1.117	0.899	0.438	2.050
20°	0.349	0.342	0.940	0.364	65°	1.134	0.906	0.423	2.145
21°	0.367	0.358	0.934	0.384	66°	1.152	0.914	0.407	2.246
22°	0.384	0.375	0.927	0.404	67°	1.169	0.921	0.391	2.356
23°	0.401	0.391	0.921	0.424	68°	1.187	0.927	0.375	2.475
24°	0.419	0.407	0.914	0.445	69°	1.204	0.934	0.358	2.605
25°	0.436	0.423	0.906	0.466	70°	1.222	0.940	0.342	2.748
26°	0.454	0.438	0.899	0.488	71°	1.239	0.946	0.326	2.904
27°	0.471	0.454	0.891	0.510	72°	1.257	0.951	0.309	3.078
28°	0.489	0.469	0.883	0.532	73°	1.274	0.956	0.292	3.271
29°	0.506	0.485	0.875	0.554	74°	1.292	0.961	0.276	3.487
30°	0.524	0.500	0.866	0.577	75°	1.309	0.966	0.259	3.732
31°	0.541	0.515	0.857	0.601	76°	1.326	0.970	0.242	4.011
32°	0.559	0.530	0.848	0.625	77°	1.344	0.974	0.225	4.332
33°	0.576	0.545	0.839	0.649	78°	1.361	0.978	0.208	4.705
34°	0.593	0.559	0.829	0.675	79°	1.379	0.982	0.191	5.145
35°	0.611	0.574	0.819	0.700	80°	1.396	0.985	0.174	5.671
36°	0.628	0.588	0.809	0.727	81°	1.414	0.988	0.156	6.314
37°	0.646	0.602	0.799	0.754	82°	1.431	0.990	0.139	7.115
38°	0.663	0.616	0.788	0.781	83°	1.449	0.993	0.122	8.144
39°	0.681	0.629	0.777	0.810	84°	1.466	0.995	0.105	9.514
40°	0.698	0.643	0.766	0.839	85°	1.484	0.996	0.087	11.43
41°	0.716	0.656	0.755	0.869	86°	1.501	0.998	0.070	14.30
42°	0.733	0.669	0.743	0.900	87°	1.518	0.999	0.052	19.08
43°	0.750	0.682	0.731	0.933	88°	1.536	0.999	0.035	28.64
44°	0.768	0.695	0.719	0.966	89°	1.553	1.000	0.017	57.29
45°	0.785	0.707	0.707	1.000	90°	1.571	1.000	0.000	

Table 2 Exponential functions

x	e^x	e^{-x}	x	e^x	e^{-x}
0.00	1.0000	1.0000	2.5	12.182	0.0821
0.05	1.0513	0.9512	2.6	13.464	0.0743
0.10	1.1052	0.9048	2.7	14.880	0.0672
0.15	1.1618	0.8607	2.8	16.445	0.0608
0.20	1.2214	0.8187	2.9	18.174	0.0550
0.25	1.2840	0.7788	3.0	20.086	0.0498
0.30	1.3499	0.7408	3.1	22.198	0.0450
0.35	1.4191	0.7047	3.2	24.533	0.0408
0.40	1.4918	0.6703	3.3	27.113	0.0369
0.45	1.5683	0.6376	3.4	29.964	0.0334
0.50	1.6487	0.6065	3.5	33.115	0.0302
0.55	1.7333	0.5769	3.6	36.598	0.0273
0.60	1.8221	0.5488	3.7	40.447	0.0247
0.65	1.9155	0.5220	3.8	44.701	0.0224
0.70	2.0138	0.4966	3.9	49.402	0.0202
0.75	2.1170	0.4724	4.0	54.598	0.0183
0.80	2.2255	0.4493	4.1	60.340	0.0166
0.85	2.3396	0.4274	4.2	66.686	0.0150
0.90	2.4596	0.4066	4.3	73.700	0.0136
0.95	2.5857	0.3867	4.4	81.451	0.0123
1.0	2.7183	0.3679	4.5	90.017	0.0111
1.1	3.0042	0.3329	4.6	99.484	0.0101
1.2	3.3201	0.3012	4.7	109.95	0.0091
1.3	3.6693	0.2725	4.8	121.51	0.0082
1.4	4.0552	0.2466	4.9	134.29	0.0074
1.5	4.4817	0.2231	5	148.41	0.0067
1.6	4.9530	0.2019	6	403.43	0.0025
1.7	5.4739	0.1827	7	1096.6	0.0009
1.8	6.0496	0.1653	8	2981.0	0.0003
1.9	6.6859	0.1496	9	8103.1	0.0001
2.0	7.3891	0.1353	10	22026	0.00005
2.1	8.1662	0.1225			
2.2	9.0250	0.1108			
2.3	9.9742	0.1003			
2.4	11.023	0.0907			

Table 3 Natural logarithms of numbers

n	$\log_e n$	n	$\log_e n$	n	$\log_e n$	n	$\log_e n$
0.0	*	3.5	1.2528	7.0	1.9459	15	2.7081
0.1	7.6974	3.6	1.2809	7.1	1.9601	16	2.7726
0.2	8.3906	3.7	1.3083	7.2	1.9741	17	2.8332
0.3	8.7960	3.8	1.3350	7.3	1.9879	18	2.8904
0.4	9.0837	3.9	1.3610	7.4	2.0015	19	2.9444
0.5	9.3069	4.0	1.3863	7.5	2.0149	20	2.9957
0.6	9.4892	4.1	1.4110	7.6	2.0281	25	3.2189
0.7	9.6433	4.2	1.4351	7.7	2.0412	30	3.4012
0.8	9.7769	4.3	1.4586	7.8	2.0541	35	3.5553
0.9	9.8946	4.4	1.4816	7.9	2.0669	40	3.6889
1.0	0.0000	4.5	1.5041	8.0	2.0794	45	3.8067
1.1	0.0953	4.6	1.5261	8.1	2.0919	50	3.9120
1.2	0.1823	4.7	1.5476	8.2	2.1041	55	4.0073
1.3	0.2624	4.8	1.5686	8.3	2.1163	60	4.0943
1.4	0.3365	4.9	1.5892	8.4	2.1282	65	4.1744
1.5	0.4055	5.0	1.6094	8.5	2.1401	70	4.2485
1.6	0.4700	5.1	1.6292	8.6	2.1518	75	4.3175
1.7	0.5306	5.2	1.6487	8.7	2.1633	80	4.3820
1.8	0.5878	5.3	1.6677	8.8	2.1748	85	4.4427
1.9	0.6419	5.4	1.6864	8.9	2.1861	90	4.4998
2.0	0.6931	5.5	1.7047	9.0	2.1972	95	4.5539
2.1	0.7419	5.6	1.7228	9.1	2.2083	100	4.6052
2.2	0.7885	5.7	1.7405	9.2	2.2192		
2.3	0.8329	5.8	1.7579	9.3	2.2300		
2.4	0.8755	5.9	1.7750	9.4	2.2407		
2.5	0.9163	6.0	1.7918	9.5	2.2513		
2.6	0.9555	6.1	1.8083	9.6	2.2618		
2.7	0.9933	6.2	1.8245	9.7	2.2721		
2.8	1.0296	6.3	1.8405	9.8	2.2824		
2.9	1.0647	6.4	1.8563	9.9	2.2925		
3.0	1.0986	6.5	1.8718	10	2.3026		
3.1	1.1314	6.6	1.8871	11	2.3979		
3.2	1.1632	6.7	1.9021	12	2.4849		
3.3	1.1939	6.8	1.9169	13	2.5649		
3.4	1.2238	6.9	1.9315	14	2.6391		

*Subtract 10 from $\log_e n$ entries for $n < 1.0$.

***Toolkit* programs**

Function Evaluator

A.10 The Distributive Law for Vector Cross Products

In this appendix we prove the distributive law

$$\mathbf{A} \times (\mathbf{B} + \mathbf{C}) = \mathbf{A} \times \mathbf{B} + \mathbf{A} \times \mathbf{C} \tag{1}$$

from Article 13.7.

Proof. To see that Eq. (1) is valid, we interpret the cross product $\mathbf{A} \times \mathbf{B}$ in a slightly different way. The vectors **A** and **B** are drawn from the common point O and a plane M is constructed perpendicular to **A** at O (Fig. A.14). Vector **B** is now projected orthogonally onto M, yielding a vector $\mathbf{B}'$ whose length is $|\mathbf{B}| \sin\theta$. The vector $\mathbf{B}'$ is then rotated 90° about **A** in the positive sense to produce a vector $\mathbf{B}''$. Finally, $\mathbf{B}''$ is multiplied by the length of **A**. The resulting vector $|\mathbf{A}|\mathbf{B}''$ is equal to $\mathbf{A} \times \mathbf{B}$ since $\mathbf{B}''$ has the same direction as $\mathbf{A} \times \mathbf{B}$ by its construction (Fig. A.14) and

$$|\mathbf{A}||\mathbf{B}''| = |\mathbf{A}||\mathbf{B}'| = |\mathbf{A}||\mathbf{B}| \sin\theta = |\mathbf{A} \times \mathbf{B}|.$$

Now each of these three operations, namely,

1. projection onto M,
2. rotation about **A** through 90°,
3. multiplication by the scalar $|\mathbf{A}|$,

when applied to a triangle whose plane is not parallel to **A**, will produce another triangle. If we start with the triangle whose sides are **B**, **C**, and $\mathbf{B} + \mathbf{C}$ (Fig. A.15) and apply these three steps, we successively obtain:

1. a triangle whose sides are $\mathbf{B}'$, $\mathbf{C}'$, and $(\mathbf{B} + \mathbf{C})'$ satisfying the vector equation
$$\mathbf{B}' + \mathbf{C}' = (\mathbf{B} + \mathbf{C})';$$
2. a triangle whose sides are $\mathbf{B}''$, $\mathbf{C}''$, and $(\mathbf{B} + \mathbf{C})''$ satisfying the vector equation
$$\mathbf{B}'' + \mathbf{C}'' = (\mathbf{B} + \mathbf{C})'';$$

A.14 For reasons explained above. $\mathbf{A} \times \mathbf{B} = |\mathbf{A}|\mathbf{B}''$.

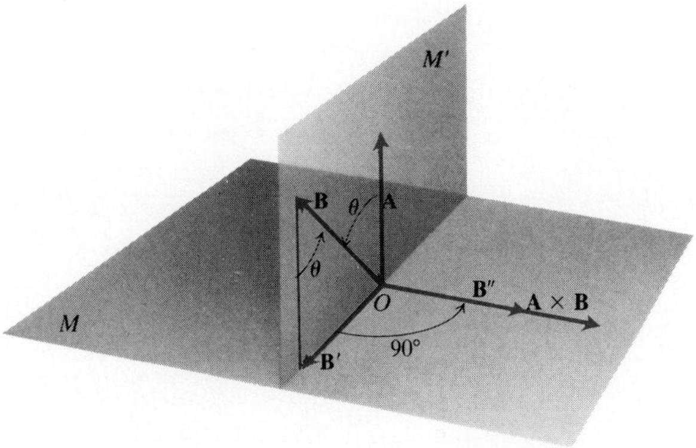

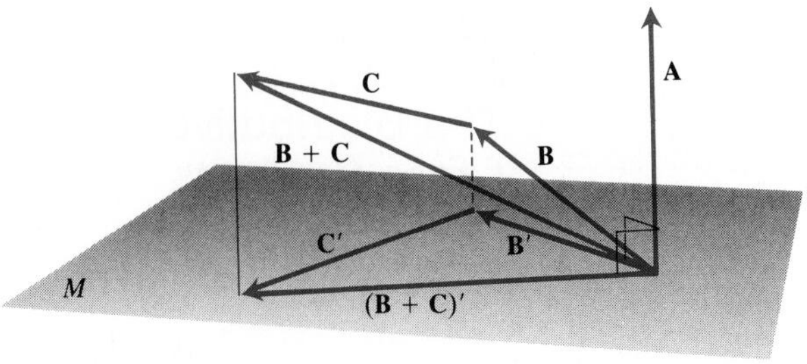

A.15 The vectors **B, C,** and **B** + **C** projected onto a plane perpendicular to **A.**

(the double-prime on each vector has the same meaning as in Fig. A.14); and finally,

3. a triangle whose sides are $|\mathbf{A}|\mathbf{B}''$, $|\mathbf{A}|\mathbf{C}''$, and $|\mathbf{A}|(\mathbf{B}+\mathbf{C})''$ satisfying the vector equation

$$|\mathbf{A}|\mathbf{B}'' + |\mathbf{A}|\mathbf{C}'' = |\mathbf{A}|(\mathbf{B}+\mathbf{C})''. \tag{2}$$

When we use the equations $|\mathbf{A}|\mathbf{B}'' = \mathbf{A}\times\mathbf{B}$, $|\mathbf{A}|\mathbf{C}'' = \mathbf{A}\times\mathbf{C}$ and $|\mathbf{A}|(\mathbf{B}+\mathbf{C})'' = \mathbf{A}\times(\mathbf{B}+\mathbf{C})$, which result from our discussion above, Eq. (2) becomes

$$\mathbf{A}\times\mathbf{B} + \mathbf{A}\times\mathbf{C} = \mathbf{A}\times(\mathbf{B}+\mathbf{C}),$$

which is the distributive law, (1), that we wanted to establish. ■

ANSWERS

CHAPTER 11

Article 11.1, p. 595

1. $\frac{3}{2}, \frac{7}{4}, \frac{15}{8}, \frac{31}{16}, \frac{63}{32}, \frac{127}{64}$ **3.** $1, \frac{1}{2}, \frac{1}{4}, \frac{1}{8}, \frac{1}{16}, \frac{1}{32}$ **5.** 1, 2, 3, 5, 8, 13

Article 11.2, pp. 603–604

1. $0, -\frac{1}{4}, -\frac{2}{9}, -\frac{3}{16}$; converges to 0 **3.** $\frac{1}{3}, \frac{1}{9}, \frac{1}{27}, \frac{1}{81}$; converges to 0 **5.** $1, -\frac{1}{3}, \frac{1}{5}, -\frac{1}{7}$; converges to 0 **7.** 0, −1, 0, 1; diverges **9.** $1, -1/\sqrt{2}, 1/\sqrt{3}, -\frac{1}{2}$; converges to 0 **11.** Converges to 0 **13.** Converges to 1 **15.** Diverges **17.** Converges to $-\frac{2}{3}$ **19.** Converges to $\sqrt{2}$ **21.** Converges to 0 **23.** Diverges **25.** Converges to 1 **27.** Converges to −5 **29.** Converges to 1 **31.** Converges to 1 **33.** Converges to 5 **35.** Converges to 0 **37.** Converges to 0. **39.** Converges to $\sqrt{2}$ **41.** Diverges **43.** Converges to 0 **45.** Diverges **47.** Converges to 0 **49.** Converges to $\pi/2$ **51.** Converges to 1 **53.** Converges to $\frac{1}{2}$ **55.** Use $0 < n!/n^n \leq 1/n$ **57.** Yes. No; the limit is 1 for any $x > 0$. **61.** 1

Article 11.3, pp. 607–608

1. Converges to 0 **3.** Converges to 0 **5.** Converges to 0 **7.** Converges to e^7 **9.** Converges to 0 **11.** Diverges **13.** Converges to 1 **15.** Converges to 0 **17.** Converges to $1/e$ **19.** Converges to 0 **21.** Converges to x **23.** Converges to 1 **25.** Converges to 1 **27.** Converges to $1/e$ **29.** Converges to 0 **31.** Diverges **33.** Converges to $1/(p-1)$ **35.** $N \geq 9124$ **37.** 10^{75}

Article 11.4, pp. 618–619

1. $s_n = 3(1 - (\frac{1}{3})^n)$; $\lim_{n\to\infty} s_n = 3$ **3.** $s_n = (1 - (1/e)^n)/(1 - (1/e))$; $\lim_{n\to\infty} s_n = e/(e-1)$ **5.** $s_n = \frac{1}{3}(1 - (-2)^n)$, diverges **7.** $s_n = -\ln(n+1)$, diverges **9.** a) $\sum_{n=-2}^{\infty} 1/[(n+4)(n+5)]$ b) $\sum_{n=0}^{\infty} 1/[(n+2)(n+3)]$ c) $\sum_{n=5}^{\infty} 1/[(n-3)(n-2)]$ **11.** $\frac{4}{3}$ **13.** $\frac{7}{3}$ **15.** 11.5 **17.** $\frac{5}{3}$ **19.** 1 **21.** $\frac{1}{9}$ **23.** a) $0.234234\ldots = 234 \sum_{n=1}^{\infty} (\frac{1}{1000})^n = \frac{26}{111}$ b) Yes. If $x = 0.a_1a_2 \ldots a_n \overline{a_1a_2 \ldots a_n}$, then $(10^n - 1)x = a_1a_2 \ldots a_n = p$ is an integer and $x = p/q$ with $q = 10^n - 1$. **25.** $2 + \sqrt{2}$ **27.** 1 **29.** Diverges **31.** $e^2/(e^2 - 1)$ **33.** Diverges **35.** $\frac{3}{2}$ **37.** Diverges **39.** $a = 1$, $r = -x$ **41.** 8

Article 11.5, pp. 639–641

Note. The tests mentioned in Problems 1–30 may not be the only ones that apply.

1. Converges; ratio test; geometric series, $r < 1$ **3.** Converges; compare with $\Sigma\,(1/2^n)$ **5.** Converges; ratio test **7.** Diverges; compare with $\Sigma\,(1/n)$ **9.** Converges; ratio test **11.** Diverges; compare with $\Sigma\,(1/n)$ **13.** Diverges; ratio test **15.** Converges; ratio test, root test **17.** Converges; compare with $\Sigma\,(1/n^{3/2})$ **19.** Converges; ratio test **21.** Diverges; integral test **23.** Diverges; n^{th}-term test, $a_n \to e$ **25.** Converges, ratio test **27.** Converges; ratio test or comparison with $\Sigma\,(1/n!)$ **29.** Converges; ratio test **31.** $\Sigma\,(1/(2n-1)) > \Sigma\,(1/2n) = \frac{1}{2}\Sigma\,(1/n)$; the last series diverges **33.** $x^2 \leq 1$ U **35.** 1.649767731 **37.** $n = 27$, $s_n + n^{-3}/3 = 1.082324151$; $\pi^2/90 = 1.082323234$; difference $\approx 0.9 \times 10^{-6}$ **39.** $\Sigma\,(1/nx) = (1/x)(\Sigma\,(1/n))$. Since $\Sigma\,(1/n)$ diverges, its nonzero multiple $(1/x)(\Sigma\,(1/n))$ also diverges.

Article 11.6, pp. 647–648

Note. In the answers provided for this article, "Yes" means that the series converges absolutely. "No" means that the series does not converge absolutely. (The series may still converge.) The reasons given may not be the only appropriate ones.

1. Yes; p-series, $p = 2$ **3.** No; harmonic series, divergence **5.** Yes; compare with p-series, $p = 2$ **7.** Yes; compare with p-series, $p = \frac{3}{2}$ **9.** Yes; comparison test **11.** No; n^{th} term test **13.** Yes; geometric series, $r < 1$ **15.** Yes; ratio test **17.** Yes; compare with $(2+n)/n^3$

Article 11.7, pp. 654–655

1. Converges **3.** Diverges **5.** Converges **7.** Converges **9.** Converges **11.** Absolutely convergent **13.** Absolutely convergent **15.** Conditionally convergent **17.** Divergent **19.** Conditionally convergent **21.** Absolutely convergent **23.** Absolutely convergent **25.** Divergent **27.** Conditionally convergent **29.** $\frac{1}{5} = 0.2$ **31.** $(0.01)^5/5 = 2 \times 10^{-11}$ **33.** 0.54030 (actual: $\cos 1 = 0.540302306\ldots$) **35.** a) The a_n are not decreasing (condition 2) b) $-\frac{1}{2}$

Miscellaneous Problems Chapter 11, pp. 655–656

1. $s_n = \ln[(n+1)/(2n)]$; series converges to $-\ln 2$. **7.** Diverges, since n^{th} term doesn't approach 0 **9.** Diverges **11.** Converges **13.** Diverges **15.** Converges **17.** Diverges **19.** Converges

CHAPTER 12

Article 12.1, p. 662

1. $1 - x + (x^2/2!) - (x^3/3!)$; $1 - x + (x^2/2!) - (x^3/3!) + (x^4/4!)$ **3.** $1 - (x^2/2!)$; $1 - (x^2/2!) + (x^4/4!)$ **5.** $x + (x^3/3!)$; $x + (x^3/3!)$ **7.** $-2x + 1$; $x^4 - 2x + 1$ **9.** $x^2 - 2x + 1$; $x^2 - 2x + 1$ **11.** x^2 **13.** $1 + [3/2]x + [3/(4 \cdot 2!)]x^2 - [3/(8 \cdot 3!)]x^3 + [(3 \cdot 3)/(16 \cdot 4!)]x^4 - [(3 \cdot 3 \cdot 5)/(32 \cdot 5!)]x^5 + \cdots + [(\frac{3}{2})(\frac{1}{2})(-\frac{1}{2}) \cdots (\frac{3}{2} - n + 1)/n!]x^n + \cdots$ **15.** $e^{10} + e^{10}(x-10) + (e^{10}/2!)(x-10)^2 + (e^{10}/3!)(x-10)^3 + \cdots = e^{10} \sum_{n=0}^{\infty} [(x-10)^n/n!]$ **17.** $(x-1) - [(x-1)^2/2] + [(x-1)^3/3] - [(x-1)^4/4] + \cdots \sum_{n=1}^{\infty} [(-1)^{n+1}(x-1)^n]/n$ **19.** $-1 - (x+1) - (x+1)^2 - (x+1)^3 - \cdots = -\sum_{n=0}^{\infty} (x+1)^n$ **21.** $1 + 2(x - (\pi/4)) + 2(x - (\pi/4))^2$ **23.** 1.0100 with error less than 0.0001 (use two terms)

Article 12.2, pp. 671–672

1. $1 + (x/2) + \frac{1}{4}(x^2/2!) + \frac{1}{8}(x^3/3!) + \frac{1}{16}(x^4/4!) + \cdots$ **3.** $5 - (5/\pi^2)(x^2/2!) + (5/\pi^4)(x^4/4!) - (5/\pi^6)(x^6/6!) + \cdots$ **5.** $(x^4/4!) - (x^6/6!) + (x^8/8!) - (x^{10}/10!) + \cdots$ **9.** $1 - x + x^2 + R_2(x, 0)$ **11.** $1 + (x/2) - (x^2/8) + R_2(x, 0)$ **13.** $|x| < \sqrt[5]{0.06} \approx 0.57$ **15.** $|R| < (10^{-9}/6)$; $x < \sin x$ for $x < 0$ **17.** $|R| < (e^{0.1}/6000) \approx 1.84 \times 10^{-4}$ **19.** $R_4(x, 0) < 0.0003$ **23.** a) $-1 + 0i$ b) $(\sqrt{2}/2) + (i\sqrt{2}/2)$ c) $-i = 0 - i$ d) $-1 + i$ **27.** $\int e^{ax} \cos bx\, dx = [e^{ax}/(a^2 + b^2)](a \cos bx + b \sin bx) + \text{const.}$; $\int e^{ax} \sin bx\, dx = [e^{ax}/(a^2 + b^2)](a \sin bx - b \cos bx) + \text{const.}$

Article 12.3, pp. 678–679

1. Use $\cos x \approx (\sqrt{3}/2) - \frac{1}{2}(x - (\pi/6))$; $\cos 31° \approx 0.857299$. Actual value: $\cos 31° = 0.857167\ldots$ **3.** Use $\sin x \approx x - 2\pi$; $\sin 6.3 \approx 0.0168147$. Actual value: $\sin 6.3 = 0.0168139\ldots$ **5.** Use $\ln x \approx (x-1) - [(x-1)^2/2] + [(x-1)^3/3] - [(x-1)^4/4]$; $\ln 1.25 \approx 0.222982$. Actual value: $\ln 1.25 = 0.223143\ldots$ **7.** $\ln(1 + 2x) = 2x - [(2x)^2/2] + [(2x)^3/3] - [(2x)^4/4] + \cdots = \sum_{n=1}^{\infty} [(-1)^{n+1}(2x)^n/n]$; converges for $|x| < \frac{1}{2}$ **9.** $\int_0^{0.1} [(\sin x)/x]\, dx \approx x - (x^3/18)\big|_0^{0.1} = 0.0999\overline{4}$ **13.** $\ln 1.5 \approx 0.405465$ **17.** $c_3 = 3.141592665$ **21.** c) 6 d) $1/q$

Article 12.4, p. 681

1. 1 **3.** $-\frac{1}{24}$ **5.** -2 **7.** 0 **9.** $-\frac{1}{3}$ **11.** $\frac{1}{120}$ **13.** $\frac{1}{3}$ **15.** -1 **17.** 3 **19.** 0 **21.** b) $\frac{1}{2}$ **23.** This approximation gives better results than the approximation $\sin x \approx x$. See table.

x	±1.0	±0.1	±0.01
$6x/(6 + x^2)$	±0.857142857	±0.099833611	±0.009999833
sin x	±0.841470985	±0.099833417	±0.009999833

Article 12.5, pp. 693–694

1. $|x| < 1$. Diverges at $x = \pm 1$ **3.** $|x| < 2$. Diverges at $x = \pm 2$ **5.** $-\infty < x < \infty$ **7.** $-4 < x < 0$. Diverges at $x = -4, 0$ **9.** $|x| < 1$. Converges at $x = \pm 1$ **11.** $-\infty < x < \infty$ **13.** $|x| < (1/e)$. Converges at $x = -(1/e)$; diverges at $x = (1/e)$ **15.** $x = 0$ **17.** $\frac{1}{2} < x < \frac{3}{2}$. Diverges at $x = \frac{1}{2}, \frac{3}{2}$ **19.** $|x| < \sqrt{3}$. Diverges at $x = \pm\sqrt{3}$ **21.** e^{3x+6} **25.** $\sum_{n=1}^{\infty} 2nx^{2n-1}$ **27.** 0.002666 **29.** 0.000331 **31.** 0.363305 **33.** 0.099999 **35.** a) $\sinh^{-1} x = x - \frac{1}{2}(x^3/3!) + [(1 \cdot 3)/(2 \cdot 4)](x^5/5!) - [(1 \cdot 3 \cdot 5)/(2 \cdot 4 \cdot 6)](x^7/7!) + \cdots$ b) $\sinh^{-1} 0.25 \approx 0.247$. Actual value: $\sinh^{-1} 0.25 = 0.247466\ldots$ **37.** $\sum_{n=1}^{\infty} n^n x^n$ **39.** By Theorem 1, the series converges absolutely for $|x| < d$, for all positive $d < r$. This means that the series converges absolutely for all $|x| < r$. [Take $d = \frac{1}{2}(r + |x|)$.] **41.** $x + x^2 + (x^3/3) - (x^5/30) + \cdots$ **43.** $\ln(\sec x) = (x^2/2) + (x^4/12) + (x^6/45) + \cdots$

Miscellaneous Problems Chapter 12, pp. 694–696

1. a) $\sum_{n=2}^{\infty} (-1)^n x^n$ b) No, because it will converge in an interval symmetric about $x = 0$, and it cannot converge when $x = -1$. **3.** $e^{\sin x} = 1 + x + (x^2/2!) - (3x^4/4!)$ **5.** a) $\ln(\cos x) = -(x^2/2) - (x^4/12) - (x^6/45) - \cdots$ b) -0.00017 **7.** 0.747 **9.** $\sum_{n=0}^{\infty} (-1)^{n+1}(x-2)^n$, $1 < x < 3$ **11.** $\cos x = \frac{1}{2}\sum_{n=0}^{\infty} (-1)^n[(1/(2n)!)(x - (\pi/3))^{2n} + (\sqrt{3}/(2n+1)!)(x - (\pi/3))^{2n+1}]$ **13.** 1.543 **15.** $e^{(e^x)} = e(1 + x + (2x^2/2!) + (5x^3/3!) + \cdots)$ **17.** -0.0011 **19.** $e^{-1/6}$ **21.** $-5 < x < 1$; converges at $x = -5$, diverges at $x = 1$ **23.** $-\infty < x < \infty$ **25.** $1 < x < 5$; diverges at $x = 1, 5$ **27.** $0 \le x \le 2$ **29.** $-\infty < x < \infty$ **31.** $x \ge \frac{1}{2}$ **37.** $\tan^{-1} x/(1 - x) = x + x^2 + \frac{2}{3}x^3 + \frac{2}{3}x^4 + \frac{13}{15}x^5 + \cdots$

CHAPTER 13

Article 13.1, p. 703

1. a) b) c) d) e)

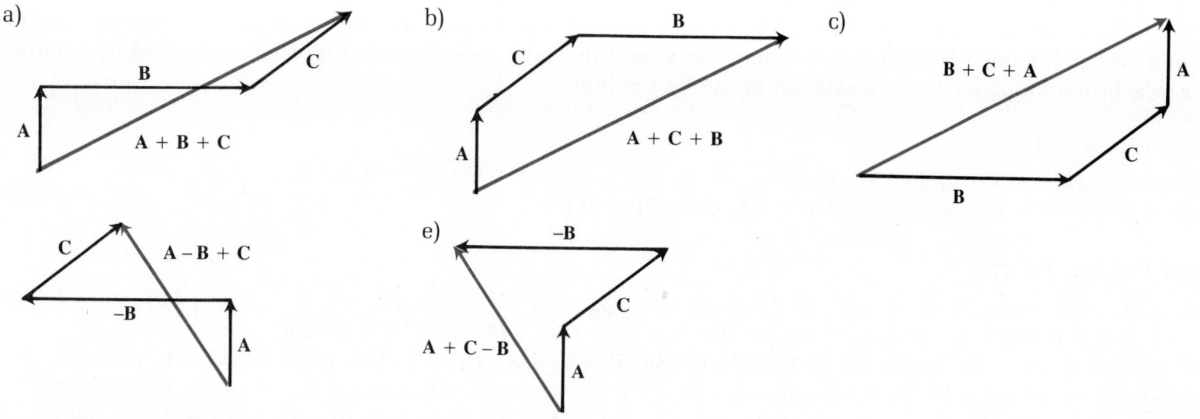

3. $-\mathbf{i} + \mathbf{j}$ **5.** $\mathbf{0}$ **7.** $\frac{1}{2}\sqrt{3}\mathbf{i} - \frac{1}{2}\mathbf{j}$ **9.** $\pm(\mathbf{i} + 4\mathbf{j})/\sqrt{17}$ (either one) **11.** a) $-(\mathbf{i} + 4\mathbf{j})/\sqrt{17}$ b) $(4\mathbf{i} - \mathbf{j})/\sqrt{17}$ **13.** $\sqrt{2}$, $(\mathbf{i} + \mathbf{j})/\sqrt{2}$, $\pi/4$ **15.** 2, $(\sqrt{3}\mathbf{i} + \mathbf{j})/2$, $\pi/6$ **17.** 13, $(5\mathbf{i} + 12\mathbf{j})/13$, $\tan^{-1}(12/5) \approx 67.4°$ **19.** $2\sqrt{5}$

Article 13.2, p. 708

1. $\alpha_1 = \frac{1}{2}\sin^{-1} 0.8 = 26°34'$, flight time 28.0 seconds; $\alpha_2 = 90° - \alpha_1 = 63°26'$ flight time 55.9 seconds **5.** Yes; if $\alpha = \frac{1}{2}\sin^{-1}(\frac{9}{10}) \approx 32°5'$, y max = 31.3 ft and range = 200 ft **7.** $x = e^t$, $y = 2e^{-t}$ **9.** $x = -\ln(1 - t)$, $y = 1/(1 - t)$ **11.** a) $x = \sin t$, $y = -\cos t$ b) $x^2 + y^2 = 1$ c) (1, 0), (0, 1)

Article 13.3, pp. 713–715

1. $x^2 + y^2 = 1$ **3.** $(x/4)^2 + (y/2)^2 = 1$ **5.** $x = 1 - 2y^2$ **7.** Right-hand branch of the hyperbola $x^2 - y^2 = 1$ (right-hand branch because $\sec t > 0$ for $-\pi/2 < t < \pi/2$). **9.** One arch of the cycloid above the interval $0 \le x \le 2\pi$ (see Fig. 13.16). The points (x, y) on this curve satisfy the equation $x = \cos^{-1}(1 - y) - \sqrt{2y - y^2}$. **11.** $y = x^{2/3}$ **13.** $4x^2 - y^2 = 4$ **15.** $y = x^2 - 2x + 5$; $x \ge 1$ **17.** $[(x - 3)^2/4] + [(y - 4)^2/9] = 1$; $x > 3$ **19.** $x = a \tanh \theta$; $y = a \operatorname{sech} \theta$ **21.** $x = a \sinh^{-1}(s/a)$; $y = \sqrt{a^2 + s^2}$ **23.** $x = (a + b)\cos\theta - b\cos([(a + b)/b]\theta)$; $y = (a + b)\sin\theta - b\sin([(a + b)/b]\theta)$ **25.** $8a$ **29.** 4 **31.** $16\pi/3$ **33.** $x = 2\cos t + 1$; $y = \sin t$; ellipse, center at (1, 0), vertices at (−1, 0) and (3, 0), semimajor axis 2, semiminor axis 1.

Article 13.4, pp. 718–719

1. Line parallel to z-axis through point (1, 1, 0) **3.** Straight line, 5 units above and parallel to the line $y = x$ in the xy-plane **5.** Circle of radius 2 with center (0, 0, −2) and parallel to the xy-plane **7.** Circle of radius 1 in yz-plane with center at the origin **9.** Straight line in yz-plane that passes through the origin and forms an angle of $\pi/4$ radians with y and z axes

Note. In Problems 11–19, entries that are completely arbitrary are denoted by "*."

	(x, y, z)	**(r, θ, z)**	**(ρ, φ, θ)**
11.	(1, 0, 0)	(1, 0, 0)	$(1, \pi/2, 0)$
13.	(0, 0, 1)	(0, *, 1)	(1, 0, *)
15.	$(\sqrt{2}, 0, 1)$	$(\sqrt{2}, 0, 1)$	$(\sqrt{3}, \tan^{-1}\sqrt{2}, 0)$
17.	$(0, -\frac{3}{2}, \sqrt{3}/2)$	$(\frac{3}{2}, -\pi/2, \sqrt{3}/2)$	$(\sqrt{3}, \pi/3, -\pi/2)$
19.	$(0, 0, -\sqrt{2})$	$(0, *, -\sqrt{2})$	$(\sqrt{2}, \pi, 3\pi/2)$

21. Straight line in the plane $\theta = \pi/6$ forming an angle of 45° with the z-axis **23.** Two spirals emanating from the origin and situated on the cone formed by rotating the line $z = \frac{3}{2}x$ about the z-axis **25.** The intersection of the sphere $\rho = 5$ and the cone $\phi = \pi/4$ is a circle, radius $5/\sqrt{2}$ centered at $(0, 0, 5/\sqrt{2})$ and parallel to xy-plane **27.** A semicircle in the yz-plane, center at (0, 0, 2), radius 2 **29.** $\rho = 2$; $r^2 + z^2 = 4$ **31.** $\phi = \pi/4$ or $\phi = \frac{3}{4}\pi$; $x^2 + y^2 = z^2$ **33.** $\theta = \pi/4$; $\theta = \pi/4$ **35.** As an equation in cylindrical coordinates, $\theta = 0$ defines the cartesian plane $y = 0$. To get this plane in spherical coordinates, we piece together the two half-planes $\theta = 0$ and $\theta = \pi$. As an equation in spherical coordinates, the equation $\theta = 0$ defines the cartesian half-plane $y = 0$, $x \geq 0$. To describe this half-plane in cylindrical coordinates we may write $\theta = 0$, $r \geq 0$. **37.** $x^2 + y^2 + z^2 = 1$; $r^2 + z^2 = 1$ **39.** $x = 2$, $r \cos\theta = 2$ **41.** Shell and its boundaries formed between spheres of radii 3 and 5 centered at the origin. **43.** Intersection of the solid wedge bounded by planes $y = 0$ and $y = x$, and the solid cone bounded by surface formed by rotating line $z = x$ in the xz-plane about z-axis, all in the first octant.

Article 13.5, p. 722

1. $(-2, 0, 2)$, $\sqrt{8}$ **3.** $(0, 0, a)$, $|a|$ **5.** a) $\sqrt{y^2 + z^2}$ b) $\sqrt{z^2 + x^2}$ c) $\sqrt{x^2 + y^2}$ d) $|z|$ **7.** 3 **9.** 9 **11.** $(2^2 + 3^2 + 7^2)^{1/2}/\sqrt{62} = \sqrt{62}/\sqrt{62} = 1$ **13.** $(2\mathbf{i} + 2\mathbf{j} + 4\mathbf{k})/3$

Article 13.6, pp. 728–729

1. 10, $\sqrt{13}$, $\sqrt{26}$, $(5\sqrt{2})/13$ **3.** 4, $\sqrt{14}$, 2, $\sqrt{\frac{2}{7}}$ **5.** 2, $\sqrt{34}$, $\sqrt{3}$, $\sqrt{\frac{2}{51}}$ **7.** $\sqrt{3} - \sqrt{2}$, $\sqrt{2}$, 3, $(\sqrt{3} - \sqrt{2})/(3\sqrt{2})$ **9.** -25, 5, 5, -1 **11.** $B = \cos^{-1}(\frac{15}{19}) \approx 37.9°$, $A = C = \cos^{-1}(\sqrt{\frac{2}{19}}) \approx 71.1°$ **13.** $\cos^{-1}(\sqrt{\frac{2}{3}}) \approx 35.3°$ **15.** a) $(\frac{7}{45})(10\mathbf{i} + 11\mathbf{j} - 2\mathbf{k}) = \frac{7}{45}\mathbf{A}$ b) $\frac{7}{3}$ **17.** $(3, 3, 0)$ **19.** $\mathbf{B} = [(\frac{3}{2})\mathbf{i} + (\frac{3}{2})\mathbf{j}] + [-(\frac{3}{2})\mathbf{i} + (\frac{3}{2})\mathbf{j} + 4\mathbf{k}]$ **21.** a) $2\sqrt{10}$ b) $6\sqrt{10}$ **23.** a) $\frac{1}{3}(\mathbf{i} - \mathbf{j} + \mathbf{k})$ b) $\mathbf{j} + \mathbf{k}$ c) $-\frac{1}{3}(-2\mathbf{i} - \mathbf{j} + \mathbf{k})$ **25.** $\cos\alpha = 1$, $\cos\beta = 0$, $\cos\gamma = 0$ **27.** $\cos\alpha = -2/\sqrt{53}$, $\cos\beta = 7/\sqrt{53}$, $\cos\gamma = 0$ **29.** $\cos\alpha = 1/\sqrt{2}$, $\cos\beta = 1/\sqrt{3}$, $\cos\gamma = 1/\sqrt{6}$ **31.** $\cos\alpha = 1/\sqrt{2}$, $\cos\beta = 0$, $\cos\gamma = 1/\sqrt{2}$ **33.** -5 newton-meters **35.** $2000\sqrt{3}$ newton-meters **37.** If $|\mathbf{v}_1| = |\mathbf{v}_2|$, then $(\mathbf{v}_1 + \mathbf{v}_2)$ and $(\mathbf{v}_1 - \mathbf{v}_2)$ are orthogonol. **39.** $z^2 = x^2 + y^2$

Article 13.7, pp. 731–732

1. $-\mathbf{i}$ w $3\mathbf{j} + 4\mathbf{k}$ **3.** $c(2\mathbf{i} + \mathbf{j} + \mathbf{k})$, c = scalar **5.** $\sqrt{6}/2$ **7.** 3 **9.** $2\mathbf{i} + 4\mathbf{j} + 7\mathbf{k}$; 0; 0 **11.** a) No pairs b) **A** and **C** **13.** a) $\cos^{-1}(2\sqrt{2}/\sqrt{33}) \approx 60.6°$ b) Yes; the normal vectors are not parallel, so the planes are not parallel. c) $\mathbf{A} \times \mathbf{B} = -5\mathbf{j} - 5\mathbf{k}$, or $\mathbf{j} + \mathbf{k}$ **15.** $\pm[2\sqrt{2}/3, 1/(3\sqrt{2}), 1/(3\sqrt{2})]$ **17.** $12/\sqrt{62}$

Article 13.8, pp. 738–739

1. $(9, -5, 12)$ **3.** $x = 1$, $y = 1 + t$, $z = t - 1$ **5.** $(5, 15, 10)$ **7.** $3x + y + z = 5$ **9.** $x + 3y - z = 9$ **11.** $x + 3y + 4z = 34$ **13.** $x - 2y + z = 6$ **15.** $x + 6y + z = 16$ **17.** a) $6x - 2y + 3z = 23$ b) $23/7$ c) $25/7$ d) $(x - 4)/6 = (2 - y)/2 = (z + 1)/3$ **19.** a) $z = -1$, $y = 2 - x$ b) $(x - 1)/14 = y/2 = z/15$ c) $x = 4$, $y = 2z + 1$ **21.** a) The complement of the angle between the line and a normal to the plane b) $\sin^{-1}(\frac{8}{21}) \approx 22.4°$ **23.** 3 **25.** $(\cos\alpha)(x - x_0) + (\cos\beta)(y - y_0) + (\cos\gamma)(z - z_0) = 0$ **31.** Both lines contain the point $(2, 7, 5)$, so they intersect. $5x - 7y - 4z = -59$

Article 13.9, p. 745

1. $-2\mathbf{C} = -6\mathbf{i} + 8\mathbf{j} - 24\mathbf{k}$; $-22\mathbf{A} = -88\mathbf{i} + 176\mathbf{j} - 22\mathbf{k}$ **3.** a) $\mathbf{A} \times \mathbf{B} = 15\mathbf{i} + 10\mathbf{j} + 20\mathbf{k}$; $(\mathbf{A} \times \mathbf{B}) \times \mathbf{C} = 200\mathbf{i} - 120\mathbf{j} - 90\mathbf{k}$ b) $(\mathbf{A} \cdot \mathbf{C})\mathbf{B} - (\mathbf{B} \cdot \mathbf{C})\mathbf{A} = 56\mathbf{B} + 22\mathbf{A} = 200\mathbf{i} - 120\mathbf{j} - 90\mathbf{k}$ **5.** $\frac{2}{3}$ **7.** $\frac{4}{3}$ **9.** a) 0 b) 2 c) $2\mathbf{A}$ **11.** $a = (\mathbf{A} \times \mathbf{B}) \cdot \mathbf{D}$, $b = -(\mathbf{A} \times \mathbf{B}) \cdot \mathbf{C}$ **15.** $7 - x = (3y - 19)/12 = (4 - 3z)/30$

Article 13.10, p. 748

1. Right circular cylinder of radius 1, axis along y-axis. **3.** Parabolic cylinder, containing the z-axis, and meeting the xy-plane in the parabola $x = -y^2$. **5.** Plane determined by the x-axis and the line $z = -y$ in the yz-plane. **7.** Hyperbolic cylinder, axis along the y-axis and meeting the xz-plane in the hyperbola $x^2 - z^2 = 1$. **9.** Right circular cylinder, radius 4, axis along the z-axis. **11.** Right circular cylinder, radius $\frac{1}{2}$, axis is the line $x = \frac{1}{2}$, $y = 0$. **13.** Right circular cylinder, radius a, axis along z-axis. **15.** Elliptic cylinder $x^2 + 4(z - \frac{1}{2})^2 = 1$; axis is the line $x = 0$, $z = \frac{1}{2}$ parallel to the y-axis. The cylinder lies on top of the y-axis.

Article 13.11, p. 753

1. Paraboloid of revolution, vertex $(0, 0, 1)$ opening upward. **3.** Ellipsoid, center $(0, 0, 0)$, semiaxes $a = 2$, $b = 1$, $c = 2$. **5.** Sphere, center $(0, 1, 0)$, radius 1. **7.** Two-sheeted hyperboloid of revolution, axis parallel to x-axis, center $(-2, -3, 0)$ **9.** Rotate the xy-plane through 45° and have $z^2 + 2y'^2 = 2x'^2$; elliptic cone with vertex O, axis along x′-axis. **11.** Paraboloid of revolution obtained by rotating the parabola $z = x^2$ about the z-axis. **13.** Rotate $z^2 = x$ about the z-axis. **15.** Elliptic cone, vertex O, axis is the x-axis. **17.** $(x - 1)^2 + 4(y + 1)^2 - (z - 2)^2 = 1$. One-sheeted hyperboloid, center $(1, -1, 2)$, axis $x - 1 = y + 1 = 0$. **19.** Elliptic cylinder, axis along y-axis. **21.** Plane, $z = x$. **23.** "Propeller"-shaped surface generated by rotating about the z-axis from $\theta = 0$ to $\theta = \pi/2$ a line through the z-axis and parallel to the xy-plane whose height above the xy-plane is $z = \sin\theta$. **25.** a) $A(z_1) = \pi ab(1 - (z_1^2/c^2))$ b) $V = \frac{4}{3}\pi abc$ **27.** a) $\pi abh(1 + (h^2/3c^2))$ b) $A_0 = \pi ab$, $A_h = \pi ab(1 + (h^2/c^2))$, $V = (h/3)(2A_0 + A_h)$ **29.** When $\theta = 0$, $\rho = F(\phi)$ describes a curve in the xz-plane. When this curve is rotated about the z-axis, it generates the surface $\rho = F(\phi)$. **31.** A cardioid of revolution.

Miscellaneous Problems Chapter 13, pp. 754–757

1. $x = (1-t)^{-1}$, $y = 1 - \ln(1-t)$, $t < 1$ **3.** $x = e^t$, $y = e^{(e^t)} - e$ **5.** $x = t - \sin t$, $y = 1 - \cos t$ **7.** $x = \sinh^{-1} t$, $y = t\sinh^{-1} t - \sqrt{1+t^2} + 1$ **9.** $x = 4\sin t$, $y = 4\cos t$ **13.** $x = 2a\sin^2\theta\tan\theta$, $y = 2a\sin^2\theta$ **15.** a) $3\pi a^2$ b) $8a$ c) $64\pi a^2/3$ d) $(\pi a, 5a/6)$ **17.** 0 **25.** $-\frac{1}{10}$ **27.** $7\sqrt{2}/10$ **29.** $\mathbf{j} - \mathbf{k}$ **31.** $(\mathbf{j} + \mathbf{k})/\sqrt{2}$ **33.** $(10\mathbf{i} - 2\mathbf{j} - 6\mathbf{k})/\sqrt{35}$ **35.** $\cos^{-1}(3/\sqrt{35})$ **39.** $(1, -2, -1)$; $(x-1)/-5 = (y+2)/3 = (z+1)/4$ **43.** $\frac{1}{3}\sqrt{78}$ **45.** a) $2x + 7y + 2z + 10 = 0$ b) $9/(5\sqrt{57})$ **49.** a) 1 b) $-10\mathbf{i} - 2\mathbf{j} - 12\mathbf{k}$ **57.** $(x-1)^2 + (y-2)^2 + (z-3)^2 = 9$

CHAPTER 14

Article 14.1, pp. 762–763

1. $\mathbf{i}$ **3.** $\mathbf{j} - \mathbf{k}$ **5.** $\mathbf{i} + \mathbf{k}$ **7.** All t **9.** $t \neq -1$ **11.** All t **13.** $2e^{2t}\mathbf{i} + (1 - \varepsilon)e^{-t}\mathbf{j}$; domain is all t. **15.** 0; domain is all t. **17.** $[1/(|x|\sqrt{9x^2 - 1})]\mathbf{i} + (2\sinh 2x)\mathbf{j} + (4\,\mathrm{sech}\,4x)\mathbf{k}$, for $|3x| > 1$. **19.** $\mathbf{F}'(s) = -\mathbf{i}/4(s-1)^{5/4} + \mathbf{j}(\cos\sqrt{s-1})/2\sqrt{s-1} + \mathbf{k}/s$ **21.** For $x + 1 > 0$, $\mathbf{F}'(x) = -\mathbf{i}\sin(x+1) + \mathbf{j}(2x+2)/[1 + (x+1)^4] + \mathbf{k}/(x+2)^2$ **23.** $\mathbf{F}'(s) = -\sqrt{2}(\mathbf{i}\cos t + \mathbf{j}\sin t)$ **25.** $\mathbf{F}'(t) = 4t(\mathbf{i}\cos 2\theta - \mathbf{j}\sin 2\theta)$

Article 14.2, pp. 766–767

1. $\mathbf{v} = (-a\omega\sin\omega t)\mathbf{i} + (a\omega\cos\omega t)\mathbf{j}$, $\mathbf{a} = -(a\omega^2\cos\omega t)\mathbf{i} + -(a\omega^2\sin\omega t)\mathbf{j} = -\omega^2\mathbf{R}$; when $\omega t = \pi/3$, $\mathbf{v} = a\omega(-\frac{1}{2}\sqrt{3}\mathbf{i} + \frac{1}{2}\mathbf{j})$, speed $= a\omega$, $\mathbf{a} = -a\omega^2(\frac{1}{2}\mathbf{i} + \frac{1}{2}\sqrt{3}\mathbf{j})$ **3.** $\mathbf{v} = \mathbf{i} + 2t\mathbf{j}$, $\mathbf{a} = 2\mathbf{j}$; at $t = 2$, $\mathbf{v} = \mathbf{i} + 4\mathbf{j}$, $\mathbf{a} = 2\mathbf{j}$, speed $= \sqrt{17}$ **5.** $\mathbf{v} = e^t\mathbf{i} - 2e^{-2t}\mathbf{j}$, $\mathbf{a} = e^t\mathbf{i} + 4e^{-2t}\mathbf{j}$; at $t = \ln 3$, $\mathbf{v} = 3\mathbf{i} - \frac{2}{9}\mathbf{j}$, $\mathbf{a} = 3\mathbf{i} + \frac{4}{9}\mathbf{j}$ speed $= \frac{1}{9}\sqrt{733}$ **7.** $\mathbf{v} = (3\sinh 3t)\mathbf{i} + (2\cosh t)\mathbf{j}$, $\mathbf{a} = (9\cosh 3t)\mathbf{i} + (2\sinh t)\mathbf{j}$; at $t = 0$, $\mathbf{v} = 2\mathbf{j}$, $\mathbf{a} = 9\mathbf{i}$, speed $= 2$ **9.** $\mathbf{v} = -e^{-t}\mathbf{i} - (6\sin 3t)\mathbf{j} + (6\cos 3t)\mathbf{k}$, $\mathbf{a} = e^{-t}\mathbf{i} - 18(\cos 3t)\mathbf{j} - 18(\sin 3t)\mathbf{k}$; at $t = 0$, $\mathbf{v} = -\mathbf{i} + 6\mathbf{k}$, $\mathbf{a} = \mathbf{i} - 18\mathbf{j}$, speed $= \sqrt{37}$ **11.** $\mathbf{v} = -(4\sin t)\mathbf{i} - (3\cos t)\mathbf{j} + 2\mathbf{k}$, $\mathbf{a} = (-4\cos t)\mathbf{i} + (3\sin t)\mathbf{j}$; at $t = \pi/3$, $\mathbf{v} = -2\sqrt{3}\mathbf{i} - \frac{3}{2}\mathbf{j} + 2\mathbf{k}$, $\mathbf{a} = -2\mathbf{i} + (3\sqrt{3}/2)\mathbf{j}$, speed $= \sqrt{73}/2$ **13.** $t = 1/\sqrt[6]{72}$ **15.** All t **17.** For $t = 0$, $\theta = \pi/2$ **19.** a) $\mathbf{v} = \mathbf{i} + 2t\mathbf{j}$, $\mathbf{a} = 2\mathbf{j}$ b) At $t = 2$, $\mathbf{v} = \mathbf{i} + 4\mathbf{j}$, $\mathbf{a} = 2\mathbf{j}$ **21.** No, it would take about 38.7 seconds to reach the hill, longer than the 31 seconds the round is airborne. **23.** $\mathbf{v} = (\frac{3}{2}t^2 + 4)\mathbf{i} + 4t\mathbf{j} + t\mathbf{k}$; $\mathbf{R} = (\frac{1}{2}t^3 + 4t)\mathbf{i} + (2t^2 + 5)\mathbf{j} + (t^2/2)\mathbf{k}$

Article 14.3, p. 771

1. $-\mathbf{i}\sin t + \mathbf{j}\cos t$ **3.** $-\mathbf{i}\cos t + \mathbf{j}\sin t$ **5.** $-(2\mathbf{i}\cos t + \mathbf{j})/\sqrt{1 + 4\cos^2 t}$ **7.** $[(\cos t - \sin t)/\sqrt{2}]\mathbf{i} + [(\cos t + \sin t)/\sqrt{2}]\mathbf{j}$ **9.** $\frac{52}{3}$ **11.** $\mathbf{j}$ **13.** $\mathbf{T} = \sqrt{\frac{1}{3}}[(\cos t - \sin t)\mathbf{i} + (\cos t + \sin t)\mathbf{j} + \mathbf{k}]$; $|\mathbf{v}| = \sqrt{3}e^t$; $\sqrt{3}(e^\pi - 1)$ **15.** $\mathbf{T} = (9t^2 + 25)^{(-1/2)}[3(\cos t - t\sin t)\mathbf{i} + 3(\sin t + t\cos t)\mathbf{j} + 4\mathbf{k}]$; $|\mathbf{v}| = [9t^2 + 25]^{1/2}$; $[(\pi/2)\sqrt{9\pi^2 + 25} + (\frac{25}{6})\sinh^{-1}(3\pi/5)]$ **17.** $\mathbf{v}(0) = 6\mathbf{j}$; $\mathbf{T} = \mathbf{j}$; speed $= 6$; $\mathbf{a} = 6\mathbf{i}$ **19.** $\frac{1}{4}[2\sqrt{5} + \sinh^{-1} 2]$

Article 14.4, pp. 778–779

1. $|\cos x|$ **3.** $|\cos y|$ **5.** $48y^5/[(4y^6 + 1)^2]$ **7.** $1/[2(1 + t^2)^{3/2}]$ **9.** $\mathbf{T} = (\cos t)\mathbf{i} + (\sin t)\mathbf{j}$; $\kappa = 1/t$; $\mathbf{N} = -\sin t\mathbf{i} + \cos t\mathbf{j}$ **11.** $\mathbf{T} = \sqrt{\frac{1}{2}}[(\cos t - \sin t)\mathbf{i} + (\sin t + \cos t)\mathbf{j}]$; $\kappa = \sqrt{\frac{1}{2}}e^{-t}$; $\mathbf{N} = \sqrt{\frac{1}{2}}[-(\sin t + \cos t)\mathbf{i} + (\cos t - \sin t)\mathbf{j}]$ **13.** $\mathbf{T} = \frac{1}{13}[(12\cos 2t)\mathbf{i} - (12\sin 2t)\mathbf{j} + 5\mathbf{k}]$; $\kappa = \frac{24}{169}$; $\mathbf{N} = -(\sin 2t)\mathbf{i} - (\cos 2t)\mathbf{j}$; $\mathbf{B} = \frac{1}{13}[(5\cos 2t)\mathbf{i} - (5\sin 2t)\mathbf{j} - 12\mathbf{k}]$ **15.** $\mathbf{T} = (1/\sqrt{3})[(1+t)^{1/2}\mathbf{i} - (1-t)^{1/2}\mathbf{j} + \mathbf{k}]$; $\kappa = (\sqrt{2}/3)(1 - t^2)^{-1/2}$; $\mathbf{N} = (1/\sqrt{2})[(1-t)^{1/2}\mathbf{i} + (1+t)^{1/2}\mathbf{j}]$; $\mathbf{B} = (\sqrt{2}/2)[-(1+t)^{1/2}\mathbf{i} + (1-t)^{1/2}\mathbf{j} + (2\sqrt{3}/3)\mathbf{k}]$ **17.** $\mathbf{T} = (\cos t)\mathbf{i} + (\sin t)\mathbf{j}$; $\kappa = 1/t$; $\mathbf{N} = -(\sin t)\mathbf{i} + (\cos t)\mathbf{j}$; $\mathbf{B} = \mathbf{k}$. Equations for tangent line: $x = \pi/2$, $z = 3$. **19.** $\mathbf{T} = (1/\sqrt{5})(-2(\sin t)\mathbf{i} + 2(\cos t)\mathbf{j} + \mathbf{k})$; $\kappa = \frac{2}{5}$; $\mathbf{N} = -(\mathbf{i}\cos t + \mathbf{j}\sin t)$; $\mathbf{B} = (1/\sqrt{5})[\mathbf{i}\sin t - \mathbf{j}\cos t + 2\mathbf{k}]$; $s(t) = \sqrt{5}t$ **21.** a) $\mathbf{R}(t) = (1 + (2/\sqrt{17})t)\mathbf{i} + (-2 + (2/\sqrt{17})t)\mathbf{j} + (3 - (3/\sqrt{17})t)\mathbf{k}$ b) $\mathbf{v}(t) = (1/\sqrt{17})(2\mathbf{i} + 2\mathbf{j} - 3\mathbf{k})$ c) $\mathbf{T}(t) = (1/\sqrt{17})(2\mathbf{i} + 2\mathbf{j} - 3\mathbf{k})$ d) $\mathbf{a}(t) = 0$ e) $\kappa = 0$ **23.** $[x - x_c(t)]^2 + [y - y_c(t)]^2 = \frac{1}{4}(t + t^{-1})^4$ where $x_c(t) = 2\ln t - \frac{1}{2}(t^2 - t^{-2})$, $y_c(t) = -2(t + t^{-1})$ **25.** $(x+2)^2 + (y-3)^2 = 8$ for the circle; $y' = -(x+2)/(y-3)$, $y'' = -8/(y-3)^3$. Both of these $= +1$ at (0, 1). **29.** $d\mathbf{R}/dt = \mathbf{T}(ds/dt)$, $d^2\mathbf{R}/dt^2 = \mathbf{T}(d^2s/dt^2) + \mathbf{N}\kappa(ds/dt)^2$

Article 14.5, pp. 784–785

1. $\mathbf{v} = 2\mathbf{i} + 2t\mathbf{j}$; $\mathbf{a} = 2\mathbf{j}$; $ds/dt = 2(1 + t^2)^{1/2}$; $a_T = 2t/(1 + t^2)^{1/2}$; $a_N = 2[1 - t^2/(1 + t^2)]^{1/2} = 2/(1 + t^2)^{1/2}$; $\mathbf{a} = [2t/(1 + t^2)^{1/2}]\mathbf{T} + 2/(1 + t^2)^{1/2}\mathbf{N}$ **3.** $\mathbf{v} = (-2\sin t)\mathbf{i} + (2\cos t)\mathbf{j}$; $ds/dt = 2$; $a_T = 0$; $a_N = 2$;

$\mathbf{a} = -(2\cos t)\mathbf{i} - (2\sin t)\mathbf{j} = 2\mathbf{N}$ **5.** $\mathbf{v} = e^t[(\cos t - \sin t)\mathbf{i} + (\sin t + \cos t)\mathbf{j}]$; $ds/dt = \sqrt{2}e^t$; $a_T = a_N = \sqrt{2}e^t$; $\mathbf{a} = \sqrt{2}e^t(\mathbf{T} + \mathbf{N})$ **7.** a) $\mathbf{v} = -(\sin t)\mathbf{i} + (\cos t)\mathbf{j} + b\mathbf{k}$; $ds/dt = [1 + b^2]^{1/2}$; $a_T = 0$; $a_N = 1$, $\mathbf{a} = \mathbf{N}$ b) $\kappa = (b^2 + 1)^{-1}$; increasing b reduces curvature **9.** $\mathbf{v} = \mathbf{i} + 2\mathbf{j} + 2\mathbf{k}$; $\mathbf{a} = 2\mathbf{k}$; $ds/dt = 3$; $a_T = \frac{4}{3}$; $a_N = 2\sqrt{5}/3$ **11.** $\mathbf{v} = \mathbf{j} + \mathbf{k}$; $\mathbf{a} = 2\mathbf{i}$; $ds/dt = \sqrt{2}$; $a_T = 0$; $a_N = 2$ **13.** $\mathbf{v} = 3\mathbf{i} + 4\mathbf{j}$; $\mathbf{a} = 6\mathbf{i} + 8\mathbf{j} + 6\mathbf{k}$; $ds/dt = 5$; $a_T = 10$; $a_N = 6$ **15.** a) $\sqrt{3}\ 1\ k$ b) $z = 1$, $y = x - 2$ c) $x - y + 2z = 4$ **17.** If $ds/dt =$ constant, then $d^2s/dt^2 = a_T = 0$ so there is no tangential acceleration. Thus all acceleration and hence all force is directed along the normal.

Article 14.6, pp. 789–790

1. $\mathbf{v} = (3a\sin\theta)\mathbf{u}_r + 3a(1 - \cos\theta)\mathbf{u}_\theta$, $a = 9a(2\cos\theta - 1)\mathbf{u}_r + 18a(\sin\theta)\mathbf{u}_\theta$ **3.** $\mathbf{v} = 2ae^{a\theta}\mathbf{u}_r + 2e^{a\theta}\mathbf{u}_\theta$; $\mathbf{a} = 4e^{a\theta}(a^2 - 1)\mathbf{u}_r + 8ae^{a\theta}\mathbf{u}_\theta$ **5.** $\mathbf{v} = -(8\sin 4t)\mathbf{u}_r + (4\cos 4t)\mathbf{u}_\theta$; $\mathbf{a} = -(40\cos 4t)\mathbf{u}_r - (32\sin 4t)\mathbf{u}_\theta$ **7.** a) $\mathbf{v} = (\cos t - t\sin t)\mathbf{i} + (\sin t + t\cos t)\mathbf{j} + \mathbf{k}$; $\mathbf{a} = -(2\sin t + t\cos t)\mathbf{i} + (2\cos t - t\sin t)\mathbf{j}$ b) $\mathbf{v} = \sqrt{2}\mathbf{u}_r + r\mathbf{u}_\theta = \sqrt{2}\mathbf{u}_r + \sqrt{2}\,t\,\mathbf{u}_\theta$; $\mathbf{a} = -r\mathbf{u}_r + 2\sqrt{2}\mathbf{u}_\theta = -\sqrt{2}t\mathbf{u}_r + 2\sqrt{2}\mathbf{u}_\theta$ **11.** $2a = 13.5046 \times 10^8$ cm $=$ 8391.3 mi **17.** $v_0 = 3.01 \times 10^6$ cm/sec $= 18.7$ mi/sec

Miscellaneous Problems Chapter 14, pp. 790–793

1. a) $\mathbf{v} = 2^{-(3/2)}(-\mathbf{i} + \mathbf{j})$, $\mathbf{a} = 2^{-(5/2)}(\mathbf{i} - 3\mathbf{j})$ b) $t = 0$ **3.** Speed $= a\sqrt{1 + t^2}$, $a_T = at/\sqrt{1 + t^2}$, $a_N = a(t^2 + 2)/\sqrt{1 + t^2}$ **5.** b) $\pi/2$ **7.** $x = a\theta + a\sin\theta$, $y = -a(1 - \cos\theta)$ **9.** b) πab, πa^2 **11.** $y = \pm\sqrt{1 - x^2} \pm \ln[(1 - \sqrt{1 - x^2})/x] + C$. Set $C = 0$, then $x = e^{-s}[s = -\ln x$, measured from $(1, 0)]$; $y = \pm\sqrt{1 - e^{-2s}} \pm \ln(e^s - \sqrt{e^{2s} - 1})$ **13.** a) $320\sqrt{10}$ b) $16[\sqrt{2} + \ln(\sqrt{2} + 1)]$ **15.** a) $a_T = 0$, $a_N = 4$ b) 1 c) $r = 2\cos\theta$ **17.** $\kappa = |f^2 + 2f'^2 - ff''|/(f^2 + f'^2)^{3/2}$ **19.** $\sqrt{2}\mathbf{u}_r$ **21.** a) $\mathbf{v} = -\mathbf{u}_r + 3\mathbf{u}_\theta$, $\mathbf{a} = -9\mathbf{u}_r - 6\mathbf{u}_\theta$ b) $\sqrt{37} + \frac{1}{6}\ln(6 + \sqrt{37})$ **23.** a) $2\sqrt{\pi bg/(a^2 + b^2)}$ b) $\theta = [b/(a^2 + b^2)](\frac{1}{2}gt^2)$, $z = [b^2/(a^2 + b^2)]\frac{1}{2}gt^2$ c) $d\mathbf{R}/dt = gt[b/\sqrt{a^2 + b^2}]\mathbf{T}$; $d^2\mathbf{R}/dt^2 = g[b/\sqrt{a^2 + b^2}]\mathbf{T} + (gt)^2[ab^2/(a^2 + b^2)^2]\mathbf{N}$. There is never any component of acceleration in the direction of the binormal. **25.** a) $\mathbf{u}_\rho = \mathbf{i}\sin\phi\cos\theta + \mathbf{j}\sin\phi\sin\theta + \mathbf{k}\cos\phi$; $\mathbf{u}_\phi = \mathbf{i}\cos\phi\cos\theta + \mathbf{j}\cos\phi\sin\theta - \mathbf{k}\sin\phi$; $\mathbf{u}_\theta = -\mathbf{i}\sin\theta + \mathbf{j}\cos\theta$ d) Yes, they form a right-handed system of mutually orthogonal vectors because of (b) and (c). **27.** a) $ds^2 = dr^2 + r^2d\theta^2 + dz^2$ b) $ds^2 = d\rho^2 + \rho^2d\phi^2 + \rho^2\sin^2\phi\,d\theta^2$ **29.** $x = (a\cos\theta)/\sqrt{1 + \sin^2\theta}$, $y = (a\sin\theta/\sqrt{1 + \sin^2\theta}$, $z = -(a\sin\theta)/\sqrt{1 + \sin^2\theta}$; $L = 2\pi a$ **31.** $2\sqrt{3}(\mathbf{i} + \mathbf{j} - 2\mathbf{k})$ **33.** a) $\mathbf{T}(\frac{2}{3}, \frac{1}{3}, \frac{2}{3})$, $\mathbf{N}(1/\sqrt{5}, -2/\sqrt{5}, 0)$, $\mathbf{B}(4/3\sqrt{5}, 2/3\sqrt{5}, -\sqrt{5}/3)$ b) $2\sqrt{5}/9$ **35.** $\frac{1}{2}(t^2 + 1)^{-2}$ **37.** $x^2/4 + y^2/9 + z^2 = 1$, an ellipsoid

CHAPTER 15

Article 15.1, p. 799

1. Domain: $\{(x, y): xy \neq 0\}$. The xy-plane excluding the axes. Range: $(-\infty, 0) \cup (0, \infty)$. Level curves: hyperbolas $xy = k$, $k \neq 0$. **3.** Domain: $\{(x, y): y > -x\}$. Points above the line $y = -x$. Range: $(0, \infty)$. Level curves: lines $x + y = k$, $k > 0$. **5.** Domain: $\{(x, y) \neq (0, 0)\}$. xy-plane excluding the origin. Range: $(-\infty, \infty)$. Level curves: circles $x^2 + y^2 = k$. **7.** Domain: $\{(x, y): x \neq 0, y \geq 0\}$. Upper half-plane excluding y-axis. Range: $(-\infty, \infty)$. Level curves: the "punctured" parabolas $y = kx^2$, $k > 0$, $x \neq 0$, and the "punctured" x-axis, $y = 0$, $x \neq 0$. **9.** Domain: $\{(x, y): x \neq 0\}$. xy-plane excluding y-axis. Range: $(-\pi/2, \pi/2)$. Level curves: "punctured" lines through the origin $y = mx$, $x \neq 0$. **11.** Domain: xy-plane. Range: $[-1, 1]$. Level curves: hyperbolas $xy = k$. **13.** Domain: $\{(x, y): y \neq \pm x\}$. xy-plane excluding the lines $y = \pm x$. Range: $(-\infty, \infty)$. **15.** Domain: $\{(x, y, z): z \geq 0\}$. Range: $[0, \infty)$. **17.** Domain: All (x, y, z) except the origin. Range: $[0, 1]$. **19.** Spheres centered at the origin. **21.** Spheres centered at the origin. **23.** Planes $x =$ constant, parallel to yz-plane. **25.** Planes $x + y + z =$ constant. **27.** Cylinders, with axis the x-axis. **29.** Paraboloids $z = x^2 + y^2 + k$. **31.** Parabolic cylinders $y = x^2 + k$. **33.** The surface $z = x$ is the plane containing the y-axis that makes a 45° angle with the positive x-axis. The level curves are the lines $x =$ const. in the xy-plane. **35.** The surface $z = x^2 + y^2$ is the circular paraboloid whose axis is the z-axis and that opens upward from the origin above the xy-plane. The level curves are circles in the xy-plane centered at the origin. **37.** The surface $z = \sqrt{y - x^2}$ consists of the upper half $(z \geq 0)$ of the circular paraboloid $z^2 = y - x^2$ or $y = x^2 + z^2$ opening to the right of the origin and lying on and above the xy-plane. The level curves are the curves $\sqrt{y - x^2} = k$ or $y = x^2 + k^2$, the family of parabolas in the xy-plane, with axis the y-axis, opening upward away from the x-axis, and congruent to $y = x^2$. These level curves nest to fill the domain $y \geq x^2$ shown in Fig. 15.1.

39. Setting $f(x, y) = k$ gives $-(x-1)^2 - y^2 + 1 = k$, or $1 - k = (x-1)^2 + y^2$. From this we see that the level curves of f are the circles in the xy-plane centered at (1, 0) and with radius $\sqrt{1-k}$, $k \leq 1$.

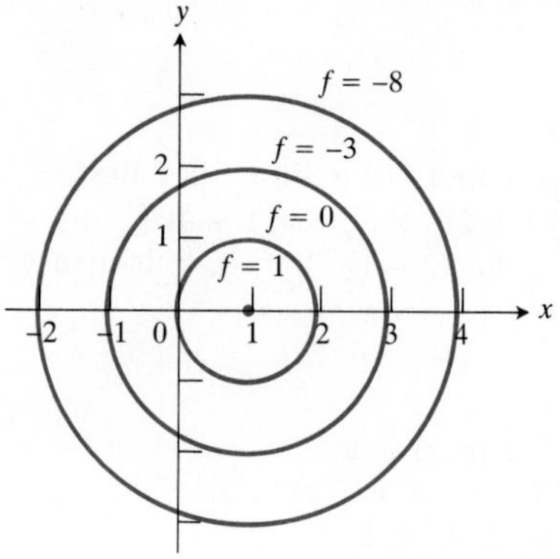

41. $x^2 + y^2 = 10$ **43.** The domain of $z = \sqrt{\sin(\pi xy)}$, $xy \geq 0$ consists of the set of points in the first and third quadrants for which $\sin(\pi xy) \geq 0$. These points make up the closed region between the coordinate axes and the hyperbola $xy = 1$, together with the alternate bands between the hyperbolas $xy = n$, n a positive integer: $2 \leq xy \leq 3$, $4 \leq xy \leq 5$, $\cdots$, $2k \leq xy \leq 2k+1$, $\cdots$. See the figure.

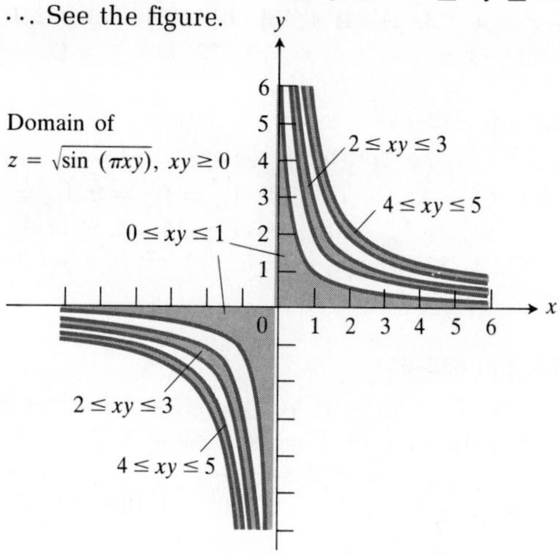

Article 15.2, p. 805

1. $\frac{5}{2}$ **3.** $\frac{1}{2}$ **5.** 5 **7.** 1 **9.** $\pi/2$ **11.** 1 **13.** No limit **15.** 0 **17.** 2 **25.** $\sqrt{x^2+y^2} < 0.1$ (inside the disc of radius 0.1 around (0, 0)) **27.** $|x| < 0.005$, $|y| < 0.005$ works **29.** $\sqrt{x^2+y^2+z^2} < 0.1$ (inside the ball of radius 0.1, centered at origin) **31.** $|x| < 0.005$, $|y| < 0.005$, $|z| < 0.005$ works **33.** Yes.

Article 15.3, pp. 810–811

1. $f_x = f_y = 1$ **3.** $f_x = ye^x$, $f_y = e^x$ **5.** $f_x = e^x \sin y$, $f_y = e^x \cos y$ **7.** $f_x = x/(x^2+y^2)$, $f_y = y/(x^2+y^2)$ **9.** $f_x = 1$, $f_y = 0$ **11.** $f_x = f_y = 0$ **13.** $f_x = 2x - y$, $f_y = 2y - x$ **15.** $f_x = f_y = -1/(x+y)^2$ **17.** $f_x = 1/(y+3)$, $f_y = -(x+2)/(y+3)^2$ **19.** $f_x = (-1-y^2)/(xy-1)^2$, $f_y = (-1-x^2)/(xy-1)^2$ **21.** $f_x = f_y = \sec(x+y)\tan(x+y)$ **23.** $f_x = ex^{e-1}y^e$, $f_y = ex^e y^{e-1}$ **25.** $f_x = 2xe^{2y+3z}\cos(4w)$, $f_y = 2x^2e^{2y+3z}\cos(4w)$, $f_z = 3x^2e^{2y+3z}\cos(4w)$, $f_w = -4x^2e^{2y+3z}\sin(4w)$ **27.** $f_u = 2u/(v^2+w^2)$, $f_v = -2v(u^2+w^2)/(v^2+w^2)^2$, $f_w = -2w(u^2-v^2)/(v^2+w^2)^2$ **29.** $f_x = 2x/(u^2+v^2)$, $f_y = 2y/(u^2+v^2)$, $f_u = -2u(x^2+y^2)/(u^2+v^2)^2$, $f_v = -2v(x^2+y^2)/(u^2+v^2)^2$ **31.** $f_x = y + z$, $f_y = z + x$, $f_z = x + y$ **33.** $f_x = 0$, $f_y = 2y$, $f_z = 4z$ **35.** $f_P = QR$, $f_Q = PR$, $f_R = PQ$ **37.** By the law of cosines, $\cos A = (b^2+c^2-a^2)/2bc$; $\partial A/\partial a = a/bc \sin A$, $\partial A/\partial b = (c^2-a^2-b^2)/2b^2c \sin A$ **39.** $x/\rho = x/\sqrt{x^2+y^2+z^2}$ **41.** $x/(x^2+y^2)$ **43.** $xz/\rho^3 \sin\phi = xz/(x^2+y^2+z^2)\sqrt{x^2+y^2}$ **45.** $\mathbf{i}\sin\phi\cos\theta + \mathbf{j}\sin\phi\sin\theta + \mathbf{k}\cos\phi$ **47.** $\rho\sin\phi(-\mathbf{i}\sin\theta + \mathbf{j}\cos\theta)$

Article 15.4, pp. 817–818

1. $4e^{2t}$ **3.** $2t(t^2+1)^2[(4t^2+1)\cos 4t - 2t(t^2+1)\sin 4t]$ **5.** a, b **11.** $2(x^2 - xy - y^2)$ **13.** $e^x + y^2(1+2x)$ **15.** 2 **17.** $\partial w/\partial r = e^{2r}/\sqrt{e^{2r}+e^{2s}}$; $\partial w/\partial s = e^{2s}/\sqrt{e^{2r}+e^{2s}}$ **19.** 0

Article 15.5, p. 822

1. a) 0 b) $1 + 2z$ c) $1 + 2z$ **3.** a) $\partial U/\partial p + (v/nR)(\partial U/\partial T)$ b) $\partial U/\partial T + (nR/v)(\partial U/\partial p)$ **5.** $\partial w/\partial x = \cos\theta(\partial w/\partial r) - (\sin\theta/r)(\partial w/\partial\theta)$; $\partial w/\partial y = \sin\theta(\partial w/\partial r) + (\cos\theta/r)(\partial w/\partial\theta)$ **9.** Use $(\partial x/\partial y)_z = -f_y/f_x$, etc.

Article 15.6, pp. 834–837

1. $\mathbf{E} = -2\mathbf{i} - 2\mathbf{j} + 4\mathbf{k}$ **3.** $\mathbf{E} = 2\mathbf{i}$ **5.** $\mathbf{E} = (\mathbf{i} + 2\mathbf{j} - 2\mathbf{k})/27$ **7.** $\mathbf{E} = 5\mathbf{k}$ **9.** $-\pi/\sqrt{2}$ **11.** $\frac{2}{13}$ **13.** $(\pi - 3)/2\sqrt{5} \approx 0.032$ **15.** $4\sqrt{3}$ **17.** $\frac{2}{3}$ **19.** a) $[1/\sqrt{1+(2-\pi/2)^2}][(2-(\pi/2))\mathbf{i} - \mathbf{j}]$ b) $\sqrt{1+(2-(\pi/2))^2}$ **21.** a) $(1/\sqrt{10})(-3\mathbf{i} + \mathbf{j})$ b) $8\sqrt{10}$ **23.** a) $-(10\mathbf{i} + 4\mathbf{j} + 10\mathbf{k})$ b) $-6\sqrt{6}$ **25.** $0.3/\sqrt{2} = 0.15\sqrt{2}$ **27.** $0.2/3$ **29.** $\pm(-\frac{12}{13}\mathbf{i} + \frac{5}{13}\mathbf{j})$ **31.** $\mathbf{n} = 4\mathbf{i} + 2\mathbf{j}$ **33.** $\mathbf{n} = 2\sqrt{2}\mathbf{i} - \mathbf{j}$ **35.** $\mathbf{n} = 4\mathbf{i} - 4\mathbf{j}$ **37.** a) $6x + 10y + 8z = 36$ b) $(x-3)/6 = (y-5)/10 = (z+4)/8$ **39.** a) $z - 2x = -2$ b) $(x-2)/-2 = z - 2$, $y = 0$ **41.** a) $3x + 3y + 4z = 25$

b) $(x - 1)/3 = (y - 2)/3 = (x - 4)/4$ **43.** a) $2x + 2y + z = 4$ b) $x/2 = (y - 1)/2 = z - 2$ **45.** a) $x - 2y + 2z = 9$ b) $2x = -y = z$ **47.** a) $x - y + 2z = \frac{\pi}{2}$ b) $x - 1 = 1 - y = (z - (\pi/4))/2$ **49.** a) $z = 0$ b) $x = y = 0$ **51.** a) $x + y + z = 1$ b) $x = y - 1 = z$ **53.** a) $9x - 7y - z = 21$ b) $(x - 2)/9 = (y + 3)/-7 = (z - 18)/-1$ **55.** a) $x + 2z = 2$ b) $x = (z - 1)/2,\ y = 0$ **57.** a) $x - y = 0$ b) $x = -y,\ z = 0$ **59.** $\mathbf{n} = \mathbf{i} + 2\mathbf{j} + \mathbf{k}$ **61.** $\mathbf{n} = 2\mathbf{j} + \mathbf{k}$ **63.** $\mathbf{n} = \mathbf{i} + 2\mathbf{j} + 2\mathbf{k}$ **65.** $\mathbf{n} = 2\mathbf{i} + 4\mathbf{j} - \mathbf{k}$ **67.** a) $2\mathbf{i} + 2\mathbf{j} - 2\mathbf{k}$ b) $2\sqrt{2}$ **69.** Two lines: $x = z = -y \pm 4$ **73.** a) $24\mathbf{i} + 13\mathbf{j}$ b) $\sqrt{745}$ c) $-24\mathbf{i} - 13\mathbf{j}$ d) $\pm(13\mathbf{i} - 24\mathbf{j})$ e) $24x + 13y - z = 42$ **75.** $(x - \sqrt{2})/-1 = y - \sqrt{2},\ z = 4$ **77.** $x - 1 = (z - \frac{1}{2})/-1,\ y = 1$ **79.** $y - 1 = 1 - z,\ x = 1$

Article 15.7, pp. 842–844

1. $f_{xx} = 2(y^2 - x^2)/(x^2 + y^2)^2$, $f_{yy} = 2(x^2 - y^2)/(x^2 + y^2)^2$, $f_{xy} = f_{yx} = -4xy/(x^2 + y^2)^2$ **3.** $f_{xx} = 2y - y \sin x$, $f_{yy} = -\cos y$, $f_{yx} = 2x + \cos x$ **5.** $f_{xx} = f_{yy} = 0$, $f_{xy} = z$ **11.** a) x first b) y first c) x first d) x first e) y first f) y first **13.** $w_{xy} = f_{uu} + uf_{uv} + vf_{vv}$ **15.** $2f_x + 8rsf_{xy} + 4s^2f_{xx} + 4r^2f_{yy}$ **17.** $-f_x \sin t + 2f_y + f_{xx} \cos^2 t + 4t^2f_{yy} + 4t \cos t f_{xy}$ **19.** 2 **31.** a) $f(x, y) = bxy + c^2$; b, c arbitrary b) $f(x, y) = ax^3 + bx^2y - 3axy^2 - (b/3)y^3$; a, b arbitrary **39.** $w = e^{-c^2\pi^2 t} \sin \pi x$

Article 15.8, pp. 852–854

1. a) $T = 1$ b) $T = 2(x - 1) + 2(y - 1) + 3 = 2x + 2y - 1$ **3.** a) $T = 1 + x$ b) $T = (\pi/2) - y$ **5.** a) $T = 3x - 4y + 5$ b) $T = 3x - 4y + 5$ **7.** a) $T = x$ b) $T = (x + y)/\sqrt{2}$ c) $T = (x + y + z)/\sqrt{3}$ **9.** a) $T = 1 + x$ b) $T = (\pi/2) - y - z$ **11.** $T = 7x - 3$; $|E| \leq 0.02$ **13.** $T = x + y - 2$; $|E| \leq \frac{1}{8}$ **15.** $T = 4x + 5y - 8$; $|E| \leq \frac{1}{2}(5.2)(0.2)^2 = 0.104$ **17.** a) $T = 3x + 3y + 2z - 5$, $|E| \leq 0.01$ b) $T = x + y - z - 1$, $|E| \leq (3/2)(0.03)^2 = 0.00135$ c) $T = y + z + 1 - \pi/4$, $|E| \leq (\sqrt{2}/2)(0.03)^2 < 0.00064$ **19.** a) $S_0[\Delta p/100 + \Delta x - 5\,\Delta w - 30\,\Delta h]$ b) A 1-cm increase in height decreases sag 6 times as much as a 1-cm increase in width. **21.** Approximately 340 ft^2 **23.** 0.14 **25.** Approximately 5% **27.** $|x - 1| \leq \frac{1}{70}$, $|y - 1| \leq \frac{1}{70}$. **29.** a) r: 0.014, θ: 0.003 b) r more sensitive to changes in y, θ more sensitive to changes in x **33.** a) More sensitive to changes in x b) $\Delta y/\Delta x = -2$ **35.** $h = f_0/(f_x^2 + f_y^2 + f_z^2)_0$. (Smaller values of h should probably be used in calculations.) For the three equations given, we might use $h = 0.001$. One approximate solution is $(x, y, z) = (1.059, 1.944, 3.886)$. For any solution (a, b, c), there are five other solutions: (c, b, a), $(-b, -a, c)$, $(c, -a, -b)$, $(a, -c, -b)$, and $(-b, -c, a)$.

Article 15.9, pp. 864–865

1. Minimum at $(-3, 3, -5)$ **3.** Saddle point at $(\frac{6}{5}, \frac{69}{25}, -\frac{112}{25})$ **5.** Saddle point at $(-2, 1, 3)$ **7.** Maximum at $(\frac{4}{9}, \frac{2}{9}, -3\frac{1}{9})$ **9.** Minimum at $(1, -2, 0)$ **11.** Minimum at $(2, -1, -6)$ **13.** Saddle point at $(\frac{1}{6}, \frac{4}{3}, -\frac{11}{12})$ **15.** Saddle point at $(1, 2, 9)$ **17.** Saddle point at $(0, 0, 0)$ **19.** Minimum at $(-2, 3, -2)$ **21.** Minimum at $(1, -2, -3)$ **23.** Saddle point at $(0, 0, 6)$; saddle point at $(\frac{2}{3}, \frac{2}{3}, \frac{46}{9})$ **25.** Minimum at $(0, 0, 0)$; saddle point at $(1, -1, 1)$ **27.** Saddle point at $(0, 0, 15)$; minimum at $(1, 1, 14)$ **29.** Maximum at $(1, 1, 2)$; maximum at $(-1, -1, 2)$; saddle point at $(0, 0, 0)$ **31.** Minimum of 1 at $(0, 0)$; minimum of -5 at $(1, 2)$ **33.** Minimum of 0 at $(0, 0)$; maximum of 4 at $(0, 2)$ **35.** Maximum of 11 at $(0, -3)$; minimum of $-\frac{37}{4}$ at $(5, -\frac{5}{2})$ and at $(\frac{9}{2}, -3)$ **37.** Minimum of -4 at $(2, 0)$; maximum of -3 at $(1, 0)$ and $(3, 0)$ **39.** Minimum of $-\frac{1}{4}$ at $(\frac{1}{2}, 0)$; maximum of $\frac{9}{4}$ at $(-\frac{1}{2}, \sqrt{3}/2)$ and $(-\frac{1}{2}, -\sqrt{3}/2)$ **41.** Absolute maximum of 4 at $(1, 1)$; no minima **43.** Low point: $(0, 0, 0)$; high points at the corners: $(\pm 1, \pm 1, 1)$; $\partial z/\partial x$, $\partial z/\partial y$ do not exist at $(0, 0, 0)$ but do exist at the other points and equal (in absolute value) $1/\sqrt{2}$.

Article 15.10, pp. 874–876

1. Maximum at $\pm(1/\sqrt{2}, \frac{1}{2})$; minimum at $(1/\sqrt{2}, -\frac{1}{2})$ and $(-1/\sqrt{2}, \frac{1}{2})$ **3.** Maximum value of -5.4 at $(1.2, 3.6)$ **5.** Local minimum of 0 at $(0, 3)$; local maximum of 4 at $(2, 1)$; no absolute minimum, maximum **7.** a) Minimum value -8 at $(-4, -4)$ b) Maximum value 64 at $(8, 8)$ **9.** $r = 2$ cm, $h = 4$ cm **11.** Maximum $T = 125$ at $\pm(2\sqrt{5}, -\sqrt{5})$; minimum $T = 0$ at $\pm(\sqrt{5}, 2\sqrt{5})$ **13.** Maximum of 20 at $(2, 4)$; minimum of 0 at $(0, 0)$ **15.** $(\frac{3}{2}, 2, \frac{5}{2})$ **17.** Minimum distance is 1 **19.** $(2, -2, 0)$ and $(-2, 2, 0)$ **21.** 50 **23.** $4096\sqrt{5}/125$ **25.** Minimum value $-\frac{1}{3}$ at $(0, -\frac{1}{3})$; maximum value 5 at $(0, 1)$ **27.** Hottest points are $(\frac{4}{3}, -\frac{4}{3}, -\frac{4}{3})$ and $(-\frac{4}{3}, -\frac{4}{3}, -\frac{4}{3})$ **29.** $\sqrt{a_1^2 + \cdots + a_n^2}$ **31.** $\frac{4}{3}$ **33.** Cone $|z| = r$, plane $z = 1 + r(\cos\theta + \sin\theta) = 1 + r\sqrt{2}\sin(\theta + \pi/4)$; curve of intersection lies on the hyperbolic cylinder $r = g(\theta) = 1/[\pm 1 - \sqrt{2}\sin(\theta + \pi/4)]$, $|\overrightarrow{OP}| = \sqrt{2}\,r$ with $r = g(\theta)$. The minimum value of $|g(\theta)|$ is $1/(1 + \sqrt{2}) = \sqrt{2} - 1$, and $|\overrightarrow{OB}| = \sqrt{2}(\sqrt{2} - 1) = 2 - \sqrt{2} = \sqrt{(6 - 4\sqrt{2})}$. B is the point on the cone nearest the origin; A is the point nearest the origin on the other branch of the hyperbola; there is no point farthest from the origin. **35.** a) $2xy + 2x + 2y + 1 = 0$ is the hyperbola $2(x + 1)(y + 1) = 1$ in the xy-plane. b) In space, $\{(x, y, z): 2xy + 2x + 2y + 1 = 0\}$ is a hyperbolic cylinder. Minimum for $y = x = -1 + \sqrt{\frac{1}{2}}$; no maximum.

Article 15.11, pp. 883–884

1. Yes; $f = \frac{2}{5}x^5 + x^2y^3 + \frac{3}{5}y^5 + C$ **3.** Yes; $f = x^2 + xy + y^2 + C$ **5.** No **7.** No **9.** Yes; $f = y \sin x + x \sin y + \cos y + C$ **11.** Yes; $f = 3x^2y^5 + 6xy - 5y^2 + C$ **13.** Yes; $f = 5x^2 + e^{xy} + C$ **15.** General solution: $6x^2y + 2xy^2 = C$; solution through $(1, -3)$: $6x^2y + 2xy^2 = 0$ **17.** $y \ln x + xe^y + y^2 = C$; solution through $(1, 0)$: $y \ln x + xe^y + y^2 = 1$ **19.** $x^2 - e^{-y} \cos x + e^y = 0$ **21.** a) $\alpha = 2$ b) $g = xy^5 + 2x^{3/2}y^3 + y^2 + 2$

Article 15.12, pp. 887–888

3. $y = \frac{3}{4}x + \frac{5}{3}$; $y = \frac{14}{3}$ at $x = 4$ **5.** $y = \frac{5}{14}x + \frac{15}{14}$; $y = \frac{5}{2}$ at $x = 4$ **7.** a) $y = -0.369x$ b) $I/I_0 = e^{-0.369x}$ **9.** $H \approx 17$

Miscellaneous Problems Chapter 15, pp. 889–893

1. No, $\lim_{x\to 0} f(x, 0) = 1$, $\lim_{x\to 0} f(x, x) = 0$ **5.** a) $(\partial/\partial x)(\sin xy)^2 = 2y \sin xy \cos xy = y \sin 2xy$, $(\partial/\partial y)(\sin xy)^2 = 2x \sin xy \cos xy = x \sin 2xy$ b) $(\partial/\partial x) \sin (xy)^2 = 2xy^2 \cos (xy)^2$, $(\partial/\partial y) \sin (xy)^2 = 2x^2y \cos (xy)^2$ **9.** a) Hyperboloid of one sheet b) $2\mathbf{i} + 3\mathbf{j} + 3\mathbf{k}$ c) $2(x - 2) + 3(y + 3) + 3(z - 3) = 0$, $(x - 2)/2 = (y + 3)/3 = (z - 3)/3$ **11.** $\partial f/\partial x = 1$, $\partial f/\partial y = 3$. The derivative of f at $P_0(1, 2)$ in the direction of $\mathbf{u} = \frac{3}{5}\mathbf{i} + \frac{4}{5}\mathbf{j}$ is 3. **15.** $\sqrt{3}$, maximum $= \sqrt{3}$ **17.** 7 **19.** $-\sqrt{2/3}$
21. a) $\mathbf{N}(x, y, z) = (x^2 + y^2)^{-1/2}[x(1 + 3x^2 + 3y^2)\mathbf{i} + y(1 + 3x^2 + 3y^2)\mathbf{j} - \sqrt{x^2 + y^2}\mathbf{k}]$ b) $[(1 + 3x^2 + 3y^2)^2 + 1]^{-1/2}$, $1/\sqrt{2}$ **23.** a) $4\mathbf{j} + 6\mathbf{k}$ b) $4(y + 1) + 6(z - 3) = 0$ **31.** Direction makes an angle of $\theta + \pi/2$ with the positive x-axis; magnitude is $1/r$ **35.** $h'(x) = f_x(x, y) + f_y(x, y)[-g_x(x, y)/g_y(x, y)]$
37. $dx/dz = (\sin y + \sin z - 2y^2 \cos z)(\sin y + \sin z)^{-1}(\sin x + x \cos x)^{-1}$ **39.** $\partial F/\partial x = \frac{1}{2}(\partial f/\partial u + \partial f/\partial v)$, $\partial F/\partial y = \frac{1}{2}(\partial f/\partial v - \partial f/\partial u)$ **43.** $a = b = 1$ **47.** $d^2y/dx^2 = -(1/f_y^3)[f_{xx}f_y^2 - 2f_{xy}f_xf_y + f_{yy}f_x^2]$ **49.** $x_t = x - v + ax_{vv}/x_v^2$, $0 \le v \le 1$, $t \ge 0$, $x(0, t) = 0$, $x(1, t) = 1$ **55.** $(1, 1, 1)$; $(1, -1, -1)$; $(-1, 1, -1)$; $(-1, -1, 1)$ **57.** $e^{-2}/6$ **59.** $\sqrt{3}abc/2$
61. y^2 **65.** Yes; $f = \frac{2}{5}x^5 + x^2y^3 + \frac{3}{5}y^5 + C$

CHAPTER 16

Article 16.2, pp. 903–904

1. 16

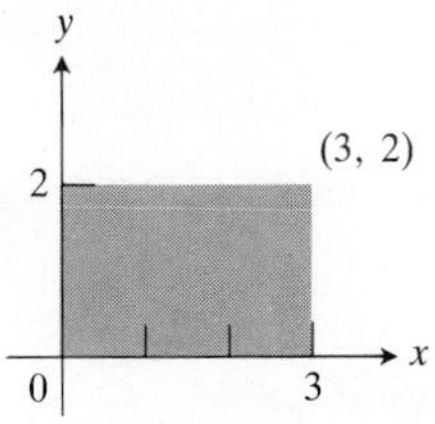

3. 0

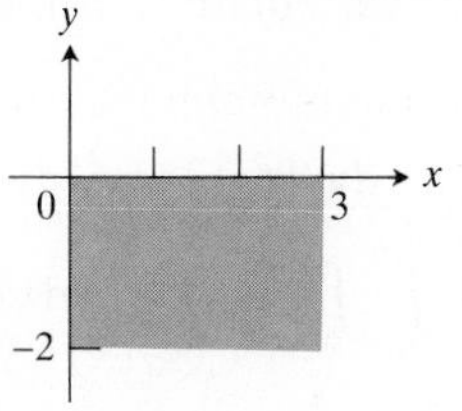

5. $(4 + \pi^2)/2$

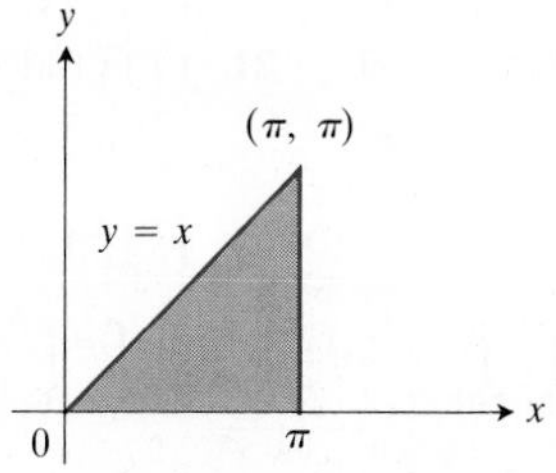

7. $\pi/4$

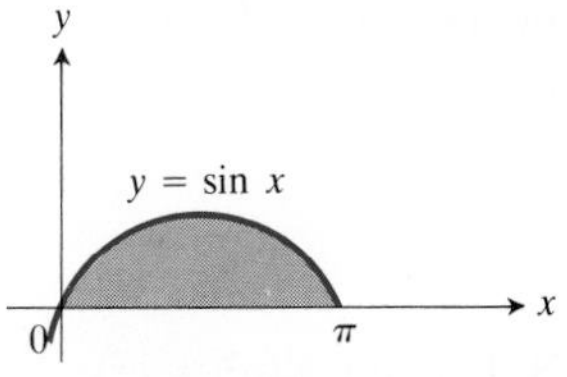

9. $9 - 9e$ **11.** $(3 \ln 2)/2$ **13.** $-\frac{1}{10}$ **15.** $(\ln 2)^2$ **17.** $\int_1^{e^2}\int_{\ln y}^{2} dx\, dy = e^2 - 3$

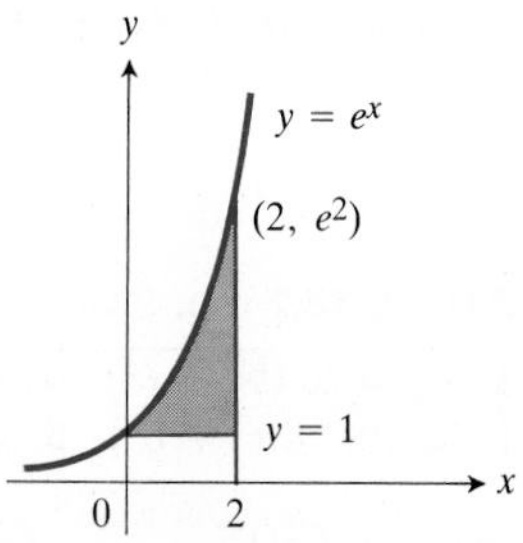

19. $\int_{-2}^{2}\int_0^{\sqrt{(4-x^2)/2}} y\, dy\, dx = \frac{8}{3}$

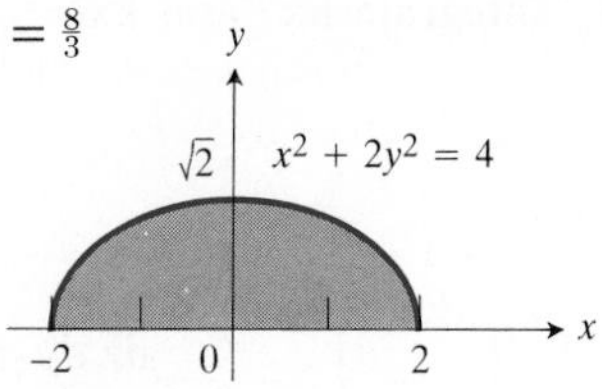

21. $\int_0^1\int_y^{\sqrt{y}} f(x, y)\, dx\, dy$ **23.** $\int_1^e\int_{\ln y}^{1} dx\, dy$ **25.** $\int_0^8\int_{\sqrt[3]{y}}^{2} f(x, y)\, dx\, dy$ **27.** 2 **29.** $(e - 2)/2$ **31.** $\frac{1}{4}\ln 17$ **33.** $\frac{625}{12}$ **35.** $(9\pi - 8)/3$ **37.** 32 **39.** $e - 1$ **41.** $[(\pi + 1)/2\pi] \ln 5 - (1/2\pi) \ln [(1 + 4\pi^2)/5] + 2 [\tan^{-1} 2\pi - \tan^{-1} 2]$

Article 16.3, pp. 905–906

1. $a^2/2$ **3.** 4 **5.** $\frac{1}{3}$ **7.** $\frac{4}{3}$ **9.**

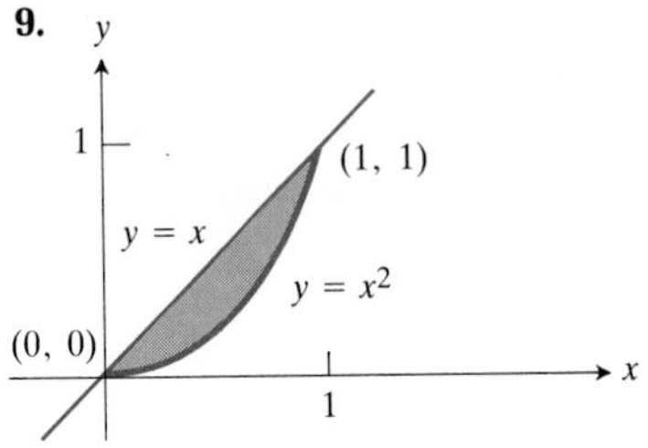

11.

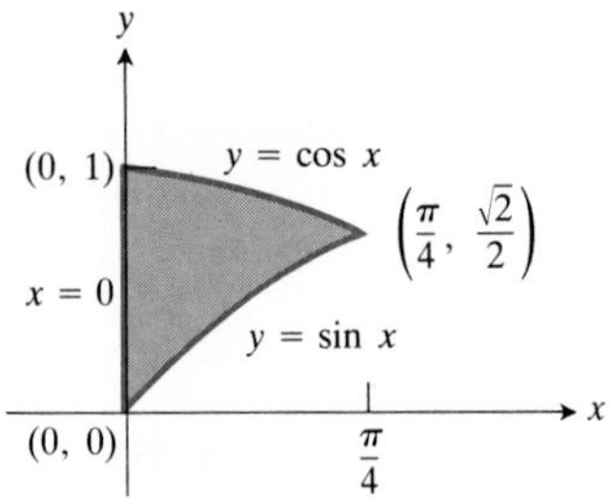

13.

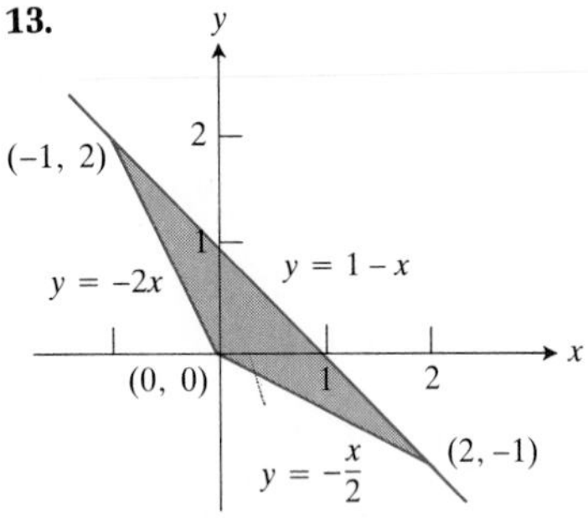

Article 16.4, pp. 909–910

1. $\frac{5}{12}$ **3.** $(a/3, a/3)$ **5.** $(0, a(8 + 3\pi)/(40 + 12\pi))$ **7.** $\bar{x} = \frac{3}{2}$ **9.** $(4a/3\pi, 4a/3\pi)$ **11.** $\frac{1}{9}[e^3 - 1]$ **13.** $\frac{64}{105}$ **15.** $\frac{104}{3}$
17. a) $I_x = bh^3/3$, $R_x = h/\sqrt{3}$ b) $I_x = bh^3/12$, $R_x = h/\sqrt{6}$ c) $I_x = \pi a^4/4$, $R_x = a/2$ d) $I_x = \pi a^4/16$, $R_x = a/2$
e) $I_x = \pi ab^3/4$, $R_x = b/2$ **19.** 1 **21.** $1000/3r\ \text{lb/ft}^2$

Article 16.5, p. 916

1. $\int_0^{2\pi}\int_0^a r\,dr\,d\theta = \pi a^2$ **3.** $\int_0^{\pi/4}\int_0^a r^2\cos\theta\,dr\,d\theta = a^3\sqrt{2}/6$ **5.** $\int_0^{\pi/4}\int_0^{2\sec\theta} r^2\sin\theta\,dr\,d\theta = \frac{4}{3}$
7. $\int_0^{\pi/6}\int_0^{3\sec\theta} dr\,d\theta = \frac{3}{2}\ln 3$ **9.** $(3\pi + 8)/8$ **11.** $(\pi - 1)/2$ **13.** $\frac{4}{3}$ **15.** $((32 + 15\pi)a/(48 + 6\pi), 0)$ **17.** $(a^3/24)[32 + 15\pi]$
19. $(3\pi + 20 - 16\sqrt{2})2\sqrt{2}\,a^3/9$ **21.** $\int_1^3\int_1^2 (2u/v)\,dv\,du = 8\ln 2$ **23.** $\pi ab(a^2 + b^2)/4$

Article 16.6, pp. 920–921

1. $\int_0^1\int_0^2\int_0^3 dz\,dy\,dx$, $\int_0^2\int_0^1\int_0^3 dz\,dx\,dy$, $\int_0^1\int_0^3\int_0^2 dy\,dz\,dx$, $\int_0^3\int_0^2\int_0^1 dx\,dy\,dz$, $\int_0^3\int_0^1\int_0^2 dy\,dx\,dz$. The value of each integral is 6. **3.** $\int_0^3\int_0^2\int_0^{\sqrt{4-x^2}} dz\,dx\,dy$, $\int_0^2\int_0^3\int_0^{\sqrt{4-x^2}} dz\,dy\,dx$, $\int_0^2\int_0^{\sqrt{4-x^2}}\int_0^3 dy\,dz\,dx$, $\int_0^2\int_0^{\sqrt{4-z^2}}\int_0^3 dy\,dx\,dz$, $\int_0^3\int_0^2\int_0^{\sqrt{4-z^2}} dx\,dz\,dy$, $\int_0^2\int_0^3\int_0^{\sqrt{4-x^2}} dx\,dy\,dz$. The value of each integral is 12π. **5.** $4\int_0^2\int_0^{\sqrt{1-x^2}}\int_0^{x^2+y^2} dz\,dy\,dx$

7. $\frac{1}{6}|abc|$ **9.** $\frac{2}{3}$ **11.** $\frac{2}{3}$ **13.** $\frac{2}{7}$ **15.** 32π **17.** $16a^3/3$ **19.** $\frac{4}{3}\pi abc$ **21.** a) $\int_{-1}^1\int_0^{1-x^2}\int_{x^2}^{1-z} dy\,dz\,dx$ b) $\int_0^1\int_{-1}^1\int_{x^2}^{1-z} dy\,dx\,dz$

c) $\int_0^1\int_0^{1-z}\int_{-\sqrt{y}}^{\sqrt{y}} dx\,dy\,dz$ d) $\int_0^1\int_0^{1-y}\int_{-\sqrt{y}}^{\sqrt{y}} dx\,dz\,dy$ e) $\int_0^1\int_{-\sqrt{y}}^{\sqrt{y}}\int_0^{1-y} dz\,dx\,dy$

Article 16.7, pp. 923–924

1. $R_x = \sqrt{(b^2 + c^2)/12}$, $R_y = \sqrt{(a^2 + c^2)/12}$, $R_z = \sqrt{(a^2 + b^2)/12}$ **3.** $I_x = (m/3)(b^2 + c^2)$, $I_y = (m/3)(a^2 + c^2)$, $I_z = (m/3)(a^2 + b^2)$ **5.** Center of mass: $(0, 0, \frac{12}{5})$; $I_x = 3952/105$; $I_y = 2416/63$; $I_z = 128/45$ **7.** $32k/15$

9. $m = \int_{-2}^{2}\int_{-\sqrt{9-(9/4)x^2}}^{\sqrt{9-(9/4)x^2}}\int_{-\sqrt{1-(x^2/4)-(y^2/9)}}^{\sqrt{1-(x^2/4)-(y^2/9)}} kx\,dz\,dy\,dx$. To get M_{yz} and I_y, integrate kx^2 and $kx(x^2 + z^2)$, respectively, with the same limits and order of integration. **11.** $m(3a^2 + 4b^2)/4$

Article 16.8, pp. 928–930

1. a) $\int_0^{2\pi}\int_0^{\pi}\int_0^2 \rho^2\sin\phi\,d\rho\,d\phi\,d\theta$ b) $\int_0^{2\pi}\int_0^2\int_{-\sqrt{4-r^2}}^{\sqrt{4-r^2}} r\,dz\,dr\,d\theta$ c) $\int_{-2}^2\int_{-\sqrt{4-x^2}}^{\sqrt{4-x^2}}\int_{-\sqrt{4-x^2-y^2}}^{\sqrt{4-x^2-y^2}} dz\,dy\,dx$

3. a) $\int_{-1}^{1}\int_{-\sqrt{1-x^2}}^{\sqrt{1-x^2}}\int_{0}^{\sqrt{4-x^2-y^2}} dz\, dy\, dx$ b) $\int_0^{2\pi}\int_0^{\pi/6}\int_0^{2} \rho^2 \sin\phi\, d\rho\, d\phi\, d\theta$ **5.** $\int_{-\pi/2}^{\pi/2}\int_0^{\cos\theta}\int_0^{3r^2} f(r,\theta,z)\cdot r\, dz\, dr\, d\theta$

7. a) $\int_0^{2\pi}\int_0^{1}\int_0^{\sqrt{1-r^2}} r^3\, dz\, dr\, d\theta$ b) $\int_0^{2\pi}\int_0^{\pi/2}\int_0^{1} \rho^4 \sin^3\phi\, dp\, d\phi\, d\theta$ **9.** $\pi/2$ **11.** 8π **13.** $5\pi/2$ **15.** $(4\pi a^3/3)(8 - 3\sqrt{3})$
17. $(8\sqrt{2} - 7)\pi a^3/6$ **19.** $100\pi/3$ **21.** $(2\pi a^5/15)(128 - 51\sqrt{3})$ **23.** $\bar{x} = \bar{y} = 0, \bar{z} = 7a/(16\sqrt{2} - 14) \approx 0.8114a$
25. $\frac{2}{5}ma^2$ **27.** $64\pi a^5/35$ **29.** $\frac{7}{5}ma^2$ **31.** $\frac{15}{8}$ **33.** $\pi a^3/3$ **35.** $\pi a^3/9$ **37.** $(16 - 10\sqrt{2})\pi/3$ **39.** 8π **41.** $(0, 0, 3a/8)$
43. $a\sqrt{1270/651} \approx 1.4a$ **45.** $\pi r^4 h/10$

Article 16.9, pp. 933–934

1. $(\pi/6)[17\sqrt{17} - 1]$ **3.** $(\pi/6)[17\sqrt{17} - 1]$ **5.** $8\sqrt{3}/3$ **7.** $\frac{1}{3}(2\sqrt{2} - 1)$ **9.** $(\pi/6)(37\sqrt{37} - 5\sqrt{5})$
11. $\int_{-3}^{3}\int_{-\sqrt{(9-x^2)/4}}^{\sqrt{(9-x^2)/4}} 3\, dy\, dx/\sqrt{9 - x^2 - y^2}$ **15.** $\frac{1}{2}\sqrt{a^2b^2 + a^2c^2 + b^2c^2}$ **17.** $\pi a^2(1 + c^2)$ **19.** $(\pi a^2/6)(5\sqrt{5} - 1)$
21. a) $\pi a^2/2$ b) $\pi a^2/2$

Miscellaneous Problems Chapter 16, pp. 934–937

1. $\int_{-2}^{0}\int_{2x+4}^{4-x^2} dy\, dx = \frac{4}{3}$ **3.** $\int_0^1\int_{-\sqrt{y}}^{\sqrt{y}} dx\, dy$ **5.** $\int_0^1\int_x^{2-x} (x^2 + y^2)\, dy\, dx = \frac{4}{3}$ **9.** $(4a/3\pi, 4a/3\pi)$
11. $(28\sqrt{2}/9\pi, 28(2 - \sqrt{2})/9\pi)$ **17.** $b\sqrt{\frac{2}{5}}$ **21.** $\int_{-\pi/2}^{\pi/2}\int_0^{a} r^2\cos\theta\, dr\, d\theta$ **25.** $A = 2a^2, I_y = (3\pi + 8)a^4/12$
27. $K(a) = \int_0^{a\cos\beta}\int_0^{x\tan\beta} \ln(x^2 + y^2)\, dy\, dx + \int_{a\cos\beta}^{a}\int_0^{\sqrt{a^2-x^2}} \ln(x^2 + y^2)\, dy\, dx$ **29.** $(0, 0, \frac{1}{4})$ **31.** $(\pi/4)$ **33.** $49\pi a^5\delta/15$
35. $a^2b^2c^2/6$ **37.** $4\pi a^3/3 + \frac{16}{3}[(b/2)(3a^2 - b^2)\sin^{-1}(b/\sqrt{a^2 - b^2}) + (b^2/2)\sqrt{a^2 - 2b^2} - a^3\tan^{-1}(a/\sqrt{a^2 - 2b^2})]$

39. $\int_0^{1}\int_{\sqrt{1-x^2}}^{\sqrt{3-x^2}}\int_1^{\sqrt{4-x^2-y^2}} xyz^2\, dz\, dy\, dx + \int_1^{\sqrt{3}}\int_0^{\sqrt{3-x^2}}\int_1^{\sqrt{4-x^2-y^2}} xyz^2\, dz\, dy\, dx$ **41.** $2a^3(3\pi - 4)/9$ **43.** $\pi^2a^3/4$
47. $2\pi(\sqrt{3} - 1)$ **49.** $\pi\sqrt{2}\,a^2$ **51.** $4\int_0^{a}\sqrt{a^2 - y^2}\sqrt{4y^2 + 1}\, dy$ **53.** $8\pi^2$ **55.** $8r^2$ **57.** $(6M/a^2)[1 - h/\sqrt{a^2 + h^2}]$

CHAPTER 17

Article 17.1, p. 942

1. a) At the center of the cylinder b) At the outer surface **5.** Replace x, y, z by $x - x_0$, $y - y_0$, $z - z_0$
7. $\vec{\mathbf{F}} = 2(x^2 + y^2 + z^2)^{-1}[x\mathbf{i} + y\mathbf{j} + z\mathbf{k}]$ **9.** $\mathbf{F} = 2\mathbf{i} - 3\mathbf{j} + 5\mathbf{k}$ **11.** $\pi a^4\delta/2$

Article 17.2, pp. 947–948

1. 9 **3.** $\frac{1}{4}abc(ab + ac + bc)$ **5.** $-\pi a^3$ **7.** $\frac{28}{3}\sqrt{14}$ **9.** $(0, 0, a/2)$ **11.** $3\pi\sqrt{2}/2$ **13.** $\pi a^2/2$ **15.** $\pi a^3/6$ **17.** $\pi a^2/4$
19. $-4(e - 1)$ **21.** $\frac{14}{3}$ **23.** πa^3

Article 17.3, pp. 958–960

1. $\frac{5}{6}\sqrt{2}$ **3.** π **5.** a) 0 b) $-\frac{1}{3}$. The endpoints are the same, yet the integral along different paths is different.
7. a) $\frac{9}{2}$ b) $\frac{9}{2}$ c) $\frac{9}{2}$ **9.** a) $\pi/4$ b) ln 2 c) $\frac{1}{2}$ **11.** a) 1 b) $\frac{17}{18}$ c) $\frac{1}{2}$ **13.** a) 2.143 b) 2.538 c) 2 + sin 1 **15.** a) e^3
b) e^3 c) e^3 **17.** $\frac{1}{2}$ **19.** $-\pi$ **21.** 48 **23.** $f = x^2 + \frac{3}{2}y^2 + 2z^2$ **25.** $f = xe^{(y+2z)}$ **29.** 24 **31.** -1 **33.** 16 **35.** -2
37. 8 + ln 2 **43.** $n = -1$

Article 17.4, pp. 966–967

3. $2\pi\delta$ **5.** a) 0 b) 4π **7.** $\int_C \mathbf{F}\cdot\mathbf{n}\, ds$ represents flux, $\int_C \mathbf{G}\cdot\mathbf{T}\, ds$ represents work.

Article 17.5, pp. 974–975

1. 0 **3.** $\pi a^4/2$ **5.** $-\pi a^2$ **7.** $-\frac{1}{12}$ **9.** $\frac{3}{8}\pi a^2$ **11.** 0 **13.** -16π **15.** $\frac{2}{33}$ **21.** $x^2/4 + y^2 = 1$

Article 17.6, pp. 982–984

1. 0 **3.** 0 **5.** 0 **7.** 48π **9.** $4\pi a^3$ **11.** 32π **13.** $128 + 8\pi$ **15.** 0 **17.** 5 **21.** Flux = 76π **25.** $\pi a/2$

Article 17.7, pp. 991–992

1. a) 0 b) $-\pi$ **3.** $-\frac{5}{6}$ **5.** 0 **7.** $-\pi/4$ **9.** 4π **11.** 0

Miscellaneous Problems Chapter 17, pp. 993–995

5. 0 **7.** $(\pi/60)(17^{5/2} - 41)$ **9.** $8\pi a^3$ **11.** $-\frac{25}{3}$ **13.** 0 **15.** 2π **17.** 0 **19.** 0 **31.** a) $C_1 = -100$, $C_2 = 200$ b) $C_1 = 100$, $C_2 = 0$

CHAPTER 18

Article 18.3, p. 1000

1. $(x^2 + 1)(2y - 3) = C$ **3.** $e^y = e^x + C$ **5.** $\sinh 2y - 2\cos x = C$ **7.** $(y - 1)e^y + (x^2/2) + \ln|x| = C$ **9.** $\cosh^{-1} y + \sinh^{-1} x = C$ **11.** 10 ln 10 years **13.** 48.2 minutes

Article 18.4, p. 1002

1. $x^2(x^2 + 2y^2) = C$ **3.** $\ln|x| + e^{-y/x} = C$ **5.** $\ln|x| - \ln|\sec(y/x - 1) + \tan(y/x - 1)| = C$ **7.** $\ln[(x + 2)^2 + (y - 1)^2] + 2\tan^{-1}[(x + 2)/(y - 1)] = C$, $a = -2$, $b = 1$ **9.** a) $y = kx^2$ b) $x^2 + 2y^2 = C$ **11.** $x^2 - y^2 = C$

Article 18.5, p. 1004

1. $y = e^{-x} + Ce^{-2x}$ **3.** $x^3y = C - \cos x$ **5.** $x = y/2 + C/y$ **7.** $y\cosh x = C - e^x$ **9.** $xy = y^2 + C$ **11.** a) $C = G/100\,kV + (C_0 - G/100\,kV)e^{-kt}$ b) $G/100\,kV$

Article 18.6, p. 1006

1. $y = Cx^2 - x$ **3.** $y = Cx$ **5.** $e^x + e^y(x^2 + 1) = C$ **7.** $x^2 + 2x\sqrt{y^2 + 1} - y^2 = C$ **9.** $y = x\tan(x + C)$ **11.** $e^x + x\ln y + y\ln x - \cos y = C$ **13.** $xy + \cos x = C$ **15.** $c = -2b$, a is arbitrary; $ax - b(y^2/x) = \text{const.}$

Article 18.7, p. 1008

1. $y = C_1e^{-x} + C_2$ **3.** $y = C_1\int e^{-x^2/2}\,dx + C_2$. (The integral cannot be calculated explicitly as a finite combination of elementary functions.) **5.** $y = C_1\sinh(x + C_2)$ **7.** $y = C_1x^4 + C_2x + C_3$ **9.** $x = A\sin(t\sqrt{k/m} + \pi/2) = A\cos(t\sqrt{k/m})$

Article 18.9, p. 1012

1. $y = C_1 + C_2e^{-2x}$ **3.** $y = C_1e^{-x} + C_2e^{-5x}$ **5.** $y = e^{-x/2}[C_1\cos(\sqrt{3}x/2) + C_2\sin(\sqrt{3}x/2)]$ **7.** $y = (C_1 + C_2x)e^{-3x}$ **9.** $y = e^x(C_1\cos(\sqrt{3}x) + C_2\sin(\sqrt{3}x))$

Article 18.10, pp. 1020–1021

1. $y = C_1 + C_2e^{-x} + \frac{1}{2}x^2 - x$ **3.** $y = C_1\cos x + C_2\sin x - \frac{1}{2}x\cos x$ **5.** $y = (C_1 + C_2x)e^{-x} + \frac{1}{2}x^2e^{-x}$ **7.** $y = C_1e^x + C_2e^{-x} + \frac{1}{2}xe^x$ **9.** $y = e^{-2x}(C_1\cos x + C_2\sin x) + 2$ **11.** $y = A\cos x + B\sin x + x\sin x + \cos x\ln(\cos x)$ **13.** $y = C_1e^{5x} + C_2e^{-2x} + \frac{3}{10}$ **15.** $y = C_1 + C_2e^x + \frac{1}{2}\cos x - \frac{1}{2}\sin x$ **17.** $y = C_1\cos x + C_2\sin x - \frac{1}{8}\cos 3x$ **19.** $y = C_1e^{2x} + C_2e^{-x} - 6\cos x - 2\sin x$ **21.** $y = C_1e^x + C_2e^{-x} - x^2 - 2 + \frac{1}{2}xe^x$ **23.** $y = C_1e^{3x} + C_2e^{-2x} - (e^{-x}/4) + \frac{49}{50}\cos x + \frac{7}{50}\sin x$ **25.** $y = C_1 + C_2e^{-5x} + x^3 + \frac{3}{5}x^2 - \frac{6}{25}x$ **27.** $y = C_1 + C_2e^{3x} + 2x^2 + \frac{4}{3}x + \frac{1}{3}xe^{3x}$ **29.** $y = C_1 + C_2e^{5x} + \frac{1}{10}x^2e^{5x} - \frac{1}{25}xe^{5x}$ **31.** $y = C_1\cos x + C_2\sin x - \frac{1}{2}x\cos x + x\sin x$ **33.** $y = C_1 + C_2e^x + \frac{1}{2}e^{-x} + xe^x$ **35.** $y = C_1e^{5x} + C_2e^{-x} - \frac{1}{8}e^x - \frac{4}{5}$ **37.** $y = C_1\cos x + C_2\sin x - (\sin x)[\ln(\csc x + \cot x)]$ **39.** $y = C_1 + C_2e^{8x} + \frac{1}{8}xe^{8x}$ **41.** $y = C_1 + C_2e^x - x^4/4 - x^3 - 3x^2 - 6x$ **43.** $y = C_1 + C_2e^{-2x} - \frac{1}{3}e^x + x^3/6 - x^2/4 + x/4$ **45.** $y = C_1\cos x + C_2\sin x + (x - \tan x)\cos x - \sin x\ln(\cos x) = C_1\cos x + C_2'\sin x + x\cos x - (\sin x)\ln(\cos x)$ **47.** $y = Ce^{3x} - \frac{1}{2}e^x$ **49.** $y = Ce^{3x} + 5xe^{3x}$ **51.** $y = 2\cos x + \sin x - 1 + \sin x\ln(\sec x + \tan x)$ **53.** $y = 1 - e^{-x}$

Article 18.11, p. 1023

1. $y = C_1 + C_2e^x + C_3e^{2x}$ **3.** $y = (C_1 + C_2x)e^{x\sqrt{2}} + (C_3 + C_4x)e^{-x\sqrt{2}}$ **5.** $y = e^{x\sqrt{2}}(C_1\cos\sqrt{2}x + C_2\sin\sqrt{2}x) + e^{-x\sqrt{2}}(C_3\cos\sqrt{2}x + C_4\sin\sqrt{2}x)$ **7.** $y = (C_1 + C_2x + C_3x^2 + C_4x^3)e^x + 7$

Article 18.12, p. 1027

1. $x = x_0\cos\omega t + (v_0/\omega)\sin\omega t = \sqrt{x_0^2 + v_0^2/\omega^2}\sin(\omega t + \phi)$, where $\omega = \sqrt{k/m}$, $\phi = \tan^{-1}(\omega x_0/v_0)$ **3.** a) $I = C_1\cos\omega t + C_2\sin\omega t$ b) $I = C_1\cos\omega t + C_2\sin\omega t + (A\alpha/L(\omega^2 - \alpha^2))\cos\alpha t$ c) $I = C_1\cos\omega t + C_2\sin\omega t + (A/2L)t\sin\omega t$ d) $I = e^{-5t}(C_1\cos 149t + C_2\sin 149t)$ **5.** $\theta = \theta_0\cos\omega t + (v_0/\omega)\sin\omega t$, where $\omega = \sqrt{2k/mr^2}$ **7.** a) $x = C_1\cos\omega t + C_2\sin\omega t + (\omega^2A/(\omega^2 - \alpha^2))\sin\alpha t$ where $\omega = \sqrt{k/m}$ b) $x = C_1\cos\omega t + C_2\sin\omega t - (\omega A/2)t\cos\omega t$ where $\omega = \sqrt{k/m}$

Article 18.13, p. 1029

1. $y = C\sum_{n=0}^{\infty}[x^n/n!]$ **3.** $y = C\sum_{n=0}^{\infty}[(2x)^n/n!]$ **5.** $y = C_1\sum_{n=0}^{\infty}[x^n/n!] + C_2\sum_{n=0}^{\infty}[(-1)^n x^n/n!]$ **7.** $y = x + C_1\sum_{n=0}^{\infty}[(-1)^n x^{2n}/(2n)!] + C_2\sum_{n=0}^{\infty}[(-1)^n x^{2n+1}/(2n+1)!]$ **9.** $y = x + C_1\sum_{n=0}^{\infty}[x^n/n!] + C_2\sum_{n=0}^{\infty}[(-1)^n x^n/n!]$ **11.** $y = C_1(1 - (x^2/2) + (x^4/24) - (x^6/720) + \cdots) + C_2(x - (x^5/120) - (x^9/362{,}880) + \cdots)$ **13.** $y = a_0(1 - (x^4/12) + (x^8/672) - (x^{12}/88{,}704) + \cdots) + a_1(x - (x^5/20) + (x^9/1440) - (x^{13}/224{,}640) + \cdots) + (x^3 - (x^7/42) + (x^{11}/4620) - (x^{15}/970{,}200) + \cdots)$

Article 18.14, pp. 1033–1034

7. $\tan^{-1}(x + y) = x + C$. The solution passes through (0, 0) when $C = 0$. **9.** a) $y = x - 1$ b) Concave upward for $y > x - 1$; concave downward for $y < x - 1$ **11.** $y_0(x) = 1$; $y_1(x) = 1 + x$; $y_2(x) = 1 + x + (x^2/2)$; $y_3(x) = 1 + x + (x^2/2) + (x^3/6)$ **13.** $y_0(x) = 0$; $y_1(x) = x^2/2$; $y_2(x) = (x^2/2) + (x^3/6)$; $y_3(x) = (x^2/2) + (x^3/6) + (x^4/24)$ **15.** $y_0(x) = 1$; $y_1(x) = -1 - x + x^2$; $y_2(x) = \frac{1}{6} + x + \frac{3}{2}x^2 - (x^3/3)$; $y_3(x) = -\frac{1}{4} - (x/6) + (x^2/2) - (x^3/2) + (x^4/12)$

Article 18.15, pp. 1038–1039

1. $y(1) \approx 2.488320$; exact value is $e = 2.718281$ **3.** $y(1) \approx 2.702708$ **5.** $dy/dx = y$; $y(0) = 1$ **7.** $dy/dx = 1/x$; $y(1) = 0$ **9.** $y = y_0 + \int_{x_0}^{x} f(t)\,dt$ **11.** The general solution of $y' = a^2 + y^2$ is $(1/a)\tan^{-1}(y/a) = x + C$. If $y(a) = f(a)$, then $x = a + (1/a)\tan^{-1}(y/a) - (1/a)\tan^{-1}(f(a)/a)$. **13.** $y = 1 + x + x^2 + \frac{4}{3}x^3 + \frac{5}{6}x^4$

Miscellaneous Problems Chapter 18, pp. 1040–1041

1. $\ln(C\ln y) = -\tan^{-1}x$ **3.** $y = \ln(C - \frac{1}{3}e^{-3x})$ **5.** $y^2 + \sqrt{y^4 + 1} = Ce^{2x}$ **7.** $y^2 = x^2 + Cx$ **9.** $y = x\tan^{-1}(\ln Cx)$ **11.** $y = (x^2 + 2)/4 + C/x^2$ **13.** $y = \cosh x \ln(C\cosh x)$ **15.** $y^3 + 3xy^2 + 3(x + y) = C$ **17.** $x^3 + 3xy^2 + 3\sinh y = C$ **19.** $x^2 + e^y(x^2 + y^2 - 2y + 2) = C$ **21.** $y = C$, or $y = a\tan(ax + C)$, or $y = -a\tanh(ax + C)$, or $y = -1/(x + C)$ **23.** $y = -\ln|\cos(x + C)| + D$ **25.** $y = \frac{1}{2}(\ln Cx)^2 + D$ **27.** $y = C_1 + (C_2x + C_3)e^x$ **29.** $y = C_1e^{-x} + (C_2 + x/3)e^{2x}$ **31.** $y = C_1\sqrt{x} + C_2/\sqrt{x}$ **33.** $\frac{3}{2}x^2 + y^2 = D$ **35.** $y^2 = D^2 + 2Dx$ **37.** $y = e^x - x - 1$; $y = \sum_{n=2}^{\infty}(x^n/n!)$ **39.** a) $y = (\pi/2) + x + (x^2/2!) - (x^3/3!) - (3x^4/4!)$ b) $y = -(\pi/2) - x + (x^2/2!) + (x^3/3!) - (3x^4/4!)$

APPENDIXES

Appendix A.1, p. A-8

1. −5 **3.** 1 **5.** a) −7 b) −7 **7.** a) 38 b) 38 **9.** (−4, 1) **11.** (3, 2) **13.** (3, −2, 2) **15.** (2, 0, −1)

Appendix A.2, pp. A-16–A-17

1. a) $\begin{bmatrix} 2 & -3 & 4 \\ 6 & 4 & -2 \\ 1 & 5 & 4 \end{bmatrix}\begin{bmatrix} x \\ y \\ z \end{bmatrix} = \begin{bmatrix} -19 \\ 8 \\ 23 \end{bmatrix}$ **5.** (−4, 1) **7.** (3, 2) **9.** (3, −2, 2) **11.** (2, 0, −1) **15.** (−4, 1)

17. a) $\begin{bmatrix} 1 & -2 & 1 \\ 0 & \frac{1}{4} & -\frac{1}{2} \\ 0 & 0 & \frac{1}{3} \end{bmatrix}$ b) $(-20, \frac{15}{2}, -\frac{10}{3})$ **19.** a) $\begin{bmatrix} 1 & 1 \\ -r_2 & -r_1 \end{bmatrix}\begin{bmatrix} C \\ D \end{bmatrix} = \begin{bmatrix} a \\ b \end{bmatrix}$ b) Whenever $r_1 \neq r_2$

Appendix A.3, pp. A-21–A-22

1. Generalization: If $F_1(t) \to L_1$, $F_2(t) \to L_2, \ldots, F_n(t) \to L_n$ as $t \to c$, then $F_1(t) + F_2(t) + \cdots + F_n(t) \to L_1 + L_2 + \cdots + L_n$ as $t \to c$, for any positive integer n.

Appendix A.8, p. A-45

1. a) (14, 8) b) (−1, 8) c) (0, −5) **11.** a) Circle, radius 2, center at origin b) Interior of circle in (a) c) Exterior of circle in (a) **13.** Circle, radius 1, center at (−1, 0) **15.** Points on line $y = -x$ **17.** $e^{(\pi/2)i}$ **19.** $\sqrt{65}e^{i\tan^{-1}(-0.125)}$ **21.** $1, \frac{1}{2}(-1 + i\sqrt{3}), -\frac{1}{2}(1 + i\sqrt{3})$ **23.** $-(\sqrt{3} + i), (\sqrt{3} - i), 2i$ **25.** $\pm(1/\sqrt{2})(\sqrt{3} + i), \pm(1/\sqrt{2})(\sqrt{3} - i)$ **27.** $\pm(1 + i\sqrt{3}), \pm(1 - i\sqrt{3})$

INDEX

Note. Numbers in parentheses refer to problems on the pages indicated.

A • B, 723
A × B, 729, 731
(A × B) • C, 740, 741
ABRAMOWITZ, M., 691
Absolute convergence, 641
  rearrangement theorem, 643
Absolute value(s)
  of complex number, A-41
  of vector, 699, 720
Acceleration
  in space, 766
  tangential and normal components, 782
  vector, 764
Addition formulas, cos $(A \pm B)$, sin $(A \pm B)$, A-30
Adjoint matrix, A-15
AGNESI, MARIA GAETNA, 714(28)
Alfalfa yield, 887(8)
Algebra, fundamental theorem, A-45
Alternating current, 1027(3)
Alternating series, 648
  estimation theorem, 651
  Leibniz's convergence theorem, 649
Alternating harmonic series, 650
American Mathematical Monthly, 628
Amplitude, 1025
Approximation(s)
  $\csc x \approx (1/x) + (x/6)$, 681
  linear, 845
ARCHIMEDES (287–212 B.C.), 674
Arc length
  space curves, 768
  spherical and cylindrical coordinates, 792(27)
Area, 931
  corollary to Green's theorem, 969
  by double integrals, 904
  surface of revolution, other surfaces, 931
Argand diagram, A-41
Astronomy, 887(9), 888
Atmosphere, dynamics of, xiii
Attraction, 937(57)

BARROW, ISAAC (1630–1677), xiv
Beams, deflection of, 851, 908
BERNOULLI, JAMES (1654–1705), xiv, 711
BERNOULLI, JOHN (1667–1748), xiv, 711
Bernoulli's equation of order 2, 1021(52)
Binomial series, 677
Binomial theorem, A-32
Binormal vector, 777
Biochemistry, 841
Blood sugar, concentration, 1004(11)
BOAS, RALPH P., JR., 628
BOLZANO, BERNHARD (1781–1848), xv
BOYER, MARTINE, 676
Brachistochrones, 711
BROUNCKER, LORD, 675
BUDENHOLZER, F. E., 887(7)
Butterfly, 833

Calculus, xiii
  of variations, 711
Carbon-14 dating, xiii
Cartesian coordinates, 715
Catenary, 713(21)
CAUCHY, AUGUSTIN LOUIS (1789–1857), xv
  condensation test, 640(43)
CAVALIERI, BONAVENTURA (1598–1647), xiv
Center of mass, 922
Central force field, 787
Chain rule, for $f(x, y)$, 812
  for $f(x, y, z)$, 813
  for f(many variables), 817
  for functions on paths, 811
  for functions on surfaces, 814
  for vector functions, 761
Chain rules, remembering them, 816
Change
  absolute, relative, percent, 848
  of variables in double integrals, 913
Characteristic equation, 1010, 1022
Chemistry, 841
Circle(s), of curvature, 774
  involute of, 713(22)
  osculating, 774
Circular helix, 767
Circular paraboloid, 749
Circulation of a field, 989
Closed and bounded regions, 804
Cofactors and minors, A-3
Comparison test, 622
  limit form, 629
Completing the square, A-32
Complex conjugate, A-40
Complex numbers, 668, A-36 ff.
Components of a vector, 697
Computations using series, 672
  of pi, 674
Conditional convergence, 648
Conditionally convergent series, rearrangements of, 653
Cone, elliptic, 750
Conjugate complex numbers, A-40
Conservative fields, 956
Continuity, 801
  of motion, 938
  theorem of hydrodynamics, 981
  of vector functions, 760
Continuous functions
  maximum/minimum of, 805
  of a sequence, 601
Contour lines (level curves), 796
Convergence, absolute, 641
  of alternating series, 649
  conditional, 648
  of geometric series, 611
  of Picard's method, 646
  of power series, 634, 666, 682
  tests for, comparison, 622
    integral, 624
    limit comparison, 629
    ratio, 632
    root, 637
Coordinates
  cartesian (rectangular), 715, 928

Coordinates, (*continued*)
cylindrical, 716, 928
polar, 570
and cylindrical, 785
rectangular, cylindrical, and spherical, 928
spherical, 717, 928
Coulomb's law, 961
COURANT, R., and F. JOHN, 662
Cramer's rule, A-6
Crater Lake, xiii
Craters of Mars, 888(10)
Critical points, 854
Cross product of vectors, 729
associative and distributive laws, 730, A-49
Curl of vector field, 960(44), 973, 983(22)
paddle wheel model, 990
Curvature, 772 ff.
circle of, 774
of a circle, 773
radius of, 774, 778
of space curves, 776, 778
and unit normal vector, 775, 778
vector formula for, 773, 783
and vector formulas, 778
Curve(s)
curvature of, 772, 778, 783
length of, 768
level, 795, 836(82)
space, 767
useful formulas for, 784
Cycloid, 710, 714(25, 26)
Cyclotrons, xiii
Cylinders, 745
Cylindrical coordinates, 716, 924, 928

D, the operator d/dx, 1008
Damped vibrations, 1026
Decreasing sequence, 648
Deflection of beams, 851
del operator (∇), 973
de Moivre's theorem, A-43
Dependent variable, 794
Derivatives. *See also* Differential
chain rules for, 761, 812, 813, 817
partial, 794, 806
of power series, 686
of products of vectors, 779
of vector functions, 761
of vectors of constant length, 781
of vector products, 779
DESCARTES, RENE (1596–1650), xiv
Determinants, A-1
formulas that use, A-7
and inverse of matrix, A-15
properties of, A-14
Dice, 679(21)
Differentiable functions and continuity, 853(30)
Differentiable vector functions, 761
Differential(s)
exact, 876, 957
forms, 879
of $f(x, y)$, 876
Differential analyzers, 706
Differential calculus, xiii
Differential equations, 996
approximate solutions, 1028
degree, 996
first order, exact, 879, 1005
homogeneous, 1000
linear, 1002
separable, 999
linear with constant coefficients, 1008, 1022
as mathematical models, 997
numerical methods, 1034
order, 996
second order, homogeneous, 1009
nonhomogeneous, 1013
special types, 1006
solutions, 998
by Euler's methods, 1034, 1035
by Picard's method, 1031
Runge-Kutta method, 1035
by power series, 1028
by undetermined coefficients, 1016
by variation of parameters, 1014
type, 996
Differential geometry, 778
Diffusion equation, 841
Direction
cosines, 727
fields, 1029
of a vector, 721, 727
Directional derivative(s), 823, 834
and gradient, 824
properties of, 826
Discriminant of $f(x, y)$, 855
Distance formulas, 719
from point to plane, 735
Divergence of series, 614
Divergence theorem, 977
for more complex regions, 979
Divergence of vector field, 974, 982
Division of series, 691
Domain of a function, 794
restriction on, 795
Dot (scalar) product, 723
Double integrals, 894
and area, 904
bounded regions, 899
determining limits, 902
Fubini's theorem, 898
physical applications, 906
in polar coordinates, 912
finding limits, 912
properties of, 895
over rectangles, 895
e, 604, 622, 658, 666, 669
$e^{(a + ib)x}$, 672(26, 27)
Earth, mass of, 789
$e^{i\theta}$ (Euler's formula), 668
Electric circuits, 1027(3)
Electric fields, 939, 961
Electrical engineering, 841
Elementary mathematical formulas, A-31 ff.
Elementary row operations, A-11
Ellipsoid, 748, 753(25)
Elliptic cone, 750
Elliptic cylinder, 747
Elliptic paraboloid, 749, 753(26)
Encyclopaedia Britannica, 790(12–14)
Epicycloid, 713(23)
Epsilons and deltas, A-18
Equations of lines, 733
Error estimates
in Picard's scheme, 646, 647
in Taylor's series, 666
Estimation of increments, 847
EULER, LEONHARD (1707–1783), xiv
Euler's constant, γ, 640(45)
formula ($e^{i\theta}$), 668
$\pi^2/6$, 640(36)
identities, 672(24)
methods for differential equations, 1034
partial differential equation, 938
e^x, 605, 658, 666, 668
Exact differentials, 876, 957
test for, 882
Expansion of the universe, 887(9)
Exponential function. *See* e^x
Expected value of a random variable, 679(22)
Exponential growth, 1000

Factor theorem, A-32
FERMAT, PIERRE DE (1601–1665), xiv
Fields
electrical, 939, 961
fluid flow, 938
force, 940
central, 787
conservative, 956
gradient, 941
gravitational, 798, 940
two-dimensional, 960
vector, 938
Fitting lines and curves
least squares, 886
Taylor polynomials, 658
Fluid flow, 938
Flux, 960, 964
Force field, 940
central, 787
conservative, 956
Formulas from elementary mathematics, A-31

Fortran IV, A–7
FUBINI, GUIDO (1879–1943), 898
Fubini's theorem, 898, 900
Function(s), 794, 804, 808
 contours (level curves), 796
 domain of, 794
 graphs of, 795
 level curves, 795
 Maclaurin series for, 659
 of more than one variable, 794, 804, 808
 natural domain of, 795
 range, 794
 Taylor polynomials for, 658
 Taylor series for, 660
 vector, 758
Fundamental theorem, of algebra, A–45

Galaxies, 887(9)
General solution, 998
Geoid, 798
Geoidal height, 798
Geometric series, 611
Geometry formulas, A–33
GISLASON, E. A., 887(7)
Gradient(s), 822, 824, 834, 973
 algebraic properties, 833
 and curl, 972
 and directional derivative, 824
 and level surfaces, 827
  normal to, 828
Gradient field, 941
Graphs (graphing), 795
Gravitational attraction, 937(45, 57)
Gravitational field, 940
 Earth's, 798
Green's theorem, 967, 973
 area corollary, 969
GREGORY, JAMES (1638–1675), xiv
Growth. *See* Exponential growth
GUILLOUD, JEAN, 676
Gyration, radius of, 909

Handbook of Mathematical Functions, 691
Harmonic functions, 944(28), 983(23, 24)
Harmonic series, 623
 alternating, 650
Heat equation, 841, 844(39, 40)
Heat flow, 993(21–23), 994(30, 31)
Helix, 767
 curvature of, 776
 tangent vector, 770
Homogeneous differential equations, 1000, 1009, 1022
Hooke's law, 1008(9), 1023
Horner's method, A–33
HUBBLE, EDWIN POWELL (1889–1953), 887
Hubble's, constant, 887(9)
 law, 887(9)
HUYGENS, CHRISTIAN (1629–1695), xiv
Hydrodynamics, 938
Hyperbolic paraboloid, 752
Hyperboloids, 751
Hypocycloid, 714(24)

Ideal gas law, 819
Identity, function, matrix, A–13
Imaginary, exponents, 668
 numbers, A–36
Incompressible fluid, 982
Increasing sequence(s), 620
 theorem, 621
Increment(s), 809, 847
 theorems, 809, 810, A–22
Independent variables, 794
Indeterminate forms, 679
Index of permutation, A–2
Inertial navigation, xiii
Infinite product, 695(35)
Infinite series, 609. *See also* Series
 convergence or divergence, 609
 *n*th term of, 609
 partial sums of, 609
 seven tests for convergence, 638
Ingenious devices, 1005
Initial condition(s), 998
Inner product, 723
Integers, A–37
Integral(s). *See also* Integration; and tables on endpapers
 approximations, by series, 678(9, 10)
 calculus, xiii
 definite, approximating, 678(9, 10)
 double, 894
 $e^{ax}\cos bx\,dx$, 672(27)
 $e^{ax}\sin bx\,dx$, 672(27)
 formulas (*see* endpapers)
 iterated, or repeated, 898
 line, 949
 multiple, 894 ff.
 in polar coordinates, 910
 surface, 931, 942
Integral calculus, xiii
Integral test, 624
Integrating factor, 1003
Integration. *See also* Integral
 in cylindrical coordinates, 924
 formulas (*see* endpapers)
 and length of curves, 768
 in polar coordinates, 910
 of power series, 687
 in spherical coordinates, 926
 and surface area, 930
 and volume, 896, 918, 925, 927
Interest, compound, 1000(11), 1004(12)
Intermolecular potential, 887(7)
Intervals of convergence, 683
Inverses of matrices, A–13
Inversions in permutations, A–2
Investment, 1000(11), 1004(12)
Involute of circle, 713(22), 783
Isoclines, 1029
Iterated integral, 898
Iteration method for differential equations, 1031

Jacobian determinant, 913
JASTROW, ROBERT, 887(9)
JOHN, FRITZ, 662
JORGENSEN, A. D., 887(7)
Journal of Chemical Physics, 887(7)

KEPLER, JOHANNES (1571–1630), xiii
Kepler's laws, xiv, 788, 789
Kinetic energy and moments, 907
KLINE, MORRIS, xv

LAGRANGE, JOSEPH-LOUIS (1736–1813), xiv
Lagrange multipliers, 865
 with two constraints, 872
Laplace equation, 842(20–31)
Laplace operator, 994(23–28)
Law of cosines, 1071, 1078
Law of sines, A–36
Least squares, 886
LEIBNIZ, GOTTFRIED WILHELM (1646–1716), xiv
Leibniz's formula for $\pi/4$, 675
Leibniz's theorem for alternating series, 649
Lemniscate, 747
Length
 of space curve, 768
 of a vector, 699, 720
Level curves, 795, 836(82)
Level surfaces, 796, 828
 and gradients, 827
 and normal line, 828
 and tangent plane, 828
l'Hôpital's rule, 602
Limit(s), A–18
 that arise frequently, 604
 comparison test, for series, 629
 and continuity, 799
 of functions $f(x, y)$, 800
 of sequences, 598
 theorems, proofs of, A–18
 of vector functions, 760
Line(s) and planes in space, 732 ff.
 equations for, 733, 734
Line integrals, 949
 algebraic properties, 950
 path independence, 951
 and work, 948
Linear approximations, 845, 851
 for $f(x, y)$, error estimate, 861
 for $f(x, y, z)$, 851
Linear differential operator, 1009

Linear equations and matrices, A–9
Linearization, 847
Logarithms, 677

Maclaurin series, 659
  for $(1 + x)^m$, 661, 662
  for cos x, 661, 667
  for cosh x, 667
  for $e^{i\theta}$, 668
  for e^x, 666
  for sin x, 666
  for tan x, 680
  frequently used, 677
Magnetic force fields, 939
Mars, craters of, 888(10)
Mass, 922
  center of, 922
  conservation of, 982
  of thin shells, 947
  transport of, 962
  variable (rocket), 1039(7)
Mathematical induction, A–27
Mathematical Thought from Ancient to Modern Times, xv
Mathematics of Computation, 678(17)
Mathematics Magazine, 628
Matrix (matrices), A–1
  adjoint of, A–15
  inverse of, A–13
  and linear equations, A–9
  row reduction, A–11
MATYC Journal, A–7
Maxima and minima, 854
  constrained, 865
  Lagrange multiplier method, 865
  of $f(x, y)$, 854
    or saddle points of, 854
    second derivative test, 855, 859
  tests, summary of, 863
Mercator map of Earth, 798
Method of steepest descent, 853(35)
Midpoint, position vector for, 721
MILLER, E., A–7
Minima. *See* Maxima and minima
Minimax, 753
Minors and cofactors, A–3
Mixed partial derivative, 837
  $f_{xy}(0, 0) \neq f_{yx}(0, 0)$ example, 893(64)
  proof of theorem, A–25
  theorem, 838
Möbius strip, 986
Moment(s), 906
  and center of mass, 906
  first and second, 906
  of inertia, 922
  and kinetic energy, 907
Motion, 758
  in a plane, 772
  of a projectile, 703
Multiple integrals, 894
Multiplication of series, 644, 691

NEUMANN, JOHN VON (1903–1957), xv
NEWTON, SIR ISAAC (1642–1727), xiii
Newton's law of cooling, 1000(13)
Newton's second law of motion, 764
Nonnegative series, 620
Normal component of acceleration, 782
Normal line, 702, 828, 831, 834
Normal plane to a space curve, 793(34)
Normal vector, principal, 774, 778
Normalize to a unit vector (direction), 721
*n*th root test, 637
*n*th roots of $re^{i\theta}$, A–44
Number(s)
  complex, 668, A–36
  systems, A–36
Numerical methods of solving differential equations, 1034

Ocean currents, xiii
One-dimensional heat equation, 841
Orbits of planets, 788
Orbits of satellites, xiii
Orthogonal trajectories, 1002(9–11)
Orthogonal vectors, 725
Osculating circle, 774
Osculating plane, 793(34)

Parabolic cylinder, 746
Paraboloid, 749, 796
  elliptic, 749, 753(25)
  hyperbolic, 752
  of revolution, 749
Parallelogram area, 729
Parameters, variation of, 1014, 1022
Parametric equations, 704, 707
  for circle, 709
  for circles, 709
  for cycloids, 710
  for ellipses, 709
  for hyperbolas, 709
  for involute of a circle, 713(22)
  for parabolic arch, 710
  for trochoids, 710
Partial derivatives, 794, 806
  chain rules, 811 ff., 816
  functions not given explicitly, 820
  $f_{xy}(0, 0) \neq f_{yx}(0, 0)$ example, 893(64)
  higher order, 840
  mixed derivative theorem, 838
    proof of, A–25
  nonindependent variables, 818
Partial differential equations for physics, 837
PASCAL, BLAISE (1623–1662), xiv
Peacock, 803
Pendulum, 1027(4)
Period, 1025
Permutation
  index of, A–2
  inversions in, A–2
Perpendicular lines, 11. *See also the* Orthogonal *entries*
Phase angle, 1025
Physical applications, in three dimensions, 921
Pi (π), computation of, 604(58), 674, 676, 678(17)
  $\pi^2/6$ (Euler), 640(36)
Picard method, convergence of, 646
  for differential equations, 1031
Picard's theorem, 1031
Planes, 734, 787
Planet(s) and orbits, xiv, 785, 788
Polar coordinates
    double integrals, 912
  unit vectors u_r and u_θ, 785
  velocity and acceleration vectors, 786
Population growth, 999
Position vector, 706
  for midpoints, 721
Positive definite, 728(38)
Positive series (nonnegative), 620
Potassium ions and xenon, 887(7)
Potential, 937(56), 958
Power series, 657
  convergence, 682
  differentiation of, 686
  division of, 691
  integration of, 687
  multiplication of, 691
  radius of convergence, 683
  for solving differential equations, 1028
Principal normal vector, 775
Products, infinite, 695(35)
  of vectors, 722, 729
    three or more, 740
Projectile motion, 703 ff.
  angle of elevation, 705
  height and range, 705

Quadratic formula, A–33
Quadric surfaces, 748
Quality control, 679(21)
Quotients using series, 679

Radar, xiii
Radius
  of convergence, 683
  of curvature, 774
  of gyration, 909
Random variable, expected value, 679(22)
Range of a function, 794
Range tables, 706
Ratio test, 632
Rational numbers, A–37
Real numbers, A–38

Rearrangements of series, 643, 653
Regions, connected, 881
Regression line, 887
Remainder estimation in series, 626, 651. *See also* Taylor's theorem
Remainder theorem, A-32
Ridge surface, 832
R-L-C series circuits, 1027(3)
Rocket motion, 1039(7)
ROGUES, ALBAN J., A-7
Root test, 637
Round-off errors, 647, 677
Runge–Kutta method, 1035

Saddle, 752
  point, 753, 854, 855, 857, 861
SALAMIN, EUGENE, 678(17)
Sandwich theorem, 601
  proof of, A-20
Satellite(s), 785
Scalar product, 722 ff.
  triple, 740
SEARS, FRANCIS W. (1898–1975). *See* Sears, Zemansky, and Young
Sears, Zemansky, and Young, *University Physics*, 740
Second derivatives
  test for max/min, 855, 859
Separation of variables, 999
Sequences, 594 ff.
  bounded above, 620
  convergence of, 598
  decreasing, 648
  divergence of, 599
  increasing, 620
  least upper bound, 620
  limit theorems, 600
  sandwich theorem, 601
  tail of, 599
Series, 608. *See also* Infinite series
  absolute convergence, 641
  alternating, 648
  alternating harmonic, 650
  binomial, 677
  comparison test, 622
  computations, 672 ff.
  convergence of, 609
    absolute, 641
    conditional, 648
    tests for, 619, 622, 624, 629, 632, 637
  for cos x, 661, 667
  for cosh x, 667
  differentiation of, 686
  divergence, 609
    *n*th term test for, 614
  division of, 691
  for e^x, 666
  geometric, 611
  harmonic, 623
  indexing of, 617
  integral test, 624
  limit comparison test, 629
  logarithmic *p*-series, 641(46)
  for logarithms, 674, 677
  Maclaurin's, 659
    frequently used, 677
  multiplication of, 644
  of nonnegative terms, 619
  partial sums, 609
  p-series, 625
  power, 657
    radius of convergence, 683
  ratio test, 632
  remainder (or truncation error), 626
    estimated, by integrals, 626
    estimated by ratio test, 625
  remainder estimation theorem, 666
  root test, 637
  for sin x, 667, 677
  sums and multiples of, 616
  for tan x, 691
  for $\tan^{-1} x$, 676, 677
  Taylor's, with remainder, 663
Shell, mass and moments, 947
Simply connected, 881
$\sin A \pm \sin B$, A-35
SMITH, DAVID A., 676
Snow plow problem, 1000(12)
Space coordinates, 715
  relations among, 718
Space curves, 767
  arc length, 768
Sphere, 748
Spherical coordinates, 717
  surface integrals in, 948(23)
  triple integrals in, 927
Spring, vibration of, 1027(7)
STARR NORTON, 893
Steady-state flows, 938
Steepest descent method, 853(35)
STEGUN, I. A., 691
Stokes's theorem, 987
  for surfaces with holes, 989
Straight lines in space, 732, 733
Sun, 788
  mass of, 789
Surfaces, max/min, 854 ff.
  quadric, 748
  $z = f(x, y)$, 796
Surface area, 931, 948(23)
Surface integrals, 942, 948(23, 24)
  for flux, 945

Tables
  exponential, A-47
  natural logarithms, A-48
  natural trigonometric, A-46
Tail of a sequence, 599
Tangent planes, 828, 831, 834
Tangent/normal components of velocity and acceleration, 781, 782
Tangential vectors, 763 ff.
Tautochrones, 711
TAYLOR, BROOK (1685–1741), 663
Taylor polynomials, 658, 665
Taylor series, 660
  remainder, 663, 665
Taylor's theorem, 663
  application to max/min, 859
  for $f(x, y)$, 863
Telegraph equation, 841, 842
Tests for convergence/divergence of series, 638, 639
Tests for max/min, 863
Thin shells, 947
THOMPSON, M. H., 888
Torsion, 1027(5)
Torus, 937(53)
Tractrix, 791(11)
Trajectories, orthogonal, 1002(9–11)
Trigonometric definitions and identities, A-35, A-36
Trigonometry, A-35
Triple integral, 918
  in cartesian coordinates, 916
  in cylindrical coordinates, 924
  in spherical coordinates, 927
Triple scalar product, 741
  as volume of box, 740
Triple vector products, 743
Trochoids, 710
Truncation error in series, 626
  estimation of, 635
TURNBULL, HERBERT W., 675
Twisted butterfly, 833
Two-Year College Mathematics Journal, A-7

Undetermined coefficients, 1016
Unit vectors, 700
  binormal, 777
  and direction cosines, 727
  normal, 775
  polar (u_r, u_θ), 785
  tangent, 770
University of California Experimental Station, 887(8)

Variable(s), 794
Variation of parameters, 1014, 1022
Vector(s), 697
  acceleration, 707, 764, 766, 781, 786
  addition, 698
  binormal, 777
  components, 697
  cross product, 729
    distributive law, proof, A-49
  curvature formula, 773
  derivatives of, 758
  direction, 701

Vector(s), (*continued*)
  and distance, 719
  dot (inner or scalar) product, 723
  equality of, 698
  i and **j**, 698
  i, **j**, and **k**, 719
  inner product, 723
  length of, 699, 720
  multiplication by scalars, 699
  negative of, 699
  normal, 775
  orthogonal, 725
  and parametric equations, 704, 709
  perpendicular to line $ax + by = c$, 726
  polar coordinates, 785
  position, 706
  products of three or more, 740
  projection of **B** onto **A**, 723
  scalar product, 723
  in space, 719
  subtraction, 699
  tangent and normal, 702
    and binormal, 777
  triple scalar product, 741
  triple vector product, 743
  unit, 698, 700, 702, 719, 777, 785
  vector product, 729, 731
  velocity, 707, 759, 764, 766, 781, 786
  zero, 700
Vector analysis, 938
Vector field(s), 938, 939
  conservative, 956
  curl of, 973
  divergence of, 974
Vector functions, 758
Vibrations, 1023
  damped, 1026
VIÉTA (VIETE), FRANCOIS (1540–1603), 675
Viéta's formula, 675
Volume(s)
  as a double integral, 896
  as a triple integral, 918
    in cylindrical coordinates, 925
    in spherical coordinates, 927

WALLIS, JOHN (1616–1703), xiv, 675
Wallis's formula, 675
Wave equation, 843(32–37), 844(38)
WEIERSTRASS, KARL (1815–1897), xv
Whales, 1040
Witch of Maria Agnesi, 714(28)
Work, 950
  and line integrals, 948
  path independence, 951
World of Mathematics, xv

YOUNG, H. D. *See* Sears, Zemansky, and Young

ZEMANSKY, M. W. *See* Sears, Zemansky, and Young
Zero vector, 700
"Zipper" theorem, 654(38)

65. $\displaystyle\int \frac{\cos ax}{\sin ax}\,dx = \frac{1}{a}\ln|\sin ax| + C$

66. $\displaystyle\int \cos^n ax \sin ax\,dx = -\frac{\cos^{n+1} ax}{(n+1)a} + C, \qquad n \neq -1$

67. $\displaystyle\int \frac{\sin ax}{\cos ax}\,dx = -\frac{1}{a}\ln|\cos ax| + C$

68. $\displaystyle\int \sin^n ax \cos^m ax\,dx = -\frac{\sin^{n-1} ax \cos^{m+1} ax}{a(m+n)} + \frac{n-1}{m+n}\int \sin^{n-2} ax \cos^m ax\,dx,$

$n \neq -m$ (If $n = -m$, use No. 86.)

69. $\displaystyle\int \sin^n ax \cos^m ax\,dx = \frac{\sin^{n+1} ax \cos^{m-1} ax}{a(m+n)} + \frac{m-1}{m+n}\int \sin^n ax \cos^{m-2} ax\,dx,$

$m \neq -n$ (If $m = -n$, use No. 87.)

70. $\displaystyle\int \frac{dx}{b+c\sin ax} = \frac{-2}{a\sqrt{b^2-c^2}}\tan^{-1}\left[\sqrt{\frac{b-c}{b+c}}\tan\left(\frac{\pi}{4}-\frac{ax}{2}\right)\right] + C, \qquad b^2 > c^2$

71. $\displaystyle\int \frac{dx}{b+c\sin ax} = \frac{-1}{a\sqrt{c^2-b^2}}\ln\left|\frac{c+b\sin ax+\sqrt{c^2-b^2}\cos ax}{b+c\sin ax}\right| + C, \qquad b^2 < c^2$

72. $\displaystyle\int \frac{dx}{1+\sin ax} = -\frac{1}{a}\tan\left(\frac{\pi}{4}-\frac{ax}{2}\right) + C$

73. $\displaystyle\int \frac{dx}{1-\sin ax} = \frac{1}{a}\tan\left(\frac{\pi}{4}+\frac{ax}{2}\right) + C$

74. $\displaystyle\int \frac{dx}{b+c\cos ax} = \frac{2}{a\sqrt{b^2-c^2}}\tan^{-1}\left[\sqrt{\frac{b-c}{b+c}}\tan\frac{ax}{2}\right] + C, \qquad b^2 > c^2$

75. $\displaystyle\int \frac{dx}{b+c\cos ax} = \frac{1}{a\sqrt{c^2-b^2}}\ln\left|\frac{c+b\cos ax+\sqrt{c^2-b^2}\sin ax}{b+c\cos ax}\right| + C, \qquad b^2 < c^2$

76. $\displaystyle\int \frac{dx}{1+\cos ax} = \frac{1}{a}\tan\frac{ax}{2} + C$

77. $\displaystyle\int \frac{dx}{1-\cos ax} = -\frac{1}{a}\cot\frac{ax}{2} + C$

78. $\displaystyle\int x\sin ax\,dx = \frac{1}{a^2}\sin ax - \frac{x}{a}\cos ax + C$

79. $\displaystyle\int x\cos ax\,dx = \frac{1}{a^2}\cos ax + \frac{x}{a}\sin ax + C$

80. $\displaystyle\int x^n \sin ax\,dx = -\frac{x^n}{a}\cos ax + \frac{n}{a}\int x^{n-1}\cos ax\,dx$

81. $\displaystyle\int x^n \cos ax\,dx = \frac{x^n}{a}\sin ax - \frac{n}{a}\int x^{n-1}\sin ax\,dx$

82. $\displaystyle\int \tan ax\,dx = -\frac{1}{a}\ln|\cos ax| + C$

83. $\displaystyle\int \cot ax\,dx = \frac{1}{a}\ln|\sin ax| + C$

84. $\displaystyle\int \tan^2 ax\,dx = \frac{1}{a}\tan ax - x + C$

85. $\displaystyle\int \cot^2 ax\,dx = -\frac{1}{a}\cot ax - x + C$

86. $\displaystyle\int \tan^n ax\,dx = \frac{\tan^{n-1} ax}{a(n-1)} - \int \tan^{n-2} ax\,dx, \qquad n \neq 1$

87. $\displaystyle\int \cot^n ax\,dx = -\frac{\cot^{n-1} ax}{a(n-1)} - \int \cot^{n-2} ax\,dx, \qquad n \neq 1$

88. $\displaystyle\int \sec ax\,dx = \frac{1}{a}\ln|\sec ax + \tan ax| + C$

89. $\displaystyle\int \csc ax\,dx = -\frac{1}{a}\ln|\csc ax + \cot ax| + C$

Continued overleaf.

90. $\int \sec^2 ax\, dx = \frac{1}{a}\tan ax + C$

91. $\int \csc^2 ax\, dx = -\frac{1}{a}\cot ax + C$

92. $\int \sec^n ax\, dx = \frac{\sec^{n-2} ax \tan ax}{a(n-1)} + \frac{n-2}{n-1}\int \sec^{n-2} ax\, dx, \qquad n \neq 1$

93. $\int \csc^n ax\, dx = -\frac{\csc^{n-2} ax \cot ax}{a(n-1)} + \frac{n-2}{n-1}\int \csc^{n-2} ax\, dx, \qquad n \neq 1$

94. $\int \sec^n ax \tan ax\, dx = \frac{\sec^n ax}{na} + C, \qquad n \neq 0$

95. $\int \csc^n ax \cot ax\, dx = -\frac{\csc^n ax}{na} + C, \qquad n \neq 0$

96. $\int \sin^{-1} ax\, dx = x \sin^{-1} ax + \frac{1}{a}\sqrt{1-a^2x^2} + C$

97. $\int \cos^{-1} ax\, dx = x \cos^{-1} ax - \frac{1}{a}\sqrt{1-a^2x^2} + C$

98. $\int \tan^{-1} ax\, dx = x \tan^{-1} ax - \frac{1}{2a}\ln(1+a^2x^2) + C$

99. $\int x^n \sin^{-1} ax\, dx = \frac{x^{n+1}}{n+1}\sin^{-1} ax - \frac{a}{n+1}\int \frac{x^{n+1}\, dx}{\sqrt{1-a^2x^2}}, \qquad n \neq -1$

100. $\int x^n \cos^{-1} ax\, dx = \frac{x^{n+1}}{n+1}\cos^{-1} ax + \frac{a}{n+1}\int \frac{x^{n+1}\, dx}{\sqrt{1-a^2x^2}}, \qquad n \neq -1$

101. $\int x^n \tan^{-1} ax\, dx = \frac{x^{n+1}}{n+1}\tan^{-1} ax - \frac{a}{n+1}\int \frac{x^{n+1}\, dx}{1+a^2x^2}, \qquad n \neq -1$

102. $\int e^{ax}\, dx = \frac{1}{a}e^{ax} + C$

103. $\int b^{ax}\, dx = \frac{1}{a}\frac{b^{ax}}{\ln b} + C, \qquad b > 0,\ b \neq 1$

104. $\int xe^{ax}\, dx = \frac{e^{ax}}{a^2}(ax - 1) + C$

105. $\int x^n e^{ax}\, dx = \frac{1}{a}x^n e^{ax} - \frac{n}{a}\int x^{n-1}e^{ax}\, dx$

106. $\int x^n b^{ax}\, dx = \frac{x^n b^{ax}}{a \ln b} - \frac{n}{a \ln b}\int x^{n-1} b^{ax}\, dx, \qquad b > 0,\ b \neq 1$

107. $\int e^{ax}\sin bx\, dx = \frac{e^{ax}}{a^2+b^2}(a \sin bx - b \cos bx) + C$

108. $\int e^{ax}\cos bx\, dx = \frac{e^{ax}}{a^2+b^2}(a \cos bx + b \sin bx) + C$

109. $\int \ln ax\, dx = x \ln ax - x + C$

110. $\int x^n \ln ax\, dx = \frac{x^{n+1}}{n+1}\ln ax - \frac{x^{n+1}}{(n+1)^2} + C, \qquad n \neq -1$

111. $\int x^{-1}\ln ax\, dx = \frac{1}{2}(\ln ax)^2 + C$

112. $\int \frac{dx}{x \ln ax} = \ln|\ln ax| + C$

113. $\int \sinh ax\, dx = \frac{1}{a}\cosh ax + C$

114. $\int \cosh ax\, dx = \frac{1}{a}\sinh ax + C$

115. $\int \sinh^2 ax\, dx = \frac{\sinh 2ax}{4a} - \frac{x}{2} + C$

116. $\int \cosh^2 ax\, dx = \frac{\sinh 2ax}{4a} + \frac{x}{2} + C$

117. $\int \sinh^n ax\, dx = \frac{\sinh^{n-1} ax \cosh ax}{na} - \frac{n-1}{n}\int \sinh^{n-2} ax\, dx, \qquad n \neq 0$